Pulling Rabbits Out of Hats

Advances in Biochemistry and Biophysics

This monograph series offers expert summaries of cutting edge topics across all areas of biological physics. Individual titles address such topics as molecular biophysics, statistical biophysics, molecular modeling, single-molecule biophysics, and chemical biophysics. The goal of the series is to facilitate interdisciplinary research by training biologists and biochemists in quantitative aspects of modern biomedical research and to teach key biological principles to advanced students in physical sciences and engineering.

Pulling Rabbits Out of Hats

Using Mathematical Modeling in the Material, Biophysical, Fluid Mechanical, and Chemical Sciences
David Wollkind, Bonni J. Dichone

For more information about this series, please visit: https://www.routledge.com/Advances-in-Biochemistry-and-Biophysics/book-series/CRCBIOPHY

Pulling Rabbits Out of Hats

Using Mathematical Modeling in the Material, Biophysical, Fluid Mechanical, and Chemical Sciences

David Wollkind
Washington State University
with
Bonni Dichone
Gonzaga University

CRC Press
Taylor & Francis Group
Boca Raton London New York

CRC Press is an imprint of the
Taylor & Francis Group, an **informa** business

First edition published 2022

by CRC Press
6000 Broken Sound Parkway NW, Suite 300, Boca Raton, FL 33487-2742

and by CRC Press
2 Park Square, Milton Park, Abingdon, Oxon, OX14 4RN

© 2022 David Wollkind, Bonni J. Dichone

CRC Press is an imprint of Taylor & Francis Group, LLC

Library of Congress Cataloging-in-Publication Data

Names: Wollkind, David J., 1942- author. | Dichone, Bonni, author.
Title: Pulling rabbits out of hats : using mathematical modeling in the material, biophysical, fluid mechanical, and chemical sciences/David Wollkind with Bonni Dichone.
Description: Boca Raton : CRC Press, 2021. | Includes bibliographical references and index.
Identifiers: LCCN 2021021531 (print) | LCCN 2021021532 (ebook) | ISBN 9781032047874 (hardback) | ISBN 9781032050072 (paperback) | ISBN 9781003195603 (ebook)
Subjects: LCSH: Mathematical models. | Mathematical analysis.
Classification: LCC QA401 .W74 2021 (print) | LCC QA401 (ebook) | DDC 511/.8--dc23
LC record available at https://lccn.loc.gov/2021021531
LC ebook record available at https://lccn.loc.gov/2021021532

ISBN: 9781032047874 (hbk)
ISBN: 9781032050072 (pbk)
ISBN: 9781003195603 (ebk)

DOI: 10.1201/9781003195603

Typeset in CMR10 font
by KnowledgeWorks Global Ltd.

For Juan, Natasha, and Olli; Niki, Jonathan, and Jeremy; and in memory of Edward Pate and thesis advisor Lee Segel.
David J. Wollkind

For my husband Paulo, daughter Georgina, and mother Gloria Kealy.
Bonni Dichone

Contents

Foreword

It was Fall of 2007. I was at graduate school orientation at Washington State University (WSU). I already had a Master's degree in Mathematics from Eastern Washington University and had spent the last two years lecturing at other universities, but colleagues had convinced me to go back to school and pursue a Ph.D. I was hesitant to do so, because I thought I was not "good enough" at mathematics and had been out of that kind of in-depth study for too long. But, somehow I found myself at WSU, terrified of what lay ahead of me and certain I would fail.

It was during this orientation that I first met Dr. David J. Wollkind or just "Wollkind," as I still call him. He came into orientation and gave a presentation on his research and the courses he taught. He was extremely intense, talked very quickly, and disseminated information at a rapid speed, while pacing and clicking through slides being projected (from one of those old-fashioned projectors that had a wheel of literal film slides) at an alarming and aggressive rate. Being in the state of mind that I was, you would have thought such a presentation from this clearly intelligent and entirely formidable mathematician would have intimidated me but instead I was intrigued and fascinated. I approached him after the session and he said, in the most matter of a fact way, "Bonni, you should sign up for my Continuum Mechanics course this Fall." Which I did and every course he taught, thereafter.

Lectures with Wollkind were unique. He used a very specific set of colored chalk, in addition to white chalk, along with a yard stick, all of which he used to draw very detailed pictures and graphs up on the board, well before class would begin. I learned very quickly that one needed to arrive at least 10 minutes before class to copy said figures down, as there was no time during his lecture to copy them when he referenced them. He would lecture rapidly and his handwriting, while quite legible compared to other professors, degraded as he would go, with t's, x's, and z's all looking nearly the same by the end of class, not to speak of his use of ξ's , ζ's, and the like. Each lecture was jam-packed with mathematical technique and its applications, relayed in the most carefully planned out and truly beautiful and artistic way. Every class felt like a master class in the topic being presented, from *the* master, himself. Sprinkled throughout the complex mathematics, he would often relay these anecdotal stories of when and how he learned the mathematics or discovered the techniques he was presenting. I always looked forward to these accounts, because he would get a twinkle in his eye and usually include some kind of sarcastic joke or comment, which was always funny and also gave me a chance to catch up in my note taking.

One semester in, I asked him to be my thesis advisor and the rest, as they say, is history (see Chapter 12 and onward). Throughout my time at WSU, we formed a very close working relationship and a great friendship, which has continued to this day. He shepherded me through my qualifying exams, doctoral exams, dissertation defense, publications, and presentations. In so doing, he built my confidence as a mathematician and I successfully graduated with my Ph.D. and, after a Post-Doctoral position, was hired in a tenure-track position, as one of the first Applied Mathematicians, at Gonzaga University (GU). Our continued collaboration has contributed greatly to not only my hiring but also promotion and tenure at GU, where I am now an Associate Professor. I am forever indebted to him, my "mathematical father," if you will. He is my "Lion" (see the conclusion of Chapter 18).

When we were initially working on our first textbook, Comprehensive Applied Mathematical Modeling in the Natural and Engineering Sciences ([284]), at one point, I suggested that maybe, instead of writing a textbook, we should write some kind of hybrid book that included all the great anecdotes around his mathematical research and career that he would share in class. I was always amazed at the seemingly "magical" techniques he used (that I didn't see in any other textbook) and fascinated by the narrative around his career and how so many of those techniques and results were discovered. We tabled that idea, as the textbook was actually the thing that needed to be done at the time, but I never quit gently reminding him of this proposition and different kind of book that could be written. He listened and thus, it is my great honor to have helped him bring to life this book, Pulling Rabbits Out of Hats, and in so doing, preserve his life's work and career's story.

Bonni J. Dichone, Ph.D.

Preface

Scope

Fifty years ago, I attempted to provide a better understanding of the natural and engineering sciences by employing a variety of mathematical techniques to quantify a number of phenomenological problems. The term "comprehensive applied mathematical modeling" was coined by C.C. Lin, my thesis advisor's thesis advisor, to categorize such quantification since its resulting theoretical predictions are to be compared with observational or experimental evidence from the phenomenon under investigation. The purpose of this book is to catalogue my endeavors in that regard. It represents a personal odyssey of discovery using mathematical modeling in the material, biophysical, fluid mechanical, and chemical sciences and hence its subtitle.

Many people, upon hearing about such mathematical modeling, assume the applied mathematician has only to analyze an already formulated model and then compare the results obtained with existing observational or experimental data sets for validation purposes. However, with virtually all the phenomena to be discussed in this book, the formulation and validation steps described above have never been that straight-forward. Paradoxically those examinations of these phenomena required both the addition of extra effects not considered in the original model formulation and the introduction of simplifications to make that modified model more tractable for mathematical analysis. Once the resulting theoretical predictions were obtained subtle interpretations within the framework of that model of the data, sometimes also generated, were often necessary in order to validate it. I and my co-author, Bonni Dichone, began referring collectively to these interventions as Pulling Rabbits Out of Hats, the descriptions of which are the primary focus of this book, and hence its title. Indeed, the whole process is really more of an art than a science.

Another popular misconception which this book tries to remedy is that the conducting of applied mathematical modeling research and its dissemination are cut-and-dried processes. They are not. No affair involving personal interactions can be, since human beings do not always act rationally even when working in a scientific area that should require objectivity rather than subjective behavior. Some pains are taken to put a human face on the research presented by profiling a fairly wide range of people including pioneers in the field, mentors, collaborators, colleagues, competitors, administrators, and political figures when appropriate. Further, it is almost impossible to describe applied mathematical modeling research without analyzing the requisite governing systems for which some people have an aversion. The relevant mathematical terms are developed graphically and from first principles as much as possible. My thesis advisor Lee Segel would have labeled this effort as "carrying the gospel to the savages" although Lee was referring to integrating applied mathematicians into formerly pure mathematics departments when making that remark to me about my initial employment at Washington State University in the Fall of 1970. Developments of this sort are intended to provide the general readership with enough background information to aid them in following these descriptions but may be skimmed over by those with more expertise in deciphering technical writing.

Except for the first and last chapters, each intervening one concentrates on particular phenomenological topics covering specific chosen problems from those areas treated throughout my whole research career with this material being presented in chronological order for the most part. Each chapter title includes both the problems described and the mathematical modeling techniques employed, which would allow readers to rearrange these chapters by the latter topic ordered from the simplest to the most complex methodologies, *i.e.*, dynamical systems; linear stability; and longitudinal, square, rhombic, and hexagonal planform nonlinear stability analyses; *e.g.* – Chapters 4, 16, 5, 17, 6, 15, 3, 2, 8, 7, 12, 9, 10, 11, 14, and 13 – should that layout be more attractive to them. Hence, this book may be considered both an annotated semi-autobiographical account of that research's completion and a monograph on comprehensive applied mathematical modeling in which the phenomenological data and the model predictions are self-consistent. Thus it also serves as a companion volume to Wollkind and Dichone ([284]), our Springer textbook on the subject of such comparisons in both the natural and engineering sciences.

Content

This book consists of 18 chapters. Chapter 1 presents background information about the book and its co-authors, as well as describing its philosophy and organization. Chapter 2, which serves as a template for comprehensive applied mathematical modeling, deals with interfacial morphological stability of the solid-liquid interface during dilute binary alloy solidification and melting while introducing my thesis advisor, Lee Segel, whose influence plays a prominent role throughout this book, and my Ph.D. students Sassan Raissi, David Oulton, and Rukmini Sriranganathan. Chapter 3 concentrates on my chemically driven convection postdoctoral research with supervisor Harry Frisch and colleague John Bdzil. Whereas the subject of those and most of the other chapters in the book is pattern formation of governing partial differential equations (PDEs), the subject area of Chapter 4 concerns the various dynamical systems of ordinary differential equations (ODEs) for temperature-dependent predator-prey mite interactions on fruit trees analyzed in collaboration with my Ph.D. students Jesse Logan and John Collings, the former from entomology. Instead of the global behavior of the governing systems of equations considered in most of the chapters of this book, Chapter 5 deals with the local behavior of the systems of PDEs appropriate for modeling the multi-layer fluid phenomena of Rayleigh-Bénard-Marangoni convection and Kelvin-Helmholtz rock folding which served as the thesis topics for my Ph.D. students Eric Ferm and Iwan Alexander, the latter from geology. Chapter 6, which returns to nonlinear stability theory, are longitudinal planform analyses of two-phase fluid flow of aerosols and convection in planetary atmospheres, both of which I conducted with my Ph.D. student Limin Zhang, while Chapter 7 is an analysis of hexagonal chemical Turing pattern formation done in collaboration with my Ph.D. student Laura Stephenson. The relevant models of Chapters 8-11 can each be reduced to a single spatio-temporal partial differential evolution equation: Namely, a Lennard-Jones lubrication equation for thin liquid layer instabilities, a damped Kuramoto-Sivashinsky equation for ion-sputtered erosion of thin solid layers, a modified Swift-Hohenberg equation for nonlinear optical patterns, and a fourth order logistic equation for nonlinear vegetative distributions in dryland environments analyzed in collaboration with my Ph.D. students Mei Tian, Adoon Pansawan, Dean Edmeade and

Francisco Alvarado, and Francisco and Nichaphat Boonkorkuea, respectively. The subject of Chapters 12 and 13 is diffusive versus differential flow instabilities of dryland and mussel-bed ecological Turing pattern formation analyzed in collaboration with my Ph.D. student and co-author Bonni Dichone and my Ph.D. student and her colleague Richard Cangelosi, respectively. Chapter 14 deals with root suction driven rhombic vegetative dryland pattern formation analyzed in collaboration with my Ph.D. student Inthira Chaiya and Chapter 15 done in collaboration with my undergraduate student Mitch Davis is concerned with the subcritical behavior of a model interaction-dispersion equation, as opposed to the other analyses in the book which are primarily resticted to supercritical behavior. Finally, Chapter 16, done in collaboration with Richard Cangelosi and my colleague Elissa Schwartz, analyzes the topical subject of a non-cytopathic viral-target cell dynamical system and Chapter 17, done in collaboration with my MS student Kohl Gill, deduces Jeans' criterion for a linear gravitational instability in the presence of uniform rotation. The book concludes with Chapter 18 which synthesizes a number of topics introduced previously before closing by presenting the famous fable for Ph.D. graduate students and their thesis advisors entitled "The Rabbit, the Fox, and the Wolf."

Target Audience

The target audience for this book consists predominantly of people who have an interest in how scientific contributions are actually made. Such people tend to come from the following two groups: Individuals possessing some knowledge of science or engineering but who have never worked professionally in either capacity and those who have. The first group includes the rather large number of individuals having educational credentials nominally related to STEM (Science, Technology, Engineering, and Mathematics), but for a variety of reasons chose to take positions in other fields after completing that training. These come from both the private and public sector, including commercial, financial, personnel, actuarial, or legal positions that require a good deal of creativity and intellectual acumen but of a somewhat different nature. The second group consists of individuals usually from academia, industry, or government laboratories who are scientists or engineers. Both groups, having concentrated on the specific specialties inherent to their professional duties, may not be intimately familiar with all the particular phenomena and methodologies introduced in this book, but tend to retain an innate curiosity about the procedure employed to make contributions in these areas never-the-less. It is such curiosity that this book has been written to assuage.

They are being presented from first principles upon their initial employment for the purpose of review or preview depending on the background information of them possessed by a particular reader. For this reason special effort has been taken during the development of the book's required mathematical concepts from the subdisciplines of advanced calculus and differential equations. As mentioned in the scope section of the preface, the theme of this book is the various "Pulling of Rabbits Out of Hats" that occurred during the research presented in it. The genesis of these unexpected occurrences are carefully explained and each one is labeled by that identification in capital letters when they appear to alert prospective readers to their introduction.

Also the book is organized as much as possible about the figures contained in each chapter in the spirit of the old English adage "a picture is worth a thousand words." Indeed for each of these chapters the figures were selected first as a guide to help that reader follow its material and then the text was composed in essence around them. These figures, including their captions, illustrate the phenomena under investigation, basic mathematical concepts, and theoretical research results as well as those observational or experimental data and numerical simulations to which the latter are subsequently compared. All of these figures have been created to allow readers more easily to understand those theoretical analytical developments and make comparisons of them to relevant observational, experimental, or numerical results. This is another point that distinguishes applied from pure mathematics. Applied mathematics especially in modeling-type research tends to use multiple figures whereas many pure mathematicians may not have produced a single figure in all of their publications combined. When asked by "the person in the street" about what they do, comprehensive applied mathematical modelers are able to explain that research by means of a set of figures depicting the phenomena under examination, whereas most of their purer brethren are unable to offer an analogous explanation which can be understood by such a layperson.

The ultimate goal of this book is to make the process of quantification of scientific and engineering phenomena as transparent as possible while at the same time serving as a monograph on comprehensive applied mathematical modeling for its intended audience of interested amateurs and professionals.

David J. Wollkind, Ph.D.

Acknowledgements

We would like to acknowledge our appreciation to the following individuals and societies for kindly giving us their permission to publish the third-party figures listed below:

Qi Ouyang
Figs. 1.2, 7.1, 7.12, 9.12, 13.18

Mustapha Tlidi
Figs. 11.12, 11.15, 14.12d

Günter Reiter
Fig. 8.1

Ehud Meron
Fig. 11.20

Jack Douglas
Fig. 8.6

Jonathan Sherratt
Figs. 12.6, 12.12

American Physical Society
Fig. 8.7

Rene Lefever
Fig. 12.8

Rodolfo Cuerno
Figs. 9.1, 9.7

Vincent Deblauwe
Figs. 12.18, 12.19

Stefan Facsko
Figs. 9.8, 9.9

Heather Silverman
Fig. 13.3

Stefano Rusponi
9.13a, 9.13b, 9.14

Quan-Xing Liu
Figs. 13.17, 13.20

American Institute of Physics
Fig. 9.15

Karna Gowda
Figs. 14.11, 14.12a, 14.12b, 14.12c

Thomas Michely
Figs. 9.13c, 9.13d

American Astronomical Society
Fig. 17.5

We would also like to thank our Taylor & Francis Group CRC Press handling and production editors, Carolina Antunes and Lara Spieker, respectively, for the encouragement and guidance they provided during the publication process including the conversion of the manuscript text and its figures to the appropriate form for the particular required template and the obtaining of the permissions listed above.

1

Introduction

Scientific research is currently disseminated in a manner analogous to the origin of the Greek Goddess of Love and Beauty, Aphrodite, who sprang fully formed from the foam of the sea (the name Aphrodite is derived from the Greek word for foam "aphros" and is related to this story of her origin). This precludes the average journal reader from gaining any insight into the evolutionary iterative process often required to achieve publishable results. As an example of the fact that this has not always been the case, consider James Clerk Maxwell's ([138]) paper on governors for engines, published in the *Proceedings of the Royal Society of London*. This seminal paper on control theory or cybernetics is incomplete, in that Maxwell only carried it out as far as he could before posing an open question, the answer to which was required for its completion. That led to Routh and Hurwitz independently deducing the criterion, now bearing their name (see Chapter 3), needed to resolve this open question and thus allow the completion of the control theory problem for governors under investigation (Bellman and Kalaba, [11]). Today Maxwell's paper could not be published until after that Rabbit had been Pulled Out of the Hat, and anyone reading the paper in its final form would have no idea an open question ever existed.

To alleviate this deficiency for a particular paper would require an identification of its open questions and an explanation of the thought processes involved in resolving them, which is often more of an art than a science. Given that this is a subjective procedure, it could only be accomplished by the authors themselves. Hence specific individuals would be restricted to their own research when discussing such endeavors. Toward that end, it is the purpose of this book to offer examples of comprehensive applied mathematical models in the natural and engineering sciences from my own university research career that required the Pulling of Rabbits Out of Hats during their formulation, analysis, or validation. To do so, I have adopted a style of presentation using diagrammatic representations whenever possible, popularized by Kip Thorne ([247]) in his book on general relativity entitled *Black Holes and Time Warps: Einstein's Outrageous Legacy*. Although many scientists are loath to admit publicly that such occurrences play a role in their creative activities, sometimes human interactions were involved in these endeavors and those incidents will be discussed as well, when appropriate. To place this exposition in its proper context, it is helpful for me to provide the following definitions and background information.

Comprehensive applied mathematical modeling starts with a particular scientific phenomenon, abstracts to a mathematical model using fundamental first principles, introduces whatever simplifications are necessary to make the model tractable but still preserve its salient features, analyzes that modified model by employing a variety of analytical or numerical techniques, and then compares the resulting theoretical predictions of that analysis with existing observational or experimental data relevant to the phenomenon under examination. Hence these phenomenological observables are used both to formulate and validate the model. The optimal situation is that the theoretical modeling predictions provide very good qualitative and quantitative agreement with those observables. In this context, it is important to realize that comprehensive applied mathematical modelers are individuals who

DOI: 10.1201/9781003195603-1

have a thorough knowledge, understanding, and appreciation of the proper relationship between mathematics and all the sciences and who professionally devote themselves to the totality of such phenomenological modeling. These people are neither pure mathematicians with an interest in certain specializations of the theoretical sciences, nor are they theoretical natural or engineering scientists with an acquired expertise in a given subdiscipline of mathematics closely related to their particular field. Rather their goal is to elucidate scientific concepts and describe specific phenomena through the use of mathematics.

The phrase **"Pull (or bring) a Rabbit Out of the (or a) Hat"** is defined by the Oxford Dictionary as "to do something unexpected but ingeniously effective in response to a problem" and by the Macmillan Dictionary as "to do something very clever to solve a problem," with the synonym of "to deal successfully with a problem or difficulty." Its origin is the magician's trick of pulling some unexpected object, often a rabbit, out of a hat and hence is often used to mean "get magical results." Indeed, when presenting examples in my modeling classes of Pulling Rabbits Out of Hats to obtain results, I would often tell the students "It's Magic, You Dope!," with apologies to Jack Sharkey whose novelette of the same title about an alternate universe where magical laws superseded natural ones originally appeared as a two-part serial in the November and December 1962 issues of the science fiction magazine *Fantastic: Stories of Imagination.*

At the time of its publication, I was in the first semester of my junior year at Rensselaer Polytechnic Institute (RPI), a university located in Troy, New York, specializing in STEM (an acronym for Science Technology Engineering and Mathematics) subjects, as it is referred to today. My major was mathematics to which I had switched the semester before from my original choice of chemistry. The trouble was that neither the physical sciences nor the mathematics, to which I had been exposed, appealed to me; the former being too empirical and the latter, too pure (abstract analysis and algebra with no applications) for my taste. Then, as a first semester senior, I enrolled in George H. Handelman's FOAM (Foundations of Applied Mathematics) course. Professor Handelman, who was the chair of the Department of Mathematics, showed me that applied mathematics could be used effectively to describe real-world phenomena, particularly in the field of elasticity. I had, in effect, finally found something I liked to do and accepted the NASA (National Aeronautics and Space Administration) Traineeship he subsequently offered me to work for a Ph.D. (Doctor of Philosophy degree) in Mathematics. Four years later, I received that degree under the direction of Lee A. Segel, spent the next two years as a postdoctoral research associate of Harry L. Frisch in the Department of Chemistry at the State University of New York Albany (SUNYA), and then became a faculty member in the Department of Mathematics at Washington State University (WSU), the land grant institution of that state, which exclusive of its experiment stations consisted of the Pullman campus alone in those days.

Lee Segel was renowned as a leading scholar in Comprehensive Applied Mathematical Modeling. Originally trained as an applied mathematical modeler in the physical and engineering sciences with an emphasis on fluid mechanics and materials science, he was also a pioneer in modern Mathematical Biology. Segel was one of the earliest promoters of the need for close collaborative contact between theoretical and experimental biology and fostered the appreciation of modeling by biologists and of biology by mathematicians. His research on pattern formation and morphogenesis has been seminal in launching a burgeoning field at the intersection of developmental biology, applied mathematics, and numerical computation. Lee Segel was a forefather of the field now known as theoretical immunology and as the editor of the *Bulletin of Mathematical Biology* introduced the idea that theoretical mathematical modeling predictions of biological phenomena should be compared with relevant experimental or observational data in order to be eligible for publication. After starting at RPI, Segel moved to the Weizmann Institute in Rehovot, Israel. Harry Frisch, who was a theoretical physical chemist of some repute at Bell Laboratories in Murray Hill, New Jersey, when SUNYA hired him, demonstrated to me the power of employing a variety of mathematical techniques to understand physical chemistry from first principles and how to work in a research group, which included my postdoctoral colleague, John Bdzil, on chemically driven convection. In the process he also weaned me from my dissertation topic of metallurgical dilute binary alloy solidification. Perhaps of equal importance, his letter of recommendation convinced B. Roger Ray, the Dean of Sciences at WSU in 1970 and a Professor of Physical Chemistry, to hire me as an Assistant Professor of Mathematics.

Influenced by those mentors, my research activities have been primarily concerned with modeling natural and engineering science phenomena by means of mathematical methods. These phenomena include the solidification of metals and alloys, convection of fluids and aerosols, folding of sedimentary rock multilayers, development of stars in the outer arms of spiral galaxies, dynamics of predator-prey mite and noncytopathic viral-target cell interactions, and the occurrence of Turing structures in chemical reaction-diffusion systems, of morphological phase separation in thin liquid films, of nonlinear optical patterns in sodium vapor ring cavities, of nanoscale etchings on solid surfaces during ion-sputtered erosion, of vegetation distributions in arid environments, and of self-organized patterning in young beds of mussels feeding on algae. The methodologies employed have been both analytical and numerical, with stability analyses and perturbation techniques, examples of the former and differential equation solvers and computer-assisted bifurcation algorithms examples of the latter. A common theme in all these investigations is the comparison of theoretical predictions, deduced from the relevant continuum mechanical study, with experimental or observational evidence characteristic of the phenomenon under investigation. Virtually all of the comprehensive applied mathematical modeling research catalogued above was done in collaboration with either those mentors or my own Ph.D. graduate students. I have supervized the dissertation research of 25 such students, 20 in Mathematics, 1 each in Entomology and Geology at WSU, and 3 from Mahidol University in Bangkok, Thailand, with which we have a Memorandum of Agreement (MOA) that allows them to spend a year at WSU as Visiting Scholars and me, to be the dissertation advisor for their Ph.D.'s in Applied Mathematics from Mahidol University. This research was collaborative in nature resulting in joint publications since I have always regarded my Ph.D. graduate students as co-authors and saved my best problems for their dissertations. Many of these students made significant contributions to the Pulling of Rabbits Out of Hats aspect of my research and will be acknowledged individually on a case by case basis along with my thesis adviser, postdoctoral supervisor, and other professional colleagues in this exposition.

Bonni J. Dichone (nee Kealy) was one of my last Ph.D. students at WSU (see Fig. 1.1). I met her in the Fall Semester of 2007 when she entered our doctoral program and enrolled in my two-semester graduate-level continuum mechanics course. Particle mechanics deals with the equilibrium and motion of systems of point masses. In continuum mechanics one deals with continuous media in which smoothly varying properties such as density (ρ), temperature (T), and velocity ($\mathbf{v}$, a *vector* quantity) must be assigned at each point (x, y, z) in space instantaneously occupied by a material body under examination at the time (t) in question, using the so-called continuum hypothesis. There exist a large number of phenomena to which this hypothesis can be applied and hence may be modeled by the various techniques of continuum mechanics. Such techniques usually associated with postulated balance-type conservation laws produce governing systems of partial differential equations (PDEs). Often consistent with experimental, thermodynamic, or empirical evidence, it

FIGURE 1.1

Photograph of David Wollkind presenting his Ph.D. student Bonni Dichone (nee Kealy) with the Sidney and Eleanor Hacker Graduate Teaching Award for the third consecutive year. Dichone is holding the award certificate and Martin Gardner's annotated edition of *The Hunting of the Snark* given to her by Wollkind during his presentation in commemoration of Bonni's having fulfilled the Bellman's Rule of Three "What I tell you three times is true," contained in the first stanza of that book, a nonsense poem by Lewis Carroll (pen name of Charles Dodgson), the author of *Alice's Adventures in Wonderland* and its sequel *Through the Looking-Glass*.

becomes necessary to adopt particular constitutive relations, equations of state, or some other relationship between the dependent variables. All of these equations along with appropriate boundary (space) and initial (time) conditions constitute the mathematical formulation of the problem. In January of 2008, Bonni asked me if I would be her Ph.D. adviser and we have been working closely together ever since. She took all five of my graduate modeling and applied analysis courses, as well as my undergraduate differential equations class at WSU, did her dissertation with me on vegetative pattern formation in arid flat environments, and served as my postdoctoral associate for the next two years upon its completion. Bonni, currently an Associate Professor of Mathematics, is starting her ninth year at Gonzaga University where she has been implementing a comprehensive applied mathematics program. During that time Bonni saw me make a number of assumptions and deduce a variety of results, which she thought to be reasonably clever, in order to complete a particular modeling investigation. We began labeling such endeavors "Pulling Rabbits Out of Hats." Further more, the courses she took with me included slide show presentations highlighting my comprehensive applied mathematical modeling research results, many of which also entailed Pulling such a Rabbit Out of a Hat. Bonni collaborated with me on a comprehensive applied mathematical modeling textbook (Wollkind and Dichone [284]) as my co-author. When I started planning that textbook she said, "You know what I'd really like to see you write is a popular science book about all the Rabbits you've Pulled Out of Hats during your research career. In fact, I'll even help you with its development if you decide to do it." Without that conversation I would never have even thought of the idea of writing a book of this sort, so its genesis is owed totally to Bonni Dichone's suggestion and its completion to her collaboration. She has helped me with the selection of the Rabbits to be Pulled Out of Hats, the text, the figures, and the preparation of the manuscript: Hence, my author listing of "with Bonni J. Dichone" which is usually reserved for crediting the contributions of ghost writers to autobiographies.

There is also some merit in pointing out at this time the role collaboration plays in applied mathematical modeling. Unlike in pure mathematics where single-authored papers are somehow valued more than those produced by joint authorship, collaborative research is encouraged in interdisciplinary applied mathematical modeling. Indeed, as will become clear in later chapters, one of the advantages of such collaboration is that it gives each author the opportunity to use their co-authors as sounding boards when attempting to Pull the proverbial Rabbit Out of a Hat. Another difference between pure and applied mathematicians is that, whereas the former often make their major contributions at a relatively early age professionally, experience is a significant part of the latter's ability to model phenomena successfully which tends to be correlated with how long one has been modeling.

This book is organized in chronological order about the various scientific phenomena investigated and methodologies employed by me that required the Pulling of a Rabbit Out of a Hat, with an emphasis on these processes and the people involved in the successful completion of those comprehensive applied mathematical modeling analyses. As catalogued above, many of these investigations were pattern formation studies and such spatial patterns consisting, for example, of parallel stripes and close-packed hexagonal arrays of either spots or honeycombs, as depicted in Fig. 1.2, can only occur if the model under examination is nonlinear. Thorne, in his general relativity book ([247]), described nonlinearity by stating that a quantity is called linear if its total size is equal to the sum of its parts; otherwise, it is nonlinear. Using an algebraic analogy, the operation of scalar multiplication, defined by $L[a] \equiv \alpha a$, is linear since

$$L[a + b] = \alpha(a + b) = \alpha a + \alpha b = L[a] + L[b]$$

Stripes	Spots	Honeycombs

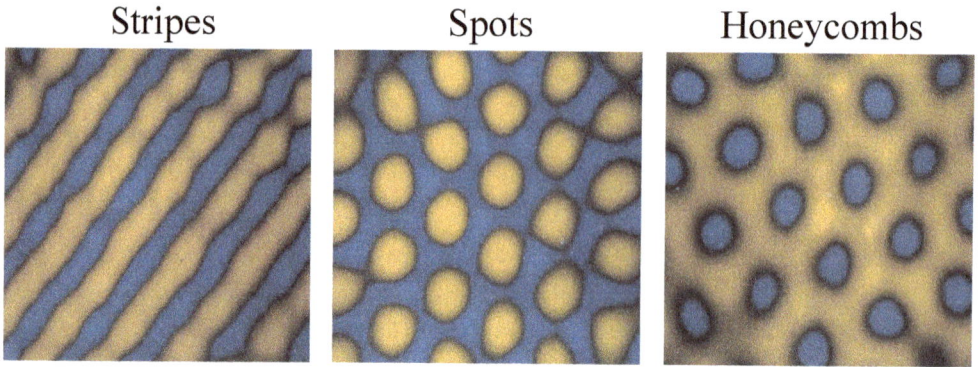

FIGURE 1.2
Photographic images of the chemical Turing patterns of stripes (bands), spots (dots), and honeycombs (nets) observed during the CDIMA (an acronym for *C*hlorine *D*ioxide-*I*odine-*M*alonic *A*cid)/starch reaction by Ouyang and Swinney ([163]) and predicted by a nonlinear stability analysis of the appropriate model system of PDEs for this phenomenon (see Chapter 7). Here the yellow portions of the images represent the designated patterns.

while that of squaring, defined by $N_2[a] \equiv a^2$, is nonlinear since

$$N_2[a + b] = (a + b)^2 = a^2 + 2ab + b^2 \neq a^2 + b^2 = N_2[a] + N_2[b].$$

In addition, these analyses used perturbation methods also explained by Thorne employing this square power law. Extending that explanation to a quartic power, consider the binomial formula

$$N_4[a + b] = (a + b)^4 = a^4 + 4a^3b + 6a^2b^2 + 4ab^3 + b^4.$$

Then a linear perturbation of b to a yields the approximation that

$$N_4[a + b] \cong a^4 + 4a^3b$$

by neglecting all the higher-order terms in the perturbation quantity b, while a nonlinear one, through third-order terms in b, yields, upon the neglect of b^4, the approximate cubic expansion

$$N_4[a + b] \cong a^4 + 4a^3b + 6a^2b^2 + 4ab^3.$$

2

Solidification and Melting of Dilute Binary Alloys: Longitudinal and Hexagonal Planform Nonlinear Stability Analyses

Since this chapter includes my dissertation topic, I shall begin it by presenting the historical details of how Lee Segel became my advisor. I first met him in the Spring Semester of 1966. At the time, I was taking a graduate fluid mechanics course from Richard C. DiPrima who said to me one day that "Professor Segel is looking for a Ph.D. student and I suggested you. Please meet with him tomorrow afternoon." I was already familiar with Lee since he had spent a week as a guest lecturer in this course developing the Rayleigh-Bénard problem of a buoyancy-driven convection layer heated from below. I went to his office the next afternoon. Segel told me about the two problems on which he was currently interested in working. One was a highly theoretical treatment of an iteration scheme for analyzing nonlinear stability problems and the other, a phenomenological problem dealing with the stability of the shape of the solid-liquid interface during the controlled plane-front solidification of a dilute binary alloy. For me that choice was no contest since I much preferred modeling this metallurgical phenomenon to concentrating on abstract nonlinear stability theory.

In preparing for my candidacy examination and getting ready to work on the alloy solidification problem, I purchased the recently published University of Chicago book entitled "Nonequilibrium thermodynamics, variational techniques, and stability" edited by Russell Donnelly, Robert Herman, and Ilya Prigogine, which was the proceedings of a conference held there the year before. Segel had a chapter in this book about the application of nonlinear hydrodynamical stability theory to the Rayleigh-Bénard and Taylor-Couette (flow in the annulus between two rotating concentric cylinders, DiPrima's area of expertise) problems. DiPrima had distributed a preprint of that chapter (Segel, [212]) to us as a prelude to Segel's guest lectures. This material made a great impression on me and I reread it before my meeting with Segel described above. Another contribution in that book by James Kirkaldy concentrated on the interfacial morphological patterns observed during dilute-binary alloy solidification and this was how Segel became aware of what ultimately became my thesis problem.

That dissertation dealt with the stability of the shape of a moving initially planar interface between the liquid and solid phases during the unidirectional freezing of a dilute binary alloy, which is a two-component metallic mixture, one component of which is of much lower concentration than the other such as a tin-lead alloy with 0.01% by weight of its lead impurity. Lee Segel and I considered a controlled *solidification* situation in the absence of liquid-phase convection which experimentally might be accomplished by use of the thermal valve technique, represented schematically in Fig 2.1.

In this technique of constrained crystal growth, a metal rod with a cooling coil soldered to one end, is positioned between two heating coils. These heating coils are raised to a

DOI: 10.1201/9781003195603-2

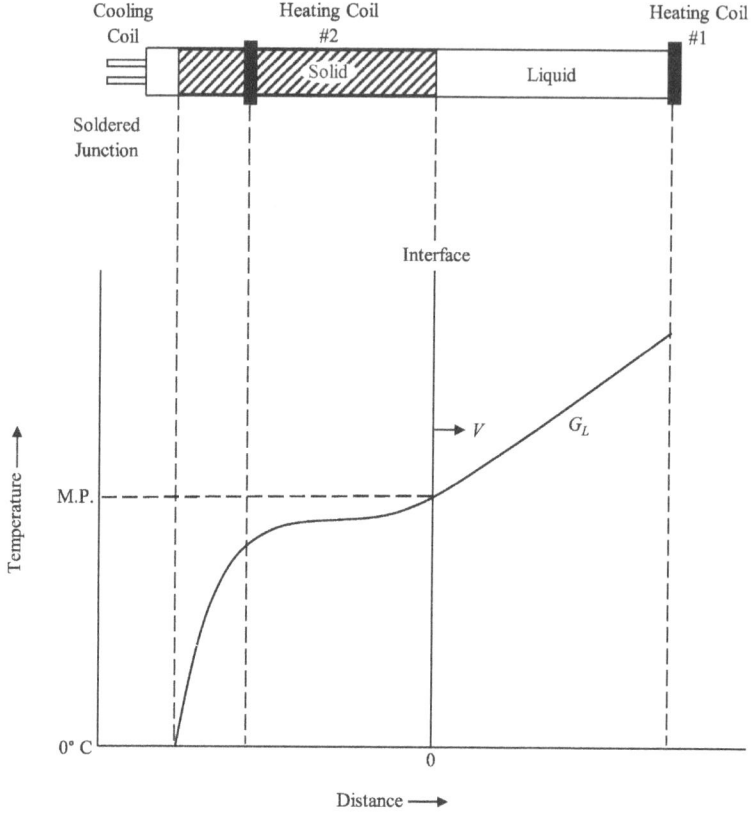

FIGURE 2.1

A schematic representation of the controlled alloy solidification situation and the relevant temperature distribution under consideration. Once a particular alloy of a fixed solute concentration C_P in the liquid at a flat interface is selected, this experimental situation can be completely prescribed by the two parameters V and G_L, its constant specified solidification speed and imposed liquid temperature gradient at the interface, respectively. Here M.P. $\equiv$ melting point.

temperature high enough to melt the metal between them and cause a temperature distribution as depicted in Fig. 2.1 with gradients G_S and G_L in the solid and liquid phases, respectively. The interface between the phases is then moved forward by lowering the temperature of the two heating coils at the appropriate rates calculated, so that its mean position will be advanced into the *liquid* phase with a constant specified speed V while minimizing liquid-phase convection. The imposed temperature gradients and specified speed of solidification are related by the conservation of heat condition at a flat interface $\kappa_S G_S - \kappa_L G_L = \mathcal{L}V$ where $\kappa_S(\kappa_L) \equiv$ thermal diffusivity in the solid (liquid) phase and $\mathcal{L} \equiv$ latent heat of fusion. As solidification conditions intensify the initially planar solid-liquid interface can become unstable and form a variety of steady-state periodic spatial patterns consisting of parallel bands and hexagonal arrays of either nodes (circular depressions) or elevated cells (see Fig. 2.2).

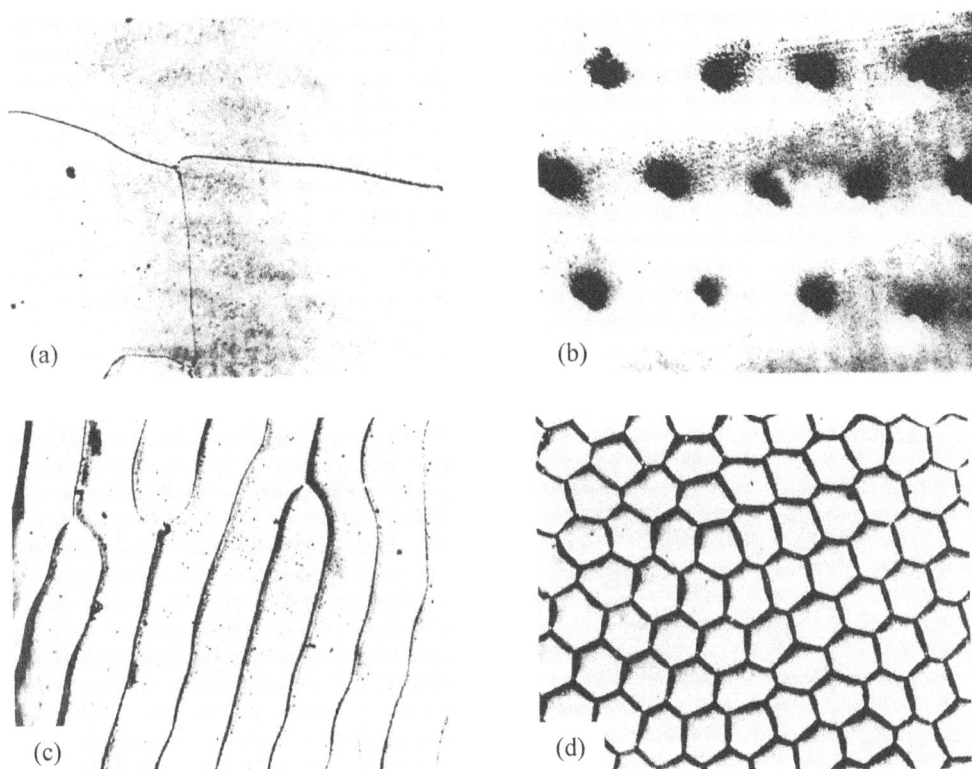

FIGURE 2.2
Interfacial patterns occurring during lead-tin dilute binary alloy solidification. Here photograph (a) is of a planar interface; (b), of nodes; (c), of bands; and (d), of cells; which develop sequentially as solidification conditions intensify.

The main purpose in the solidification portions of this chapter is to explain how stability methods were employed to examine the morphology of the solid-liquid interface, first by using a two-dimensional longitudinal-planform (my thesis problem) and next, a three-dimensional hexagonal-planform (my Ph.D. student Rukmini Sriranganathan's thesis problem) analysis. Also the companion situation of alloy *melting* where the interface is constrained to advance into the *solid* phase (my first Ph.D. student Sassan Raissi's thesis problem) will be considered. The emphasis here is on the methodologies employed and the subsequent results obtained in the spirit of Pulling Rabbits Out of Hats.

To develop a feel for the periodicity of these interfacial patterns it is helpful to observe the periodic behavior of the circular functions presented pictorially in Fig. 2.3. Here $\cos(\theta)$ and $\sin(\theta)$ are defined as the x- and y- coordinates of a point on the unit circle with central angle θ measured in the counter-clockwise direction from the x-axis. These periodic functions are plotted versus x in that figure as well. Further $\tan(\theta)$ is defined as their ratio $\sin(\theta)/\cos(\theta)$.

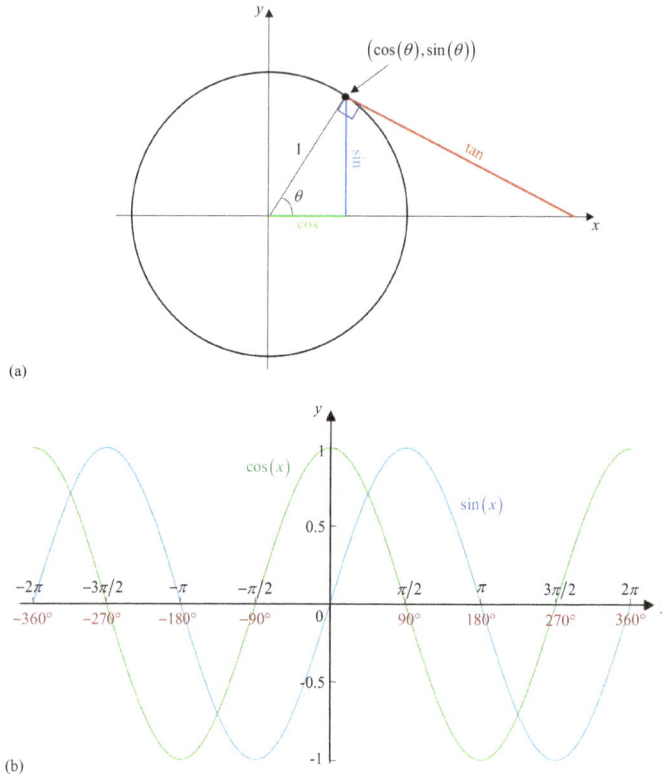

(a)

(b)

FIGURE 2.3
(a) Diagrammatic representation of the sine, cosine, and tangent as functions of the central angle θ in a unit circle. (b) Plots of the cosine and sine as functions of x in both radian (black) and degree (red) measure where each has period 2π in radians or $360°$ in degrees.

Returning to the solidification problem diagrammed schematically in the x-z plane of Fig. 2.4, it can be seen that the interface satisfies the equation

$$z = Vt + \zeta(x, t)$$

where Vt is the mean position of the planar interface and $\zeta(x,t)$, its deviation from that position; while $z > Vt + \zeta(x,t)$ represents the liquid phase and $z < Vt + \zeta(x,t)$, the solid one.

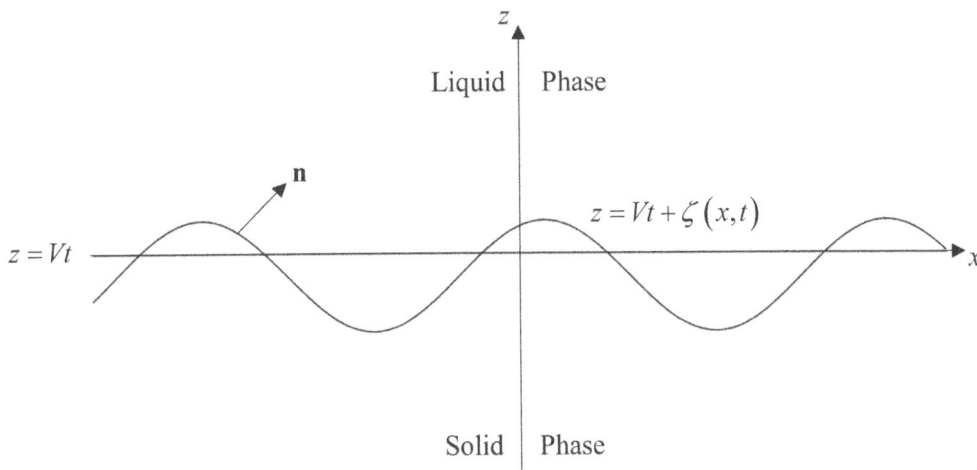

FIGURE 2.4
Schematic diagram of the two-dimensional alloy solidification coordinate system.

Lee and I modeled this solidification situation by posing balance-type conservation laws for the impurity solute concentration and heat in the solid and liquid phases. In particular, we assumed that the diffusion of solute D_L in the liquid phase satisfied $0 < D_L \ll \kappa_{L,S}$ and its diffusion D_S in the solid phase was negligible or $D_S = 0$. These assumptions yielded a diffusion equation depending on both temporal and spatial partial derivatives for the solute concentration C_L in the liquid phase

$$\frac{\partial C_L}{\partial t} = D_L \nabla^2 C_L \text{ where } \nabla^2 \equiv \frac{\partial^2}{\partial x^2} + \frac{\partial^2}{\partial z^2}$$

but a Laplace's equation depending only on such spatial derivatives for the temperature T_L or T_S

$$\nabla^2 T_L = 0, \ \nabla^2 T_S = 0,$$

in either the liquid or solid phases.

To understand the concept of derivative as introduced above consider Fig. 2.5a which is a plot of the function $y = f(x)$. Let $P(a, f(a))$ be a reference point and $Q(x, f(x))$, a nearby point with $x \neq a$. Compute the slope of the line $\overline{PQ}$ joining P to Q:

$$m_{\overline{PQ}} = \frac{\Delta y}{\Delta x} = \frac{f(x) - f(a)}{x - a}.$$

Let Q approach P along the curve C with equation $y = f(x)$ by letting x approach a or $x \to a$. If in this limit $m_{\overline{PQ}}$ approaches a number m or $m_{\overline{PQ}} \to m$, then the tangent line τ to the curve C at P has slope $m = \tan(\theta)$ where θ is the angle of inclination that line makes with the positive x-axis (see Fig. 2.5b). Under this condition we define the derivative of $f(x)$ at $x = a$ or $f'(a)$ by

$$f'(a) = \lim_{x \to a} \frac{f(x) - f(a)}{x - a},$$

if that limit exists, which geometrically represents the slope of the tangent line τ to C at

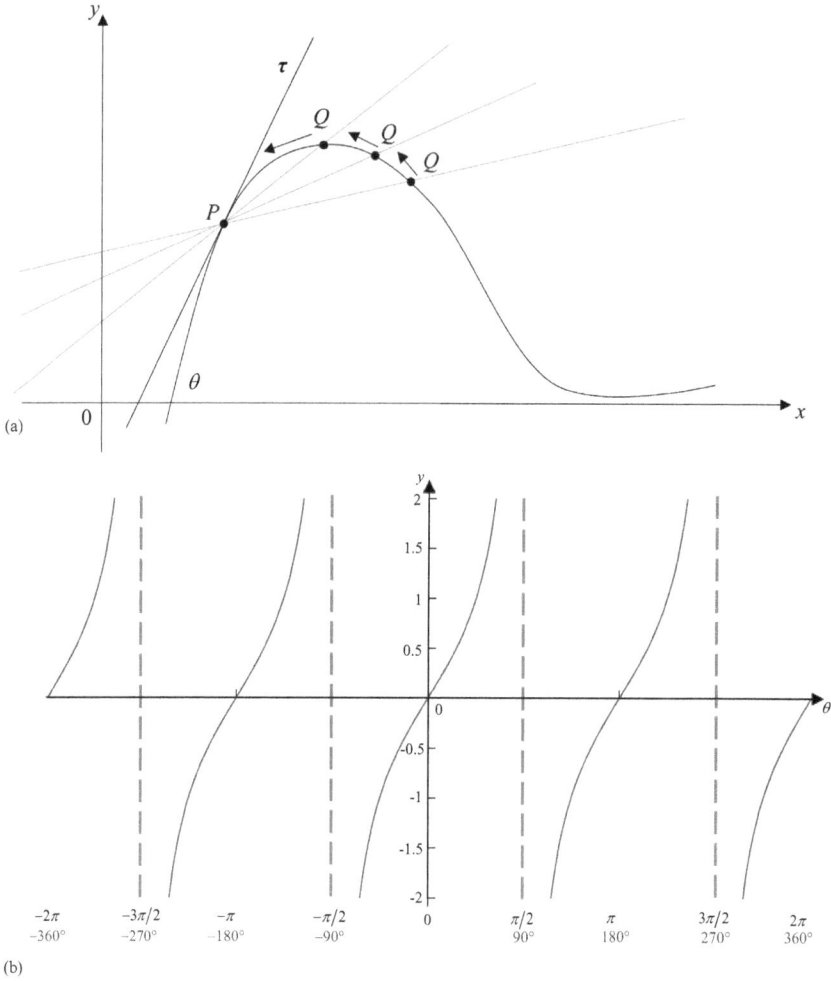

FIGURE 2.5
(a) Diagrammatic representation of the derivative $f'(x) = \tan(\theta)$ where θ is the angle of inclination the tangent line $\boldsymbol{\tau}$ to $y = f(x)$ at the point P makes with the positive x-axis. (b) Plot of the tangent as a function of θ in radian (black) and degree (red) measure which has a period of π in radians or $180°$ in degrees.

P. The derivative of $f(x)$ can be defined for any point x by the equivalent expression

$$\frac{df}{dx}(x) \equiv f'(x) = \lim_{h \to 0} \frac{f(x+h) - f(x)}{h}.$$

using the so-called "double d" notation due to Leibniz. The *chain rule of differentiation* for a composite function $f[g(x)]$ employing this notation may be written in the form

$$\frac{d}{dx} f[g(x)] = \frac{df}{dg}[g(x)]\frac{dg}{dx}(x) = f'[g(x)]g'(x).$$

This concept can be generalized to the derivative of a point on the surface $z = f(x,y)$ in the direction of a unit vector $\boldsymbol{b} = (b_1, b_2)$ of length one, -*i.e.*, $b_1^2 + b_2^2 = 1$, by

$$\frac{\partial f}{\partial b}(x,y) = \lim_{h \to 0} \frac{f(x+hb_1, y+hb_2) - f(x,y)}{h}.$$

Defining $F(h) = f(x+hb_1, y+hb_2)$, this *directional derivative* can be represented as

$$\frac{\partial f}{\partial b}(x,y) = \lim_{h \to 0} \frac{F(h) - F(0)}{h} = F'(0) = \frac{\partial f}{\partial x}(x,y)b_1 + \frac{\partial f}{\partial y}(x,y)b_2,$$

by extending the chain-rule of differentiation to *partial derivatives*. Note that for the special cases of $\boldsymbol{b} = \boldsymbol{i} = (1,0)$ or $\boldsymbol{b} = \boldsymbol{j} = (0,1)$ lying along the coordinate axes in the x- or y-directions, the directional derivative reduces to the standard partial derivatives

$$\frac{\partial f}{\partial i}(x,y) = \frac{\partial f}{\partial x}(x,y) \text{ or } \frac{\partial f}{\partial j}(x,y) = \frac{\partial f}{\partial y}(x,y),$$

respectively. In other words, the partial derivatives of f with respect to x or y are just special cases of the directional derivative while the higher derivative $\partial^2 f/\partial x^2 \equiv (\partial/\partial x)(\partial f/\partial x)$.

Some phenomena when modeled by the various techniques of continuum mechanics give rise to inherent surfaces at which the dependent variables exhibit discontinuities. These are usually referred to as surfaces of discontinuity and careful application of the postulated conservation or balance laws to them produce jump-type boundary conditions satisfied by the dependent variables across those surfaces. The solid-liquid interface during the freezing of the dilute binary alloys under investigation represents such a surface of discontinuity, since among other things the solute concentration in the solid satisfies $C_S = kC_L$ at that interface where the solute segregation coefficient $0 < k < 1$, although the temperature is assumed to be continuous there or $T_S = T_L$. The jump-type boundary conditions for solute and heat at the solidification interface are given by

$$D_L \frac{\partial C_L}{\partial n} = (k-1)C_L w_n \text{ and } \kappa_S \frac{\partial T_S}{\partial n} - \kappa_L \frac{\partial T_L}{\partial n} = \mathcal{L} w_n,$$

where $\boldsymbol{n} \equiv$ interfacial unit normal pointing into the liquid (see Fig. 2.4) and hence the directional derivative $\partial/\partial n = (\partial/\partial z - \zeta_x \partial/\partial x)/\sqrt{1+\zeta_x^2}$ while $w_n \equiv$ interfacial normal speed. Here $\zeta_x \equiv \partial\zeta/\partial x$. Further it is necessary to impose the Gibbs-Thomson equation at the interface

$$T_L = T_M[C_L] + \Delta T$$

which describes the alteration, ΔT, of the interfacial temperature from that of the equilibrium melting temperature $T_M[C_L]$ at a planar interface as a function of C_L due to the curvature η of the interface itself.

For a dilute alloy using a linear approximation

$$T_M[C_L] \cong T_M[0] + \frac{dT_M}{dC_L}[0]C_L = T_M + m_0 C_L$$

and taking $\Delta T = T_M \Gamma \eta$ with $\Gamma \equiv$ interfacial capillarity, the Gibbs-Thomson equation becomes

$$T_L = T_M + m_0 C_L + T_M \Gamma \eta \text{ where } \eta = \frac{\zeta_{xx}}{(1 + \zeta_x^2)^{3/2}}.$$

Given that heat diffuses so much faster than solute it has now become standard operating procedure to neglect the time derivatives in the heat diffusion equations for temperature and hence replace them by Laplace's equations as indicated earlier. Lee and I employed a systematic scaling argument to justify this reduction procedure and in so doing Pulled my first Rabbit Out of a Hat. Using the scale factors of T_M, V, and D_L/V for temperature, interfacial normal speed, and length, respectively, the boundary condition for heat was converted into the dimensionless form

$$\kappa \frac{\partial T_S}{\partial n} - \frac{\partial T_L}{\partial n} = \varepsilon \ell w_n$$

where $\kappa = \kappa_S/\kappa_L$, $\varepsilon = D_L/\kappa_L$, and $\ell = \mathcal{L}/T_M$. Here for ease of exposition the same notation is being employed to denote these nondimensional variables as was used to denote the original ones. Since ε has the typically very small value of $10^{-6} = 0.000001$ while $\kappa \cong 2.0$ and $\ell \cong 0.4$, we replaced the original boundary condition by

$$\kappa_S G_S \cong \kappa_L G_L.$$

Such a replacement dramatically simplified the required stability analyses examining the long-time behavior of the *planar interface solution* of this model system (see below) since

$$w_n = \frac{V + \zeta_t}{\sqrt{1 + \zeta_x^2}}$$

contains a time derivative that is then also eliminated from the system and this approximation did not alter the fundamental results of these analyses.

To complete the formulation of that system, far field boundary conditions must be imposed which depend on the distance $z_{new} = z_{old} - Vt$ from the mean position of the interface, with $C_L \equiv C_P$ at $z_{new} = 0$. Introducing this Galilean transformation, these far field conditions satisfy

$$C_L \to C_0(z), \ T_L \to T_0(z) \text{ as } z \to \infty; \ T_S \to \Theta_0(z) \text{ as } z \to -\infty;$$

where $C_0(z), T_0(z)$, and $\Theta_0(z)$ represent the planar interface solution for $\zeta(x,t) \equiv 0$ depicted in Fig. 2.6 and the simplification that the phases are infinite in thickness has been implicitly adopted. Further that Galilean transformation converted the diffusion equation for C_L into

$$\frac{\partial C_L}{\partial t} - V \frac{\partial C_L}{\partial z} = D_L \nabla^2 C_L \text{ for } z > \zeta(x,t)$$

by employing the chain rule for partial differentiation with respect to time.

Lee and I examined the stability of the planar interface solution to a particular class of perturbations using a two-dimensional longitudinal-planform analysis by assuming that, to third order terms in an *amplitude function* $A(\tau)$, the interface satisfied the equation

$$z = \zeta(x,t) = A(\tau)\cos(qx) + A^2(\tau)\zeta_2(x) + A^3(\tau)\zeta_3(x) + O(A^4)$$

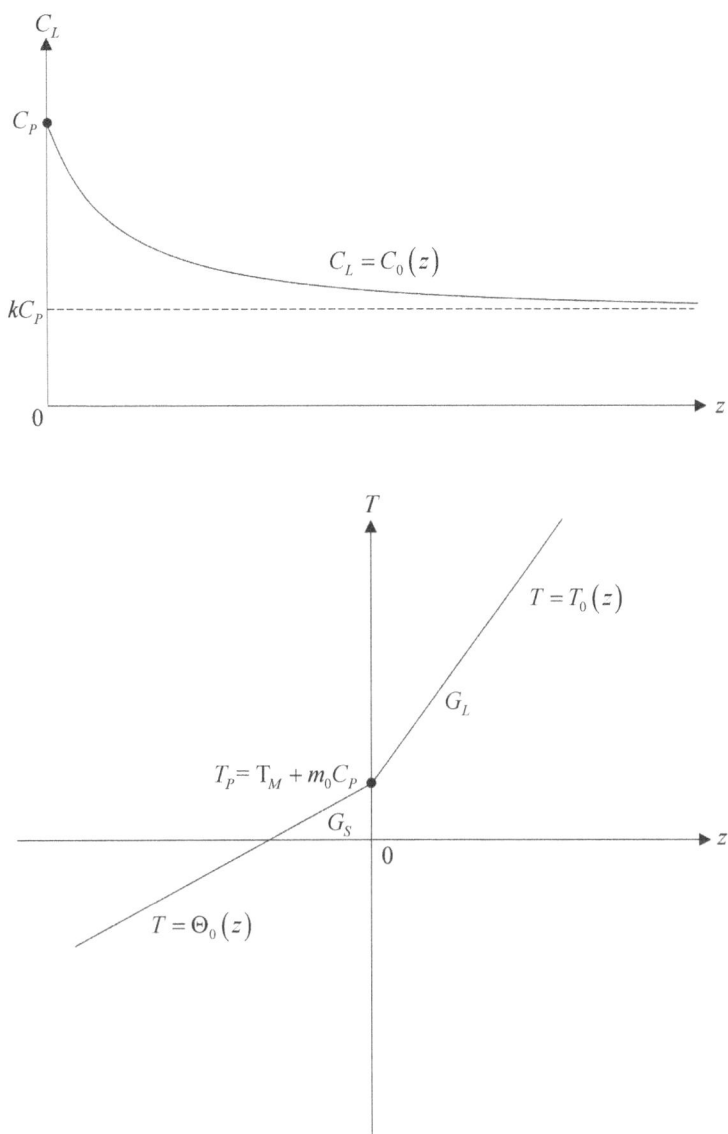

FIGURE 2.6
The planar interface steady-state solution corresponding to $\zeta(x,t) \equiv 0$. Here G_L and G_S represent the slopes or gradients of the straight line temperature distributions in the liquid and solid phases, respectively, with common intercept $T_P = T_M + m_0 C_P$ while the solute concentration component in the liquid has intercept C_P and horizontal asymptote kC_P where the solute segregation coefficient $0 < k < 1$ for the alloys under investigation.

where $O(A^4)$ denotes terms of order A^4 which are being neglected. Here, we introduced an interfacial z-scale such that $D_L/V \equiv 1$ unit of deviation, $\tau = V^2 t/D_L \equiv$ nondimensional time, and $q = 2\pi/\lambda$, λ being the dimensional wavelength of the periodic disturbance.

The forms of $\zeta_{2,3}(x)$ were motivated as follows: To lowest order the interface satisfies the approximation

$$z \sim \zeta_1(x,\tau) = A(\tau)\cos(qx).$$

Upon comparing the forms of

$$\zeta_1^2(x,\tau) = A^2(\tau)\cos^2(qx) = A^2(\tau)\frac{1+\cos(2qx)}{2},$$

$$\zeta_1^3(x,\tau) = A^3(\tau)\cos^3(qx) = A^3(\tau)\frac{3\cos(qx)+\cos(3qx)}{4};$$

where relevant trigonometric identities have been employed to derive the above relationships, with the $O(A^2)$ and $O(A^3)$ terms in the interfacial equation, respectively, we deduced that the proper choices for these quantities should be

$$\zeta_2(x) = \zeta_{20} + \zeta_{22}\cos(2qx) \text{ and } \zeta_3(x) = \zeta_{31}\cos(qx) + \zeta_{33}\cos(3qx).$$

Here the ζ_{nm} are constants, such that each is a coefficient of a term in the interfacial equation of the form $A^n(\tau)\cos(mqx)$ and $\zeta_{11} = 1$ while $\zeta_{00} = 0$.

We next examined that interfacial equation for terms proportional to $\cos(qx)$ and found them to contain the amplitude functions $A(\tau)$ and $A^3(\tau)$. Hence, we concluded that $A(\tau)$ satisfied the following amplitude equation to third-order

$$\frac{dA}{d\tau} = \sigma A - a_1 A^3 + O(A^5)$$

where σ, the growth rate of linear stability theory, and a_1, the Landau constant, are to be determined. We also assumed analogous expansions for $C_L(x,z,t)$, $T_L(x,z,t)$, and $T_S(x,z,t)$ consistent with that for $\zeta(x,t)$ where now the coefficient of each term corresponding to a ζ_{nm} of the latter was considered to be a function of the z-variable denoted by $C_{nm}(z)$, $T_{nm}(z)$, and $\Theta_{nm}(z)$, respectively, with $n = m = 0$ being the planar interface solution in that variable.

Finally, one says a function $f(A)$ is expandable in a Taylor series about $A = 0$ provided

$$f(A) = f(0) + f'(0)A + f''(0)\frac{A^2}{2!} + f'''(0)\frac{A^3}{3!} + O(A^4),$$

where the higher derivatives are defined by $f''(A) \equiv d^2 f(A)/dA^2$ and $f'''(A) \equiv d^3 f(A)/dA^3$ and the factorials, by $2! \equiv 2 \cdot 1 = 2$ and $3! \equiv 3 \cdot 2 \cdot 1 = 6$.

Then, after substituting our two-dimensional longitudinal-planform solution into the governing system of equations and expanding the boundary conditions in such a Taylor series about $A = 0$ while employing the relevant identities in the resulting products of its trigonometric functions, we obtained a set of systems of ordinary differential equations for $C_{nm}(z)$, $T_{nm}(z)$, and $\Theta_{nm}(z)$, with boundary conditions evaluated at $z = 0$ and as $|z| \to \infty$, one for each pair of n and m values corresponding to a term of the form $A^n(\tau)\cos(mqx)$ appearing explicitly in that solution.

In particular, the $n = m = 1$ system is equivalent to the linear stability problem from which it was possible to deduce that σ satisfied the semi-quadratic secular equation or dispersion relation

$$\sigma + a = 2b\sqrt{\sigma + c} \text{ such that } \text{Re}(\sqrt{\sigma + c}) \geq \alpha \geq 0$$

for

$$2a = 1 + (2k - 1)(Q + \beta\omega^2), \ 2b = 1 - (Q + \beta\omega^2), \ c = \omega^2 + \frac{1}{4}, \ \alpha = 0,$$

where the nondimensional *wavenumber*, speed, and temperature gradient were defined by

$$\omega = q\frac{D_L}{V}, \ \beta = \frac{VT_P\Gamma}{D_LC_Pm_0(k-1)} \text{ with } T_P = T_M + m_0C_P,$$

$$Q = \frac{2G_LD_L}{VC_P(1 + \kappa)m_0(k-1)}.$$

Here, given a complex quantity $w = u + iv$ with u and v being real numbers and $i = \sqrt{-1}$, we are employing the notation $\text{Re}(w) = u$ and $\text{Im}(w) = v$. Let $\sigma = \sigma_0$ denote that root of our semi-quadratic with the largest real part. If one considers $\sigma_0 = \text{Re}(\sigma_0) + i\text{Im}(\sigma_0)$ and examines the locus $\text{Re}(\sigma_0) = 0$ in our semi-quadratic this implies $\text{Im}(\sigma_0) = 0$ as well and hence that locus can be characterized by $\sigma_0 = 0$. Since Segel had never encountered a problem for which such a so-called exchange of stabilities was valid and σ_0 was not real, we implicitly assumed in our nonlinear analysis that it was real everywhere over the relevant parameter space. Then, because our amplitude equation to this order yields a solution of the form $A(\tau) = A(0)e^{\sigma_0\tau}$, where the natural exponential e^x satisfies $(e^x)' = e^x$ with $e^0 = 1$ which implies

$$e \equiv \lim_{\delta \to 0}(1 + \delta)^{1/\delta} = 2.718281828459\cdots \equiv \text{Euler's number},$$

and

$$e^{\sigma_0\tau} \to \begin{cases} 0 & \sigma_0 < 0 \\ 1 \quad \text{for} & \sigma_0 = 0 \quad \text{as } \tau \to \infty, \\ \infty & \sigma_0 > 0 \end{cases}$$

we say that our planar interface solution is stable, neutrally stable, or unstable to initially infinitesimal disturbances depending upon whether σ_0 is less than, equal to, or greater than zero, respectively. Note that $y = a^x \Leftrightarrow x = \log_a(y)$ while $\log_e(y) \equiv \ln(y)$ and $\log_{10}(y) \equiv \log(y)$.

We shall ultimately be interested in plotting various σ_0-loci in the V-G_L plane for a particular alloy of fixed concentration C_P and hence proceed to solve our semi-quadratic secular equation for Q, obtaining

$$Q_{\sigma_0}(\omega^2) = 1 - \beta\omega^2 - \frac{\sigma_0 + k}{k - 1/2 + \sqrt{\sigma_0 + \omega^2 + 1/4}}.$$

This curve has a maximum at $\omega^2 = \omega_{\sigma_0}^2$ such that

$$\frac{dQ_{\sigma_0}}{d\omega^2}(\omega_{\sigma_0}^2) = 0 \ \Rightarrow \ \beta = \frac{\sigma_0 + k}{2\sqrt{\sigma_0 + \omega_{\sigma_0}^2 + 1/4}\left[k - 1/2 + \sqrt{\sigma_0 + \omega_{\sigma_0}^2 + 1/4}\right]^2}$$

and this maximum is given by $Q_{\sigma_0} = Q_{\sigma_0}(\omega_{\sigma_0}^2)$.

For $\sigma_0 = 0$, Lee and I represented this curve by $Q_0(\omega^2)$ and its maximum point by (ω_c^2, Q_c) where the critical wavenumber squared satisfied

$$\beta = \frac{4k}{\sqrt{4\omega_c^2 + 1}[2k - 1 + \sqrt{4\omega_c^2 + 1}]^2} \text{ if } 0 \leq \beta \leq \frac{1}{k}$$

or

$$\omega_c^2 = 0 \text{ if } \beta > \frac{1}{k}$$

and

$$Q_c = Q_0(\omega_c^2) = 1 - \beta\omega_c^2 - \frac{2k}{2k - 1 + \sqrt{4\omega_c^2 + 1}}.$$

That curve $Q = Q_0(\omega^2)$ on which $\sigma_0 = 0$ is plotted in Fig. 2.7a. Observe from this figure that it is a marginal stability curve separating the unstable region where $\sigma_0 > 0$ from the stable region where $\sigma_0 < 0$. Note from Figs. 2.7a,b that for $Q > Q_c$ there exists no ω^2 such that $\sigma_0 > 0$, while for $Q < Q_c$ there exists an interval of such ω^2 centered about ω_c^2. Further, as is standard operating procedure for nonlinear stability analyses, we identified the q contained in our two-dimensional longitudinal-planform expansions with this critical wavenumber of linear stability theory and hence took $q \equiv q_c = \omega_c V / D_L$. Then, since $\omega_c^2 = \omega_c^2(V)$ and $Q_c = Q_c(V)$ for a particular alloy of fixed impurity concentration C_P, the curve corresponding to $\sigma_0 = 0$ in V-G_L space given by

$$G_L = G_c(V) = \frac{C_P m_0(k-1)(1+\kappa)}{2D_L} V Q_c(V)$$

is marginal in the sense that $\sigma_0 > 0$ if and only if $G_c(V) < G_L$ (see Fig. 2.8). With σ_0 so determined it was possible to solve for nonzero $C_{11}(z)$, $T_{11}(z)$, and $\Theta_{11}(z)$.

There were two second-order systems corresponding to $n = 2$ and $m = 0$ or 2 which could be solved in a straight-forward manner to yield ζ_{2m}, $C_{2m}(z)$, $T_{2m}(z)$, and $\Theta_{2m}(z)$; $m = 0$ and 2. As an aside, to satisfy the far field boundary conditions most easily, we had wanted

$$C_{nm}(z), \ T_{nm}(z) \to 0 \text{ as } z \to \infty \text{ and } \Theta_{nm}(z) \to 0 \text{ as } z \to -\infty,$$

which, in particular, required the alteration of the mean temperature to vanish identically or

$$T_{20}(z) = \Theta_{20}(z) \equiv 0.$$

In extending the original linear stability analysis for this problem by Mullins and Sekerka ([152]), we had at first mistakenly retained their partial derivatives with respect to z in the interfacial boundary conditions instead of replacing them by the normal directional derivatives, which is the proper procedure when nonlinearities are being considered. With such a formulation, this requirement proved impossible to satisfy. Lee realized that error during the Christmas recess of the first year we had been working on the problem. After it was corrected the whole thing "fell into place like a champ," as my Ph.D. student David Oulton would say, and I was able to take the alteration of the mean temperature equal to zero identically. Segel then said that once you had its proper formulation, everything else for a problem would follow accordingly and this truism has guided much of my subsequent modeling research especially when Pulling Rabbits Out of Hats.

Although there were also two third-order systems, only the one corresponding to $n = 3$ and $m = 1$, which contained the Landau constant a_1, was of interest to us. The determination of this constant can be reduced to an examination of an equation of the form

$$a_1 - 2\sigma_0\zeta_{31}(\sigma_0) = r_{31}(\sigma_0)$$

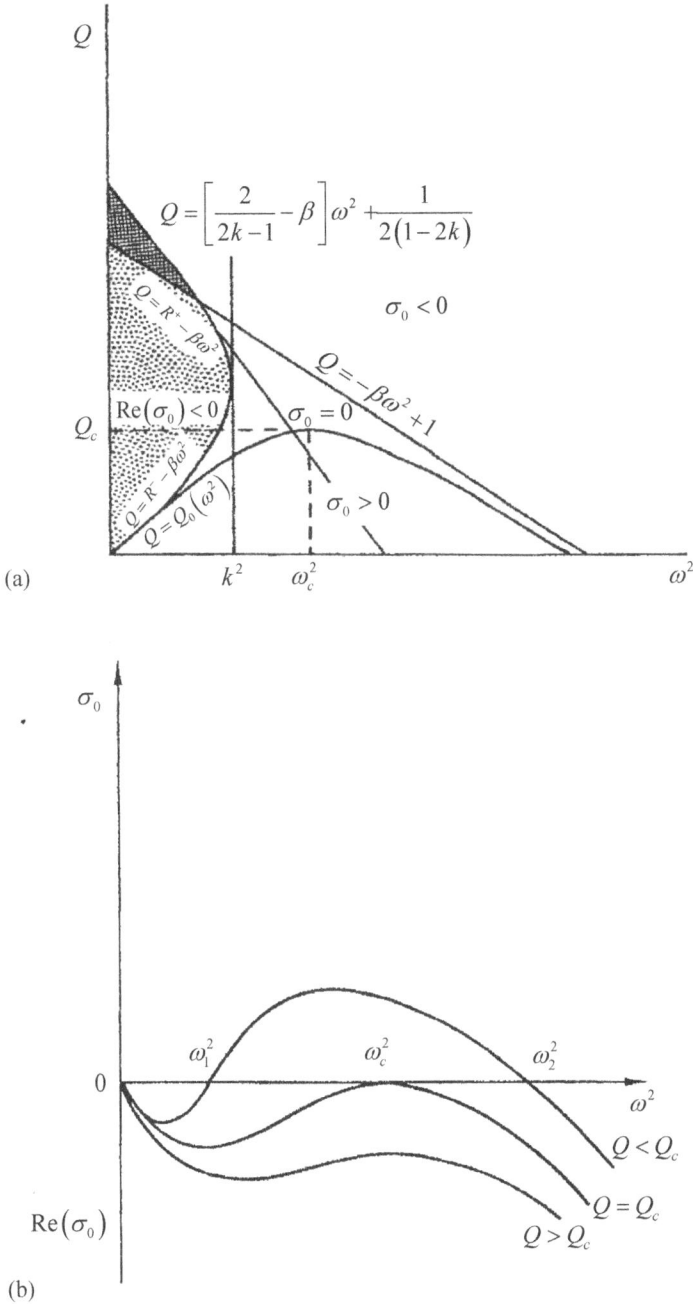

FIGURE 2.7

(a) Plot of the marginal stability curve $Q = Q_0(\omega^2)$ on which $\sigma_0 = 0$ separating the unstable region where $\sigma_0 > 0$ from the stable region that also includes a subregion of complex σ_0 designated by stippling. If $k > 1/4$ there also exists a part of that subregion designated by shading corresponding to identical stability where solutions to the semi-quadratic are extraneous. This figure has been drawn for $1/4 < k < 1/2$ and $\beta < 1/k$. (b) Plots of $\mathrm{Re}(\sigma_0)$ versus ω^2 for various values of Q associated with part (a). Note that $\mathrm{Re}(\sigma_0) = \sigma_0$ where $\mathrm{Re}(\sigma_0) \geq 0$.

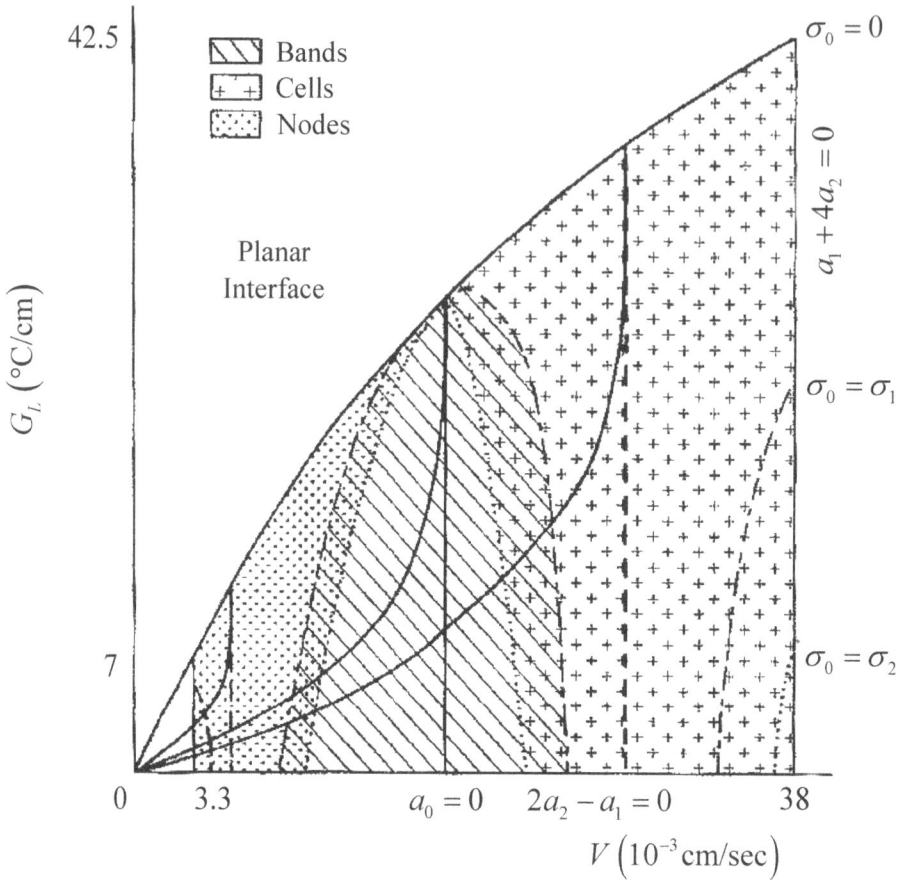

FIGURE 2.8

Plots in the V-G_L plane of the marginal stability curves $G_L = G_{\sigma_0}(V)$ for $\sigma_0 = 0$, σ_1, σ_2; denoting the predicted interfacial morphologies for the parameter values of Table 2.2. Recall, in the two-dimensional analysis, $G_{\sigma_0=0}(V)$ was represented by $G_c(V)$. Note that $a_0 = 0$ for $V = V_c$; $2a_2 - a_1 = 0$ for $V = V_2, V_3$; and $a_1 + 4a_2 = 0$ for $V = V_1, V_4$; with $a_1 > 0$ for $V_1 \leq V \leq V_4$. The hexagonal-planform stability method while incorporating the nonlinearities of the solidification model system basically pivots a perturbation procedure about the critical point of linear stability theory. The advantage of such an approach over strictly numerical procedures is that it allows one to deduce quantitative relationships between system parameters and stable patterns which are valuable for comparison with experimental evidence and difficult to accomplish using simulation alone. Here, the curves starting at the origin and asymptotically approaching the vertical lines $V = V_2, V_c, V_3$ are schematic representations into what those lines might be transformed should the Landau coefficients be considered functions of σ_0 instead of constants.

where $r_{31}(\sigma_0)$ is a known function of σ_0 depending on the planar interface solution and the solutions to the linear and second-order systems evaluated at $z = 0$. Here a_1 is independent of σ_0 and hence called the Landau constant, after the great Soviet physicist Lev Landau ([106]), who first proposed an equivalent form of our amplitude equation in his work on the transition from laminar to turbulent flow. Taking the limit of this equation as $G_L \to G_c(V)$, noting from Fig. 2.8

$$\lim_{G_L \to G_c(V)} \sigma_0 = 0,$$

and assuming $\zeta_{31}(\sigma_0)$ is well behaved as $\sigma_0 \to 0$, we found the so-called solvability condition

$$a_1 = r_{31}(0).$$

In order to evaluate this quantity for a particular alloy of fixed concentration C_P, we next assigned the following typical values to its relevant parameters implicitly employed in Fig. 2.8

$$C_P = 0.01\% \text{ by wt. of solute, } m_0 = -2°C/\% \text{ by wt. of solute, } T_P = 500K,$$

$$\kappa = 2, \ \Gamma = 10^{-8} \text{cm}, \ D_L = 10^{-5} \text{cm}^2/\text{sec}.$$

Under these conditions a_1 was a function of w_c^2 and k. Hence, given the previous discussion, it could ultimately be thought of as a function of V and k.

I began our examination of this Landau constant by writing a computer program to evaluate a_1 for different w_c and for a number of k-values from 0 to 1. In particular, this program was designed to compute its values as w_c ranged from 10 to 0.1 by tenths and from 0.1 to 0.01 by hundredths and as k ranged from 0.1 to 0.9 by tenths or 981 total values in all. Since numerical computation by means of digital computing was in its infancy and this represented my first attempt at using it, Lee had me calculate a_1 directly by hand for the particular values of $w_c = 1.48$, $k = 0.2$ and then check that answer with the one obtained from the computer program restricted to those values before running the whole program. I did the hand calculation arriving at an answer of $a_1 = -0.739$ as compared to the result of $a_1 = -0.737$ yielded by my computer program. Segel, whose seminal work on nonlinear stability theory ([219]) analyzed the Rayleigh-Bénard problem of buoyancy-driven natural convection in a fluid layer heated uniformly from below, for which the Landau constant is identically positive (see the next chapter), was unpleasantly surprised by this negative value and originally doubted its validity.

The main fruits of linear stability theory applied to an equilibrium state of a system are the prediction of the critical conditions for the onset of instability to initially infinitesimal disturbances and the wavelength characteristic of those growing disturbances as demonstrated above. Its principal deficiency is that it cannot determine the long-time behavior of such growing disturbances and hence does not predict the stationary pattern formational aspects of this system. In order to ascertain that behavior which is the primary goal of our longitudinal-planform analysis it was necessary to take the nonlinear terms into account. When the Landau constant is positive the system undergoes a re-equilibration that results in a stationary banded pattern in the region of parameter space where the linear analysis predicted instability, but when that constant is negative no such saturation can occur at third-order (see Fig. 2.9).

What concerned Segel the most in this instance was the possibility that our Landau constant might be identically negative for all values of w_c and k which would require the expansions to be extended to so-called quintic (fifth)-order and thus could entail an enormous amount

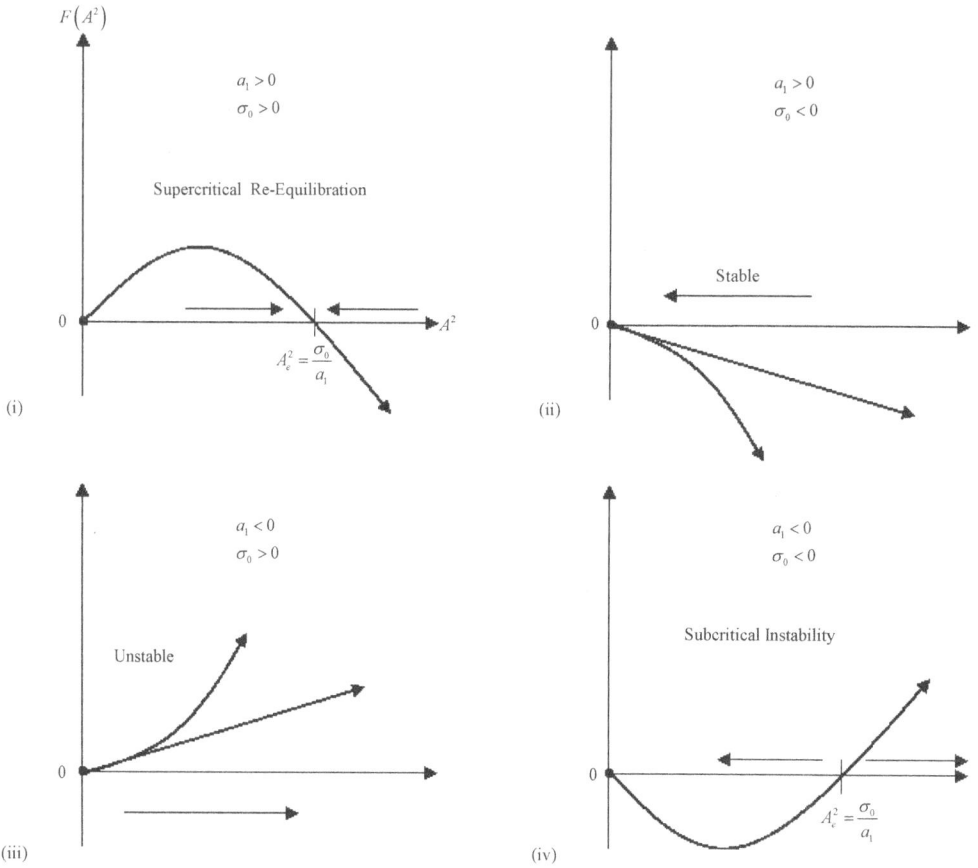

FIGURE 2.9

Plots of the qualitative behavior of the truncated Landau equation in the four cases representing the possible choices for the signs of σ_0 and a_1. Note in this context that a function increases if its derivative is positive and decreases if it is negative. See Chapter 15 for a further discussion of the behavior of this Landau equation when $a_1 < 0$.

of very tedious extra calculations (see Chapter 15). Then I ran the complete program for $k = 0.2$ and found that although, unlike the Rayleigh-Bénard problem, a_1 was indeed negative when $\omega_c > \omega_c^* = 0.55$ or $V < V^* = 21.9 \times 10^{-3}$cm/sec, consistent with our computation for $\omega_c = 1.48$, and zero at $\omega_c = \omega_c^*$ or $V = V^*$, it became positive, to the immense relief of both of us, when $\omega_c < \omega_c^*$ or $V > V^*$. The same qualitative behavior occurred for all the other values of $0 < k < 1$. Quantitatively, $\omega_c^* = 0.35$ or $V^* = 11.3 \times 10^{-3}$cm/sec for $k = 0.1$, since ω_c^* increased, or equivalently, V^* decreased with increasing k becoming $\omega_c^* = 1.55$ or $V^* = 2.7 \times 10^{-4}$cm/sec for $k = 0.9$. Thus, for $V > V^*$ where $a_1 = r_{31}(0) > 0$ and in the neighborhood of the marginal stability curve $G_L = G_c(V)$, the Landau amplitude equation could be truncated through terms of third-order to obtain

$$\frac{dA}{d\tau} \sim \sigma_0 A - a_1 A^3$$

or

$$A \frac{dA}{d\tau} = \frac{1}{2} \frac{d}{d\tau} (A^2) \sim \sigma_0 A^2 - a_1 A^4 = \sigma_0 A^2 \left[1 - \frac{A^2}{\sigma_0/a_1} \right] = F(A^2),$$

which has two equilibrium points satisfying $F(A^2) = 0$. Hence, we concluded from Fig. 2.9 that

(i) For $G_L < G_c(V)$ where $\sigma_0 > 0$, $A^2 = A_e^2 = \sigma_0/a_1$ was stable, yielding a periodic one-dimensional interfacial pattern consisting of stationary parallel bands

$$\zeta(x,t) \to \zeta_e(x) \sim A_e \cos\left(\frac{2\pi x}{\lambda_c}\right) \text{ as } t \to \infty$$

of characteristic wavelength $\lambda_c = (D_L/V)(2\pi/\omega_c)$.

(ii) For $G_L > G_c(V)$ where $\sigma_0 < 0$, the undisturbed state $A^2 = 0$ was stable, yielding a planar interface

$$\zeta(x,t) \to 0 \text{ as } t \to \infty.$$

This, in essence, completed my dissertation research and I was ready to defend my thesis. The thesis defense was scheduled for the morning of June 21, 1968. My committee consisted of Segel, DiPrima, Lester Rubenfeld who was a replacement for an original member unable to be present, and Lemuel Tarshis, a metallurgist from General Electric in Schenectady, who was the nondepartmental examiner. About an hour before the defense was to start Les came to me with a question about my thesis that he wished to have clarified. His question was how did we know that σ_0 was real since my methodology depended on this being true. Having gone to Segel with that question, only one look at his face told me all I needed to know. We did a quick back of the envelope calculation and showed that σ_0 was real at least for a particular choice of parameter values in the banded interface region. Its general behavior still being an open question I went through with the thesis defense. When it was over the committee decided that a decision on my thesis would have to be deferred until that behavior was decided conclusively and let Segel make it for them by proxy. I started work on the answer to this question immediately. It entailed an analysis of the semi-quadratic secular equation introduced earlier. Squaring both sides of that equation and using the quadratic formula (see Chapter 6), I found that σ_0 could be represented by

$$\sigma_0 = (b + \sqrt{\Delta})^2 - c \text{ with } \Delta = b^2 + c - a.$$

This would be complex only where $\Delta < 0$ or, employing the specific values of a, b, and c, where

$$R^- < Q + \beta\omega^2 < R^+ \text{ with } R^{\pm} = 2(k \pm \sqrt{k^2 - \omega^2}) \text{ for } \omega^2 < k^2.$$

The boundary curves of that region which have a vertical tangent at their point of intersection where $\omega^2 = k^2$ are plotted in the ω^2-Q plane of Fig. 2.7a and within this region $\mathrm{Re}(\sigma_0) < 0$, as indicated in that figure. Hence, one could guarantee that σ_0 would remain real simply by requiring that $\omega_c > k$ or equivalently $V < V_k$ which was satisfied identically for the low speeds under investigation. With this Rabbit having been Pulled Out of a Hat and appended to my dissertation, Segel added his signature to the thesis defense approval form and I had my Ph.D.

These seminal results were published in our paper D.J. Wollkind and L.A. Segel, A nonlinear stability analysis of the freezing of a dilute binary alloy, *Philosophical Transactions of the Royal Society of London Series A*, volume 268, pages 351-380, which appeared on December 24, 1970, while Oulton *et al.* ([161]) provided the definitive linear stability analysis of semi-quadratic secular equations. Ten years after Wollkind and Segel ([294]), my graduate student Rukmini Sriranganathan and I, in collaboration with David Oulton, then at Old Dominion University, under a WSU subcontract to his National Science Foundation (NSF) grant, refined those *two*-dimensional predictions so that they could be compared with all of the experimental morphologies depicted in Fig. 2.2 and not just bands. To do so we examined the stability of the planar interface solution of the appropriate *three*-dimensional version of our system to a particular class of perturbations using a hexagonal-planform analysis by assuming to lowest order that the interface now satisfied the equation

$$z = \zeta(x,y,t) \sim A_1(\tau)\cos[q_c x + \varphi_1(\tau)] + A_2(\tau)\cos\left[\frac{q_c(x - \sqrt{3}y)}{2} - \varphi_2(\tau)\right]$$

$$+ A^3(\tau)\cos\left[\frac{q_c(x + \sqrt{3}y)}{2} - \varphi_3(\tau)\right] = f(x,y,\tau)$$

where, for $(j,k,\ell) \equiv$ even permutation of $(1,2,3)$,

$$\frac{dA_j}{d\tau} \sim \sigma A_j - 4a_0 A_k A_\ell \cos(\varphi_j + \varphi_k + \varphi_\ell) - A_j[a_1 A_j^2 + 2a_2(A_k^2 + A_\ell^2)],$$

$$A_j\frac{d\varphi_j}{d\tau} \sim 4a_0 A_k A_\ell \sin(\varphi_j + \varphi_k + \varphi_\ell),$$

with analogous expansions for $C_L(x,y,z,t)$, $T_L(x,y,z,t)$, and $T_S(x,y,z,t)$ as described in the two-dimensional analysis. Here, by even permutations of $(1,2,3)$ we mean the three natural order permutations $(1,2,3)$, $(2,3,1)$, and $(3,1,2)$. Hence, there are three amplitude equations involving the derivatives of A_j for $j = 1,2,3$ and three phase equations involving the derivatives of φ_j for $j = 1,2,3$. This hexagonal-planform expansion, originally developed by Segel ([211]) to study Rayleigh-Bénard cellular convection, consisted, at lowest order, of three sets of modes the wavenumber vectors of which are $120°$ apart in the x-y plane and each of overall magnitude $q_c = V\omega_c/D_L$. Since we were interested in the long-time asymptotic behavior of these modes and performed our analysis in the neighborhood of the maximum point (ω_c, Q_c) of the marginal stability curve, all other modes decayed exponentially in time ([105]). Further by the appropriate three-dimensional version of our system we mean one in which the $\partial/\partial n$, ∇^2, w_n, η, and $\Gamma \equiv \gamma/\mathcal{L}$ of the two-dimensional system were replaced by

$$\frac{\partial}{\partial n} = \frac{\left(\frac{\partial}{\partial z} - \zeta_x \frac{\partial}{\partial x} - \zeta_y \frac{\partial}{\partial y}\right)}{(1 + \zeta_x^2 + \zeta_y^2)^{1/2}},$$

$$\nabla^2 = \frac{\partial^2}{\partial x^2} + \frac{\partial^2}{\partial y^2} + \frac{\partial^2}{\partial z^2}, \quad w_n = \frac{V + \zeta_t}{(1 + \zeta_x^2 + \zeta_y^2)^{1/2}},$$

$$\eta = \frac{\zeta_{xx}(1 + \zeta_y^2) - 2\zeta_{xy}\zeta_x\zeta_y + \zeta_{yy}(1 + \zeta_x^2)}{(1 + \zeta_x^2 + \zeta_y^2)^{3/2}},$$

and $\gamma = \gamma_P - s(T_L - T_P) + \mu(C_L - C_P)$ for $\gamma_P = \gamma_0 - sT_P + \mu C_P$, where $\gamma \equiv$ interfacial surface free energy, $\gamma_0 \equiv$ interfacial surface energy, $s \equiv$ interfacial surface entropy, and $\mu \equiv$ coefficient of solutal capillarity. Unlike the previous formulation where $\gamma \equiv \gamma_P$ or $\Gamma \equiv \Gamma_P = \gamma_P/\mathcal{L}$, we allowed for the possibility of the variation of it with T_L and C_L. In so doing, we implicitly assumed that the tangential components of momentum at the solid-liquid interface caused by this variation were balanced identically by the imposed elastic surface shear stress vector of the solid.

The nonlinear stability behavior of the amplitude-phase equations to be described below depended only on the values of their growth rate and Landau constants. We determined that growth rate and the solvability conditions for these Landau constants by letting

$$A_2(\tau) = A_3(\tau) = \frac{B_1(\tau)}{2}, \quad \varphi_1(\tau) = \varphi_2(\tau) = \varphi_3(\tau) \equiv 0,$$

which reduced our hexagonal-planform expansion to

$$\zeta(x, y, t) \sim f(x, y, t) = A_1(\tau)\cos(q_c x) + B_1(\tau)\cos\left(\frac{q_c x}{2}\right)\cos\left(\frac{\sqrt{3}q_c y}{2}\right)$$

where

$$\frac{dA_1}{d\tau} \sim \sigma A_1 - a_0 B_1^2 - A_1(a_1 A_1^2 + a_2 B_1^2),$$

$$\frac{dB_1}{d\tau} \sim \sigma B_1 - 4a_0 A_1 B_1 - B_1\left[2a_2 A_1^2 + (a_1 + 2a_2)\frac{B_1^2}{4}\right].$$

We note that the forms of the second- and third-order terms in this expansion can be deduced by examining the functional dependence of $f^2(x, y, t)$ and $f^3(x, y, t)$, respectively. Observe in addition that the terms in its amplitude equations are then those with coefficients containing components proportional to $\cos(q_c x)$ and $\cos(q_c x/2)\cos(\sqrt{3}q_c y/2)$, respectively. Further, note in this context that by taking $B_1(\tau) \equiv 0$ and $A_1(\tau) = A(\tau)$, our expansion can be reduced to that of Wollkind and Segel ([294]) once we make the identification $\zeta_{nm} = \zeta_{n0M0}$ for $M = 2m$ where the notation ζ_{njMk} is being employed to represent the coefficient of each higher-order term appearing in this expansion of the form $A_1^n(\tau)B_1^j(\tau)\cos(Mq_c x/2)\cos(k\sqrt{3}q_c y/2)$. Thus, $\sigma = \sigma_0$ but since $\Gamma \neq \Gamma_P$ we still needed to re-evaluate a_1 and compute the two other Landau constants a_0 and a_2. Proceeding in the same manner as we did with the two-dimensional analysis, the determination of these constants could be reduced to an examination of equations of the form

$$a_0 - \sigma_0 \zeta_{0220}(\sigma_0) = r_0(\sigma_0), \quad a_1 - 2\sigma_0 \zeta_{3020}(\sigma_0) = r_1(\sigma_0),$$

$$\text{and } a_2 - 2\sigma_0 \zeta_{1220}(\sigma_0) = r_2(\sigma_0),$$

respectively, where $r_0(\sigma_0)$ is a known function of σ_0 depending on the planar interface solution and the solutions to the first-order systems evaluated at $z = 0$ while $r_{1,2}(\sigma_0)$ are similar functions of σ_0 depending on the planar interface solution and the solutions to the first and second-order systems evaluated at $z = 0$. Hence, by taking the limit of these equations as $\sigma_0 \to 0$, we obtained

$$a_j = r_j(0) \text{ for } j = 0, 1, \text{ and } 2.$$

Having determined formulae for their growth rate and Landau constants, we returned to the six-disturbance hexagonal-planform amplitude phase-equations. In cataloguing the critical points of these equations and summarizing their orbital stability behavior, it is necessary to employ the quantities

$$\sigma_{-1} = \frac{-4a_0^2}{a_1 + 4a_2}, \ \sigma_1 = \frac{16a_1a_0^2}{(2a_2 - a_1)^2}, \ \sigma_2 = \frac{32(a_1 + a_2)a_0^2}{(2a_2 - a_1)^2}.$$

There exist equivalence classes of critical points of these equations of the form $(A_0, B_0, B_0, 0, 0, 0)$ corresponding to $\phi_1 = \phi_2 = \phi_3 = 0$ with

$$\text{I:} A_0 = B_0 = 0; \ \text{II:} A_0^2 = \frac{\sigma}{a_1}, \ B_0 = 0;$$

$$\text{III}^{\pm}\text{:} A_0 = B_0 = A_0^{\pm} = \frac{-2a_0 \pm [4a_0^2 + (a_1 + 4a_2)\sigma]^{1/2}}{a_1 + 4a_2};$$

$$\text{IV:} A_0 = \frac{-4a_0}{2a_2 - a_1}, \ B_0^2 = \frac{\sigma - \sigma_1}{a_1 + 2a_2},$$

where it is assumed that a_1, $a_1 + 4a_2 > 0$ (see below). The orbital stability conditions for these critical points can be posed in terms of σ. Thus, critical point I is stable in this sense for $\sigma < 0$, while the stability behavior of II and III$^{\pm}$, which depends upon the signs of a_0 and $2a_2 - a_1$, has been catalogued in Table 2.1 under the further assumption that $a_1 + a_2 > 0$ ([277]). In this parameter range, $A_0^+ > 0$ and $A_0^- < 0$. Finally, critical IV, which reduces to II for $\sigma = \sigma_1$ and to III$^{\pm}$ for $\sigma = \sigma_2$, and hence, called a generalized cell by Kuske and Matkowsky ([105]), is not stable for any value of σ (Wollkind, [277]). Here, we use the term orbital stability of pattern formation to mean a family of solutions in the plane that may interchange with each other but do not grow or decay into a solution type from a different family. Such an interpretation depends upon the translational and rotational symmetries inherent to the amplitude-phase equations, these invariancies also limiting each equivalence class of critical points to a single member that must be considered explicitly. Also, note that by stability in the third row of this table, we merely mean neutral stability.

TABLE 2.1
Orbital stability behavior for critical points II and III$^{\pm}$ with a_0 and $2a_2 - a_1$.

a_0	$2a_2 - a_1$	Stable Structures
$+$	$-, 0$	III$^-$ for $\sigma > \sigma_{-1}$
$+$	$+$	III$^-$ for $\sigma_{-1} < \sigma < \sigma_2$, II for $\sigma > \sigma_1$
0	$-$	III$^{\pm}$ for $\sigma > 0$
0	$+$	II for $\sigma > 0$
$-$	$+$	III$^+$ for $\sigma_{-1} < \sigma < \sigma_2$, II for $\sigma > \sigma_1$
$-$	$-, 0$	III$^+$ for $\sigma > \sigma_{-1}$

We next offered a morphological interpretation of the potentially stable critical points catalogued above relative to the interfacial patterns under investigation. Then critical points

I and II represented the planar and banded states, respectively, described in the two-dimensional analysis. Observing that, to lowest order, the equilibrium function associated with critical points $III^{\pm}$ satisfied

$$\zeta(x,y,t) \to \zeta_e(x,y) \sim A_0^{\pm} f_0(x,y) \text{ as } t \to \infty \text{ where}$$

$$f_0(x,y) = \cos\left(\frac{2\pi x}{\lambda_c}\right) + 2\cos\left(\frac{\pi x}{\lambda_c}\right)\cos\left(\frac{\sqrt{3}\pi y}{\lambda_c}\right),$$

and, noting from Fig. 2.10 that $f_0(x,y)$ exhibited hexagonal symmetry, we deduced that these critical points gave rise to an interface possessing hexagonal structure with circular elevations at the centers of the hexagons for III^+ since $A_0^+ > 0$ and depressions for III^- since $A_0^- < 0$ with both the diameter of these regular hexagons and the distance between adjacent centers equal to $2\lambda_c/\sqrt{3}$. Hence, we concluded that III^+ represented dome-shaped regular hexagonal cells and III^-, a hexagonal close-packed array of circular depressions of liquid into the solid or nodes. Note, from these identifications and Table 2.1 that cells can only be stable if $a_0 < 0$ and nodes if $a_0 > 0$.

Having summarized those stability criteria and morphological identifications, we now returned to our expressions for the Landau constants. Once a particular alloy of fixed C_P was selected these constants could be calculated as functions of V alone as in our two-dimensional analysis. Employing the same parameter values as in that analysis with $\Gamma_P = 1 \times 10^{-8}$cm, we evaluated those constants versus V for various k, $\partial\gamma/\partial T_L$, and $\partial\gamma/\partial C_L$. We observed from these calculations that for $\gamma \equiv \gamma_P$, $a_0 > 0$ identically and thus, given the necessary condition for the stability of cells, this guaranteed the instability of such structures. We also observed that the sign of a_0 was virtually independent of our choice for $\partial\gamma/\partial T_L$ but depended crucially on the parameters k and $\partial\gamma/\partial C_L$. In particular, for $0 < k < 1$, stable cells could occur for $\partial\gamma/\partial C_L > 0$. Thus, we considered an alloy for which $k = 0.1$, $\partial\gamma/\partial C_L = 2.7\gamma_P/C_P$, and, for the sake of definiteness, $\partial\gamma/\partial T_L = -10\gamma_P/T_P$ (note, since $\gamma_P > 0$ this implied $\gamma_0 > 7.3\gamma_P$). We next examined the signs of a_0, a_1, $2a_2 - a_1$, and $a_1 + 4a_2$ as functions of V and found besides V^* and V_k defined earlier there also existed the following other significant values of V (see Table 2.2)

$$V^* < V_1 < V_2 < V_c < V_3 < V_4 < V_k$$

such that

$$a_1 + 4a_2 = 0 \text{ for } V = V_1 \text{ or } V_4, \ a_1 + 4a_2 > 0 \text{ for } V_1 < V < V_4;$$
$$2a_2 - a_1 = 0 \text{ for } V = V_2 \text{ or } V_3, \ 2a_2 - a_1 > 0 \text{ for } V_2 < V < V_3$$
$$2a_2 - a_1 < 0 \text{ for } V < V_2 \text{ or } V < V_3$$
$$a_0 = 0 \text{ for } V = V_c, \ a_0 > 0 \text{ for } V < V_c, a_0 < 0 \text{ for } V > V_c.$$

TABLE 2.2
Significant values of V, for $k = 0.1$ and $\gamma/\gamma_P = 1 - 10(T_L/T_P - 1) + 2.7(C_L/C_P - 1)$, as measured in the units of 10^{-3}cm/sec

V_1	V_2	V_c	V_3	V_4
3.3	5.5	18	30	38

In order to identify the various interfacial morphologies in the V-G_L plane of Fig. 2.8 as

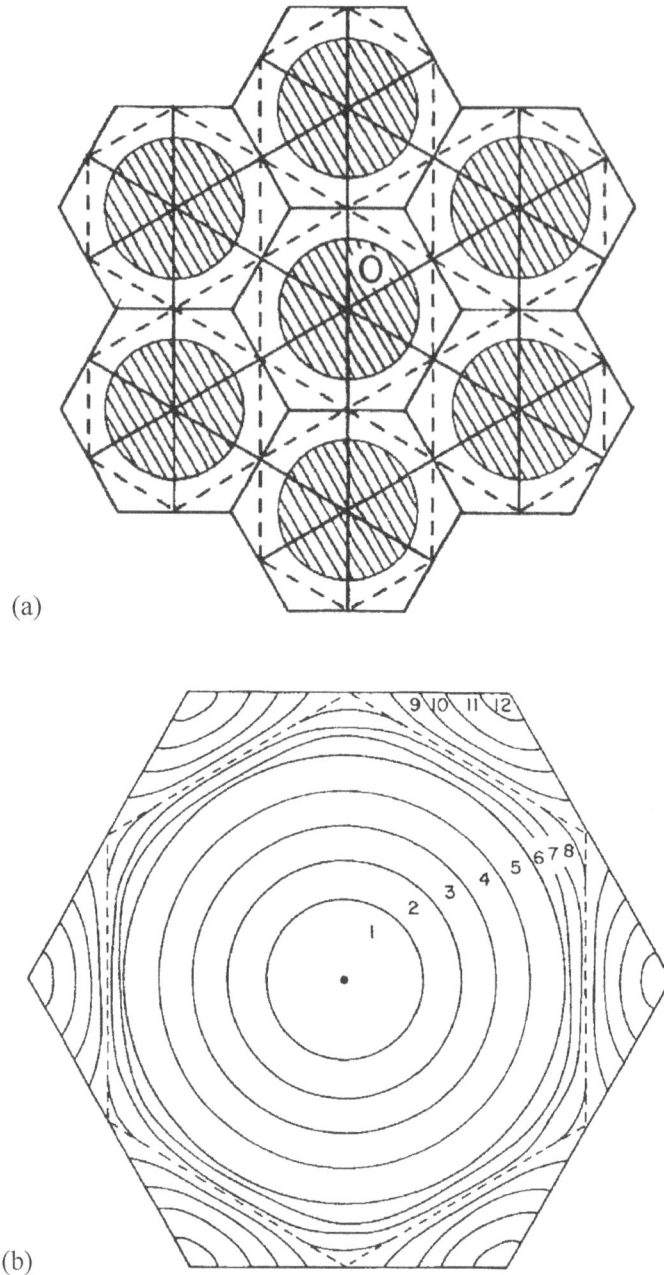

(a)

(b)

FIGURE 2.10
(a) Contour plot of $z = f_0(x, y)$ relevant to hexagonal symmetry. The shaded regions lie above the $z = 0$ plane and the unshaded ones, below it. Here the point O corresponds to the origin. The solid parallel straight lines are characterized by $z = 1$ and the dashed ones, $z = -1$. (b) An enlargement of the hexagonal cell in part (a) containing the origin (the point at its center) with representative level curves. Here the origin corresponds to $z = 3$ while the curves labeled 1, 2, ..., 11, and 12 are for values of $z = 2.25$, 1.50, 0.75, 0, -0.60, -0.75, -0.90, -0.975, -1.08, -1.20, -1.32, and -1.44, respectively. The value of z on the inscribed hexagon, shown by dashed lines as in part (a), is -1 while at the vertices of the principal hexagon $z = -1.50$.

predicted by Table 2.1 we needed to define

$$G_{\sigma_0}(V) = \frac{C_P m_0 (k-1)(1+\kappa)}{2D_L} V Q_{\sigma_0}(V)$$

and, in an analogous manner to the marginal curve of linear stability theory $G_L = G_c(V)$ on which $\sigma_0 = 0$, plot their generalizations (see Chapter 11 for more details on this procedure)

$$G_L = G_{\sigma_j(V)}(V) \equiv G_j(V) \text{ for } j = -1, 1, \text{ and } 2$$

in this plane where σ_{-1}, σ_1, and σ_2 as defined earlier only depended on the Landau constants, themselves functions of V. Observing that $G_{-1}(V) \cong G_c(V)$ we characterized the three loci $G_L = G_c(V)$, $G_1(V)$, and $G_2(V)$ by $\sigma_0 = 0$, σ_1, and σ_2, respectively, in Fig. 2.8 and represented the regions corresponding to the relevant interfacial morphologies as predicted by Table 2.1 graphically in the V-G_L plane of that figure. In this context we noted that hexagonal structures can only exist for $a_1 + 4a_2 > 0$ or where $V_1 < V < V_4$ and in this region $a_1 > 0$ since now $V^* < V_1$. Further we observed that as $\partial\gamma/\partial C_L > 0$ decreased the vertical lines in Fig. 2.8 would be shifted to the right. This is consistent with the fact that $r_1(0)$ of our three-dimensional analysis reduces to $r_{31}(0)$ of our two-dimensional one for $\gamma \equiv \gamma_P$.

Finally, to justify the truncation procedure inherent to the asymptotic expansions of the hexagonal planform analysis it was necessary that the Landau constants satisfied the size constraint

$$|a_0| \leqq (a_1 + 4a_2)^2$$

which was valid for the parameter range of interest. Hence, we concluded that our theoretical predictions were conclusive rather than merely suggestive as would be the case for such analyses should that condition be violated.

As pointed out in our introductory Chapter 1 the basic feature of comprehensive applied mathematical modeling is the comparison of its theoretical predictions with experimental or observable data for the phenomenon under investigation. Toward that end we next made an interpretation of the morphological predictions summarized in Fig. 2.8 with respect to the experimental observations of Morris and Winegard ([150]). In particular, the latter performed a very careful study of the sequence of morphologies from which regular cells developed as a function of V for various values of G_L and the microsegregation of those structures during dilute binary lead (Pb)-antimony (Sb) alloy solidification having its solute with $0 < k < 1$. As a prelude to this discussion we first re-examined our morphological predictions in some detail. The level curves of the contour plot $z = f_0(x, y)$ contained in Fig. 2.10 for z in the range $0 < z < 3$ are circular in shape (the shaded regions of Fig. 2.10a); for $-1 < z < 0$, oval to hexagonal; and, finally, for $-1.5 < z < -1$, they are centered about the vertices. The principal hexagon which forms the boundary of the cells is not itself a level curve but is tessellated varying in depth from $z = -1.5$ at its vertices to $z = -1$ at the midpoints of each edge. We then determined that $C_e(x, y)$, the equilibrium C_L at an interface exhibiting hexagonal structure, satisfied to first order

$$C_L(x, y, \zeta, t) \to C_e(x, y)$$

$$\sim C_P \left[1 + (k-1) \frac{\sqrt{4\omega_c^2 + 1} - 1}{2k - 1 + \sqrt{4\omega_c^2 + 1}} A_0^{\pm} f_0(x, y) \right] \text{ as } t \to \infty.$$

Given this behavior, we deduced that the deviation of the hexagonal interface from a plane and its solutal impurity concentration were inversely related when $0 < k < 1$. Hence, we concluded that for cells, since $A_0^+ > 0$ the highest solute concentration should occur at its vertices with the lowest concentrations occurring at its circular or sometimes oval caps, while for nodes, since $A_0^- < 0$ this should be reversed. The latter is consistent with the photomicrograph of nodes reproduced from Morris and Winegard ([150]) in Fig. 2.11a. Here, due to the lighting conditions the most solute rich regions appear dark, the next richest appear white, and the lowest, gray. In the case of nodes, which Morris and Winegard ([150]) described as circular wells of liquid penetrating the solid, the shaded regions of Fig. 2.10a represent depressions. Since these regions are rich in solute, they each appear as dark circles surrounded by a white annulus in Fig. 2.11a, while the unshaded regions of Fig. 2.10a which are depleted of solute appear as a uniformly gray background dramatically demonstrating the fact that only regions of high solute concentration are delineated in such photographs. Thus, these nodes are of uniform circular cross-section, arranged in a close-packed formation lacking any demarcation at the vertices of the hexagons.

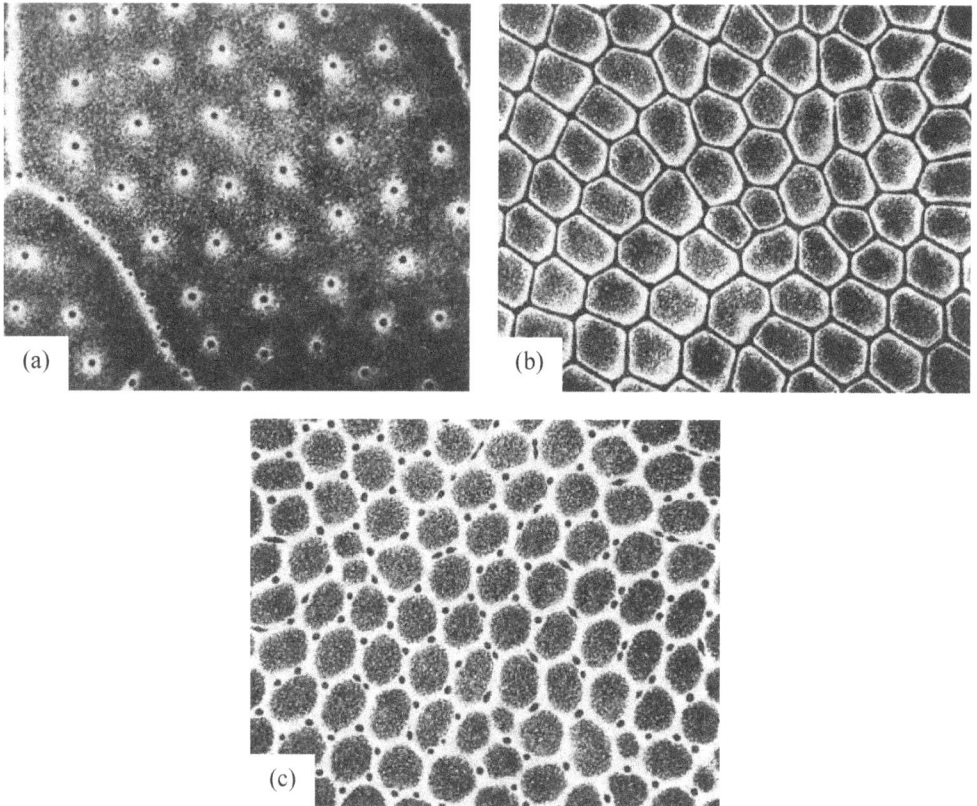

FIGURE 2.11
Photomicrographs of interfacial cross-sections for a dilute lead-antimony alloy of (a) nodes; (b) cells taken at a quenched interface; and (c) cells taken 1 mm before the quench. Here the dark regions are those of highest solute concentration; the white regions, those of next highest concentration; and the gray regions, those of lowest concentration.

Now that our identification of the III^- critical point with nodes had been confirmed, we ironically became concerned about our identification of the III^+ critical point with cells. When Morris and Winegard ([150]) photographed regular cells in a cross-section taken near the position where the growth front was quenched (rapidly cooled) reproduced in Fig. 2.11b, they found the cells had straight boundaries forming regular hexagons as was also the case for the shape of the cells in Fig. 2.2d involving a decanted (liquid drained) interface. Given the appearance of the nodes described above we would have expected the cells to have more circular or oval gray caps rather than hexagonal ones with its vertices marked by dark circles and its edges by white regions rather than uniformly dark boundaries. In order to resolve this discrepancy, I telephoned Larry Morris at Alcan Corporation in Kingston. He told me it was their conjecture that the hexagonal shape of the regular cells (and by inference the decanted interface cellular shapes) had been produced by the lateral growth of the cell caps during quenching and that these caps were more circular in cross-section during steady-state growth. Further, Larry stated that if the result of such growth conditions were desired, it was necessary to shave 1 mm off the quenched interface. He said they had photomicrographs of this sort of cell and asked me if I would be interested in receiving them. Having answered that I would be, Larry sent me copies of these photographs one of which is reproduced in Fig. 2.11c. Observe from the latter that the cells now appear exactly as predicted above confirming our identification of the III^+ critical point with a cellular morphology. This whole process represents, in my opinion, the ultimate Rabbit being Pulled Out of a Hat.

Further, another Rabbit being Pulled Out of a Hat was the assumption that the interfacial capillarity in the Gibbs-Thomson relation was a function of liquid-phase temperature and solute concentration because without this assumption no cells could be predicted. Given the importance of that morphological prediction it seems appropriate to explain our rationale for making this assumption in the first place. With very thin fluid layers the convection cells such as those originally observed by Henri Bénard ([12]) in his experiments involving whale oil are surface-tension rather than buoyancy-driven, as is the case with deeper layers. The so-called Marangoni-Bénard model for this situation depends crucially upon considering the surface tension at the liquid-air interface to be a linear function of interfacial temperature. When extrapolated to thermo-haline convection which includes both temperature and salinity gradients in a liquid, such as sea water, this surface-tension variation must be extended to include the interfacial salt concentration as well. Motivated by that behavior, it was natural to introduce the linear variation of interfacial capillarity with liquid-phase temperature and solute concentration for our alloy solidification model. If after the nonlinear stability analysis was completed that variation proved to be unnecessary in order to make the proper morphological predictions in comparison with experimental data, then we would merely have set it equal to zero. If, however, we had not made that assumption and the nonlinear stability analysis with $\gamma \equiv \gamma_P$ failed to yield the appropriate predictions, as in our case, then we would have had to redo the whole analysis after introducing this variation in an attempt to obtain those predictions. For situations of this sort I have always found it far better to include such extra effects from the outset rather than introduce them later if needed, since they can always be set equal to zero if not required to yield the appropriate results.

Having examined the morphology and solute segregation of our predicted hexagonal structures, we returned to an interpretation of Fig. 2.8 within the framework of the experiments of Morris and Winegard ([150]). They found that for a fixed value of G_L, in the 10 to $12°C/cm$ range, the sequence of morphologies observed during the transition from a planar to a cellular interface structure, as V was increased, depended strongly on the crystallographic orientation of the planar interface. In this context, one says a plane has Miller indices $(hk\ell)$, named after the British mineralogist William Hallowes Miller who introduced them in 1839, if it can be represented in the form $h\boldsymbol{a}_1 + k\boldsymbol{a}_2 + \ell\boldsymbol{a}_3$ for the set of basis vectors $\{\boldsymbol{a}_1, \boldsymbol{a}_2, \boldsymbol{a}_3\}$ while then $\langle hk\ell \rangle$ denotes the direction perpendicular to that plane (see Fig. 2.12). In particular, for their face-centered cubic lattice, a cubic array with an element in the center of each of its six sides, Morris and Winegard ([150]) found that when growth was in a $\langle 100 \rangle$ direction this sequence consisted of a planar interface, a regularly spaced formation of nodes, curved elongated cells the boundaries of which were random in orientation, and finally hexagonal cells; while for growth in a $\langle 110 \rangle$ direction the sequence was planar interface, longitudinal cells or parallel bands, and then regular cells. Taking $G_L = 11°\text{C/cm}$ in Fig. 2.8, we predicted a morphological sequence of planar interface, nodes, bands, and then cells. We note that in the overlap regions, where bands and nodes or bands and cells, respectively, can both be stable the initial conditions determine which of these possible morphologies is actually observed. Here where our analysis predicted stable parallel banded patterns, the equivalence class designated as II earlier in this section actually contains three solutions ([211])

$$A_j^2 = \frac{\sigma_0}{a_1}, \ A_k = A_\ell = 0, \ (j, k, \ell) \equiv \text{ even permutation of } (1, 2, 3).$$

These represent a family of bands aligned parallel to the y-axis, as per our original identification, plus two similar families of bands making angles of $\pm 60°$ with them (see Fig.2.10a, where the three sets of parallel lines intersecting at angles of $60°$ are representative of these families of bands), for which stable coexistence with either the original family or one another is impossible ([211]). Then, as initial conditions vary from point to point on the interfacial surface (see Fig. 2.12), we deduced that such families of bands could give rise to polygonal arcs which would appear quite random in orientation. Hence, we next identified our band region of Fig. 2.8 with those patterns described by Morris and Winegard ([150]) as "curved" elongated cells and concluded that our sequence of morphologies predicted for $G_L = 11°\text{C/cm}$ corresponded to growth in the $\langle 100 \rangle$ direction. More precisely, our choice for γ_P, and hence Γ_P, corresponded to growth in that direction and these qualitative results would hold for any constant G_L up to that value $G_C \equiv G_c(V_c) = G_1(V_c) = G_2(V_c)$ where all three of these marginal curves intersected at their point of common tangency in Fig. 2.8.

In order to determine how a different choice for γ_P would have affected our results quantitatively, it was necessary to understand the manner Fig. 2.8 had been deduced in the first place. Sriranganathan *et al.*'s ([236]) corresponding results depended only on the values of two nondimensional quantities H, a dimensionless liquid temperature gradient, and U, a dimensionless solidification speed, which were related to G_L and V, respectively, by

$$G_L = \frac{T_P}{\Gamma_P} H \text{ and } V = \frac{D_L}{\Gamma_P} U.$$

If Γ_P were increased then the plot of G_L versus V in Fig. 2.8 would concomitantly be reduced in scale. Thus $G_L = 11°\text{C/cm}$ could now correspond to G_C in Fig. 2.8 and the sequence of morphologies, given the proper initial conditions (see Fig. 2.12), would be planar interface, parallel bands, and regular cells. This was the sequence Morris and Winegard ([150]) found to be associated with growth in the $\langle 110 \rangle$ direction. Such an interpretation was consistent with the conjecture of Sharp and Hallewell ([224]) that

$$\gamma_{P_{\langle 110 \rangle}} > \gamma_{P_{\langle 100 \rangle}},$$

and is the most simplistic way to introduce anisotropy into an isotropic model which assumes all properties are independent of orientation.

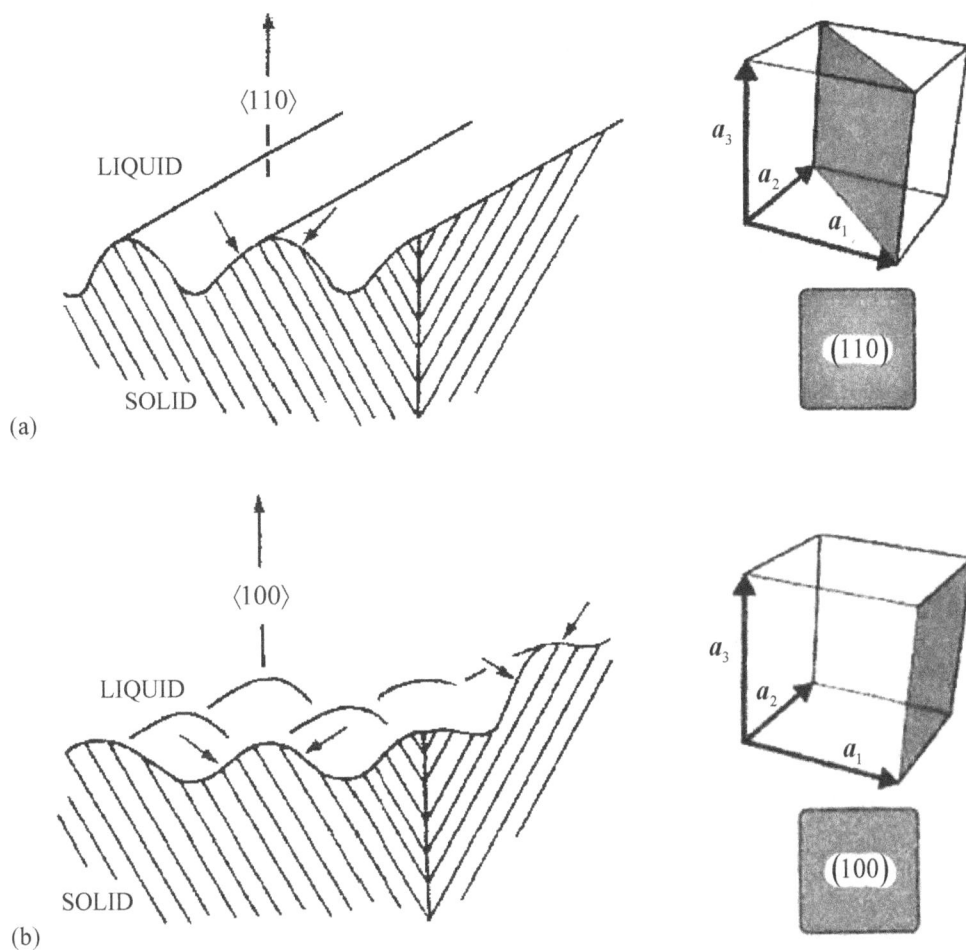

FIGURE 2.12
Schematic representation of developing perturbations from a planar interface for (a) the
⟨110⟩ growth direction which is perpendicular to the (110) plane and bands develop; and
(b) the ⟨100⟩ direction which is perpendicular to the (100) plane and nodes develop. It is
plausible that bands developing from a uniform planar interface would all be aligned parallel
as is the case for (a) while those developing from the nodes of (b) would give rise to all
three possible band types due to the nonuniformity of initial conditions as one moves from
point to point on the interface.

We also noted that if V were held constant between 10×10^{-3} and 22×10^{-3} cm/sec which are the intersection points on the V-axis of $G_L = G_1(V)$ in Fig. 2.8 and G_L decreased along such a vertical line, our sequence of morphologies would be planar interface, nodes, nodes or bands, and bands for $V < V_c = 18 \times 10^{-3}$ cm/sec, while it would be planar interface, cells, cells or bands, and bands for $V > V_c$. This was reminiscent of the behavior of the Rayleigh-Bénard problem as the Rayleigh number, a nondimensional temperature gradient, increased where the comparable transition states were that of pure conduction, hexagons, hexagons or rolls, and rolls (see Fig. 2.13, a more detailed explanation of which appears in Chapter 3).

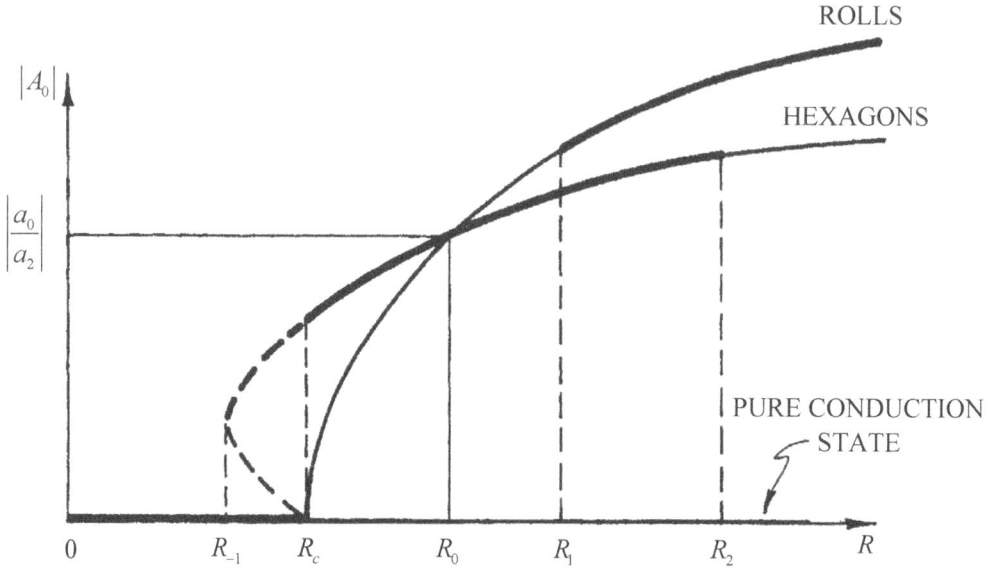

FIGURE 2.13
Stability bifurcation diagram in the R-$|A_0|$ plane for the Rayleigh-Bénard problem where R is the Rayleigh number and A_0, the equilibrium amplitude. Here, a_1, $2a_2 - a_1 > 0$ and $0 < |a_0| \ll 1$ while R_{-1}, R_c, R_0, R_1, and R_2 correspond to $\sigma_0 = \sigma_{-1}$, 0, $a_1 a_0^2 / a_2^2$, σ_1, and σ_2, respectively. Further, heavy curves denote stable states and light curves, unstable ones, with subcritical portions represented by broken curves. The designations "hexagons" and "rolls" refer to the states of hexagonal cellular and rolling longitudinal convection patterns, respectively.

As a final conclusion we calculated explicitly the range of spacing characteristic of the various interfacial morphologies predicted by our model. Recall that the width of a band was given by λ_c while both the distance between adjacent nodes and the diameter of a cell, by $d = 2\lambda_c/\sqrt{3}$ whereas defined earlier

$$\lambda_c = \frac{2\pi}{\omega_c} \frac{D_L}{V}.$$

Since ω_c was defined implicitly as a function of V, we were able to calculate the desired spacings from these relations for the predicted interfacial morphologies over the relevant intervals of solidification speed depicted in Fig. 2.8. Thus, we determined that the distance between adjacent nodes ranged from 213 to 72 μm as V varied from 3.5×10^{-3} to

17.4×10^{-3}cm/sec; the width of bands, from 137 to 47 μm as V varied from 5.5×10^{-3} to 26.7×10^{-3}cm/sec; and the diameter of cells, from 67 to 44 μm as V varied from 19.3×10^{-3} to 37.8×10^{-3}cm/sec; where μm $\equiv 10^{-4}$ cm. This was in qualitative agreement with the experimental observations of Morris and Winegard ([150]) who found the spacing characteristic of all three morphologies to decrease with increasing V, but noted that cells seemed to be a much more stable structure in this regard than either nodes or bands. It was also in quantitative agreement with their observation that the nodes appearing at the onset of instability of the planar interface were typically spaced about 200 μm apart and Tiller and Rutter's ([250]) measurements of the range of diameters of cells. Further Morris and Winegard's ([150]) observation that the transition from bands to cells proceeded by the folding of the interface down the center of the original band producing cells of diameter of one-half their width, which fold laterally as V increases to form regular cells, was consistent with our results. In fact, we selected the numerical value of $\partial\gamma/\partial C_L$, which has been employed in this chapter, precisely because that selection yielded results in good quantitative agreement with these experimentally observed spacings of interfacial morphologies. Hence, our model was predictive in the sense that it could be used to determine the proper value for $\partial\gamma/\partial C_L$, a heretofore difficult quantity to measure experimentally.

We complete the solidification portion of the chapter by suggesting an extension of the analysis just presented. Although our predictions are in both qualitative and quantitative agreement with the experimental results of Morris and Winegard ([150]) for a fixed value of G_L, they are only in qualitative agreement with those of Takahashi $et\ al.$ ([244]) in which both G_L and V were varied, as depicted in Fig. 2.14. Given the truncation procedure of our analysis through terms of third-order, it is only strictly valid in a neighborhood of the marginal stability curve $G_L = G_c(V)$. Subsequent to our research I devised a method in conjunction with my Ph.D. student Michael Vislocky that would allow those results to be extended for a considerable distance from this curve. The method for doing so employed by Wollkind and Vislocky ([298]) in their liquid phase electro-epitaxy solidification model equation analysis, required a departure from those procedures used in Landau-type nonlinear stability theory. When solving for Landau-type coefficients in the amplitude equations represented by a, for ease of exposition, one always obtains a relation involving them written in the generic form

$$a(\sigma) = (n-1)\sigma\zeta(\sigma) + r(\sigma)$$

where n is the order of the term, -$i.e.$, $n = 2$ at second-order or $n = 3$ at third-order; $\zeta(\sigma)$ is the relevant interfacial perturbation, and $r(\sigma)$ is a known function. We, as traditionally has been done, took

$$a(\sigma) \equiv a(0) = r(0),$$

$i.e.$assumed a to be $constant$ - and then solved for $\zeta(\sigma)$, obtaining

$$\zeta(\sigma) = \begin{cases} \frac{r(0)-r(\sigma)}{(n-1)\sigma}, & \sigma \neq 0 \\ -\frac{1}{n-1}\frac{dr}{d\sigma}(0) & \sigma = 0 \end{cases}.$$

In order to extend these results to those $\sigma > 0$ reasonably far from $G_L = G_c(V)$ in the V-G_L plane, Wollkind and Vislocky ([298]) instead considered

$$\zeta(\sigma) \equiv \zeta(0) = -\frac{1}{n-1}\frac{dr}{d\sigma}(0)$$

and then calculated

$$a(\sigma) = r(\sigma) - \frac{1}{n-1}\frac{dr}{d\sigma}(0)\sigma,$$

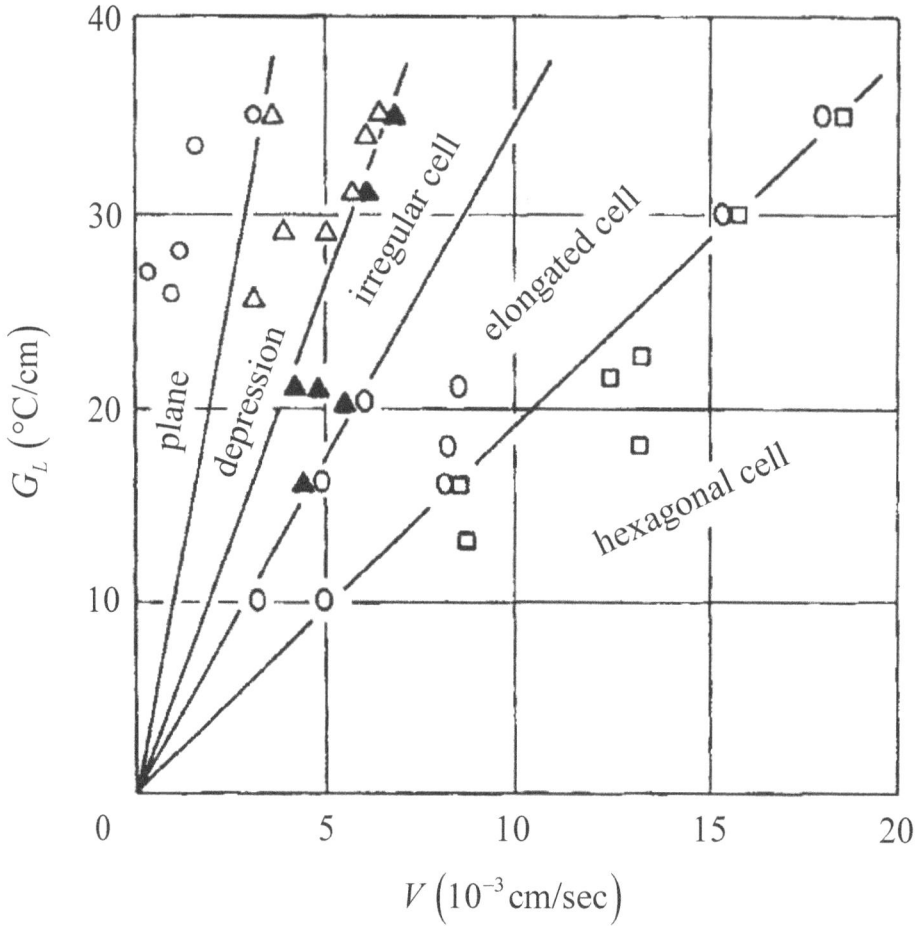

FIGURE 2.14

The morphological transition of the solid-liquid interface from a plane to hexagonal cells in a 99.997% aluminum dilute binary alloy. The slopes of the lines measured in units of $10^{3\circ}$ C sec/cm^3 dividing these regions are 10, 5.1, 3.4, and 1.9, in order of transition.

i.e., after DiPrima ([50]) assumed the Landau-type coefficients were not constant. Note this relation reduces to our previous result on the marginal stability curve where $\sigma = 0$ since $a(0) = r(0)$. Then, for those σ_0 such that $\sigma_0^{1/2}$ and $|a_0|$ are both small with respect to a_1 and $a_1 + 4a_2$, it will still have been permissible to have truncated the amplitude equations through third-order. Determining the Landau coefficients in this manner, the vertical lines in Fig. 2.8 should be transformed into curves passing through the origin having those lines as vertical asymptotes, as hypothetically depicted in that figure, with its other curves transformed accordingly, which would then compare quite favorably to the observations of Takahashi *et al.* ([244]). In this context, the curves $G_L = G_1(V)$ and $G_L = G_2(V)$ of Fig. 2.8 were calculated by determining how the problem behaved as σ_1 and σ_2, respectively, varied through relevant positive values evaluated when $\sigma_0 = 0$. Now, we would be able to obtain a similar behavior when $\sigma_0 > 0$.

Controlled solidification of dilute binary alloys is used to grow single crystals, refine materials, and obtain uniform or nonuniform composition within such substances. The most important industrial applications of this methodology are growth of crystals for metal oxide semiconductors and of oxides for optical devices in laser (an acronym for *l*ight *a*mplification by *s*timulated *e*mission of *r*adiation) systems. That technology can also be used to make crystals for synthetic jewels and similar procedures arise in the steel and glass industries. Quantitative predictions of solid-liquid interfacial morphologies during solidification, including information on spacing and solute redistribution as just summarized in this chapter and published in Wollkind *et al.* ([295]), are therefore extremely valuable to industrial researchers.

Somewhat analogous and also of considerable interest industrially, due to the process of zone refining, was the companion problem of the investigation of the stability of the solid-liquid interface during the controlled melting of a dilute binary alloy, which had received much less theoretical attention until my first Ph.D. graduate student Sassan Raissi and I began examining it to explain more fully the experimental results of Verhoeven and Gibson ([263]). We first modified the Wollkind and Segel ([285]) two-dimensional solidification formulation so that it could be applied to melting instead. We retained the same equations in the liquid and solid phases which now satisfied $z < \zeta(x,t)$ and $z > \zeta(x,t)$, respectively, as adopted for the solidification problem, while adding the extra equation for solute diffusion in the solid phase given by

$$\frac{\partial C_S}{\partial t} - V\frac{\partial C_S}{\partial z} = D_S\nabla^2 C_S \text{ for } z > \zeta(x,t).$$

Unlike the case for solidification, a significant amount of redistribution of solute in the solid phase was to be expected during melting, since the interface now advanced into that phase. Thus, whereas in solidification only solute diffusion in the liquid phase was significant, in the case of melting such diffusion must be accounted for in both phases. We retained the interfacial boundary conditions of the solidification problem at the solid-liquid interface while explicitly considering

$$C_S = kC_L \text{ at } z = \zeta(x,t),$$

changing the sign of the curvature term in the Gibbs-Thompson equation, and introducing an extra flux term in the conservation of solute jump-type boundary condition obtaining

$$D_L\frac{\partial C_L}{\partial n} - D_S\frac{\partial C_S}{\partial n} = (k-1)C_L w_n \text{ at } z = \zeta(x,t)$$

to account for the facts that now C_S was a dynamical variable, the interface advanced into the solid phase, and solute diffusion in this phase $D_S \neq 0$, respectively. Although, Lee and I, following Mullins and Sekerka ([152]), had taken $D_S = 0$ as a first approximation in modeling the solidification process, Tiller and Sekerka ([251]) considered that problem for the case in which the solute diffusion coefficients of the solid and liquid phases were assumed equal. Sassan and I, adopting the same approach, considered the melting problem under this assumption that $D_S = D_L$. To complete our formulation we imposed the far field boundary conditions that

$$C_L \to C_0(z),\ T_L \to T_0(z) \text{ as } z \to -\infty;\ C_S \to c_0(z),\ T_S \to \Theta_0(z) \text{ as } z \to \infty;$$

where the melting problem's planar interface solution's solute concentration components now satisfied

$$C_0(z) \equiv C_P,\ c_0(z) = C_P[1 + (k - 1)e^{-Vz/D_L}];$$

while its temperature components retained the identical form as that appearing in Fig. 2.6 for the solidification problem except their gradients changed sign since $G_S, G_L < 0$ during melting.

We next performed the same two-dimensional nonlinear stability analysis on our melting model system as Wollkind and Segel ([285]) applied to their solidification problem and found that this analysis yielded qualitatively similar results to those obtained for solidification. In particular, the marginal stability curve of linear theory satisfied

$$Q = Q_0(\omega^2) = -\beta\omega^2 + \frac{\sqrt{4\omega^2 + 1} - 1}{1 - k + (1 + k)\sqrt{4\omega^2 + 1}}.$$

This curve had a maximum point at (ω_c^2, Q_c) where the critical wavenumber squared satisfied

$$\beta = \frac{4}{\sqrt{4\omega_c^2 + 1}[1 - k + (1 + k)\sqrt{4\omega_c^2 + 1}]^2} \text{ if } 0 \le \beta \le 1,$$
$$\omega_c^2 = 0 \text{ if } \beta > 1.$$

Here we are using the same notation as employed for the solidification problem, except that now G_L has been replaced by its absolute value. For the experiments of Verhoeven and Gibson ([263]), which involved the melting of a tin-bismuth alloy with the relevant parameter values

$$C_P = 6\% \text{ by wt. of bismuth},\ k = 0.3,\ m_0 = -1.32°C/\% \text{ by wt. of bismuth},$$
$$T_P = 504K,\ \kappa = 1.78,\ \Gamma_P = 1.44 \times 10^{-8} \text{ cm},$$
$$D_S = 4.1 \times 10^{-10} \text{ cm}^2/\text{sec},\ D_L = 10^{-5} \text{ cm}^2/\text{sec},$$

the Landau constant satisfied

$$a_1 > 0 \text{ for } 0 < V < V^* = 4.7 \text{ mm/sec} < V_{k=0.3},$$
$$a_1 = 0 \text{ for } V = V^*,\ a_1 < 0 \text{ for } V > V^*.$$

Verhoeven and Gibson ([263]) found the interface to be essentially flat (planar) for a solid temperature gradient of $-168°C/\text{cm}$ and speeds of melting from 0.2 to 2.0 $\mu\text{m/sec}$ while for melting conditions of $-168°C/\text{cm}$, 20 $\mu\text{m/sec}$ and $-45°C/\text{cm}$, 2 $\mu\text{m/sec}$ that planar interface became unstable. This instability took the form of a cellular structure similar to that observed in the solidification experiments of Morris and Winegard ([150]). Since our predicted condition for $a_1 < 0$ required speeds of melting far in excess of what was experimentally realizable ($V > V^* = 4,700\ \mu\text{m/sec}$), we could restrict ourselves to the two cases of Fig. 2.9 for which $a_1 > 0$. Recalling that $G_L = \kappa G_S$, Sassan and I defined the dimensionless measures of the solid temperature gradient and speed of melting

$$H_{|G_S|} = \frac{\Gamma_P |G_L|}{T_P} = \frac{\Gamma_P \kappa |G_S|}{T_P} \text{ and } U_V = \frac{\Gamma_P V}{D_L},$$

respectively, analogous to the nondimensional parameters introduced for the solidification problem, and plotted these experimentally determined points in the U-H plane of Fig. 2.15, which includes the marginal stability curve of linear theory $H = H_c(U)$ corresponding to $G_L = G_c(V)$. In accordance with our nonlinear stability predictions of Fig. 2.9, we denoted the planar interface region of that plane where $H > H_c(U)$ by (ii) and the patterned region where $H < H_c(U)$ by (i). Since we dealt with two-dimensional disturbances to the planar interface solution, only stable bands could conclusively be predicted in this region. Noting that for the Rayleigh-Bénard problem, as depicted in Fig. 2.13, a three-dimensional analysis had shown that stable hexagonal cells occurred where a two-dimensional analysis predicted rolls, we tentatively referred to this region as the cellular regime. Retroactively, given the three-dimensional morphological stability results for the solidification problem summarized earlier, we can with more confidence label this region the cellular regime. Hence, from Fig. 2.15 it can be seen that our theoretical morphological stability predictions are in both qualitative and quantitative agreement with the experimental observations of Verhoeven and Gibson ([263]). Further, in this set of experiments whenever the interface was essentially flat, they observed that a network of liquid channels and wells was present in a small band of solid ahead of the melting interface. Such an aberration ahead of the interface into the solid was in agreement with our predicted nonlinear effect of $\zeta_{20} > 0$ which implied that the mean position of the melting interface should lead its original position.

As can be seen from the previous results, there is a high correlation between the experimental observations of Verhoeven and Gibson ([263]) and the theoretical interfacial morphologies as predicted by our model. It seems appropriate at this point to amplify somewhat the description of comprehensive applied mathematical modeling included in Chapter 1: Namely, starting with a particular scientific phenomenon, one abstracts to a mathematical model, introduces whatever approximations are needed to employ a desired method of analysis on it, and arrives at certain predictions upon application of this methodology to that modified model. The justification of these approximations is determined by the adherence of those theoretical predictions to observables of the scientific phenomenon under examination. In other words, approximations of this sort are only tolerable when they are necessary and only permissible when they work, in the sense that theoretical predictions deduced within the framework of these assumptions agree with phenomenological observations. Here, I am paraphrasing the actor Will Geer's statement on presidential assassination schemes from the 1973 movie "Executive Action," a fictionalized account commemorating the tenth anniversary of John F. Kennedy's death, which made enough of an impression on me to be incorporated in the form reproduced above into the paper on this research Wollkind and Raissi ([293]) published at that time.

Specifically, our model with $D_S = D_L$ gave an amazingly good prediction of interface morphology for experiments that did not satisfy this condition provided the common value was taken to be D_L. Although it was possible that such a correlation might simply have been fortuitous, the experimental results of Chen and Jackson ([32]) offered some further evidence for the justification of our having set the solute diffusion coefficients equal and taken their common value to be D_L. They observed ripples appearing on the solid-liquid interface originally planar in morphology during the controlled melting of transparent organic binary mixtures of low entropies of fusion which behave much like metallic alloys. Chen and Jackson

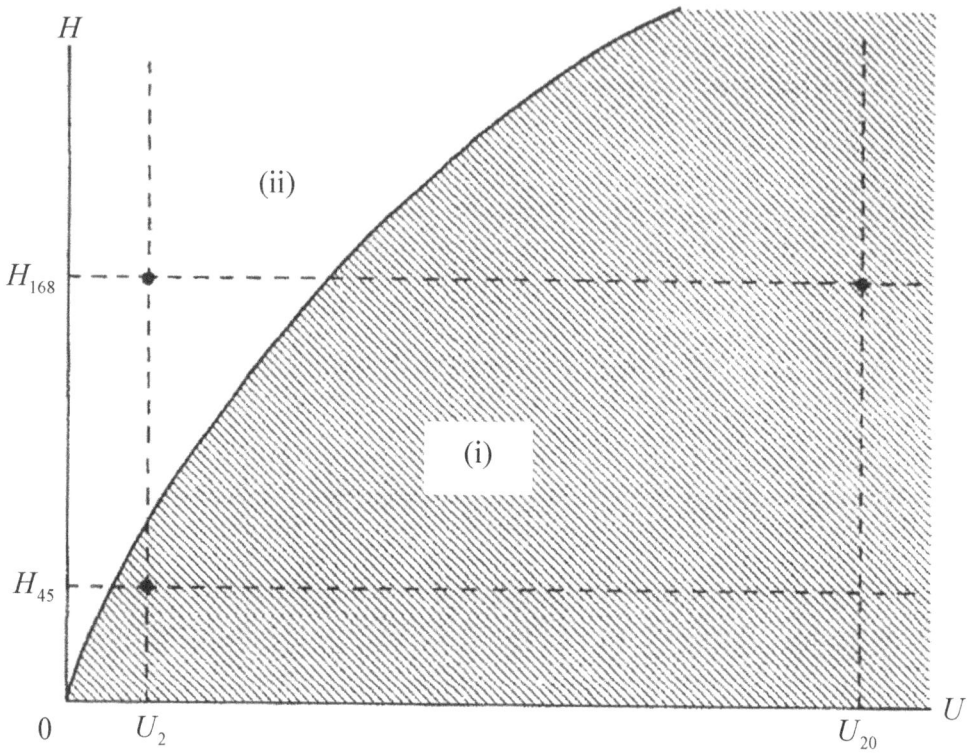

FIGURE 2.15

Plot in the U-H plane relevant to the morphologies at the melting solid-liquid interface observed by Verhoeven and Gibson ([263]) comparing their experimental results (•) with the theoretical predictions of nonlinear stability theory where the delineated regions correspond to those cases indicated in Fig. 2.9.

([32]) found the wavelength of these ripples ranged from 20 to 40 μm and was larger by an order of magnitude than the diffusion length in the solid phase $D_S/V \approx \mu$m. Since for those experiments $D_L/D_S = 50$, they concluded that, just as during solidification, the instability depended crucially on solute diffusion in the liquid phase because it was so much faster than diffusion in the solid. That is the wavelength λ of a ripple was of the order of magnitude

$$\lambda \approx \left(\frac{D_L}{D_S}\right)\left(\frac{D_S}{V}\right) = \frac{D_L}{V}.$$

Comparing λ with the formula λ_c for solidification it seems only natural to have taken the common value of the solute diffusion coefficient to be D_L. This whole procedure represented yet again another example of our Pulling a Rabbit Out of a Hat. Segel had perhaps as good a description as any for what it takes to make such a contribution when, while we were working on my dissertation, he said to me "You don't have to be so smart to obtain a result like that. You just have to be a little less dumb than the other guys."

I shall close this chapter with a commentary on a topic that is seldom discussed in the scientific literature: Namely, who gets ultimate credit for such a major contribution. The average person would assume that credit of this sort goes to the first researchers who publish the result in question. The average person would be wrong. It actually goes instead to the last researchers who publish that result. In a perfect world those would of course be one and the same, but unfortunately the world is far from perfect. It is natural to ask how they can possibly be different.

All that takes is a second set of researchers reproducing the results of the first set by publishing them without the former giving proper attribution to the latter and the reviewers of their manuscript not catching this sin of commission or omission depending upon whether or not these authors were aware of the original publication. Then, if subsequent papers referring to that result cite the secondary rather than the primary reference, since it was published later than the first and presumably must contain something not included in the earlier work, the appropriation has been completed. A clever way of accomplishing this if the second set of researchers were actually aware of the original publication is to rename the result with a newly created label and then anyone using that label in the future must cite this secondary reference, depriving the first set of researchers of the ultimate credit for their contribution. A classic example of a sin of omission in this regard in the scientific literature is the so-called McKendrick-von Foerster first order linear partial differential equation governing the dynamical behavior of an age-structured population. It was first posed by McKendrick ([140]) in the context of epidemiology and rediscovered by von Foerster ([265]) for describing cell proliferation. Known for years as the von Foerster equation, it was finally relabeled the McKendrick-von Foerster equation after this oversight had been corrected but, although sometimes called the McKendrick equation today, it is still being referred to as the von Foerster equation in numerous recent publications.

An example of a sin of commission in this regard involving my two-dimensional longitudinal-planform pattern formation analyses of alloy solidification with Segel and melting with Raissi concerns James S. Langer, then of Carnegie-Mellon University and from 1982 of the University of California Santa Barbara. I first became aware of Langer when he sent me two preprints of his alloy solidification research in the middle 1970s. The next I heard about him was from Segel after Lee had attended a Gordon Conference on stability in fluid mechanical and materials science phenomena. Langer had presented a paper on the morphological stability of a solid-liquid interface during dilute binary alloy solidification for a solute concentration model equation, to which our system reduces provided one employs the simplification of equal thermal diffusivities $\kappa_S = \kappa_L$ in both phases. During the discussion period following Langer's talk Segel introduced himself and asked, "How was it that you failed to mention the earlier and far more complete work on the same subject by Wollkind and Segel?" to which Langer replied, "I didn't know you were that Segel!" I subsequently met Langer at International Conference on Crystal Growth-5 and having asked him about his lack of providing citations when giving talks was informed that, "I never worry about including references, but just let people like you and Segel tell me what should have been cited." Thus I was not particularly surprised when a few years later his review article Langer ([108]) entitled "Instabilities and pattern formation in crystal growth" appeared, which only peripherally cited our work in its introduction by the indirect employment of "see Wollkind and Raissi ([293]) and references contained therein." Further, Langer using a renaming procedure called his solute concentration solidification system with $D_S = 0$ and $D_S = D_L$ the one- and two-sided models, respectively, without any proper attribution and hence appropriated these formulations to himself. Therefore, a number of later phase change papers cite Langer ([108]) as the primary reference for these model systems, rather than say Mullins and Sekerka ([152]) and Tiller and Sekerka ([251]).

Although this commentary might seem to be a fitting place to end the chapter, I would feel somewhat remiss in not including my final interactions with Langer during two 1983 scientific conferences to which we were both invited, especially since they have a bearing on the dissemination of the solidification results described herein. Let me begin by stating that, as just amply demonstrated, publishing significant results in professional journals is not always the best way to ensure other researchers become aware of them. This is one of the reasons that scientists and engineers also consider it important to attend professional meetings and present those results in person. During the 1982-1983 academic year, the NSF-funded Institute for Mathematics and its Applications (IMA) began operation at the University of Minnesota under the leadership of its director Hans Weinberger and assistant director George Sell. Each academic year the IMA selected a particular area of application upon which to concentrate and that first academic year it was mathematical modeling in materials science. They organized various workshops related to a topic in the selected area and invited prominent researchers in the field to visit the IMA on either a long- or short-term basis. The workshop held during the first two weeks in January of 1983 was called the Winter Session on Phase Transformations and they invited me (at Segel's suggestion), Langer, and John W. Cahn of the National Institute of Standards and Technology (NIST) as primary participants. Our responsibility was to give a series of plenary lectures on the subject and interact with the other visitors that were attending the IMA at this time. We were given offices on the fourth floor of Vincent Hall, the building where the Department of Mathematics was housed at the University of Minnesota. Jim Langer and I shared an office. We actually got along famously. In appearance he bore an uncanny resemblance to the actor John Zacherley who recently died on October 27, 2016, at the age of 98 (see their photographs in Fig. 2.16 and, given the prominent role Segel played in my research career, I have also included his photograph in Fig. 2.17). Zacherley hosted a television show both in Philadelphia and New York City from 1958-1963 which featured old horror movies with cut-ins of him dressed in his characteristic undertaker's costume seamlessly spliced into the film. After that Langer seemed much more human to me. Those two weeks were profitable on a number of levels. I gave my talks on solidification theory, tutored one of the visitors Jean Taylor, then of Rutgers University, in that subject, and got the opportunity to interact with John Cahn, perhaps the most famous theoretician in the field. Further I wrote up my solidification model equation analysis paper Wollkind *et al.* ([292]) with Oulton and Sriranganathan, which showed the equivalence of the Landau constant formula determined by the so-called adjoint operator method employed in Wollkind and Segel ([285]) and the more direct method, actually deduced by Oulton, used in Wollkind *et al.* ([295]) and sketched in this chapter which ironically made the most mathematically significant aspect of my thesis irrelevant. This publication also appeared in the IMA preprint series which served as a means of accountability to NSF and thus, by deferring its write up to those two weeks, I fulfilled my obligation to Weinberger and Sell who strongly encouraged such participant contributions.

FIGURE 2.16
Photographs of James S. Langer (on the left) and John Zacherley (on the right). Note that the title *Lady Dracula's Lover* of the book held by "Zach," as his fans called him, was a play on words relating to D. H. Lawrence's *Lady Chatterley's Lover*, the subject of a 1960 American obscenity trial, while Dracula is the fictional vampire Bram Stoker created in his 1897 novel of the same name, based loosely on the historical figure Vlad III known alternatively as Vlad Tepes "the Impaler" or Vlad Dracula "the son of the Dragon," a fifteenth century Wallachian ruler. In this context, one of the old movies that Zacherley parodied was Bela Lugosi's Dracula of 1931.

Of equal importance, the IMA workshop led directly to my invitation to participate in a special session on crystal growth at the May 1983 conference on Fronts, Interfaces, and Patterns hosted by the Center for Nonlinear Studies affiliated with the Los Alamos National Laboratory. They had originally invited John Cahn for that special session but he, being unable to attend, suggested me as his replacement. The session consisted of four thirty minute talks with the other speakers being Langer, Herbert Levine of Schlumberger-Doll Research, and Geoffrey B. McFadden, a colleague of Cahn at NIST. It was scheduled in the evening, with Robert F. Sekerka of Carnegie-Mellon University giving an hour invited address earlier

FIGURE 2.17

Photograph of Lee A. Segel presenting a talk at a professional meeting. Segel was an exceptional speaker and, during his time at RPI and the Weizmann and Santa Fe Institutes, gave a great many invited addresses and tutorial lectures at conferences, workshops, and colloquia. One of his favorite opening slides for such presentations was entitled "Quest for Understanding" and reproduced the Albert Einstein quote: "The grand aim of all science is to cover the greatest number of empirical facts by logical deduction from the smallest number of hypotheses or axioms." Such a statement adequately summed up Segel's principle of employing the minimum number of slides for these presentations in order to get the maximal number of ideas across to his audience without burdening them unnecessarily with cumbersome details. It also provides another good description for the philosophy of comprehensive applied mathematical modeling.

that same day in order to introduce the subject. Langer also claimed responsibility for my invitation to both the IMA and this special session, which he had actually organized. The session was chaired by Michael Nauenberg of the University of California Santa Cruz, a visitor during that academic year at the University of California Santa Barbara's Institute for Theoretical Physics (ITP) of which Langer later served as director. Just before we were scheduled to start, Jim approached Geoff and me saying that he and Levine would like to add Eshel Ben-Jacob, a postdoctoral fellow at ITP, as a fifth participant, which they felt would strengthen the session. In order to accommodate this extra speaker Langer proposed that all three of them restrict their talks to 20 minutes each, while we retained our 30 minute time slots. Jim asked if this arrangement were agreeable to us and we answered that it was. In the event, Nauenberg as chair of the session allowed them to take a total of 80 minutes for their presentations rather than the agreed upon 60 and hence mine started 20 minutes late. Fifteen minutes, into my talk he informed me that I only had 5 minutes left. I told him that there were still 15 minutes to go. Nauenberg then stated that because of Los Alamos security regulations the session had to end at its originally scheduled completion time and therefore my talk needed to be truncated. I told him that he should have thought of that earlier and the fact that the first three speakers ran over their allotted time was not going to come out of my or Geoff's hide, as it were. McFadden graciously said that his presentation only required about 13 minutes in all and hence I proceeded to take 12 more minutes to complete the talk. Since Martin D. Kruskal of Princeton University, a good friend of Segel's, when giving an overview of all the talks in his closing address labeled my presentation, published as Wollkind *et al.* ([295]), the tour de force of the whole conference, I would say it worked out pretty well. Nigel D. Goldenfeld, another of Langer's postdoctoral fellows, liked the talk so much that he suggested I present it to his group. I replied "Have your boss invite me." I am still waiting for that invitation to the ITP from Langer.

One of the other good things to come out of this conference was that it led in turn to an invitation for Bob Sekerka and me to give a joint presentation of our solidification research at the NATO workshop on the Structure and Dynamics of Partially Solidified Systems held at Fallen Leaf Lake in May of 1986. After our hour-long presentation was finished, an attendee from Cambridge University during the discussion period wanted to know why Newell-Whitehead-Segel instabilities associated with other wavelengths in an interval centered about the critical one were not introduced. This side-band instability theory had been developed independently by Segel ([213]) and Newell and Whitehead ([156]) for Rayleigh-Bénard convection. A Woods Hole attendee pointed out that these instabilities were implicit to our results since they would always occur where my three-dimensional nonlinear stability analysis predicted parallel banded patterns. The Cambridge man was still unconvinced and demanded, "How can you be so sure of that?," to which this individual then responded "I'm Whitehead." Talk about Pulling a Rabbit Out of a Hat!

3

Chemically Driven Convection of Dissociating Gases: Longitudinal Planform Nonlinear Stability Analysis

FIGURE 3.1
Photograph of Nelson A. Rockefeller, New York State Governor from 1958 to 1974.

Since this chapter deals with my postdoctoral research topic, I shall begin it by presenting the historical details of how Harry Frisch became my postdoc-supervisor. Nelson Rockefeller (see Fig. 3.1), as governor of New York state from 1958 to 1974, was extremely interested in upgrading its higher educational system. At the time of his initial election to that governorship, California had the best state system of higher education. This consisted of three tiers: The University of California system of comprehensive colleges offering both undergraduate and graduate degrees, the California State system of four-year colleges offering only undergraduate degrees, and the California Community College system of two-year colleges offering so-called associate degrees. Rockefeller wished to create a New York state system of higher education based on that structure. He did so by developing a similar three-tier system consisting of the State Universities of New York, the State of New York Colleges, and the New York Community Colleges, respectively. The highest tier was composed of the four University centers located at Albany, Binghamton, Buffalo, and Stonybrook, the first two of which had been smaller colleges. The State University of New York at Albany, for instance, was a 15,000-student campus housed in a new academic podium constructed on Washington Avenue that had been superposed upon the existing 1,500 student Albany State Teachers College, originally housed farther downtown. In order to organize that University center, prestigious faculty members were lured away from leading academic and industrial positions by sparing no expense. Harry Frisch (see Fig. 3.2), one of best theoretical physical chemists in the country, was hired from the Bell Laboratories unit at Murray Hill, New Jersey. He immediately obtained a National Science Foundation Grant from the Division of Chemistry to study chemically-driven convective instabilities in gases. That grant included support for several postdoctoral scholars and Ph.D. students. In the Spring semester of 1968, while I was completing my Ph.D. thesis, Harry began running a seminar on nonlinear problems at SUNYA and invited Lee to give a talk about his current research. I accompanied him to that seminar and after his presentation, Harry asked me to explain my dissertation topic, which had been highlighted during the talk, more completely. Upon the conclusion of that discussion, he offered me a two-year postdoctoral position to work with him on his NSF grant. I had applied for several university appointments and, without knowing the outcome of those applications, hesitated to accept Harry's offer immediately. He gave me a month to consider it and, when that time had elapsed with no other offers forthcoming, I gratefully accepted the position. As an addendum to this exposition, let me recount my own meeting with Nelsen Rockefeller. RPI held one graduation ceremony during the year in June and all students who obtained degrees in January of that year or August of the previous year went through it as well. Since my Ph.D. date was August of 1968, I went through the ceremony in June of 1969 at the start of the second year of my postdoctoral studies. We were lined up outside the RPI fieldhouse and I was last in line. Feeling a tap on my shoulder, I turned around and faced Nelsen Rockefeller, the commencement speaker, who had just arrived by helicopter. He asked me what degree I was receiving and what job I had accepted. When told about the SUNYA postdoctoral position, Rockefeller said he was especially pleased that I had remained in state with my employment at one of his SUNY centers.

I actually started working with Harry three weeks before my thesis defense described in the last chapter (see Chapter 7 for how that impacted my publication of those thesis results). He first asked me to investigate the linear stability of an infinite expanse of a dissociating diatomic gas using a chemical quasi-equilibrium model for such a situation due to M.J. Lighthill ([118]). In particular, that required the consideration of a gas undergoing the reversible dissociation reaction $A_2 \rightleftharpoons 2A$, and the introduction of the mass fraction of dissociation

$$\alpha = \alpha(x, y, z, t) = \frac{N_A}{N_A + 2N_{A_2}}$$

where $N_A \equiv$ number of A atoms and $N_{A_2} \equiv$ number of A_2 molecules. To explain that model, it is necessary to define a few more mathematical vector operations: Namely, the gradient of a scalar field and the divergence of a vector one. Given a scalar field $f(\boldsymbol{r}, t)$ and a vector one $\boldsymbol{v}(\boldsymbol{r}, t) = (v_1, v_2, v_3)[\boldsymbol{r}, t]$ with $\boldsymbol{r} = (x_1, x_2, x_3) \equiv$ position vector then the gradient of the former which is a vector and divergence of the later which is a scalar are defined by

$$\boldsymbol{\nabla} f = \left(\frac{\partial f}{\partial x_1}, \frac{\partial f}{\partial x_2}, \frac{\partial f}{\partial x_3}\right) \text{ and } \boldsymbol{\nabla} \cdot \boldsymbol{v} = \frac{\partial v_1}{\partial x_1} + \frac{\partial v_2}{\partial x_2} + \frac{\partial v_3}{\partial x_3},$$

respectively. Further, the substantial or total derivative of these quantities with respect to

FIGURE 3.2
Photograph of Harry L. Frisch, Professor of Chemistry at SUNYA from 1967 to 2007.

time t, which can be derived by an application of the chain rule of partial differentiation, are given by

$$\frac{Df}{Dt} = \frac{\partial f}{\partial t} + \boldsymbol{\nabla} f \cdot \boldsymbol{v} \text{ and } \frac{D\boldsymbol{v}}{Dt} = \left(\frac{Dv_1}{Dt}, \frac{Dv_2}{Dt}, \frac{Dv_3}{Dt} \right)$$

where $\boldsymbol{v}$ is to be interpreted as a velocity vector and its scalar product with another vector field $\boldsymbol{w} = (w_1, w_2, w_3)$ are defined by

$$\boldsymbol{w} \cdot \boldsymbol{v} = w_1 v_1 + w_2 v_2 + w_3 v_3 \equiv \sum_{i=1}^{3} w_i v_i.$$

Observe that, unlike the metallurgical models introduced in the previous chapter for which convection in the liquid phase was ignored, here we are dealing with a moving continuum having characteristic velocity vector $\boldsymbol{v}$.

The general governing equations for this situation were given by ([118])

$$\frac{D\rho}{Dt} + \rho\theta = 0 \text{ where } \theta = \boldsymbol{\nabla} \cdot \boldsymbol{v},$$

$$\rho\frac{D\boldsymbol{v}}{Dt} = -\boldsymbol{\nabla}p + \mu_0 \left(\nabla^2 \boldsymbol{v} + \frac{\boldsymbol{\nabla}\theta}{3} \right)$$

$$\text{where } p = \rho \left(\frac{\mathcal{R}}{M} \right) T(1+\alpha) \equiv \text{ ideal gas pressure,}$$

$$\rho \left(C_p\frac{DT}{Dt} + c_0\frac{D\alpha}{Dt} \right) + \left(p + 2\mu_0\frac{\theta}{3} \right)\theta - 2\mu_0\boldsymbol{\varepsilon} \cdot\cdot \boldsymbol{\varepsilon} = k_0\nabla^2 T + \rho_0 k_\alpha D_0 \nabla^2\alpha$$

$$\text{where } [\boldsymbol{\varepsilon}]_{ij} = \frac{\frac{\partial v_i}{\partial x_j} + \frac{\partial v_j}{\partial x_i}}{2} = \varepsilon_{ij} \equiv \text{ rate of strain and } \boldsymbol{\varepsilon} \cdot\cdot \boldsymbol{\varepsilon} \equiv \sum_{i,j=1}^{3} \varepsilon_{ij}\varepsilon_{ij},$$

$$\rho \left[\frac{D\alpha}{Dt} + \frac{\alpha - \alpha_e(T_0)}{\tau_d} \right] = \rho_0 D_0 \nabla^2\alpha,$$

which represent balance of mass, momentum, energy, and species, respectively. Here, $\rho \equiv$ density with reference value ρ_0, $T \equiv$ temperature with reference value T_0, $\mathcal{R} \equiv$ the universal gas constant $= 8.314$ joules/(mole K), $M \equiv$ gram-molecular mass of the gas, $\mu_0 \equiv$ shear viscosity coefficient, $C_p \equiv$ specific heat at constant pressure, $c_0 \equiv$ specific dissociation energy, $k_0 \equiv$ thermal conductivity, $k_\alpha \equiv$ cross species thermal flux coefficient, $D_0 \equiv$ Fickian diffusion coefficient of the binary mixture, $\tau_d \equiv$ dissociation relaxation time, and $\alpha_e(T_0) \equiv$ equilibrium dissociation as an explicit function of the reference temperature.

Lighthill ([118]) made a major simplification to this formulation by assuming that the system under investigation was in a state of chemical quasi-equilibrium, *i.e.*, there is only a slight departure from a dissociating gas in chemical equilibrium, replacing the Laplacian of α on the right-hand sides of the energy and species equations by

$$\nabla^2\alpha = \varphi_0 \nabla^2 T$$

where $\varphi_0 \equiv$ chemical quasi-equilibrium proportionality constant > 0, and then reducing those right-hand sides to (note that $\nabla^2 \equiv \boldsymbol{\nabla} \cdot \boldsymbol{\nabla}$)

$$k_e\nabla^2 T \text{ where } k_e = k_0 + \rho_0 k_\alpha D_0\varphi_0 \text{ and } \rho_0 D_0\varphi_0\nabla^2 T,$$

respectively. There exists a static homogeneous solution

$$\boldsymbol{v} \equiv \boldsymbol{0}, \ \rho \equiv \rho_0, \ T \equiv T_0, \ \alpha \equiv \alpha_0 = \alpha_e(T_0)$$

satisfying these chemical quasi-equilibrium governing equations. It was the (linear) stability of this state to initially infinitesimal normal mode disturbances with which we were concerned and hence considered a solution to that governing system of equations of the form

$$\boldsymbol{v} = \varepsilon_1\boldsymbol{v}_1 + \boldsymbol{O}(\varepsilon_1^2), \ \rho = \rho_0 + \varepsilon_1\rho_1 + O(\varepsilon_1^2),$$
$$T = T_0 + \varepsilon_1 T_1 + O(\varepsilon_1^2), \ \alpha = \alpha_0 + \varepsilon_1\alpha_1 + O(\varepsilon_1^2),$$

where $\boldsymbol{v}_1 = (u_1, v_1, w_1)$ and $|\varepsilon_1| \ll 1$, such that for dimensionless time $\tau = t/\tau_d$

$$[u_1, v_1, w_1, \rho_1, T_1, \alpha_1](\boldsymbol{r}, t) = [U, V, W, R, \Theta, A] \exp(i\boldsymbol{k} \cdot \boldsymbol{r} + \sigma\tau)$$

with

$$|U|^2 + |V|^2 + |W|^2 + |R|^2 + |\Theta|^2 + |A|^2 \neq 0,$$
$$\exp(\ldots) \equiv e^{\cdots}, \; i = \sqrt{-1}, \; \boldsymbol{k} = (k_1, k_2, k_3),$$

the components of the wavenumber vector $\boldsymbol{k}$ being real, while the growth rate σ may be complex.

We next defined the diffusion ratios G and S, called the Lewis and Schmidt numbers, by

$$G \equiv \frac{\kappa_e}{D_0} \text{ and } S \equiv \frac{\nu_0}{D_0} \text{ where } \kappa_e = \frac{k_e}{\rho_0 C_p} \text{ and } \nu_0 = \frac{\mu_0}{\rho_0 C_p},$$

and introduced the nondimensional wavenumber

$$q = \sqrt{D_0 \tau_d} k \text{ where } k^2 = \boldsymbol{k} \cdot \boldsymbol{k}.$$

We further defined $0 < \alpha_0 = \alpha_e(T_0) < 1$, implicitly as a function of T_0, by the relation ([118])

$$\frac{\alpha_0^2}{1 - \alpha_0} = \left(\frac{\rho_d}{\rho_0}\right) \exp\left(\frac{-T_d}{T_0}\right) \text{ where } \rho_d \text{ and } T_d \text{ are characteristic values.}$$

Substituting this solution into our governing system of chemical quasi-equilibrium equations, neglecting terms of $O(\varepsilon_1^2)$, cancelling the resulting common ε_1 exponential factor, and imposing the nontriviality condition, we obtained the fourth-order characteristic polynomial in σ

$$\mathfrak{L}_4(\sigma) = q_4\sigma^4 + q_3\sigma^3 + q_2\sigma^2 + q_1\sigma + q_0 = 0,$$

where q_0 and q_4 were positive being explicitly given by

$$q_0 = \beta_0 G q^4 \text{ with } \beta_0 = \tau_d\left(\frac{\mathcal{R}}{M}\right) T_0 \frac{1 + \alpha_0}{D_0} \text{ and } q_4 = \frac{c_0}{C_p T_0},$$

while the other q_j's as defined in Wollkind and Frisch ([285]) were also functions of q^2 and G.

Recalling that the perturbation time dependence is of the form $\exp(\sigma\tau)$, to determine the linear stability of the static homogeneous solution we need to examine the behavior of that exponential as $\tau \to \infty$. Toward this end, since it is may be complex, consider $\sigma = \sigma_r + i\sigma_i$ where $\sigma_{r,i}$ are real and define the magnitude of any complex function $f(z) = u(x,y) + iv(x,y)$, where $z = x + iy$ and x, y, u, v are real, by $|f(z)|^2 = u^2(x,y) + v^2(x,y)$. Further, note that

$$\exp(\sigma\tau) = \exp(\sigma_r\tau)\exp(i\sigma_i\tau)$$

while by Euler's Formula

$$\exp(i\sigma_i\tau) = \cos(\sigma_i\tau) + i\sin(\sigma_i\tau).$$

Thus,

$$\exp(\sigma\tau) = \exp(\sigma_r\tau)\cos(\sigma_i\tau) + i\exp(\sigma_r\tau)\sin(\sigma_i\tau).$$

Hence,

$$|\exp(\sigma\tau)|^2 = [\exp(\sigma_r\tau)\cos(\sigma_i\tau)]^2 + [\exp(\sigma_r\tau)\sin(\sigma_i\tau)]^2$$
$$= [\exp(\sigma_r\tau)]^2[\cos^2(\sigma_i\tau) + \sin^2(\sigma_i\tau)] = [\exp(\sigma_r\tau)]^2$$

which implies that

$$|\exp(\sigma\tau)| = \exp(\sigma_r\tau) \rightarrow \begin{cases} 0 & \sigma_r < 0 \\ 1 & \text{for} \quad \sigma_r = 0 \quad \text{as } \tau \rightarrow \infty. \\ \infty & \sigma_r > 0 \end{cases}$$

Hence, we say that our static homogeneous solution is stable, neutrally stable, or unstable to initially infinitesimal disturbances depending upon whether σ_r is less than, equal to, or greater than zero, respectively. Here the neutral stability curve in any parameter space can be characterized by setting $\sigma_r = 0$ or $\sigma = i\sigma_i$. Should $\sigma_i = 0$ on such a curve and thus $\sigma = 0$, we say there is an exchange of stabilities; while if $\sigma_i \neq 0$, we say there is overstability. Note that for σ real or $\sigma_i = 0$ and $\sigma_r = \sigma$, this reduces to our results of the previous chapter. Recall in this context that Segel had not originally considered the possibility of σ_0 being complex for my thesis, discussed in Chapter 2, since he had never before encountered a problem having an exchange of stabilities on the marginal curve where the growth rate was not also real in the rest of the space.

Let us first examine the neutral stability curve $\sigma_r = 0$ or $\sigma = i\sigma_i$ for $\mathfrak{L}_4(\sigma) = 0$. That is, noting

$$i^2 = -1, \ i^3 = -i, \ \text{and} \ i^4 = 1,$$

consider

$$\mathfrak{L}_4(i\sigma_i) = q_4\sigma_i^4 - iq_3\sigma_i^3 - q_2\sigma_i^2 + iq_1\sigma_i + q_0 = 0 = 0 + 0i.$$

Then setting its real and imaginary parts equal to 0, we obtain

$$q_4\sigma_i^4 - q_2\sigma_i^2 + q_0 = 0 \ \text{and} \ \sigma_i(q_1 - q_3\sigma_i^2) = 0.$$

Solving these equations, we can deduce that either

$$\sigma_i = 0 \Rightarrow q_0 = 0$$

or

$$\sigma_i^2 = \frac{q_1}{q_3} > 0 \Rightarrow D_3 = q_1q_2q_3 - q_1^2q_4 - q_0q_3^2 = 0.$$

Since $q_0 > 0$, as defined, the first possibility must be rejected and hence we can conclude that the neutral stability curve in question satisfied $D_3 = 0$ where $q_1/q_3 > 0$. Employing the representations for the q_j's, we deduced that D_3 then took the form

$$D_3 = q^4[a_3(q^2)G^3 + a_2(q^2)G^2 + a_1(q^2)G + a_0(q^2)],$$

which upon cancellation of q^4 implied that the neutral stability curve in the q^2-G plane was given by

$$f_3(q^2, G) = a_3(q^2)G^3 + a_2(q^2)G^2 + a_1(q^2)G + a_0(q^2) = 0 \ \text{where} \ \frac{q_1}{q_3} > 0.$$

This represented an algebraic function, in that it was a cubic polynomial in G with co-efficients a_j for $j = 0, 1, 2, 3$, and 4, which were cubic polynomials in q^2 also defined in Wollkind and Frisch ([285]). Here, for ease of exposition, we have suppressed the dependence of these coefficients on all parameters except q^2.

Although at this time the neutral stability curve on which $\sigma_r = 0$ had been deduced, I did not know definitively that it was marginal, in the sense of separating stable regions where $\sigma_r < 0$ from unstable ones where $\sigma_r > 0$. Having reported my progress on our problem to

Frisch before he left on a trip to give an invited address at a professional meeting, Harry said he expected me to figure it out during the two weeks before his return to Albany. I despondently went back to my office and randomly pulled the Theory of Equations by J. V. Uspensky ([258]) from the bookcase. There, lo and behold, I found that book contained an Appendix III entitled *On Equations Whose Roots Have Negative Real Parts* which developed the Routh-Hurwitz criterion, mentioned in the first paragraph of Chapter 1, in conjunction with Maxwell's paper on governors, which incidentally involved the same question for his fifth-order polynomial. Applying this criterion to my fourth-order polynomial with q_0 and q_4 positive, I deduced the following three conditions on its coefficients to guarantee that $\sigma_r < 0$:

$$q_1 > 0, q_3 > 0, D_3 = q_1 q_2 q_3 - q_1^2 q_4 - q_0 q_3^2 > 0.$$

This demonstrated that the neutral stability curve $D_3 = 0$ where $q_1/q_3 > 0$ was indeed marginal and also gave me the region of stability in the q^2-G plane. Talk about Pulling a Rabbit Out of a Hat! Harry, having returned from his trip, was suitably impressed assuming that I had spent the full two weeks working out this result rather than the five minutes that it actually took me which occurred before he had even left Albany. Naturally, I never disabused him of his misconception.

Now it remained to plot $f_3(q^2, G) = 0$ as well as the $q_j(q^2, G) = 0$ for $j = 1, 3$ in the q^2-G plane and identify the stability region deduced above. Although these loci for the $q_j = 0$ were so-called rectangular hyperbolae of the form $G = g_j + C_j/q^2$ with the g_j's and C_j's being constants, with respect to q^2 and G, determining loci for the cubic algebraic function $f_3 = 0$ were a much more complicated matter. During that first year of postdoctoral research in the Chemistry Department at SUNYA, I also held the position of Visiting Assistant Professor in the Mathematics Department teaching two sections of freshman Calculus, since both Harry and Lee thought such an appointment would be useful for my future career opportunities. Except for me, the faculty members in that department were all pure mathematicians, as described in Chapter 1, and had no interest in applications. Nevertheless, since one of them was an expert in algebraic function theory, Harry felt it might be beneficial for me to ask him about my cubic algebraic function. It turned out that he had never studied an actual example of such a function before, given that his sole interest lay in their abstract theoretical analysis, and hence could provide me little help in the solving of my problem. Since RPI had no pure mathematicians (as Dick DiPrima liked to say: "We hired a pure mathematician once; we never made that mistake again."), this was my first experience dealing with a pure mathematician and I was unimpressed. Not only the faculty but even some of the Ph.D. students in that department had quite an attitude. A graduate student from the Sudan, upon asking on what was I currently working and having been provided with my answer, haughtily informed me "That's not mathematics. That's mathematical physics."

This being the case, I decided it was necessary to solve my cubic algebraic function numerically. For each $q^2 > 0$ and $0 < \alpha_0 < 1$ from discrete enumerated sets (just as in my thesis calculations), I used the SUNYA Univac mainframe digital computer to solve for its three roots in G by means of Cardan's formulae listed in Chapter V of Uspensky ([258]) after employing the relationships (Lighthill, [118]):

$$\nu_0 = 5.4 \times 10^{-6} \left(\frac{T_0}{K}\right)^{1.75} \frac{\text{cm}^2}{\text{sec}}, \ T_0 = \frac{T_d}{\ln\left(\frac{\rho_d}{\rho_0} \frac{1-\alpha_0}{\alpha_0^2}\right)};$$

and assigning the following parameter values appropriate for nitrogen ([118]):

$$T_d = 113,000 \text{ K}, \ \rho_d = 130\frac{\text{g}}{\text{cm}^3}, \ \rho_0 = 0.00125\frac{\text{g}}{\text{cm}^3},$$

$$c_0 = 3.35 \times 10^{11}\frac{\text{cm}^2}{\text{sec}^2}, \ M = 28 \text{ g}, \ \tau_d = 10^{-5} \text{ sec}, S = 0.8.$$

Like all numerical computations, however, it took a while to get the program compiling correctly. The Univac in operation at SUNYA, during the Fall Semester of 1968 when I was running these programs, could not do complex arithmetic so all computations had to have separate entries for their real and imaginary parts. Each morning between my 8 and 10 AM classes, I would go to the computer center to pick up my program which at first would fail to run. Most of the errors were easy to correct, but the last one was really perplexing. That perplexing error message read: "Cannot take the logarithm of a negative number." To my knowledge there were no logarithms of negative numbers in my program. Finally, I figured it out. Cardan's formula did contain cube roots of the form $\sqrt[3]{x}$. Cube roots should have no trouble with negative entries in that if $x < 0$ then $\sqrt[3]{x} = -\sqrt[3]{|x|}$ where $|x| = -x > 0$ for $x < 0$, while $|x| = x$ for $x \geq 0$. The trouble was the Univac was calibrated to convert all real powers r of x or x^r to

$$x^r \equiv e^{r \ln(x)} \text{ since } e^{\ln(x)} = x.$$

Such an equivalence is required for powers that are real but not rational. Thus, for a cube root or $r = 1/3$, it converted $\sqrt[3]{x}$ to

$$\sqrt[3]{x} = x^{1/3} \equiv e^{\ln(x)/3}$$

and then for negative x tried to take the natural logarithm of a negative number. Hence, the error message. To fix things up, I merely replaced my cube root by

$$\sqrt[3]{x} \equiv \text{sgm}(x)\sqrt[3]{|x|} \text{ where } \text{sgm}(x) \equiv \begin{cases} +1 & \text{for} & x \geq 0 \\ -1 & \text{for} & x < 0 \end{cases}.$$

With that Rabbit Being Pulled Out of a Hat and the cube roots so defined, the program ran perfectly and I had the information needed to plot the three relevant roots of G in the q^2-G plane for each value of α_0 which was the only remaining parameter in the problem. Figure 3.3 represents a plot of this sort for $\alpha_0 = 0.5$ where the roots are designated by $G_{1,2,3}$ when real and the loci $q_{1,3} = 0$ appear as well. The region of stability over which $\text{Re}(\sigma) = \sigma_r < 0$ is also identified in this figure.

Before examining that region in detail, we tabulate the temperatures T_0 corresponding to the values α_0 of interest from the appropriate relationship and parameters listed above for nitrogen.

TABLE 3.1
The values of $T_0 = T_d / \ln\left(\frac{\rho_d}{\rho_0}\frac{1-\alpha_0}{\alpha_0^2}\right)$ versus α_0 for $T_d = 113,000$ K, $\rho_d = 130$ g/cm^3, and $\rho_0 = 0.00125$ g/cm^3.

α_0	0.10	0.33	0.50	0.67	0.90
T_0 (K)	7,040	8,452	9,228	10,049	11,945

The shaded region in Fig. 3.3 is the locus of points for which $D_3, q_1, q_3 > 0$ and hence includes all those (q^2, G) corresponding to stability. The critical point (q_c^2, G_c) on the marginal stability curve $D_3 = 0$ is such that G_c represents the maximum value of G on this curve.

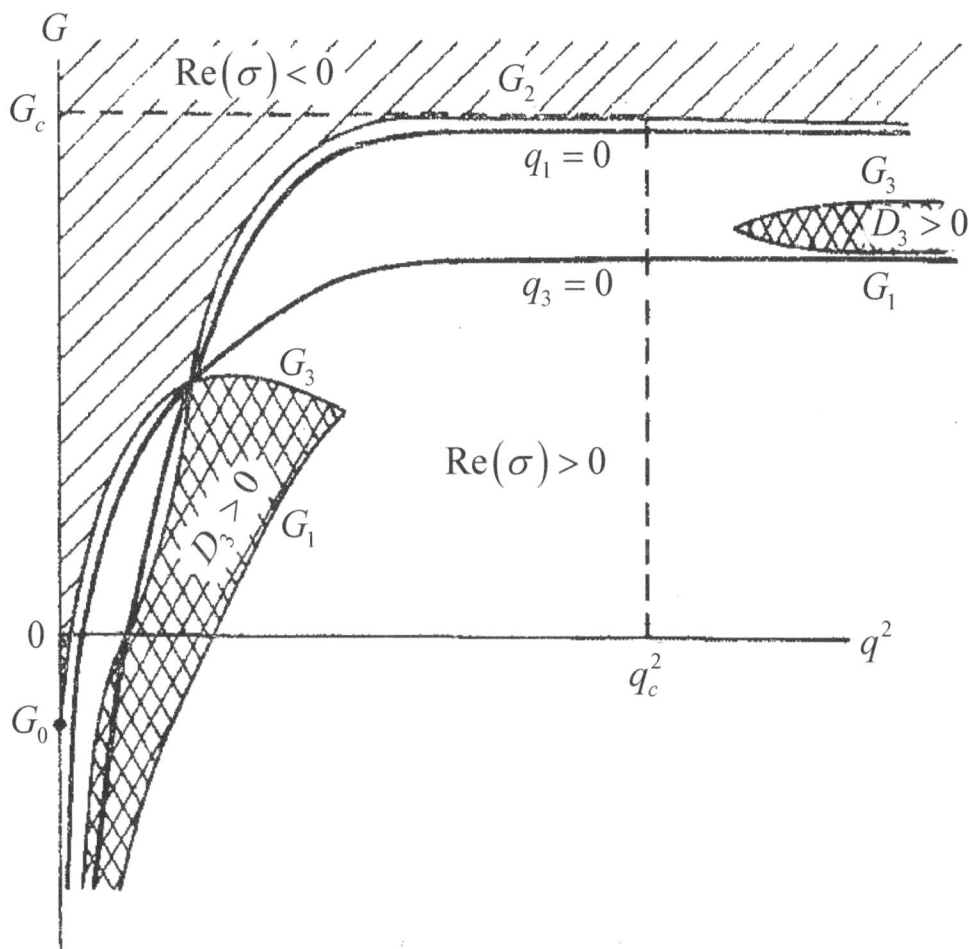

FIGURE 3.3
Stability diagram in the q^2-G wavenumber squared-Lewis number plane. Here those portions of the roots G_2 and G_3 of $D_3 = 0$ where $q_1, q_3 > 0$ comprise the marginal stability curve.

Note that for $G > G_c$, $\sigma_r < 0$ for all wavenumbers q^2, while for $G < G_c$ there exists an interval of wavenumbers centered about q_c^2 such that in this interval $\sigma_r > 0$.

For the case represented in Fig. 3.3, $G_c = 0.8777$; however, once the proper representations for k_0, k_α, C_p, D_0, and φ_0 as defined by Lighthill ([118]) had been substituted into it, the actual Lewis number for this problem with its parameters assigned those values listed above was a function of α_0 alone or $G(\alpha_0)$ and the same was true for $G_c(\alpha_0)$. Thus, we had stability or instability for each choice of α_0 depending on whether $G(\alpha_0) > G_c(\alpha_0)$ or $G(\alpha_0) < G_c(\alpha_0)$, respectively. For the case depicted in Fig. 3.3,

$$G(0.5) = 1.0203 > G_c(0.5) = 0.8777;$$

and hence we had stability. The qualitative features of Fig. 3.3 were the same regardless of the value for α_0. Therefore, G_c was closely approximated by the horizontal asymptote of the rectangular hyperbolic curve $q_1 = 0$ as $q^2 \to \infty$ or $G_c(\alpha_0) \cong g_1(\alpha_0)$. If one used this approximation then the criterion for stability took the form $G(\alpha_0) > g_1(\alpha_0)$ which yielded the satisfaction of the following inequality ([285])

$$x + a > \frac{x}{1+b}$$

where

$$x = 0.5\alpha_0(1-\alpha_0^2)\left(\frac{T_d}{T_0+1}\right)^2, \quad a = 1.5S[2.5(1+\alpha_0)+1.25(1-\alpha_0)],$$

$$b = \left(\frac{4}{3}\right)\left(\frac{T_d}{T_0}\right)\frac{\nu_0 T_0}{[(1+\alpha_0)c_0\tau_d]}.$$

Since $a, b > 0$, this was satisfied identically for all $x > 0$ and hence by all $0 < \alpha_0 < 1$. Thus, our problem did not predict any chemical instabilities for the prototype dissociating gas of nitrogen.

Harry was less than pleased with this negative result, even though it ultimately appeared in the *Physics of Fluids* as my first publication, and asked me if weakly nonlinear stability theory might not yield the chemical instabilities that he had anticipated and now desired. Since nonlinear analyses of this sort, as sketched in the previous chapter, are basically pivoted about the critical point of linear stability theory, I told him that lacking an instability to infinitesimal disturbances this analysis could not be conducted. Harry's NSF proposal predicted a chemical instability for this situation that could then be extrapolated to various parallel-flow experiments and without it he felt his research program was jeopardized. In my opinion, the reason our problem had not predicted any chemical instability was that it contained no inherent mechanism to produce such an occurrence. I suggested that we should instead consider an entirely different scenario: Namely, the Rayleigh-Bénard problem of a Boussinesq convection layer involving a chemical quasi-equilibrium gas being heated from either below or *above*. Given that in the latter case there was gravitational stability for normal buoyancy-driven convection, if it could become unstable for a dissociating gas then Harry's proposed chemically driven instability would, in fact, be a reality.

So, in the middle of the 1969 Spring semester, I started working on that problem. The first order of business was to formulate the governing equations in two-dimensions. I considered a layer of dissociating gas in chemical quasi-equilibrium confined between two horizontal pure-conducting stress-free surfaces on which the tangential component of stress was zero that were maintained at the constant temperatures of $T = T_0$ and T_1 on the lower ($z = 0$)

and upper ($z = d$) surfaces, respectively. Starting with the Lighthill ([118]) model from the previous problem, I added a gravitational body-force term of the form $\boldsymbol{f} = -\rho g \boldsymbol{e}_3$, where $\boldsymbol{e}_3 = (0, 1)$ was a unit vector in the vertical direction and $g \equiv$ the acceleration due to gravity, to the momentum equations; took $k_\alpha = c_0$ for the sake of definiteness in the energy equation; and replaced the equilibrium term in the species equation with its generalized form $\alpha_e(T)$ which was implicitly defined by

$$\frac{\alpha_e^2(T)}{1 - \alpha_e(T)} = \frac{\rho_d}{\rho_0} \exp\left(-\frac{T_d}{T}\right).$$

Next, I introduced the Boussinesq approximation which consisted of assuming that the density satisfied the linear equation of state

$$\rho = \rho_0[1 - a(T - T_0) - b(\alpha - \alpha_0)]$$

and then neglecting this density variation from $\rho \equiv \rho_0$ everywhere except in the body-force term of the momentum equations. Finally, as a simplification, I retained only the first two terms of the Taylor series expansion for $\alpha_e(T)$ about the point $T = T_0$, yielding (Wollkind and Bdzil, [280])

$$\alpha_e(T) = \alpha_e(T_0) + \alpha_e'(T_0)(T - T_0) = \alpha_0 + m_0(T - T_0) \text{ where } m_0 = \alpha_e'(T_0) \equiv \varphi_0,$$

consistent with both the Boussinesq approximation and the chemical quasi-equilibrium assumption.

Under these conditions the two-dimensional governing Boussinesq-Lighthill equations of motion for $\boldsymbol{v} = (u, w)$, $\boldsymbol{\nabla}_2 \equiv (\partial/\partial x, \partial/\partial z)$, $D/Dt = \partial/\partial t + \boldsymbol{v} \cdot \boldsymbol{\nabla}_2$, and $\boldsymbol{\nabla}_2 \cdot \boldsymbol{\nabla}_2 \equiv \nabla_2^2$ were given by

$$\boldsymbol{\nabla}_2 \cdot \boldsymbol{v} = 0; \quad \frac{D\boldsymbol{v}}{Dt} = -\left(\frac{1}{\rho_0}\right)\boldsymbol{\nabla}_2 p - g[1 - a(T - T_0) - b(\alpha - \alpha_0)]\boldsymbol{e}_3 + \nu_0\nabla_2^2\boldsymbol{v};$$

with the energy and species equations

$$\rho_0\left(C_p\frac{DT}{Dt} + c_0\frac{D\alpha}{Dt}\right) = \Phi + k_e\nabla_2^2 T$$

$$\text{where } \Phi \equiv 2\mu_0 \sum_{i,j=1}^{2} \varepsilon_{ij}\varepsilon_{ij} \text{ and } k_e = k_0 + \rho_0 c_0 D_0 m_0;$$

$$\frac{D\alpha}{Dt} + \frac{\alpha - \alpha_0}{\tau_d} = m_0\left[\frac{T - T_0}{\tau_d} + D_0\nabla_2^2 T\right];$$

having the following boundary conditions adopted at the stress-free surfaces:

$$w = \frac{\partial u}{\partial z} + \frac{\partial w}{\partial x} = 0 \text{ at } z = 0, d;$$

and

$$T = T_0, \; \alpha = \alpha_e(T_0) = \alpha_0 \text{ at } z = 0; \; T = T_1, \; \alpha = \alpha_e(T_1) \text{ at } z = d.$$

Finally, I further simplified the energy equation by neglecting its $\Phi \sim \mu_0(\partial v_i/\partial x_j)^2$ dissipation term relative to its $k_e\nabla_2^2 T \sim k_0\partial^2 T/\partial z^2$ flux term, which was justified by making the following comparison between their sizes: From the momentum equation for w, I deduced that

$$w\frac{\partial w}{\partial z} \sim ag\Delta T \Rightarrow [w]^2 = ag(\Delta T)d \Rightarrow [\Phi] = \frac{\mu_0[w]^2}{d^2} = \frac{\mu_0 ag\Delta T}{d}$$

and

$$[k_e \nabla_2^2 T] = \frac{k_0 \Delta T}{d^2} \Rightarrow \frac{[\Phi]}{[k_e \nabla_2^2 T]} = \frac{\mu_0 a g \Delta T / d}{k_0 \Delta T / d^2} = \frac{\mu_0 a g d}{k_0} \ll 1,$$

since, when $d \cong 1$ cm, $\mu_0 a g d / k_0$ is typically of the order 10^{-7} consistent with

$$\frac{\mu_0 a g d}{k_0} = \frac{\mu_0}{\rho_0} \frac{\Delta T}{T_0} \frac{g}{C_p} \frac{\rho_0 C_p}{k_0} \frac{d}{\Delta T} = \frac{\nu_0}{\kappa_0} \frac{\Gamma_{ad}}{\beta_0} \frac{\Delta T}{T_0} = Pr D_s \frac{\Delta T}{T_0},$$

where for gases $Pr \equiv$ Prandtl number $= \nu_0 / \kappa_0 \cong 0.7$ with $\kappa_0 = k_0 / (\rho_0 C_p)$ and $\nu_0 = \mu_0 / \rho_0$; $D_s \equiv$ dissipation number $= \Gamma_{ad} / \beta_0$ with $\Gamma_{ad} = g / C_p$ and $\beta_0 = \Delta T / d$ is extremely small; and $a = 1/T_0$ for an ideal gas, while in general $\Delta T = T_0 - T_1$ is relatively small with respect to T_0. That neglect reduced the energy equation to

$$\rho_0 \left(C_p \frac{DT}{Dt} + c_0 \frac{D\alpha}{Dt} \right) = k_e \nabla_2^2 T.$$

This was the exact methodology Segel employed when simplifying the energy equation during his development of the Rayleigh-Bénard model for DiPrima's fluids class mentioned earlier in my description of our first meeting.

Having formulated the mathematical model for this problem, I sought a static $(u = w \equiv 0)$ stratified (dependent on z alone) exact solution of it of the form

$$u = w \equiv 0, \ T = T_0(z), \ \alpha = \alpha_0(z), \ p = p_0(z),$$

and found that

$$T_0(z) = T_0 - \beta_0 z, \ \alpha_0(z) = \alpha_0 - m_0 \beta_0 z, \ p_0(z) = p_A - \rho_0 g \left[1 + \frac{(a + m_0 b) \beta_0 z}{2} \right] z,$$

where $p_A \equiv$ atmospheric pressure.

Next, I began to investigate the linear stability of that state by considering a normal mode solution of the governing equations and boundary conditions of the form

$$[u, w, T, \alpha, p](x, z, t) = [0, 0, T_0(z), \alpha_0(z), p_0(z)]$$

$$+ \varepsilon_1 [u_1, w_1, T_1, \alpha_1, p_1](x, z, t) + \boldsymbol{O}(\varepsilon_1^2)$$

where, for the dimensionless time $\tau = \nu_0 t / d^2$,

$$[u_1, w_1, T_1, \alpha_1, p_1](x, z, t) = [U, W, \Theta, A, P](z) \exp\left(i q \frac{x}{d} + \sigma \tau \right).$$

Substituting this solution into my model system, neglecting terms of $O(\varepsilon_1^2)$, and cancelling the resulting common exponential factor times ε_1, an eigenvalue problem for the growth rate σ would be produced, that when solved yielded a cubic secular equation, the analysis of which, using the relevant Routh-Hurwitz conditions, determined its linear stability. In this context, the corresponding secular equation for the regular Rayleigh-Bénard problem was a quadratic, which is much easier to analyze. Before I even had a chance to start that more difficult analysis, however, an intervening development occurred which made its investigation no longer required.

To explain this intervention, it is necessary for me to back track a little to the last month of my first postdoctoral semester at SUNYA. During December of 1969, Harry told me that he had finally found a candidate to fill the other postdoctoral position on his NSF grant. The individual he had chosen was John Bdzil, a recent Ph.D. graduate of the University of Minnesota, who had done his dissertation in physical chemistry under the direction of Stephen Prager. The spelling of John's name caused me no end of fun over the years whenever I was trying to find a new telephone phone number for him either in Albany or Los Alamos, New Mexico, after his hiring by its national laboratory LANL in 1974. When one was trying to find such a number in those days, it had to be done by communicating with an information operator. That was accomplished by dialing the information number 1-(area code for the place in question)-555-1212. I had this exchange down to a science and it always went the same way:

Me: "I'd like the number of a John Bdzil. That is B as in boy, D as in dog, Z as in zero, I as in idiot, and L as in love."

Operator: "Do you have an address for this name, sir?"

Me: "Do you think there will be more than one?"

Operator: "The number is...."

It was my job, during the first month of John's postdoctoral appointment, to get him up to speed in chemical quasi-equilibrium fluid mechanics and stability theory as applied to governing systems of partial differential equations. He grasped basic concepts very quickly and in no time began working on his own aspects of these problems.

In the summer of 1969, Steve Prager came to SUNYA to work with Harry. He was the son of William Prager, who started the US renaissance in applied mathematics at Brown University after the Second World War. George Handelman, one of Prager's first Ph.D. students at Brown and department chair at RPI, said there was a time when all applied mathematics departments in the US could trace their lineage back to him. During that summer, Harry had me discuss the problem, on which I was working, with Steve. He immediately made the following suggestion: Why not extend the chemical quasi-equilibrium approximation employed by Lighthill ([118]) only for the Laplacian to the substantial derivative, as well and make the replacement $D\alpha/Dt = m_0 DT/Dt$. That Rabbit, having been Pulled Out of a Hat, allowed the transformation of the energy and species equations into

$$\frac{DT}{Dt} = \kappa_e \nabla_2^2 T = \frac{\alpha_0 - \alpha}{m_0 \tau_d} + \frac{T - T_0}{\tau_d} + D_0 \nabla_2^2 T,$$

where $\kappa_e = k_e/(\rho_0 c_e)$ with k_e as defined above and $c_e = C_p + c_0 m_0$ were the effective thermometric conductivity, thermal conductivity, and specific heat of the binary mixture, respectively. Now, using that new species equation, it was possible to eliminate this variable from the problem by converting the momentum vector equation into

$$\frac{D\boldsymbol{v}}{Dt} = -\left(\frac{1}{\rho_0}\right)\boldsymbol{\nabla}_2 p - g[1 - (a + bm_0)(T - T_0) - bm_0\tau_d(D_0 - \kappa_e)\nabla_2^2 T]\boldsymbol{e}_3 + \nu_0 \nabla_2^2 \boldsymbol{v}.$$

That equation in addition to the simplified continuity and energy equations, *i.e.*,

$$\boldsymbol{\nabla}_2 \cdot \boldsymbol{v} = 0 \text{ and } \frac{DT}{Dt} = \kappa_e \nabla_2^2 T,$$

respectively, - along with the appropriate kinematic, dynamical, and thermal boundary conditions, *i.e.*,

$$w = \frac{\partial u}{\partial z} + \frac{\partial w}{\partial x} = 0 \text{ at } z = 0, d; \text{ and } T = T_0 \text{ at } z = 0, \ T = T_1 \text{ at } z = d;$$

respectively, - as already developed, constituted the mathematical formulation for the problem. Note, by setting $b = 0$, this formulation can be reduced to that of the regular Rayleigh-Bénard problem. For an ideal gas, $b = 1/(1 + \alpha_0)$.

As a final detail let us deduce the proper formula for $m_0 = \alpha'_e(T_0)$ from the implicit relationship satisfied by $\alpha_e(T)$; namely

$$\frac{\alpha_e^2(T)}{1 - \alpha_e(T)} = \frac{\rho_d}{\rho_0} \exp\left(-\frac{T_d}{T}\right).$$

This can be accomplished by taking the derivative of both sides of this relationship and then evaluating the result at $T = T_0$. In order to proceed, we need to catalogue a few more properties of the derivative: $(x^r)' = r x^{r-1}$, $(f - g)' = f' - g'$, and $(f/g)' = (f'g - fg')/g^2$. Further, recall from the definition of the natural exponential and the chain rule of differentiation, both introduced in Chapter 2, that $(e^g)' = e^g g'$. Now, taking the derivative of the left-hand side of this relationship with respect to T, we obtain, from the quotient and chain rules, that

$$\frac{2\alpha_e(T)[1 - \alpha_e(T)] + \alpha_e^2(T)}{[1 - \alpha_e(T)]^2} \alpha'_e(T) = \frac{[2 - \alpha_e(T)]\alpha_e(T)}{[1 - \alpha_e(T)]^2} \alpha'_e(T),$$

while taking the derivative of its right-hand side with respect to T, we obtain, from the differentiation rule for $e^{-T_d/T}$ and $(-T_d/T)' = T_d(-T^{-1})' = T_d T^{-2} = T_d/T^2$, that

$$\frac{\rho_d}{\rho_0} \exp\left(-\frac{T_d}{T}\right) \frac{T_d}{T^2} = \frac{\alpha_e^2(T)}{1 - \alpha_e(T)} \frac{T_d}{T^2}.$$

Thus, evaluating these two derivatives at $T = T_0$, setting them equal, solving for $\alpha'_e(T_0)$, and recalling that $\alpha_e(T_0) = \alpha_0$, yields the desired formula ([285])

$$\alpha'_e(T_0) = m_0 = \frac{\alpha_0(1 - \alpha_0)}{2 - \alpha_0} \frac{T_d}{T_0^2}.$$

Considering the perturbation system for our exact solution relevant to these finalized governing equations in conjunction with the appropriate BC's and nondimensionalizing position, time, velocity, pressure, and temperature by introducing the scale factors d, d^2/ν_0, ν_0/d, $\rho_0 \nu_0^2/d^2$, and $\beta_0 d$, respectively, we obtained

$$\nabla_2 \cdot \boldsymbol{v}_1 = 0 \text{ with } \boldsymbol{v}_1 = (u_1, w_1),$$

$$\frac{D\boldsymbol{v}_1}{Dt} = -\nabla_2 p_1 + R Pr^{-1}[(1 + \delta)T_1 + \varepsilon B \nabla_2^2 T_1]\boldsymbol{e}_3 + \nabla_2^2 \boldsymbol{v}_1,$$

$$\frac{DT_1}{Dt} - w_1 = Pr^{-1} \nabla_2^2 T_1;$$

$$w_1 = \frac{\partial u_1}{\partial z} + \frac{\partial w_1}{\partial x} = T_1 = 0 \text{ at } z = 0, 1;$$

where

$$Pr = \frac{\nu_0}{\kappa_e}, \quad R = \frac{g a \beta_0 d^4}{\kappa_e \nu_0}, \quad \delta = \frac{b m_0}{a}, \quad S = \frac{\nu_0}{D_0}, \quad \varepsilon = \frac{T_d \nu_0 \delta}{d^2}, \quad B = S^{-1} - Pr^{-1};$$

and the same nomenclature has been employed for the dimensionless variables as that used for the original ones. It is instructive to point out that the thermal expansivity coefficient

a contained in the Rayleigh number R has been traditionally denoted by α (Koschmeider, [103]), which we have reserved for our species variable now eliminated from the formulation of the problem. This helped cause a typographical error to go undetected in my textbook (Wollkind and Dichone, [284]) on comprehensive applied mathematical modeling with Bonni mentioned in the Introduction. Specifically, an α was employed instead of an a in the Rayleigh number for a chemically driven convection capstone problem of its concluding chapter. That just proved one of Lee Segel's favorite adages: Namely: "Although there can only be a finite number of typographical errors in the proofs of any publication, it takes an infinite amount of time to find them all."

We are now ready to return to the linear stability problem which resulted from Steve Prager's suggestion. Writing this system in component form, we obtained from our general perturbation equations

$$\frac{\partial u_1}{\partial x} + \frac{\partial w_1}{\partial z} = 0,$$

$$\frac{\partial u_1}{\partial t} = -\frac{\partial p_1}{\partial x} + \nabla_2^2 u_1,$$

$$\frac{\partial w_1}{\partial t} = -\frac{\partial p_1}{\partial z} + RPr^{-1}[(1+\delta)T_1 + \varepsilon B \nabla_2^2 T_1] + \nabla_2^2 w_1,$$

$$\frac{\partial T_1}{\partial t} - w_1 = Pr^{-1}\nabla_2^2 T_1;$$

$$w_1 = \frac{\partial u_1}{\partial z} + \frac{\partial w_1}{\partial x} = T_1 = 0 \text{ at } z = 0,1;$$

upon neglect of its nonlinear terms which reduces the substantial derivatives D/Dt in those equations to $\partial/\partial t$.

In order to motivate the form of the normal mode solution to this system, which shall be introduced below, it is necessary to resume the examination of the properties of the circular functions begun in the last chapter by determining their derivatives. Toward that end we consider the triangles depicted in Fig. 3.4. Then, recalling that the area of a triangle is equal to one-half the product of the lengths of its base times its altitude, we can deduce that the areas of $\Delta 0P1$ and $\Delta 0T1$ are given by $\sin(\theta)/2$ and $\tan(\theta)/2$, respectively We can further deduce that the area of a sector subtending an angle θ in radian measure of a circle of radius r satisfies the ratio

$$\frac{\text{area of a sector}}{\text{area of the circle}} = \frac{\theta}{2\pi}$$

or

$$\text{area of a sector} = \frac{\theta}{2\pi}\pi r^2 = \frac{\theta r^2}{2}.$$

Thus, the area of sector $0P1$ in Fig. 3.4 is given by $\theta/2$, since in this instance $r = 1$. Observing from this figure that

$$\text{area of } \Delta 0P1 < \text{area of sector } 0P1 < \text{area of } \Delta 0T1,$$

we obtain

$$\sin(\theta) < \theta < \tan(\theta) = \frac{\sin(\theta)}{\cos(\theta)}$$

upon cancellation of the common factor 2 and employment of the definition of $\tan(\theta)$. Now, assuming without loss of generality, that $0 < \theta < \pi/2$ and dividing this inequality by $\sin(\theta)$ yields

$$1 < \frac{\theta}{\sin(\theta)} < \frac{1}{\cos(\theta)}.$$

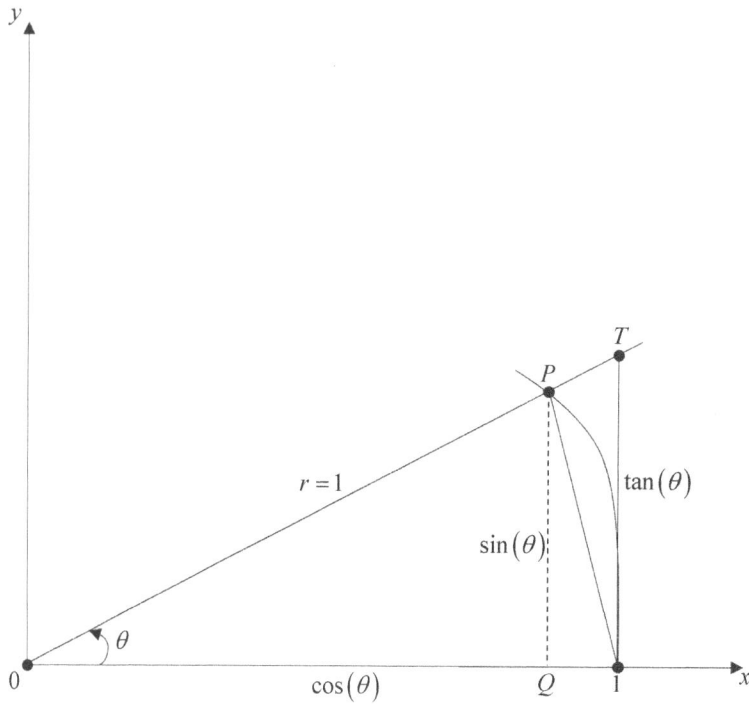

FIGURE 3.4
The geometric scenario used to deduce that the limit of $\sin(\theta)/\theta \to 1$ as $\theta \to 0$.

Then, taking the reciprocal of each entry in this expression and noting that such a procedure reverses the sense of the inequalities, we obtain

$$\cos(\theta) < \frac{\sin(\theta)}{\theta} < 1.$$

Finally, given that $\cos(\theta) \to 0$ in the limit as $\theta \to 0$, we have deduced the fundamental result

$$\lim_{\theta \to 0} \frac{\sin(\theta)}{\theta} = 1.$$

To compute the derivatives of the sine and cosine functions from the definition of derivative, we next need to demonstrate one more limiting result

$$\lim_{\theta \to 0} \frac{\cos(\theta) - 1}{\theta} = 0,$$

which can be derived from the fundamental result as follows: First examine

$$f(\theta) = \frac{\cos(\theta) - 1}{\theta} = \frac{[\cos(\theta) - 1][\cos(\theta) + 1]}{\theta[\cos(\theta) + 1]}$$

$$= \frac{\cos^2(\theta) - 1}{\theta[\cos(\theta) + 1]} = -\frac{\sin(\theta)}{\theta} \frac{\sin(\theta)}{\cos(\theta) + 1},$$

where use has been made in the above derivation of the relationship

$$\cos^2(\theta) + \sin^2(\theta) = 1,$$

given that these functions were defined as coordinates of a point on the unit circle. Then employing the fact that the limit of a product is the product of the limits of its factors if such limits exist, we obtain the desired result

$$\lim_{\theta \to 0} f(\theta) = -\left[\lim_{\theta \to 0} \frac{\sin(\theta)}{\theta}\right]\left[\frac{\sin(\theta)}{\cos(\theta) + 1}\right] = -1 \cdot \frac{0}{2} = -1 \cdot 0 = 0.$$

Further, to compute the derivatives in question, we must also make use of the standard trigonometric identities involving the sum of angles:

$$\sin(x + \theta) = \sin(x)\cos(\theta) + \cos(x)\sin(\theta)$$

and

$$\cos(x + \theta) = \cos(x)\cos(\theta) - \sin(x)\sin(\theta).$$

We are finally ready to compute the derivatives of the $\sin(x)$ and the $\cos(x)$:

$$\sin'(x) = \lim_{\theta \to 0} \frac{\sin(x + \theta) - \sin(x)}{\theta} = \lim_{\theta \to 0} \frac{\sin(x)\cos(\theta) + \cos(x)\sin(\theta) - \sin(x)}{\theta}$$

$$= \sin(x)\lim_{\theta \to 0} \frac{\cos(\theta) - 1}{\theta} + \cos(x)\lim_{\theta \to 0} \frac{\sin(\theta)}{\theta}$$

$$= \sin(x) \cdot 0 + \cos(x) \cdot 1 = \cos(x)$$

and

$$\cos'(x) = \lim_{\theta \to 0} \frac{\cos(x + \theta) - \cos(x)}{\theta} = \lim_{\theta \to 0} \frac{\cos(x)\cos(\theta) - \sin(x)\sin(\theta) - \cos(x)}{\theta}$$

$$= \cos(x)\lim_{\theta \to 0} \frac{\cos(\theta) - 1}{\theta} - \sin(x)\lim_{\theta \to 0} \frac{\sin(\theta)}{\theta}$$

$$= \cos(x) \cdot 0 - \sin(x) \cdot 1 = -\sin(x).$$

Returning to the linear perturbation system, we looked for a real valued version of our previous normal mode solution of the form

$$u_1(x, z, t) = u_{11} \sin(qx) \cos(\pi z) e^{\sigma t},$$
$$[w_1, T_1]x, z, t = [w_{11}, T_{11}] \cos(qx) \sin(\pi z) e^{\sigma t},$$
$$p_1(x, z, t) = p_{11} \cos(qx) \cos(\pi z) e^{\sigma t};$$

where $u_{11}^2 + w_{11}^2 + T_{11}^2 + p_{11}^2 \neq 0$ and $q \geq 0$. Since using the chain rule

$$[\sin(qx)]' = q \cos(qx), \quad [\cos(qx)]' = -q \sin(qx),$$
$$[\sin(\pi z)]' = \pi \cos(\pi z), \quad [\cos(\pi z)]' = -\pi \sin(\pi z),$$

and

$$\nabla_2^2 [u_{11}, w_{11}, T_{11}] = -k^2 [u_{11}, w_{11}, T_{11}] \text{ where } k^2 = q^2 + \pi^2,$$

we deduced that the first and second linear perturbation equations were proportional to $\cos(qx) \cos(\pi z)$ and $\sin(qx) \cos(\pi z)$, respectively; the third and fourth equations and the first and third B.C.'s, to $\cos(qx) \sin(\pi z)$; and the second B.C., to $\sin(qx) \sin(\pi z)$. Note, that given $\sin(\pi z) = 0$ at $z = 0, 1$, this solution satisfied the B.C.'s identically while the four perturbation equations yielded an algebraic linear homogeneous system in the four constants $u_{11}, w_{11}, T_{11}, p_{11}$. Then, in order to satisfy the nontriviality condition, the determinant of the coefficients of these constants must vanish from which the following quadratic secular equation was obtained

$$Prk^2 \sigma^2 + k^4 (Pr + 1) \sigma + k^6 - Rq^2 (1 + \delta - \varepsilon Bk^2) = 0.$$

In particular, this linear homogeneous system represented an eigenvalue problem for σ which admitted the trivial solution while nontrivial ones could only occur provided that secular equation was satisfied.

As an aside, given its relevance to the procedure just employed to obtain this equation, let me offer a personal experience that took place in one of Dick DiPrima's lectures during his differential equations course at RPI in the Spring Semester of 1962. DiPrima in introducing such matrix eigenvalue problems asked Pat Munroe, the only woman in the class: "A linear homogeneous system of equations only has a nontrivial solution provided the determinant of its coefficients is nonzero; right, Miss Munroe?" Of course, she answered "Yes." to which DiPrima then responded: "Wrong, Miss Munroe! If this determinant were nonzero then by Cramer's rule there would be a unique solution and since zero is a solution that would be the only solution. So, in order for there to be a nontrivial solution this determinant must be zero." Ever since that time I have always referred to this nontriviality condition for matrix eigenvalue problems as the Pat Munroe algorithm.

There is another exchange in that class this time between DiPrima and me which is relevant to the distinction between the complex and real forms of the normal mode solution presented earlier in our development. It occurred when DiPrima began a class by writing the simple second order ordinary differential equation $y''(x) + y(x) = 0$ on the board and then asked me "What are the two linearly independent solutions to this equation? You!" So, I answered "$e^{\pm ix}$," having mentally let $y(x) = e^{mx}$ and calculated $m^2 = -1$ or $m_{1,2} = \pm i$. That wasn't good enough for DiPrima, however, who yelled at me "I don't want that answer! I want an answer that any God damned high school student could understand!" Recalling Euler's Formula introduced earlier in the chapter

$$e^{ix} = \cos(x) + i \sin(x) = g(x; i),$$

which incidentally can be most easily demonstrated by using the Maclaurin series

$$e^z = \sum_{n=0}^{\infty} \frac{z^n}{n!}, \quad \cos(x) = \sum_{k=0}^{\infty} \frac{(-1)^k x^{2k}}{(2k)!}, \quad \sin(x) = \sum_{k=0}^{\infty} \frac{(-1)^k x^{2k+1}}{(2k+1)!}$$

for $z \in \mathbb{C}$ and $x \in \mathbb{R}$, taking $z = ix$ where $i^2 = -1$, and considering

$$e^{ix} = \sum_{n=0}^{\infty} \frac{(ix)^n}{n!} = \sum_{k=0}^{\infty} \frac{i^{2k} x^{2k}}{(2k)!} + \sum_{k=0}^{\infty} \frac{i^{2k+1} x^{2k+1}}{(2k+1)!}$$

$$= \sum_{k=0}^{\infty} \frac{(-1)^k x^{2k}}{(2k)!} + i \sum_{k=0}^{\infty} \frac{(-1)^k x^{2k+1}}{(2k+1)!} = g(x; i);$$

I responded with: "Alright, Professor DiPrima. Then it's the damned $\cos(x)$ and the God damned $\sin(x)$."

We have examined the stability behavior of a quartic polynomial satisfied by σ earlier in this chapter by employing the Routh-Hurwitz criterian. We shall now do so for the just deduced quadratic secular equation which is of the form

$$q_2 \sigma^2 + q_1 \sigma + q_0 = 0$$

where

$$q_2 = Prk^2 > 0, \quad q_1 = k^4(Pr + 1) > 0, \quad q_0 = k^6 - Rq^2(1 + \delta - \varepsilon Bk^2)$$

or, upon division by q_2,

$$\sigma^2 + 2P_1\sigma + P_0 = (\sigma - \sigma^+)(\sigma - \sigma^-) = 0$$

where

$$2P_1 = \frac{q_1}{q_2}, \quad P_0 = \frac{q_0}{q_2}.$$

We shall demonstrate for such quadratics by direct means that $\mathrm{Re}(\sigma^{\pm}) < 0$, if and only if (iff) $P_{1,0} > 0$, which is equivalent to the Routh-Hurwitz criterion for second-order polynomials, as follows:

Here from the quadratic formula $\sigma^{\pm} = -P_1 \pm \sqrt{P_1^2 - P_0}$ and there are two cases to consider:

(i) $P_1^2 - P_0 < 0$:
$$\mathrm{Re}(\sigma^{\pm}) = -P_1 < 0 \text{ iff } P_1 > 0 \Rightarrow P_0 > P_1^2 > 0;$$

(ii) $P_1^2 - P_0 \geq 0$:

$$\mathrm{Re}(\sigma^{\pm}) = \sigma^{\pm} \text{ where } \sigma^+ + \sigma^- = -2P_1, \ \sigma^+\sigma^- = P_0 \Rightarrow \sigma^{\pm} < 0 \text{ iff } P_{1,0} > 0.$$

Since, for our quadratic $P_1 > 0$, this reduces to the stability criterion $P_0 > 0$ or equivalently

$$R < R_0(q^2; \varepsilon B) = \frac{(q^2 + \pi^2)^3}{q^2[1 + \delta - \varepsilon B(q^2 + \pi^2)]}.$$

In what follows, it will be assumed that $0 < \varepsilon B \ll 1$, with a detailed examination of this assumption being deferred until later in the chapter. Under this assumption we concluded that the marginal stability curve on which $\sigma = 0$

$$R = R_0(q^2; \varepsilon B) = \frac{(q^2 + \pi^2)^3}{q^2[1 + \delta - \varepsilon B(q^2 + \pi^2)]}$$

had a vertical asymptote at $q_0^2 = (1+\delta)/(\varepsilon B) - \pi^2 > 0$, with a normal convective instability occurring for $R > R_0 > 0$ when $0 < q^2 < q_0^2$ and a chemically driven one occurring for $R < R_0 < 0$ when $q^2 > q_0^2$. These two branches are denoted by "+" and "−," respectively, in Fig. 3.5, which is a schematic plot of $R = R_0(Q; \varepsilon B)$ in the Q-R plane where $Q = q^2$. Here $Q_0 = q_0^2$ and linearly stable or unstable regions are identified by $P_0 > 0$ or $P_0 < 0$, respectively.

As can be seen from this figure that marginal stability curve has a minimum point at (Q_c^+, R_c^+) on its "+" branch and a maximum point at (Q_c^-, R_c^-) on its "−" branch. Denoting the square of these critical wavenumbers generically by Q_c, they satisfy $dR_0(Q_c; \varepsilon B)/dQ = 0$ which implies that

$$\overbrace{\varepsilon B Q_c^2}^{①} - \overbrace{2(1+\delta)Q_c}^{②} + \overbrace{\pi^2(1+\delta - \varepsilon B \pi^2)}^{③} = 0.$$

Although one could use the quadratic formula and then expand those exact solutions, we employed the following simpler procedure to obtain asymptotic representations for its roots $Q_c^{\pm}$: Sequentially retain two of the terms, as denoted above by the encircled numbers, solve the truncated equation, and compare the resulting size of the neglected term with one of those retained. If it is small for $0 < \varepsilon B \ll 1$, our result will be valid but, if it is not, then that result must be rejected. There are three possible cases:

(i) Retain ② & ③ and neglect ①:

$$③ - ② = 0 \Rightarrow Q_c \sim \frac{\pi^2}{2} \Rightarrow ① \sim \frac{\pi^4 \varepsilon B}{4} \ll ③ \text{ as } \varepsilon B \to 0.$$

(ii) Retain ① & ② and neglect ③:

$$① - ② = 0 \Rightarrow Q_c \sim \frac{2(1+\delta)}{\varepsilon B} \Rightarrow ③ \sim \pi^2(1+\delta) \ll ② \text{ as } \varepsilon B \to 0.$$

(iii) Retain ① & ③ and neglect ②:

$$① + ③ = 0 \Rightarrow Q_c \sim \pm \frac{i\pi(1+\delta)^{1/2}}{(\varepsilon B)^{1/2}}$$

$$\Rightarrow |②| \sim \frac{2\pi(1+\delta)^{3/2}}{(\varepsilon B)^{1/2}} \gg ③ \text{ as } \varepsilon B \to 0.$$

Here recall that $i \equiv \sqrt{-1}$ and the notation $|\ldots|$ is to be interpreted as the absolute value of a complex quantity where, for $z = x + iy$ with $x, y \in \mathbb{R} \equiv$ reals, $|z| \equiv \sqrt{x^2 + y^2}$.

Observe that cases (i,ii) are valid corresponding to the lead terms of $Q_c^{\pm}$, respectively, or

$$Q_c^+ \sim \frac{\pi^2}{2} \text{ and } Q_c^- \sim \frac{2(1+\delta)}{\varepsilon B} \text{ as } \varepsilon B \to 0,$$

while case (iii) must be rejected. In this context, note that the $Q_c = 0$ root for case (ii) corresponds to the root of case (i).

Observe from Fig. 3.5 that

$$Q_c^+ \sim \frac{\pi^2}{2} < Q_0 \sim \frac{1+\delta}{\varepsilon B} < Q_c^- \sim \frac{2(1+\delta)}{\varepsilon B} \text{ as } \varepsilon B \to 0.$$

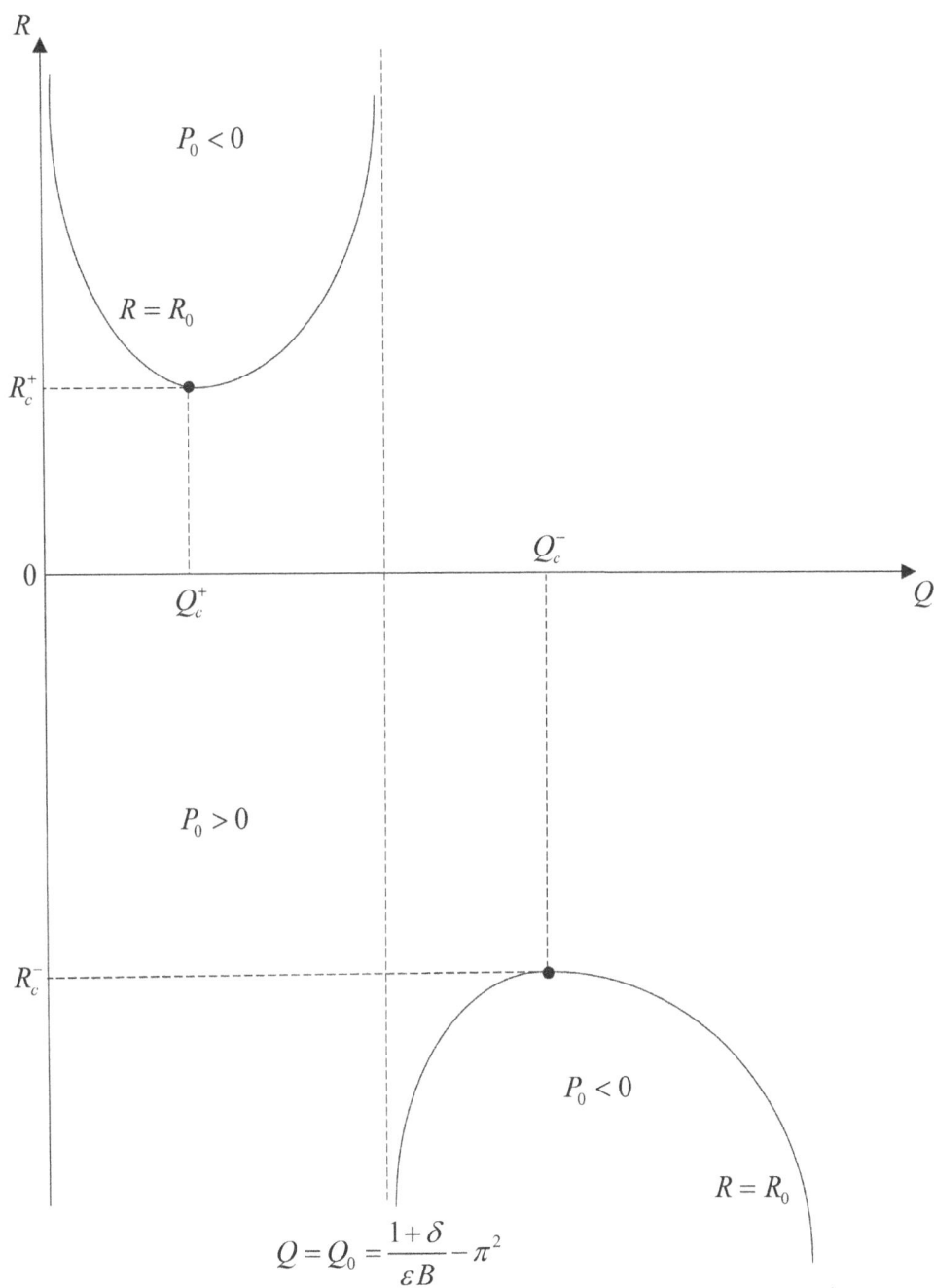

FIGURE 3.5
A schematic plot of R_0 vs. Q for $B > 0$ and $0 < \varepsilon B \ll 1$.

Now defining $R_c^\pm = R_0(Q_c^\pm; \varepsilon B)$, we deduced in a straight-forward manner given these results that

$$R_c^+ \sim \frac{27\pi^4}{4(1+\delta)} \text{ and } R_c^- \sim \frac{-4(1+\delta)}{(\varepsilon B)^2} \text{ as } \varepsilon B \to 0.$$

For $R_c^- < R < R_c^+$, we had $P_0 > 0$ and hence predicted linear stability, while for $R > R_c^+ > 0$ or $R < R_c^- < 0$, we had $P_0 < 0$ and hence predicted there would be a band of wavenumbers squared centered about Q_c^+ or Q_c^-, respectively, characterized by the occurrence of linearly unstable disturbances. Note, in this context, that $R > 0$ corresponds to $T_0 > T_1$ and thus represents heating from below or cooling from above, while $R < 0$ corresponds to $T_0 < T_1$ and thus represents heating from above or cooling from below. Therefore, upon identification of the linearly unstable disturbance associated with the "−" branch of our marginal stability curve with the chemical instability being sought for heating from above, my prediction of the possibility of such an occurrence, as described by Harry in his research proposal, became a reality and, needless to say, he was ecstatic about it.

The next obvious project to pursue was an investigation of the nonlinear stability behavior of this chemical instability but, unlike Lee with my thesis research, Harry wanted to publish the linear result separately. His reason for proceeding in this manner was that he just heard the renowned Belgian chemist Ilya Prigogine had been working on a similar problem where a fluid convective instability could be produced upon heating from above. Harry was afraid of being scooped if we waited to publish our results until after the nonlinear stability analysis had been completed. Indeed, I did not actually start that nonlinear analysis until the Fall of 1969 taking the month of August to write up my thesis results which appeared as Wollkind and Segel ([285]), discussed in the previous chapter. Since I began working for Harry even before defending my thesis, there had been no time to do so until that month. In those days, the NSF would only support a postdoctoral scholar for 11 out of 12 months of the academic year, so this was my month "off," which was when he let me work with Lee on my thesis publication. So, Harry and I wrote up the linear problem as quickly as possible in July of 1969, had that manuscript typed with its figure corresponding to Fig. 3.5 completed by our graphic artist (whom John always referred to as "the Peacock," after her last name), and submitted the whole thing to the *Physics of Fluids* in October of 1969. Ironically, we drew Lee as our reviewer and he requested that an examination of the parameter range for which our assumption of $0 < \varepsilon B \ll 1$ was valid be included in a revised version of the original manuscript. In compliance with this request, we added an appendix, consisting of that examination, to our manuscript which was then accepted in revised form for publication and thus this linear chemical instability paper appeared as Wollkind and Frisch ([286]) in volume 14 of that journal. This examination can be summarized as follows:

Since ε, as defined, was a positive parameter, our chemically driven linear instabilities depended upon the parameter $B > 0$ and $0 < \varepsilon B \ll 1$. Given the definition of B, the former inequality meant that the effective Prandtl number Pr must exceed the Schmidt number S. We derived a necessary and sufficient condition for that occurrence in this case to be

$$S < \frac{8(4+\alpha_0)}{15(3+\alpha_0)}.$$

Recalling that Pr typically has a value of 0.7 for gases, the plausibility of this condition was reinforced by the fact that the right-hand side of this inequality ranged from 0.7 to 2/3 as α_0 varied over its domain of definition $0 < \alpha_0 < 1$. Since S is known to range from 0.5 to 1.0 as α_0 varies over its domain ([118]), we assumed a linear interpolation for S; namely,

$$S = \frac{1+\alpha_0}{2};$$

and showed this inequality then held when $0 < \alpha_0 < \alpha_0^+ = [\sqrt{769} - 22]/15 \cong 0.38$. Further recalling that $a = 1/T_0$ and $b = 1/(1 + \alpha_0)$, we demonstrated in this context, that

$$\delta = \frac{bm_0}{a} = \frac{\alpha_0(1 - \alpha_0)}{(2 - \alpha_0)(1 + \alpha_0)} \frac{T_d}{T_0} \text{ and } \varepsilon = \frac{\tau_d \nu_0 \delta}{d^2}$$

had the typical values of 1.33 and 0.535×10^{-3}, respectively, for nitrogen with $\alpha_0 = 0.33 < \alpha_0^+$, $T_0 = 8{,}452$ K, $T_d = 113{,}000$ K, $\tau_d = 10^{-5}$ sec, $d = 1.0$ cm, and $\nu_0 = 40.23$ cm^2/sec $= 5.4 \times 10^{-6}(T_0/K)^{1.75}$ cm^2/sec (see Table 3.1 where, in addition, $\rho_0 = 0.00125$ g/cm^3 and $\rho_d = 130$ g/cm^3).

In the event, it turned out that Harry had good reason to be worried. Prigogine's paper with Robert Schechter and J.R. Hamm of the University of Texas at Austin entitled "Thermal diffusion and convective stability," although involving a thermohaline fluid, such as a heated horizontal layer of saltwater, rather than our dissociating gas, appeared as Schechter *et al.* ([201]) in volume 15 of the same journal, the *Physics of Fluids*, with the following abstract:

> The stability of a two-component fluid layer subjected to a temperature gradient has been studied, and the associated thermal diffusion separation has been found to exert a large influence even when the separations are small. The most unexpected and perhaps important result is that an instability has been found which can give rise to convection currents even though the density gradient is not adverse. Thus, a system heated from above can become unstable even when the fluid is less dense at the top of the system provided the denser substance rises to the upper plate.

They did not reference our result, in spite of the fact that Harry provided them with a preprint, and left the lasting impression that this was the first time a buoyancy-driven as opposed to a surface tension-driven convective instability had been produced upon heating from above. Schechter claimed in a correspondence that it was a different scenario because their density increased with the salt concentration while ours decreased with the dissociated species concentration. I pointed out to him that, had we introduced the *associated* species instead, so would ours with no change in the results. This is another example of how one can lose credit for a research accomplishment, as described in the last chapter. That was not helped when the Royal Swedish Academy of Sciences decided to award Prigogine the 1977 Nobel Prize in Chemistry "for his contributions to non-equilibrium thermodynamics, particularly the theory of dissipative structures," while noting that "the most well-known dissipative structure is perhaps the so-called Bénard instability."

So, in September of 1969, I began a nonlinear stability analysis of our chemical instability problem by considering a solution to its perturbation system of the normal mode form

$$\begin{bmatrix} w_1 \\ T_1 \end{bmatrix} \sim A(t)\mathbf{g}_{11}(z)\cos(q_c x) + A^2(t)[\mathbf{g}_{20}(z) + \mathbf{g}_{22}(z)\cos(2q_c x)]$$

$$+ A^3(t)[\mathbf{g}_{31}(z)\cos(q_c x) + \mathbf{g}_{33}(z)\cos(3q_c x)] \text{ with } \mathbf{g}_{nm}(z) = \begin{bmatrix} w_{nm} \\ T_{nm} \end{bmatrix}(z),$$

$$p_1(x,z,t) \sim A(t)p_{11}\cos(q_c x) + A^2(t)[p_{20}(z) + p_{22}(z)\cos(2q_c x)]$$

$$+ A^3(t)[p_{31}(z)\cos(q_c x) + p_{33}(z)\cos(3q_c x)],$$

and

$$u_1(x,z,t) \sim A(t)u_{11}(z)\sin(q_c x) + A^2(t)[u_{20}(z) + u_{22}(z)\sin(2q_c x)]$$

$$+ A^3(t)[u_{31}(z)\sin(q_c x) + u_{33}(z)\sin(3q_c x)]$$

where, as usual, $A(t)$ satisfies the amplitude equation

$$\frac{dA}{dt} \sim \sigma A(t) - a_1 A^3(t).$$

Here, q_c generically denoted $q_c^{\pm}$, since we wanted to investigate the nonlinear behavior of both the instabilities predicted by linear theory. Substituting this solution into our perturbation system, equating coefficients of terms proportional to $A^n(t) \cos(mq_c x)$ or $A^n(t) \sin(mq_c x)$, and eliminating the p_{nm}'s or the u_{nm}'s by using the momentum or continuity equations, respectively, we obtained a system of ordinary differential equations and B.C.'s for each $g_{nm}(z)$ appearing explicitly in that solution. Defining the 2×2 matrix ordinary differential operators

$$L_m = \begin{pmatrix} [D^2 - m^2 q_c^2]^2 & -m^2 q_c^2 R P r^{-1}\{1 + \delta + \varepsilon B[D^2 - m^2 q_c^2]\} \\ 1 & P r^{-1}[D^2 - m^2 q_c^2] \end{pmatrix}$$

and

$$M_m = \begin{pmatrix} [D^2 - m^2 q_c^2] & 0 \\ 0 & 1 \end{pmatrix} \quad \text{where } D \equiv \frac{d}{dz},$$

we catalogued those systems as follows:

Terms of $O(A)$ gave:

$$(L_1 - \sigma M_1)g_{11}(z) = \mathbf{0} \equiv \begin{bmatrix} 0 \\ 0 \end{bmatrix}, \quad g_{11}(z) = \mathbf{0} \text{ at } z = 0, 1.$$

Here, for

$$L = \begin{pmatrix} L_{11} & L_{12} \\ L_{21} & L_{22} \end{pmatrix} \text{ and } g(z) = \begin{bmatrix} W \\ \theta \end{bmatrix}(z),$$

$$Lg(z) = \begin{bmatrix} L_{11}W(z) + L_{12}\theta(z) \\ L_{21}W(z) + L_{22}\theta(z) \end{bmatrix};$$

while the B.C.-notation is to be interpreted as

$$g_{nm}(z) = \mathbf{0} \text{ at } z = 0, 1 \Rightarrow w_{nm} = D^2 w_{nm} = T_{nm} = 0 \text{ at } z = 0, 1.$$

Terms of $O(A^2)$ gave:

$$w_{20}(z) \equiv 0; \quad (Pr^{-1}D^2 - 2\sigma)T_{20}(z) = S_{20}(z), \quad T_{20}(z) = 0 \text{ at } z = 0, 1;$$

and

$$(L_2 - 2\sigma M_2)g_{22}(z) = \boldsymbol{S}_{22}(z), \quad g_{22}(z) = \mathbf{0} \text{ at } z = 0, 1;$$

where S_{20} and $\boldsymbol{S}_{22}$ are quadratic scalar and vector functions, respectively, of w_{11} and T_{11}.

Although there were also two systems of $O(A^3)$, for our purpose we only needed to consider that one proportional to $\cos(q_c x)$, which was given by

$$(L_1 - 3\sigma M_1)g_{31}(z) + a_1 M_1 g_{11}(z) = \boldsymbol{S}_{31}(z), \quad g_{31}(z) = \mathbf{0} \text{ at } z = 0, 1;$$

where $\boldsymbol{S}_{31}$ is a quadratic vector function of w_{11}, T_{11}, T_{20}, w_{22}, and T_{22}.

The $O(A)$ system is equivalent to the linear eigenvalue problem for σ with $q = q_c$. Wishing to examine the long-time behavior of its so-called most dangerous mode, we took σ to satisfy

$$\sigma = \sigma_0(R; q_c) \text{ where } 2k_c^2 Pr\sigma_0 = \sqrt{\mathfrak{D}} - (Pr+1)k_c^4 \text{ with}$$

$$k_c^2 = q_c^2 + \pi^2 \text{ and } \mathfrak{D} = (Pr-1)^2 k_c^8 + 4PrRq_c^2 K_c \text{ for } K_c = 1 + \delta - \varepsilon Bk_c^2.$$

Here, σ_0 represented the largest root of our quadratic secular equation and was real since $\mathfrak{D} \geq 0$ by virtue of $R, K_c > 0$ for $q_c = q_c^+$ and of $R, K_c < 0$ for $q_c = q_c^-$. Indeed, σ_0 being real is implicit to our perturbation expansion procedure as in the last chapter and hence, given the debacle that this caused for my thesis problem, I was much more careful to demonstrate that assumption from the outset this time. Further, the corresponding eigenvector was then given by

$$\boldsymbol{g}_{11}(z) = \begin{bmatrix} 1 \\ \gamma_{11} \end{bmatrix} \sin(\pi z) \text{ with } \gamma_{11} = \frac{Pr}{k_c^2 + Pr\sigma_0}.$$

Recall, the main goal of any nonlinear stability analysis is to evaluate the Landau constants. In the last chapter, what Wollkind *et al.* ([292]) termed a direct method for doing so, was sketched. Since that methodology had not yet been developed at the time of this analysis, we employed the adjoint operator method instead to determine the Landau constant a_1 of our amplitude equation. Hence, the next order of business for us was to pose and solve the so-called adjoint operator linear eigenvalue problem. We proceeded by considering

$$(L_1^+ - \sigma^+ M_1^+)\boldsymbol{g}_{11}^+(z) = \boldsymbol{0}, \ \boldsymbol{g}_{11}^+(z) = \begin{bmatrix} w_{11}^+ \\ T_{11}^+ \end{bmatrix}(z) = \boldsymbol{0} \text{ at } z = 0, 1;$$

such that

$$\langle L_1 \boldsymbol{g}_{11}(z), \boldsymbol{g}_{11}^+(z) \rangle = \langle \boldsymbol{g}_{11}(z), L_1^+ \boldsymbol{g}_{11}^+(z) \rangle$$

and

$$\langle M_1 \boldsymbol{g}_{11}(z), \boldsymbol{g}_{11}^+(z) \rangle = \langle \boldsymbol{g}_{11}(z), M_1^+ \boldsymbol{g}_{11}^+(z) \rangle$$

with the *inner product* of any two vectors $\boldsymbol{g}(z) = \begin{bmatrix} W \\ \theta \end{bmatrix}(z)$ and $\boldsymbol{g}^+(z) = \begin{bmatrix} W^+ \\ \theta^+ \end{bmatrix}(z)$ defined by

$$\langle \boldsymbol{g}(z), \boldsymbol{g}^+(z) \rangle = \int_0^1 [W(z)W^+(z) + \theta(z)\theta^+(z)]\, dz$$

where that *definite integral* (see Fig 3.6)

$$\int_0^1 f(z)\, dz = \lim_{\substack{\Delta \to 0 \\ n \to \infty}} \sum_{j=1}^n f(\bar{z}_j)\Delta z_j;$$

can be evaluated by employing the *Fundamental Theorem of the Calculus* (see Stewart, [240])

$$\int_0^1 f(z)\, dz = F(z)|_{z=0}^{z=1} \equiv F(1) - F(0) \text{ for } F'(z) = f(z),$$

which relates integration to anti-differentiation and represents the area under $y = f(z) \geq 0$ from $z = 0$ to 1.

Using those inner-product identities, we deduced that $L_1^+ = L_1^T$ and $M_1^+ = M_1^T = M_1$ in a straight-forward manner where for the matrix L as defined above its transpose

$$L^T = \begin{pmatrix} L_{11} & L_{21} \\ L_{12} & L_{22} \end{pmatrix}.$$

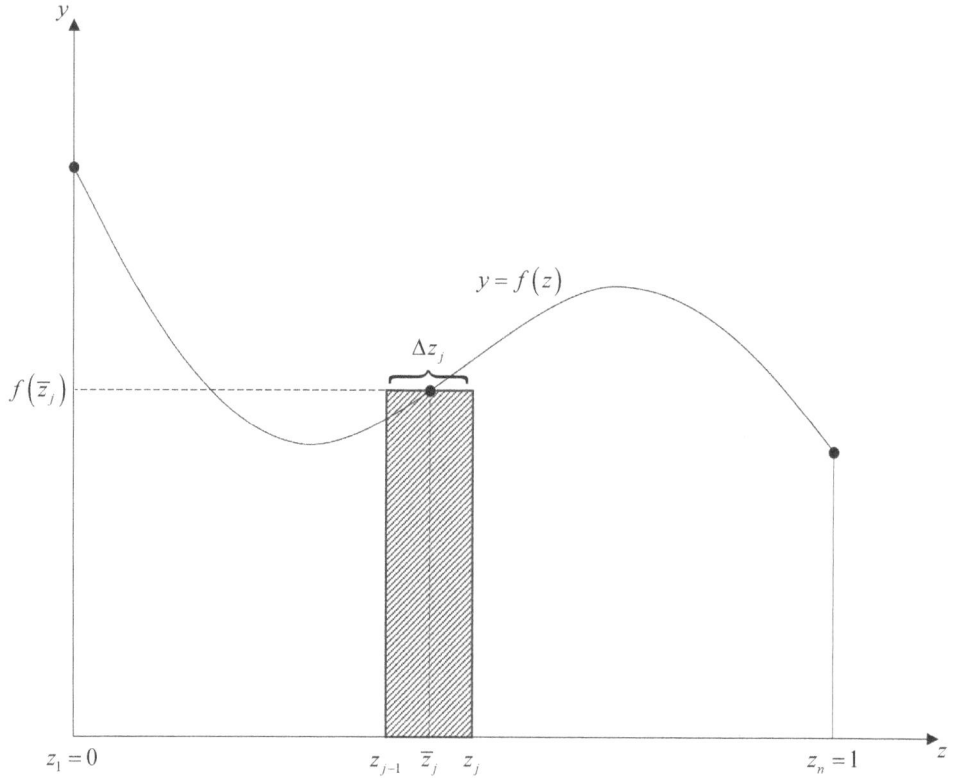

FIGURE 3.6
Plot of the curve $y = f(z) \geq 0$ where its definite integral from $z = 0$ to 1, with $\Delta z_j = z_j - z_{j-1}$, $j = 1, 2, \ldots, n$; $z_1 = 0$; $z_n = 1$; $\bar{z}_j \in \Delta z_j$; and $\Delta = \max_j \Delta z_j$; is independent of the choice for both the partition Δz_j and the intermediate points $\bar{z}_j$. That integral can be shown to equal $F(1) - F(0)$ for $F'(z) = f(z)$ by the Fundamental Theorem of the Calculus and represents the area under this curve (see Stewart, [240]).

Now, employing these results and interpreting the B.C.'s in a similar sense to that for the linear eigenvalue problem, we determined the adjoint operator eigenvalue $\sigma^+ = \sigma$ since it satisfied exactly the same secular equation. Hence, wishing to examine the fate of the most-dangerous mode, we again took $\sigma^+ = \sigma_0$ and then the corresponding adjoint eigenvector was given by

$$\boldsymbol{g}_{11}^+(z) = \begin{bmatrix} 1 \\ \gamma_{11}^+ \end{bmatrix} \sin(\pi z) \text{ with } \gamma_{11}^+ = -(k_c^4 + k_c^2 \sigma_0).$$

Turning our attention to the $O(A^2)$ problems, we determined that for our choice of $\boldsymbol{g}_{11}(z)$

$$S_{20}(z) = \frac{\pi \gamma_{11}}{2} \sin(2\pi z) \text{ and } \boldsymbol{S}_{22}(z) \equiv \boldsymbol{0}.$$

Then substituting those functions into these problems, we solved them to obtain

$$T_{20}(z) = C_{20} \sin(2\pi z) \text{ where } C_{20} = -\frac{\pi \gamma_{11}}{2(4\pi^2 Pr^{-1} + 2\sigma_0)} \text{ and } \boldsymbol{g}_{22}(z) \equiv \boldsymbol{0}.$$

Examining the $O(A^3)$ system proportional to $\cos(q_c x)$, we noted that except for the presence of a "3" instead of a "1" in its operator, $\boldsymbol{g}_{31}(z)$ satisfied a nonhomogeneous problem of the same form as the linear. In this context, we determined that for our solutions of the $O(A)$ and $O(A^2)$ systems

$$\boldsymbol{S}_{31}(z) = \pi C_{20} \begin{bmatrix} 0 \\ \sin(3\pi z) - \sin(\pi z) \end{bmatrix}.$$

In order to apply the usual Fredholm solvability condition (Ince, [85]) to this system, we took the inner product of it with $\boldsymbol{g}_{11}^+(z)$ and obtained

$$\langle L_1 \boldsymbol{g}_{31}(z), \boldsymbol{g}_{11}^+(z) \rangle - 3\sigma_0 \langle M_1 \boldsymbol{g}_{31}(z), \boldsymbol{g}_{11}^+(z) \rangle + a_1 \langle M_1 \boldsymbol{g}_{11}(z), \boldsymbol{g}_{11}^+(z) \rangle$$
$$= \langle \boldsymbol{S}_{31}(z), \boldsymbol{g}_{11}^+(z) \rangle.$$

Then making use of the adjoint properties of the operators and the eigenvalues, we deduced that

$$\langle L_1 \boldsymbol{g}_{31}(z), \boldsymbol{g}_{11}^+(z) \rangle = \langle \boldsymbol{g}_{31}(z), L_1^T \boldsymbol{g}_{11}^+(z) \rangle$$
$$= \langle \boldsymbol{g}_{31}(z), \sigma_0 M_1 \boldsymbol{g}_{11}^+(z) \rangle = \sigma_0 \langle M_1 \boldsymbol{g}_{31}(z), \boldsymbol{g}_{11}^+(z) \rangle.$$

Finally, employing this result, noting that

$$\sigma_0(R_c^\pm; q_c^\pm) = 0,$$

and taking the limit of the previous expression, we found the Fredholm solvability condition

$$a_1 = \lim_{R \to R_c^\pm} \left[\frac{\langle \boldsymbol{S}_{31}, \boldsymbol{g}_{11}^+(z) \rangle}{\langle M_1 \boldsymbol{g}_{11}(z), \boldsymbol{g}_{11}^+(z) \rangle} \bigg|_{q_c = q_c^\pm} \right].$$

To evaluate the inner products contained in this representation for a_1 it was necessary to employ the so-called orthogonality property for the functions $\sin(m\pi z)$ where m is a positive integer

$$2 \int_0^1 \sin(m\pi z) \sin(n\pi z) \, dz = \delta_{mn} = \begin{cases} 1 \text{ for } m = n \\ 0 \text{ for } m \neq n \end{cases},$$

which can be demonstrated from the trigonometric identity

$$2 \sin(\theta) \sin(\varphi) = \cos(\theta - \varphi) - \cos(\theta + \varphi)$$

by taking $\theta = m\pi z$ and $\varphi = n\pi z$ in that identity. Then

$$2\int_0^1 \sin(m\pi z)\sin(n\pi z)\, dz = \int_0^1 \cos([m-n]\pi z)\, dz - \int_0^1 \cos([m+n]\pi z)\, dz.$$

Since

$$\int_0^1 \cos([m-n]\pi z)\, dz = \begin{cases} 1 & \text{for } m = n \\ \left.\frac{\sin([m-n]\pi z)}{(m-n)\pi}\right|_{z=0}^{z=1} & \text{for } m \neq n \end{cases} = \delta_{mn}$$

and

$$\int_0^1 \cos([m+n]\pi z)\, dz = \left.\frac{\sin([m+n]\pi z)}{(m+n)\pi}\right|_{z=0}^{z=1} = 0,$$

by virtue of the fact that $\sin([m \pm n]\pi z) = 0$ at $z = 0, 1$, the result follows immediately.

Given that

$$M_1 \boldsymbol{g}_{11}(z) = \begin{bmatrix} -k_c^2 \\ \gamma_{11} \end{bmatrix} \sin(\pi z),$$

we employed our previous results to evaluate the inner products contained in the expression for a_1 and obtained

$$\langle M_1 \boldsymbol{g}_{11}(z), \boldsymbol{g}_{11}^+(z) \rangle = (-k_c^2 + \gamma_{11}\gamma_{11}^+) \int_0^1 \sin^2(\pi z)\, dz = \frac{-k_c^2 + \gamma_{11}\gamma_{11}^+}{2}$$

and

$$\langle \boldsymbol{S}_{31}(z), \boldsymbol{g}_{11}^+(z) \rangle = \pi C_{20}\gamma_{11}^+ \left[\int_0^1 \sin(3\pi z)\sin(\pi z)\, dz - \int_0^1 \sin^2(\pi z)\, dz \right]$$

$$= \frac{-\pi C_{20}\gamma_{11}^+}{2}.$$

Thus

$$a_1 = \lim_{R \to R_c^\pm} \left[\frac{\pi^2 \gamma_{11}\gamma_{11}^+}{2(4\pi^2 Pr^{-1} + 2\sigma_0)(-k_c^2 + \gamma_{11}\gamma_{11}^+)} \bigg|_{q_c = q_c^\pm} \right]$$

where

$$\gamma_{11}\gamma_{11}^+ = \frac{-Pr(k_c^4 + k_c^2\sigma_0)}{k_c^2 + Pr\sigma_0}.$$

Since

$$\lim_{R \to R_c^\pm} \left[\gamma_{11}\gamma_{11}^+ \big|_{q_c = q_c^\pm} \right] = -Pr k_c^{\pm 2} \quad \text{where } k_c^{\pm 2} = q_c^{\pm 2} + \pi^2$$

and

$$\lim_{R \to R_c^\pm} \sigma_0(R; q_c^\pm) = \sigma_0(R_c^\pm; q_c^\pm) = 0,$$

$$a_1 = \frac{\pi^2 Pr k_c^{\pm 2}}{8\pi^2 Pr^{-1} k_c^{\pm 2}(1 + Pr)} = \frac{Pr^2}{8(Pr + 1)} > 0.$$

Note that this positive Landau constant was identical for both our two cases of instability and independent of δ or εB. Indeed, given that setting $\delta = 0$, which implies $\varepsilon = 0$ as well, since the latter parameter is proportional to the former, reduces our system to the classical Rayleigh-Bénard problem, this meant that the Landau constant is unaltered by the introduction of a dissociating gas in chemical quasi-equilibrium. This is a direct demonstration of our assertion in the last chapter that such a Rayleigh-Bénard problem produced an identically positive Landau constant which caused Segel such consternation when the Landau constant associated with my thesis research became negative for certain ranges of parameter values. Examining our amplitude equation as in Chapter 2, we concluded that for $\sigma_0 > 0$

$$\lim_{t \to \infty} A(t) = \left(\frac{\sigma_0}{a_1} \right)^{1/2} = A_e.$$

Hence, for $R > R_c^+$ or $R < R_c^-$ where $\sigma_0 > 0$, we obtained the equilibrium amplitude

$$A_e = \frac{2}{Pr} (2\sigma_0 [Pr + 1])^{1/2}$$

and, therefore, predicted stable one-dimensional periodic finite-amplitude solutions or *rolls*

$$\left[\begin{array}{c} w_1 \\ T_1 \end{array} \right] (x, z, t) \to \left[\begin{array}{c} w_e \\ T_e \end{array} \right] (x, z) \sim A_e g_{11}(z) \cos(q_c x)$$

$$\text{as } t \to \infty \text{ where } q_c = \left\{ \begin{array}{ll} q_c^+ & \text{for } R > R_c^+ \\ q_c^- & \text{for } R < R_c^- \end{array} \right. ,$$

with characteristic wavelengths $\lambda_c^{\pm} = 2\pi/q_c^{\pm}$ and $\lambda_c^{*\pm} = \lambda_c^{\pm} d$ in dimensionless and dimensional variables, respectively. Employing our previously deduced asymptotic results for $q_c^{\pm} = (Q_c^2)^{1/2}$ this yielded

$$\lambda_c^{*+} \sim 2\sqrt{2}d \text{ and } \lambda_c^{*-} \sim \sqrt{2}\pi d \left(\frac{\varepsilon B}{1 + \delta} \right)^{1/2} \text{ as } \varepsilon B \to 0.$$

Now, using the typical parameter values introduced earlier for nitrogen with $\alpha_0 = 0.33$, we found

$$B = S^{-1} - Pr^{-1} \cong \frac{2}{1 + \alpha_0} - (0.7)^{-1} = 0.075$$

and obtained $\lambda_c^{*+} \cong 2\sqrt{2}$ cm and $\lambda_c^{*-} \cong 0.018$ cm. Thus, in these parameter ranges where linear theory predicted instability to infinitesimal perturbations, our nonlinear analysis showed such disturbances actually re-equilibrated into stationary finite amplitude roll solutions.

Further, for $R_c^+ < R < R_c^-$ where $\sigma_0 < 0$, the pure conduction solution was stable to both infinitesimal and finite-amplitude disturbances. Thus, here the nonlinear stability analysis enhanced the predictions of linear theory.

This completed the nonlinear analysis of our chemical instability so we again submitted the paper to the *Physics of Fluids* and drew Segel as its reviewer. Lee, who was always concerned with physical mechanisms responsible for the occurrence of instabilities, wanted us to suggest one for our chemical instability in a slightly revised version of the manuscript. We did so as follows: From the body force term in the momentum equation, it can be seen that the change in the "effective density" $\Delta\rho$ caused by a small perturbation in the temperature ΔT_1 is given by

$$\Delta\rho = -a\rho_0 (1 + \delta - k_c^2 \varepsilon B) \Delta T_1 \text{ where } a = \frac{1}{T_0} > 0.$$

When the fluid is heated from below, we can think of this as an introduction of a fluctuation

$\Delta T_1 > 0$ at $z = 0$. Then, since the inequalities $B > 0$ and $k_c^2 = k_c^{+2} < (1 + \delta)/(\varepsilon B)$ are satisfied for the occurrence of the normal convective instability, $\Delta \rho < 0$, and we shall have an "effective density" inversion, which causes this buoyancy-driven instability given that the "lighter" fluid is at the bottom while when the fluid is heated from above under the conditions that $B > 0$, but now $k_c^2 = (k_c^-)^2 > (1 + \delta)/(\varepsilon B)$, as in the case for the chemical instability, then the introduction of the fluctuation $\Delta T_1 > 0$ at $z = d$ results in $\Delta \rho > 0$ and the "effective density" inversion will be preserved, given that the "heavier" fluid is still at the top.

Let us pose a physical argument for this chemical instability. Consider a small fluid element in the bulk which is displaced upward into a region of somewhat higher temperature. In accordance with our chemical quasi-equilibrium assumption, some of the molecules of this element will dissociate by absorbing heat from the surrounding fluid in that region. Since for $B > 0$ species diffusivity effects outweigh thermal diffusivity ones, local changes of physical density due to dissociation will be quickly compensated for by species diffusion, while the temperature changes produced by these chemical reactions will be dissipated more slowly by virtue of the relatively lower thermometric conductivity. Thus, the small fluid element will be locally hotter than the surrounding fluid and hence rise. Similarly, a small fluid element displaced downward will have a tendency to sink.

Segel also stated that, in the classical Rayleigh-Bénard problem, only rolls were stable for the three-dimensional nonproperty varying model. If the variation of the shear viscosity with temperature were taken into account, hexagonal patterns would be stable for certain parameter ranges. Due to the similarity of the mathematical formulation, he expected analogous results for our problem and, since experiments had shown that the cell size of these hexagonal patterns for the classical Bénard problem were of the order of the wavelength of the critical disturbance, cell sizes approximately equal to $\lambda_c^{*\pm}$ could be anticipated. In his opinion, the length scale λ_c^{*-} of the interactions responsible for our chemical instability being small, lent plausibility to this result.

To comply with Segel's suggestions, we added an extra section to a revised version of the manuscript both including the physical explanation for our instabilities and incorporating his statement given in the previous paragraph with an acknowledgement to him. This was accepted upon resubmission and appeared as Wollkind and Frisch ([287]) in volume 14 of the *Physics of Fluids*. Observe that this was published before Schechter *et al.* ([201]) as well, so we really did not have to break the paper up into two parts. These linear and nonlinear results were entitled Chemical Instabilities I and III, respectively, since in the interim John had submitted his first joint paper with Harry to the *Physics of Fluids*, which was also reviewed by Segel, and appeared as Chemical Instabilities II. In his review of our second paper, Lee added an addendum requesting that I teach John not to split infinitives in his expositions. An example of such a split infinitive, incidentally, is if I were to rewrite the previous phrase as "that I teach John to not split infinitives." Segel was always a stickler for proper grammatical usage in scientific papers. For instance, when working on Wollkind and Segel ([285]) in August of 1969, I had used the phrase "Motivated by the five-component vectors." Segel objected to this and said to me, "What did they do? Whisper in your ear? It should be 'Motivated by *the form of* the five-component vectors' instead." He also insisted upon "we shall" rather than the "we will" often found in the literature.

Eventually, about ten years later, John and Harry (Bdzil and Frisch, [10]) performed both the linear and longitudinal planform nonlinear stability analyses of a model system for this problem that did not include either the quasi-equilibrium assumptions employed by

Lighthill ([118]) or suggested by Steve Prager. This gave rise to a cubic secular equation, as previously described for the Lighthill model, and that required numerical computation to complete these analyses which then reproduced our result of the occurrence of a chemical instability upon heating from above.

We shall conclude this chapter by examining the results of the nonlinear stability analysis of the classical Rayleigh-Bénard problem discussed earlier and summarized by Fig. 2.13, given that it has been a subject of historical importance in both the present as well as the previous one. Our formulation can be converted into this three-dimensional model, that Stuart and Segel ([219]) and Segel ([211]) analyzed, if we consider the velocity vector $\boldsymbol{v}_1 = (u_1, v_1, w_1)$, take $\delta = 0$, adopt the constitutive relation for the nondimensional shear viscosity with scale factor μ_0

$$\mu(T_1) = 1 + \gamma \cos(\pi[T_1 - z]) \cong 1 + \gamma[\cos(\pi z) + \pi \sin(\pi z)T_1] \text{ where } |\gamma| \ll 1,$$

and replace $\boldsymbol{\nabla}_2$ and $\nabla_2^2 \boldsymbol{v}_1$ by $\boldsymbol{\nabla} = (\partial/\partial x, \partial/\partial y, \partial/\partial z)$ and $\boldsymbol{\nabla} \cdot [\mu(T_1)\{\boldsymbol{\nabla} \boldsymbol{v}_1 + (\boldsymbol{\nabla} \boldsymbol{v}_1)^T\}]$, respectively.

They sought a solution of their basic equations that, to lowest order, consisted of the z-component of velocity
$$w_1(x, y, z, t) \sim f(x, y, t)\sin(\pi z),$$

where $f(x, y, t)$ represented the hexagonal planform function involving the amplitudes $A_j(t)$ and phases $\varphi_j(t)$ for $j = 1, 2, 3$, satisfying the six amplitude-phase differential equations containing growth rate σ_0 and Landau constants $a_{0,1,2}$, as defined in the interfacial alloy solidification pattern formation analysis of Chapter 2, with corresponding expansions for its other dependent variables consistent to this. Substituting that solution into their governing equations, they obtained values for these coefficients of the amplitude-phase equations in the same manner as we determined those in our amplitude equation. In particular, for this classical Rayleigh-Bénard problem, the growth rate was given by

$$2Prk_c^2\sigma_0(R; q_c) = \sqrt{\mathfrak{D}} - (Pr + 1)k_c^4$$

where

$$k_c^2 = q_c^2 + \pi^2, \quad \mathfrak{D} = (Pr + 1)^2 k_c^8 + 4Prk_c^2 q_c^2(R - R_c), \quad q_c^2 = \frac{\pi^2}{2}, \quad R_c = \frac{27\pi^4}{4},$$

which is equivalent to our representation when $\delta = 0$, while the Landau constants satisfied

$$a_0 = \gamma b_0 \text{ with } b_0 = O(1) > 0$$

and

$$2a_2 > a_1 > 0,$$

the second of the above inequalities following by virtue of the fact that, as stated earlier, their determination for a_1 was identical to ours. Observe, from this representation for $\sigma_0(R; q_c)$ it is straightforward to deduce that $\sigma_0(R; q_c) < 0$ for $R < R_c$, $\sigma_0(R_c; q_c) = 0$, and $\sigma_0(R; q_c) > 0$ for $R > R_c$.

As in Chapter 2, the potentially stable critical points of the amplitude-phase equations can be catalogued by considering the following member of each relevant equivalent class corresponding to $A_1 = A_0$, $A_2 = A_3 = B_0$, $\varphi_1 = \varphi_2 = \varphi_3 = 0$ where

$$\text{I}: A_0 = B_0 = 0; \quad \text{II}: A_0^2 = \frac{\sigma_0}{a_1}, \ B_0 = 0;$$

$$\text{III}^\pm : A_0 = B_0 = A_0^\pm = \frac{-2a_0 \pm [4a_0^2 + (a_1 + 4a_2)\sigma_0]^{1/2}}{a_1 + 4a_2} \quad \text{or}$$

$$(a_1 + 4a_2)A_0^{\pm^2} + 4a_0 A_0^\pm - \sigma_0 = 0.$$

Here critical point I represents the pure conduction solution while II and III$^\pm$ correspond to roll- and hexagonal-type convection cells, respectively. Again, as in Chapter 2, the orbital stability behavior of these potentially stable critical points can be posed in terms of σ. Thus, critical point I is stable in this sense for $\sigma < 0$, while the stability behavior of II and III$^\pm$, which now depends only on the sign of a_0 since $2a_2 - a_1 > 0$ for the Rayleigh-Bénard problem, has been summarized in Table 3.2 that contains the relevant entries of Table 2.1 under this additional condition where, as before,

$$\sigma_{-1} = -\frac{4a_0^2}{a_1 + 4a_2}, \quad \sigma_1 = \frac{16a_1 a_0^2}{(2a_2 - a_1)^2}, \quad \sigma_2 = \frac{32(a_1 + a_2)a_0^2}{(2a_2 - a_1)^2},$$

while over the parameter range in question, $A_0^+ > 0$ and $A_0^- < 0$.

TABLE 3.2
Orbital stability behavior of critical points II and III$^\pm$ with a_0 when $2a_2 - a_1 > 0$.

a_0	Stable structures
$+$	III$^-$ for $\sigma_{-1} < \sigma < \sigma_2$, II for $\sigma > \sigma_1$
0	II for $\sigma > 0$
$-$	III$^+$ for $\sigma_{-1} < \sigma < \sigma_2$, II for $\sigma > \sigma_1$

Note, from Table 3.2, that stable hexagonal patterns satisfy the necessary condition $a_0 A_0^\pm < 0$. Thus, III$^+$ patterns which have amplitude $A_0^+ > 0$ are only stable provided $a_0 < 0$ while III$^-$ patterns, which have amplitude $A_0^- < 0$, are only stable provided $a_0 > 0$. Observe from the a_0-relationship for the classical Rayleigh-Bénard problem that the sign of the second-order Landau constant is the same as the sign of γ. Since

$$\frac{d\mu(T_1)}{dT_1} = \gamma[\pi \sin(\pi z)] \text{ and } \pi \sin(\pi z) > 0 \text{ for } 0 < z < 1,$$

this means that motion in hexagonal cellular fluid flow will be ascending in the cell centers when $d\mu/dT_1 < 0$, which is true for most liquids and descending when $d\mu/dT_1 > 0$, true for most gases. Further, the condition for truncation

$$|a_0| \ll (a_1 + 4a_2)^2,$$

would be satisfied identically by virtue of $|\gamma| \ll 1$. Finally, note from Table 3.2, that should the variation of shear viscosity with temperature not be taken into account by assuming $\mu(T_1) \equiv 1$ then $\gamma = 0$, which implies $a_0 = 0$ as well, and hence only stable rolls would be predicted for the classical Rayleigh-Bénard problem.

We complete this examination by a detailed explanation of the bifurcation diagram in the R-$|A_0|$ plane that comprises Fig. 2.13. Here, $\sigma_0(R_j; q_c) = \sigma_j$ for $j = -1, 1$, and 2, while $\sigma_0(R_c; q_c) = 0$, as defined in its caption. The three branches in that figure correspond to plots of the absolute value of its amplitude versus Rayleigh number for the three critical points I, II, and III$^\pm$ of the problem, which have been labeled "pure conduction state," "rolls," and "hexagons," respectively. This diagram represents the stability results of the Rayleigh-Bénard problem graphically. Here, heavy black solid curves denote stability and light solid ones, instability. On the hexagonal branch

$$|A_0| = \begin{cases} A_0^+ \text{ for } \gamma < 0 \\ -A_0^- \text{ for } \gamma > 0 \end{cases}.$$

That portion of this branch, denoted by a broken curve, corresponds to a subcritical state occurring when $\sigma_0 < 0$: The heavy segment representing subcritically stable hexagons; the light segment, that unstable continuation which would be defined for the opposite sign of γ, i.e., $\gamma > 0$ for A_0^+ and $\gamma < 0$ for A_0^-. In order to determine the stable states predicted for a given value of R in such a diagram, one needs only draw a vertical line through that value and examine its intersection(s) with those three branches. Doing so yields the following predictions catalogued in Table 3.3.

TABLE 3.3
Morphological predictions versus R for the Rayleigh-Bénard problem from Fig. 2.13.

R-Interval	Stable states
$0 < R < R_{-1}$	Pure conduction
$R_{-1} < R < R_c$	Pure conduction and Hexagons
$R_c < R < R_1$	Hexagons
$R_1 < R < R_2$	Hexagons and Rolls
$R > R_2$	Rolls

In the intervals where there is bistability, such as

$$R_{-1} < R < R_c \text{ or } R_1 < R < R_2,$$

the initial conditions will determine which of the two potentially stable states is selected.

There remains only one final detail of Fig 2.13 to explain: Namely, a determination of that value of σ_0, denoted by σ_e in what follows, corresponding to the point of intersection of the "hexagon" and "roll" branches at $R = R_0$. Here, the equilibrium values of these two amplitudes will coincide and, using the notation A_e to designate that value, we shall have

$$(a_1 + 4a_2)A_e^2 = \sigma_e - 4a_0 A_e \text{ and } A_e^2 = \frac{\sigma_e}{a_1}.$$

Hence

$$(a_1 + 4a_2)\frac{\sigma_e}{a_1} = \sigma_e + 4a_2\frac{\sigma_e}{a_1} = \sigma_e - 4a_0 A_e$$

or

$$\sigma_e = -a_1 a_0 \frac{A_e}{a_2} > 0.$$

Since $a_{1,2} > 0$ while on the stable portion of the hexagonal branch

$$a_0 A_e < 0,$$

this relationship is self-consistent. We note that it implicitly assumes the adoption of a similar convention for rolls involving the solutions

$$A_e^\pm = \pm\left(\frac{\sigma_e}{a_1}\right)^{1/2},$$

in the spirit of Pulling a Rabbit Out of a Hat, each of which generates the same morphological pattern. Now, squaring both sides of this relationship, we find that

$$\sigma_e^2 = \frac{a_1^2 a_0^2 A_e^2}{a_2^2} = \frac{a_1^2 a_0^2 \left(\frac{\sigma_e}{a_1}\right)}{a_2^2} = \frac{a_1 a_0^2 \sigma_e}{a_2^2}$$

or, since $\sigma_e > 0$, this yields the desired result

$$\sigma_e = \frac{a_1 a_0^2}{a_2^2}.$$

Hence, given that this intersection point corresponds to $R = R_0$, we have deduced that

$$\sigma_0(R_0; q_c) = \sigma_e = \frac{a_1 a_0^2}{a_2^2}.$$

Finally, synthesizing these results, we obtain

$$A_e^2 = \frac{\sigma_e}{a_1} = \frac{a_0^2}{a_2^2},$$

which, upon taking the square root of both sides of this expression, implies that

$$|A_e| = \sqrt{A_e^2} = \sqrt{\left(\frac{a_0}{a_2}\right)^2} = \left|\frac{a_0}{a_2}\right|$$

as indicated in Fig. 2.13, and thus completing our explanation of that figure and its caption.

We shall encounter bifurcation diagrams of this sort again in the next chapter, which concentrates upon the ecological agro-economic phenomenon of a temperature-dependent predator-prey biological-control mite interaction on apple tree foliage located in the orchards of the central valleys of Washington state.

4

Temperature-Dependent Predator-Prey Mite Interaction on Apple Tree Foliage: Dynamical Systems Analysis

When Lee Segel became interested in biomathematics and life science modeling, he first concentrated his research efforts on developmental biological phenomena, such as the life cycle of the cellular slime mold, and then on biomedical phenomena, such as the response of the human immune system. Lee felt that these areas represented those frontiers of mathematical biology where significant new advances could be made. At this time he thought that all of the most important contributions to modeling in mathematical ecology had already been made by its pioneers such as Alfred Lotka ([127]). Since natural history always fascinated me and many of the books still in my library, such as *Wildlife in Color* (1951) and *Nature's Ways* (1951) by the famous naturalists Roger Tory Peterson and Roy Chapman Andrews, respectively, dealt with this subject, I did not agree with that assertion and this difference of opinion led to a number of discussions between us in the summer of 1971. I thought that predator-prey population dynamics offered a fertile area for quantification, especially with all the new data being collected relevant to the biological control of certain arthropod pest species by their natural enemies. Lee felt that working on such agro-ecosystems would not produce anything worthwhile. I was particularly influenced in this regard by the paper Rosenzweig ([190]), which appeared in the journal *Science* earlier that year and interpreted the results of a set of laboratory experiments involving predator-prey mite interactions on oranges conducted by Huffaker *et al.* ([84]).

Robert MacArthur, who played a fundamental role in the development of theoretical ecology by changing it from a primarily descriptive field to an experimental one, was Michael Rosenzweig's Ph.D. advisor at the University of Pennsylvania. Michael followed in his mentor's footsteps as the reference cited above and its sequel (Rosenzweig, [191]) indicate. My initial research in this area, Wollkind ([275]) was an extension, allowing intraspecific carnivore interaction, of one of his subsequent papers Rosenzweig ([192]) that introduced a graphical representation of the linear stability analysis of the community equilibrium point for a carnivore-herbivore-plant exploitation interaction, both of which were published in the journal *American Naturalist*. Although Rosenzweig was actually a Biology Department faculty member at SUNY Albany in my last year of postdoctoral study there, we did not meet until I invited him to be our keynote speaker at a conference hosted by the Department of Mathematics at WSU, while my three-trophic level paper was in press. By then he was a University of New Mexico faculty member before assuming his current position at the University of Arizona. During the course of that visit, Michael admired my new McGraw-Hill book *Insects of the World* (1972) by Walter Linsenmaier, translated from the German by Leigh Chadwick, which was unique in its presentation since Linsenmaier was trained as a painter and not an entomologist. Possessing two copies of this extraordinarily illustrated volume, I presented him with one of them as a gift. Princeton University Press in its monographs in population biology, edited by Robert MacArthur who started this series

DOI: 10.1201/9781003195603-4

upon moving to Princeton, had just published *Stability and Complexity in Model Ecosystems* [139] by Robert May, which was to influence my future modeling endeavors greatly, and we also discussed the potential impact of that publication on theoretical ecology.

FIGURE 4.1
The artist W. G. Wiles' impression of the attack by the predacious mite *Metaselius occidentalis* on the spider mite pest *Tetranychus mcdanieli*, occurring upon an apple tree leaf. This oil painting hung on the wall of Stan Hoyt's office at the WSU Tree Fruit Research Center in Wenatchee, Washington.

Shortly thereafter Jesse Logan, a Ph.D. student in the Department of Entomology at WSU, enrolled in my two-semester continuum mechanics course. As luck would have it, he was doing his dissertation research Logan ([123]) on the association of the McDaniel spider mite pest, *Tetranychus mcdanieli* McGregor, with its control agent, the predacious soil mite *Metasieulus occidentalis* Nesbitt, in the apple agro-ecosystem (see Fig. 4.1), which was a predator-prey mite interaction very similar to the one studied by Huffaker *et al.* ([84]). This work was being conducted under the direction of Stan Hoyt, the head of the Fred L. Overly Laboratory at the Tree Fruit Research Center, a WSU agricultural station located in Wenatchee, Washington. The major pest on apple trees in the interior valleys of the Wenatchee area of Washington state is the codling moth whose larvae feed on the apple itself. Its economic impact is so severe that it must be treated by pesticides. The comprehensive control program established in these orchards of the central valleys of Washington is one that integrates the chemical control of this insect pest with the biological control of the McDaniel spider mite. That integrated pest management program was introduced by Jesse's Entomology advisor, an early proponent of such a strategy (Hoyt, [83]).

Jesse Logan began his collaboration with me on modeling temperature-rate phenomena in arthropods after I had completed presenting boundary layer theory, during the second semester of my continuum mechanics class. In order to explain exactly how that transpired it is necessary for me to develop the concept of boundary-layer behavior by introducing a canonical model equation. Consider the singularly perturbed linear boundary value problem for $y(x; \varepsilon)$:

$$L[y] = \varepsilon \frac{d^2 y}{dx^2} + \frac{dy}{dx} + b_0 y = 0, \ 0 < \varepsilon \ll 1, \ b_0 > 0, \ 0 < x < 1;$$

with boundary conditions (B.C.'s) : $y(0; \varepsilon) = 0, \ y(1; \varepsilon) = 1$.

Since this is a constant-coefficient ordinary differential equation, we let $y(x; \varepsilon) = e^{m(\varepsilon)x}$. Thus,

$$L[y] = p(m)e^{m(\varepsilon)x} = 0 \text{ where } p(m) = \varepsilon m^2 + m + b_0,$$

which follows from $[e^{mx}]' = me^{mx}$. Given that exponentials are always positive, we can conclude $p(m) = 0$. Hence, this equation has a general solution of the form ([262])

$$y(x; \varepsilon) = c_1(\varepsilon)e^{m^{(1)}(\varepsilon)x} + c_2(\varepsilon)e^{m^{(2)}(\varepsilon)x}$$

where $m^{(1,2)}(\varepsilon)$ are the two unequal roots of that quadratic. Then applying the B.C.'s we find that

$$c_1(\varepsilon) = -c_2(\varepsilon) = \frac{1}{e^{m^{(1)}(\varepsilon)} - e^{m^{(2)}(\varepsilon)}}.$$

Therefore our exact solution is given by

$$y(x; \varepsilon) = \frac{e^{m^{(1)}(\varepsilon)x} - e^{m^{(2)}(\varepsilon)x}}{e^{m^{(1)}(\varepsilon)} - e^{m^{(2)}(\varepsilon)}}.$$

Employing the same method as applied to the quadratic in the last chapter, we obtain

$$m^{(1)}(\varepsilon) \sim -b_0, \ m^{(2)}(\varepsilon) \sim \frac{-1}{\varepsilon} \text{ as } \varepsilon \to 0;$$

and hence, can deduce that

$$y(x; \varepsilon) \sim \frac{e^{-b_0 x} - e^{-x/\varepsilon}}{e^{-b_0} - e^{-1/\varepsilon}} \sim e^{b_0}(e^{-b_0 x} - e^{-x/\varepsilon}) = y_u^{(0)}\left(x, \frac{x}{\varepsilon}\right) \text{ as } \varepsilon \to 0,$$

since $e^{-1/\varepsilon} \to 0$ as $\varepsilon \to 0$ and $1/e^{-b_0} = e^{b_0}$. An examination of this solution for x in the two intervals of interest $x = O(1)$ and $x = O(\varepsilon)$ as $\varepsilon \to 0$, yields the asymptotic behavior

$$y(x; \varepsilon) \sim \begin{cases} y_0(x) = e^{b_0(1-x)} & \text{for } x = O(1) \\ Y_0\left(\frac{x}{\varepsilon}\right) = e^{b_0}(1 - e^{-x/\varepsilon}) & \text{for } x = O(\varepsilon) \end{cases} \text{ as } \varepsilon \to 0.$$

since, in addition, $e^{-b_0 \varepsilon} \to 1$ as $\varepsilon \to 0$. Here $y_0(x)$ represents the so-called outer solution and $Y_0(\xi)$ for $\xi = x/\varepsilon$, the so-called boundary-layer or inner solution. All of these solutions are plotted in Fig. 4.2. Further, note from this figure that the outer and boundary-layer solutions satisfy the one-term matching rule as it was termed by Milton Van Dyke ([262])

$$y_0(0) = \lim_{\xi \to \infty} Y_0(\xi) = e^{b_0}$$

which can be deduced by employing the so-called intermediate limit technique of matching.

The obvious question is: How does one find this behavior for a nonconstant coefficient equation the exact solution of which cannot be determined in closed form? The answer is to use the method of matched asymptotic expansions or the method of matched asymptotes, as Jesse liked to call it, by proceeding as follows:

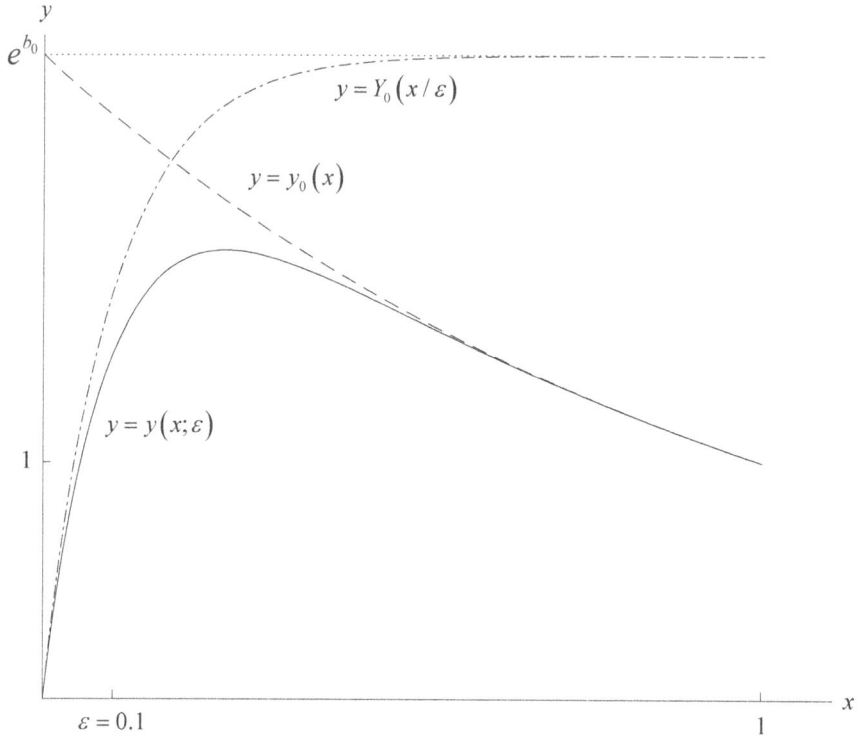

FIGURE 4.2
A schematic representation of the asymptotic behavior of $y(x;\varepsilon) \sim y_u^{(0)}(x, x/\varepsilon)$ as $\varepsilon \to 0$ with $\varepsilon = 0.1$ where $y_0(x)$ and $Y_0(x/\varepsilon)$ denote its outer and inner solutions, respectively.

For $x = O(1)$: Assume $y(x; \varepsilon)$ and all its derivatives are bounded as $\varepsilon \to 0$ and take that limit of the differential equation and its B.C. at $x = 1$, obtaining

$$y_0'(x) + b_0 y_0(x) = 0 \text{ and } y_0(1) = 1 \text{ where } y_0(x) = \lim_{\varepsilon \to 0} y(x; \varepsilon),$$

which again yields the outer solution

$$y_0(x) = e^{b_0(1-x)}.$$

For $x = O(\delta)$ where $\delta = \delta(\varepsilon) > 0$ such that $\lim_{\varepsilon \to 0} \delta(\varepsilon) = 0$: Define the boundary-layer variables,

$$\xi = \frac{x}{\delta(\varepsilon)}, \ Y(\xi; \delta) = y(x; \varepsilon);$$

such that Y and all its derivatives are bounded which, upon substitution into the differential equation and its B.C. at $x = 0$, yields the boundary-layer thickness $\delta(\varepsilon) = \varepsilon$ and the problem

$$\frac{d^2 Y}{d\xi^2} + \frac{dY}{d\xi} + \varepsilon b_0 Y = 0 \text{ and } Y(0; \varepsilon) = 0 \text{ where } Y = Y(\xi; \varepsilon) \text{ for } \xi = \frac{x}{\varepsilon}.$$

Now taking the limit of that boundary-layer equation and its B.C. as $\varepsilon \to 0$, we obtain

$$Y_0''(\xi) + Y_0'(\xi) = 0 \text{ and } Y_0(0) = 0 \text{ where } Y_0(\xi) = \lim_{\varepsilon \to 0} Y(\xi; \varepsilon),$$

which has a one-parameter family of solutions

$$Y_0(\xi) = C_0(1 - e^{-\xi})$$

with the parameter C_0 determined by the Van Dyke matching rule

$$\lim_{\xi \to \infty} Y_0(\xi) = C_0 = e^{b_0} = y_0(0)$$

or

$$Y_0\left(\frac{x}{\varepsilon}\right) = e^{b_0}(1 - e^{-x/\varepsilon}),$$

reproducing our previous result for the boundary-layer or inner solution.

It only remains to deduce that function denoted by $y_u^{(0)}(x, x/\varepsilon)$ earlier. We do so by defining the so-called uniformly-valid additive composite of the outer and inner solutions (Van Dyke, [262])

$$y_u^{(0)}\left(x, \frac{x}{\varepsilon}\right) = y_0(x) + Y_0\left(\frac{x}{\varepsilon}\right) - y_0(0)$$
$$= e^{b_0(1-x)} + e^{b_0}(1 - e^{-x/\varepsilon}) - e^{b_0} = e^{b_0}(e^{-b_0 x} - e^{-x/\varepsilon}),$$

which completes the derivation.

Here the boundary layer is located at $x = 0$ since the coefficient of the dy/dx term in $L[y]$ is positive. Should that coefficient be negative as in the problem for $u(z; \varepsilon)$:

$$\varepsilon \frac{d^2 u}{dz^2} - \frac{du}{dz} + b_0 u = 0, \ 0 < \varepsilon \ll 1, \ b_0 > 0, \ 0 < z < 1;$$
$$u(0; \varepsilon) = 1, \ u(1; \varepsilon) = 0;$$

then the boundary layer would occur at $z = 1$, instead. Indeed, the latter case can be transformed into the former one by introducing the change of variables

$$x = 1 - z, \ y(x; \varepsilon) = u(z; \varepsilon);$$

since then

$$\frac{dy}{dx} = \frac{-du}{dz} \text{ while } \frac{d^2y}{dx^2} = \frac{d^2u}{dz^2}$$

and

$$z = 0 \text{ corresponds to } x = 1 \text{ while } z = 1 \text{ corresponds to } x = 0.$$

Once this material had been presented in the continuum mechanics class, Jesse paid a visit to my office to ask me to become a member of his Ph.D. committee and help him formulate an analytic model for the description of temperature dependent rate phenomena in arthropods, which was fundamental to his thesis work and the eventual title of our seminal paper Logan *et al.* ([126]) on the subject. He felt that the mere knowledge of the singular perturbation results just developed in class could be used to accomplish this determination of the appropriate analytic expression for rate-temperature phenomena in arthropods.

Mites and ticks are arthropods in the taxon *Acarina* that possess one body part and eight legs (see Fig. 4.1). They are insect relatives, often classified as a sub-order of *Arachnids* which predominantly include spiders, and have been typically studied by entomologists. Of all exogenous variables affecting mite predator-prey interactions on apple tree leaves, temperature is by far the most important, say, as compared to humidity. From the outset, the model is restricted to the female component of both species. This restriction implicitly assumes that there are enough males present in the population to assure mating and that males play a minor role in comparison to females with respect to economic impact on the apple crop or as biological control agents. At one time it was thought that mites were parthenogenetic and did not need to have their eggs fertilized. In point of fact, mites are haploid-diploid, in the sense that males develop from unfertilized eggs and are haploid, while females develop from fertilized eggs and are diploid. Hence a fertilized female will produce female offspring until she has used up all her fertilized eggs. Then she will produce male offspring and mate with her progeny, starting the cycle all over again.

The critical importance of temperature to the rate of development of these populations has long been recognized and entomologists have expended a great deal of experimental effort determining the temperature dependence of their life history parameters. In particular, the effect of temperature on the rate of development for these arthropod species measured in the number of adult female mites produced per day can be divided into two phases: Namely, Phase I, which occurs from some base to optimum temperature T_0 and is characterized by a monotone increasing slope of constant rate ρ; and Phase II, which occurs over a temperature interval of width ΔT, should this optimum temperature ever be exceeded and is characterized by a rapid, often precipitous, decline in growth rate to zero at its lethal maximum temperature T_M, *i.e.*, a temperature at or above which life processes cannot be maintained for any appreciable length of time. This standard arthropod behavior is represented schematically in Fig. 4.3.

Although one could deduce empirical formulations to describe the rate of development in these two phases separately, it is advantageous to determine a single analytical expression for rate of development over the whole relevant temperature domain involving biologically meaningful and hence measurable parameters. In order to convince me that the mere knowledge of the form of the uniformly valid additive composite for our singularly perturbed

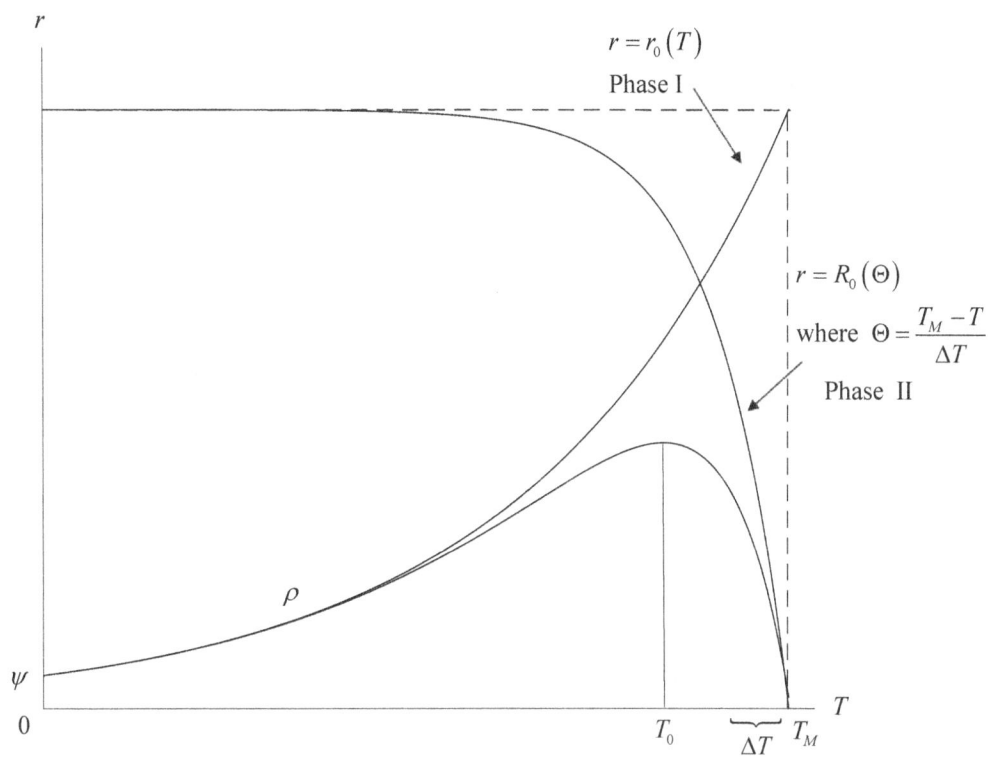

FIGURE 4.3
A schematic representation of rate-temperature phenomena in arthropods with $r_0(T)$ for Phase I and $R_0(\Theta)$ with $\Theta = (T_M - T)/\Delta T$ for Phase II, where $T = {}^\circ C - 10^\circ C \in [0, T_M]$.

boundary value problem could be used to derive the required relationship, Jesse brought the diagram depicted in Fig. 4.3 to his meeting with me. It was his contention that, in this context, Phase II was an outer solution and Phase I, a boundary-layer solution. I Pulled my first Rabbit Out of a Hat in this regard by realizing it was actually the other way around: Phase II represented the boundary-layer solution and Phase I, the outer solution, which has already been anticipated in selecting the notation of Fig. 4.3; in other words, the outer solution is inside and the inner solution, outside, as was true with the latter case of our singularly perturbed boundary value problem for $u(z; \varepsilon)$ posed above.

In particular, considering Phase I to be an outer solution and Phase II to be a boundary layer one and denoting the intrinsic growth rate of either mite species by $r(T)$ where T is temperature measured in °C above the given base temperature of 10°C, we defined a boundary layer variable $\Theta = (T_M - T)/\Delta T$ and assumed a representation of the form

$$r(T) \sim \begin{cases} r_0(T) = \psi e^{\rho T} & \text{for } T \text{ in Phase I} \\ R_0(\Theta) = \psi e^{\rho T_M}(1 - e^{-\Theta}) & \text{for } T \text{ in Phase II} \end{cases} \quad \text{as} \quad \frac{\Delta T}{T_M} \to 0,$$

which followed from that latter case upon introduction of the change of variables and parameters

$$z = \frac{T}{T_M}, \ u(z; \varepsilon) = \frac{r(T)}{\psi}, \ \varepsilon = \frac{\Delta T}{T_M}, \ b_0 = \rho T_M,$$

where $\psi \equiv$ the rate of population growth at the base temperature $T = 0$, as depicted in Fig. 4.3. Implicit to this formulation was the satisfaction of the one-term matching rule

$$r_0(T_M) = \lim_{\Theta \to \infty} R_0(\Theta) = \psi e^{\rho T_M}$$

denoted by the dashed lines in Fig. 4.3. Seeking a representation that would be uniformly valid for all $T \in [0, T_M]$, we formed in the usual way its additive composite defined by

$$r(T) = r_0(T) + R_0\left(\frac{T_M - T}{\Delta T}\right) - r_0(T_M) = \psi[e^{\rho T} - e^{\rho T_M - (T_M - T)/\Delta T}],$$

after noting that $e^a e^b = e^{a+b}$, and finally obtained

$$r(T) = \psi[e^{\rho T} - e^{T/\Delta T - T_M(1/\Delta T - \rho)}].$$

This representation had several desirable characteristics. It is analytic over its entire temperature domain and described by biologically meaningful parameters. Here the relevance of ψ and T_M in this context follows directly from their definitions, while ρ and ΔT can be interpreted as a composite temperature coefficient for the critical biochemical reaction associated with population growth in Phase I and the width of the temperature interval for which thermal breakdown due to desiccation becomes the overriding influence in Phase II, respectively. From it, recalling that $1/e^a = e^{-a}$ and $\ln(e^a) = a$ while $[e^{aT}]' = ae^{aT}$, the optimal temperature T_0 such that $r'(T_0) = 0$ can be shown to satisfy

$$[e^{\rho T}]'(T_0) = [e^{T/\Delta T - T_M(1/\Delta T - \rho)}]'(T_0)$$

$$\Rightarrow \rho e^{\rho T_0} = \left(\frac{1}{\Delta T}\right) e^{T_0/\Delta T - T_M(1/\Delta T - \rho)} \Rightarrow \rho \Delta T = e^{(T_0 - T_M)(1/\Delta T - \rho)}$$

$$\Rightarrow \ln(\rho \Delta T) = (T_0 - T_M)\left(\frac{1}{\Delta T} - \rho\right) \Rightarrow T_0 = T_M + \Delta T \frac{\ln(\rho \Delta T)}{1 - \rho \Delta T}$$

$$\Rightarrow T_0 = T_M\left[1 + \left(\frac{\Delta T}{T_M}\right)\frac{\ln(\rho \Delta T)}{1 - \rho \Delta T}\right]$$

$$\text{or } T_0 = T_M\left[1 + \varepsilon\frac{\ln(\rho \Delta T)}{1 - \rho \Delta T}\right].$$

One of the most perplexing problems encountered in the natural and engineering sciences concerns itself with the deduction of an optimal closed form representation, which provides the best fit for a given set of data points. In this instance, it was first necessary to decide exactly what data to use to obtain the most effective growth rate of each mite species for modeling purposes and then to deduce the appropriate identification of the parameters involved, so that our derived representation would provide this best fit to those data. Although it was possible to have determined these growth rates using observed data for the ovipositional adult stages, this approach, besides being difficult to accomplish, suffered from an additional basic deficiency in that it did not take the complete life history of the mite species into account. In order to overcome this deficiency, Jesse came up with the clever idea of incorporating the fecundity and life history information for each mite species into our formulation by generating the data points to be used for the adult parameter identifications through the employment of a discrete time dynamical systems approach at a fixed set of temperatures $\{T_n\}_{n=1}^N$ with the various life-stages of mite metamorphosis as state variables updated at each stage using our temperature-rate relation and the age-specific deterministic transfer mechanisms from a specific such stage to the next as information flows, while allowing the adult growth rate to converge to a set of corresponding maximums $\{R_n\}_{n=1}^N$. In particular, given this generated data set $\{(T_n, R_n)\}_{n=1}^N$ and our analytic representation for mite growth rate as a function of temperature denoted by $r(T; \psi, \rho, T_M, \Delta T)$, a parameter identification residual least squares fit to that data can be determined by defining

$$E(\psi, \rho, T_M, \Delta T) = \sum_{n=1}^N [r(T_n; \psi, \rho, T_M, \Delta T) - R_n]^2$$

and then minimizing this function by solving for ψ, ρ, T_M, ΔT such that

$$\frac{\partial E}{\partial \psi}(\psi, \rho, T_M, \Delta T) = \frac{\partial E}{\partial \rho}(\psi, \rho, T_M, \Delta T)$$

$$= \frac{\partial E}{\partial T_M}(\psi, \rho, T_M, \Delta T) = \frac{\partial E}{\partial \Delta T}(\psi, \rho, T_M, \Delta T) = 0$$

employing the appropriate algorithm. This procedure can be accomplished much more efficiently if one has a closed form representation for $r(T; \psi, \rho, T_M, \Delta T)$ over the whole temperature domain, as in our case, rather than merely a separate expression for each of the phases. In particular, the results of such a procedure for the McDaniel spider mite then yielded

$$\psi = 0.048, \quad \rho = 0.103, \quad T_M = 28.033, \quad \Delta T = 2.710;$$

and, incorporating this parameter identification into $r(T)$, we obtained the following explicit temperature-dependent per capita growth rate per day for $T.\ mcdanieli$ given by

$$r_1(T) = 0.048(e^{0.103T} - e^{0.369T - 7.457})$$

where $T = {}^\circ C - 10^\circ C$, which is plotted in Fig. 4.4. Employing the same procedure for the predacious mite $M.\ occidentalis$, then yielded

$$\psi = 0.089, \quad \rho = 0.055, \quad T_M = 27.215, \quad \Delta T = 2.070;$$

$$r_2(T) = 0.089(e^{0.055T} - e^{0.483T - 11.648});$$

which is plotted in Fig. 4.5 along with $r_1(T)$ for comparison purposes.

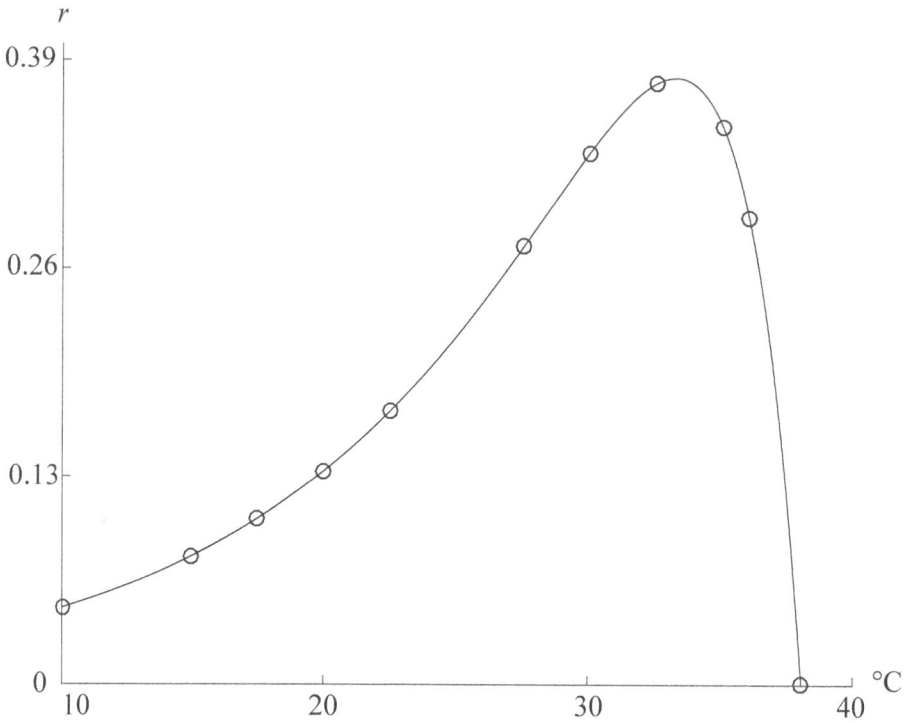

FIGURE 4.4
A plot of $r = r_1(T)$, the intrinsic growth rate per day for the McDaniel spider mite versus temperature measured in °C. The circled points represent the generated data with $N = 11$ while the curve is the graph obtained from our general formulation $r(T)$ by a least-squares parameter identification of that data. Here the base temperature is 10°C and the optimal temperature measured in °C above it, determined from our derived formula, satisfies $T_0 = 23.24$ and $r_1(T_0) = 0.38$. Note the very high degree of accuracy between these generated data points and this graph which, given that there are 11 data points and only 4 parameters to be identified in the graph, most certainly qualifies as another Rabbit having been Pulled Out of a Hat.

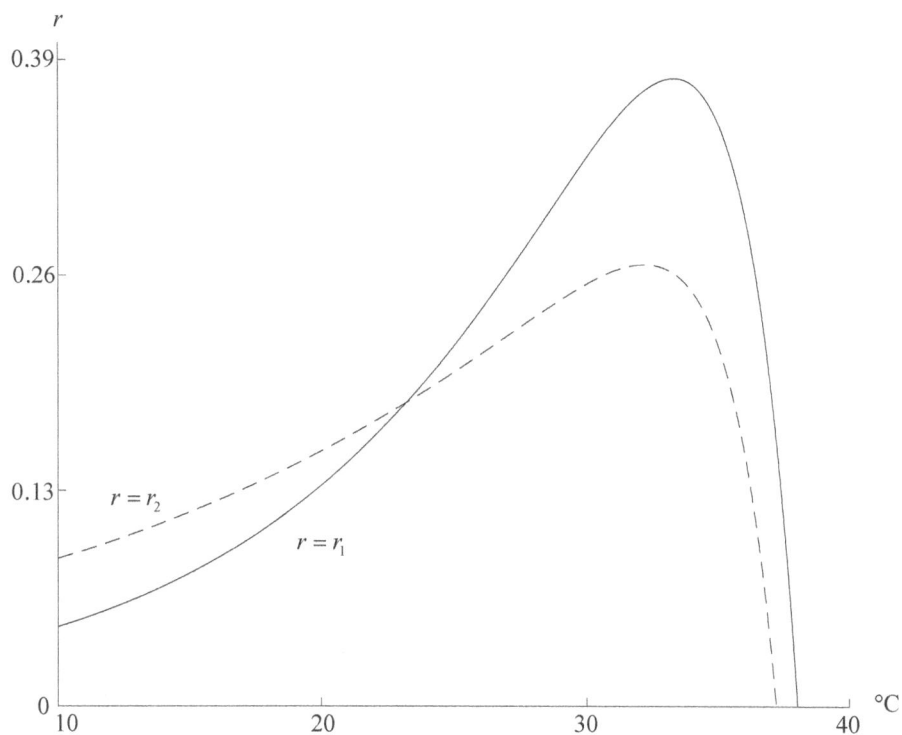

FIGURE 4.5
A comparison of the simulated $r_{1,2}$ values as functions of temperature for the two mite species. The solid line is $r_1(T)$ for *T. mcdanieli* as given in Fig. 4.4 while the broken line is $r_2(T)$ for *M. occidentalis* where $T = °C -10°C$.

It was our assertion that a procedure of the sort illustrated here can provide further biological insights. Specifically, the problem for $u(z; \varepsilon)$ under the transformation introduced above had dimensional form

$$r'(T) = \rho r(T) + \Delta T r''(T), \ 0 < T < T_M;$$
$$r(0) = \psi, \ r(T_M) = 0;$$

which has the biological interpretation that the change of growth rate with temperature is composed of a term (ρr) due to the effect on critical enzyme-catalyzed biochemical reactions and another $(\Delta T r'')$ due to the dissociation of the epicuticle monomolecular wax layer resulting in subsequent desiccation. Fundamentally, it is our contention that such a process can be used to deduce governing biological laws where only empiricism existed before. We note that should the results of this procedure prove fruitful for a particular natural phenomenon, the usual order of steps standardly taken when one quantifies such phenomena mathematically would have been exactly reversed and the ultimate Rabbit Pulled Out of a Hat.

We knew from the outset that this use of boundary-layer methods to model ecological processes was a significant contribution to the field of theoretical ecology and should be disseminated in such a way as to maximize its exposure by publishing journal papers and presenting talks at professional meetings related to its development. We first published the paper Logan *et al.* ([126]), referenced earlier with co-authors Stan Hoyt and the WSU Wenatchee entomologist Lynell Tanigoshi, who provided much of the mite data required for the curve-fitting procedure, in the journal *Environmental Entomology* to reach that community. This seven page paper, which featured our temperature-rate representation for arthropods, had a two and one-half page appendix deriving that formulation from first principals written by me. Then I presented a paper entitled the use of singular perturbation techniques to model ecosystems at a special AMS (American Mathematical Society)-SIAM (Society for Industrial and Applied Mathematics) Joint Symposium on Asymptotic Methods and Singular Perturbations, organized by Robert O'Malley. This talk also featured our temperature-rate representation for mites, an extended abstract of which was published as Wollkind and Logan ([288]) in the proceedings of that symposium, edited by O'Malley. Further, I published a full-length paper of this talk Wollkind ([276]) in *SIAM Review* which, while highlighting the temperature-rate representation, compared asymptotic results obtained by the method of matched asymptotic expansions with those deduced by the so-called method of multiple scales. When applied to our prototype singularly perturbed boundary value problem, that latter method yielded the asymptotic representation

$$y(x; \varepsilon) \sim f_0\left(x, \frac{x}{\varepsilon}\right) = e^{b_0}(e^{-b_0 x} - e^{b_0 x} e^{-x/\varepsilon}) \text{ as } \varepsilon \to 0,$$

which would result in the following refinement of our temperature-rate relation

$$r(t) = \psi[e^{\rho T} - e^{\rho T_M - (T_M - T)/\Delta_1 T}] \text{ where } \frac{1}{\Delta_1 T} = \frac{1}{\Delta T} - \rho.$$

Since this is of the same form as our original representation with the only change being the replacement of ΔT by $\Delta_1 T$, the mite parameter identifications would be unaltered upon its employment instead. During my talk's discussion period, Milton Van Dyke asked me why it was that I found a different representation for the method of multiple scales from that of matched asymptotic expansions, when he found them to be identical in his book *Perturbation Methods in Fluid Mechanics* (Van Dyke, [262]). I first said that having learned singular perturbation theory from the first edition of his book, it seemed strange to be telling him about this. He responded with, "Be that as it may, I am unfamiliar with why I did not find a difference." I told him that he had picked as an illustrative example, the only case where they yielded the same asymptotic representations and then extrapolated that result as a general conclusion. I really liked Van Dyke. Academic Press had published the first edition of his book in 1964 and when they wanted to produce a second edition and sell it at what he considered an inflated price, Milton purchased the copyright from them, republished it under his own in-house Parabolic Press, and sold that book at cost. A few years later when he was the banquet speaker at a conference I was attending, Milton began his after-dinner talk with the following story about an individual who was to be sacrificed to a lion at the Coliseum in ancient Rome. As the lion strides toward him that individual whispers something in its ear whereupon the lion retreats with its tail between its legs and disappears from the scene. The Roman emperor tells him that his life will be spared if only he will relate what he whispered in the lion's ear to make it behave in this manner. The individual in question then responds, "I hope you realize that you are expected to make an after-dinner speech."

In addition, we published Wollkind and Logan ([289]) that again developed our temperature-rate relationship for the mites as a prelude to the analysis of a temperature-dependent predator-prey differential equation model, extended from that introduced in Jesse's thesis, by employing boundary-layer methods to reformulate its so-called predator functional response in terms of a prey-predator density ratio. Finally, we published Wollkind *et al.* ([290]), with WSU entomologist Alan Berryman as our co-author, that developed boundary-layer asymptotic methods for modeling a variety of ecological phenomena and then applied them to obtain our temperature-dependent rate representations for mites, survival relationships for bark beetles, and density-dependent competition functions for single populations satisfying difference equations.

In the meantime, WSU had hired Alan Hastings, Simon Levin's latest Ph.D. graduate from Cornell University, who was an expert in environmental and evolutionary biology. Alan and I began working together, not only turning out a series of papers on age-structure in predator-prey systems but also bouncing our research ideas off each other, as it were. One of the things I discussed with him while sketching its development on his blackboard was my just-completed application of the boundary-layer method to obtain density-dependent competition functions for a discrete-time single population model from Wollkind *et al.* ([290]). After spending two years at WSU, Alan left for the greener pastures of the University of California at Davis's Department of Mathematics. During his second academic year at UC Davis, he sent me the galley proofs of his paper Hastings *et al.* ([78]) entitled "Boundary-layer model for the population of single species," with Francisco Ayala and the latter's Ph.D. student Juan Serradilla, then in press in the journal *PNAS* (*Proceedings of the National Academy of Sciences*) which had the following abstract:

We develop a new discrete-time model, called the boundary-layer model, to describe the dynamics of single species that have a capacity for fast growth at very low population densities. The model explicitly separates the dynamics of the population at very low densities (within the "boundary layer") and at high densities. The boundary-layer model provides a better fit than other models such as the logistic or the θ model to data from experimental populations of *Drosophila willistoni* and *D. pseudoobscura*.

Their new discrete-time model in nondimensional form for a single population n_t at time t was

$$n_{t+1} = e^b(e^{-bn_t} - e^{-n_t/\varepsilon}) \text{ with } n = \frac{N}{K} \text{ and } 0 < \varepsilon \ll 1;$$

while my corresponding scramble-type competition density-dependent difference equation model from Wollkind *et al.* ([290]), describing the size n_g of a similar population in generation g, was

$$n_{g+1} = f(n_g) \text{ where } f(n) = e^b(e^{-bn} - e^{-n/\varepsilon}) \text{ with } 0 < \varepsilon \ll 1;$$

their Eqs. 6, 2 and my Eqs.15, 20c, respectively. Thus, nothing was new about their new model.

Besides redeveloping and then employing my existing model equation without the proper attribution, they claimed the form of their "new discrete-time model, called the boundary-layer model" was suggested by the approximate solution of the same singularly-perturbed boundary value problem analyzed earlier in this chapter that appeared in both Wollkind and Logan ([289]), which they referenced only in that capacity, and the appendix to the unreferenced Logan *et al.* ([126]), where its complete asymptotic behavior had been treated. This claim left the distinct impression that their work represented the first time the knowledge of the asymptotic solution to such singularly-perturbed boundary value problems was used to model ecological processes. In point of fact, my scrambled competition function of Wollkind *et al.* ([290]) was developed in the identical manner from a uniformly valid additive composite

$$f(n) = F_0\left(\frac{n}{\varepsilon}\right) + f_0(n) - f_0(0),$$

where $F_0(n/\varepsilon) = e^b(1 - e^{-n/\varepsilon})$ and $f_0(n) = e^{b(1-n)}$ were an inner and an outer solution, respectively, and represented the limiting case of a *generalized* competition function defined by

$$f(n) = F_0\left(\frac{n}{\varepsilon}\right) + c[f_0(n) - f_0(0)]$$

or

$$f(n) = e^b(1 - c + ce^{-n} - e^{-n/\varepsilon}) \text{ for } 0 \le c \le 1$$

at $c = 1$. Here its other limiting case of $c = 0$ or $f(n) = F_0(n/\varepsilon)$ corresponded to a state of contest competition, while, for $0 < c < 1$, $f(n)$ represented an intermediate competitive state.

Just as with Langer and Prigogine recounted in Chapters 2 and 3, respectively, I felt that if this *PNAS* paper were allowed to go into print unchallenged there was an excellent chance my credit for making the original contribution in yet another research area would be lost. Since journals tend to require authors to return corrected galley proofs within 48 hours, there was not much time to act. I still did not understand why Alan sent me these galley proofs in the first place. If he had only approached me earlier, I would have suggested that they use my generalized competition function to fit their *Drosophila* fruit fly data instead of its scramble-competition limiting case. No matter how good the fit was with $c = 1$, it could only be improved or at worst remain the same if their parameter identification scheme were extended to include c as well and doing so would then have resulted in a joint publication with me removing any possible dispute. Given this situation it seemed to me that my best strategy would be to deal with the editorial office of *PNAS* directly. Toward that end, I mailed a packet containing a letter of explanation to its editor Daniel Koshland, the eminent biochemist from UC Berkeley, and including copies of all the relevant papers as supporting evidence. This resulted in my receiving a phone call from Koshland during which he stated that, in the most generous interpretation, it represented a case of subliminal plagiarism, *i.e.*, the individuals involved having forgotten exactly how materials in question had been encountered adopted them as their own using precisely the same notation.

The problem was that there were two types of submissions to *PNAS*: One, a direct submission and the other, a contributed paper from a member of the National Academy of Sciences (NAS). Hastings *et al.* ([78]) was of the latter variety, in that it had been contributed by one of its co-authors Francisco J. Ayala, a NAS member in genetics. In the event of the contributing member being a co-author of such a paper, it underwent no further review by *PNAS* and simply appeared in print as is, unless the contributing member agreed to a requested revision. Koshland said that, although the paper clearly required my requested revision, because of *PNAS* rules this could only be accomplished if Ayala would agree to it and suggested that the optimal way to obtain this agreement would be for me to have a conversation with him about the revision. This resulted in a phone call from Ayala, who at first claimed that their final dimensional boundary-layer model given by Eq. 7 of Hastings *et al.* ([78])

$$N_{t+1} = K[e^{-bN_t/K} - e^{-N_t/(K\varepsilon)}]$$

was indeed new, since my formulation was nondimensional to which I responded, "Come on now. That's a distinction without a difference. Such nondimensional populations are always implicitly scaled by their carrying capacity K in the ecological literature as was true for your Eq. 6." In making that claim he simultaneously employed both Langer's tactic of relabeling an old model and Schechter's of inventing a specious difference between models to defend his contention of devising a new model, which I considered quite a feat. Then Ayala stated that he did not know about my previous work because of its appearance in an obscure journal but would rectify *Alan*'s failure to give it the proper attribution by appending the following paragraph to their paper:

> **Note Added in Proof**. David J. Wollkind has kindly pointed out to us that our nondimensional Eq. **6** has previously been presented as a model for population growth—namely, equation **20c** in ref.11. In that paper, the authors proposed asymptotic models for various biological processes.

Here the additional ref. 11 was Wollkind *et al.* ([290]), whereas the original preprint had only 10. Ayala closed by asserting that such a paragraph appearing at the end of a *PNAS* paper would give my work more exposure than it previously had being published in the journal *Researches on Population Ecology*. That was the best I could accomplish in the way of a requested revision. In point of fact, this did not actually satisfy Koshland who considered the note to be incongruous in tone to that of the rest of the paper and in a follow up phone call informed me that, although nothing more could be done, he had used my case to convince his editorial board to change the rules of *PNAS* so that in the future all submitted papers, even those co-authored and contributed by NAS members, would have to undergo further review, possibly preventing such an occurrence from happening again. As a postscript to my relating of this incident, a year later my colleague the mathematical geneticist Michael Moody was hosting a pizza party to which he had invited Mike Kahn from the WSU Institute of Bio-Chemistry. I, overhearing Kahn discussing the use of boundary-layer techniques to model ecological systems, interjected, "You mean the Wollkind-Logan method?" to which he replied "What's that? I meant the Hastings-Ayala method!"

Needless to say relations between me and Alan Hastings were a bit strained following what I now, in retrospect, consider to be this over reaction on my part. By the start of 1985, I decided that it was up to me to repair the damage. So I organized a mini-symposium, entitled Bifurcation Problems in the Life Sciences, for that Fall's SIAM meeting to be held at Arizona State University in Tempe, Arizona, and invited Alan, Hans Othmer of the University of Utah, and Peter Antonelli of the University of Alberta to be my three other speakers. Alan graciously accepted, as did Hans and Peter. Having spent the intervening years working predominantly on the materials science and fluid mechanical problems featured in Chapter 2 and the next chapter, respectively, I was resuming my modeling of the temperature-dependent mite predator-prey interaction on apple tree foliage and, for that purpose, had selected a differential equation exploitation model system due to May ([139]), assembled from dynamical components originally developed by Holling ([81]) and Leslie ([116]), respectively. My talk was on bifurcation theory and the paradox of enrichment (Rosenzweig, [190]) in the context of this May ([139]) model. Once the mini-symposium was concluded, Alan warned me that he and Si Levin had reservations about the stability behavior of this model as interpreted by May ([139]). John Collings and I bore that admonishment in mind as we began work on our May mite model.

John Collings entered the WSU Applied Mathematics Ph.D. program in the Fall Semester of 1984. After obtaining a computer science BS and MS from UC Irvine and San Diego, he had worked for five years as a trouble shooter for General Telephone Company in southern California. John said that his GTE job was becoming repetitive and he was looking for new challenges. Most of our supported Ph.D. students had teaching or research assistantships but given John's unique qualifications and the needs of the department at this time, he served as the director of our newly established departmental computer center in that first year of his graduate studies. As with most of my Ph.D. students, John took all four of my graduate courses during his first two years and started his thesis research in the second year of study. That research entailed both formulating and analyzing the temperature-dependent May mite model described above.

In particular we defined $H(t) \equiv$ population density of the prey mite species *T. mcdanieli* and $P(t) \equiv$ population density of its mite predator *M. occidentalis*, measured in numbers of mites per apple tree leaf, where $t \equiv$ time is measured in days. These populations were assumed to be differential functions of a continuous time variable and distributed uniformly over each apple tree leaf. One of the modeling techniques of biological population dynamics for a scenario of this sort is to represent the rate of growth of such populations by the following system of coupled nonlinear autonomous ordinary differential equations of the form

$$\frac{dH}{dt} = HF(H, P), \ t > 0; \ H(0) = H_0; \ \frac{dP}{dt} = PG(H, P), \ t > 0; \ P(0) = P_0;$$

where the specific per capita growth rate functions F and G satisfy the exploitation conditions

$$\frac{\partial F}{\partial P} < 0, \ \frac{\partial G}{\partial H} > 0,$$

and are defined on a case by case basis. Toward that end, we adopted those particular per capita growth rate functions employed in the May ([139]) model, a Kolmogorov ([99])-type system,

$$F(H, P) = r_1\left(1 - \frac{H}{K}\right) - \frac{aP}{H + b}, \ G(H, P) = r_2\left[1 - \frac{P}{\gamma H}\right].$$

Here $r_1, r_2 \equiv$ prey and predator growth rates; $a \equiv$ maximal predator per capita consumption rate; $b, K \equiv$ prey densities necessary to achieve one-half that rate and to be at carrying capacity where $b < K$; and $\gamma \equiv$ food quality coefficient of the prey for conversion into predator births. Note that this formulation employed a Holling ([81]) type II functional response $f_2(H) = aH/(H + b)$ and a variable predator carrying capacity of γH, as introduced by Leslie ([116]).

When Caughley ([27]) employed this system to model the biological control of prickly-pear cactus by the moth *Cactoblastis cactotum*, he generated the theoretical prediction that within two years initially extensive prickly-pear stands of density 500 plants per acre would be reduced to a highly *stable equilibrium value* of 11 plants per acre in impressive agreement with actual field data. This was the reason I decided to employ it to model the predator-prey mite interaction under investigation. For other parameter ranges, the May model is capable of generating stable limit cycle behavior possessing both maximum amplitudes and periods independent of initial conditions (May, [139]). Perhaps the most salient feature that can be discovered upon examination of population data collected during laboratory, greenhouse, or field experiments involving a single predacious species feeding upon a single herbivorous one is the occurrence of periodic oscillations or limit cycles. Such population cycles can be anticipated by arguing that once the predator has overexploited its prey, the density level of the former will decline drastically because of lack of food. Then, if this, in turn, gives the prey population a sufficient chance to recover so that the surviving predators can be provided with food soon enough to enable them to increase until they overexploit their prey again, a self-perpetuating predator-prey cycle will have been established. Some of the most compelling evidence in support of such an interpretation for population fluctuations was obtained by Huffaker *et al.* ([84]), who studied the interaction between *M. occidentalis* and its prey, the six-spotted spider mite, feeding on oranges in a carefully controlled environment, which was an ecosystem very similar to the one under examination and mentioned earlier in conjunction with the *Science* paper of Rosenzweig ([190]). Another occurrence for which predator-prey model systems have been developed to describe is that of outbreak phenomenon wherein population levels can typically increase by several orders of magnitude. Since mean temperature is a biologically meaningful parameter for mite predator-prey systems in seasonal climates and outbreaks also occur naturally in such communities (Hoyt, [83]), we wished to develop a temperature-dependent so-called composite (see below) May model that was unified in the sense it could be used to represent both periodic oscillations and outbreak occurrences in populations simultaneously.

This predator-prey mite ecosystem represented one of the few two-species interactions between terrestrial arthropod populations that can be effectively modeled by differential equations. The generally high biotic potential of mite populations allows them quite rapidly to achieve a stable age distribution with overlapping generations, the latter being a necessary prerequisite for the employment of a differential equations representation (May, [139]). If, in addition, these populations are relatively numerous with the mean distance of separation between individuals relatively small when compared to the characteristic length scale of their habitat, then they may be represented by differential functions of continuous variables. Further, should those populations be distributed uniformly in space, then they may be considered to depend upon the time variable alone. Difference equations models as employed by Hastings *et al.* ([78]) for their *Drosophila* populations are more appropriate to situations where the generations are nonoverlapping (Hassell, [77]).

As a prelude to the discussion of how temperature dependence was incorporated into our May mite model, we first nondimensionalize it and then examine the linear stability of its community equilibrium point. In order to produce a nondimensionalized version of the May model system, we introduced the following dimensionless variables and parameters

$$\tau = r_1 t, \ \mathcal{H} = \frac{H}{K}, \ \mathcal{P} = \frac{P}{\gamma K}, \ D = \frac{b}{K}, \ \theta = \frac{r_2}{r_1}, \ \phi = \frac{a\gamma}{r_1},$$

which transformed that system into

$$\frac{d\mathcal{H}}{d\tau} = \mathcal{H}\mathcal{F}(\mathcal{H},\mathcal{P}), \ \frac{d\mathcal{P}}{d\tau} = \mathcal{P}\mathcal{G}(\mathcal{H},\mathcal{P}), \ \tau > 0;$$
$$\mathcal{H}(0) = \mathcal{H}_0, \ \mathcal{P}(0) = \mathcal{P}_0;$$

where

$$\mathcal{F}(\mathcal{H},\mathcal{P}) = 1 - \mathcal{H} - \frac{\phi\mathcal{P}}{\mathcal{H} + D}, \ \mathcal{G}(\mathcal{H},\mathcal{P}) = \theta\left(1 - \frac{\mathcal{P}}{\mathcal{H}}\right);$$

with

$$\mathcal{H}_0 = \frac{H_0}{K}, \ \mathcal{P}_0 = \frac{P_0}{\gamma K}.$$

We note that the parameters $\theta, \phi > 0$ while $0 < D < 1$. Further, observe that this particular nondimensionalization differed from the one introduced by May ([139]), in that the latter employed b rather than K as the basic population density scale factor. Since one of the primary purposes in scaling is to introduce new dependent variables the magnitudes of which are bounded by unity (Segel, [214]), we shall see from our results, to be presented later in the chapter, that these density variables contain the proper scale factors to achieve this goal. Our nondimensional system had a *community equilibrium point* such that

$$\mathcal{H}(\tau) \equiv \mathcal{H}_e, \ \mathcal{P}(\tau) \equiv \mathcal{P}_e$$

where $\mathcal{H}_e, \mathcal{P}_e > 0$ satisfy $\mathcal{F}(\mathcal{H}_e, \mathcal{P}_e) = \mathcal{G}(\mathcal{H}_e, \mathcal{P}_e) = 0$ given by

$$\mathcal{H}_e = \mathcal{P}_e = \frac{1}{2}(\Delta + 1 - \phi - D) > 0 \text{ for } \Delta = [(1 - \phi - D)^2 + 4D]^{1/2}.$$

In order to examine the linear stability of this community equilibrium state we sought a solution to our system of the form

$$\begin{bmatrix} \mathcal{H} \\ \mathcal{P} \end{bmatrix}(\tau) = \mathcal{H}_e \begin{bmatrix} 1 \\ 2 \end{bmatrix} + \varepsilon \begin{bmatrix} \mathcal{H}_1 \\ \mathcal{P}_1 \end{bmatrix} e^{\sigma\tau} + \mathbf{O}(\varepsilon^2)$$

for $0 < |\varepsilon| \ll 1$ and $|\mathcal{H}_1|^2 + |\mathcal{P}_1|^2 \neq 0$. Then substitution of this solution into that system, neglecting terms of order ε^2, cancelation of the resulting common ε factor, and imposition of the Pat Munroe algorithm yielded the following quadratic secular equation

$$\sigma^2 + (\theta - \theta_c)\sigma + \frac{2\theta\Delta}{1 + \Delta + D + \phi} = 0 \text{ where } \theta_c = \frac{2(\phi - \Delta)}{1 + \Delta + D + \phi};$$

from which we concluded, as usual (see Chapter 3), that there would be linear stability, provided both its coefficients were positive. Since $2\theta\Delta/(1 + \Delta + D + \phi) > 0$, we have deduced the linear stability criterion

$$\theta > \theta_c = \frac{2(\phi - \Delta)}{1 + \Delta + D + \phi}$$

in agreement with that obtained by May ([139]) for his nondimensionalized system.

The neutral stability curve $\theta = \theta_c$ associated with this criterion is plotted in the ϕ-θ plane of Fig. 4.6 for the typical values of (Franz, [62])

$$b = 1 \text{ mite per every 25 leaves and } K = 300 \text{ mites per leaf}$$

which corresponds to $D = b/K \cong 0.00013$. Here we have treated D as a parameter and

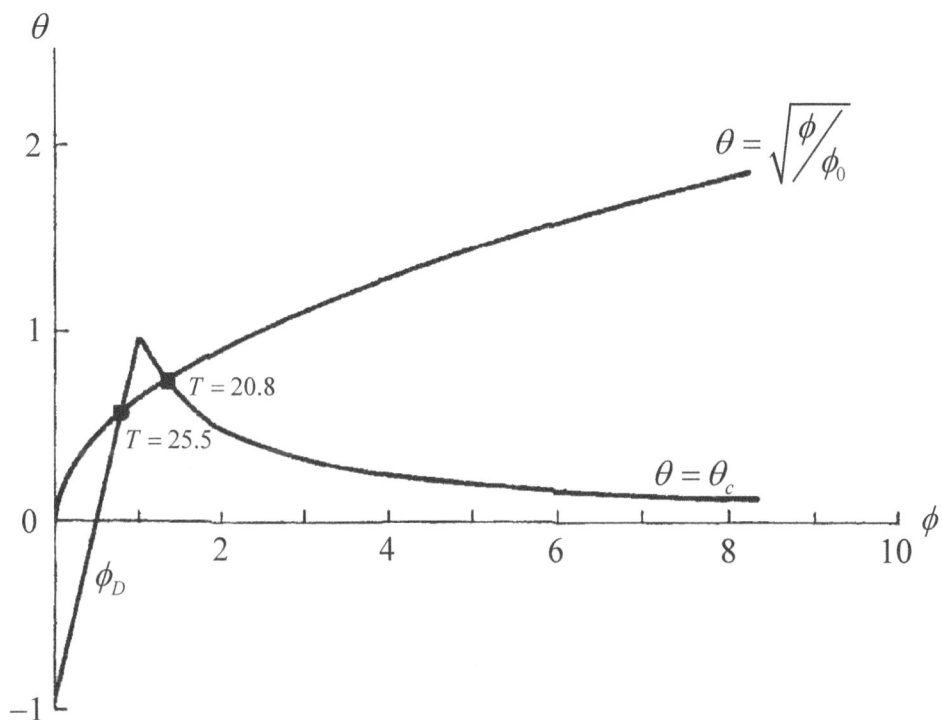

FIGURE 4.6
A plot of $\theta = r_2/r_1$ versus ϕ comparing the linear marginal stability curve $\theta = \theta_c$ with its actual value $\theta = \sqrt{\phi/\phi_0}$ given parametrically in terms of $T = °C - 10°C$ for $\phi_0 = 2.4$ and $D = 0.00013$. Note that $\phi = \phi_D = (1 + D)^2/[2(1 - D)] \cong 1/2$ corresponds to $\theta_c = 0$.

hence considered θ_c to be a function of ϕ alone or $\theta_c = \theta_c(\phi; D)$. We note that θ_c then achieved a maximum value of Θ_c at $\phi = \phi_c$ where

$$\phi_c = \frac{1+D}{2(1-D)}\{2 + [2D(D+1)]^{1/2}\} \cong 1 \text{ and } \Theta_c = 1 + 3D - 2[2D(D+1)]^{1/2} \cong 1.$$

This locus served as the marginal stability curve in that plot separating linearly stable states lying above it from unstable ones lying below.

It only remained to graph the actual temperature-dependent values for θ and ϕ on these same set of axes. To accomplish this we naturally began by introducing the parametric representation

$$\theta = \Theta(T) = \frac{r_2(T)}{r_1(T)} \text{ for } 0 \leq T \leq T_M = 27.22 \text{ where } \Theta(T_M) = 0.$$

Here $r_{1,2}(T)$ are given in Fig. 4.5. As an aside, at that time only the formulation of $r_1(T)$ for the McDaniel mite actually appeared in the open literature. Jesse retained proprietary rights to the formulation of $r_2(T)$ for the predaceous mite developed earlier in this chapter and only divulged it to his future collaborators. Since our research on the May mite model was being performed in conjunction with him, this posed no problem for us. The question was: What should we take to be the corresponding representation for $\phi = \Phi(T)$? One day on my office blackboard, where we had graphed the marginal stability curve under investigation, I plotted another locus of the generic form of the second curve appearing in Fig. 4.6, but that so far was undetermined. Denoting the temperature values of its intersection points with that marginal curve by $T = T_{1,2}$ where $T_2 > T_1$ and traversing this curve in the direction of increasing T, limit cycle behavior would be implied for $T_1 < T < T_2$ since in Kolmogorov-type systems the instability of the equilibrium solution is a sufficient condition for such behavior (May, [139]), a result demonstrated directly for the May model by Arrowsmith and Place ([8]). That the temperature interval $T_1 < T < T_2$ could be characterized by limit cycle behavior was analogous in a general ecological context to the variation of larch bud moth populations with altitude in Switzerland if one assumed a simple inverse relationship between temperature and altitude. More precisely the larch bud moth exhibits stable oscillations at altitudes between 1700 and 1900 m, whereas no such phenomena occurs above or below this range (May, [139]). Hence, we felt an occurrence of this sort was desirable. John said our prospective generic curve looked like a logarithmic plot to him. I responded that it looked more like a parabolic one of the form $\phi = \phi_0\theta^2$ or $\theta = \sqrt{\phi/\phi_0}$ to me. Since $\phi = a\gamma/r_1$, the trick was to select an a representation to yield the desired parabolic form.

That is

$$\phi = \frac{a\gamma}{r_1} = \phi_0\frac{r_2^2}{r_1^2} = \phi_0\theta^2,$$

which when solved for a yielded

$$a = \frac{\phi_0}{\gamma}\frac{r_2^2}{r_1} = a_0\frac{r_2^2}{r_1}$$

where a_0 and ϕ_0 were related by

$$a_0 = \frac{\phi_0}{\gamma} \text{ or } \phi_0 = a_0\gamma.$$

We chose the appropriate value for the proportionality constant a_0 to provide the maximal

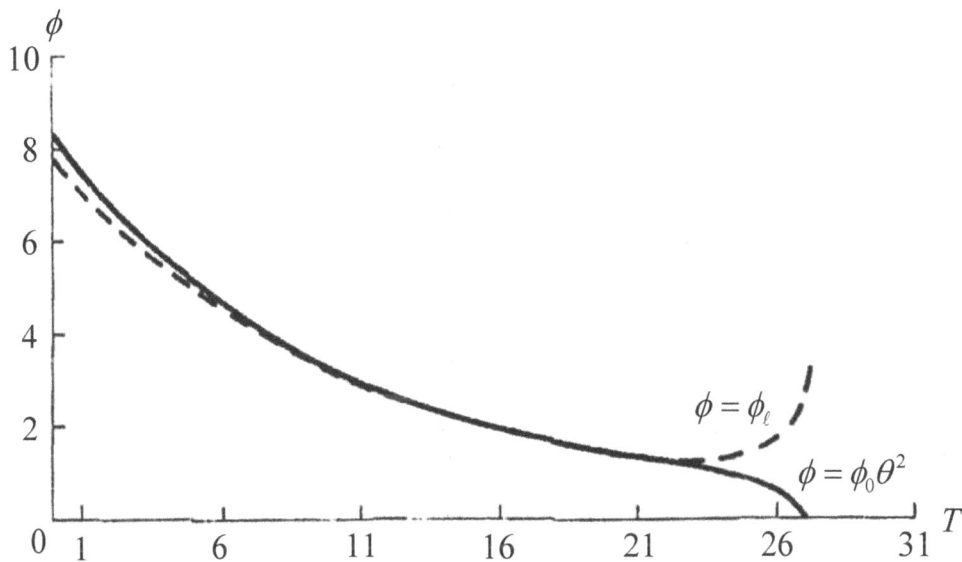

FIGURE 4.7

A plot of $\phi = a\gamma/r_1$ versus $T = {}^\circ C - 10{}^\circ C$ comparing $\phi = \phi_\ell$ based upon $a = a_\ell(T)$ with $\phi = \phi_0\theta^2$ based upon the more realistic $a = a_0 r_2^2/r_1$ for $\gamma = 0.15$, $\phi_0 = 2.4$, and $r_{1,2}(T)$ as given in Fig. 4.5. Observe that these two representations are virtually coincident over the exact temperature domain $11.1 \leq T \leq 22.2$ for which the linear interpolation may be presumed valid.

amount of correlation between $\phi = \phi_0\theta^2$ and $\phi = \phi_\ell = a_\ell\gamma/r_1$ as depicted in the temperature plots of Fig. 4.7. Here a_ℓ represented the linear interpolation for a versus T employed by Logan ([123])

$$a_\ell(T) = 0.29(8.520 + 0.125T)$$

over the latter's domain of validity, $11.1 \leq T \leq 22.2$, where the two end points in question were determined by growth chamber experiments at those fixed temperatures and the factor 0.29 was a conversion of egg-to-adult consumption rate, while γ, after Logan and Hilbert ([125]), was assigned the typical value of 0.15. We can observe from this figure that these two representations for $a_0 = 16$, or equivalently $\phi_0 = 2.4$, and $\theta = \Theta(T)$ are virtually coincident for the temperature domain over which the linear interpolation may be presumed valid. If that were not the epitome of Pulling a Rabbit Out of a Hat, then nothing would ever be. Thus, to complete our temperature-dependent parametric representation of the May mite model, we took $\phi = \Phi(T) = \phi_0\Theta^2(T)$. Under these conditions, $T_{1,2}$ were explicitly given by $T_1 = 20.85$ and $T_2 = 25.54$ (see Fig. 4.6).

Interpreting ϕ as a nondimensional predator maximal consumption rate, we note from Fig. 4.7 that $\phi = \phi_0\theta^2$ based upon $a = a_0 r_2^2/r_1$ is more realistic than $\phi = \phi_\ell$ based upon $a = a_\ell(T)$, since the former disappears at $T = T_M \equiv$ predator lethal maximum temperature, while the latter becomes large at this temperature, implying the predacious mites would in essence be feasting on their deathbed as it were. One of the criticisms frequently levelled against systems, such as the May model, is that of being overly simplistic. In order to help mitigate this concern, we have completed our formulation by selecting the May mite model parameters in such a way as to add biological credibility to our system. Since the generated data points used for the parameter identifications in the formulations for $r_{1,2}(T)$ included fecundity and life history information, while a, as defined, depended only on these formulations, our model was actually a composite or hybrid one consisting of an analytical framework containing simulation components (Logan, [124]). The idea behind such a composite model is to capture the descriptive capabilities of an analytical model and the predictive capabilities of a simulation one, by combining the best features of these two methods through the linkage just described (Plant and Mangel, [174]).

Hence, from the linear stability analysis of our temperature-dependent composite May mite model system, we concluded that its community equilibrium point was locally stable for both $0 \leq T < T_1$ and $T_2 < T < T_M$, while it was locally unstable for $T_1 < T < T_2$, which implied the occurrence of limit cycle behavior. Here, the temperature values of the end points of that interval $T = T_{1,2}$ corresponded to so-called Hopf bifurcation points, denoted by squares in Fig. 4.6. We now examine May's ([139]) interpretation of these results in more detail. Appealing to the work of Kolmogorov ([99]), May ([139]) asserted that this model system possessed either a globally stable equilibrium point *or* a globally stable limit cycle. In particular, he concluded that its linear stability criterion revealed whether the community equilibrium point was stable, whereupon the complete global character of this system was laid bare in that the associated marginal stability curve then separated the region in parameter space where these predator-prey equations had an asymptotically stable equilibrium point from the region where they exhibited stable limit cycle behavior. As per Alan Hastings' admonishment, this was an oversimplification of Kolmogorov's ([99]) result, which actually stated that the *single* community equilibrium point for such systems was *either* globally stable *or* was surrounded by a closed curve which was globally stable from the outside, with various outcomes possible inside this curve including the occurrence of a bilaterally stable limit cycle. Thus, of all possible behaviors inside this curve, the simplest and most practically interesting was that of a unique globally stable limit cycle which was the situation anticipated by May ([139]). Since Kolmogorov ([99]) was unable to find a necessary condition for the occurrence of these periodic limit cycles, but concluded that the linear instability of its equilibrium point was a sufficient condition as mentioned earlier, May ([139]) pointed out its occurrence could not necessarily be precluded for any given choice of parameter values in his model system.

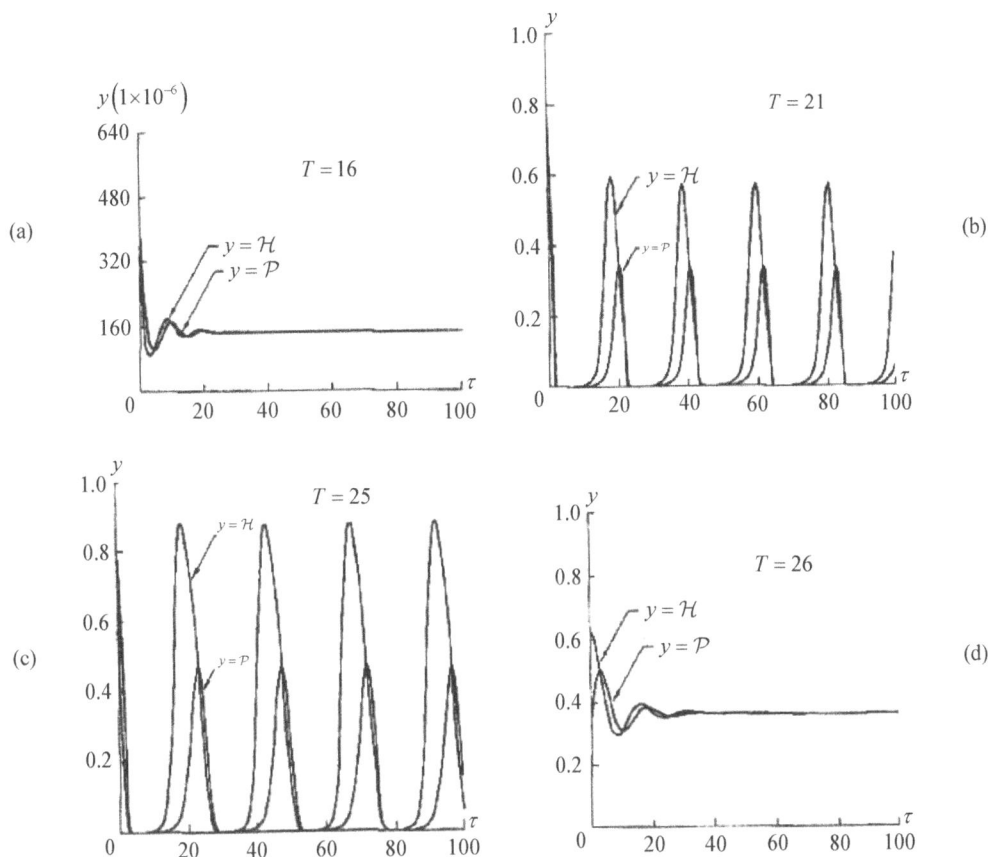

FIGURE 4.8
Plots of $\mathcal{H}$ and $\mathcal{P}$ versus τ for $T = 19$, 21, 25 and 26 in parts (a), (b), (c), and (d), respectively, with $\gamma = 0.15$ and $D = 0.00013$.

As an initial check on the validity of May's assertion, John applied a numerical differential-equation solver to our composite May mite model system for various representative temperatures in the specific intervals delineated above and for a number of different possible initial conditions. A partial compilation of these results is presented in the four parts of Fig. 4.8. From that figure, we observed that there typically occurred an extremely low-population stable node-type equilibrium for $0 \leq T < T_1$ (1 spider mite per every 23 leaves when $T = 16$) and a relatively high-population such equilibrium for $T_2 < T \leq T_M$ (109 spider mites per leaf when $T = 26$), while for the intermediate temperature interval of $T_1 < T < T_2$ stable limit cycle behavior occurred characterized by spider mite population maximums ranging in amplitude from 60% of its carrying capacity when $T = 21$ to 90% when $T = 25$. Further, the long-time asymptotic properties of our system were independent of the choice of initial conditions for the temperatures exhibited in Fig. 4.8. Thus, the asymptotic stability predictions inherent to these numerical computations would seem to have corroborated May's speculative assertion concerning the apparent correlation between the linear stability results and the global stability behavior of this predator-prey model system.

As a final conclusive test of May's ([139]) global stability hypothesis involving our composite May mite model, we next applied AUTO 86, the most recently updated version of Doedel's ([52]) computer-assisted continuation and bifurcation software package for autonomous ordinary differential equations, to that system. The code AUTO automatically traces a bifurcation diagram for such systems as a given parameter is varied. It computes equilibrium points and determines their stability for both the cases of stationary and Hopf bifurcations, in the latter case following branches of periodic solutions, investigating the orbital stability properties of these limit cycles, and measuring their periods and maximum amplitudes of oscillation. Further, AUTO also has the capability of identifying regions of interest containing loci of so-called limit (defined below) and Hopf bifurcation points in a two-parameter space.

We began with a one-parameter bifurcation analysis of our composite May mite model system in which all parameters in that system except temperature T were fixed by taking γ and D to have their typical assigned values of $\gamma = 0.15$ and $D = 0.00013$. Using AUTO 86 John then determined how the solution structure of our composite May mite model system depended on T. The results of this computation are summarized in the bifurcation diagram of Fig. 4.9. In this diagram, T is on the horizontal axis while the vertical axis represents the maximum value of the prey population and has been designated by max $\mathcal{H}$. Here

$$\max \mathcal{H} = \left\{ \begin{array}{ll} \mathcal{H}_e & \text{for stationary solutions} \\ \displaystyle\max_{\tau \in \Omega} \mathcal{H}(\tau) & \text{for periodic solutions} \end{array} \right.$$

where Ω is the period of oscillation. There are two such branches, one of each type in Fig. 4.9. Here, lines are being used to denote steady-state behavior on the stationary solution branch while the periodic solution branch consists of circles designating limit cycle behavior. Solid squares represent Hopf bifurcation points. Stable stationary solutions are drawn with solid lines and unstable ones with dashed. Solid circles represent stable periodic orbits while open ones denote unstable oscillations. That particular circle, designated by a half-moon symbol in Fig. 4.9, serves as a limit point for the periodic solution branch where it undergoes a change of direction which can, in this instance, be characterized by the occurrence of a vertical tangent.

Although Fig. 4.9 has only been drawn for $19 \leq T \leq 26$, our AUTO calculations were actually carried out for the same interval $0 \leq T \leq T_M = 27.22$, as employed in Fig. 4.6. We can recover our linear theory results completely from the stationary solution branch of Fig. 4.9 in regard to both the stability and steady-state values of the community equilibrium point. Hence, we see that our Hopf bifurcations occur at $T = T_1 = 20.85$ and $T = T_2 = 25.54$, which are represented by solid squares consistent with Fig. 4.6. In addition, there is a low temperature-low population stable equilibrium for $19 \leq T < T_1$ ($0 \leq T < T_1$ from our extended results) and a high temperature-high population stable equilibrium for $T_2 < T \leq 26$ ($T_2 < T \leq T_M$ from our extended results) while there is instability of the stationary solution branch for $T_1 < T < T_2$. We now turn our attention to the periodic solution branch of Fig. 4.9. The limit point on that branch occurs at $T = T_{-1} = 19.86 \equiv$ the lowest temperature at which the system exhibits periodic solutions and separates its upper portion

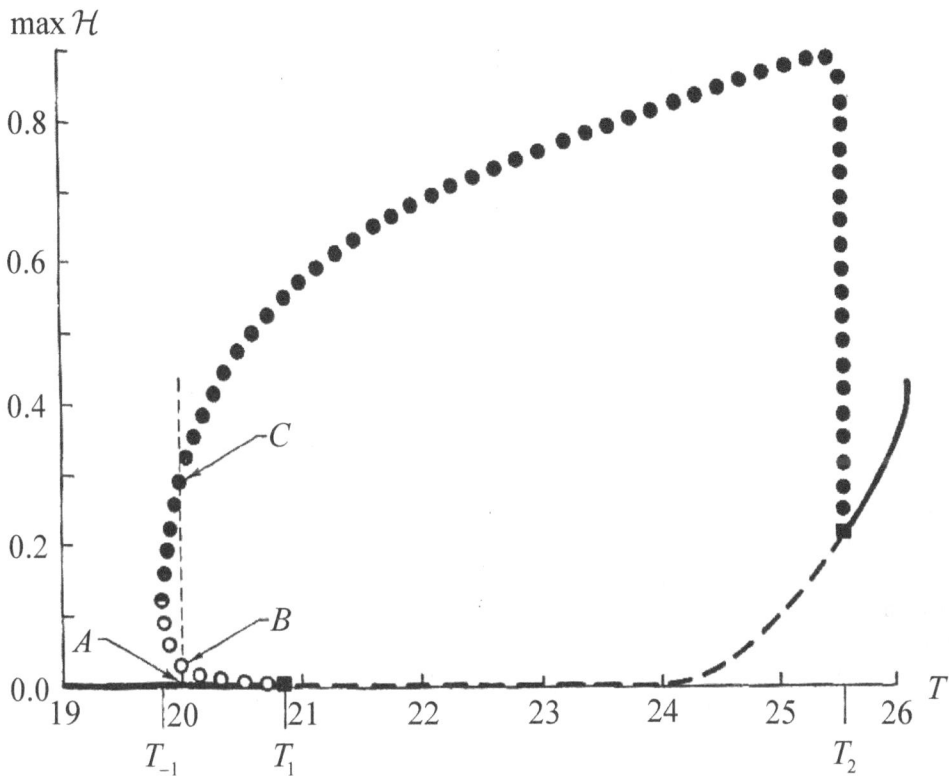

FIGURE 4.9
A max $\mathcal{H}$ versus T one-parameter bifurcation diagram with $\gamma = 0.15$ and $D = 0.00013$.
Here, the vertical dotted line $T = 20$ intersects the stationary solution branch at point A
and the periodic solution branch at points B and C.

consisting of stable limit cycles from its lower portion consisting of unstable ones. The limit point itself corresponds to a semi-stable limit cycle. Unlike the bilaterally symmetric behavior characteristic of stable (attracting) or unstable (repelling) limit cycles trajectories spiral into this semi-stable limit cycle from the outside and away from it on the inside.

Let us examine Fig. 4.9 in some detail. As mentioned above, it is a truncation of our original AUTO results with the region of paramount interest to us enlarged for the purpose of emphasis. Since the deleted intervals possessed the same qualitative bifurcation properties as those associated with the stable portions of the stationary-solution branch included in that figure, our presentation preserved the important features of this analysis. Observe that the individual stability properties of each branch just described are only local in nature. For any fixed value of temperature, in the same manner as that introduced at the end of the last chapter to interpret Fig. 2.13, the global stability behavior of the system can be deduced from the bifurcation diagram of Fig. 4.9 by examining the intersection of its solution branches with a vertical line drawn through that T value. Doing this, we found that low-population stationary equilibria were the only stable states which could exist for $0 \leq T < 19.86$ and thus concluded that they were globally stable in that temperature interval. Similarly, we concluded that high-population stationary equilibria were globally stable for $25.54 < T \leq 27.22$. Further, we also concluded that globally stable limit cycles occurred for $20.85 < T < 25.54$. All of these conclusions were in accordance with May's ([139]) global stability hypothesis and agreed with our numerical results summarized in Fig. 4.8.

For $19.86 < T < 20.85$, however, our AUTO predictions seemed to be in direct violation of that hypothesis. This was the first thing I noticed about Fig. 4.9 when John took that diagram to me after he had generated it. To my comment of, "So this means that May was wrong, doesn't it?" John responded with, "It sure looks like it." In particular, for this interval we could conclude that there actually existed multiple stable states. As a check on this prediction, we again applied a numerical differential equation solver to our system for a variety of different possible combinations of initial conditions, but now with $T = 20$. The results of these computations are summarized in Fig. 4.10. The schematic phase plane plot of Fig. 4.10a showed that these states consisted of a locally stable equilibrium solution (point A in Fig. 4.9) separated in the phase plane from a locally stable limit cycle (point C in Fig. 4.10) by an unstable limit cycle (point B in Fig. 4.9). The latter served as a separatrix dividing the phase plane into two basins of attraction. Trajectories starting at initial points inside it terminate as a focus at the equilibrium point (see Fig. 4.10b), while those starting outside it converge to the stable limit cycle (see Fig. 4.10c). The stable equilibrium point, unstable limit cycle, and stable limit cycle correspond to the three points of intersection of the solution branches with the vertical dotted line through $T = 20$ in the bifurcation diagram of Fig. 4.9. We note that should $T = 19.86$ the two limit cycles coincide and become semi-stable.

Given that linear theory predicted stability of the community equilibrium solution to initially infinitesimal disturbances in these instances, while nonlinear theory indicates that this solution could become unstable provided the perturbations exceeded a certain critical threshold, we have a prototypical case of what biologists commonly term metastability and applied mathematicians usually refer to as subcritical instability (see Chapter 2) occurring. Further, by considering states possessing amplitudes in the neighborhood of the low-temperature stable portion of the stationary solution branch in the bifurcation diagram of Fig. 4.9 and gradually increasing temperature, we see that limit cycle behavior will not spontaneously occur until $T > 20.85$, while if we start in the neighborhood of the stable portion of the periodic solution branch in that figure and gradually decrease temperature, limit

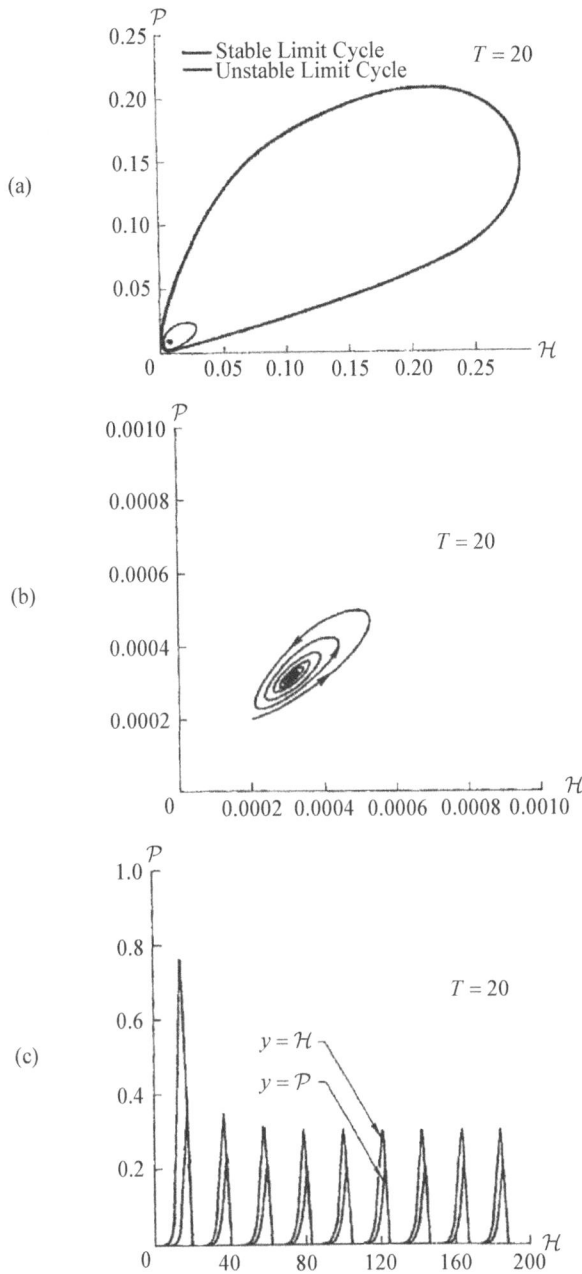

FIGURE 4.10

Phase plane and dynamical plots exhibiting the metastable behavior characteristic of a representative temperature in the hysteresis region of Fig. 4.9. These graphs depict all three possible solutions that may occur for $T = 20$ plotted schematically in (a) and both the multiple stable states which can result at that temperature upon the appropriate choice for the initial conditions, $e.g.$, (b) a stable focus or spiral equilibrium point when $\mathcal{H}_0 = \mathcal{P}_0 = 0.2 \times 10^{-3}$ and (c) a stable limit cycle when $\mathcal{H}_0 = 1 \times 10^{-5}$ and $\mathcal{P}_0 = 0.5 \times 10^{-5}$.

cycle behavior will not spontaneously cease occurring until $T < 19.86$. Such a hysteresis effect is often associated with metastable phenomena. Finally, we observe that although the occurrence of the multiple stable states just described is contrary to May's ([139]) nonlinear interpretation of his predator-prey model, it is in complete agreement with the possibilities allowed for such Kolmogorov-type systems as catalogued earlier. I think our discovery of this occurrence for the temperature-dependent composite May mite model using AUTO 86 definitely qualifies as another prime example of the Pulling of a Rabbit Out of a Hat.

The dilemma facing us was exactly how should we report this oversimplification appearing in May's ([139]) influential Princeton University Press monograph and perpetuated in many subsequent publications. A number of people, when informed of our discovery, warned us that due to the reputation of that monograph in the ecological literature we had better make a very strong case for our interpretation when disseminating these results. We took that advice to heart and reported our findings along the lines already described in this chapter when submitting the paper Wollkind *et al.* ([282]) to the *Bulletin of Mathematical Biology* only to have its two reviewers object to its tone. One reviewer was Robert May who identified himself and said he wished it had been some other author's oversimplification being corrected by this very nice paper, but wished we had not hit him so hard. The other reviewer who remained anonymous said that all references to the fact that we were right and May was wrong must be removed from the paper. Lee Segel, the editor of the journal, asked me to rewrite the paper by being a little less "Wollkind" like in its revision. I did so by introducing the AUTO 86 analysis of our composite May mite model system directly following the linear stability analysis and lumping all the numerical computations for the representative temperatures together without mentioning anything specifically about May's ([139]) interpretation. The only mention of that discrepancy was included tangentially in the first paragraph of the concluding discussion section as follows:

> "We begin this section by noting that previous researchers who analyzed the May model apparently either overlooked or did not recognize the importance of the metastable behavior demonstrated in the last section. In their defense, its occurrence eluded us for more than a year until we applied AUTO to our model, a computer program that did not even exist at the time the May model was initially being investigated. Further, had we not been fortunate enough to have reduced our system to one containing the single parameter temperature, that metastability might still have gone undiscovered in spite of our application of AUTO to the system."

So far, all the results obtained for our composite May mite model have been with $\gamma = 0.15$ and $D = 0.00013$. In order to examine the sensitivity of our system to other choices for these quantities, we next used AUTO to produce bifurcation diagrams as two parameters are varied. In such two-parameter bifurcation diagrams curves of Hopf bifurcation and limit points are plotted in the plane of the varying parameters. Three diagrams of this sort are presented in Fig. 4.11. Here Fig. 4.11a contain plots in the T-γ plane with $D = 0.00013$; Fig. 4.11b, plots in the T-D plane with $\gamma = 0.15$; and Fig. 4.11c, a plot in the D-γ plane with $T = 16.7$. In Figs. 4.11a,b the solid curves represent Hopf bifurcations, and the dashed ones, limit points. From the latter, we can observe that there is no measurable interval of metastability unless $\gamma > 0.09$ in Fig. 4.11a or $D < 0.005$ in Fig. 4.11b. Further, since these curves delineate the instability region in their respective spaces, we can see that there exists a point (T_c, γ_c) in Fig. 4.11a, with $T_c = 14.02$ and $\gamma_c = 0.0672$, such that for $\gamma < \gamma_c$ or $T > T_c$ the community equilibrium point would be globally stable and a value $D_c = 0.0177$ in Fig. 4.11b such that the same thing would be true for $D > D_c$.

109

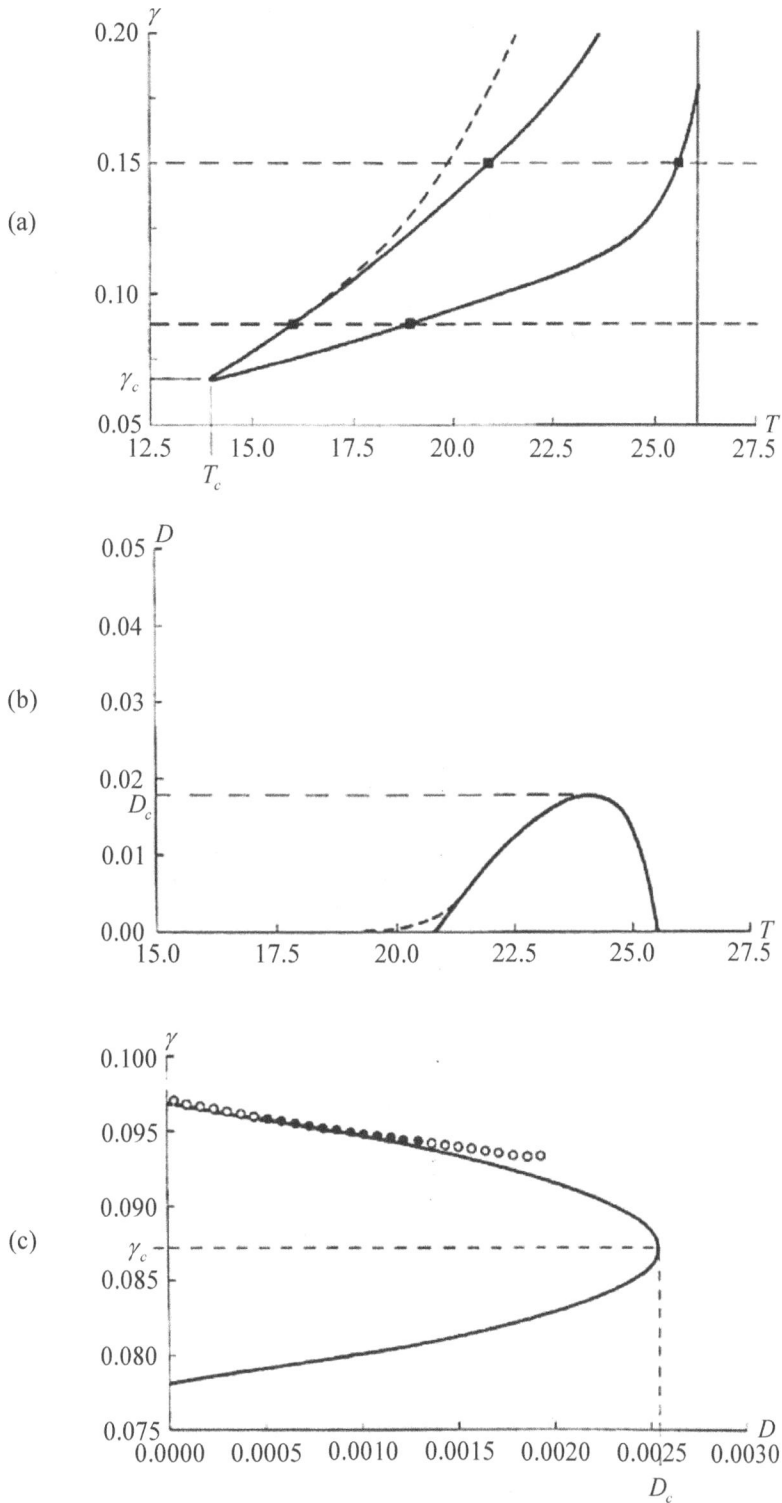

FIGURE 4.11
Two-parameter bifurcation diagrams for the composite May mite model: (a) γ versus T with $D = 0.00013$; (b) D versus T with $\gamma = 0.15$; (c) γ versus D with $T = 16.7$.

Finally, for any fixed temperature, the region in D-γ space characterized by globally stable limit cycle oscillatory behavior can be represented with a plot of the Hopf bifurcation curves as we have done in Fig. 4.11c when $T = 16.7$. Specifically, the point (D_0, γ_c) in that figure is explicitly given by $D_0 = 0.00254$ and $\gamma_c = 0.0872$. The actual region of oscillations is slightly larger than that depicted once metastability is taken into account. For instance, consistent with Fig. 4.11a when $T = 16.7$, the Hopf bifurcations occur at $\gamma_1 = 0.074$ and $\gamma_2 = 0.0966$, while the limit point occurs at $\gamma_{-1} = 0.0980$ should $D = 0.00013$. Once a particular choice of D and γ has been found that yields a desired period at a given temperature, it is possible for AUTO to trace a curve of solutions with this fixed period in the D-γ plane. Such a curve is also shown in Fig. 4.11c where the locus of solutions of period $\Omega = 18.8$ (corresponding at $T = 16.7$ to 73.75 days) appears as a sequence of circles, some of which occur in the region of metastability. The choice of temperature and period for this sequence are related to the laboratory experiments of Huffaker *et al.* ([84]) mentioned earlier and to be discussed in greater detail below.

Recall that our goal in developing this temperature-dependent composite May mite model was to produce a unified predator-prey system capable of representing both periodic oscillations and outbreak occurrences in populations simultaneously. From Fig. 4.11, we can conclude that increasing γ or decreasing D has a destabilizing influence upon our model, which ecologically corresponds to triggering outbreaks through food quality or inducing the paradox of enrichment by carrying capacity, respectively. Specifically γ_c of Fig. 4.11a serves as a threshold for outbreak behavior, which can occur provided $\gamma > \gamma_c$ as temperature increases. In a more general ecological context, we note that the existence of such a threshold is reminiscent of the behavior of the grasshopper *Austroicetes cruiciata*, whose outbreaks are triggered by a rise in food quality (Caughley, [27]). In a similar more general ecological vein, we note that Fig. 4.11b can be used to represent the paradox of enrichment (Rosenzweig, [190]) graphically. May ([139]) defined this term as a tendency for a given population to shift its dynamics from a stable equilibrium point to a stable limit cycle as life gets better for it, in the sense that its environmental carrying capacity increases. Taking any constant $T_1 < T < T_2$ and decreasing D along such a line in Fig. 4.11b, we are able to demonstrate the desired result since D is inversely proportional to this carrying capacity. Specifically for large enough $D \cong 1$, the stable stationary equilibrium point is a focus which converges to the equilibrium point after a single oscillation, while for very small values of D the oscillations associated with the limit cycle behavior may become sufficiently drastic to drive the prey density low enough so that stochastic effects cause extinction of the system to occur. This was the case with the experiments of Huffaker *et al.* ([84]), in which an originally stable periodic oscillatory situation was destabilized by tripling the food supply available to the herbivorous six-spotted spider mite. The interaction in the enriched universe was brought to a halt after only a single cycle because predator overexploitation had reduced the prey population to such an extremely low level that the predator did not stand an appreciable chance of survival. Hence, in the universes considered by Huffaker *et al.* ([84]) during their laboratory experiments, this paradox was manifested by the occurrence of periodic oscillatory behavior becoming so drastic upon enrichment that extinction eventually resulted.

In order to simplify our analysis, we have assumed from the outset that the environment being considered is spatially homogeneous with mite population densities being distributed uniformly. Such an assumption oversimplifies the ecological situation not only for field environments but also for some laboratory experiments as well including those of Huffaker *et al.* ([84]). They studied the interaction of *M. occidentalis* with its prey, the six-spotted spider mite *Eotetranychus sexmaculatus* Riley, in a carefully controlled laboratory environment which was maintained at a temperature of 26.7°C and consisted of cabinets containing arrays of oranges, replenished at regular intervals, that served as the food source for the prey with various pathways and barriers between the oranges. Huffaker *et al.* ([84]) were able to construct systems in which the populations of the two mite species did not converge to an equilibrium value, but rather oscillated with a relatively constant period and somewhat variable amplitude as shown in Fig. 4.12a and these oscillations tended to persist in ecological time. Indeed, that experiment lasted 60 weeks and was terminated not by predator overexploitation, but only when the six-spotted spider mite contracted a viral infection which caused its population density to crash. The limit cycle oscillations characteristic of our model summarized in Fig. 4.10 show this constant period and those dramatic differences between minimum and maximum amplitudes exhibited by these data but not their variation in amplitude.

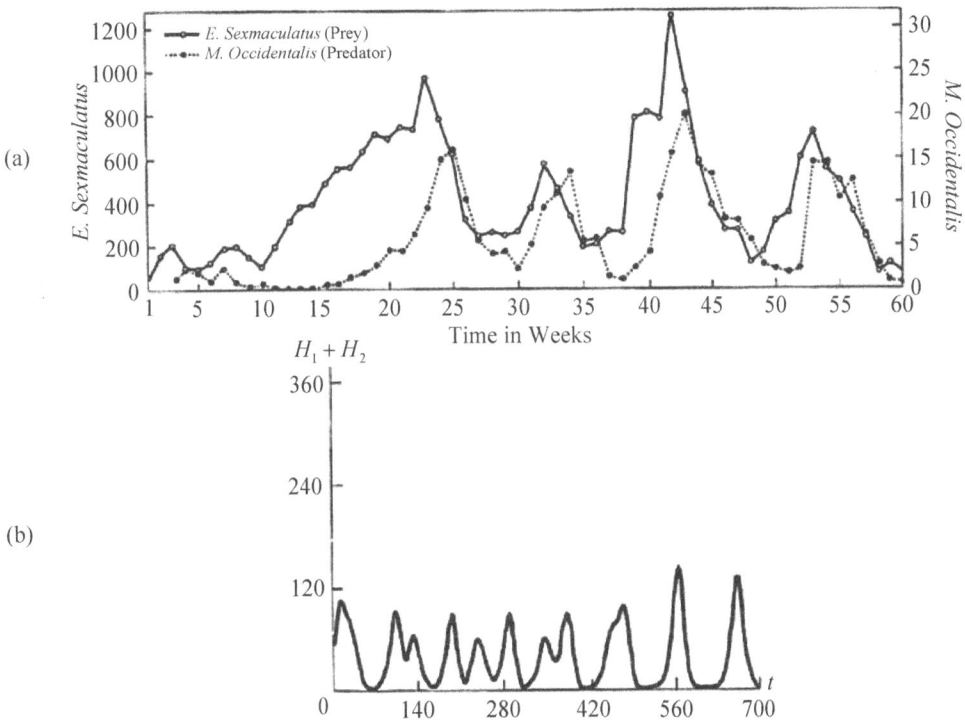

FIGURE 4.12

Plots of population versus time at $T = 16.7$: (a) as observed by Huffaker *et al.* ([84]) in one of their experimental universes where the densities are measured in number per whole orange equivalents and time in weeks. (b) as predicted by applying the composite May mite model to two independent subpopulation pairs with $\gamma_1 = 0.09$, $D_1 = 0.00013$ and $\gamma_2 = 0.084$, $D_2 = 0.00020$.

Variations in amplitude of that sort can be achieved if we relaxed this assumption of spatial uniformity. The easiest way to introduce spatial heterogeneity into a previously homogeneous model system is to postulate the existence of nonmigrating subpopulations. Since Huffaker *et al.* ([84]) inadvertently used oranges of an inferior quality for the prey food source during part of their experiment, while only one seventh of the oranges were replaced in each cabinet per week, there is a very real possibility that the prey mites were divided into subpopulations with different carrying capacities and inherent food qualities. If we assume, for simplicity, the existence of just two independent prey subpopulations with corresponding predator subpopulations, such that each subpopulation pair obeys the dynamics of our composite May mite model with $T = 16.7$, we can produce variable amplitudes in the oscillations of the total population densities obtained by summing these subpopulations, as illustrated in Fig. 4.12b. The amplitude fluctuations in the total population densities are due to the different values of carrying capacity and food quality selected for each of the two prey subpopulations, as measured by their (D, γ) pairs chosen appropriately from Fig. 4.11c. In particular, these parameters and the relevant initial densities for those two subpopulation pairs of Fig. 4.12b are given by

$$\gamma_1 = 0.09, \ D_1 = 0.00013 \text{ or}$$
$$K_1 = 300 \text{ mites per leaf}, \ (\mathcal{H}_1)_0 = 0.1370, \ (\mathcal{P}_1)_0 = 0.0517;$$
$$\gamma_2 = 0.08, \ D_2 = 0.00020 \text{ or}$$
$$K_2 = 200 \text{ mites per leaf}, \ (\mathcal{H}_2)_0 = 0.0117, \ (\mathcal{P}_2)_0 = 0.0047;$$

respectively. Further, $H_1 = K_1\mathcal{H}_1$, $H_2 = K_2\mathcal{H}_2$, $t = \tau/r_1(16.7) = 3.92\tau$ days in that figure. Upon comparison of the two parts of Fig. 4.12, it can be seen that our theoretical predictions are in good qualitative agreement with the experimental data.

This approach is entirely analogous to the situation of a patchy environment with D and γ varying between the patches in the absence of interpatch migration. This idea was suggested to us by my colleague Michael Moody, an expert in mathematical genetics, who, upon being shown the experimental data of Huffaker *et al.* ([84]) depicted in Fig. 4.12a, commented that it looked like the sum of subpopulations to him. For Pulling that particular Rabbit Out of a Hat by merely making his suggestion, I added Mike as a co-author on the paper Collings *et al.* ([37]) reporting this result. A natural extension of such an analysis can be accomplished by introducing dispersal effects into our model through the consideration of either passive diffusion between discrete patches allowing for colonization and invasion or active diffusion in space incorporating the flux mechanisms of mite motility and predator aggregation. In the latter case, our model would consist of a governing system of interaction-dispersion partial differential equations (Wollkind *et al.*, [281]). We shall consider biological phenomena more properly modeled by systems of this sort in later chapters.

As one final detail of our temperature-dependent May mite model system, note, in particular, that the intersection points of the dashed line $\gamma = 0.15$ in Fig. 4.11a with its relevant plots reproduce T_{-1}, T_1, and T_2 of Fig. 4.9. Observe that small deviations of γ from that value yield identical qualitative behavior and hence, our system is resilient to such parameter variation, which as a general rule should be true for realistic models of biological systems. That resiliency of this sort involving γ as just described may not always be the case for all choices of temperature dependence in the coefficients of our model is graphically demonstrated by Fig 4.13. This figure illustrates the results of a two-parameter T-γ Hopf bifurcation analysis with $D = 0.00013$ as in Fig. 4.11a, but for $a = a_\ell(T)$ instead. Upon comparison of the $\gamma = 0.09$ and 0.15 lines in Fig. 4.13, we observe that the qualitative behavior of the solution profile with temperature is dramatically altered upon variations in γ which is quite unlike the situation for the corresponding lines in Fig. 4.11a. Since such resiliency is to be expected in biological systems (Holling, [82]), our model would seem preferable to any in which a was implicitly taken equal to $a_\ell(T)$.

This completed the research phase of the analysis of our temperature-dependent composite May mite model system. Next came its dissemination phase, which, as always, included

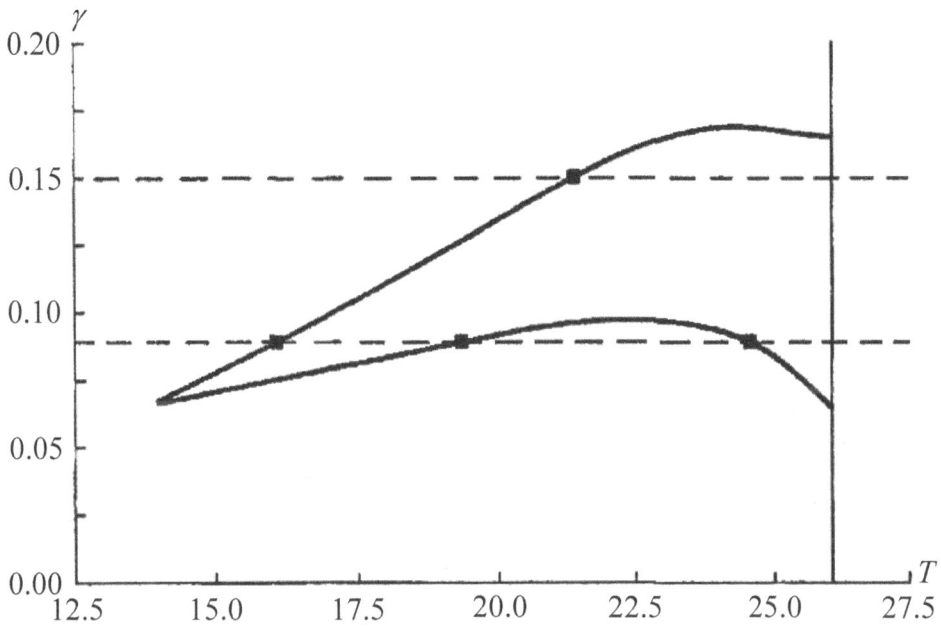

FIGURE 4.13
Two-parameter Hopf bifurcation diagram analogous to Fig. 4.11a but for a model with $a = a_\ell(T)$.

publication in appropriate journals and presentations at scholarly meetings or academic colloquia. I previously alluded to its original submission to the *Bulletin of Mathematical Biology*, edited by Lee Segel, and its subsequent publication in revised form as Wollkind *et al.* ([282]). Simultaneously, Maurice Sabelis and Wim Helle organized a symposium entitled Population Dynamics of Spider Mites and Predatory Mites to be held at the Tropenmuseum in Amsterdam from 5-10 July 1987 and invited Jesse to be one of their plenary speakers. Since he had made a prior conflicting commitment, I went instead and presented our joint work at their symposium. All the invited addresses were published in a special issue of the Elsevier journal *Experimental and Applied Acarology* and ours appeared as Wollkind *et al.* ([283]). I finished that talk by discussing the implications of our theoretical model predictions for the planning of control strategies relevant to agricultural pest management. In particular, regarding T as mean daily temperature and considering the mid- to late-growing season as a scenario during which it is high enough to be in the interval of hysteresis, the question arises when should predacious mites be added to the orchard if the system is already undergoing limit-cycle behavior. The results of Fig. 4.10 suggest the best biological control strategy would be the introduction of predators when the population density of both species is low rather than when the spider mite population is high, since the tendency in the former case is to return the system to its equilibrium point while in the latter case the limit-cycle behavior would be preserved. In other words, I pointed out, one should employ the American version of the British sense of fair play and kick them when they are down. The Brits in the audience enjoyed that comment immensely although many of the Americans were not amused.

Dana Wrench, a geneticist who specialized in acarology, was one of the Americans who was amused and she invited me to give a series of talks that Fall at her institution The Ohio State University, as it calls itself. I gave two of them; the first, in Columbus for the Department of Zoology, which included their entomologists and the second, at Ohio State's Wooster experiment station. After the first talk, I had a meeting with the eminent ecologist Peter Chesson who was then at Ohio State. To put it in no uncertain terms, he did not much care for the Leslie predator equation of the May model. Peter said that its per capita growth rate depended only on the ratio of H/P and did not have its numerical response (see below) proportional to the functional response of the prey equation, characteristics of which he strongly disapproved. To support his first argument, Chesson claimed that if two predators encountered six prey they did not then decide to consume three each. Upon my return to WSU, I related my conversation with Peter to John Collings, who became angry over the objections raised by Chesson. In point of fact, John was starting an ONR (Office of Naval Research) postdoctoral fellowship with me that Fall and spent a good part of the next three years analyzing the two models we devised by introducing modifications into our original system for the express purpose of refuting both of Chesson's claims. This actually resulted in two more papers we might never have produced had I not given those talks at The Ohio State University and met Peter Chesson in the process.

This chapter will conclude with a sketch of our formulation of these two systems which we called the generalized May and Bazykin mite models, respectively, followed by a summary of the results of their bifurcation analyses. We shall proceed to treat each of them individually:

The generalized May mite model: The presence of the ratio of the predator to the prey in the Leslie equation of the May model system was reminiscent of the predator-prey mite laboratory experiments on bean plants conducted by Eveleigh and Chant ([56]), who found the fecundity of their predaceous phytoseiid mite *Amblyseius degenerans* Berlese depended mainly on the number of its prey *Tetranicus pacificus* McGregor available per predator.

Indeed, it was primarily to accommodate this and several of their other experimental results reported in Eveleigh and Chant ([55]) that we introduced our extensions to the model of Wollkind *et al.* ([282]) to be developed. In particular, they determined the fecundity of adult mated females measured in number of eggs per female per day for various densities of prey and predator per bean leaf, as reproduced in Fig. 4.14a. We also incorporated the idea that prey density can affect the predator death rate, since Eveleigh and Chant ([56]) observed that some predaceous mites engaged in cannibalism when underfed. Thus, we replaced the Leslie predator equation by

$$\frac{1}{P}\frac{dP}{dt} = n(H, P) - d(H, P)$$

where the numerical response and the per capita death rate satisfied

$$n(H, P) = \begin{cases} 0 \\ r_2\left(1 - \alpha\frac{P}{H}\right) \end{cases} \text{when} \quad \begin{matrix} H < \alpha P \\ H > \alpha P \end{matrix} \text{and} \quad d(H, P) = d_1\frac{P}{H}$$

with $\alpha, d_1 > 0$. Since Eveleigh and Chant's ([55, 56]) experiments were conducted at 25°C this numerical response is plotted in Fig. 4.14b for $T = 15$ and $\alpha = 3$. Upon comparison with Fig. 4.14a, observe that this plot is in qualitative agreement with the experimental data and note that it is actually in quantitative agreement as well, should the developmental factor for the conversion of an ovipositional to an adult production rate be taken into account. In honor of Peter Chesson, we took α, the threshold ratio for reproduction to occur, equal to three, given that Eveleigh and Chant's ([56]) experiments indicated that if two predators encountered six prey they would indeed consume three each. In addition, we made the identification that

$$d_1 = \left(\frac{1}{\gamma} - \alpha\right) r_2 > 0$$

which required $\gamma\alpha < 1$ and was automatically satisfied for $\gamma = 0.15$ and $\alpha = 3$. The other experiments of Eveleigh and Chant ([55]) also indicated the functional response depended on the predator density, although to a lesser extent than did the numerical response. Hence, we modified our prey equation to reflect this fact by including the extra factor $1/(1 + \beta P)$ in its Holling type-II functional response. Then, nondimensionalizing this system employing the same variables and parameters as introduced for the May mite model, we obtained its generalization given by

$$\frac{d\mathcal{H}}{d\tau} = \mathcal{H}(1 - \mathcal{H}) - \frac{\phi\mathcal{P}\mathcal{H}}{(\mathcal{H} + D)(1 + c_1\mathcal{P})}$$

and

$$\frac{d\mathcal{P}}{d\tau} = \begin{cases} \frac{-\theta(1-c_2)\mathcal{P}^2}{\mathcal{H}} & \mathcal{H} < c_2\mathcal{P} \\ \theta\mathcal{P}\left(1 - \frac{\mathcal{P}}{\mathcal{H}}\right) & \mathcal{H} > c_2\mathcal{P} \end{cases},$$

where $c_1 = \gamma K\beta \geq 0$ and $0 \leq c_2 = \gamma\alpha < 1$, which reduced to it in the limiting case $c_1 = c_2 = 0$.

This system had community equilibrium points $\mathcal{P}_e = \mathcal{H}_e > 0$ satisfying the cubic equation

$$(\mathcal{H}_e - 1)(\mathcal{H}_e + D)(c_1\mathcal{H}_e + 1) + \phi\mathcal{H}_e = 0$$

and hence those points were determined by the positive roots of that equation. We found that when

$$0 < c_1 < c_{crit} = \left(\frac{1}{1 - D^{1/3}}\right)^3$$

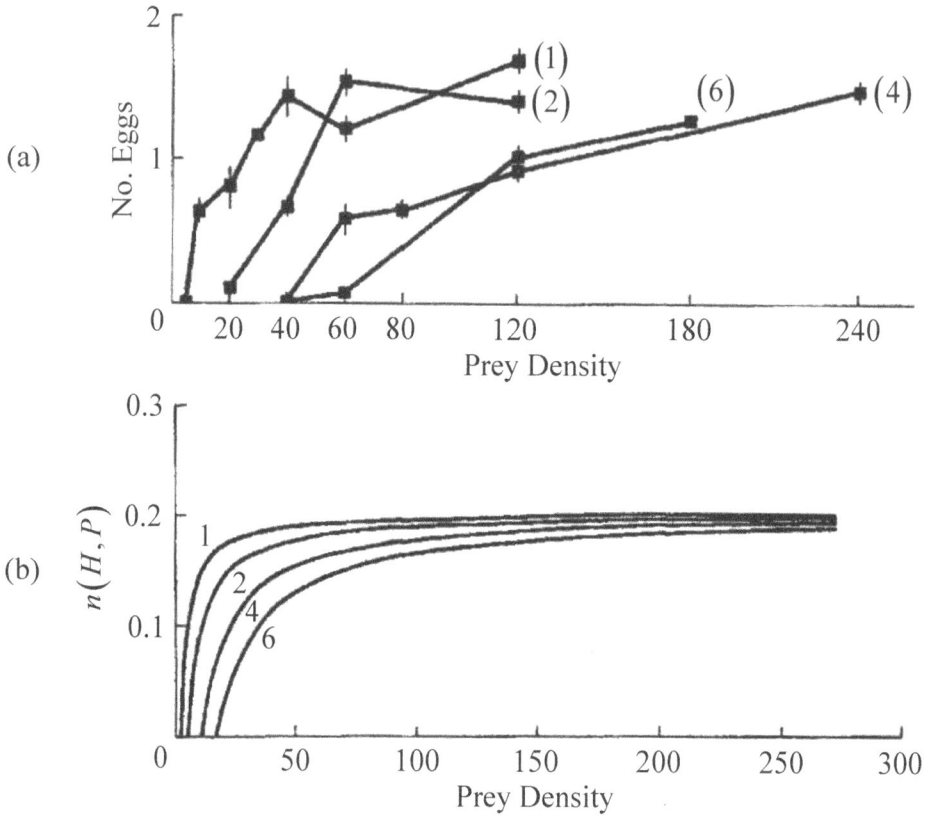

FIGURE 4.14

The effect of predator density on numerical response (a) as determined experimentally for the fecundity of *A. degenerans* by Everleigh and Chant ([56]) and (b) as modeled by $n(H, P)$ of the generalized May mite model with $\alpha = 3$ and $T = 15$ where the numerals used to designate each curve represent the number of predators per leaf.

it possessed exactly one community equilibrium point for all $0 < T < T_M$, while when $c_1 > c_{crit}$ three community equilibria actually occurred over part of that temperature domain, with two occurring at the end points and one outside of this subinterval. Note in that context, $c_{crit} \cong 1.2$ for $D = 0.00013$.

We next performed a one-parameter bifurcation analysis on this system using AUTO with $D = 0.00013$ for $\gamma = 0.15$ and $\alpha = 3$ or $c_2 = 0.45$. Fig. 4.15 summarizes the results of that analysis by plotting the bifurcation diagrams with temperature for the four representative values of c_1, yielding qualitatively different behavior. If $c_1 = 1 < c_{crit} \cong 1.2$, there was a single community equilibrium point and the global behavior of the system is similar to that of the composite May mite model, as suggested by the typical bifurcation diagram of Fig. 4.15a upon comparison with Fig. 4.9. Here, the main difference between these figures was the presence of a second interval of metastability above T_2 caused by a sufficiently large value of the birth threshold $c_2 > 0.45$. This is indicated by the two-parameter bifurcation diagram for c_2 versus T of Fig. 4.16a. It is interesting that the critical limit for the occurrence of this second interval of metastability actually coincided with its value, as selected, in honor of Peter Chesson.

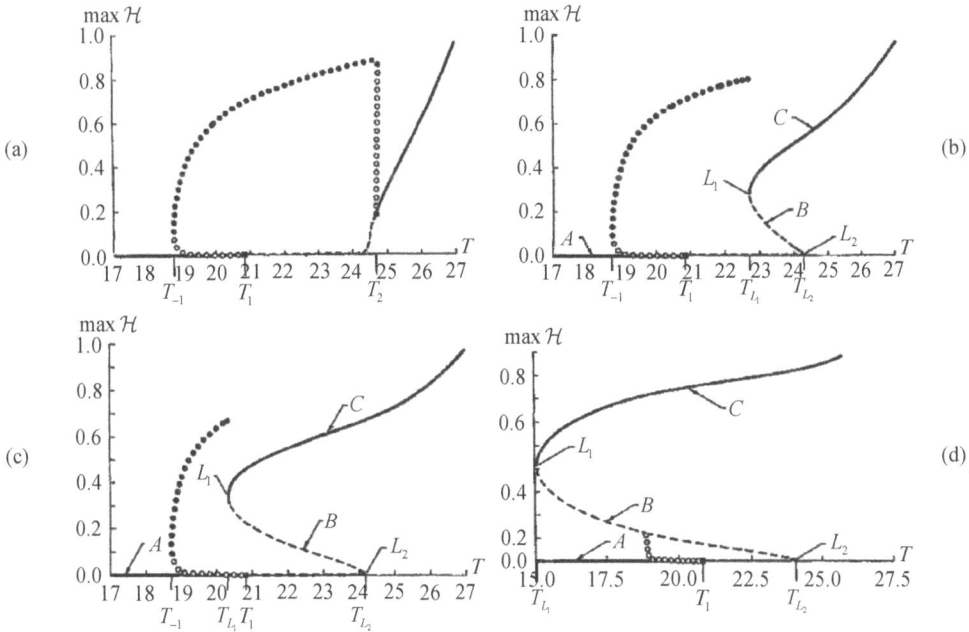

FIGURE 4.15
A sequence of max $\mathcal{H}$ versus T one-parameter bifurcation diagrams for the generalized composite May mite model with $\gamma = 0.15$, $D = 0.00013$, and $c_2 = 0.45$ when (a) $c_1 = 1$, (b) $c_1 = 2.3$, (c) $c_1 = 3.2$, and (d) $c_1 = 6$.

At this time, it is appropriate to relate an anecdote involving John's application of AUTO to the May mite model which resulted in Fig. 4.9. At first, he had trouble generating that one-parameter bifurcation diagram by increasing temperature due to the presence of the metastable region below T_1, so John merely generated it by decreasing temperature instead.

When I asked him what he would have done if there were a second region of metastability above T_2 as in this case, John replied, starting at an intermediate temperature in the stable limit cycle interval, he would then have both decreased T past T_1 and increased it past T_2. John Collings was nothing if not extremely inventive, which allowed us to be successful in Pulling that Rabbit Out of a Hat.

If $c_1 > c_{crit} \cong 1.2$ then, besides the community equilibrium point on the branch labeled A, two extra community equilibria would occur associated with the branches labeled B and C in Figs. 4.15b,c,d which correspond to $c_1 = 2.3, 3.2, 6.0$, respectively. The latter two branches give rise to two more limit points labeled $L_{1,2}$ occurring at $T = T_{L_{1,2}}$ in that figure, while no T_2 exists. The behavior of these two limit points with c_1 is plotted in the two-parameter bifurcation diagram of Fig. 4.16b. Here Figs. 4.15b,c,d are characteristic of c_1 such that $T_{-1}(c_1) < T_1(c_1) < T_{L_1}(c_1)$, $T_{-1}(c_1) < T_{L_1}(c_1) < T_1(c_1)$, and $T_{L_1}(c_1) < T_1(c_1)$ with T_{-1} ceasing to exist for those c_1 greater than that where $T_{-1}(c_1) = T_{L_1}(c_1)$, respectively. In Figs. 4.15b,c the periodic limit cycles terminate at T_{L_1} with a so-called homoclinic orbit of infinite period that starts and ends on the branch B saddle point (see below) represented by a dashed line. In particular for Fig. 4.15b, $T_{L_1} = 20.36465$. From Fig. 4.15, we can see that in this instance the system possesses multiple stable states consisting of two stable fixed points or one stable fixed point and a stable limit cycle and hence is unified in the sense it can model outbreaks with or without subsequent oscillations.

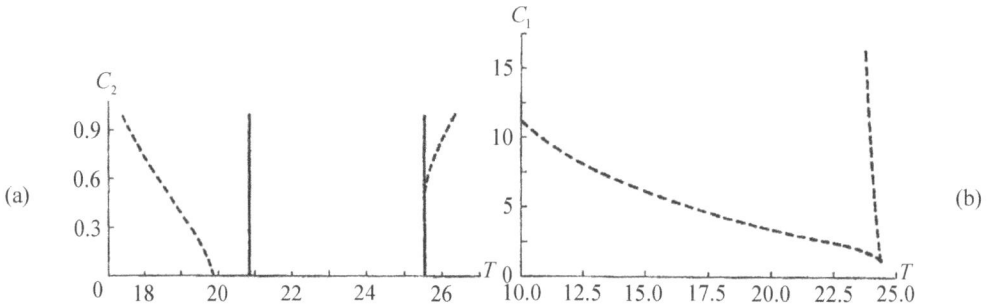

FIGURE 4.16
Two-parameter bifurcation diagrams for the generalized composite May mite model with $\gamma = 0.15$ and $D = 0.00013$ showing how (a) the Hopf bifurcation and limit points on the branch of periodic solutions vary in T-c_2 space (with $c_1 = 0$) and (b) the limit points $L_{1,2}$ on the stationary solution branch vary in T-c_1 space (with $c_2 = 0.45$). Here, the c_1 coordinate at which the two curves intersect is given by $c_1 \cong 1.2$, the value of c_{crit} when $D = 0.00013$.

The Bazykin mite model: We began by showing that the application of a Taylor series expansion technique, retaining only its linear terms, to the predator per capita growth rate function rewritten in terms of the predator functional response in the May mite model led to a Bazykin-type predator-prey system as follows. Denoting the predator functional response in the nondimensionalized May mite model by

$$\mathcal{Y} = f(\mathcal{H}) = \frac{\phi \mathcal{H}}{\mathcal{H} + D},$$

solving for

$$\frac{1}{\mathcal{H}} = \frac{1}{D}\left(\frac{\phi}{\mathcal{Y}} - 1\right),$$

and substituting that result into its predator equation, we obtained

$$\frac{d\mathcal{P}}{d\tau} = \theta\mathcal{P}g(\mathcal{Y},\mathcal{P})$$

where

$$g(\mathcal{Y},\mathcal{P}) = 1 + \frac{\mathcal{P}}{D}\left(1 - \frac{\phi}{\mathcal{Y}}\right).$$

Now expanding $g(\mathcal{Y},\mathcal{P})$ in a two-variable Taylor series about the May mite model equilibrium point $\mathcal{P}_e = \mathcal{H}_e$, with $\mathcal{Y}_e = f(\mathcal{H}_e)$, and retaining only its linear terms, we deduced that

$$g(\mathcal{Y},\mathcal{P}) \cong g(\mathcal{Y}_e,\mathcal{P}_e) + \frac{\partial g}{\partial\mathcal{Y}}(\mathcal{Y}_e,\mathcal{P}_e)(\mathcal{Y} - \mathcal{Y}_e) + \frac{\partial g}{\partial\mathcal{P}}(\mathcal{Y}_e,\mathcal{P}_e)(\mathcal{P} - \mathcal{P}_e)$$

where

$$g(\mathcal{Y}_e,\mathcal{P}_e) = 0, \quad \frac{\partial g}{\partial\mathcal{Y}}(\mathcal{Y}_e,\mathcal{P}_e) = \frac{\phi\mathcal{P}_e}{D\mathcal{Y}_e^2}, \quad \frac{\partial g}{\partial\mathcal{P}}(\mathcal{Y}_e,\mathcal{P}_e) = -\frac{1}{\mathcal{P}_e}.$$

Then introducing this linear expansion through terms of first order into our predator equation we approximated it by

$$\frac{d\mathcal{P}}{d\tau} = -\alpha\mathcal{P} - \beta\mathcal{P}^2 + \rho\frac{\mathcal{P}\mathcal{H}}{\mathcal{H} + D}$$

where

$$\alpha = \frac{\theta\mathcal{H}_e}{D}, \quad \beta = \frac{\theta}{\mathcal{H}_e}, \quad \rho = \frac{\theta(\mathcal{H}_e + D)^2}{D\mathcal{H}_e},$$

which is Bazykin ([9]) in form. Here, α, β, and ρ represent the predator per capita rates of death, intraspecific competition for nonprey resources, and birth conversion efficiency for prey consumed, respectively. Note that, the predator numerical response of this system, which retains our May mite model prey equation, is directly proportional to its predator functional response in accordance with Peter Chesson's preference for such predator-prey systems, which was why John developed our approximation in the first place. John Collings was nothing if not extremely clever.

The community equilibrium points of this system corresponded to the intersection of the prey zero-isocline

$$\mathcal{P} = \frac{1}{\phi}(1 - \mathcal{H})(\mathcal{H} + D)$$

with the predator zero-isocline

$$\mathcal{P} = \frac{(2\mathcal{H}_e + D)\mathcal{H} - \mathcal{H}_e^2}{\mathcal{H} + D}$$

and thus its $\mathcal{H}$-component was again found to satisfy a cubic equation. Given the way this system had been derived, we immediately deduced one of its community equilibrium points $(\mathcal{H}_A,\mathcal{P}_A)$ was identical to that of the May mite model or

$$\mathcal{H}_A = \mathcal{H}_e = \mathcal{P}_e = \mathcal{P}_A.$$

Then, factoring $\mathcal{H} - \mathcal{H}_e$ from this cubic yielded a quadratic equation satisfied by the $\mathcal{H}$-components of the other two possible equilibrium points and solving it we obtained their coordinates $(\mathcal{H}_B, \mathcal{P}_B)$ and $(\mathcal{H}_C, \mathcal{P}_C)$ where $\mathcal{P}_{B,C}$ were determined from either of the two zero-isoclines once $\mathcal{H}_{B,C}$ had been evaluated. Next, we performed a one-parameter bifurcation analysis on this system's composite version using AUTO with $D = 0.00013$ and $\gamma = 0.15$.

Fig. 4.17 summarizes the results of that analysis by plotting its bifurcation diagram in the T-log(max $\mathcal{H}$) plane. In this diagram the branches labeled A, B, C correspond to the community equilibrium point with $\mathcal{H}$-component designated by $\mathcal{H}_{A,B,C}$, respectively. Here log(max $\mathcal{H}$) has been employed due to the extremely small distance separating the two solutions $\mathcal{H}_A$ and $\mathcal{H}_B$. As before, the solid and dashed lines represent stable and unstable solutions, while in this instance the saddle points on branch B are represented by a dotted line. Again, Hopf bifurcation points are denoted by solid squares and occur at the same temperatures $T_1 = 20.85$ and $T_2 = 25.54$ on branch A as they did for the May mite model system. The branches of stable periodic solutions which emanate from those points are depicted by solid circles in the insets to that figure and terminate in homoclinic orbits at $T_{H_1} = 21.2$ and $T_{H_2} = 25.4$. As can be seen from those insets, the oscillations characteristic of these limit cycles are quite small with amplitudes limited by the population saddle point values on branch B. Finally, $T_{L_2} = 25.49$, where branches B and C intersect, represents both a limit point for this system and the maximum temperature for which these branches exist.

The global behavior of this system can be described by dividing T into five intervals. On $T < T_{H_1}$ or $T_{H_2} < T < T_{L_2}$, it possesses multiple stable states. In both instances one of these states is the stable community equilibrium point on branch C, while the other state is that associated with branch A being either the stable community equilibrium point on that branch for the first interval or a stable limit cycle surrounding its unstable community equilibrium point for the second. Fig. 4.18 is a phase plane plot representative of the latter case for $T = 25.45$. Here the stable states are separated by an unstable saddle point on branch B, the typical trajectories of which are depicted in that figure. The community equilibrium point on branch C is the only stable state for $T_{H_1} < T < T_{H_2}$, while for $T > T_{L_2}$ the system possesses the single community equilibrium point on branch A, unstable and surrounded by a stable limit cycle for $T_{L_2} < T < T_2$ but globally stable for $T > T_2$. The metastability indicated above allows this system to model outbreak behavior although the extremely small basin of attraction associated with the solutions on branch A for $T < T_{H_1}$ and the fact that solutions on branch C show the prey near its carrying capacity indicate that in most situations the predator is ineffective in controlling the prey. Given these limitations involving oscillations and biological control, we concluded that this composite Bazykin mite predator-prey system would not be a very good choice for modeling the interaction between *M. occidentalis* and *T. McDanielli* on apple tree foliage.

We close by relating our experiences during the dissemination of these results for the generalized May and Bazykin mite models involving publications in academic journals and presentations at professional meetings. John and I published a paper with Michael Moody highlighting the results of the generalized May mite model system that, in reviewing those of the May mite model, included Mike's subpopulation suggestion alluded to earlier and, in previewing future work, concluded with the Bazykin mite model formulation. Our paper was actually submitted to the *American Naturalist*, but accepted for publication by *Theoretical Population Biology*, this having transpired because the reviewers for the *American Naturalist* did not think that journal the proper place for it, but its editor Marty Feldman who was also the *Theoretical Population Biology* editor accepted it unilaterally without revision for the latter journal. During the publication process I got a phone call from the

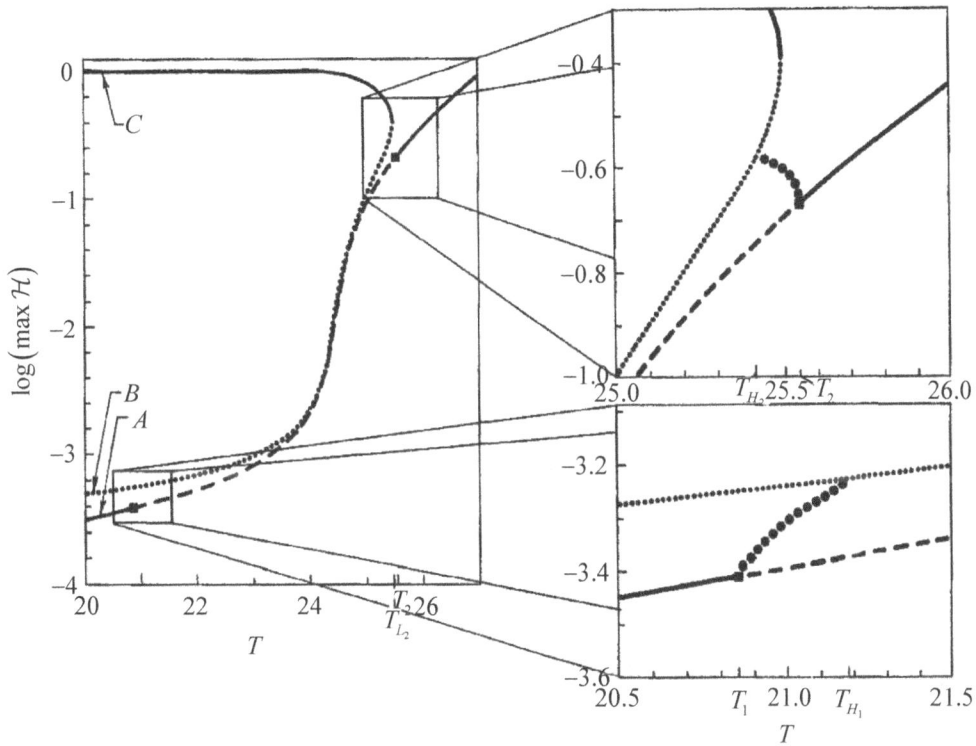

FIGURE 4.17
A one-parameter bifurcation diagram for the Bazykin mite model system with $\gamma = 0.15$ and $D = 0.00013$ plotting $\log(\max \mathcal{H})$ versus T. The insets show the branches of the periodic solutions originating at the Hopf bifurcation points corresponding to $T_{1,2}$ and terminating in homoclinic orbits at T_{H_1, H_2} where $T_1 = 20.85$, $T_2 = 25.54$ and $T_{H_1} = 21.2$, $T_{H_2} = 25.4$, while $T_{L_2} = 25.49$.

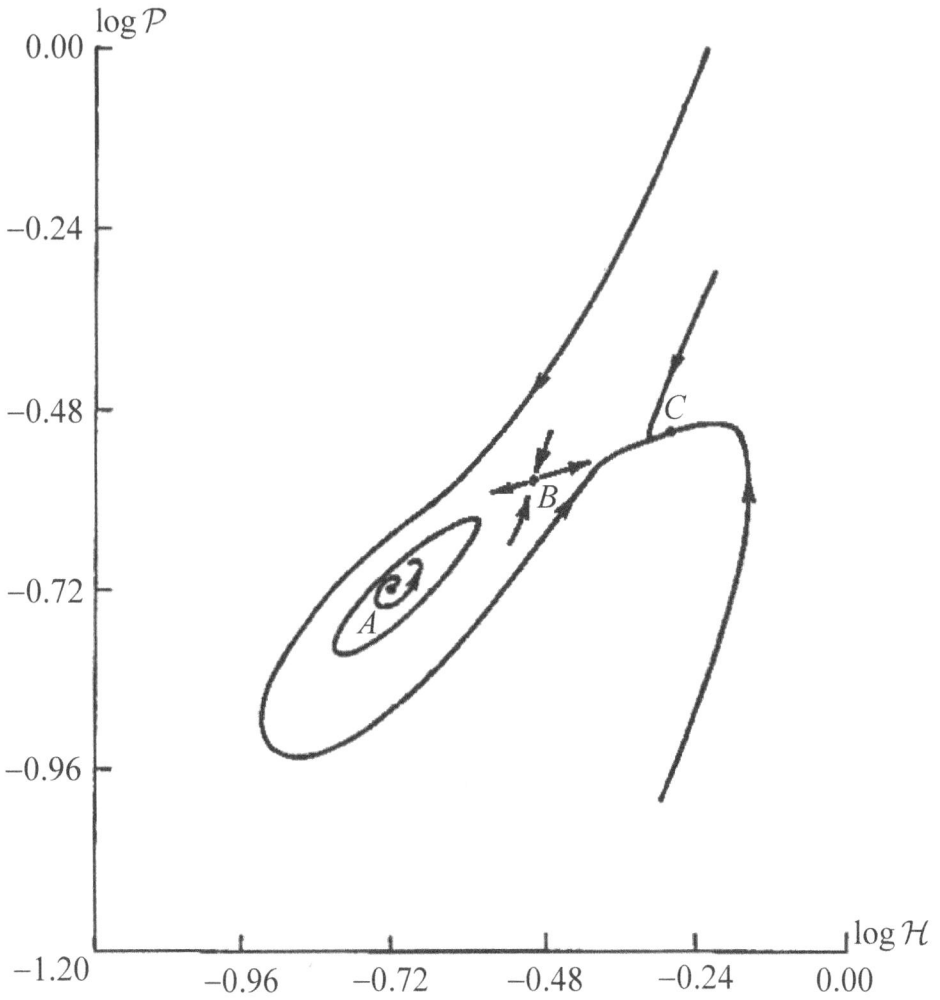

FIGURE 4.18
A phase portrait in the log $\mathcal{H}$-log $\mathcal{P}$ plane for the Bazykin mite model system with $\gamma = 0.15$ and $D = 0.00013$ for $T = 25.45$.

copy editor for *Theoretical Population Biology* complaining about the paper not complying with *TPB*'s stated format. I responded to her by replying, "In our defense, madam, this paper was never submitted to *Theoretical Population Biology*, but rather to the *American Naturalist* and Marty made the editorial decision to publish it as is in your journal." She, of course, was not amused. John and I then published our results for the Bazykin mite model as Collings and Wollkind([35]) in the *SIAM Journal for Applied Mathematics*, there being no question about its appropriateness for that journal since the definitive paper Hainzl ([76]) on the behavior of Bazykin predator-prey systems had recently appeared in it.

We presented talks at a variety of venues publicizing our results for these model systems, the most important of which were John's generalized May mite model presentation at the International Workshop on the Population Dynamics of Outbreaks, organized by Peter Antonelli, which was published as Collings and Wollkind ([36]), and my invited addresses featuring both of these mite model systems at the Pacific Northwest Workshop on Mathematical Biology, the Gordon Research Conference on Theoretical Biology and Biomathematics, and the 1992 Society of Mathematical Biology Annual Meeting. During those presentations, I promoted the power of AUTO to produce bifurcation diagrams, by first revealing the content of one of the negative *American Naturalist* reviews of what appeared in *TPB* as Wollkind *et al.* ([298]). In particular, when referring to its AUTO generated bifurcation diagrams, that reviewer stated, "I don't know why the authors had to use a numerical computer code to produce those plots. I could have drawn them using only a paper and a pencil!" In response to this, I would Pull a Rabbit out of a Hat by putting on the screen the slide of the complicated bifurcation diagram for the Bazykin mite model system reproduced in Fig. 4.17 while making the commentary, "Try drawing that with only a paper and a pencil!!!" Incidentally, the design of this figure using insets for the limit cycle behavior in the neighborhood of its Hopf bifurcation points was due to the creativity of my graphic artist Karen Parvin. As a cost saving measure, WSU, in all its wisdom, eventually shut down her graphics laboratory since they deemed the availability to researchers of digital software packages for producing figures of this sort made Karen's job redundant, which has become a trend nationwide. The outcome of such measures is that most of the figures in current journal papers are inferior in quality to those found in older papers published before they were instituted.

The models introduced in this chapter have provided useful information about the dynamic structure of field populations and also served as a first step toward the development of a comprehensive model for planning biological control strategies of orchard mite populations in Washington state. Once this has been completely achieved it should be possible to formulate a simulation model to predict the economic impact on those orchards inflicted by the McDaniel spider mite pests. Figure 4.19 represents a schematic flow chart devised by Lynell Tanigoshi representing such a simulation model for determining agro-economic influence of this sort. That is the same WSU entomologist who helped us gather data used for parameter identification in the Wollkind-Logan temperature-rate equation for arthropods. Entomologists invariably cite this as the Logan equation and that is just another way to lose credit for a partial research contribution.

We close this chapter by returning to a topic discussed at the end of Chapter 2: Namely, who gets credit for a major research contribution. More than ten years after Wollkind *et al.* ([282]), Sáez and González-Olivares ([199]) published a paper entitled "Dynamics of a predator-prey model" in the *SIAM Journal on Applied Mathematics* with the following abstract:

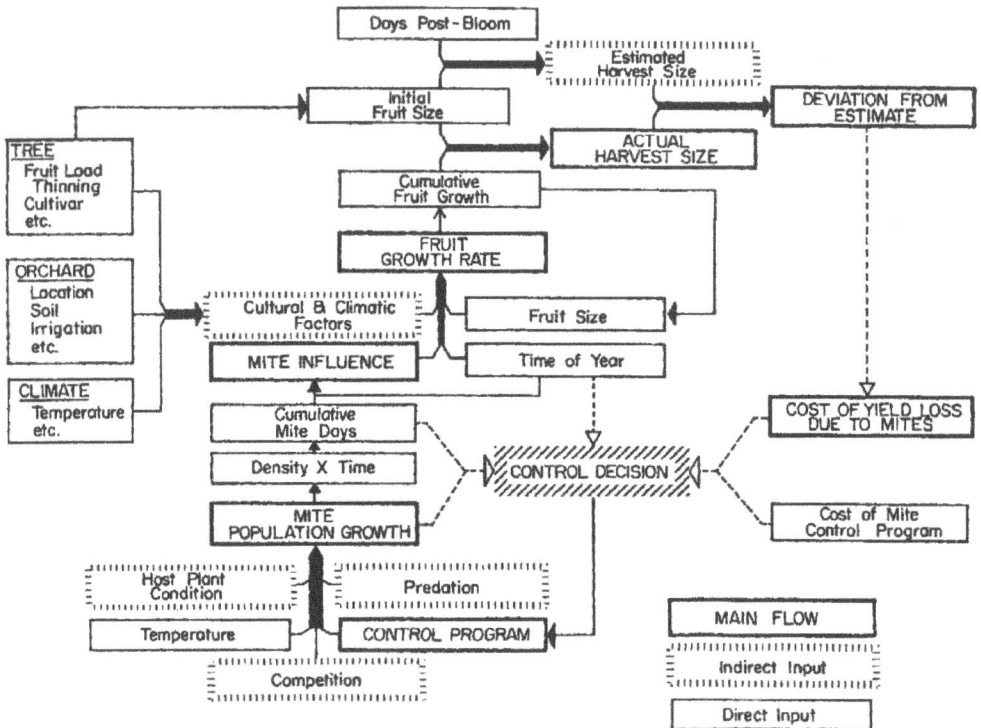

FIGURE 4.19
A schematic flow chart, due to Lionel K. Tanigoshi, representing a complete simulation model for determining the agro-economic influence of mite populations on apple orchards.

We describe the bifurcation diagram of limit cycles that appear in the first realistic quadrant of the predator-prey model proposed by R. M. May [Stability and Complexity in Model Ecosystems, Princeton University Press, Princeton, NJ, 1974]. In particular, we give a qualitative description of the bifurcation curve when two limit cycles collapse on a semi-stable limit cycle and disappear. Moreover, we show that locally asymptotic stability of a positive equilibrium point does not imply global stability for this class of predator-prey models.

In their introduction, they begin with the statements that the May model has been used by Wollkind, Collings, and Logan ([282]) to investigate numerically the dynamics of a predator-prey system under the hypothesis that the parameters depend on the temperature and that May ([139]) conjectured for predator-prey systems of this type, with a unique community equilibrium point, local and global stability are equivalent. Sáez and González-Olivares ([199]) then state:

> In this paper we show that local and global stability are not equivalent for this class of predator-prey models. In fact we prove, in the parameter space, the existence of a bifurcation manifold S of semi-stable limit cycles and the existence of an open manifold G, where the community equilibrium point is locally stable and is surrounded by two limit cycles. This implies the interesting phenomenon of coexistence of stable equilibrium and persistent oscillations. Our conclusions confirm the numerical results obtained by Wollkind, Collings, and Logan (1988a).

Sáez and González-Olivares ([199]) do this by means of a theorem-proof methodology for a class of Kolmogorov predator-prey systems including the May model and its related normal form reduction. Their main results can be catalogued as follows:

(i) The deduction that the coexistence of a focus surrounded by a limit cycle, both locally asymptotically stable, is possible. Therefore, for the May model and its normal form reduction, the conjecture that local stability of its equilibrium point implies global stability is false.

(ii) If the equilibrium point is in G, the May model and its normal form exhibit the phenomenon of bistability where there is coexistence of a locally stable limit cycle surrounding this locally stable community equilibrium point as well as an unstable limit cycle which serves as their separatrix in the phase plane. If that equilibrium point is on its boundary S, there is coexistence of this stable singularity and a semistable limit cycle which is attracting on the outside and repelling on the inside. These two cases are plotted in a figure for representative parameter values.

In essence, Sáez and González-Olivares ([199]) put our temperature-dependent AUTO-generated results for the May model on a firm mathematical foundation. To date their paper has garnered 170 citations while ours, only 136. This is also another way to lose credit at least in part for a major research contribution.

5

Multi-Layer Fluid Phenomena: Rayleigh-Bénard-Marangoni Convection and Kelvin-Helmholtz Rock Folding: Linear Stability Analyses

This chapter deals with two companion problems involving multilayer fluid models in which the interaction of gravity and interfacial surface tension effects played a key role. In addition, the linear stability analysis of each of them produced results that yielded theoretical predictions in very good qualitative and quantitative agreement with existing fluid mechanical experimental or structural geological observational data, while serving as the Ph.D. dissertation topics for Eric N. Ferm and J. Iwan D. Alexander, respectively. Given the similarity in the formulations and the subsequent mathematical analyses of these problems, it seems reasonable to group them together.

I first met Eric Ferm when he visited my office and introduced himself to me in the Fall of 1977. Eric had just completed a BS in Mathematics from Boise State University, at which two of WSU's earlier Ph.D. recipients in abstract algebra, Eugene Furuyama (class of 1972) and Marshall Sugiyama (class of 1974), were then faculty members and, given his interest in phenomenological modeling, had suggested that he enter our applied mathematics doctoral program to pursue a Ph.D. degree under my direction. At this time, I had some equations on my blackboard related to two-layer Bénard convection with interfacial deflection driven by thermal variations in surface-tension (the Marangoni effect analogous to that of surface free energy for solidification introduced in Chapter 2) and density (the Rayleigh effect introduced in Chapter 3). This was a "back-burner problem," the formulation of which I would periodically consider. After Eric had passed his preliminary doctoral examination and we were discussing a thesis topic for him, he said, "You know that problem you had on your blackboard the first day we met? That's what I want to do." Generally, I picked the thesis problems for my students, selecting only those that were both doable, in my estimation, and could be completed in a reasonable period of no more than two years. I did not select open-ended ones, nor back-burner problems, but Eric was adamant in his choice and could not be dissuaded from attempting it. Being a back burner problem, I was not as familiar with this research area as with those of the other Ph.D. students who worked with me before and after Eric. Hence, we conducted a literature search to see what had been done on Rayleigh-Bénard-Marangoni convection, both theoretically and experimentally. There existed a recent review of the subject by Normand *et al.* ([157]) that allowed us to perform an extensive examination of all the papers mentioned in it. Specifically the work of Koschmieder ([102]) and Palmer and Berg ([165]) represented two series of carefully controlled experiments for the onset of Bénard natural convection in a thin liquid layer heated from below and cooled from above by means of a gas layer (see Fig. 5.1). To compare the experimentally determined temperature gradient across the liquid layer at which the onset of convection occurred and the observed wavelength characteristic of the resulting flow with those critical conditions, as predicted by linear stability theory, it was essential to employ a

DOI: 10.1201/9781003195603-5

model including all the effects present in these experiments: Namely, the dynamics of both fluid layers, the buoyancy gravity effect due to the expansion of heated fluid, the surface tension effect due to interfacial traction on the fluid-fluid interface induced by temperature variations, and the deflection of that interface itself. Since no such theoretical analysis was mentioned in Normand *et al.* ([157]), we assumed that it had not yet been done and began completing the appropriate model for this problem as a prelude to performing a linear stability analysis of its planar interface pure conduction solution.

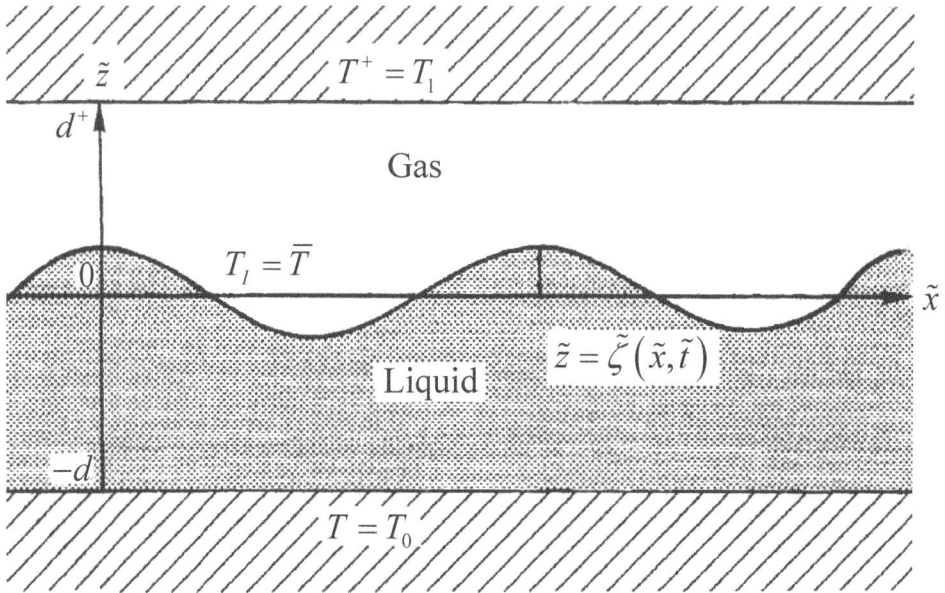

FIGURE 5.1
Schematic diagram employing dimensional variables depicting the thermal and geometric properties of our two-layer liquid-gas system with an interfacial deviation $\tilde{z} = \tilde{\zeta}(\tilde{x}, \tilde{t})$ from its mean planar position, which coincides with the $\tilde{x}$-axis and has a mean interfacial temperature of $T_I = \overline{T}$ where $T_1 < \overline{T} < T_0$. Here $\tilde{t} \equiv$ time.

Our two-dimensional governing equations of motion in the liquid and gas layers for that problem were of the same form as those employed in the Rayleigh-Bénard problem of Chapter 3 but without dissociation, while there were seven boundary conditions imposed at the liquid-gas interface between those layers; the first two, called kinematic boundary conditions, guaranteeing that this interface was a material surface separating immiscible fluids; the next three representing jump-type boundary conditions at this surface of discontinuity, arising from the balance of the tangential and normal components of momentum and the conservation of energy; and the last two being a consequence of our thermodynamic and constitutive relations of continuity of temperature and no relative tangential motion (the kinematic boundary conditions implied no relative normal motion), respectively, at that surface. Finally, three boundary conditions each were imposed at the upper and lower horizontal plates; the first two being dynamical ones of zero velocity and the last, a thermal one of being maintained at a constant temperature, due to our assumption that both of these plates represented rigid isothermal pure conductors.

Analogous to our approach of Chapter 3, we introduced the Rayleigh-Bénard Boussinesq approximation in both layers by adopting the equations of state for the densities ρ^+ and ρ, as functions of the temperatures T^+ and T

$$\rho^+ = \rho_0^+[1 - \alpha^+(T^+ - \overline{T})], \ \rho = \rho_0[1 - \alpha(T - T_0)]$$

in the body force terms of the respective momentum equations, where

$$\alpha^+, \alpha \equiv \text{ thermal expansivity } > 0, \ \Delta\rho = \rho_0 - \rho_0^+ > 0;$$

while taking (see Fig. 5.1)

$$T^+ = T_1 \text{ at } \tilde{z} = d^+, \ T^+ = T = T_I = \overline{T} \text{ at } \tilde{z} = 0, T = T_0 \text{ at } \tilde{z} = -d$$

with

$$\Delta T = T_0 - T_1 > 0, \ T_1 < \overline{T} < T_0.$$

Here, the unsuperscripted variables and parameters denoted the liquid layer $-d < \tilde{z} < \tilde{\zeta}(\tilde{x},\tilde{t})$ and the "+" superscripted ones, the gas layer $\tilde{\zeta}(\tilde{x},\tilde{t}) < \tilde{z} < d^+$, where $\tilde{z} = \tilde{\zeta}(\tilde{x},\tilde{t})$ represented the liquid-gas interfacial deviation from its mean planar interface position of $\tilde{z} = 0$. Similarly, we introduced the Marangoni constitutive relation for interfacial surface tension

$$\gamma(T) = \gamma_0 - \gamma_1(T - \overline{T}) \text{ where } \gamma_{0,1} > 0$$

in the momentum balance jump-type boundary conditions.

We next examined the thermal properties of this system for a motionless (velocity components $\tilde{u} = \tilde{w} = \tilde{u}^+ = \tilde{w}^+ \equiv 0$), steady-state ($\partial/\partial\tilde{t} \equiv 0$), stratified $[T = T_0(\tilde{z}), T^+ = T_0^+(\tilde{z})]$ planar interface ($\tilde{\zeta} \equiv 0$) situation. Since then

$$T_0^{+''}(\tilde{z}) = 0 \text{ for } 0 < \tilde{z} < d^+, \ T_0^+(d^+) = T_1;$$
$$T_0''(\tilde{z}) = 0 \text{ for } -d < \tilde{z} < 0, \ T_0(-d) = T_0;$$
$$T_0^+(0) = T_0(0) = \overline{T}, \ k^+ T_0^{+'}(0) = k T_0'(0)$$
$$\text{where } k^+, k \equiv \text{ thermal conductivity } > 0;$$
$$T_0^+(\tilde{z}) = \overline{T} - \beta^+\tilde{z} \text{ with } \beta^+ = \frac{\overline{T} - T_1}{d^+};$$
$$T_0(\tilde{z}) = \overline{T} - \beta\tilde{z} \text{ with } \beta = \frac{T_0 - \overline{T}}{d};$$

where the temperature gradients satisfied

$$k^+\beta^+ = k\beta \text{ or } \beta^+ = \frac{\beta}{m} \text{ for } m = \frac{k^+}{k}.$$

Thus,

$$\Delta T = T_0 - T_1 = (T_0 - \overline{T}) + (\overline{T} - T_1) = \beta d + \beta^+ d^+ = \beta d\left(1 + \frac{\ell}{m}\right) \text{ for } \ell = \frac{d^+}{d}.$$

Note that, although the interfacial temperature at the mean position of the interface $T_I = \overline{T}$

is an unmeasurable quantity, β can be defined independent of it by using this relation involving experimentally determined parameters.

Finally, we measured the pressures $\widetilde{p}^+$ and $\widetilde{p}$ in such a way that their zero levels corresponded to

$$p_A - \rho_0^+ g\widetilde{z} \text{ for } \widetilde{p}^+, \ p_A - \rho_0 g(1 + \alpha\beta d)\widetilde{z} \text{ for } \widetilde{p},$$

where $p_A \equiv$ atmospheric pressure and $g \equiv$ acceleration due to gravity, which are called reduced.

We nondimensionalized our system by introducing d, d^2/κ, κ/d, $\mu\kappa/d^2$, and δ as scale factors for distance, time, velocity, pressure, and deviation of the interface from its mean planar position, respectively, where $\mu \equiv$ shear viscosity and $\kappa \equiv$ thermal diffusivity, while defining dimensionless temperatures by

$$\theta^+ = \frac{\widetilde{T}^+ - \overline{T}}{\beta d}, \ \theta = \frac{T^+ - \overline{T}}{\beta d},$$

and measuring nondimensional pressures p^+ and p consistent with their dimensional zero levels. This nondimensionalization produced a number of other dimensionless ratios and parameters the most important of which for our present expository purposes were

$$a = \frac{\alpha^+}{\alpha}, \ s = \frac{\rho_0^+}{\rho_0}, \ r = \frac{\mu^+}{\mu}, \ n = \frac{\kappa^+}{\kappa}, \ \epsilon = \frac{\delta}{d};$$

$$R = \frac{g\alpha\beta d^4}{\kappa\nu} \equiv \text{Rayleigh number where } \nu = \frac{\mu}{\rho_0} \equiv \text{kinematic viscosity,}$$

$$M = \frac{\gamma_1 \beta d^2}{\kappa\mu} = \Gamma_1 R \equiv \text{Marangoni number where } \Gamma_1 = \frac{\gamma_1}{\alpha\rho_0 g d^2}.$$

The planar interface pure conduction solution in these nondimensional variables was given by

$$u^+ = w^+ \equiv 0, \ \theta^+ = \theta_0^+(z) = -\frac{z}{m}, \ p^+ = p_0^+(z) = -\frac{asRz^2}{2m} \text{ for } z > 0;$$

$$u = w \equiv 0, \ \theta = \theta_0(z) = -z, \ p = p_0(z) = -\frac{Rz^2}{2} \text{ for } z < 0;$$

$$\epsilon\zeta \equiv 0.$$

It was the linear stability of this state with which we were concerned. Hence, in order to examine that behavior we considered solutions of our basic dimensionless system of equations and boundary conditions of the usual form:

$$\epsilon\zeta(x, t; \epsilon) = 0 + \epsilon c_0 \cos(\omega x)e^{\sigma t} + O(\epsilon^2),$$

$$u(x, z, t; \epsilon) = 0 + \epsilon U(z)\sin(\omega x)e^{\sigma t} + O(\epsilon^2),$$

$$[w, p, \theta](x, z, t; \epsilon) = [0, p_0, \theta_0](z) + \epsilon[W, \Pi, \Theta](z)\cos(\omega x)e^{\sigma t} + \boldsymbol{O}(\epsilon^2);$$

with analogous expansions for u^+, w^+, p^+, and θ^+ involving U^+, W^+, Π^+, and Θ^+, respectively. Here, the parameter ϵ, a measure of the maximum deviation of the interface from its mean planar position, was assumed for the purpose of linear theory to satisfy the condition $0 < \epsilon \ll 1$; hence, terms of $O(\epsilon^2)$ were negligible in comparison with those of $O(\epsilon)$. Thus, that expansion was a superposition of the planar interface solution and an arbitrary normal-mode component of an initially infinitesimal perturbation of wavelength $2\pi/\omega$ and growth rate σ. Then, setting $\sigma = 0$, we determined the critical conditions for the onset of instability from a stationary state by substituting this solution into our basic system, expanding the boundary conditions in a Taylor series about $z = \epsilon\zeta = 0$, neglecting terms of $O(\epsilon^2)$, and, analyzing the two sixth-order ordinary differential equations with thirteen boundary conditions for $W(z)$ and $W^+(z)$, that resulted upon cancellation of the relevant common factors and elimination of all the other dependent variables. In particular, these differential equations were given by

$$[(D^2 - \omega^2)^3 + R\omega^2]W(z) = 0, \ -1 < z < 0; \ \text{where} \ D \equiv \frac{d}{dz};$$

$$[(D^2 - \omega^2)^3 + R^+\omega^2]W^+(z) = 0, \ 0 < z < \ell;$$

$$\text{where} \ R^+ = \frac{\alpha^+\beta^+gd^4}{\kappa^+\upsilon^+} = \frac{as}{nrm}R.$$

To illustrate the form of the general solution to these differential equations, we consider the special case of $\omega^6 = R\omega^2$ for the differential equation satisfied by $W(z)$. Under that condition, this differential equation reduces to $L[W] = [D^6 - 3\omega^2D^4 + 3\omega^4D^2]W(z) = 0$. Then, letting $W(z) = e^{mz}$, we find that

$$L[e^{mz}] = (m^6 - 3\omega^2m^4 + 3\omega^4m^2)e^{mz} = 0 \Rightarrow m^2(m^4 - 3\omega^2m^2 + 3\omega^4) = 0,$$

which yields the double roots $m = m_{1,2} = \pm\lambda_1$ with $\lambda_1 = 0$ while the other four roots satisfy

$$(m^2)^2 - 3\omega^2(m^2) + 3\omega^4 = 0.$$

Treating this as a quadratic equation in m^2 and applying the quadratic formula, we obtain that

$$m^2 = \frac{3\omega^2}{2} \pm \frac{\sqrt{3}}{2}\omega^2i = \sqrt{3}\omega^2\left(\frac{\sqrt{3}}{2} \pm \frac{1}{2}i\right) = \sqrt{3}\omega^2e^{\pm i\pi/6},$$

where in the above use has been made of De Moivre's Theorem introduced in Chapters 2 and 3

$$e^{\pm i\psi} = \cos(\psi) \pm i\sin(\psi)$$

for the case of $\psi = \pi/6$ since $\cos(\pi/6) = \sqrt{3}/2$ and $\sin(\pi/6) = 1/2$. Finally, taking the square root of this expression, the other four roots are given by

$$m = m_{3,4,5,6} = \pm\sqrt[4]{3}\omega e^{\pm i\pi/12} = \sqrt[4]{3}\omega\left[\pm\cos\left(\frac{\pi}{12}\right) \pm i\sin\left(\frac{\pi}{12}\right)\right].$$

In order to calculate the values of $\cos(\pi/12)$ and $\sin(\pi/12)$, we first consider the trigonometric relations $\cos(\psi \pm \varphi) = \cos(\psi)\cos(\varphi) \mp \sin(\psi)\sin(\varphi)$ with $\varphi = \psi$ or

$$\cos(2\psi) = \cos^2(\psi) - \sin^2(\psi), \ \cos(0) = 1 = \cos^2(\psi) + \sin^2(\psi),$$

from which it follows that, for $0 < \psi < \pi/2$,

$$\cos(\psi) = \sqrt{\frac{1 + \cos(2\psi)}{2}} \ \text{and} \ \sin(\psi) = \sqrt{\frac{1 - \cos(2\psi)}{2}}.$$

Then setting $\psi = \pi/12$ yields

$$\cos\left(\frac{\pi}{12}\right) = \sqrt{\frac{1 + \cos\left(\frac{\pi}{6}\right)}{2}} = \sqrt{\frac{1 + \frac{\sqrt{3}}{2}}{2}} = \frac{1}{2}\sqrt{2 + \sqrt{3}}$$

and

$$\sin\left(\frac{\pi}{12}\right) = \sqrt{\frac{1 - \cos\left(\frac{\pi}{6}\right)}{2}} = \sqrt{\frac{1 - \frac{\sqrt{3}}{2}}{2}} = \frac{1}{2}\sqrt{2 - \sqrt{3}},$$

respectively. Now, substituting these values into our expressions for those other four roots,

$$m_{3,4,5,6} = \pm\frac{\sqrt[4]{3}}{2}\sqrt{2 + \sqrt{3}}\omega \pm \frac{\sqrt[4]{3}}{2}\sqrt{2 - \sqrt{3}}\omega\, i$$

or

$$m_{3,4,5,6} = \pm\lambda_2 \pm \lambda_3 i$$

with

$$\lambda_{2,3} = \frac{\omega}{2}\sqrt{2\sqrt{3} \pm 3}$$

since $\sqrt[4]{3} = \sqrt{\sqrt{3}}$. Finally, taking the linear combination of those linearly independent real solutions corresponding to these roots $m = m_{1,2,3,4,5,6}$, we obtain (Boyce and DiPrima, [18])

$$W(z) = c_1 + c_2 z + \cosh(\lambda_2 z)[c_3 \cos(\lambda_3 z) + c_4 \sin(\lambda_3 z)]$$
$$+ \sinh(\lambda_2 z)[c_5 \cos(\lambda_3) + c_6 \sin(\lambda_3 z)]$$

where

$$\cosh(v) = \frac{e^v + e^{-v}}{2} \text{ and } \sinh(v) = \frac{e^v - e^{-v}}{2},$$

represent the so-called *hyperbolic* cosine and *hyperbolic* sine functions, respectively, given that

$$\cosh^2(v) - \sinh^2(v) = [\cosh(v) + \sinh(v)][\cosh(v) - \sinh(v)]$$
$$= e^v e^{-v} = e^{v-v} = e^0 = 1.$$

Defining

$$A = \sqrt[3]{\frac{R}{\omega^4}},$$

our special case just treated corresponds to $A = 1$. More generally, for $A \neq 1$, by similar but slightly more complicated considerations, we found that (Ferm and Wollkind, [59])

$$W(z) = c_1 W_1(z; \lambda_1) + c_2 W_2(z; \lambda_1) + \cosh(\lambda_2 z)[c_3 \cos(\lambda_3 z) + c_4 \sin(\lambda_3 z)]$$
$$+ \sinh(\lambda_2 z)[c_5 \cos(\lambda_3 z) + c_6 \sin(\lambda_3 z)]$$

where

$$\lambda_1 = \sqrt{|A - 1|}\omega, \quad \lambda_{2,3} = \frac{\omega}{2}\sqrt{2\sqrt{A^2 + A + 1} \pm (A + 2)},$$

$$W_1(z; \lambda_1) = \begin{cases} \cos(\lambda_1 z) & \text{for } A > 1 \\ \cosh(\lambda_1 z) & \text{for } A < 1 \end{cases},$$

$$W_2(z; \lambda_1) = \frac{1}{\lambda_1}\begin{cases} \sin(\lambda_1 z) & \text{for } A > 1 \\ \sinh(\lambda_1 z) & \text{for } A < 1 \end{cases}.$$

Note that this solution reduces to our special case in the limit as $A \to 1$, since

$$\lim_{\lambda_1 \to 0} W_1(z; \lambda_1) = 1, \quad \lim_{\lambda_1 \to 0} W_2(z; \lambda_1) = z;$$

the latter following by virtue of the facts that $\sin(\lambda_1 z)$, $\sinh(\lambda_1 z) \sim \lambda_1 z$ as $\lambda_1 \to 0$. The general solution for $W^+(z)$ is of the same form as that for $W(z)$, but with R replaced by R^+ and each c_i replaced by c_i^+. Substitution of these solutions into their boundary conditions yielded a system of thirteen linear homogeneous equations in the thirteen unknown constants c_0, $\{c_i\}_{i=1}^6$, and $\{c_i^+\}_{i=1}^6$.

Then application of the Pat Munroe algorithm to ensure the existence of nontrivial solutions required the determinant of the coefficient matrix of these constants to vanish, from which we obtained a relationship of the form

$$\beta = f(\beta, \omega)$$

for a given experimental situation involving particular fluids of fixed depth. For a prescribed value of ω, this was a so-called fixed point problem and we wanted to find the smallest positive value of β, such that

$$\beta = F(\omega)$$

which satisfied this relationship. Since f was a transcendental function of β, this had to be accomplished numerically. A sample of the results of such a calculation for a typical case is plotted in Fig. 5.2. As can be seen from that figure, this stationary neutral stability curve had an absolute minimum at $\omega = \omega_c$, where ω_c, the critical wavenumber, could as usual be defined by

$$F'(\omega_c) = 0.$$

Corresponding to ω_c was an associated value of β, denoted by β_c in Fig. 5.2, such that

$$\beta_c = F(\omega_c).$$

Motivated by the experimental observation that the onset of convection occurred from a stationary state, we interpreted β_c as the critical temperature gradient for the occurrence of such an instability. Hence, we considered $\beta = F(\omega)$ as a marginal stability curve in Fig. 5.2, which separated stable states lying beneath it from unstable ones lying above it. Defining R_c and M_c by

$$R_c = \frac{\rho_0 g \alpha d^4}{\kappa \mu} \beta_c \text{ and } M_c = \frac{\gamma_1 d^2}{\kappa \mu} \beta_c,$$

we thus had stability for $R < R_c$ or, equivalently, for $M < M_c$, and instability for $R > R_c$ or, equivalently, for $M > M_c$.

The values of M_c are plotted versus R_c in Fig. 5.3 for Dow Corning 200 silicone oil-air layers of oil $\nu = 1$ stoke (cm^2/sec). Each curve in that figure has been drawn for a different fixed value of ℓ, the ratio of the layer depths, while the points on a particular curve for a given such ℓ have been generated by varying d, the depth of the oil layer. Note that there is instability above such curves and stability below them. Fig. 5.4 is a log-log plot of β_c versus d with $\ell = 0.109$ from that figure. Observe from this figure that the upper part of that curve corresponding to small d has a slope of -2, while the lower part corresponding to large d has one of -4. This result, in conjunction with an examination of the β_c and d dependence of M_c and R_c, as just defined, corroborated the general observation discussed in Chapter 2, that the dominant driving mechanism for convection in shallow oil layers bounded above by an air layer was surface tension (the Marangoni effect), while that in similar deep ones was buoyancy (the Rayleigh effect). For layers of intermediate depth between these two limiting cases, we concluded that convection was driven by a combination of both mechanisms. For the situation depicted in Fig. 5.4, this transition was rather sharply defined and occurred at

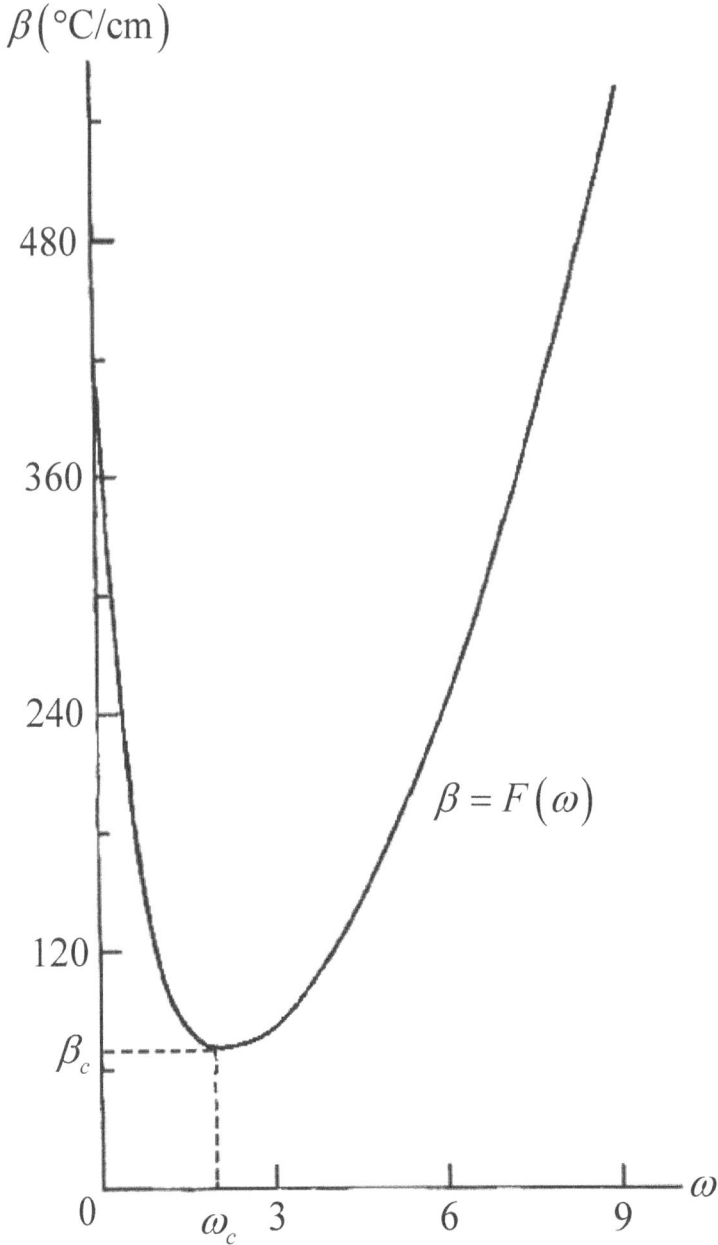

FIGURE 5.2

Plot of the marginal stability curve $\beta = F(\omega)$ for the Dow Corning 200 0.5 stoke silicone oil-air layers, with $d = 0.10$ cm and $\ell = 0.42$. The minimum point occurs at $\omega_c = 2.05$ and $\beta_c = 70.7\,°C/cm$ in this instance.

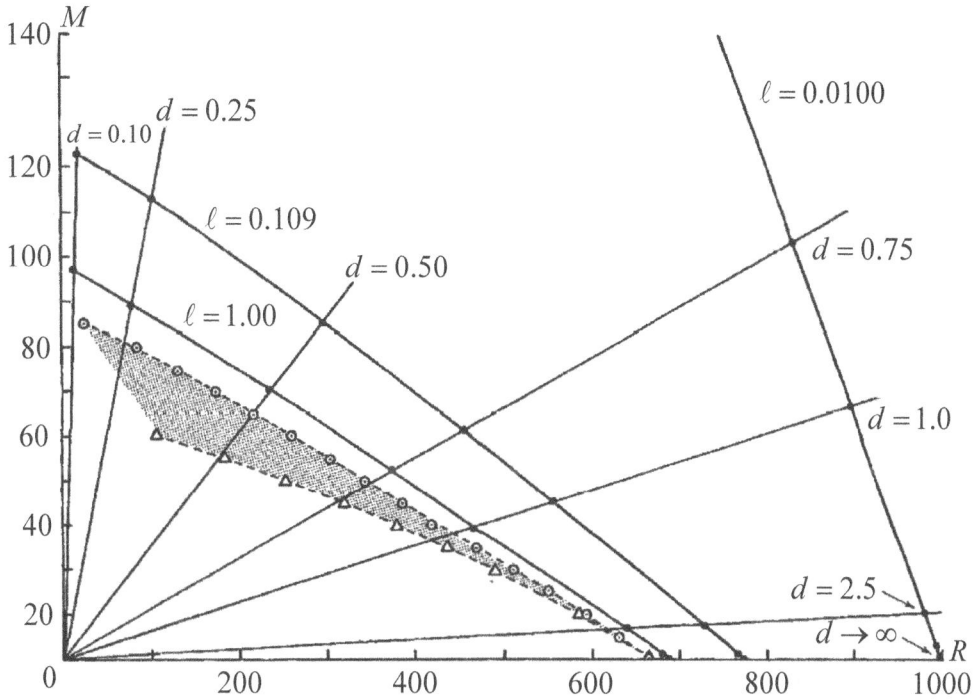

FIGURE 5.3

Plots of M_c versus R_c for Dow Corning 200 1 stoke silicone oil-air layers with ℓ values of 0.0100, 0.109, and 1.00, respectively. Since $M = \Gamma_1 R$, where $\Gamma_1 = \gamma_1/(\alpha \rho_0 g d^2)$, loci of constant d are straight lines emanating from the origin. Here the dashed curves are the graphs obtained from energy (Δ) and linear ($\odot$) theory for the passive gas layer analysis of Davis and Homsey ([47]) representing sufficient conditions for stability and instability, respectively.

a depth of $d \cong 7$ mm. That result was in good agreement with Pearson's ([170]) prediction that such a value should occur when the depths

$$d_1 = \left(\frac{M_c^{(0)} \mu \kappa}{\gamma_1 \beta_c} \right)^{1/2} \text{ and } d_2 = \left(\frac{R_c^{(0)} \mu \kappa}{\rho_0 g \alpha \beta_c} \right)^{1/4},$$

are equal, where $M_c^{(0)}$ and $R_c^{(0)}$ are the critical values of the Marangoni number in the absence of buoyancy effects and the Rayleigh number in the absence of surface tension effects, respectively. The formula for $d_e = d_1 = d_2$ can most easily be deduced by computing the relevant ratio

$$\frac{R_c^{(0)}}{M_c^{(0)}} = \frac{\rho_0 g \alpha}{\gamma_1} d_e^2$$

and solving for d_e to obtain, for the fluid situation depicted in Fig. 5.4,

$$d_e = \left(\frac{\gamma_1}{\rho_0 g \alpha} \frac{773}{116} \right)^{1/2} = 6.5 \text{ mm},$$

where $M_c^{(0)} = 116$ and $R_c^{(0)} = 773$ have been extrapolated from the $\ell = 0.109$ curve of Fig. 5.3 and the following sea level silicone oil-air layers parameter values at 293 K have been employed

$$\gamma_1 = 0.058 \frac{\text{gm}}{\text{sec}^2 \text{K}} \quad \rho_0 = 0.968 \frac{\text{gm}}{\text{cm}^3}, \quad g = 980 \frac{\text{cm}}{\text{sec}^2}, \quad \alpha = 0.96 \times 10^{-3} K.$$

Having developed the critical conditions for the onset of convective instabilities for our two-layer model, we next wished to compare these theoretical predictions with the carefully controlled experimental results mentioned earlier. We began with the series of experiments conducted by Koschmieder ([102]), who used an apparatus of fixed total depth (7.5 mm) involving silicone oil-air layers. His primary objective was the accurate determination of the wavelength of a roll in the concentric circular patterns he observed at the stationary onset of convection in the oil layer. These motions of the oil were made visible by means of aluminum powder with vertical motions appearing as dark rings while predominantly horizontal motions appeared bright. From the knowledge of the number of such rings, N, at a given depth of oil, d, it followed that the nondimensional wavelength of one roll λ^* was given by

$$\lambda^* = \frac{D_0}{2Nd} = \frac{100}{N} \frac{\text{mm}}{d},$$

where $D_0 \equiv$ the diameter of his circular plate $= 200$ mm. For example, 11 rings were found at an oil depth of 6.76 mm corresponding to $\lambda^* = 1.35$. In order to make a comparison between this wavelength and our theoretical predictions, we had to compare it to $\lambda_c = \pi/\omega_c$, which is one-half the usual critical wavelength commonly employed, given that the vertical motions in adjacent rings were in opposite directions and hence the wavelength of one roll was comparable to such a λ_c. The values of these quantities relevant to Koschmieder's ([102]) experiments are summarized in Table 5.1. Observe from this table, which are for Dow Corning 200 silicon oil-air layers of oil kinematic viscosities ν of 1 or 10 stoke (cm^2/sec), that our theoretical predictions of λ_c were in close quantitative agreement with those values of λ^* determined experimentally.

As an aside, while making these comparisons, I phoned Professor Koschmieder at the University of Texas at Austin to inquire about some aspects of his experimental procedure. He was pleasantly surprised that a theoretician wanted to compare model predictions with

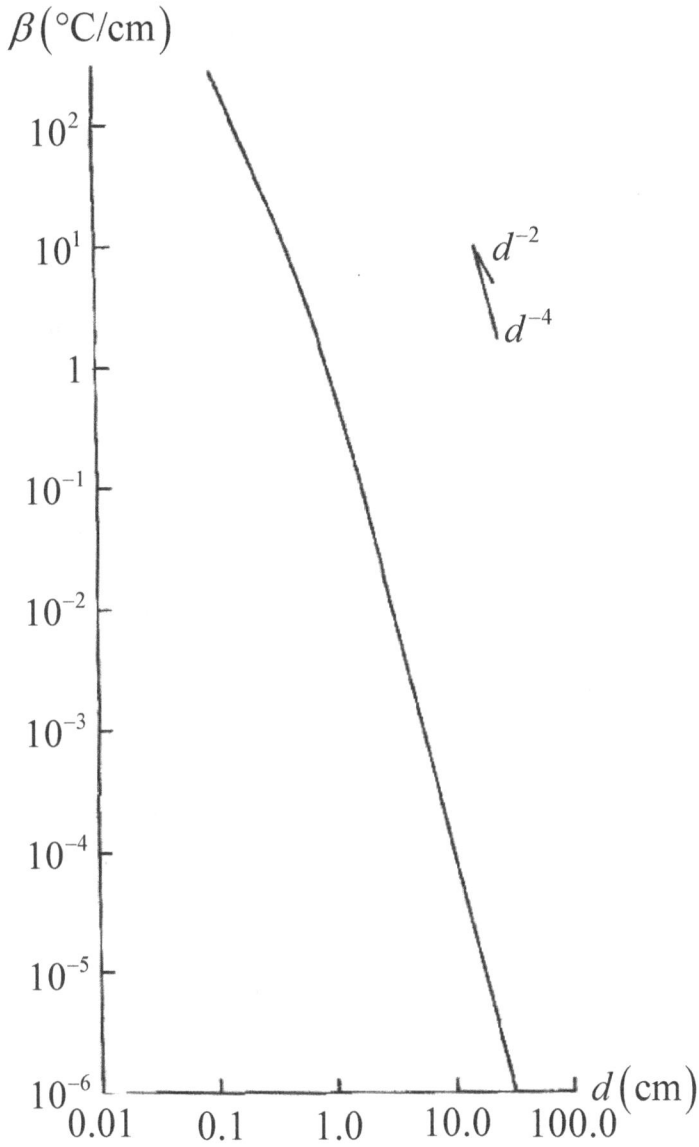

FIGURE 5.4
A log-log plot of β_c versus d corresponding to the marginal stability curve of Fig. 5.3 with $\ell = 0.109$. The insets having slopes of -2 and -4 represent the d-dependence of the Marangoni and Rayleigh effects, respectively. Figs. 5.3 and 5.4, which, according to Steve Davis, appear nowhere else in the literature, represent two Rabbits being Pulled Out of a single Hat.

TABLE 5.1
Comparison with the experimental results of Koschmieder ([102]) for Dow Corning 200
silicone oil-air layers of 7.50 mm total depth. Here $\lambda^* = 100$ mm/(Nd) and $\lambda_c = \pi/\omega_c$.

Liquid	d (mm)	N	λ^*	ℓ	ω_c	λ_c
	6.76	11	1.35	0.109	2.26	1.39
$\nu = 1$ stoke (cm^2/sec)	5.15	13	1.49	0.456	2.07	1.52
	4.28	15	1.56	0.752	1.96	1.60
$\nu = 10$ stoke (cm^2/sec)	6.44	11	1.41	0.165	2.22	1.42

experimental data. Usually, Koschmieder said, experimentalists after obtaining their results
tried to find the best existing theoretical modeling predictions to which to compare them,
but seldom in his experience did a theoretician actually act in a similar manner regarding
existing experimental results. Then he asked me who my Ph.D. adviser had been and, when
I answered, "Lee Segel," responded with, "Well, that certainly figures," complementing us
both in the process.

We continued the comparison of our theoretical predictions of the critical conditions for the
onset of convective instabilities with experimental data by considering the results of Palmer
and Berg ([165]). Their primary experimental objective was to obtain an accurate measure-
ment of the critical temperature difference $(\Delta T)_c$ between their upper and lower plates, at
which the onset of convection occurred and then to convert that to the critical drop across
the liquid layer $(\Delta T_L)_c$ by means of $(\Delta T_L)_c = (\Delta T)_c/(1 + \ell m)$ with $m = 0.113$ or 0.127,
which is an adaptation to this situation of our earlier relationship between these two differences. Here, the values of m corresponded to those of the Dow Corning 200 silicone oils
employed in their experiments. They used a series of such oils of $\nu = 0.5$ or 0.1 stoke and an
adjustable experimental apparatus that allowed them to vary the depths of both the liquid
and air layers independently, as depicted for the representative values displayed in Table 5.2.

TABLE 5.2
Comparison with the experimental results of Palmer and Berg ([165]) for Dow Corning 200
silicone oil-air layers. Here $(\Delta T_L)_c = (\Delta T)_c/(1 + \ell m)$ with $m = 0.113$ or 0.127.

Liquid	d (cm)	d^+ (cm)	$(\Delta T_L)_c$ (°C)	ℓ	β_c (°C/cm)	$\beta_c d$ (°C)	R_c	M_c
	0.381	0.160	1.54	0.420	3.95	1.51	163	78.4
	0.462	0.169	1.10	0.366	2.48	1.15	221	72.4
$\nu = 0.5$ stoke								
	0.546	0.276	0.96	0.506	1.55	0.85	269	63.1
	0.655	0.176	0.70	0.269	0.99	0.65	357	58.7
	0.152	0.147	0.67	0.967	4.42	0.67	30.9	91.8
$\nu = 0.1$ stoke								
	0.239	0.231	0.44	0.967	1.69	0.40	72.1	93.8

The top plate was maintained at a uniform temperature by means of circulating cooling water over it, while the bottom plate was heated electronically by means of a wire heater which
was separated from it by another aluminum plate and a plexiglass disk. The temperature

difference ΔT was changed by altering the heat flux to the bottom plate, which was related to the temperature drop ΔT_p across the plexiglass disk. By employing a Schmidt-Milverton like technique of plotting ΔT versus ΔT_p for both the pre- and post-convective data they were able to identify $(\Delta T)_c$, with the intersection point of those plots where its slopes changed. As can be seen from an examination of Table 5.2, there is a close quantitative correlation between the predictions of our model for $\beta_c d$ and the experimental results of Palmer and Berg ([165]) for $(\Delta T_L)_c$. Observe from that table for the 0.5 stoke oil layers, our M_c decreased with R_c, while for the 0.1 stoke layers it increased. This was a heretofore anomalous outcome of Palmer and Berg's ([165]) experiments also confirmed by our model.

Once these comparisons were completed, Eric's thesis, as I had originally envisioned it, was finished. I still had a nagging doubt however about this problem not having already been done. Hearing that Davis and Homsy ([47]) had just published a paper in the *Journal of Fluid Mechanics* entitled "Energy stability theory for free surface problems: buoyancy-thermocapillary layers," I sent Eric to the WSU Science Library to obtain a copy of it. Both of us were relieved to find that this paper, the results of which will be discussed later, considered a passive, as opposed to a dynamical, gas layer and hence was only a single layer treatment of the Rayleigh-Bénard-Marangoni convection problem. One of its references Zeren and Reynolds ([304]), entitled "Thermal instabilities in two-fluid horizontal layers," also published in the *Journal of Fluid Mechanics* but not mentioned by Normand *et al.* ([157]) gave me some concern. So I sent Eric back to the library to obtain a copy of it as well and this time we were not as lucky. When Eric returned, I said to him, "They've already done our problem, haven't they?" to which he replied, "It looks like it, chief." Indeed, it was exactly the same model and linear stability analysis. Zeren and Reynolds ([304]) numerically determined the critical conditions for the onset of convection in water-benzene layers with a total depth of 2 mm and various water depth fractions, which they then compared qualitatively to the results of such experiments conducted by them. Since they did not make any comparisons with previous silicone oil-air layers experiments, Eric's results would still be good for a review paper but not a thesis, given that such dissertations have to contain some new theoretical contribution and we had been scooped in that regard by them. Eric had, what I consider to be, the perfect temperament to do research, in that nothing seemed to faze him.

Zeren and Reynolds' ([304]) normal mode analysis differed from ours in that their z-dependence was taken in complex form while our general solution was real, *e.g.*, they took

$$W(z) = k_1 e^{i\lambda_1 z} + k_2 e^{-i\lambda_1 z} + k_3 e^{(\lambda_2+i\lambda_3)z} + k_4 e^{(\lambda_2-i\lambda_3)z}$$
$$+ k_5 e^{(-\lambda_2+i\lambda_3)z} + k_6 e^{-(\lambda_2+i\lambda_3)z}$$

for $A > 1$. This difference necessitated the use of a much more complicated numerical procedure to solve the resulting determinantial condition than we employed to solve ours, since their coefficient matrix contained complex entries. Further, this, and the fact that they only made qualitative rather than quantitative comparisons to their experimental results, led Eric to believe that perhaps there might have been errors in their numerical calculations and gave him increased devotion to redo those calculations using our approach for its model parameters assigned values, consistent with these water-benzene layers experiments. The results of our computations as well as Zeren and Reynolds' ([304]) corresponding critical conditions appear in Table 5.3.

Here and in what follows an asterisk (*) on a parameter denotes that this quantity is associated with the analysis of Zeren and Reynolds ([304]). In particular, d^* represents

TABLE 5.3
Comparison with the theoretical results of Zeren and Reynolds ([305]) for water-benzene layers of 2 mm total depth. Note $R_c^*/R_c = M_c^*/M_c = \beta_c^*/\beta_c$ and $\omega_c^* = \omega_c(1 + \ell)$.

d^*	ω_c^*	R_c^*	M_c^*	ℓ	d (mm)	ω_c	$\omega_c(1+\ell)$	R_c	M_c
0.6	2.5	36.0	703	2/3	1.20	1.49	2.48	32.1	627
0.7	2.8	24.8	357	3/7	1.40	1.98	2.82	21.4	308
0.8	3.0	35.8	394	1/4	160	2.36	2.95	29.9	329
0.9	3.2	89.7	780	1/9	1.80	2.86	3.18	72.6	631

their water depth fraction given in our notation by

$$d^* = \frac{d}{d + d^+} = \frac{1}{1 + \ell}.$$

Further, they used a wavenumber ω^* nondimensionalized by total layer depth $d + d^+$ as opposed to our ω which was nondimensionalized by d, reflecting the difference between the scale factors for length adopted in these two analyses. Thus, if $\tilde{\omega}$ is the associated dimensional wavenumber

$$\omega^* = \tilde{\omega}(d + d^+) \text{ and } \omega = \tilde{\omega}d,$$

which implies that

$$\omega^* = \omega \frac{d + d^+}{d} = \omega(1 + \ell).$$

Observe from Table 5.3, that our numerical procedure consistently yielded R_c and M_c values lower than the corresponding R_c^* and M_c^* ones obtained by Zeren and Reynolds ([305]), each set of which had been determined from the material properties of their experimental sea level water-benzene layers at 289 K, although the equivalence of both critical wavenumbers could be shown once it was realized $\omega_c^* = \omega_c(1 + \ell)$, as indicated above.

In light of the quantitatively accurate predictions of our model relevant to the experiments of Koschmieder ([102]) and Palmer and Berg ([165]), as demonstrated by the appropriate entries of Tables 5.1 and 5.2, respectively, we suspected that these discrepancies had been caused by numerical inaccuracies for their β_c^*. This supposition gained additional credibility given the qualitative assertion in Zeren and Reynolds ([305]) that their model overpredicted the values of R_c^* when compared with its experimental determination by the Schmidt-Milverton technique. They attributed this to the similarity in their liquid densities of $\rho_0^+ = 0.9$ (gm/cm^3) and $\rho_0 = 1$ (gm/cm^3). Thus, faced with the prospect of impending doom, which is exactly the way Ph.D. students and their advisers view being scooped on a dissertation we, had Pulled a Rabbit Out of a Hat, by discrediting in part the theoretical analysis of this existing equivalent model system due to the presence of these discrepancies.

So far, we had been concentrating on linear stability theory. Davis and Homsy ([47]), in their paper, mentioned earlier had formulated an energy stability theory for free-surface flows that used variational procedures to take a model's nonlinearities into account and then compared its results to those obtained by applying linear stability theory to the same problem. The specific model employed by Davis and Homsy ([47]) to illustrate this approach was a limiting case of our own, in which the gas was considered to be passive in nature and its governing system could be obtained from ours by taking $u^+ = w^+ \equiv 0$, $\theta^+ = \theta_0^+(z)$, $p^+ = p_0^+(z)$, and $s = r = 0$. Applying the methods of energy and linear stability theory to their system, they numerically calculated R_c as a function of M_c for each of those theories, in the form $R_c = R_E(M_c)$ with $0 \leq M_c \leq 50$ and $R_c = R_L(M_c)$ with $0 \leq M_c \leq 75$, respectively, which appear as dotted curves in Fig. 5.3. The area between them, indicated by shading in that figure, represents the region where subcritical instabilities due to finite amplitude disturbances are allowable.

Now that the research phase of our problem had been completed, its dissemination phase could begin. Due to the fact that I was scheduled to attend an AMS-SIAM Summer Seminar held at the University of Chicago during early July of 1981, entitled "Fluid Dynamical Problems in Astrophysics and Geophysics," and give talks on both the multi-layer fluid problems featured in this chapter, the presentation of our results at a professional meeting preceded the submission of a paper based upon them to a scholarly journal, which in the event proved extremely fortuitous. One of the attendees of that seminar was Manuel Verlarde, an expert on Bénard convection and a co-author of the review Normand *et al.* ([157]). After hearing my presentation on the Rayleigh-Bénard-Marangoni problem and being told that it had not yet been submitted for publication, he suggested that I might consider the *Journal of Non-Equilibrium Thermodynamics* for that purpose, should any difficulty be encountered with the other usual places where such a paper would normally appear. Upon my return from that seminar, Eric and I began writing the paper reporting our results. In doing so, we reversed the order of presentation of this chapter and started both the introductory and previous work comparisons sections of our paper with a discussion of the model and analysis of Zeren and Reynolds ([305]). Given that the latter appeared in the *Journal of Fluid Mechanics*, we decided this would be the best place for us to submit our paper. It was not. They rejected the paper with their referee claiming that our correction of Zeren and Reynold's ([305]) numerical inaccuracies was tantamount to pointing out a typographical error and changing a minus sign in one of their equations! I then submitted a highly condensed version of that manuscript as a letter to the *Physics of Fluids*, where all the papers from my post-doctoral work had appeared. Such letters including text, figures, and tables were limited to four journal pages and their referee requested clarifications that would have required the expansion of the manuscript back to its original length. I suspect that William Reynolds' stature in the field accounted for some of the trouble we were having in getting our paper published. Reynolds, at the time the Chair of the Mechanical Engineering Department at Stanford, was a very well respected member of the American Academy of Engineering, and it seemed clear to me that Verlarde, in offering his advice about a suitable place for the publication of our paper, anticipated that we would have trouble trying to get it published in many of the standard fluid mechanics journals. The other problem was that numerical results in papers were determined by "black box" procedures and nobody knew the suitable way to correct them if mistakes had been made. These were not equivalent to overlooked typographical errors and could totally discredit such papers in the eyes of experts, which was why reviews of the literature sometimes failed to mention them, as I assume to be the case for Normand *et al.* ([157]) with Zeren and Reynolds ([305]). In the spirit of three times being the charm, I now submitted our paper to the *Journal of Non-Equilibrium Thermodynamics*, which accepted it without revision in less than a week. I later learned Verlarde was our referee and had already read a preprint of it given to him by me. In subsequent reviews of the literature, both he and Steve Davis of Davis and Homsy ([47]) fame, who was Segel's first Ph.D. student, mentioned our paper Ferm and Wollkind ([59]), as having corrected the numerical errors of Zeren and Reynolds ([305]). Finally, while we were discussing future job plans after his thesis defense, Eric said to me, "You know your Albany colleague John Bdzil, now at LANL (Los Alamos National Laboratory)? I would like to do postdoctoral work with him." That is exactly what he did and then became a regular member of LANL. Eric is one of those people capable of planning their lives in advance from the outset.

The next year after I first met Eric Ferm, another student paid a visit to my office. His name was James Iwan Dennis Alexander, who called himself Iwan (pronounced "U win") and came from Wales. Iwan was a Ph.D. candidate in the Geology Department working on his thesis topic of rock folding under the direction of John Watkinson, who had sent him to us for some help in understanding a recent paper on that subject by Raymond C. Fletcher ([61]). Since this paper dealt with the linear stability analysis of an appropriate exact solution to a fluid dynamical governing system of partial differential equations and boundary conditions, while H. Claire Wiser was teaching our differential equations course that semester, the people in the departmental office sent him to see Claire. Wiser, a topologist by training, told Iwan that, although being able to help him with the analysis of the model proposed in the paper, he could not offer any insights into this model's formulation and suggested that I was probably the right person to provide such expertise. So Iwan came to see me.

It had long been recognized that rocks undergoing deformation over time periods, on the order of 10^8 years and at low rates of strain, could be successfully modeled for certain purposes as viscous fluids (see Ramberg, [181, 182, 182]). Such an approach had been adopted by Fletcher ([61]) to describe the structural geological phenomenon of rock folding, a process by which stratified layers of sedimentary rock subjected to lateral compression buckle to form wave-like geometries. In particular, this investigation concentrated on an isolated rock stratum that was more competent or resistant to deformation than its embedding medium and represented this situation in terms of a two-dimensional layered Newtonian fluid model, in which stress and rate of strain are linearly related, as has been implicitly assumed for all the fluids treated in Chapter 3 and this chapter. Then the occurrence of such plane folding was examined by means of a linear stability analysis of an appropriate exact solution to the governing system of partial differential equations and boundary conditions. This exact solution corresponded to both the layer and the medium undergoing a uniform parallel-layer shortening at a constant rate of compressive strain, such that the interfaces separating the two were parallel planes moving apart symmetrically.

The main results of Fletcher's ([61]) analysis were the demonstration that this exact planar interface solution was identically unstable to the fold-type perturbations under consideration and the characterization of the wave train of the subsequent plane fold by the so-called dominant wavelength associated with that disturbance having maximum perturbation growth rate. His model involved a single layer of viscous fluid embedded in an immiscible fluid medium of lower viscosity. Both this stratum and its surroundings were assumed to be isothermal homogeneous Newtonian fluids of the same constant temperature and density, while surface tension effects at the layer-medium interfaces had not even been considered. Given the similarity of that model to an isothermal version of Eric's, the first thing I noticed was this absence of surface tension and told Iwan that. I said to him, "You can't, in effect, order a layered Newtonian fluid model for rock folding from the modeling store, but say hold the surface tension at the layer-medium interfaces." I also felt that the density differences between the layer and its embedding medium should be considered as well, although the isothermal assumption seemed reasonable enough and hence could be retained. We decided to collaborate on an analysis of our reformulated single layer rock folding model and the introduction of surface tension effects and layer-medium density differences represented Rabbits being Pulled Out of Hats for that problem. This was the reason that Claire sent Iwan to see me and our subsequent collaboration proved fruitful, as will be demonstrated in what follows.

Although Fletcher's ([61]) work dealing with rock folding had been at least qualitatively applicable to some geological situations, discrepancies still existed between the theoretical predictions resulting from that model and actual observations of natural folds. For example, multilayer folds from the Castile Formation of southern New Mexico (Watkinson and Alexander, [269]) exhibit folded layers lying in relatively close proximity to planar ones (see Fig. 5.5). These layers in the Castile Formation, possessing two folded or two planar interfaces, respectively, have typical wavelengths and/or layer thicknesses ranging from 1 mm to 1 cm and, hence, are classified as small scale structures. Since, from the layer spacing inherent in this formation, it may be deduced that such planar layers interspersed among folded ones must also have been perturbed, the fact that they remained planar could not be accounted for by the linear stability analysis Fletcher ([61]) performed on his model. In particular, that some layer-medium interfaces might remain stable to folding-type perturbations was violated by the identical instability result of this analysis. In order to explain more fully the occurrence of small scale geological fold-type structures such as the Castile Formation, we wished to apply the folding analysis of Fletcher ([61]) to a single layer Newtonian fluid model which included both the density differences between the layer and the upper and lower portions of its embedding medium and the effect of interfacial surface tension at the layer-medium interfaces.

Ramberg ([181, 182]), employing stress-type arguments, concluded from his fluid models that, for single and multilayer strata being laterally compressed in a gravity field, the effect of gravitation was to damp out long wavelength perturbations, while shorter wavelength ones were unstable and resulted in the formation of folds. The failure of these earlier "gravity models" to account for such structures as the Castile may be ascribed to the absence of a stabilizing influence on the shorter wavelength perturbations. When gravity and surface tension effects were introduced at the interfaces between the fluids in the present analysis, it was our hope that gravity would stabilize the longer wavelength disturbances as in Ramberg's work, while the shorter wavelength ones, which in the absence of surface tension would have been unstable, could be stabilized as well and the interaction between the effects of gravity and surface tension produce a critical phenomenon; that is, there would exist a critical threshold value of the imposed rate of compressive strain at which folds would first begin to form. Since such anticipated behavior is somewhat reminiscent of the onset of Kelvin-Helmholtz instability for water waves being generated by the action of wind over water, under the influence of both gravity and capillarity (Yih, [303]), we have referred to it by that term in the title of this chapter. Hence, the main result of such an analysis would be the prediction that for certain geological formations there will be a rate of strain below which no folds could occur. Once this critical value of the rate of strain had been exceeded, however, folding would be initiated at some characteristic wavelength corresponding to the critical wavelength associated with that threshold strain rate on the marginal stability curve. This approach to rock folding as a marginal stability problem would offer an alternative to the dominant wavelength method used by Fletcher ([61]) to characterize folded rock layers. A final consequence of the introduction of gravity and surface tension effects into the single layer model would then be the additional prediction of a band of layer thicknesses for which folding could occur, given a rate of strain above the critical. Further, we hoped that this analysis could be used to explain why the ratio of wavelength to thickness of such small scale folds seemed to have a preference for being between 4 and 6 in agreement with observation.

That is, we wished to investigate the initiation of plane or two-dimensional folding in a single rock layer which was more competent than its embedding medium. The layer was of initial uniform thickness $\overline{H}$, while the upper and lower portions of the embedding medium were assumed to be semi-infinite in extent. We adopted the convention that a primed superscript

FIGURE 5.5
Copy of a photograph depicting folds in a multilayer sequence of anhydrite calcium sulphate (light color) and organic limestone calcium carbonate (dark color) from the Castille Formation of southern New Mexico. Note the presence of folded anhydrite layers interspersed among layers of the same composition which have remained planar. Observe also that some of these planar layers occur in close proximity to thicker folded ones.

on a bulk quantity denoted the upper portion of the medium; a double primed superscript, the lower portion; and no superscript at all, the layer itself. After Ramberg ([181, 182]) we assumed a stable gravitational stratification in that the constant densities of the layer and the upper and lower portions of the embedding medium satisfied the relation

$$\rho_0' < \rho_0 < \rho_0''$$

and defined

$$\Delta\rho' = \rho_0 - \rho_0' > 0, \ \Delta\rho'' = \rho_0'' - \rho_0 > 0.$$

We also assumed that, except for this density difference, the upper and lower portions of the embedding medium had identical material properties and, along with the single layer stratum it surrounded, were originally undergoing a uniform parallel layer compression at a constant rate of strain $\bar{e}$, such that the layer-medium interfaces were planar in shape, horizontal in orientation, and symmetrically moving apart. Our nondimensional model was developed as follows: (x, z) denoted a two-dimensional Cartesian coordinate system such that the x-axis coincided with the mean position of the center line of the single layer, while the z-axis was oriented antiparallel to the direction of gravity, and t denoted time. We defined the dependent variables and parameters: $\boldsymbol{v} = \boldsymbol{v}(x, z, t) = (u, w) \equiv$ velocity components and $p = p(x, z, t) \equiv$ reduced pressure; and $h(t) \equiv$ distance separating the mean position of either layer-medium interface from the x-axis, $\mu \equiv$ shear viscosity, $\nu = \mu/\rho_0 \equiv$ the kinematic viscosity, $\gamma_0 \equiv$ interfacial surface tension, and $g \equiv$ the acceleration due to gravity. Observe that our competency requirement implied $\mu > \mu'$.

We considered all independent and dependent variables in nondimensional form and used $\overline{H}, 1/\bar{e}, \bar{e}\overline{H}, \mu\bar{e}$, and δ as scale factors for distance, time, velocity, pressure, and the deviation of an interface from its mean planar position, respectively. In addition, we introduced the following dimensionless parameters:

$$\epsilon = \frac{\delta}{\overline{H}}, \ R_e = \frac{\bar{e}\overline{H}^2}{\nu}, \ r = \frac{\mu'}{\mu}, \ \Gamma_0 = \frac{\gamma_0}{\mu\bar{e}\overline{H}}, \ G' = \frac{g\overline{H}}{\mu\bar{e}}\Delta\rho', \ \text{and} \ G'' = \frac{g\overline{H}}{\mu\bar{e}}\Delta\rho''.$$

Then the interface between the single layer and the upper portion of the embedding medium satisfied the relation

$$z = h(t) + \epsilon\zeta(x, t) \text{ where } \lim_{\ell\to\infty} \frac{1}{\ell} \int_{-\ell}^{\ell} \epsilon\zeta(x, t)\, dx = O(\epsilon^2) \text{ as } \epsilon \to 0$$

and the one between that layer and the lower portion of the medium satisfied

$$z = -h(t) + \epsilon\zeta'(x, t) \text{ where } \lim_{\ell\to\infty} \frac{1}{\ell} \int_{-\ell}^{\ell} \epsilon\zeta'(x, t)\, dx = O(\epsilon^2) \text{ as } \epsilon \to 0$$

as depicted in Fig. 5.6; while the governing Navier-Stokes equations of motion became:

For $z > h(t) + \epsilon\zeta(x, t)$ (in the upper portion of the medium):

$$\boldsymbol{\nabla}_2 \cdot \boldsymbol{v}' = \frac{\partial u'}{\partial x} + \frac{\partial w'}{\partial z} = 0, \ 0 = -\frac{\partial p'}{\partial x} + r\nabla_2^2 u', \ 0 = -\frac{\partial p'}{\partial z} + r\nabla_2^2 w'.$$

For $-h(t) + \epsilon\zeta'(x, t) < z < h(t) + \epsilon\zeta(x, t)$ (in the layer):

$$\boldsymbol{\nabla}_2 \cdot \boldsymbol{v} = 0, \ \boldsymbol{0} = -\boldsymbol{\nabla}_2 p + \nabla_2^2 \boldsymbol{v}.$$

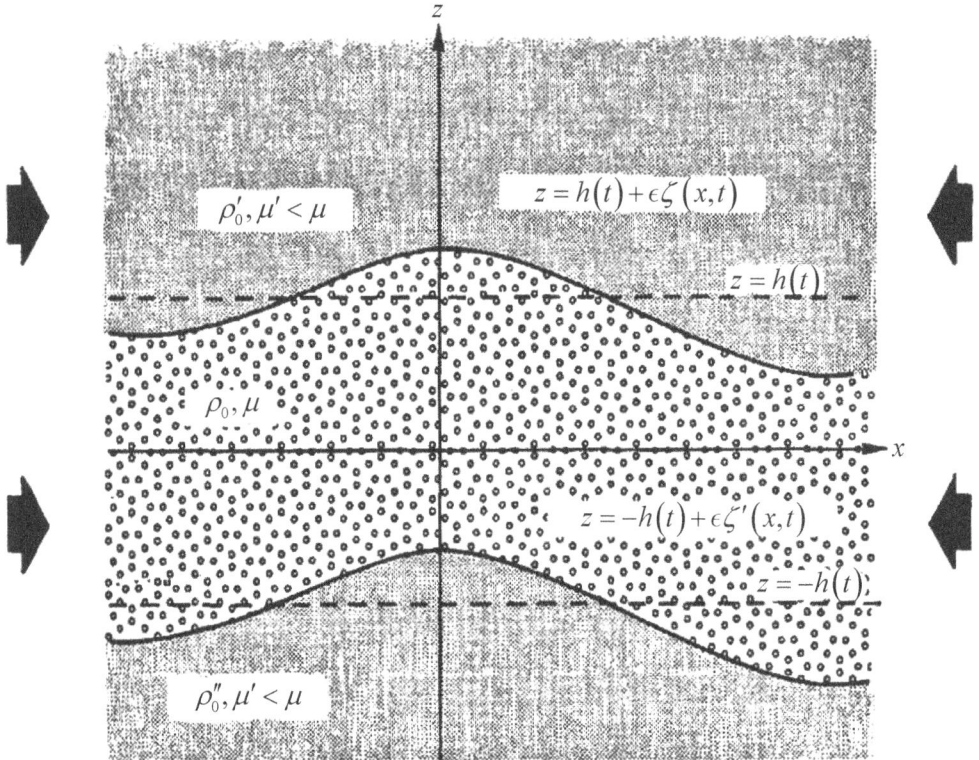

FIGURE 5.6
Schematic diagram employing dimensionless variables illustrating the single layer of initial uniform thickness $\overline{H}$ separated from the upper and lower portions of the embedding medium by the interfaces $z = h(t) + \epsilon\zeta(x,t)$ and $z = -h(t) + \epsilon\zeta'(x,t)$, respectively. When $\rho_0 = (\rho_0' + \rho_0'')/2$ with $\rho_0' < \rho_0''$, while the upper and lower portions of the embedding medium are identical in all other respects, $\zeta' = \zeta + O(\epsilon)$ for folding. The solid arrows denote the uniform-shortening parallel layer compression of constant rate of strain $\overline{e}$.

For $z < -h(t) + \epsilon\zeta'(x,t)$ (in the lower portion of the medium):

$$\boldsymbol{\nabla}_2 \cdot \boldsymbol{v}'' = 0, \ \ \mathbf{0} = -\boldsymbol{\nabla}_2 p'' + r\nabla_2^2 \boldsymbol{v}''.$$

Here $\boldsymbol{\nabla}_2 \equiv (\partial/\partial x, \partial/\partial z)$ and $\nabla_2^2 \equiv \boldsymbol{\nabla}_2 \cdot \boldsymbol{\nabla}_2 = \partial^2/\partial x^2 + \partial^2/\partial z^2$.

These equations represent conservation of mass and momentum in the upper portion of the medium, the layer, and the lower portion of the medium, respectively, for an isothermal pure substance of constant density. In those bulk equations, the Reynolds number R_e, which typically has the very small value of 10^{-38} for this problem, with $\overline{H} = 10$ cm, $\overline{e} = 10^{-14}/\text{sec}$, and $\nu = 10^{26}$ cm^2/sec (Turcotte and Oxburgh, [255]), was set equal to zero thus eliminating a term proportional to the substantial derivative of velocity from the left-hand sides of the momentum equations. Hence, we neglected inertial effects by making the quasistatic approximation due originally to William Prager ([177]). The five boundary conditions we imposed at each of the layer-medium interfaces, as in Eric's problem, were a direct consequence of the treatment of these interfaces as material surfaces of discontinuity separating two immiscible fluids. Then again there were two kinematic boundary conditions which guaranteed that each interface was a material surface and automatically satisfied conservation of mass at such a surface, while the two dynamical boundary conditions arose, in turn, from the normal and tangential components of momentum balance. Here, in particular, the surface tension effect appeared as a term $\Gamma_0\eta$ in the normal momentum balance condition where $\eta \equiv$ interfacial curvature and Γ_0 was assumed to be isotropic and uniformly constant, while the gravity effect appeared as a term $G'z$ or $G''z$ in that condition, resulting from the density difference between the layer and the upper or lower portion of the embedding medium in conjunction with our use of reduced pressure. Finally, we again adopted the constitutive relation of no relative tangential motion occurring at the layer-medium interfaces, which may be described as an adherence or no-slip type boundary condition (the kinematic boundary conditions taken together already guaranteed no such relative normal motion).

We have deferred until now the mathematical formulation of kinematic boundary conditions, but given that they played a role in the exact solution to our system and the form of the normal mode linear stability analysis of this solution, we next need to do so. Consider a material surface satisfying the condition $\varphi(x,z,t) = 0$, $e.g.$, for our interface separating the layer from the upper portion of the embedding medium $\varphi(x,z,t) = z - h(t) - \epsilon\zeta(x,t)$. Equivalent to our assertion involving the interfacial normal speed w_n of the solid-liquid interface in Chapter 2 for $h(t) = Vt$, it can more generally be shown that $w_n = -\varphi_t/\sqrt{\varphi_x^2 + \varphi_z^2}$, while the normal component of the fluid velocity $v_n = \boldsymbol{n} \cdot \boldsymbol{v} = \boldsymbol{\nabla}_2 \cdot \boldsymbol{v}/\sqrt{\varphi_x^2 + \varphi_y^2}$ (see Wollkind and Dichone, [284]). Defining the relative normal speed of the interface by $s_n = w_n - v_n$, a material surface must satisfy $s_n = 0$, which yields that the substantial derivative $D\varphi/Dt = \varphi_t + \boldsymbol{\nabla}_2\varphi \cdot \boldsymbol{v} = 0$ at $\varphi = 0$. For our layer-upper medium interface, we then obtained the kinematic boundary conditions $dh/dt + \epsilon(\zeta_t + u\zeta_x) = w$ and $dh/dt + \epsilon(\zeta_t + u'\zeta_x) = w'$, which imply $-\epsilon u\zeta_x + w = -\epsilon u'\zeta_x + w' \Leftrightarrow \boldsymbol{v} \cdot \boldsymbol{n} = \boldsymbol{v}' \cdot \boldsymbol{n}$ or no relative normal motion as stated above; while, for the layer-lower medium interface, companion conditions with dh/dt replaced by $-dh/dt$ were obtained.

There existed an exact solution of our governing system which satisfied the boundary conditions for planar interfaces located at $z = \pm h(t)$ and represented the parallel-layer compression described earlier. This uniform-shortening planar interface solution was given by

$$\zeta(x,t) = \zeta'(x,t) \equiv 0;$$
$$u' = u_0'(x) = -x, \ w' = w_0'(z) = z, \ p' = p_0'(t) \text{ for } z > h(t);$$
$$u = u_0(x) = -x, \ w = w_0(z) = z, \ p = p_0(t) \text{ for } -h(t) < z < h(t);$$
$$u'' = u_0''(x) = -x, \ w'' = w_0''(z) = z, \ p'' = p_0''(t) \text{ for } z < -h(t);$$

where

$$p_0(t) = p_0'(t) + G'h(t) + 2(1-r) = p_0''(t) + G''h(t) + 2(1-r)$$

and $h(t)$ satisfied $dh/dt = h$, which followed from the kinematic boundary conditions at either of the planar layer-medium interfaces. This differential equation, in conjunction with the initial location of each interface and the fact that $\overline{H}$ was the scale factor for distance, determined $h(t)$; hence,

$$h(t) = \frac{1}{2}e^t.$$

Thus, this uniform-shortening planar interface solution represented a case of uniform layer thickening as well.

In an actual occurrence of a single layer fold, the extent of the upper or lower embedding medium is naturally finite. A simplifying assumption in this model was that, for a fixed t-value, $z - h(t)$ extended to positive infinity in the upper medium and $z + h(t)$ extended to negative infinity in the lower one. The folding instability to be considered depends crucially upon conditions at the layer-medium interfaces, but should be virtually unaffected by conditions far from these interfaces. Thus, we can expect that far from each interface the influence of the shape of that interface on the relevant velocity and pressure fields will become negligible. This means that analogous to the far-field conditions introduced in Chapter 2

$$u' \to u_0', \ w' \to w_0', \ p' \to p_0' \text{ as } z \to \infty;$$
$$u'' \to u_0'', \ w'' \to w_0'', \ p'' \to p_0'' \text{ as } z \to -\infty;$$

where u_0', w_0', p_0' and u_0'', w_0'', p_0'' comprise the components of the uniform-shortening planar interface solution. These far-field conditions were the proper ones to impose as $|z| \to \infty$ and, along with the bulk equations and the boundary conditions, constituted the explicit mathematical formulation of the problem, once the mean interfacial positions were defined consistent with our determination of $h(t)$. In addition, x extended to positive and negative infinity, and we also adopted the implicit requirement that the dependent variables remain bounded as $|x| \to \infty$.

It was the stability of the uniform-shortening planar interface solution of our governing system to initially infinitesimal fold-type disturbances with which we were concerned. In order to investigate this stability, we considered solutions to our basic system of partial differential equations and boundary conditions of the same form as that described earlier for Eric's Rayleigh-Bénard-Marangoni problem: Namely, a superposition of this planar interface solution and a linear perturbation proportional to the parameter ϵ denoted by ζ_1, u_1, w_1, and p_1 for the unsuperscripted dependent variables, with analogous expansions for the primed and double-primed ones. Then substituting that solution into our basic system, expanding the interfacial boundary conditions in Taylor series about $z = \pm h(t)$, neglecting terms of $O(\epsilon^2)$, and cancelling the resulting common ϵ factor, we obtained a linear homogeneous system satisfied by those perturbation variables. To motivate the form Fletcher ([61])

employed for his modified normal mode analysis of such a perturbation system, consider the solution to the following prototype two-dimensional free-surface boundary-value problem:

$$\frac{\partial u}{\partial x} + \frac{\partial w}{\partial z} = 0, \ \nabla_2^2 \boldsymbol{v} = \boldsymbol{\nabla}_2 p \text{ for } z > \zeta;$$

which represents the Navier-Stokes equations with $R_e = 0$ for a material surface $z = \zeta(x,t)$, in the absence of gravitational and capillarity effects, satisfying the kinematic boundary condition

$$\frac{\partial \zeta}{\partial t} + u\frac{\partial \zeta}{\partial x} = w \text{ for } z = \zeta.$$

Taking its velocity and pressure fields to be consistent with the corresponding components of our uniform-shortening planar interface solution, we set

$$u = -x, \ w = z, \ p \equiv p_A;$$

which reduces that problem to the following initial-value one for $\zeta = \zeta(x,t)$:

$$\frac{\partial \zeta}{\partial t} - x\frac{\partial \zeta}{\partial x} = \zeta, \ t > 0;$$

since $w = z = \zeta$ for this surface which is assumed to satisfy the prescribed initial condition

$$\zeta(x,0) = A_0 \cos(\omega_0 x).$$

This is a linear first-order partial differential equation for $\zeta = \zeta(x,t)$ and can be solved by the method of characteristics which will be developed for the slightly more general so-called quasi-linear first-order partial differential equation of the form:

$$a(x,t,\zeta)\frac{\partial \zeta}{\partial x} + b(x,t,\zeta)\frac{\partial \zeta}{\partial t} = c(x,t,\zeta),$$

where ζ is prescribed to be $f(\tau)$ along the curve in the x-t plane $\boldsymbol{r}(\tau) = h(\tau)\boldsymbol{e}_x + g(\tau)\boldsymbol{e}_t$ with $\boldsymbol{e}_{x,t} \equiv$ unit vectors in the coordinate directions. Let $x = x(s,\tau)$ and $t = t(s,\tau)$ be such that

$$\frac{\partial x}{\partial s} = a, \ x(0,\tau) = h(\tau); \ \frac{\partial t}{\partial s} = b, \ t(0,\tau) = g(\tau).$$

Then defining

$$Z(s,\tau) = \zeta[x(s,\tau),t(s,\tau)],$$

we can deduce by the chain rule that

$$\frac{\partial Z}{\partial s} = \frac{\partial \zeta}{\partial x}\frac{\partial x}{\partial s} + \frac{\partial \zeta}{\partial t}\frac{\partial t}{\partial s} = a\frac{\partial \zeta}{\partial x} + b\frac{\partial \zeta}{\partial t} = c$$

and by substitution that

$$Z(0,\tau) = \zeta[x(0,\tau),t(0,\tau)] = \zeta[h(\tau),g(\tau)] = f(\tau).$$

Solving this system for $x = x(s,\tau)$, $t = t(s,\tau)$, and $Z = Z(s,\tau)$, we can invert the transformation $x = x(s,\tau)$ and $t = t(s,\tau)$, provided it is not a characteristic for the problem, obtaining $s = s(x,t)$ and $\tau = \tau(x,t)$, and finally deduce the solution to the original problem given by

$$\zeta(x,t) = Z[s(x,t),\tau(x,t)].$$

We now apply this procedure to our initial value problem for $\zeta = \zeta(x,t)$, by making the identifications that

$$a = -x, \ b = 1, \ c = \zeta = Z; \ h(\tau) = \tau, \ g(\tau) = 0, \ \text{and} \ f(\tau) = A_0 \cos(\omega_0 \tau).$$

Thus,

$$\frac{\partial x}{\partial s} = -x, \ x(0,\tau) = \tau; \ \frac{\partial t}{\partial s} = 1, \ t(0,\tau) = 0;$$

$$\frac{\partial Z}{\partial s} = Z, \ Z(0,\tau) = A_0 \cos(\omega_0 \tau).$$

Solving this system we find that

$$x = \tau e^{-s}, t = s, Z = A_0 \cos(\omega_0 \tau)e^{s}.$$

Inverting the transformation for x and t as functions of s and τ yields

$$s = t, \tau = x e^{t}.$$

Hence, we finally obtain the solution

$$\zeta(x,t) = A_0 e^{t} \cos(\omega_0 e^{t} x)$$

which is of the general form

$$\zeta(x.t) = \mathcal{A}(t) \cos(\omega x)$$

where

$$\mathcal{A}(t) = A_0 e^{t}, \ \omega = \omega(t) = \omega_0 e^{t}.$$

Returning to our perturbation system, we note that unlike most of the other such systems encountered in my research career, this one was not constant coefficient, in that its entries resulting from the kinematic boundary conditions were exactly like the prototype interfacial problem just considered. Systems of this sort, due to the presence of the coefficient $-x$ for the $\partial \zeta_1 / \partial x$ or $\partial \zeta_1' / \partial x$ term in those perturbation boundary conditions, required the modified normal-mode analysis employed by Fletcher ([61]) and suggested by the form of the solution to our prototype problem. Consistent with this approach we looked for a modified normal mode solution of our basic perturbation system of the form

$$\zeta_1(x,t) = I_0 \mathcal{A}(t) \cos(\omega x),$$
$$[w_1, p_1](x,z,t) = [W,P](z;\omega)\mathcal{A}(t)\cos(\omega x),$$
$$u_1(x,z,t) = U(z;\omega)\mathcal{A}(t)\sin(\omega x),$$

where

$$\mathcal{A}(t) = \exp\left[\int a_1(t)\,dt\right] \text{ and } \omega = \omega(t) = \omega_0 e^{t},$$

with analogous expansions for u_1', w_1', p_1', u_1'', w_1'', and p_1'' involving U', W', P', U'', W'', and P'', respectively, while taking

$$\zeta_1'(x,t) = \mathcal{C}(t)\zeta_1(x,t).$$

Solving the system of ordinary differential equations which resulted upon the substitution of our modified normal mode solution into the perturbation partial differential equations and applying the perturbation far field conditions, we found that

$$[W', U', P'](z; \omega) = -\left[C' + D'\left(z + \frac{1}{\omega}\right), C' + D'z, 2rD'\right]e^{-\omega z},$$

$$[W'', U'', P''](z; \omega) = \left[A'' + B''\left(z - \frac{1}{\omega}\right), -(A'' + B''z), 2rB''\right]e^{\omega z};$$

while the components of $[W, U, P](z; \omega)$ satisfied

$$W(z; \omega) = \left[A + B\left(z - \frac{1}{\omega}\right)\right]e^{\omega z} - \left[C + D\left(z + \frac{1}{\omega}\right)\right]e^{-\omega z},$$

$$U(z; \omega) = -[(A + Bz)e^{\omega z} + (C + Dz)e^{-\omega z}],$$

$$P(z; \omega) = 2(Be^{\omega z} - De^{-\omega z}).$$

The interfacial perturbations to the planar layer-medium interfaces can be decomposed into symmetric and antisymmetric components which, for the embedding medium under consideration, corresponded to "pinch-and-swell" and fold structures, respectively (Fletcher, [61]). This symmetry relationship was measured relative to the x-axis, and the individual interfacial components gave rise to perturbation flows characterized by a z-component of velocity, which was even in its z-variable for the case of a fold and odd for "pinch-and-swell" (Fletcher, [61]). We first determined the critical conditions for the development of folds, by imposing these symmetries characteristic of such folding, and later demonstrated that for our purposes it had been permissible to ignore the "pinch-and-swell" component in the layer shape.

Toward that end we imposed the following perturbation flow symmetries which, as mentioned above, are characteristic of the folding process under investigation (Fletcher, [61]):

$$W(z; \omega) = W(-z; \omega) \text{ for } 0 \le z \le h; \ W'(z; \omega) = W''(-z; \omega) \text{ for } z \ge h.$$

These conditions in conjunction with our solutions implied that

$$C = -A, \ D = B; \ A'' = -C', \ B'' = -D'.$$

Then substitution of those solutions into our perturbation boundary conditions at $z = h$ yielded a system of five linear homogeneous equations in the quantities A, B, C', D', and I_0. A similar substitution procedure involving the perturbation boundary conditions at $z = -h$ resulted in a companion system of equations which could be shown to be equivalent to this one provided $G'' = G'$, once it had been deduced, upon comparison with the latter under this condition, that

$$\mathcal{C}(t) \equiv 1 \Leftrightarrow \zeta_1'(x, t) = \zeta_1(x, t),$$

an equivalence often used to characterize single layer folds (Fletcher, [61]). Further, given these circumstances, it was then only necessary for us to examine that linear system of equations developed from the perturbation boundary conditions at $z = h$ in order to determine the stability behavior to fold-type disturbances of the uniform-shortening planar interface solution under investigation.

Since this occurrence depended crucially upon our adoption of $G'' = G' = G$, it was important to examine the implications of that assumption. From the definition of the gravity quantities G'' and G', this implied that the density jumps at the two interfaces were equal, *i.e.*,

$$\Delta\rho' = \rho_0 - \rho_0' = \Delta\rho'' = \rho_0'' - \rho_0 = \frac{\Delta\rho}{2} > 0 \text{ or, equivalently, } \rho_0 = \frac{\rho_0' + \rho_0''}{2},$$

from which the common value of $G = g\overline{H}\Delta\rho/(2\mu\overline{e})$ where $\Delta\rho = \rho_0'' - \rho_0' > 0$ followed.

The question was: Would our adoption of such a restrictive case be accepted as representative? That bothered me a great deal. So, similar to my perusal of Uspensky's ([258]) *Theory of Equations*, related in Chapter 3, I began going through the *Dynamics of Nonhomogeneous Fluids* by Yih ([304]), a book coincidentally ordered at the same time from the SUNYA book store. In this book I found a description of Taylor's ([245]) and Goldstein's ([68]) work on the stability of a heterogeneous fluid in a shear flow, in which they treated the identical density stratification for their three-layer configuration. Indeed, Yih ([304]), in the caption for his Figure 43 on p. 172 reproducing Goldstein's ([68]) most important results, stated that the density in the upper semi-infinite layer was $\rho - \Delta\rho/2$; that in the middle layer, ρ; and that in the lower semi-infinite layer, $\rho + \Delta\rho/2$. We are talking about Geoffrey I. Taylor and Sydney Goldstein here, two of the pioneers of modern fluid dynamics. Although the sciences do not depend on precedence to the extent jurisprudence does, an appeal to such authority carries significant weight and I knew no one was now going to dispute my assumption as being too restrictive a case; not if that were the selfsame case handled by G.I. Taylor and S. Goldstein 50 years before. After all the instances of doing so already described, it is obvious that this represents another prime example of Pulling a Rabbit Out of a Hat!

Returning to the linear system of equations involving the quantities A, B, C', D', and I_0, obtained from the substitution of our normal mode solution into the perturbation boundary conditions at $z = h(t) = e^t/2$, the coefficients of that system depended on r, a_1, k, E, and N where

$$k = 2h\omega = \omega_0 e^{2t} = \frac{2\pi}{\lambda/H}, \ E = \frac{2\mu\overline{e}H}{\gamma_0}, \ N = \frac{gH^2\Delta\rho}{2\gamma_0},$$

with

$$\lambda = \lambda_0 e^{-t} \text{ for } \lambda_0 = \frac{2\pi}{\omega_0}\overline{H}, \ H = \overline{H}e^t.$$

In particular, λ, the disturbance wavelength, and H, the layer thickness, both measured in dimensional variables, were time dependent. Thus k, the nondimensional wavenumber of the disturbance associated with the normalized wavelength λ/H, and E, a dimensionless combination of strain rate and surface tension, as well as the Bond (or Eötvös) number N, another dimensionless combination frequently arising in problems involving both gravity waves and capillary ripples (Yih, [304]), were also functions of time. We observe that, for fixed values of these functions and of the competency parameter r (note $0 < r < 1$, since $\mu' < \mu$), the linear system of equations mentioned above was an eigenvalue problem with eigenvalue a_1 and corresponding eigenvector $[A, B, C', D', I_0]$, where the growth rate a_1 plus the first four components of the eigenvector were functions of time, while I_0 was a constant. In order to ensure the existence of nontrivial solutions, we equated the determinant of the coefficient matrix of this system to zero employing the Pat Munroe algorithm and obtained the following secular equation satisfied by a_1:

$$a_1 = 1 + \frac{2(1-r)k - (1/E)(k + N/k)[r\{k + \sinh(k)\} + \cosh(k) + 1]}{(r^2 - 1)k + (1 + r^2)\sinh(k) + 2r\cosh(k)}.$$

Then observing that a_1 and $\mathcal{A}$ of our normal mode solution were related by the amplitude equation

$$\frac{d\mathcal{A}}{dt}(t) = a_1(t)\mathcal{A}(t),$$

we adopted the stability criterion that the uniform-shortening planar interface solution was stable, unstable, or neutrally stable to the type of disturbance being examined according to whether the growth rate a_1 was negative, positive, or zero identically for all time t. Given that we restricted our analysis to a period in dimensional time which was sufficiently short in comparison with the scale factor $1/\overline{e}$ so that linear theory was still applicable, it seemed reasonable to have adopted a stability criterion which treated a_1 as if it were constant, since over the corresponding period in nondimensional time, λ and H (hence k, E, and N) were such slowly varying functions of t that they could be taken as virtually constant during the interpretation of the critical conditions for the onset of instability to be developed below. This assumption was also fundamental, in an implicit sense, to the analysis of Fletcher ([61]) as well.

For fixed values of r and N, the growth rate a_1 was dependent upon E and k alone. We plotted the marginal stability curve associated with our secular equation, in the k-E plane of Fig. 5.7. This curve, characterized by $a_1 = 0$, separated the region of instability lying above it where $a_1 > 0$ from that of stability located beneath it where $a_1 < 0$ and was given by

$$E = E_c(k;r,N) = \left(k + \frac{N}{k}\right)F(k;r)$$

with

$$F(k;r) = \frac{r[k+\sinh(k)]+\cosh(k)+1}{(1-r)^2 k+(1+r^2)\sinh(k)+2r\cosh(k)}.$$

As can be seen from Fig. 5.7, this marginal curve had an absolute minimum at $k = k_c(N,r)$ and $E = E_c(N,r) = E_c[k_c(N,r);r,N]$, the values of which are tabulated in Table 5.4. For $E < E_c$, there existed no wavenumbers k such that $a_1 > 0$, while for $E > E_c$ there existed a band of such wavenumbers corresponding to growing disturbances. This meant that for $E < E_c$ one had stability, and for $E > E_c$, instability. Hence, E_c represented one measure of the critical threshold value of the imposed rate of compressive strain at which plane folding was first initiated for this single layer model.

Since the nondimensional quantity E depended on H, the dimensional thickness of our single layer, there was some merit in considering another dimensionless quantity that contained the dimensional strain rate $\overline{e}$ but was independent of H. For this purpose we defined the dimensionless rate of compressive strain:

$$\mathcal{E} = \frac{E}{\sqrt{N}} = \frac{\overline{e}}{\sqrt{g\gamma_0\Delta\rho}/(2\sqrt{2}\mu)}.$$

In Fig. 5.8, we plotted a stability diagram of $\mathcal{E}$ versus the nondimensional layer thickness

$$\mathcal{H} = \sqrt{N} = \frac{H}{\sqrt{2\gamma_0/(g\Delta\rho)}}.$$

In that figure,

$$\mathcal{E} = \mathcal{E}_c(\mathcal{H};r) = \frac{E_c(\mathcal{H}^2,r)}{\mathcal{H}}$$

represented the marginal stability curve separating unstable states lying above it from stable ones located beneath it. This curve had a minimum point at $(\mathcal{H}_c, \mathcal{E}_c)$, denoted by an

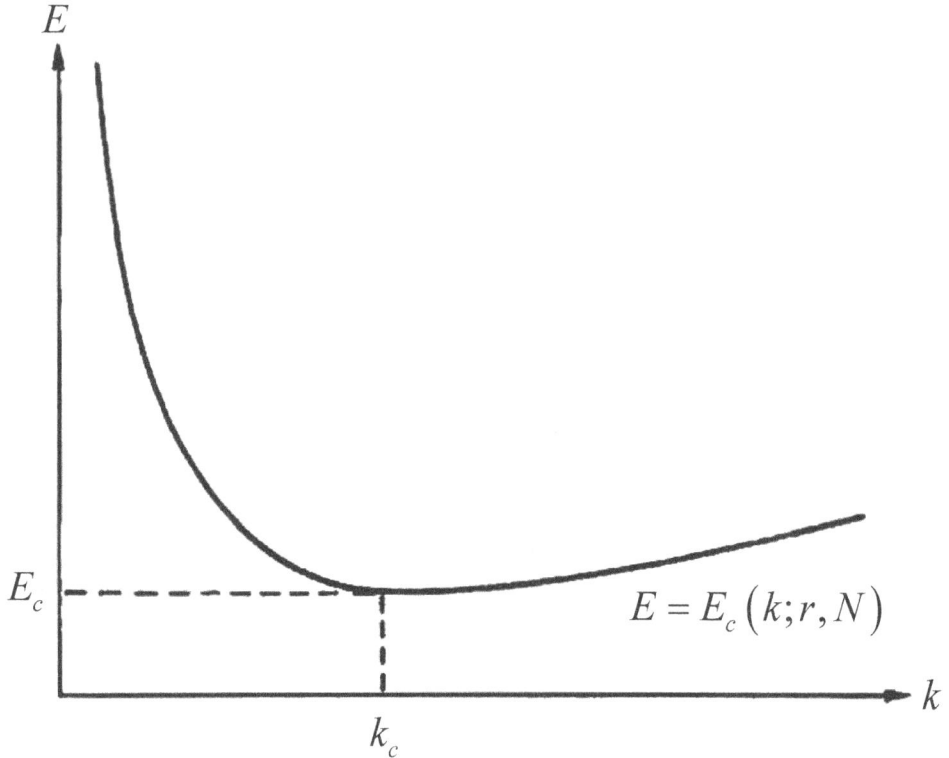

FIGURE 5.7
Plot of $E = 2\mu H\bar{e}/\gamma_0$, a dimensionless combination of strain rate and surface tension, versus k, the nondimensional disturbance wavenumber associated with the normalized wavelength λ/H. From the linear stability analysis, $E = E_c(k; r, N)$ is the marginal stability curve with minimum point (k_c, E_c) on which the disturbance growth rate $a_1 = 0$ separating the unstable region above it where $a_1 > 0$ from the stable one below it where $a_1 < 0$. Here $r = \mu'/\mu$ and $N \equiv$ Bond number $= (\Delta\rho)gH^2/(2\gamma_0)$, while $H = \overline{H}e^t$ and $\lambda = (2\pi/\omega_0)\overline{H}e^{-t}$.

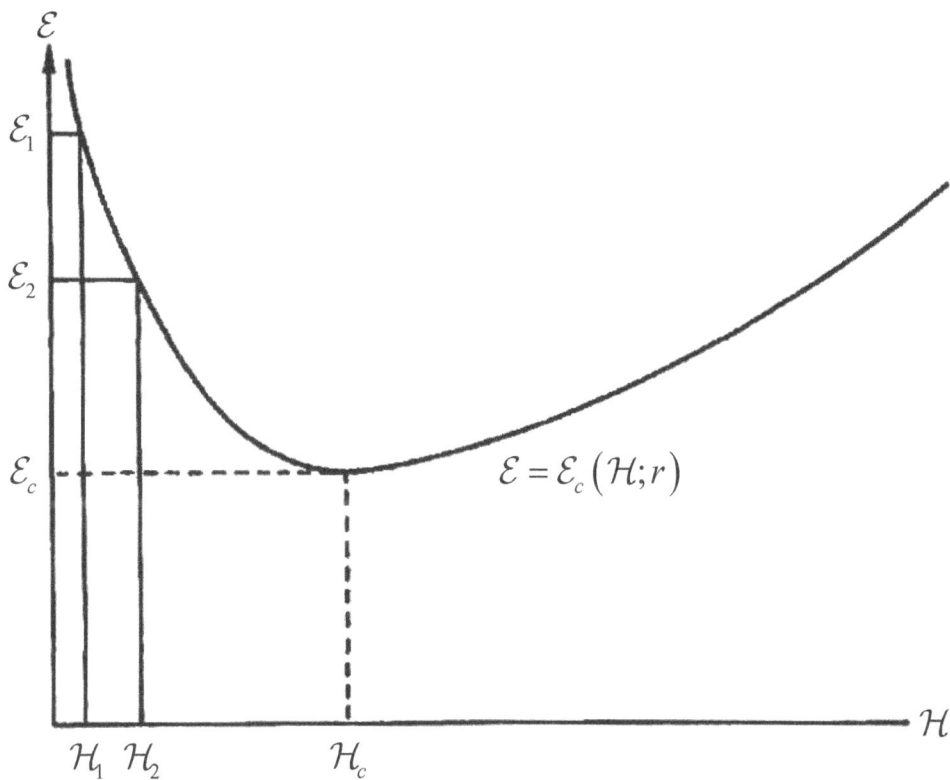

FIGURE 5.8
Stability plot of $\mathcal{E} = E_c/\sqrt{N}$, a dimensionless rate of strain independent of H, versus $\mathcal{H} = \sqrt{N}$, a dimensionless layer thickness. Here, $\mathcal{E} = \mathcal{E}_c(\mathcal{H};r)$ corresponds to that marginal stability curve with minimum point $(\mathcal{H}_c, \mathcal{E}_c)$, while the upper and lower bounds relevant to the occurrence of minor folds are designated by $\mathcal{E}_1$, $\mathcal{H}_2$, and $\mathcal{E}_2$, $\mathcal{H}_1$, respectively. Note from the asterisked quantities of Table 5.4 that $\mathcal{H}_c = k_c$ as depicted in Fig. 5.7.

Multi-Layer Fluid Phenomena

TABLE 5.4

Values relevant to the critical conditions of the plots depicted in Figs. 5.7 and 5.8 where the asterisk (*) denotes the minimum point indicated in the plot of the latter figure.

r	N	k_c	E_c	$\mathcal{H}=\sqrt{N}$	$\mathcal{E}=\frac{E_c}{\sqrt{N}}$	$\frac{\lambda_c}{H}=\frac{2\pi}{k_c}$	$\mathcal{L}=\frac{2\pi\mathcal{H}}{k_c}$
0.01	0.01	0.46	1.07	0.01	10.74	13.81	1.38
	0.10	0.86	1.28	0.32	4.05	7.31	2.31
	0.20	1.03	1.41	0.45	3.15	6.13	2.74
	0.40	1.22	1.60	0.63	2.52	5.15	3.26
	1.00	1.54	2.01	1.00	2.01	4.08	4.08
	5.86	2.42	4.04	2.42*	1.62*	2.60	2π
	10.00	2.82	5.37	3.16	1.70	2.23	7.04
0.10	0.01	0.22	0.90	0.10	9.04	29.22	2.92
	0.10	0.74	1.28	0.32	4.04	8.49	2.68
	0.20	0.94	1.43	0.45	3.21	6.72	3.00
	0.40	1.18	1.65	0.63	2.61	5.35	3.38
	1.00	1.54	2.09	1.00	2.09	4.09	4.09
	8.15	2.85	4.85	2.85*	1.697*	2.20	2π
	10.00	3.05	5.38	3.16	1.701	2.06	6.51
0.20	0.01	0.15	0.68	0.10	6.81	42.74	4.27
	0.10	0.61	1.23	0.32	3.88	10.39	3.28
	0.40	1.01	1.68	0.63	2.66	6.28	3.97
	1.00	1.52	2.15	1.00	2.15	4.13	4.13
	10.00	3.32	5.25	3.16	1.66	1.89	5.99
	16.65	4.10	6.75	4.10*	1.651*	1.53	2π
	19.89	4.40	7.43	4.50	1.652	1.43	6.43

asterisk (*) in Table 5.4, where $\mathcal{H}_c = \mathcal{H}_c(r)$ and $\mathcal{E}_c = \mathcal{E}_c(r) = \mathcal{E}_c[\mathcal{H}_c(r); r]$, as indicated in that table. Hence, for $\mathcal{E} < \mathcal{E}_c$ we had stability for all $\mathcal{H}$, while for $\mathcal{E} > \mathcal{E}_c$ there existed a band of layer thicknesses for which there was instability.

We next used the critical conditions illustrated in Fig. 5.7, Fig. 5.8, and Table 5.4 to explain some anomalous properties of naturally occurring short-wavelength folds, with particular emphasis on the Castile Formation. Fig. 5.7, a plot of E versus k for fixed values of r and N, was a schematic stability diagram for our plane folding model. We already discussed the physical significance of the critical threshold value E_c, but have deferred until now an analogous interpretation of the critical wavenumber k_c with respect to the initiation of folding.

The physical significance of k_c was that it is a measure of the nondimensional normalized wavelength λ/H of that disturbance, which is most likely to grow first, once E_c had been exceeded. Observe from Fig. 5.9, that for $E > E_c$, the band of wavenumbers $k \in (k'', k')$, corresponding to growing disturbances, is centered about k_c. Thus, k_c served as a reasonable approximation to the so-called dominant wavenumber k_d, associated with the most dangerous mode of linear theory, *i.e.*, that disturbance to the uniform-shortening planar interface solution having maximum perturbation growth rate a_1. Hence, the normalized critical wavelength associated with k_c which is given by

$$\frac{\lambda_c}{H} = \frac{2\pi}{k_c}$$

can be used to characterize the wave train of a single layer fold at the onset of plane

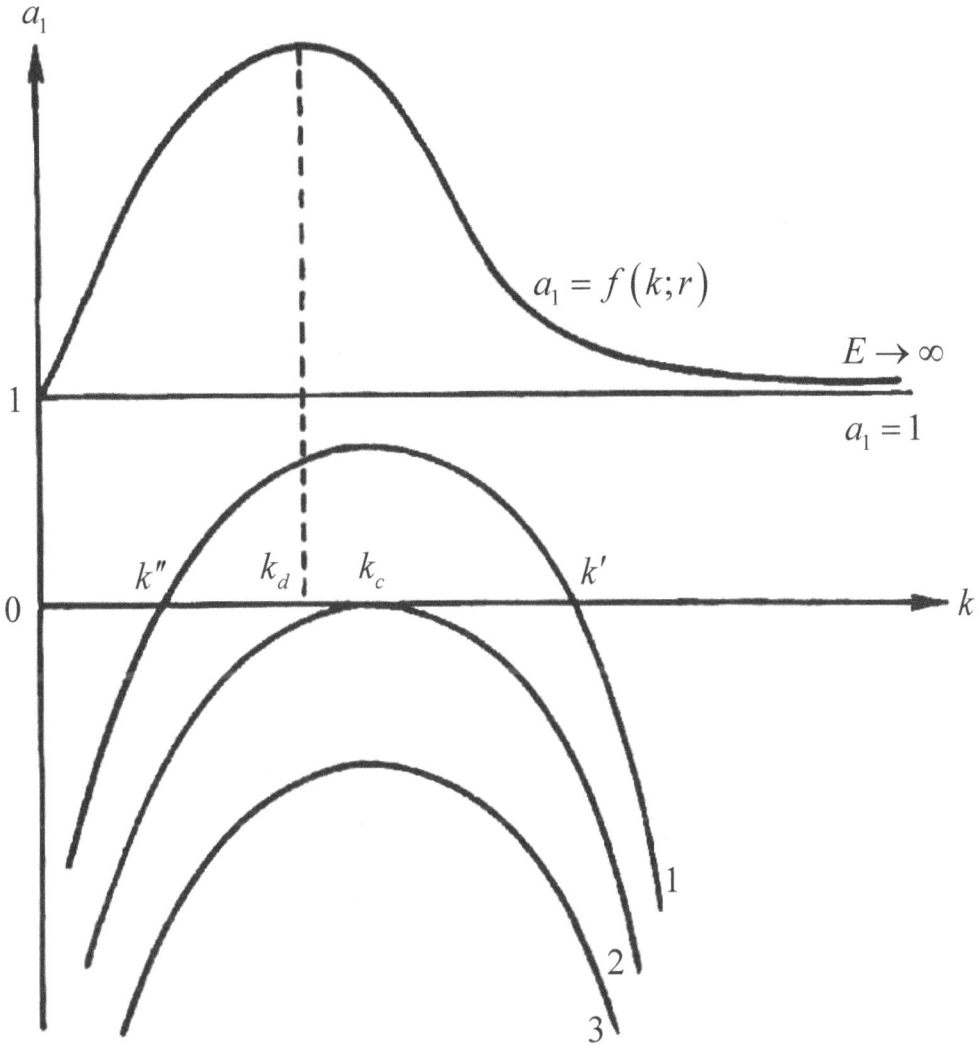

FIGURE 5.9
Schematic plots of the disturbance growth rate a_1 versus k, comparing the stability behavior of the model with or without density differences and surface tension. The curves designated by the numerals 1, 2, and 3, correspond to values of E from Fig. 5.7, such that E is greater than, equal to, or less than E_c, respectively, while the one designated by $a_1 = f(k; r)$ is associated with $N = 0$ and $E \to \infty$ or its behavior in the absence of density differences and surface tension. For the former case, curve 2 intersects the k-axis at the critical wavenumber k_c, while for the latter one, k_d is the wavenumber associated with the dominant wavelength to layer thickness ratio which corresponds to the largest value of a_1.

folding. For the case of a single layer fold, the quantity λ/H, the so-called wavelength to layer thickness ratio, is often employed by geologists to describe minor or small scale folds (Smith, [231]). Geologists have traditionally identified the dominant wavelength to layer thickness ratio,

$$\frac{\lambda_d}{H} = \frac{2\pi}{k_d},$$

as predicted by theoretical linear stability analyses, with the observed λ/H as measured in the field for naturally occurring single layer folds (Fletcher, [61]). We shall return to this discussion after a closer examination of Fig. 5.8.

We felt that Fig. 5.8, which is a plot of $\mathcal{E}$ versus $\mathcal{H}$ for a fixed value of r, should be of at least qualitative aid to field geologists, since $\mathcal{E}$ and $\mathcal{H}$ are proportional to $\bar{e}$ and H, respectively. We therefore investigated the stability predictions of this figure in more detail. For a layer of given thickness H_0 corresponding to an associated value of $\mathcal{H}$, $\mathcal{H}_0$, we found from Fig. 5.8 that the initiation of plane folding occurred at a value of $\mathcal{E}$, $\mathcal{E}_0$, satisfying

$$\mathcal{E}_0 = \mathcal{E}_c(\mathcal{H}_0; r).$$

For rates of compressive strain $\bar{e} < \bar{e}_0$, where $\bar{e}_0$ is the associated value of $\bar{e}$ corresponding to $\mathcal{E}_0$, there would be no folding, while for $\bar{e} > \bar{e}_0$ plane folding would occur. Further, such folds could be characterized by the critical wavelength to thickness ratio

$$\frac{\lambda_c^{(0)}}{H_0} = \frac{2\pi}{k_c^{(0)}} \text{ where } k_c^{(0)} = k_c(\mathcal{H}_0^2, r),$$

while the nondimensional quantity

$$\mathcal{L} = \frac{2\pi \mathcal{H}_0}{k_c^{(0)}}$$

would serve as a measure of the size of that dimensional fold wavelength. Thus, our linear stability results admitted the possibility of a layer of given thickness folding or remaining planar, depending on the relative size of the imposed rate of lateral compressive strain. This differed from the corresponding results of Fletcher ([61]) for his single-layer folding model that did not include gravity and capillary effects, which yielded the secular equation plotted in Fig. 5.9

$$a_1 = 1 + \frac{2(1-r)k}{(r^2-1)k + (1+r^2)\sinh(k) + 2r\cosh(k)} = f(k; r),$$

derivable from ours with $\Delta\rho = 0$ or $N = 0$ in the limit as $\gamma_0 \to 0$ or $E \to \infty$. Hence, since $f(k; r) \geq 1$ for $k \geq 0$, he predicted that a layer of given thickness would fold under lateral compression, no matter how small the rate of strain $\bar{e}$, as long as it were nonzero. Such a prediction could not account for the existence of any planar layers, except by making the assumption that either such layers were not subjected to lateral compression or they were subjected to a uniform such compression in the absence of all fold-type perturbations. Neither of these possibilities is very realistic and both are most likely invalid for the Castile Formation (Ramberg, [180]). If one, as a first approximation, attempts to analyze a multi-layer structure by treating it as a set of independent single layers (Ramberg, [180]), then our model went a long way toward explaining the existence of the planar Castile layers interspersed among folded ones.

Returning to our discussion of the characterization of single layer folds by their λ/H ratios, Smith ([231]) stated that minor or short wavelength folds occurring over small length scales had almost a preference for that quantity to be in the range from 4 to 6. Identifying λ/H

with Fletcher's ([61]) λ_d/H, Smith ([231]) was unable to predict any normalized wavelength in that range for a layered Newtonian fluid model, since any k_d satisfying $df(k_d;r)/dk = 0$ (see Fig. 5.9) was bounded above in such a way that $\lambda_d/H \geq 2\pi$. He therefore turned to a power law rheological non-Newtonian fluid approach, in order to obtain λ_d/H in the desired range (Smith, [231]). That being the case, we wanted to see what our model would predict in this regard.

Given the typical value of $r = 0.2$, we found from Table 5.4 that

$$\frac{\lambda_c^{(1)}}{H_1} = 6.28, \; \mathcal{E}_1 = 2.66 \text{ for } \mathcal{H}_1 = 0.63; \; \frac{\lambda_c^{(2)}}{H_2} = 4.13, \mathcal{E}_2 = 2.15 \text{ for } \mathcal{H}_2 = 1.00.$$

Thus, on our marginal stability curve in Fig. 5.8

$$4.13 \leq \frac{\lambda_c}{H} \leq 6.28, \; 2.15 \leq \mathcal{E} \leq 2.66 \text{ as } 1.00 \geq \mathcal{H} \geq 0.63.$$

Hence, we predicted normalized wavelengths in the anomalous four to six range.

Further, we deduced a plausible rationale for the preference that minor folds seem to have for any such bounded λ/H range, as follows: The lower bound corresponded to the largest value $\mathcal{H}$ could take and still have the quantity $\mathcal{L}$ remain small enough so that the fold was classified as minor, while the upper bound was associated with the largest value of $\mathcal{E}$ that could presumably be physically imposed. For our analysis $\lambda_c \cong \lambda_d$, while for the linear stability analyses of previous researchers modeling rock folding, mentioned above, there was no λ_c, since these analyses yielded identical instability. Hence, they tended to characterize their folds by λ_d/H. Although such a disturbance associated with the largest growth rate tends to predominate, it should be kept in mind that, for the case of identical instability, by the time perturbations have grown enough so that the effect of the maximum growth rate can be observed, the neglected nonlinearities may have rendered the analysis inaccurate (Segel and Stoeckly, [218]). For many nonlinear stability analyses of physical phenomena involving marginal stability curves of the form given in Fig. 5.7, however, it has been shown that the observed wavelengths are determined to a close approximation by the critical wavelength, λ_c, obtained from linear theory (see Chapter 2).

There is one final prediction we made with the aid of our single layer model. A further examination of Fig. 5.8 for $\mathcal{H}$ and $\mathcal{E}$ values, appropriate for minor folding, shows that if we chose a fixed value of $\mathcal{E}$, $\mathcal{E}_0$, such that $\mathcal{E}_2 < \mathcal{E}_0 < \mathcal{E}_1$, we then found that there was stability for $\mathcal{H}_1 \leq \mathcal{H} < \mathcal{H}_0$ and instability for $\mathcal{H}_0 < \mathcal{H} \leq \mathcal{H}_2$, where $\mathcal{E}_0 = \mathcal{E}_c(\mathcal{H}_0;r)$. The physical significance of this result was that it was possible for two layers, which are identical except for layer thickness, to undergo the same lateral compression and have the thicker layer fold, while the thinner one remained planar. This prediction, which would be impossible for an elastic model of folding, can be borne out by the Castile Formation [see Fig. 5.1, as well as Watkinson and Alexander ([269])] which contains numerous examples of thin planar layers lying in close proximity to thicker folded ones. As a partial justification for the inclusion of surface tension at the layer-medium interfaces, we cited the work of Buckmaster and Nachman ([21]), dealing with the deformation of a two-dimensional thin thread of viscous liquid, owing to the slow motion of its ends and the influence of surface tension. They determined for their viscida problem that surface tension effects would be important, provided the quantity $\gamma_0/(\mu U)$ was of the same order of magnitude as H/L, where U was the relative speed of one end of the viscida with respect to the other end, while H and L were the thickness and total arc length of that viscida, respectively. It only remained for us to show that for our range of values cited above, the Buckmaster-Nachman criterion was satisfied. That criterion could be written in the form $\gamma_0/(\mu U) = O(H/L)$, where for our problem $U = \bar{e}L$. Hence, it became $\gamma_0/(\mu \bar{e}H) = 2/E = O(1)$. Observing from Table 5.4 that for our range of values $1.68 \leq \mathcal{E}\mathcal{H} = E \leq 2.15 \Rightarrow 0.93 \leq 2/E \leq 1.19$, the Buckmaster-Nachman ([21]) criterion was satisfied for the single-layer rock folding problem under examination.

From the outset, because of the inherent symmetry of our problem about the center line of the layer (owing to the uniformity in both the imposed rate of strain field and the properties of the embedding medium), we had been able to restrict our analysis to the even z-component of the perturbation flow, which, under these circumstances, corresponded to folding. In order to show that this approach was permissible, we needed to reexamine our single layer model analysis by considering a more general class of disturbances which also included the odd z-component of the perturbation flow associated with "pinch-and-swell" interfacial structures, *i.e.*, ones categorized to lowest order by $\zeta'(x,t) = -\zeta(x,t)$. This was accomplished for our problem by redoing the analysis for such an odd solution and comparing the results obtained with the corresponding ones deduced for the even solution. Chandrasekhar ([30]) used similar reasoning to analyze the Bénard convection problem of a viscous fluid layer heated from below, for the case of rigid-rigid boundaries (see the next chapter). In that event, the "pinch-and-swell" growth rate b_1 satisfied the secular equation

$$b_1 = 1 + \frac{2(r-1)k - (1/E)(k+N/k)[r\{\sinh(k) - k\} + \cosh(k) - 1]}{(1-r^2)k + (1+r^2)\sinh(k) + 2r\cosh(k)}$$

with corresponding marginal stability curve

$$E = E_c^{(1)}(k; N, r) = \left(k + \frac{N}{k}\right) F_1(k; r)$$

where

$$F_1(k; r) = \frac{r[k\sinh(k) - k] + \cosh(k) - 1}{-(1-r)^2 k + (1+r^2)\sinh(k) + 2r\cosh(k)}.$$

We first noted that

$$\lim_{k \to 0} b_1 = 1 \text{ while } b_1 < 1 \text{ for } k > 0,$$

since $r < 1$. Hence, the critical conditions for a "pinch-and-swell" instability mode were given by

$$E_c^{(1)} = 0 \text{ and } k_c^{(1)} = k_d^{(1)} = 0$$

$$\Rightarrow (\zeta_1)_c = I_0 \exp\left[\int 1\, dt\right] \cos(0 \cdot x) = I_0 e^t = 2I_0 h(t).$$

Although these results implied that the onset of a "pinch-and-swell" instability would occur for any $E > 0$, they also predicted that the fastest-growing odd mode would be one corresponding to an infinite disturbance wavelength (Segel and Stoeckly, [218]). Since such interfacial structures located at $z = \pm h(1 + 2\epsilon I_0)$ would tend to reinforce the morphological behavior of the exact solution and thus be virtually indistinguishable from a planar layer in the context of field observation, our approach was the correct one for analyzing the observed folds in question. Hence, for our purposes, it had been possible without loss of generality to ignore the "pinch-and-swell" component of the layer shape when examining the development of folds as stated earlier.

This completed the research phase of Iwan's thesis problem, but prior to the dissemination phase two things had to happen. One of them was his thesis defense, called the Ph.D. Final Oral Examination by WSU. After John Watkinson arranged this examination, he realized that Iwan had never taken his Preliminary Doctoral Examination also required by the graduate school. So the other thing was that examination, which John scheduled the day before the Final Orals. Exactly a week prior to the date of his thesis defense, Iwan presented our research results in a geology colloquium that all his doctoral committee members attended and reduced his formal defense to a ballot meeting. Then we held his Preliminary Doctoral Examination. Besides the members of his doctoral committee, which consisted of me and John, plus two of John's geology department colleagues, a so-called representative of the graduate school was present. At WSU, this individual was a faculty member outside of the Ph.D. granting department who presided over the preliminary examination but did not vote. I was the first person to enter the examination room and that individual was the second. He immediately introduced himself to me and said, "You must be James," to which I, having no idea to what he was referring, said, "No," and introduced myself. When Iwan entered the room soon thereafter and the individual in question greeted him the same way it all became clearer. Remember that his actual first name is James and the representative of the graduate school had mistaken me for the candidate! Once Iwan had successfully completed his Preliminary Doctoral Exam, the graduate school representative said to him, "Having passed this examination, you are now ready to start your thesis work," to which Iwan replied that it was already completed and he was defending his thesis the next day. The individual in question was irate and said that things were not supposed to be done in this way, to which Iwan responded, "Be that as it may, mate, I am defending my thesis tomorrow." In fact, because of this individual's subsequent formal complaint to the graduate school, there is now a rule that a thesis defense cannot take place any earlier than three months after the preliminary doctoral examination, which I call the J. Iwan D. Alexander rule. There had always been a maximum period of time that was allowed between these two examinations, but nobody had ever thought a minimum time was required until Iwan made a mockery of the whole procedure. Incidentally, these committees at WSU no longer contain a representative of the graduate school and have the adviser presiding instead, which I certainly think is a step in the right direction.

As usual, the dissemination phase consisted of publishing our results in a scholarly journal and presenting them at professional meetings. Up until this time, I tended to publish my modeling research in specialty journals related to the scientific discipline of the phenomenon being modeled. Hence, we first submitted a paper to the *American Journal of Science*, headquartered at Yale University, in which Smith ([231]) was published and many seminal contributions to the geological literature had appeared. They declined to publish it without stating their reasons for this declination or providing us with any referees' reports, but did however return a copy of our original manuscript. In the margin next to its "partial justification for the inclusion of surface tension" entry were scrawled the words "bah," "bogus," and "bull shit" in pencil. Although John was very impressed with our work, he informed me that the geology community probably would never accept the introduction of interfacial surface tension in the rock folding model. Thus, I decided to submit an updated version of our paper along the lines outlined in this chapter to the *SIAM Journal on Applied Mathematics*, where it was refereed by geophysical fluid dynamicists, accepted after some organizational revisions, and appears as Wollkind and Alexander ([278]).

I presented this rock folding research at the 1981 SIAM Annual Meeting held at RPI in June, while Eric, who attended this meeting as well, presented our Rayleigh-Bénard-Marangoni results. These talks were scheduled back-to-back, in the same session, given the similarity of their multi-layer formulation and methodology of analysis. In point of fact, that similarity allowed Iwan and Eric to compare their thesis results, which was advantageous for dissertation proof reading. I also presented our rock folding research that summer at the "Fluid Dynamics Problems in Astrophysics and Geophysics" seminar, where my meeting with Manual Verlarde described earlier had taken place. It was generally well received, although someone during the questioning period asked me: if the stratified rock layers really acted as Newtonian fluids, why hadn't the interfaces all returned to their original planar position once the sample I exhibited was taken from the Castille Formation, thus removing the lateral compressive strain. I answered that it acted as a Newtonian fluid only under this lateral compression and upon its removal acted as an elastic solid. I thought this was a good question for a college freshman but not for a geophysical fluid dynamicist, although hopefully the tone of my response didn't reveal that attitude to the audience. Often the answering of questions of this sort required the Pulling of a Rabbit Out of a Hat.

Unlike some advisors, I think it is my job to get the students doing research with me placed in appropriate professional positions after that project has been completed. Further, should their first choice not be the right fit, I try to provide another option, which was exactly what happened in Iwan's case. He took an Assistant Professorship in the Geology Department at FIT (Florida Institute of Technology) in Melbourne, Florida, and, regretting it almost immediately, asked me if I could help him find a more suitable position, preferably a post-doctoral appointment. That next summer of 1982, David Oulton and I attended the Sixth American Conference on Crystal Growth held at Fallen Leaf Lake, California, and presented our NSF-supported joint work with Rukmini Sriranganathan on the development of interfacial cells during the solidification of a dilute binary alloy described in Chapter 2. On the morning of the last day of this conference, Bob Sekerka said he had a postdoctoral position available to work with him at Carnegie-Mellon University. He said he especially wanted someone who had experience in performing stability analyses of time-dependent solutions of governing systems of partial differential equations. I suggested Iwan and gave him a copy of Wollkind and Alexander ([278]), which had just been published and was, of course, an analysis of that sort. Bob told me to have Iwan apply for this postdoctoral position, which he did and was hired immediately. He served as Bob's postdoctoral associate for five years and they worked really well together, producing a number of significant results in physical metallurgy including a joint paper with me, Alexander *et al.* ([4]), that extended my thesis problem to the case for which the latent heat of fusion had to be retained in the energy balance condition at the solid-liquid interface. Bob pronounced Iwan's name as "E won" and apparently Iwan never informed him of the correct Welsh pronunciation. After working with Sekerka, Iwan became a senior scientist in the Marshall Space Flight Center at the University of Alabama in Huntsville doing NASA microgravity research; then, the Chair of the Department of Mechanical and Aerospace Engineering at Case-Western Reserve University in Cleveland, where he created the Great Lakes Energy Institute; and finally, the Dean of Engineering at the University of Alabama in Birmingham. All in all, I would say that J. Iwan D. Alexander's academic resume is pretty impressive, given his expertise in oceanography, structural geology, physical metallurgy, astrophysics, and mechanical, aerospace, and renewable energy and power engineering.

6

Two-Phase Fluid Flow of Aerosols and Convection in Planetary Atmospheres: Longitudinal Planform Nonlinear Stability Analyses

In Chapters 2 and 3 we examined various aspects of the classical Rayleigh-Bénard problem of a single layer of viscous Boussinesq fluid confined between two infinite, horizontal, shear-stress free pure-conducting surfaces that were being uniformly heated from below or cooled from above. Lord Rayleigh ([183]) selected these so-called free-free boundary conditions to simplify his linear stability analysis. This model has actually proven to be more useful for representing convection in closed containers the isothermal boundary planes of which are rigid (no slip) surfaces. The reasons for this are two-fold:

(i) The resulting fluid flow is then purely gravity driven as opposed to being generated by the variation of surface tension with temperature at fluid-fluid interfaces, which was the mechanism primarily responsible for the motion observed by Bénard ([12]) in his thin layers of spermaceti exposed to the air, as discussed in Chapter 2 and corroborated in Chapter 5; and

(ii) subsequent investigations of convective instability problems demonstrate that the same qualitative results are obtained for different boundary conditions (Koschmieder, [103]). Specifically, Jeffreys ([89]), realizing that the boundary conditions employed by Rayleigh ([183]) were somewhat artificial, extended the problem to take into account the possibility of rigid boundaries. The form of this rigid-rigid boundary condition model now necessitated that the critical conditions for the onset of convection be calculated numerically. Jeffreys' ([89]) numerical results, which have been progressively improved upon over the years by a series of researchers using increasingly more accurate procedures, yielded only quantitative differences when compared with Rayleigh's ([183]) analytical ones (see Table 6.1).

TABLE 6.1
Critical conditions for free-free, rigid-free, and rigid-rigid surfaces (Koschmieder, [103]). Here $R_c \equiv$ critical Rayleigh number and $q_c \equiv$ critical nondimensional wavenumber.

$z = 0\text{-}z = d$	Free-Free	Rigid-Free	Rigid-Rigid
R_c	657.511	1100.650	1707.762
q_c	2.2214	2.682	3.117
$2\pi/q_c$	2.828	2.342	2.016

Once critical conditions for the onset of thermal instability in fluids had been determined theoretically for the case of rigid-rigid boundaries, various experiments were devised to check the validity of these predictions. Those experiments, surveyed by Chandrasekhar ([30]) and, more recently, by Drazin and Reid ([53]), who referenced the modern and astonishingly accurate observations of Ahlers ([3]), employed heat transfer, optical light refraction, and

DOI: 10.1201/9781003195603-6

visual means to decide exactly when convection occurred, as relevant controllable parameters such as the temperature difference between the plates and the layer depth varied. The experimental determination of the onset of convection agreed with the theoretical prediction in liquid layers regardless of the measuring technique employed, but in gas layers there was a discrepancy when the visual method of adding an aerosol to the gas was used. The latter technique, completely analogous in concept to the liquid layer method of making motion visible in silicone oils by means of additives such as aluminum powder, resulted in an observed onset of this instability occurring at lower temperature gradients than those predicted theoretically, which were themselves in agreement with experimental evidence for clean gases as determined through heat transfer or optical means. This phenomenon was first observed by Chandra ([29]) in air layers mixed with cigarette smoke (see Fig 6.1). He found the nature of the instability to be crucially dependent upon layer depth.

For layers deeper than 10 mm, normal convection was observed at temperature gradients approximately 80% of Jeffreys' ([89]) predicted value, while if the layer depth was less than 7.5 mm, motion of an entirely different nature of a wavelength much shorter than anticipated occurred at temperature gradients a great deal lower than that prediction. Chandra termed these flow pattern convective rolls as motion of Types I and II, respectively, the latter mode being subsequently described by Sutton ([242]) as a *columnar instability*. It is somewhat ironic that an experimental technique adopted to visualize a flow (via cigarette smoke) would in fact make such major changes in the flow itself that this innocent intervention turned out to have far reaching and interesting consequences. Although there had been attempts at modeling such columnar modes theoretically by representing this aerosol (colloidal suspension of particles in a gas) as a single phase fluid, a complete survey of which can be found in the work of Scanlon and Segel ([200]), my Ph.D. student Limin Zhang and I believed that the two-phase flow approach of the latter authors seemed to promise the best chance of successfully explaining this long-standing discrepancy between theory and experiment.

Scanlon and Segel ([200]) considered the gas to be the continuous phase and the smoke particles, the dispersed phase. They then examined the continuum effect of the particles on the linear stability of a quiescent gas layer with no particle settling for their dilute particle-gas equations relevant to the Rayleigh-Bénard problem. The most significant result of their analysis was that convection occurred at temperature gradients of about 70% of Jeffreys' ([89]) predicted value, independent of layer depth. While being in substantial agreement with Chandra's observations for his relatively deep layers, this result did not actually succeed in settling the columnar instability question itself. In order to obtain even qualitative agreement with all of Chandra's experimental observations, and hence, to begin understanding this phenomenon, it was first necessary, at minimum, to devise a particle-gas model which lowered the critical temperature gradient associated with the clean gas in such a way that this reduction increased as layer depth decreased, both asymptotically approaching the Scanlon and Segel ([200]) result for relatively deep layers and becoming quite severe for shallow ones.

Before examining this model in detail, I would like to discuss the two students mentioned above. John W. Scanlon was a Ph.D. candidate in Chemical Engineering at RPI who was my immediate predecessor as a Lee Segel advisee. His thesis dealt both with the problem just referenced and a nonlinear stability analysis of a simplified Marangoni convection model for a semi-infinite fluid layer. When Jack gave a colloquium on his thesis results, it seemed to me that I would never be able to do anything so impressive. Limin Zhang, after arriving from China, began taking my graduate modeling classes upon his entrance into our doctoral

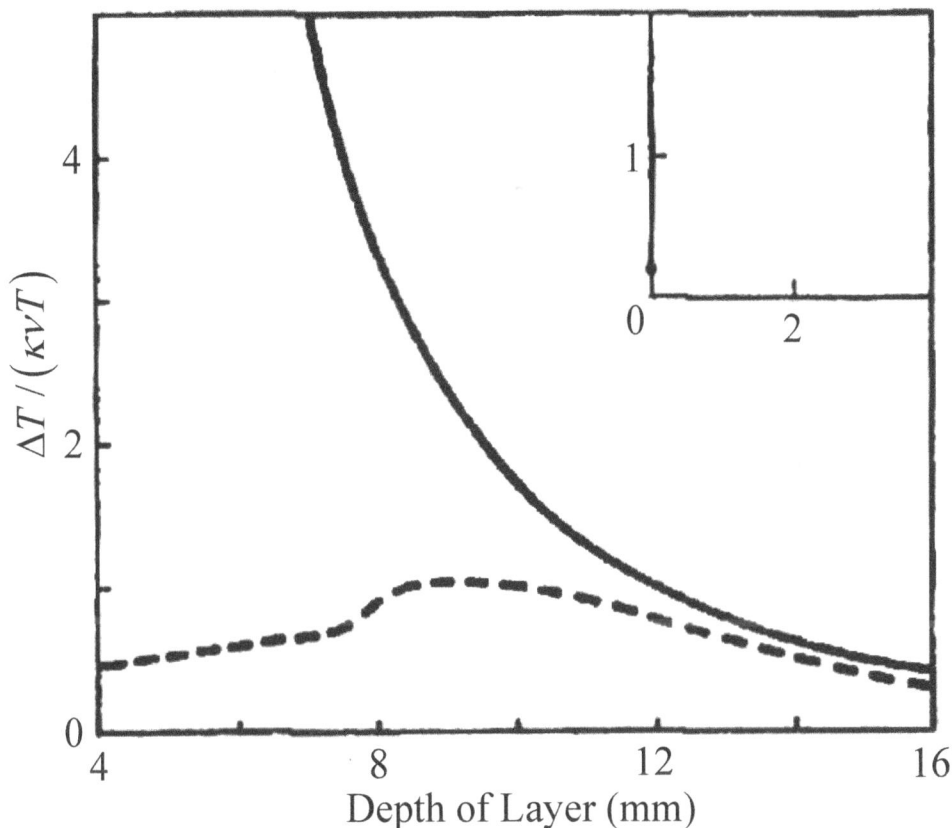

FIGURE 6.1
A reproduction of Chandra's ([29]) graphical comparison between his observed critical conditions (dashed curve) for convective instability in a smoke-air layer of variable depth involving
a rigid-rigid boundary experimental apparatus and Jeffreys' ([89]) theoretical predictions
(solid curve) for a clean gas relevant to this situation. Here, ΔT and d denote the layer temperature difference (K) and depth (mm), respectively, while T, κ, and ν are the clean gas
mean temperature (K), thermal diffusivity (cm^2/sec), and kinematic viscosity (cm^2/sec).
Chandra ([29]) employed $1709/(gd^3)$ for Jeffreys' ([89]) curve, where g is the acceleration
due to gravity at sea level (cm/sec^2), d has cm measure, and 1709 was the best available
estimate of the associated critical Rayleigh number, now determined more accurately to be
1708. In Fig. 6.5, Chandra's critical value will be adopted for comparison purposes with
his experimental data. The inset to this figure was added in order to portray explicitly the
extrapolated intercept of the columnar instability section of Chandra's curve.

program and became the research assistant on my second ONR grant in 1990 during his next year at WSU. This funding was for performing a weakly nonlinear stability analysis of a model relevant to Chandra's ([29]) and Sutton's ([242]) aerosol convection studies, which served as Limin's dissertation topic.

Limin and I analyzed that phenomenon by introducing a two-phase flow model with the gas as the continuous phase and the cigarette smoke as the discrete one. We employed a particle-gas model with the following two-dimensional discrete-phase Boussinesq-type governing equations:

Mass:
$$\boldsymbol{\nabla}_2 \cdot \boldsymbol{s} = 0;$$

Momentum:
$$\frac{D_s \boldsymbol{s}}{D_s t} = \left(\frac{\partial}{\partial t} + \boldsymbol{s} \cdot \boldsymbol{\nabla}_2 \right) \boldsymbol{s}$$
$$= -\left(\frac{1}{mN_0} \right) \boldsymbol{\nabla}_2 \mathcal{P} - g[1 - \alpha_d(\theta_d - T_0)]\boldsymbol{e}_3 + \left(\frac{K_0}{m} \right) (\boldsymbol{v} - \boldsymbol{s});$$

Energy:
$$mN_0 C_0 \frac{D_s \theta_d}{D_s t} = H_0(T - \theta_d) + k_d \nabla_2^2 \theta_d \text{ where } \nabla_2^2 = \boldsymbol{\nabla}_2 \cdot \boldsymbol{\nabla}_2;$$

Boussinesq equation of state:
$$N = N_0[1 - \alpha_d(\theta_d - T_0)] \text{ where } \alpha_d = \frac{1}{\theta_0};$$

Constitutive relations:
$$K_0 = 6\pi\mu a_0, \ H_0 = 4\pi a_0 N_0 k_0 \varepsilon_0 \text{ where } \varepsilon_0 = \frac{\rho_0}{\rho_d};$$

with boundary conditions:

Dynamical:
$$\boldsymbol{s} \cdot \boldsymbol{e}_3 = s_3 = 0 \text{ at } z = 0 \text{ and } d;$$

Thermal:
$$\theta_d = T_0 \text{ at } z = 0 \text{ and } \theta_d = T_1 \text{ at } z = d;$$

Far-Field:
$$\boldsymbol{s}, \mathcal{P}, \text{ and } \theta_d \text{ remain bounded as } x^2 \to \infty.$$

We retained the classical Rayleigh-Bénard Boussinesq governing system of equations for the gas phase with the addition of the particle drag and thermal exchange interaction terms given by
$$\left(\frac{K_0 N_0}{\rho_0} \right) (\boldsymbol{s} - \boldsymbol{v}) \text{ and } \left(\frac{H_0}{\rho_0 C_p} \right) (\theta_d - T),$$

to the right-hand sides of its momentum and energy equations, respectively. Here the spherical particles have radius $\equiv a_0$, mass $\equiv m$, material density $\equiv \rho_d$, number density $\equiv N_0$, specific heat $\equiv C_0$, and thermal conductivity $\equiv k_d$, with velocity $\equiv \boldsymbol{s} = (s_1, s_3)$, pressure $\equiv \mathcal{P}$, and temperature $\equiv \theta_d$, while mean temperature $\equiv \bar{\theta}_0$. Note in this context, it is assumed that the gas-particle flow is dilute enough to justify the neglect of viscous particle effects in its momentum equations but not so dilute as to invalidate the continuum hypothesis for these particles. It is also assumed that a_0 lies at the upper range over which the molecular process responsible for Brownian motion is of importance and hence, $\mathcal{P}$ represents the continuum effect of those particle-gas collisions. Further terms due to the substantial derivative of this pressure in the particle energy equation and Joule heating from particle drag in both energy equations have been neglected. The inclusion of the k_d term in the particle energy equation, however, accounts for the continuum mechanism by which heat can radiate directly from the hotter particles and be reabsorbed by the cooler ones without necessarily passing through the gas first as in a vacuum. Finally, we employed the stress-free boundary conditions for the gas

$$\frac{\partial u}{\partial z} + \frac{\partial w}{\partial x} = 0 \text{ at } x = 0, d$$

and, in order to compare our predictions with Chandra's ([29]) experimental observations for the rigid-rigid case, merely rescaled those critical conditions for the free-free case in accordance with Table 6.1, given that convective instability investigations of these two cases yield qualitatively similar results as mentioned in item (ii) listed earlier.

Finally, we note that this thermal disequilibrium system can be reduced to its thermal equilibrium version by solving the particle energy equation for

$$H_0(T - \theta_d) = mN_0C_0\frac{D_s\theta_d}{D_st} - k_d\nabla_2^2\theta_d,$$

substituting that value into the gas energy equation to obtain

$$\rho_0C_p\frac{DT}{Dt} + mN_0C_0\frac{D_s\theta_d}{D_st} = k_0(\nabla_2^2 T + \xi\nabla_2^2\theta_d) \text{ where } \xi = \frac{k_d}{k_0},$$

setting $\theta_d \equiv T$, and using that equation as motivation to postulate a balance of energy on the total system (Murray, [153])

$$\rho_0C_p\frac{DT}{Dt} + mN_0C_0\frac{D_sT}{D_st} = k_0(1 + \xi)\nabla_2^2 T.$$

After completing our discussion of the stability predictions relevant to the thermal disequilibrium model system, we shall compare these results to those obtained upon an analysis of such thermal equilibrium model systems including that employed by Scanlon and Segel ([200]). There exists a pure conduction solution of our thermal disequilibrium particle-gas system given by

$$\boldsymbol{v} = \boldsymbol{s} \equiv \boldsymbol{0}, \ T = \theta_d = T_0(z) = T_0 - \beta z \text{ where } \beta = \frac{\Delta T}{d} > 0,$$

$$p = p_0(z) = -\rho_0 g \int (1 + \alpha\beta z)\, dz, \ \mathcal{P} = \mathcal{P}_0(z) = -mN_0 g \int (1 + \alpha_d\beta z)\, dz.$$

That state represents the physical situation of a quiescent gas layer with no particle settling, across which an adverse affine temperature gradient is maintained. It was the stability of this solution to both two-dimensional linearly infinitesimal and weakly nonlinear finite-amplitude disturbances with which we were concerned. Given that only roll-type convection can occur in the rigid-rigid case, such stratified longitudinal-planform stability analyses were sufficient for our comparison purposes with Chandra's results. Toward that end, we first performed a normal-mode linear stability analysis of this pure conduction state for a nondimensionalized version of our particle-gas system employing scale factors, consistent with those used for Eric Ferm's problem, by seeking a solution of the same form as that introduced in the previous chapter with similar expansions of its particle variables, *e.g.*,

$$s_3(x,z,t) = 0 + \varepsilon_1 A_{11}\cos(qx)\sin(\pi z)e^{\sigma t} + O(\varepsilon_1^2) \text{ where } |\varepsilon_1| \ll 1;$$

finding critical conditions (q_c^2, R_c) for its most dangerous stationary instability mode having wavelength $2\pi/q_c$ and growth rate $\sigma = \sigma_s$, while $R = g\alpha\beta d^4/(\kappa\nu)$; and then sought a solution for the full nonlinearized system of the form introduced in Chapter 2, *e.g.*,

$$s_3(x,z,t) \sim A(t)s_{11}\cos(q_c x)\sin(\pi z) + A^2(t)[s_{20}(z) + s_{22}(z)\cos(2q_c x)]$$
$$+ A^3(t)[s_{31}(z)\cos(q_c x) + s_{33}(z)\cos(3q_c x)] \text{ where } \frac{dA}{dt} \sim \sigma_s A - a_1 A^3;$$

finding an expression for the Landau constant a_1 as a Fredholm-type solvability condition. Here, for ease of exposition, the identical notation is being used to represent its dimensionless variables as was employed to represent its dimensional ones. When the other parameters of the problem were assigned typical values, $[R_c, q_c^2, a_1] = [R_c, q_c^2, a_1](d)$. We shall also describe how k_d was introduced into our model since this is another perfect example of Pulling a Rabbit Out of a Hat.

In this context, besides the nondimensional parameters already defined for the Rayleigh-Bénard problem in this and the previous chapters,

$$r = \frac{\alpha_d}{\alpha}, \ f = \frac{mN_0}{\rho_0}, \ h = \frac{C_0 f}{C_p}, \ F = h + rf(1+h),$$

were also employed. We assigned the typical material parameter values for air at sea level with average temperature $\equiv \overline{T}_0 = 293$ K:

$$k_0 = 6.1 \times 10^{-5}\frac{\text{cal}}{\text{cm sec K}}, \ C_p = 0.242\frac{\text{cal}}{\text{gm K}}, \ \rho_0 = 0.0012\frac{\text{gm}}{\text{cm}^3}, \ \nu = 0.152\frac{\text{cm}^2}{\text{sec}};$$

and for cigarette smoke:

$$a_0 = 0.3 \times 10^{-4} \text{ cm}, \ N_0 = 2.35 \times 10^4\frac{1}{\text{cm}^3},$$
$$r = 1.0, \ \xi = 10^{-4}, \ \varepsilon_0 = \frac{1}{36}, \ f = 0.048, \ h = 4f;$$

respectively. Note that for these values of k_0, C_p, ρ_0, and ν:

$$\kappa = \frac{k_0}{\rho_0 C_p} = 0.210\frac{\text{cm}^2}{\text{sec}} \Rightarrow Pr = \frac{\nu}{\kappa} = 0.72;$$

while for those values of f, h, and r:

$$F = 0.25.$$

After assigning these parameter values, we plotted $R_c = R_c(d)$, $q_c^2 = q_c^2(d)$, and $a_1 = a_1(d)$, as reproduced in Figs. 6.2, 6.3, and 6.4, respectively. Before interpreting those plots and explaining Figs. 6.5 and 6.6 as well, in order to examine the behavior of this model with respect to layer depth, we shall describe Chandra's ([29]) experimental observations in more detail. His original apparatus consisted of a smoke-air layer of variable depth d contained

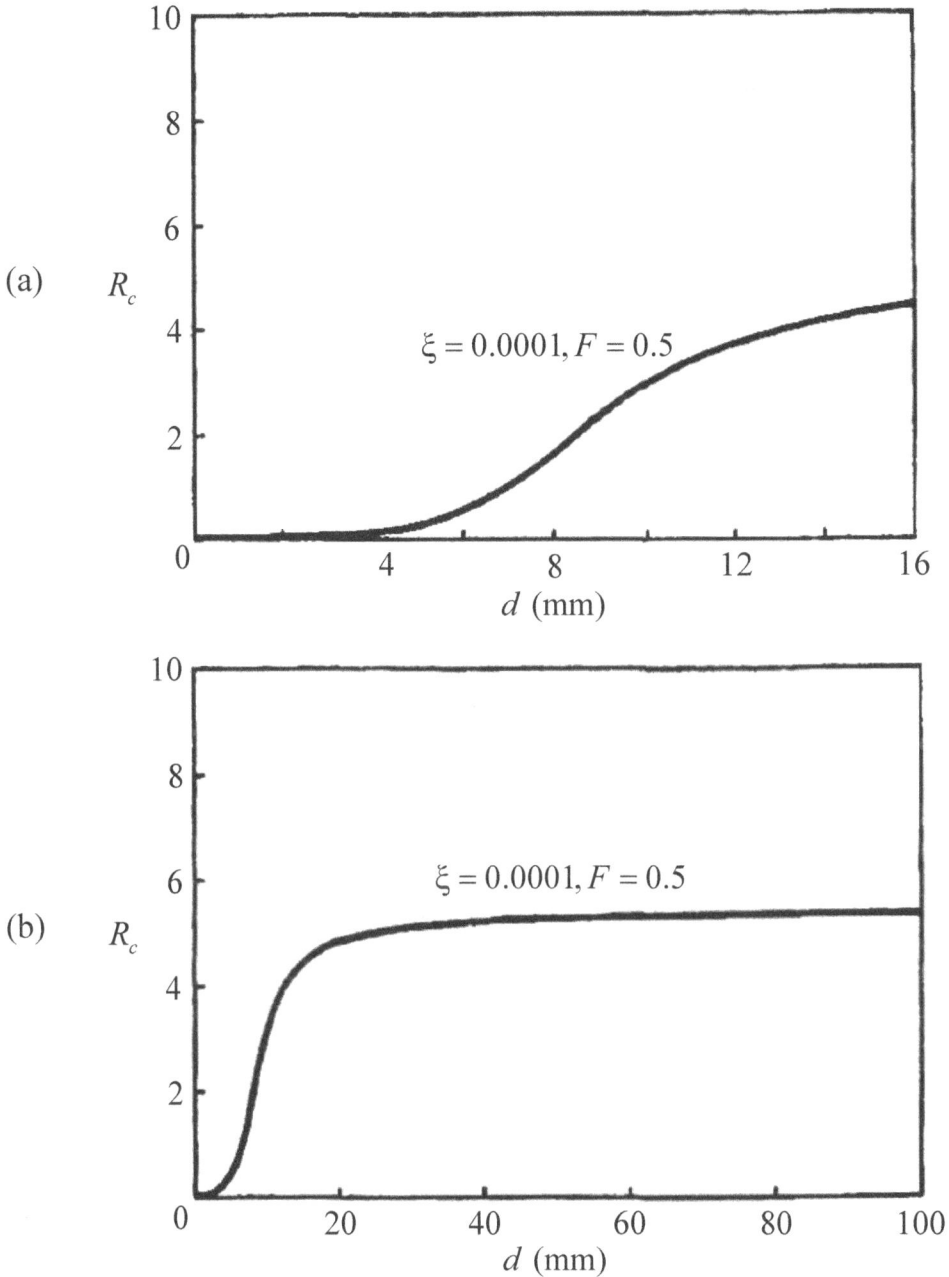

FIGURE 6.2
Plots of the stationary marginal stability curve $R = R_c(d)$ where R is measured in units of π^4 and d in mm for (a) 0 mm $\leq d \leq$ 16 mm and (b) 0 mm $\leq d \leq$ 100 mm. Here, and in the figures which follow for Chandra's experiments, the short-hand notation $\xi = 10^{-4}$ and $F = 0.25$ is being employed to represent all the particle and gas parameter value assignments.

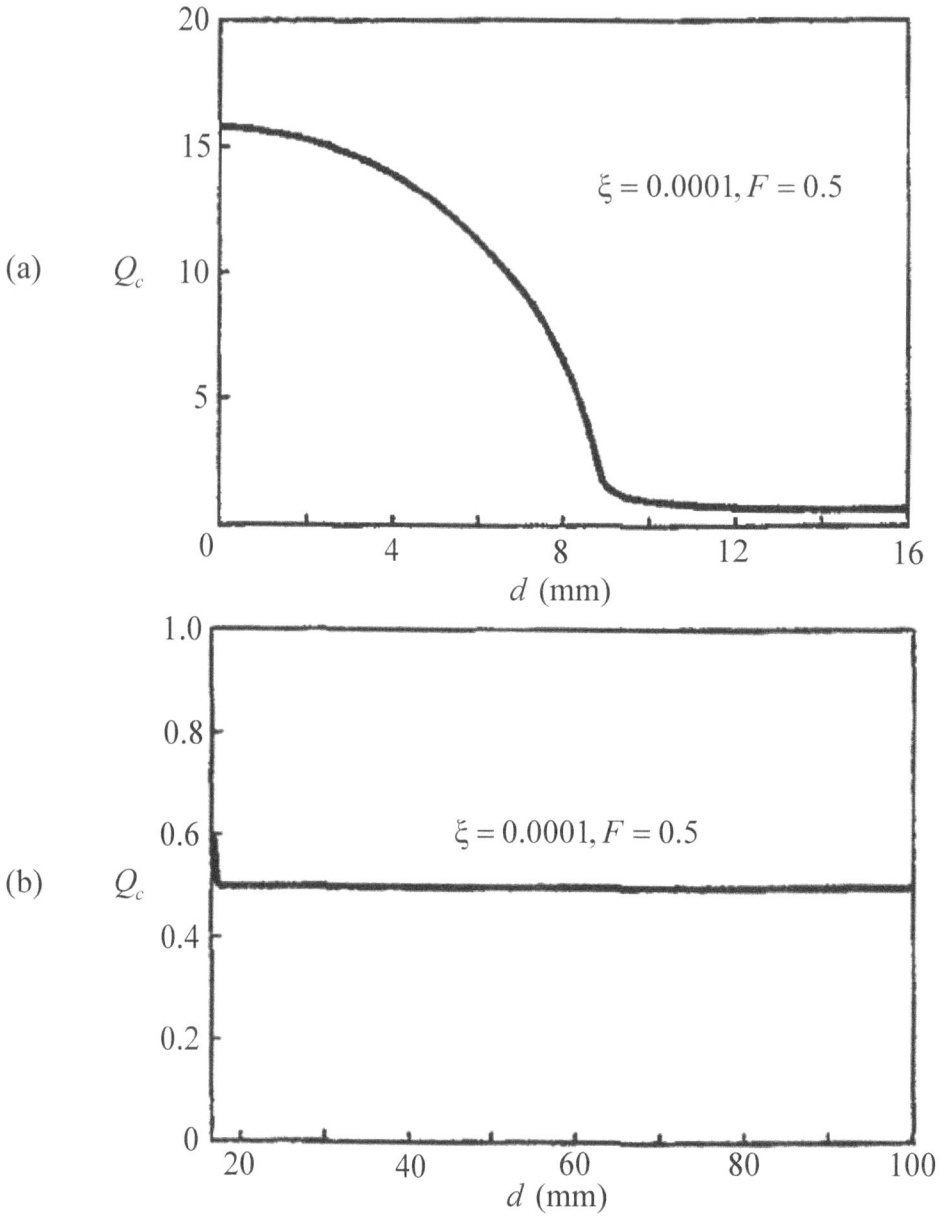

FIGURE 6.3
Plots of $Q_c = Q_c(d) \equiv q_c^2(d)$ where Q_c is measured in units of π^2 and d in mm for (a) 0 mm $\leq d \leq$ 16 mm and (b) 16 mm $\leq d \leq$ 100 mm.

between two rigid horizontal plates uniformly heated from below and cooled from above. Starting with the layer in motion at a particular depth, he lowered the temperature difference ΔT between the plates until that convective motion ceased. Chandra ([29]) reported the results of these experiments in both a tabular and graphical form, the latter by means of a plot reproduced in Fig. 6.1 of $\Delta T/(\kappa \nu T)$ versus 4 mm $\leq d \leq$ 16 mm where $T = \overline{T}_0$, the mean layer temperature. Noting that for Boussinesq gases $\alpha = 1/\overline{T}_0$, we can deduce from the definition of the Rayleigh number that

$$\frac{\Delta T}{\kappa \nu T} = \frac{R}{gd^3}.$$

Observe from this figure, that Chandra's ([29]) experimentally determined marginal stability curve (dashed line) can be divided into three parts: Namely, a columnar instability section, a normal convective section, and a transitional region linking these two sections together. In particular, the normal convective section of that graph occurs for 12 mm $\leq d \leq$ 16 mm and is closely approximated by 80% of the asymptotic curve (solid line) $1709/(gd^3)$ corresponding to Jeffreys' ([89]) theoretical prediction of the critical conditions for the onset of Rayleigh-Bénard instability in the case of rigid-rigid boundaries relevant to a clean gas, while the columnar instability section occurs for 4 mm $\leq d \leq$ 7.5 mm and is a segment of a straight line with positive slope and intercept (see the inset of Fig. 6.1). We note, in this context, that the quantity $\Delta T/(\kappa \nu T)$ has units $\sec^2/\text{cm}^4$ in Fig. 6.1, even though the depth of the layer appearing in that figure is being measured in mm. Finally, Chandra ([28]) also published photographs of the flow patterns for various depths which clearly indicated that the rolls associated with columnar or Type II instabilities were much more elongated in form when compared to those associated with normal convective or Type I instabilities than one would have anticipated from Jeffreys' ([89]) predicted critical wavelengths for such small differences in layer depth. Further, with a layer depth $d = 2$ mm, Chandra ([29]) found that stability had not yet been attained even for those ΔT which were so low as to be impossible to maintain uniformly over the whole layer. Hence, he was unable to observe the critical condition for this depth accurately.

Observe from an examination of the plot of the stationary marginal stability curve $R = R_c(d)$ in Fig. 6.2, that it becomes virtually tangent to the d-axis at $d = 2$ mm, consistent with Chandra's inability to attain a motionless state at this depth. Further, observe from Fig. 6.4, that

$$a_1(d) > 0 \text{ when } d > 0.$$

Then setting the wavenumber equal to its critical value q_c, as has been done in the nonlinear expansion, one can represent σ_s in a Taylor series about $R = R_c$

$$\sigma_s(R) = \sigma_s(R_c) + \sigma'_s(R_c)(R - R_c) + O(R - R_c)^2; \ \sigma_s(R_c) = 0, \ \sigma'_s(R_c) = \sigma_1 > 0;$$

and, taking R sufficiently close to its critical value so that $R - R_c = O(\varepsilon^2)$ where $|\varepsilon| \ll 1$, guarantee that

$$\sigma_s = O(\varepsilon^2).$$

Under these conditions, it can be shown that the implicit neglect of the fifth-order terms in the amplitude equation of the nonlinear expansion is valid (see Chapters 8 and 15), and the long-time behavior of the solution to that equation can be catalogued as follows:

(i) For $R < R_c$, the undisturbed state $A = 0$, which corresponds to the pure conduction solution, is stable to both infinitesimal and finite amplitude stationary disturbances.

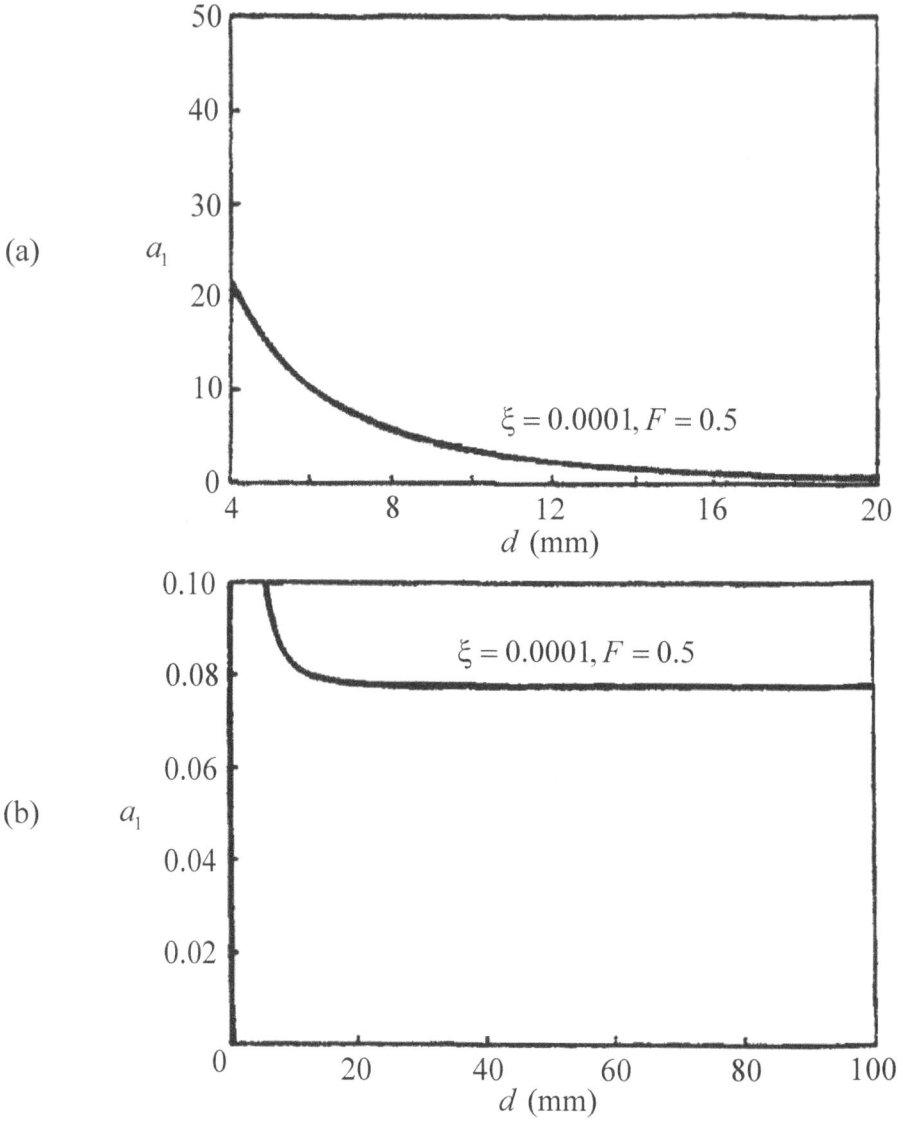

FIGURE 6.4
Plots of $a_1(d)$ for (a) 4 mm $\leq d \leq$ 20 mm and (b) 0 mm $\leq d \leq$ 100 mm with $Pr = 1$.

(ii) For $R > R_c$, there exists a stable stationary equilibrium solution $A_e^2 = \sigma_s/a_1$ such that

$$w(x, z, t) \to w_e(x, t) \sim A_e \cos(q_c x) \sin(\pi z) \text{ as } t \to \infty,$$

which physically corresponds to a stationary roll-type convection pattern of dimensional wave-length $(2\pi/q_c)d$.

Limin and I were now ready to examine the behavior of our particle-gas model with respect to layer depth and compare those results with Chandra's ([29]) experimental evidence regarding the occurrence of columnar instabilities. We wished to demonstrate that the stationary rolls predicted by our model were compatible with his observations involving smoke-air layers described above. Toward that end, we first investigated the asymptotic behavior of $R_c(d)$, $Q_c(d) \equiv q_c^2(d)$, and $a_1(d)$ for large d as depicted in Figs. 6.2b, 6.3b, and 6.4b, respectively. We deduced that

$$R_c(d) \sim \frac{1+\xi}{1+F}\frac{27}{4} \cong \frac{27}{4(1+F)} = \frac{27}{5} \text{ as } d \to \infty$$

for our choice of parameter values, which is consistent with the normal convective mode being associated with 80% of the critical value of Jeffreys' ([89]) clean gas theoretical prediction, once one realizes that the R_c of Fig. 6.2 is being measured in units of π^4. Similarly, we deduced that $Q_c(d) \sim 1/2$ as $d \to \infty$, which is consistent with the critical wavenumber squared for a clean gas Rayleigh-Bénard model having free-free boundaries, once one again realized that the Q_c of Fig. 6.3 is being measured in units of π^2. Finally, we deduced that

$$a_1(d) \sim \frac{(1+h)^2/(1+\xi)}{8[Pr^{-1}(1+f)(1+\xi)+1+h]} \cong \frac{(1+h)^2/8}{Pr^{-1}(1+f)+1+h} \text{ as } d \to \infty,$$

in comparison to $a_1 = 1/[8(Pr^{-1}+1)]$ for the clean gas Rayleigh-Bénard model having free-free boundaries, which can be obtained by setting $f = h = 0$ in the above expression. Note that this differs by a factor of Pr from the Landau constant obtained in Chapter 3, due to the fact that the scaling introduced for its dissociating gas model was based on the kinematic viscosity, while our model's scaling is based on the thermal diffusivity $\equiv$ thermometric conductivity instead.

Before examining the columnar instability behavior of our model, we shall explain how k_d was incorporated into it. Initially, after Scanlon and Segel ([200]), no such term was included in our model which can be obtained from the one defined in this chapter by taking the parameter $\xi = 0$ or equivalently setting $k_d = 0$. What occurred for this model was that then the marginal stability curve $R = R_0(q^2; d)$ had a horizontal asymptote $R = R_\infty(d) = 3.64d^4$, but only a true minimum point (q_m^2, R_m) when $d > d_c \cong 9$ mm. Here, $R_0(q^2; d) > R_\infty$ for $0 < d < d_c$, while $R_m < R_\infty$ for $d > d_c$. We originally defined

$$R_c(d) = \begin{cases} R_\infty(d) \\ R_m(d) \end{cases} \text{ for } \begin{array}{l} 0 < d < d_c \\ d > d_c \end{array}$$

and used that as a critical instability threshold but were unable to define a corresponding critical wavenumber for $0 < d < d_c$. Soon thereafter, I had the occasion to discuss the choice of thermal exchange coefficients H_0 for our model with Clayton Crowe, a two-phase flow expert in the Mechanical Engineering Department at WSU. He thought this choice to be reasonable but said the current thinking in the field was that a k_d effect should be included in such models. Two days earlier, Sam Coriell of NIST, who was visiting our department at my invitation, pointed out the deficiency of the $\xi = 0$ version of our model, in that, for

$0 < d < d_c$ it predicted instability when $R > R_\infty$ as $q^2 \to \infty$ or equivalently $\lambda \equiv 2\pi/q \to 0$, which he said was physically impossible. On Tuesday, Sam said something was missing from our model and on Thursday Clayton told me what it was, so on Saturday I included this effect and everything worked out 'like a champ' (as David Oulton would say), in that, the horizontal asymptote was converted to a linear one with a slope of $146\xi d^2$ and as a consequence there existed an absolute minimum point (q_c^2, R_c) for all $d > 0$, the coordinates of which are plotted as functions of d in Figs. 6.2 and 6.3 when $\xi = 10^{-4}$. Hence, we were finally in business. This is why I consider the introduction of k_d into our model as Pulling a Rabbit Out of a Hat.

In order to actually provide quantitative agreement with Chandra's ([29]) experimental data, it was necessary for us to take into account the intrinsic difference between the critical conditions for the onset of Rayleigh-Bénard instability involving a clean gas when the boundaries are rigid-rigid rather than free-free. As mentioned earlier, the simplest way of accomplishing this was merely to introduce the following rescaled marginal stability function appropriate for comparing our theoretical predictions with Chandra's ([29]) experimental observations summarized in Fig. 6.1:

$$1709 \left(\frac{4}{27} \right) \frac{R_c(d)}{gd^3}$$

which has the large-d asymptotic behavior

$$\left(\frac{1+\xi}{1+F} \right) \frac{1709}{gd^3}$$

where

$$\frac{1+\xi}{1+F} \cong \frac{1}{1+F} = \left\{ \begin{array}{c} 1.0 \\ 0.8 \end{array} \right. \quad \text{for } F = \left\{ \begin{array}{c} 0.00 \\ 0.25 \end{array} \right. .$$

Next, in order to examine the correlation of our model's theoretically predicted results with Chandra's columnar instabilities, we plotted this marginal stationary stability function versus d in Fig. 6.5a. Upon comparison with Chandra's columnar-instability straight-line data points of $d = 4$, 6, 7, and 7.5 mm, it can be seen from Fig. 6.5a that our marginal stability function provides an excellent fit to these points over this portion. We then plotted its asymptotic function versus d for both $F = 0$ and $F = 0.25$ in Fig. 6.5b. Note that the $F = 0$ curve, denoted by a dotted line in that figure, corresponds to Jeffreys' ([89]) theoretical prediction and the $F = 0.25$ one virtually coincides with the marginal stability function over its normal convective instability portion 12 mm $\leq d \leq$ 16 mm, as well as also providing an excellent fit to the data end points of $d = 12$ and 16 mm for this interval. Thus, this completes both the columnar and normal convective instability parts of a compatibility demonstration of our theoretical predictions with Chandra's ([29]) experimental results. In fact, even the transition region data point $d = 8$ mm in Fig. 6.5a deviates just slightly and that of $d = 10$ mm in Fig.6.5b only slightly more from that marginal stability curve, completing the final part of the desired compatibility demonstration.

It only remained for us to demonstrate that the layer depth behavior of the critical wavelength of our theoretically predicted stationary rolls was compatible with the characteristic width of the convective flow patterns photographed by Chandra ([29]). Using analogous reasoning to that employed in conjunction with the introduction of the marginal stability curve, we defined the rescaled dimensional wavelength

$$\lambda_c(d) = \frac{2.016}{2\sqrt{2}} \left(\frac{2}{\sqrt{Q_c(d)}} \right) d,$$

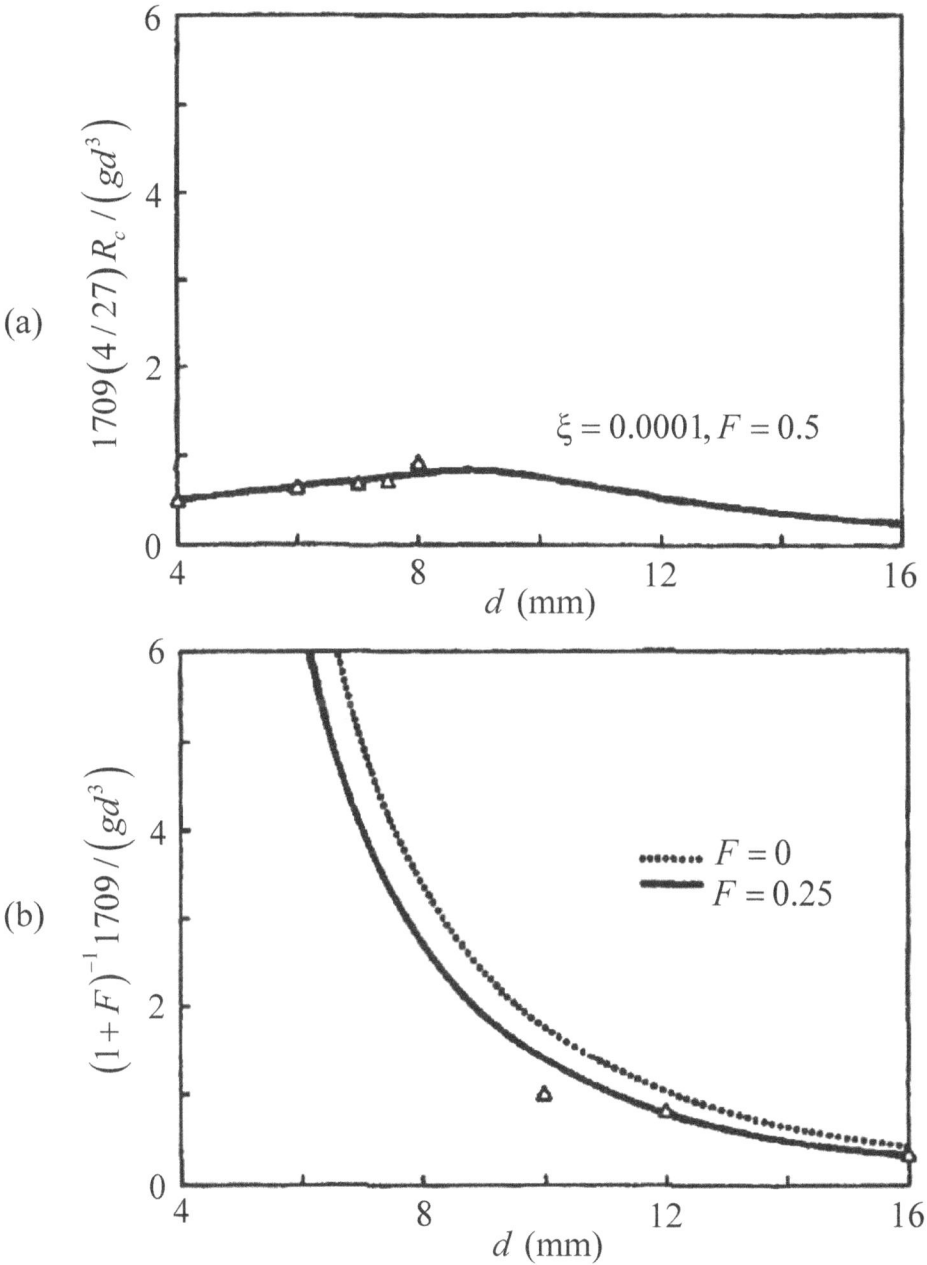

FIGURE 6.5
Plots of (a) the rescaled marginal stability function versus 4 mm $\leq d \leq$ 16 mm for the $R_c(d)$ of Fig. 6.2 and (b) its asymptotic functions versus 4 mm $\leq d \leq$ 16 mm for $F = 0$ and 0.25. Here the small triangles represent Chandra's experimentally determined data points.

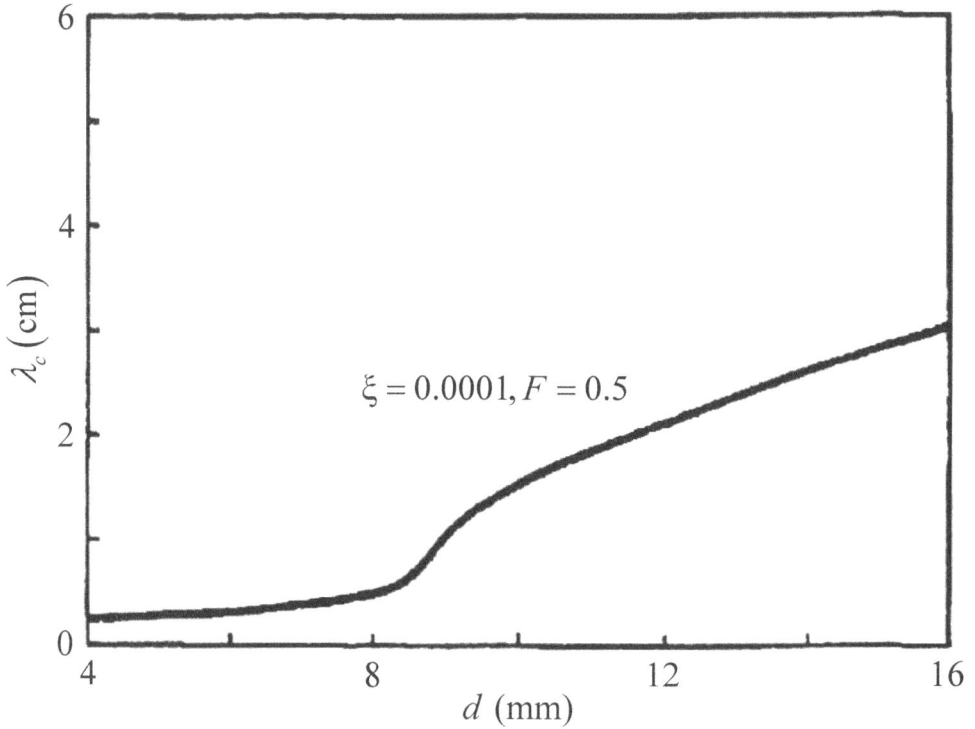

FIGURE 6.6

A plot of the rescaled critical dimensional wavelength $\lambda_c(d)$ measured in cm versus 4 mm $\leq d \leq 16$ mm for $Q_c(d)$ of Fig. 6.3.

which corresponds to a similar rescaled quantity appropriate for rigid-rigid boundaries. When we plotted this critical wavelength function versus d in Fig. 6.6, the resultant curve had an inflection point which, occurring as it did at $d \cong 9$ mm, allowed for a much more dramatic shift in characteristic roll width between those depths associated with columnar and normal convective instabilities, respectively, than could be anticipated from the linear relation for a clean gas given by

$$\lambda_c(d) = 2.016 \ d,$$

which serves as its large-d asymptote. This prediction, being in exact accordance with Chandra's ([29]) photographic experimental evidence, completed our final compatibility demonstration.

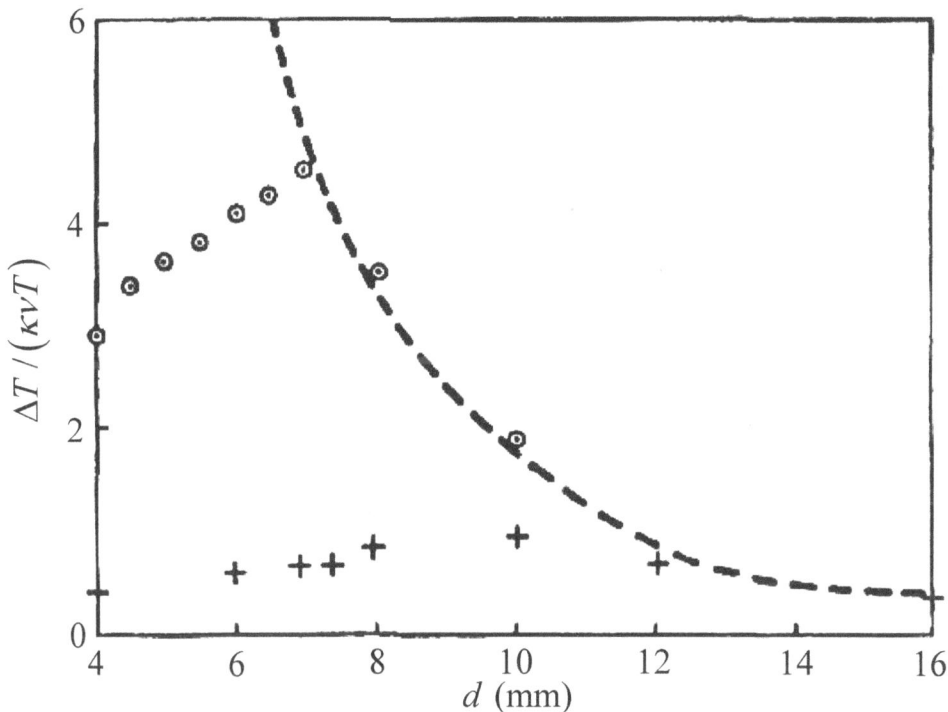

FIGURE 6.7
A reproduction of Sutton's ([242]) graphical comparison between the experimental data points of Chandra ([29]) and Dassanayake, denoted by crosses and encircled dots, respectively, relevant to the occurrence of columnar instabilities in a smoke-gas layer of variable depth involving a rigid-rigid experimental apparatus, where air was used as the carrier gas for the former series of experiments and carbon dioxide for the latter. Note that here the same notation is being employed as in Fig. 6.1, except that Jeffreys' ([89]) theoretical prediction (dashed curve) is now represented by $1708/(gd^3)$ since Sutton ([242]) adopted the more accurate 1708 determination of the associated critical Rayleigh number, as opposed to the 1709 used by Chandra ([29]). In Fig. 6.8a Chandra's critical value will be retained in order to compare our fit of Fig. 6.5 with that relevant to Dassanayake's experimental data.

Dassanayake continued Chandra's ([29]) work on columnar instabilities, employing essentially the same apparatus and methodology but replacing the air layer involved in the latter's experiments with carbon dioxide, as reported by Sutton ([242]). Dassanayake's data set, denoted by encircled dots, is reproduced from Sutton ([242]) in Fig. 6.7. This is again a plot of $\Delta T/(\kappa \nu \overline{T}_0)$ measured in $\sec^2/cm^4$ versus d in mm units, which includes Chandra's ([29]) data, denoted here by crosses, for comparison purposes. In particular, Dassanayake's marginal stability curve consists of two parts: Namely, a columnar instability section occurring for 4 mm $\leq d \leq$ 7 mm, which is a line segment with positive slope and 4 mm-intercept and a normal convective section occurring for 7 mm $\leq d \leq$ 10 mm which is asymptotic to the dashed curve $1708/(gd^3)$ in Fig. 6.7, corresponding to Jeffreys' ([89]) theoretical prediction for a clean gas.

In contrast, Chandra's marginal stability curve also contains a third part, which is transitional in nature, linking its columnar instability and normal convective sections. Upon comparing these two marginal stability curves, we see that the one for carbon dioxide has had both the 4 mm-intercept and slope of its columnar instability section's line segment so markedly increased that it now intersects its normal convective section at a cusp-type maximum point possessing a much larger ordinate value than occurs with air, while virtually eliminating the transitional region of the latter. The primary cause for this difference is that $\kappa \nu$ for air is about four and one half times greater than that for carbon dioxide as we now demonstrate. Recall, for air at $\overline{T}_0 = 293$ K that

$$\kappa = 0.210 \frac{cm^2}{\sec} \text{ and } \nu = 0.152 \frac{cm^2}{\sec} \Rightarrow \kappa \nu = 0.032 \frac{cm^4}{\sec^2}$$

while the corresponding components of κ for carbon dioxide are given by (Weast, [271])

$$k_0 = 3.8 \times 10^{-5} \frac{cal}{cm \ \sec \ K}, \ C_p = 0.200 \frac{cal}{gm \ K}, \ \rho_0 = 0.0019 \frac{gm}{cm^3}$$

$$\Rightarrow \kappa = \frac{k_0}{\rho_0 C_p} = 0.100 \frac{cm^2}{\sec}.$$

Recalling that for gases

$$Pr = \frac{\nu}{\kappa} = 0.7 \Rightarrow \nu = 0.7\kappa = 0.070 \frac{cm^2}{\sec} \Rightarrow \kappa \nu = 0.007 \frac{cm^4}{\sec^2}.$$

Hence, the ratio of this quantity for air with that for carbon dioxide is 4.57, as was to be shown. As previously summarized in this chapter, we assigned parameter values for our particle-gas system appropriate for modeling Chandra's convective experiments involving air layers at sea level, into which cigarette smoke had been mixed to provide visibility. We then provided a similar sort of correlation between our theoretical predictions and Dassanayake's data set. To do so, we varied the relevant particle parameters simultaneously about these reference values for a specific choice of the gas parameters particularized to smoke-carbon-dioxide aerosols and selected the proper set to provide the desired correlation. We defined the following ratios in addition to r and ξ:

$$\eta = \frac{\rho_{0_{CO_2}}}{\rho_{0_{air}}}, \ n = \frac{N_d}{N_0}, \ 4\zeta = \left(\frac{C_0}{C_p}\right)_{CO_2/smoke} \ ;$$

and using these ratios represented the relevant model parameters in the form

$$\varepsilon_0 = \frac{\eta}{36}, \ f = 0.048 \frac{n}{\eta}, h = 4\zeta f;$$

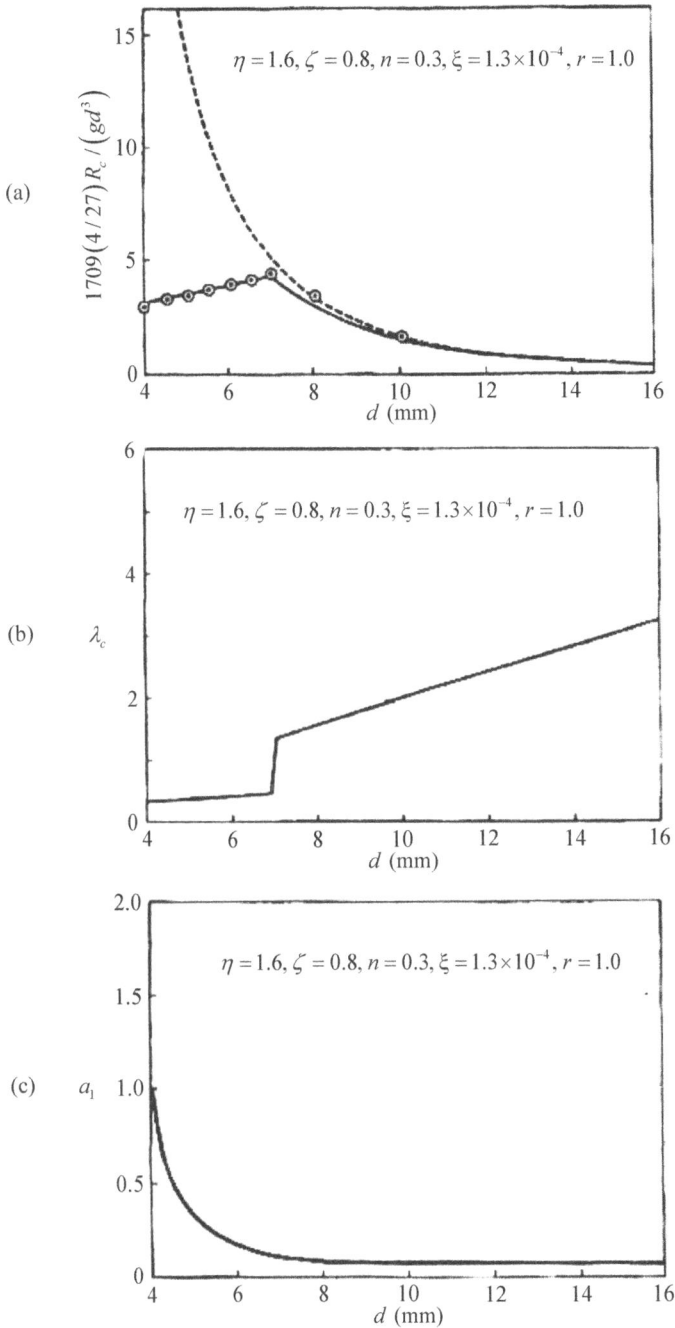

FIGURE 6.8

Plots of (a) the rescaled marginal stability function, (b) the rescaled dimensional wavelength $\lambda_c(d)$, and (c) $a_1(d)$; for 4 mm $\leq d \leq$ 16 mm with $\eta = 1.6$, $\zeta = 0.8$, $= 0.3$, $\xi = 1.3 \times 10^{-4}$, and $r = 1$. In (a), the encircled dots and dashed curve represent Dassanayake's (Sutton, [242]) experimentally determined carbon dioxide data points and Jeffreys' ([89]) theoretically predicted marginal stability locus, respectively, from Fig. 6.7. Here we have retained 1709 for comparison purposes instead of having replaced it with 1708.

which reduced to our reference values for $\eta = n = \zeta = 1$. Here, N_d represents the reference particle number density for the carbon dioxide experiments, while taking $h = 4f$, as was done for the air-smoke layers implicitly implied that $(C_0/C_p)_{\text{air/smoke}} = 4$ which, being the value of that ratio for aerosols of liquid water in air, assumes that the moisture content of air-cigarette smoke mixtures is significant enough to allow them to be treated as such aerosols. In light of the values for $\rho_{0_{CO_2}}$ and $\rho_{0_{air}}$, we assigned $\eta = 19/12 = 1.6$; retained $r = 1$; and determined the proper selection for the particle ratios ζ, n, ξ that provided the best fit to Dassanayake's data of Fig. 6.7, by varying each component of that triad about the reference point $\zeta = n = 1$, $\xi = 10^{-4}$, and making the optimal selection visually, which yielded

$$\zeta = 0.8, \ n = 0.3, \ \xi = 1.3 \times 10^{-4}.$$

The result of such a process appears in Fig. 6.8a, while the associated $\lambda_c(d)$ and $a_1(d)$ are plotted in Figs. 6.8b,c. Note that our theoretical predictions fit Dassanayake's data points extremely well, except for the 4 mm-intercept data point, which probably can be treated as an outlier due to the nonuniformity of ΔT at such a low depth, since all the other columnar instability points lie on a straight line. Observe, in addition, that the intersection of the columnar instability and normal convective segments of their marginal stability curve of Fig. 6.8a at a cusp-type maximum point results in the concomitant precipitous behavior of $\lambda_c = \lambda_c(d)$ in the neighborhood of its inflection point (see Fig. 6.8b), while, given the positivity of $a_1 = a_1(d)$, the occurrence of a supercritical stationary roll convective pattern is again predicted (see Fig. 6.8c).

We offer a final commentary on exactly what contribution each of our model system parameters made to the correlation with experiment exhibited in Figs. 6.6a or 6.8a. In particular, the slope and the 4 mm-intercept of the straight line segment columnar instability portion of our marginal stability curves depended crucially upon ε_0, ξ, and the product of a_0 with N_0 or N_d., while the reduction factor $1/(1 + F)$ characteristic of the normal convection portion depended crucially upon its constituent components r, f, and h. In this context, we note that, for the carbon dioxide-smoke marginal curve of Fig. 6.8a, $1/(1 + F) = 0.96$, as opposed to this factor's value of 0.80 for the air-smoke marginal curve of Fig. 6.6a, which was why the normal convective portion of the former virtually coincided with Jeffreys' ([89]) asymptotic curve that corresponds to $F = 0$ or $1/(1 + F) = 1$. Further, in retaining the assumption that $r = 1$, we adopted the mean temperature equilibrium condition $\overline{T}_0 = \overline{\theta}_0$, which implies that the pointwise temperature differences $T \neq \theta_d$ caused by thermal disequilibrium tend to cancel out when spatially averaged over the whole layer.

Finally, assuming a thermal equilibrium model and adopting the balance of total energy equation discussed earlier, as employed by Scanlon and Segel ([200]), we then obtained a critical Rayleigh number which had the same asymptotic behavior for large d as our thermal disequilibrium model, but only a parabolic asymptote of the form $R_2 d^2$ as $d \to 0$ where (Wollkind and Zhang, [299])

$$R_2 = \frac{4(1 + \xi)}{(1 + F)d_0^2} \text{ with } d_0 = 0.74 \text{ mm.}$$

Hence, although that result predicted a drastic lowering of the critical Rayleigh number with decreasing depth when compared to the classical model for a clean gas, this reduction did not become severe enough in the case of very thin layers to be in quantitative accord with the columnar instabilities observed by Chandra ([29]) at such shallow depths. That behavior required an asymptote of the form $R_4 d^4 + R_3 d^3$ as $d \to 0$, consistent with our determination for the thermal disequilibrium model as depicted in Figs. 6.2 and 6.5a. Note that, in comparison with the $R_\infty(d) = 3.64d^4$ of its $\xi = 0$ version, $R_4 > 3.64$ and $R_3 > 0$. Thus, the columnal instability section of the rescaled marginal stability function of Fig. 6.5a for $\xi = 10^{-4}$ has an extrapolated intercept in agreement with that depicted in Fig. 6.1 for Chandra's ([29]) data, while the corresponding function for $\xi = 0$ passed through the origin. No matter how we tried varying the other parameters of the $\xi = 0$ model, it was impossible for us to raise that intercept, but upon the introduction of a $\xi > 0$ the whole problem worked out perfectly. Like Segel always said, "Once you have the correct formulation for a model everything falls into place automatically." This is another reason why I consider introducing k_d into our model as Pulling a Rabbit Out of a Hat.

There remained only a comparison of our thermal equilibrium results with those of Scanlon and Segel ([200]), which can be obtained from ours by setting $r = \xi = 0$. Then, to first order, the particle pressure disappears identically from the problem, an occurrence that was implicit to Scanlon and Segel's ([200]) approach in their linear stability analysis of not taking that effect into account, while our critical conditions reduce to their predictions of (Wollkind and Zhang, [299])

$$Q_c = \frac{1}{2} \text{ and } R_c = \frac{27}{4H} \text{ where } H = 1 + 4f,$$

which were equivalent to our asymptotic behavior as $d \to \infty$ for $r = \xi = 0$. Taking

$$f = 0.1 \Rightarrow \frac{1}{H} = 0.71,$$

they predicted an R_c of approximately 70% of that associated with a clean gas independent of layer depth. Observe that for $\xi \cong 0$, our asymptote for R_c as $d \to \infty$ is of the same form as their R_c, but with a reduction factor of the reciprocal of

$$1 + F = (1 + rf)H,$$

which, for $r = 1$ and $f = 0.048$, becomes $1/(1+F) = 0.80$, as indicated earlier in accordance with Chandra's ([29]) experimental observations of relatively deep layers. Scanlon and Segel ([200]) felt that $f = 0.1$ was a large value for cigarette smoke and expected that its actual value would be much lower in Chandra's ([29]) experiments where smoke was mixed with air in a pump to obtain a uniform mixture. This is consistent with our value for f.

As a prelude to the analysis of both our thermal equilibrium and disequilibrium problems, we first considered them under the compressible gas version of the Boussinesq approximation, which will be developed from first principles using the following model in the absence of particles:

$$\nabla \cdot v = 0,$$

$$\rho_0 \frac{Dv}{Dt} = -\nabla p' - \rho_0(1 - \alpha T)ge_3 + \mu_0 \nabla^2 v,$$

$$\rho_0 C_p \frac{DT}{Dt} = \frac{Dp'}{Dt} + \Phi + k_0 \nabla^2 T;$$

where $v = (u, w) \equiv$ velocity, $p' \equiv$ pressure, $T \equiv$ temperature, $\rho_0 \equiv$ density, $\alpha \equiv$ thermal expansivity, $g \equiv$ gravitational acceleration at sea level, $\mu_0 \equiv$ shear viscosity, $C_p \equiv$ specific heat at constant pressure, $\Phi \equiv$ viscous dissipation, $k_0 \equiv$ thermal conductivity, $D/DT = \partial/\partial t + v \cdot \nabla$ for $\nabla \equiv (\partial/\partial x, \partial/\partial z)$, $\nabla^2 \equiv \nabla \cdot \nabla$, $e_3 \equiv (0, 1)$, $t \equiv$ time, $(x, z) \equiv$ two-dimensional position; with boundary conditions

$$w = \frac{\partial u}{\partial z} + \frac{\partial w}{\partial x} = 0 \text{ at } z = 0 \text{ and } d; \ T = 0 \text{ at } z = 0, \ T = -\Delta T < 0 \text{ at } z = d.$$

Introducing the reduced pressure

$$p = p' - p_0'(z) \text{ where } p_0'(z) = p_A - \rho_0 g z,$$

the momentum and energy equations are transformed into

$$\frac{D\boldsymbol{v}}{Dt} = -\left(\frac{1}{\rho_0}\right)\boldsymbol{\nabla}p + \alpha g T \boldsymbol{e}_3 + \nu\nabla^2\boldsymbol{v},$$

$$\frac{DT}{Dt} + \frac{gw}{C_p} = \kappa\nabla^2 T;$$

upon neglect of Dp/DT and Φ in the energy equation, where $\nu = \mu_0/\rho_0 \equiv$ kinematic viscosity and $\kappa = k_0/(\rho_0 C_p) \equiv$ thermal diffusivity since

$$\boldsymbol{\nabla}p' = \boldsymbol{\nabla}p + \boldsymbol{\nabla}p_0' = \boldsymbol{\nabla}p - \rho_0 g \boldsymbol{e}_3 \text{ and } \frac{Dp'}{Dt} = \frac{Dp}{Dt} + \frac{Dp_0'}{Dt} = \frac{Dp}{Dt} - \rho_0 g w.$$

There exists a pure conduction solution of these equations and boundary conditions given by

$$\boldsymbol{v} = \boldsymbol{0} = (0,0), \ T = T_0(z) = -\beta z \text{ for } \beta = \frac{\Delta T}{d}, \ p = p_0(z) = -\frac{\rho_0 \alpha \beta g z^2}{2}.$$

Seeking a solution of this system of the form

$$\boldsymbol{v} = \boldsymbol{0} + \boldsymbol{v}_1, \ p = p_0(z) + p_1, \ T = T_0(z) + T_1,$$

where $\boldsymbol{v}_1 = (u_1, w_1)$ and nondimensionalizing position, time, velocity, pressure, and temperature by d, d^2/κ, κ/d, $\kappa\mu_0/d^2$, and $\Delta\theta = \Delta T - gd/C_p > 0$, respectively, we obtain

$$\boldsymbol{\nabla} \cdot \boldsymbol{v}_1 = 0,$$

$$Pr^{-1}\frac{D\boldsymbol{v}_1}{Dt} = -\boldsymbol{\nabla}p_1 + RT_1\boldsymbol{e}_3 + \nabla^2\boldsymbol{v}_1,$$

$$\frac{DT_1}{Dt} = w_1 + \nabla^2 T_1;$$

$$w_1 = \frac{\partial u_1}{\partial z} + \frac{\partial w_1}{\partial x} = T_1 = 0 \text{ at } z = 0 \text{ and } 1;$$

where $Pr = \nu/\kappa$, $R = g\alpha d^3 \Delta\theta/(\kappa\nu)$, and the same nomenclature has been employed for the dimensionless variables as that used for the original ones. This set of equations and boundary conditions is equivalent to that deduced by Spiegel and Veronis ([234]), as appropriate for the study of convection in a single layer of compressible gas, and would be identical to the usual perturbation system corresponding to a Boussinesq liquid if it were not for the presence of the adiabatic gradient term $-gd/C_p$ in the $\Delta\theta$ of the Rayleigh number R, as just defined. We shall exploit this similarity in what follows by merely summarizing existing stability results for a two-dimensional situation relevant to the classical Rayleigh-Bénard problem (Segel, [213]), while using this modified R as the bifurcation parameter instead of the normal one with $\Delta\theta = \Delta T$ traditionally employed for that purpose.

Seeking a normal-mode solution of this system of the form

$$u_1(x, z, t) = A\sin(qx)\cos(\pi z)e^{\sigma t},$$

$$[w_1, T_1]x, z, t = [B, C]\cos(qx)\sin(\pi z)e^{\sigma t},$$

$$p_1(x, z, t) = E\cos(qx)\cos(\pi z)e^{\sigma t};$$

after neglecting nonlinear terms in the perturbation quantities, we obtain the secular equation

$$k^2 Pr^{-1}\sigma^2 + k^4(Pr^{-1}+1)\sigma + k^6 - Rq^2 = 0 \text{ where } k^2 = \pi^2 + q^2$$

governing the linear stability of the pure conduction state. From an analysis of this equation, similar to that employed in Chapter 2, we can conclude that the onset of convective instability occurs should

$$R > R_0(q^2) = \frac{k^6}{q^2}$$

or, since this marginal curve has a minimum point (q_c^2, R_c), where $dR_0(q_c^2)/dq^2 = 0$ given by

$$q_c^2 = \frac{\pi^2}{2} \text{ and } R_c = R_0(q_c^2) = \frac{27\pi^4}{4},$$

whenever the temperature difference in the layer exceeds the critical value (Nathenson, [155])

$$\Delta T > (\Delta T)_c(d) = \frac{gd}{C_p} + \frac{27\pi^4 \kappa\nu}{4g\alpha d^3}$$

or

$$\Delta T > (\Delta T)_c(d) = \kappa \left[\frac{\rho_0 gd}{k_0} + \frac{27\pi^4 \nu \overline{T}_0}{4gd^3} \right]$$

once one recalls that $\kappa C_p = k_0/\rho_0$ and $\alpha = 1/\overline{T}_0$, the later being a relationship satisfied by Boussinesq gases or dispersed particles, use of which has been made earlier in this chapter.

Next, seeking a weakly nonlinear solution of our perturbation system of the form

$$w_1(x,z,t) \sim A(t)w_{11}(z)\cos(q_c x) + A^2(t)[w_{20}(z) + w_{22}(z)\cos(2q_c x)]$$
$$+ A^3(t)[w_{31}(z)\cos(q_c x) + w_{33}(z)\cos(3q_c x)]$$

where $dA(t)/dt \sim \sigma A(t) - a_1 A^3(t)$, with analogous expansions for the other dependent variables, all of which are pivoted about the critical point of linear stability theory, we find that:

$$w_{11}(z) = \sin(\pi z), \ w_{20}(z) = w_{22}(z) \equiv 0;$$
$$\sigma \sim \sigma_1(R - R_c) \text{ for } \sigma_1 = \frac{2}{9\pi^2(Pr^{-1}+1)},$$
$$a_1 = \frac{1}{8(Pr^{-1}+1)}.$$

Having catalogued these nonlinear stability results, we return to the linear instability criterion and examine its behavior with respect to layer depth, especially in reference to Chandra's ([29]) experiments. When plotted schematically in Fig. 6.9 versus d, the critical curve $(\Delta T)_c(d)$ for this inequality condition exhibits a minimum value at the layer depth d_0 where $(\Delta T)_c'(d_0) = 0$, which implies that $\rho_0 g/k_0 = 81\pi^4\nu\overline{T}_0/(4gd_0^4)$ or

$$d_0 = 3\pi \left(\frac{k_0\nu\overline{T}_0}{4\rho_0 g^2} \right)^{1/4},$$

and consists of two terms which represent the adiabatic gradient and normal Rayleigh number effect, respectively, the latter predominating for the limiting case of thin layers and the former, for thick ones. The significance of d_0 in this context is that it serves as a rough line

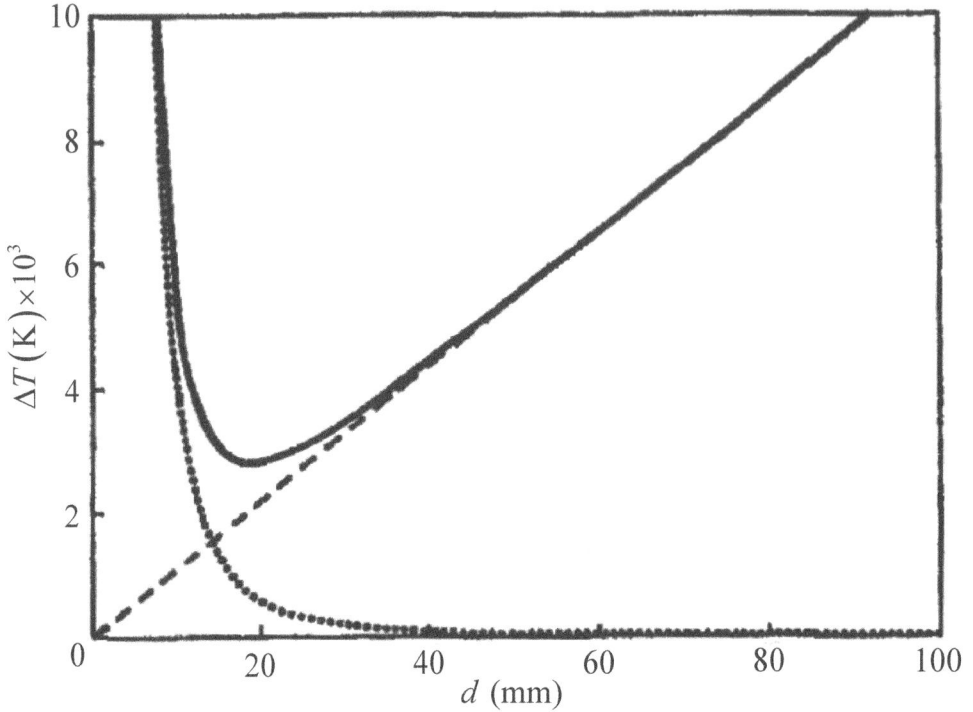

FIGURE 6.9

Schematic plot of the marginal stability curve $(\Delta T)_c(d)$ in the d-ΔT plane for the Rayleigh-Bénard model under the compressible gas version of the Boussinesq approximation. Here, the dashed and dotted asymptotes represent the adiabatic gradient and normal Rayleigh number terms, respectively. This figure has been plotted for $k_0 = 1700$ gm cm/(sec³ K) and $\kappa = \nu$ (Covey and Schubert, [41]), with all the other relevant parameters assigned the same values as employed in the text. Under these conditions, our formulae for $(\Delta T)_c(d)$ and d_0 are equal to $(\Delta T)_c(d) = 1.05 \times 10^{-4}$ (K/cm) $d + 4.54$ (K cm³)/d^3 and $d_0 = 19$ cm, respectively. The region of instability lies above this marginal stability curve and that of stability, below it.

of demarcation separating these two cases from each other. When particularized to convection experiments involving air at sea level by assigning the relevant material properties the following values appropriate for this situation (Ferm and Wollkind, [59])

$$k_0 = 2551\frac{\text{gm cm}}{\text{sec}^3\text{K}}, \ \nu = 0.152\frac{\text{cm}^2}{\text{sec}}, \ \overline{T}_0 = 293 \text{ K},$$
$$\rho_0 = 0.0012\frac{\text{gm}}{\text{cm}^3}, \ g = 980\frac{\text{cm}}{\text{sec}^2};$$

our formulae for the linear instability criterion and critical layer thickness, respectively, become

$$\Delta T > (\Delta T)_c(d) = 9.68 \times 10^{-5} \left(\frac{\text{K}}{\text{cm}}\right) d + 6.27\frac{\text{K cm}^2}{d^3} \text{ and } d_0 = 21 \text{ cm};$$

once we take the corresponding value of $\kappa = 0.210 \text{ cm}^2/\text{sec}$ (it can be shown that the k_0 value used here is equivalent to the one introduced earlier for air by employing the conversion factor of 1 cal $\equiv 4.182 \times 10^7$ gm cm^2/sec^2 at 293 K and atmospheric pressure). This instability region is depicted in the d-ΔT plane of Fig. 6.9 from which we can conclude that the adiabatic gradient effect is negligible for the sea-level air-layer convection experiments under consideration, since Chandra ([29]) conducted the latter at layer depths 4 mm $\leq d \leq$ 16 mm. Hence, we neglected the gw/C_p term in the gas energy equation for our thermal disequilibrium model, as well as the analogous adiabatic gradient term gs_3/C_0 in its particle energy equation, given that $C_0 = 4C_p$. Further, since the total energy equation employed for our thermal equilibrium model was basically a sum of these two energy equations, those terms were also neglected in that model. Finally, such neglect concomitantly reduced the $\Delta\theta$ in our Rayleigh number R to ΔT and thus, that quantity became $R = g\alpha d^3 \Delta T/(\kappa\nu)$, as usual.

Even before Limin's thesis research was completed, I presented its linear stability results, due to their significance, at the 1991 Annual Meeting of the American Physical Society Division of Fluid Dynamics held at the end of November in Scottsdale, Arizona. Recall that Scanlon and Segel ([200]), the best then existing aerosol convection result, was also only a linear stability analysis. After my talk, Russell Donnelly of the University of Oregon, perhaps the most eminent US fluid dynamicist, congratulated me on our achievement by saying it looked like we had finally resolved this long-standing discrepancy between theory and experiment. Once we had completed the research phase of Limin's thesis by finishing its weakly nonlinear stability results, the rest of the dissemination phase could begin. Given that Chandra's ([29]) and Sutton's ([242]) papers had appeared in the *Proceedings of the Royal Society of London*, I submitted our two manuscripts on the thermal equilibrium and disequilibrium models of the gas-smoke aerosol Rayleigh-Bénard problem to this journal. Due to Donnelly's endorsement, I did not anticipate having any trouble getting our results published, but that judgement proved to be overly optimistic. Reviewer number one who refereed both papers was very favorably impressed with our results and accepted them, suggesting a few minor revisions. Reviewer number two refereed the first paper alone rejecting it because the model was "wrong." That individual did not like the particle Boussinesq approximation, preferring the introduction of a conservation of mass particle continuity equation $D_s N/D_s t + N\nabla_2 \cdot s = 0$, where $N \equiv$ the particle number density is now being considered as a dynamical variable, and objected to the presence of a particle pressure term in its momentum equations. The latter felt that, instead of our pure conduction one, the exact solution should be a particle settling situation $s \equiv s_0 e_3$ of uniform density $N \equiv N_0$, which necessitated the retaining of a Joule heating term of the form $\chi_0 N(v - s) \cdot (v - s)$

on the right-hand side of the total energy equation, and anticipated a prediction of particle clustering (variations in N) to interact with drag. The managing editor requested we submit revisions of our papers which complied with the suggestions of the referees. I spoke to that managing editor by phone and asked her what she wanted us to do since one reviewer accepted both our papers, while the other rejected the first paper. She asked me if we could refute the claims of the second reviewer and when I answered, "yes," said to add a section to the first paper doing so. Then she told me to resubmit both papers, while in addition complying with the first reviewer's suggestions. It took two weeks to write that extra section of the first paper. Under the conditions postulated by the second reviewer, a linear stability analysis showed the perturbation particle number density to be identically zero, which precluded the predicted particle clustering from actually occurring. We closed the first paper by offering two possibilities of how to modify the thermal equilibrium model so that it would agree with Chandra's ([29]) observations. When the particle settling approach did not work, we suggested a thermal disequilibrium model instead as a rationale for the second paper. After resubmitting our revised papers, I got a letter from the editorial board of the journal stating that they had looked over all the correspondence on the matter and requested we consolidate the two papers into a single paper of the same length as either one, while removing most of the details of the extra section their managing editor had just asked us to add. Given all the effort we had expended in adding this section, I considered that unreasonable and retracted both our papers. I have learned a lot in the intervening years and now if faced with similar circumstances would gladly comply with those demands. Indeed the exposition summarizing aerosol convection contained in this chapter has been written in the spirit of such consolidation.

Having retracted our papers, I was faced with the problem of exactly where to resubmit them and settled on the *SIAM Journal on Applied Mathematics*. Once having made this decision, I submitted both papers to *SIAP*, as it calls itself, in exactly the same versions that had been retracted from the *Proceedings of the Royal Society of London*. The handling editor in question was Bernie Matkowsky of Northwestern University. *SIAP* also used two referees and things actually turned out worse than before. The first referee by self-identification was reviewer number 2 from my *Royal Society* experience and without even reading the manuscripts assumed they were identical to the ones previously rejected by the latter. Hence that referee's review merely reiterated in more detail the criticisms levied in the original *Royal Society* report. My rebuttal to this review was that if the referee had even bothered to read the manuscripts it would have been obvious we had shown that the suggested modifications of our original analysis led to nonclustering results for the thermal equilibrium problem and should be discarded, while the assertion that our model was wrong for the thermal disequilibrium problem flew in the face of the fact that its predictions were in very close quantitative agreement with Chandra's ([29]) experimental data. That reviewer's responses to my rebuttal were as follows: Yes, the original suggestions were indeed incorrect as had been shown, but the model was still wrong because, instead of $\mathcal{P} \equiv 0$, we should have adopted the other usual assumption of two-phase flow $\mathcal{P} = p$, while agreement of theoretical predictions with experimental or observational data was the worst way to validate a model, especially when it contained a large number of parameters to be identified. Let me offer a critique of these responses. First of all, a reviewer should have only one bite at the apple. If the original suggestions are incorrect, that reviewer does not get to keep making new ones. In other words, it is unfair to have an author try to hit a moving target in the review process. Secondly, the reviewer's assertion involving validation of a model is at variance with the basic philosophy of comprehensive applied mathematical modeling, while the thermal equilibrium model is an object lesson in the inability of a model to fit experimental data merely because it has a large number of parameters. Bernie also requested a review from a second referee, whose evaluation of our papers was confined to complaining about the complexity of the sentence structure rather than offering any opinion on scientific merit. Given these reviews, Bernie had little choice but to reject our papers for publication although subsequent events as related below will show that he was none too happy about the decision being forced upon him. In an attempt to vindicate us, I next decided on a two pronged approach by submitting both papers to the *Pergamon Journal Mathematical and Computer Modelling*, which had published some of my earlier research, while sending them, along with their complete editorial review correspondence from the previous submittals, to my friend John Ockendon, the editor of the *European Journal on Applied Mathematics*, so that he might offer me his commentary on the whole matter. *Mathl. Comput. Modelling* accepted both papers immediately and put them on the fast track for publication where they appeared consecutively as Wollkind and Zhang ([299, 300]), while John gave my total package to Alistair Fitt, judged by him to be one of the UK's leaders in two-phase flow, who, by tending to side with the authors rather than the reviewers, vindicated us. Alistair was not overly impressed by the review process stating that two-phase flow experts should know aerosol particles being charged of the same polarity would repel each other rather than clumping together and that this represented a real horror story in reviewing, while also commenting negatively on the second *SIAP* reviewer critiquing only the papers' language!

Right after these papers had been published, I got an invitation from Bernie Matkowsky to give a talk on any subject of my choice in a reaction-diffusion partial differential equations session at an Amsterdam conference honoring the retirement of Wiktor Eckhaus from Utrecht University. Professor Eckhaus was a Polish-Dutch mathematician, whose *Studies in Nonlinear Stability Theory*, published by Springer in 1965, is considered a seminal book in this field. I first met him during the summer of 1967 when he visited the Department of Mathematics at RPI to work with DiPrima and Segel. The research they started that summer eventually appeared as DiPrima et al. ([51]), an omnibus paper on weakly nonlinear stability theory published in the *Journal of Fluid Mechanics*. There is an incident which occurred that summer, involving Eckhaus, I can't help but recalling any time someone lauds an individual from their organization for a future achievement by using the phrase that concludes my following description of this event:

> On August 25, 1967, George Lincoln Rockwell was assassinated in Arlington, Virginia. The next day George Habetler, a functional analyst in the Department of Mathematics at RPI, announced this fact to an assembled group, including Eckhaus. When Wiktor asked Habetler who Rockwell was, George replied: "You've never heard of George Lincoln Rockwell, the head of the American Nazi Party? Why he was one of our boys who made good!" Eckhaus, having been living in Poland during its 1939 invasion by Nazi Germany, did not find this joke very amusing.

Returning to my choice of a talk for the Eckhaus conference, I decided to present our results on the thermal disequilibrium gas-smoke aerosol model at the session in question, because its coupled energy equations were of the reaction-diffusion type, the subject of this session, -*i.e.*,

Gas:

$$\frac{DT}{Dt} = \overbrace{\frac{H_0}{\rho_0 C_p}(\theta_d - T)}^{\text{reaction}} + \overbrace{\kappa \nabla_2^2 T}^{\text{diffusion}}$$

and

Smoke:

$$mN_0C_0\frac{D_s\theta_d}{D_st} = \underbrace{H_0(T - \theta_d)}_{\text{reaction}} + \underbrace{k_d\nabla_2^2\theta_d}_{\text{diffusion}};$$

plus, those results were of historical significance in nonlinear stability theory. Segel and, naturally Matkowsky, who along with Eckhaus' postdoctoral associate Rachel Kuske, had organized that session, attended my talk, and praised the work. Bernie said he always knew it would eventually be published and was glad things had worked out so well in that regard. I have always thought my invitation to this conference was his way of making up for the unfortunate *SIAP* experience.

A few years later I invited Bernie to give our annual T. G. Ostrom Lecture on pattern formation during the combustion of gases, his being one of the world's leading experts in this area. More recently, while J. Iwan D. Alexander was still at Case-Western University, I also invited him to give an Ostrom Lecture. Iwan spoke on renewable energy systems. Following a summary of his educational and professional background, including Iwan's research accomplishments, I finally introduced him as "one of our boys who made good," after having related the Habetler-Eckhaus incident described above. That is one particular Rabbit I never get tired of Pulling Out of a Hat!

Since Limin Zhang was being supported as a research assistant on my ONR grant, he did not have to teach classes while working on his Ph.D. thesis. Thus, Limin was able to finish that dissertation research faster than most of my other Ph.D. students and hence had some time to complete work on other problems before graduating. One of those problems was a nonlinear stability analysis of a prototype model for Rayleigh-Bénard convection in planetary atmospheres. We considered the following nondimensionalized perturbation system relevant to the pure conduction solution for Rayleigh-Bénard convection in upper planetary atmospheres under the compressible gas version of the Boussinesq approximation (Wollkind and Zhang, [301])

$$\frac{\partial u_1}{\partial x} + \frac{\partial w_1}{\partial z} = 0,$$

$$\frac{Du_1}{Dt} = -\frac{\partial p_1}{\partial x} + \left(m\frac{\partial^2}{\partial x^2} + \frac{\partial^2}{\partial z^2}\right)u_1,$$

$$\frac{Dw_1}{Dt} = -\frac{\partial p_1}{\partial z} + RT_1 + \left(m\frac{\partial^2}{\partial x^2} + \frac{\partial^2}{\partial z^2}\right)w_1,$$

$$\frac{DT_1}{Dt} + \frac{T_1 + \eta T_1^3}{\tau} = w_1 + \left(m\frac{\partial^2}{\partial x^2} + \frac{\partial^2}{\partial z^2}\right)T_1;$$

$$w_1 = \frac{\partial u_1}{\partial z} + \frac{\partial w_1}{\partial x} = T_1 = 0 \text{ at } z = 0 \text{ and } 1.$$

Here, the gas has perturbation velocity components (u_1, w_1), pressure p_1, and temperature T_1, which are functions of the spatial coordinates (x, z) and time t. Further, R is the compressible gas Rayleigh number defined by $R = gd^3\Delta\theta/(\kappa^2\overline{T}_0)$ and, to account for the role of eddies, an eddy kinematic viscosity and thermometric conductivity, which are both assumed equal, have been introduced where these diffusivities are anisotropic with $m \equiv$ the ratio of their horizontal to vertical components, the latter $\equiv \kappa$, while $\tau = \tau_0\kappa/d^2$ for $\tau_0 \equiv$ radiative cooling time.

We sought a normal-mode solution of a linearized version of this system of the same form as that employed earlier for our compressible gas Boussinesq approximation model and obtained the secular equation

$$k_1^2 \left[\sigma^2 + \left(2k_m^2 + \frac{1}{\tau} \right) \sigma + k_m^2 \left(k_m^2 + \frac{1}{\tau} \right) \right] - Rq^2 = 0$$

where

$$k_m^2 = \pi^2 + mq^2 = k_m^2(\lambda) = \pi^2 \left(1 + \frac{4m}{\lambda^2} \right) \text{ with } \lambda = \frac{2\pi}{q}.$$

Upon analyzing this quadratic, we concluded that $\sigma \in \mathbb{R}$, since its discriminant satisfied

$$\mathfrak{D} = k_1^4 \left(2k_m^2 + \frac{1}{\tau} \right)^2 - 4k_1^4 k_m^2 \left(k_m^2 + \frac{1}{\tau} \right) + 4k_1^2 q^2 R = \frac{k_1^4}{\tau^2} + 4k_1^2 q^2 R > 0$$

and the marginal curve for the onset of convective instability was given by

$$R = R_m(\lambda) = \pi^4 \left(\frac{\lambda^2}{4} + 1 \right) \left(1 + \frac{4m}{\lambda^2} \right) \left(1 + \frac{4m}{\lambda^2} + \frac{1}{\tau_1} \right)$$

with $1/\tau_1 = 1/(\pi^2 \tau)$.

Before examining these marginal stability curves, we wish to compare this model with that of Covey and Schubert ([41]), whose linear stability analysis of mesoscale convection in the clouds of Venus' atmosphere motivated our investigation. We considered their heat transfer radiation boundary conditions for $\theta_1(x, \xi, t) \equiv T_1(x, \xi + 1/2, t)$ given by

$$\frac{\partial \theta_1}{\partial \xi} \pm \gamma \theta_1 = 0 \text{ at } \xi = \pm \frac{1}{2}$$

in the limit, as its black body radiation coefficient $\gamma \to \infty$, consistent with Goody's ([71]) treatment of such black body boundaries as isothermal ($\theta_1 = 0$) pure conductors. This simplification allowed us to employ analytical rather than numerical solution methods while replicating the behavior of Covey and Schubert's ([41]) linear stability results, which were obtained numerically for $\gamma = 1$. Besides that assumption, our convection model in planetary atmospheres was a nonlinearized alteration of their model with the chief changes being the presence of D/Dt instead of $\partial/\partial t$ and of the cubic component of the radiative cooling term proportional to η. Covey and Schubert ([41]) wished to explain why the convection cells in the atmosphere of Venus are much wider than they are deep, often by a factor of almost 50, while those produced in the laboratory are always about as wide as they are high. Their linear stability analysis predicted aspect ratios in accordance with those characteristics of cells occurring in internal adiabatic layers of Venus' upper atmosphere when the anisotropic eddy diffusivity effect was sufficiently strong. They concluded by underscoring the importance of studying nonlinear effects. Since no nonlinear analysis of such atmospheric convection had yet been attempted, we wished to alleviate this deficiency, at least in part, by posing the simplest reasonable nonlinear atmospheric convection model that preserved the salient features of Covey and Schubert's ([41]) linear stability results, while allowing us to obtain analytical rather than numerical solutions for both the linear and weakly nonlinear stability problems associated with this model.

Toward that end, we now return to our marginal stability curve $R = R_m(\lambda)$ and plot it in Fig. 6.10 for the gas layer parameters (Covey and Schubert, [41])

$$\kappa = 10^6 \frac{\text{cm}^2}{\text{sec}}, \ \tau_0 = 10^7 \text{ sec}, \ d = 10^6 \text{ cm} \Rightarrow \tau = 10$$

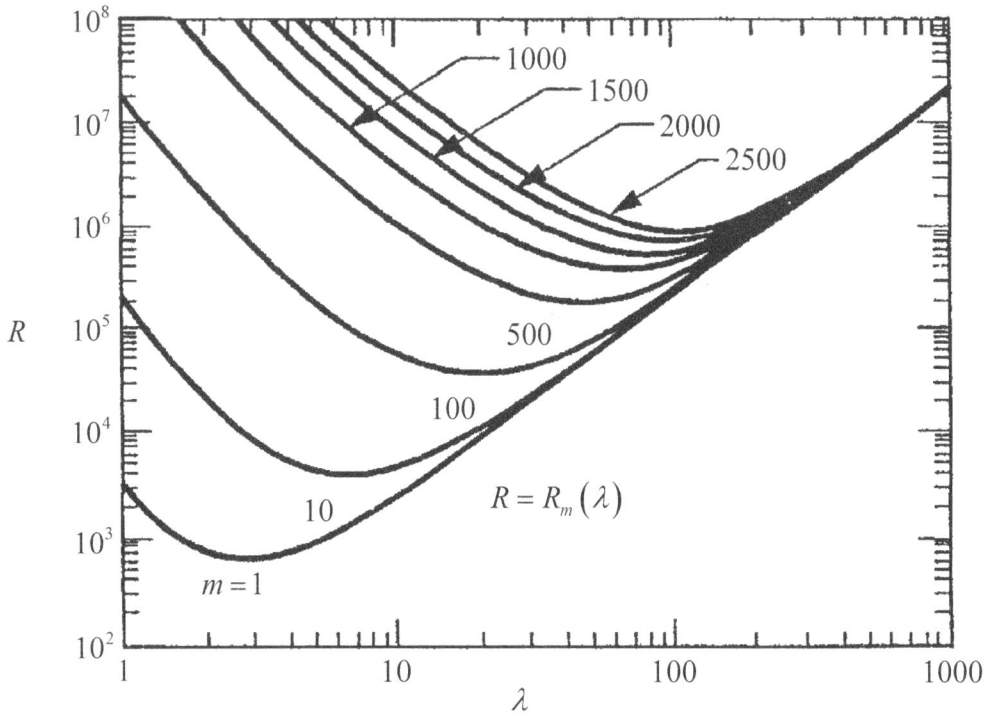

FIGURE 6.10

Plots of the marginal stability curves $R = R_m(\lambda)$ for the planetary atmosphere model system with $\tau = 10$ and $m = 1, 10, 10^2, 5 \times 10^2, 10^3, 1.5 \times 10^3, 2 \times 10^3$, and 2.5×10^3.

and for the different eddy anisotropic diffusivity ratios

$$m = 1, 10, 100, 500, 1000, 1500, 2000, \text{ and } 2500.$$

These marginal stability curves of Fig. 6.10 each have a minimum point at (λ_c, R_c), the coordinates of which $\lambda_c = \lambda_c(m)$, $R_c = R_c(m)$ are plotted in Fig. 6.11a,b, respectively. Figure 6.12a is an enlargement of the $1 \leq \lambda \leq 300$ and $10^2 \leq R \leq 10^6$ portion of Fig. 6.10 restricted to $m = 1$, 10, 100, and 1,000 in accordance with those values employed by Covey and Schubert ([41]), whose corresponding results when $\gamma = 1$ are plotted in Fig. 6.12b. Upon comparison of the two parts of this figure, we see that our pure conducting boundary simplification $(\gamma \to \infty)$ yields critical conditions involving (λ_c, R_c) of the respective curves, which are quite similar to those obtained by Covey and Schubert ([41]) for $\gamma = 1$ and this replication that allowed us to employ analytical solution methods represents yet another Rabbit being Pulled Out of the Hat!

More generally, Covey and Schubert ([41]) allowed κ to vary by considering

$$\tau = \frac{\kappa}{\left(10^5 \frac{\text{cm}^2}{\text{sec}}\right)}, \quad \gamma = \frac{\left(10^6 \frac{\text{cm}^2}{\text{sec}}\right)}{\kappa}$$

and found that $\lambda_c(m)$ increased by a factor between two to three, as κ ranged from 10^5 cm^2/sec to 10^8 cm^2/sec.

An analogous procedure involving τ applied to our problem produced a critical aspect ratio which behaved in a similar manner for 0 cm^2/sec $\leq \kappa \leq 10^5$ cm^2/sec, but remained virtually unchanged for 10^5 cm^2/sec $\leq \kappa \leq 10^8$ cm^2/sec. Given this invariancy, $\lambda_c(m)$ of Fig. 6.11a is equally valid for any value of κ in that interval, the latter range being characteristic of the atmospheric convection layers investigated by Covey and Schubert ([41]). In order to determine which of these combinations of m and κ values, represented by Fig. 6.11a, were actually physically allowable, we calculated after Covey and Schubert ([41]) the minimum heat transport due to convection predicted by our results. For instance using our parameter values of Fig. 6.10 with $m = 100$, we found that $R_c = 4 \times 10^4$ and when, in addition, taking

$$g = 862.4 \frac{\text{cm}}{\text{sec}^2}, \quad \overline{T}_0 = 270 \text{ K}, \quad \rho_0 = 0.5 \times 10^{-3} \frac{\text{gm}}{\text{cm}^3}, \quad C_p = 8.6 \times 10^6 \frac{\text{cm}^2}{\text{sec}^2\text{K}},$$

deduced that

$$\beta_c = \frac{\Delta\theta_c}{d} = \frac{\kappa \overline{T}_0 R_c}{g d^4} = 12.5 \times 10^{-9} \frac{\text{K}}{\text{cm}}$$

which is the minimum superadiabaticity required to sustain convection in this situation and implies a minimum convective heat flux of

$$\rho_0 C_p \kappa \beta_c = \frac{\rho_0 C_p \overline{T}_0 \kappa^3 R_c}{g d^4} = 5.4 \times 10^{-2} \frac{\text{W}}{\text{m}^2}$$

where $\text{W} \equiv 1 \text{ watt} = 1 \times 10^7$ gm cm^2/sec^3 and m $\equiv 1$ meter $= 100$ cm in cgs units.

Increasing κ for a fixed value of m will rapidly produce heat fluxes exceeding the downward radiative flux at the cloud levels of Venus estimated as 100 W/m^2 from Pioneer probe data (Covey and Schubert, [41]), a nonphysical result. We showed that this threshold is reached for $\kappa_c \cong 1.33 \times 10^7$ cm^2/sec, when $m = 100$ by means of Fig. 6.13, which displays the outcome of calculations analogous to that just completed, but for different m and κ. Thus,

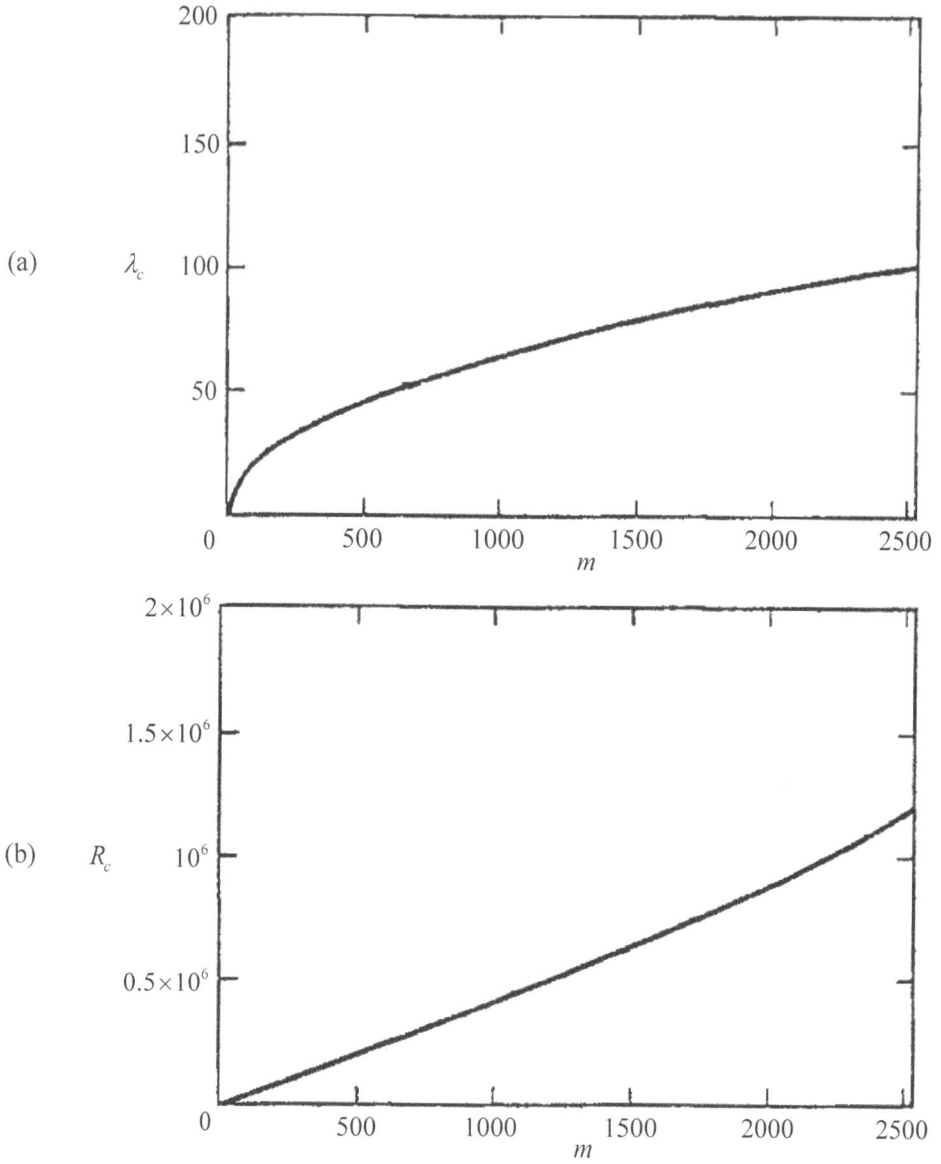

FIGURE 6.11
Plots of (a) $\lambda_c(m)$ and (b) $R_c(m)$ relevant to the coordinates of the critical points of the plots for $R_m(\lambda)$ in Fig. 6.10.

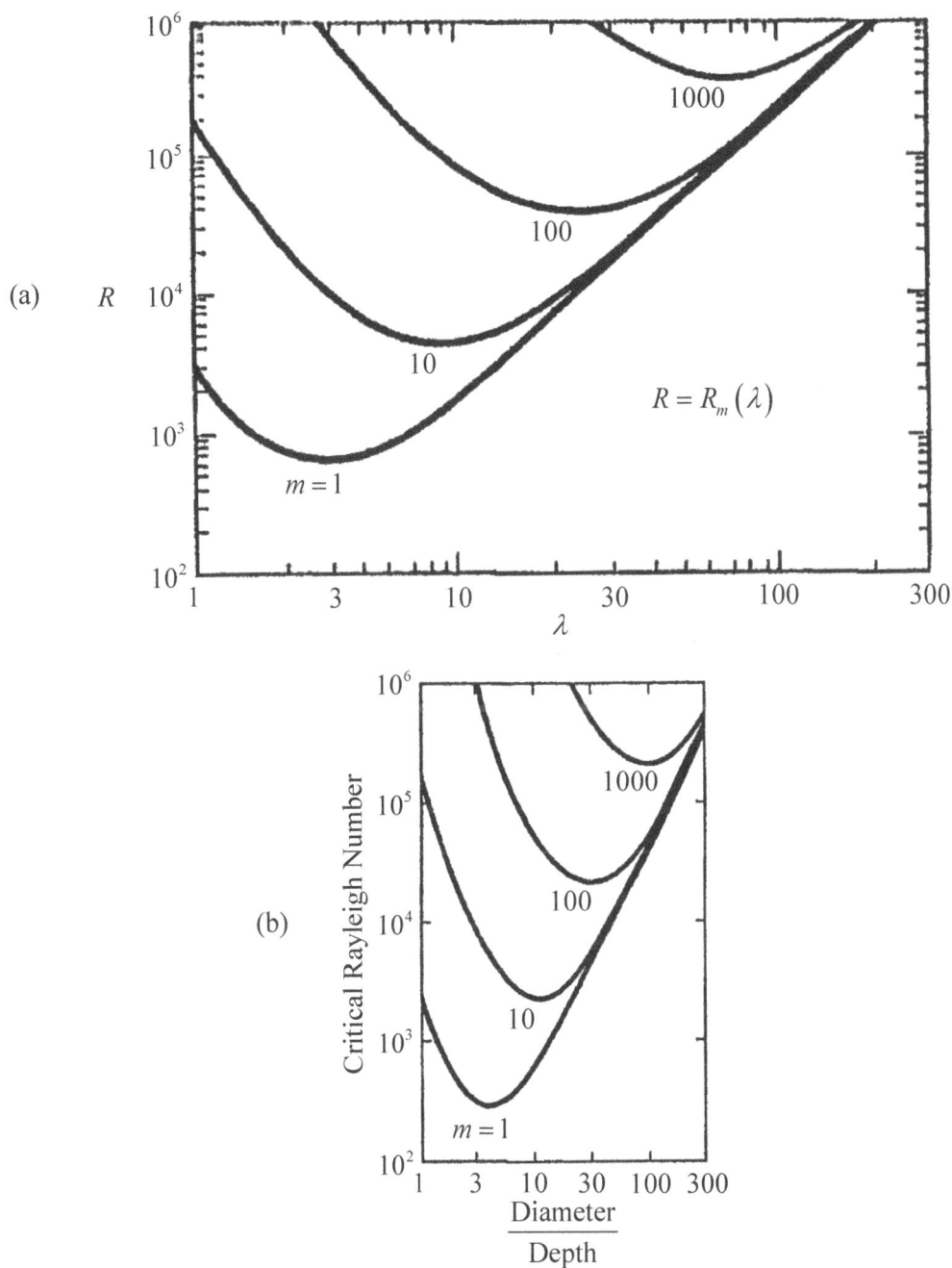

FIGURE 6.12
(a) Enlargement of the $1 \leq \lambda \leq 300$ and $10^2 \leq R \leq 10^6$ portion of Fig. 6.10 for $m = 1$, 10, 100, and 1,000. (b) Corresponding plot with $\tau = 10$ and $\gamma = 1$ from Covey and Schubert ([41]), as adapted by Wollkind and Zhang ([301]).

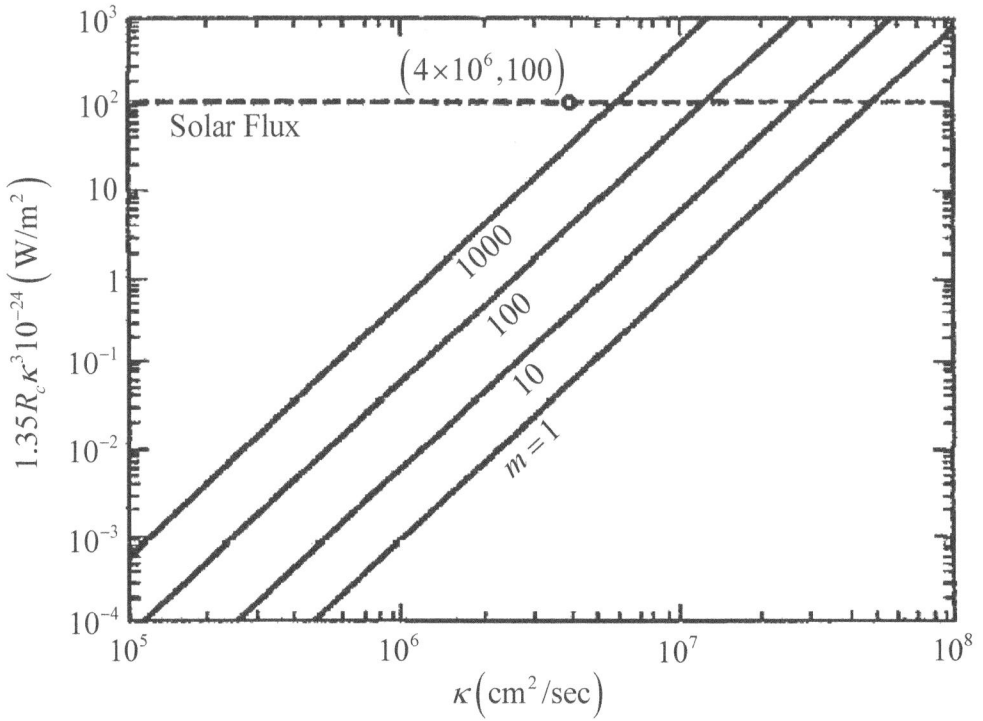

FIGURE 6.13
Plots of the critical heat flux required to maintain convection versus κ for those m-values employed in Fig. 6.12a and $\tau = \kappa/(10^5 \text{ cm}^2/\text{sec})$. Here, the point $(4 \times 10^6, 100)$ is associated with $m = 2,512$.

combinations of $m = 100$ and $\kappa > 1.33 \times 10^7$ cm^2/sec have been excluded even though they provide aspect ratios $\lambda_c \cong 19$ which are in the 10-100 range consistent with observation (Covey and Schubert, [41]). In this context, we further noted from Fig. 6.11 that

$$\lambda_c(23) = 10 \text{ and } R_c(23) = 10^4$$

while

$$\lambda_c(2, 512) = 100 \text{ and } R_c(2, 512) = 1.2 \times 10^6.$$

The corresponding thresholds $\kappa_c(m)$ for these values of m were given by

$$\kappa_c(23) = 2 \times 10^7 \frac{\text{cm}^2}{\text{sec}} \text{ and } \kappa_c(2, 512) = 4 \times 10^6 \frac{\text{cm}^2}{\text{sec}}.$$

Since our primary purpose for investigating this problem was to perform a weakly nonlinear stability analysis of its pure conduction solution, we next considered an expansion of the same form as that sought during the corresponding investigation of our compressible gas Boussinesq approximation model, where $q_c = 2\pi/\lambda_c$ and $\lambda_c = \lambda_c^*/d$, λ_c^* being the dimensional critical wavelength of the disturbance, which demonstrates conclusively that λ_c represents the latter's aspect ratio of diameter to depth. We still obtained $w_{11}(z) = \sin(\pi z)$, $w_{20}(z) = w_{22}(z) \equiv 0$; but the coefficients of the truncated amplitude equation were given in this case by

$$\sigma \sim \sigma_1(R - R_c) \text{ for } \sigma_1 = \frac{1}{(\lambda_c^2/4 + 1)[2k_m^2(\lambda_c) + 1/\tau]}$$

and

$$a_1 = k_m^2(\lambda_c) \frac{(9/16)(\eta/\tau)/[k_m^2(\lambda_c) + 1/\tau]^2 + (\pi^2/2)/(4\pi^2 + 1/\tau)}{2k_m^2(\lambda_c) + 1/\tau}.$$

We have deferred until now a discussion of the form of the radiative cooling term in the energy equation of our dimensionless perturbation system: Namely

$$\frac{r(T_1)}{\tau} \text{ where } r(T_1) = T_1 + \eta T_1^3 + O(T_1^5)$$

which contained odd powers alone, as is standard procedure for such source terms, while retaining only those through third-order, consistent with our nonlinear method of stability analysis. It remained for us to assign appropriate values for the parameter η. In making that selection, we were motivated by the form of the commonly chosen source terms catalogued below for the listed representative energy or prototype reaction-diffusion model equations:

Linear (Covey and Schubert, [41]):
$$r(T_1) = T_1.$$

Hyperbolic Sinusoidal (Wollkind and Vislocky, [298]):

$$r(T_1) = \sinh(T_1) = T_1 + \frac{T_1^3}{6} + O(T_1^5).$$

Circular Sinusoidal (Drazin and Reid, [53]):

$$r(T_1) = \sin(T_1) = T_1 - \frac{T_1^3}{6} + O(T_1^5).$$

Comparison of these functions with our radiative cooling term led us to make the assignments

$$\eta = 0$$

or

$$\eta = \pm \frac{1}{6},$$

respectively.

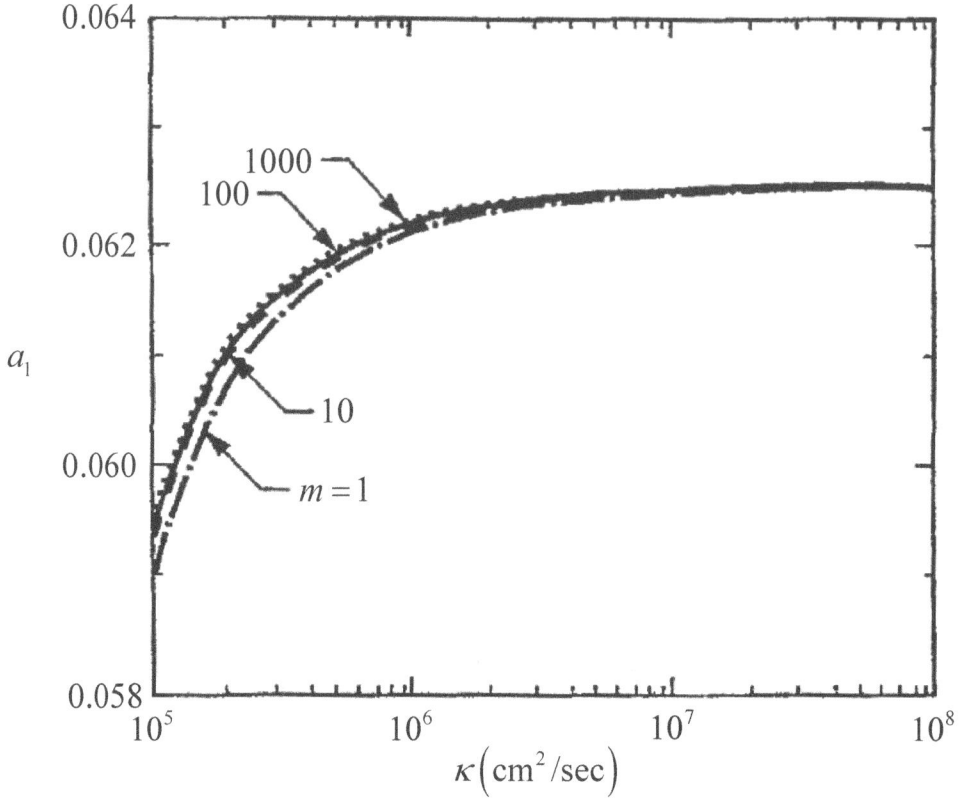

FIGURE 6.14
Plots of a_1 versus κ with $\eta = 0$ corresponding to Fig. 6.13.

Since the dynamical behavior of our truncated amplitude equation depended upon the sign of both its coefficients, it was necessary for us to analyze the expression for the Landau constant a_1 appearing above. We first examined the effect of adopting Covey and Schubert's ([41]) linear radiative cooling term constitutive relation by plotting a_1 versus κ in Fig. 6.14 for $\eta = 0$ and the other parameter values chosen consistent with those of Fig. 6.13. Under these conditions, our expression for a_1 reduced to

$$a_1 = \frac{\pi^2 k_m^2(\lambda_c)}{2(4\pi^2 + 1/\tau)[2k_m^2(\lambda_c) + 1/\tau]} \text{ with } \tau = \frac{\kappa}{10^5 \frac{\text{cm}^2}{\text{sec}}}$$

which is the function plotted in Fig. 6.14. From this figure, we can see that a_1 is a positive monotone increasing function of both κ and m which, for large values of the former, since

$$\frac{1}{\tau} \to 0 \text{ as } \tau \to \infty,$$

approaches the common asymptotic limit

$$a_1 = \frac{\pi^2 k_m^2(\lambda_c)}{16\pi^2 k_m^2(\lambda_c)} = \frac{1}{16} = 0.0625$$

independent of the latter. Next, we investigated the consequences of adopting either one of the other nonlinear constitutive relations by calculating a_1 for $\eta = \pm 1/6$ with $\kappa = 10^6$ cm^2/sec (or $\tau = 10$) and $m = 100$, which are representative values of those parameters for planetary atmospheres (Covey and Schubert, [41]). Comparison of the results of these calculations with the reference value for $\eta = 0$ compiled in Table 6.2 shows that the three virtually coincide.

TABLE 6.2
The Landau constant $a_1 = a_1(\eta)$ for $\tau = 10$ and $m = 100$.

η	a_1
$-1/6$	0.062181923
0	0.062192623
1/6	0.062203323

Given that the standard Rayleigh-Bénard problem for free-free boundaries in the absence of radiative effects (Segel, [213]) had an associated Landau constant $a_1 = 1/[8(Pr^{-1} + 1)]$, equivalent when $Pr = \nu/\kappa = 1$ to our $\eta = 0$ asymptotic value of $1/16$, and our goal of posing the simplest reasonable formulation that predicted the identical sort of re-equilibration for the planetary atmospheric problem, since such a prediction was in accordance with the observations of Covey and Schubert ([41]), we finally adopted their linear constitutive relation by taking $\eta = 0$. Observe from Table 6.2 that, for $\tau = 10$ or $d = 10^6$ cm, the $|\eta| = 1/6$ loci only deviate slightly from the reference locus of $\eta = 0$, as already pointed out, and hence this assumption may be made with no loss of generality for that depth employed by Covey and Schubert ([41]) as representative of adiabatic convection layers in the upper atmosphere of Venus.

Given the behavior of the Landau constant under the condition of $\eta = 0$, as depicted in Fig. 6.14, we were able to conclude that the main result of our two-dimensional nonlinear analysis for this pure conducting boundary atmospheric convection model was the prediction of stable roll-type patterns when $R > R_c(m)$ with aspect ratio $\lambda_c(m)$ valid, should 10^5 cm^2/sec $\leq \kappa < \kappa_c(m)$. In particular, treating this as an inverse problem, we could then anticipate aspect ratios in the desired 10-100 range when the eddy anisotropy diffusivity ratio was from the appropriate interval

$$m_1 = 23 \leq m \leq 2,512 = m_2,$$

the upper and lower bounds of which corresponded to the maximum and minimum values of that parameter required to produce theoretical pattern predictions consistent with observational data. These stable roll-type patterns are diagrammed in Fig. 6.15. Note that the fundamental unit of this pattern consists of two rolls rotating in opposite directions with a total width (diameter) of λ_c^* and an aspect ratio of diameter to depth of λ_c (see Fig. 6.12). Hence, each individual roll cell has an aspect ratio of width to depth of half this value or $\lambda_c/2$ which lies in the observed 5-50 range (Covey and Schubert, [41]) when $10 \leq \lambda_c \leq 100$, explaining why that was the desired interval. Specifically, the adiabatic cloud layers in the atmosphere of Venus tend to be situated between regions of high static stability caused by the radiative heating of their bases and cooling of their tops, are typically located in an altitude range of 50-70 km with a depth of 10 km, and have cellular features on a horizontal scale of 100-1000 km. This allowed Covey and Schubert ([41]) to impose the tangential stress-free assumption as the appropriate dynamical boundary conditions at these internal surfaces within the atmosphere. Further, they stated that the incremental heating in the perturbed state could be approximated by their linear formulation of the radiative cooling term which was valid at cloud levels in the atmosphere of Venus, an assertion directly confirmed by our previous development. Such an approach was also consistent with the introduction of the Boussinesq approximation, which only retained a linear contribution involving the perturbation temperature for density in the body force term of the momentum equations.

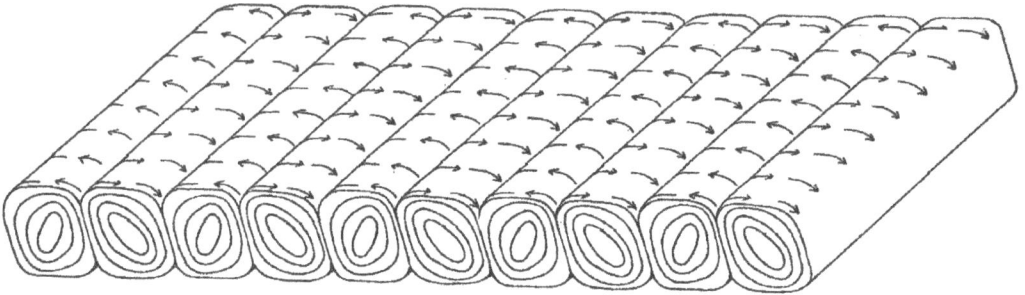

FIGURE 6.15
Schematic diagram of the predicted layer of stable roll-shaped cells. The fundamental unit of this pattern consists of two rolls that rotate in opposite directions. Hence the width of each roll corresponds to $\lambda_c^*/2$ and its depth to d. Thus, its aspect ratio of width to depth is given by $\lambda_c/2$, the same as the results for Koschmieder's ([102]) experiments summarized in Table 5.1.

Finally, this model for planetary atmospheres served as a perfect companion foil problem to Limin's aerosol thesis research, in that it represented an actual situation, where due to the relatively large value of the depth of the convection layer, the adiabatic gradient term had to be retained in its Boussinesq compressible gas Rayleigh number which now was a measure of this so-called superadiabaticity. Just when we had completed that analysis, Lokenath Debnath, whom my technical typist Dana Lohrey referred to as "Deadmath," and S. Roy Choudary of the University of Central Florida asked me to contribute a chapter to their monograph Nonlinear Stability Analysis. So we published these results in Wollkind and Zhang ([301]), which was Chapter 4 of this book. In those days, before the advent of word processors, mathematics departments actually hired technical typists which they no longer do.

At the same time, Limin also helped perform a number of calculations related to, and generated all the new figures for, our review paper with my colleague Valipuram Manoranjan on nonlinear stability analyses of prototype reaction-diffusion model equations (Wollkind *et al.*, [291]). Later, as part of Mano's student Richard Drake's Preliminary Doctoral Examination, I designed a linear stability problem for a special case of this planetary atmosphere Rayleigh-Bénard model system. Since the requisite examination of the marginal curve relevant to the linear stability analysis of this model involves a quadratic rather than a quartic equation for λ_c^2, it is instructive to present that investigation in some detail and I close my description of our planetary atmospheric research by doing so. This special case, already considered when analyzing our Landau constant formula, is for $1/\tau \to 0$, which reduces the marginal stability curve to

$$R = R_m(\lambda) = \pi^4 \left(\frac{\lambda^2}{4} + 1 \right) \left(1 + \frac{4m}{\lambda^2} \right)^2.$$

In order to find the critical point λ_c for this function, which satisfies $R'_m(\lambda_c) = 0$, we rewrite it as

$$R_m(\lambda) = \pi^4 f_m(\Lambda) \text{ where } f_m(\Lambda) = \frac{(\Lambda+1)(\Lambda+m)^2}{\Lambda^2} \text{ for } \Lambda = \frac{\lambda^2}{4}.$$

Now λ_c corresponds to $\Lambda_c = \lambda_c^2/4$ such that $f'_m(\Lambda_c) = 0$. Recalling from the quotient rule that

$$f'_m(\Lambda) = \frac{[(\Lambda+1)(\Lambda+m)^2]'\Lambda^2 - (\Lambda+1)(\Lambda+m)^2[\Lambda^2]'}{(\Lambda^2)^2}$$

and computing from the product rule, $i.e.$, $(fg)' = f'g + fg'$ - that

$$[(\Lambda+1)(\Lambda+m)^2]' = [(\Lambda+1)]'(\Lambda+m)^2 + (\Lambda+1)[(\Lambda+m)^2]'$$
$$= 1(\Lambda+m)^2 + (\Lambda+1)2(\Lambda+m) = (\Lambda+m)(3\Lambda+m+2)$$

while $[\Lambda^2]' = 2\Lambda$, we can deduce that Λ_c satisfies

$$(\Lambda_c + m)(3\Lambda_c + m + 2)\Lambda_c^2 = 2(\Lambda_c + 1)(\Lambda_c + m)^2 \Lambda_c$$

or, upon canceling the common factor $\Lambda_c(\Lambda_c + m)$,

$$3\Lambda_c^2 + (m+2)\Lambda_c = 2(\Lambda_c + 1)(\Lambda_c + m) = 2\Lambda_c^2 + 2(m+1)\Lambda_c + 2m$$

which implies that

$$\Lambda_c^2 = m(\Lambda_c + 2).$$

From this quadratic equation we can obtain the implicit relation that

$$m = \frac{\Lambda_c^2}{\Lambda_c + 2} = \frac{\Lambda_c}{1 + 2/\Lambda_c} = \frac{\lambda_c^2/4}{1 + 8/\lambda_c^2}$$

or, employing the quadratic formula and retaining the "+" root, the explicit relation

$$\Lambda_c = \frac{\lambda_c^2}{4} = \frac{m + \sqrt{m^2 + 8m}}{2}$$

which yields

$$\lambda_c^2(m) = 2m \left(1 + \sqrt{1 + \frac{8}{m}} \right) \Rightarrow \lambda_c(m) = \sqrt{2m \left(1 + \sqrt{1 + \frac{8}{m}} \right)}.$$

Note that $\lambda_c(1) = 2\sqrt{2} = 2.828\dots$ in agreement with $2\pi/q_c$ of Table 6.1 for the free-free case. Also, observe from the implicit relation that $\lambda_c = 10$ and 100 correspond to

$$m_1 = \frac{25}{1.08} = 23.1 \text{ and } m_2 = \frac{2,500}{1.0008} = 2,498,$$

respectively, closely approximating the values of $m_1 = 23$ and $m_2 = 2,512$ deduced earlier.

Limin decided that, in order to broaden his opportunity for prospective jobs, it would be best to take a teaching assistantship in the Department of Mathematics during his last year of graduate work instead of being supported as a research assistant. This actually was a very judicious decision on his part since the year Limin graduated turned out to be a difficult time for obtaining either a postdoctoral or faculty appointment at a research university. WSU has a rule that before foreign students from non-English speaking countries can teach an undergraduate class, they need the approval of the English Department. The granting of this approval is determined by a committee which consists of two members of the English Department and a faculty member from the department offering that course, typically the student's advisor, after the teaching of a typical lesson from the class in question. Limin scheduled his oral presentation during the week preceding this Fall Semester. Since he wanted to teach college algebra classes, I felt Limin should prepare a lesson involving the derivation and application of the quadratic formula. Although use of that formula is employed in both Chapters 2 and 3, as well as previously in the present chapter, its explicit formulation has been deferred until now since my intention was to introduce it in this context. Toward that end, we next describe the derivation of the quadratic formula, as presented by Limin on this occasion. We begin by making a distinction between identical and conditional equality for the quadratic equation, something not often explained in algebra classes, causing students no end of confusion when later taking the Calculus. That is we consider the two related quadratic equations

$$Ax^2 + Bx + C \equiv 0 \text{ and } ax^2 + bx + c = 0 \text{ for } a > 0;$$

the first exhibiting identical, and the second, conditional, equality. In the case of identical equality the equation must hold for all x, while in the case of conditional equality it only holds for the two roots satisfying the quadratic equation. Thus, the identical equality case implies that $A = B = C = 0$, while the conditional equality case implies that $x = x_1$ or $x = x_2$ where these roots are given by the quadratic formula

$$x_{1,2} = \frac{-b \pm \sqrt{b^2 - 4ac}}{2a},$$

which can be derived by the method of completing the square as follows:

$$ax^2 + bx + c = 0 \Rightarrow a\left(x^2 + \frac{bx}{a}\right) = -c;$$

dividing by a, taking one-half of the coefficient of the linear term on the left-hand side of this equation, squaring that quantity, and adding it to both sides of the equation, we obtain

$$x^2 + \frac{bx}{a} + \frac{b^2}{4a^2} = \frac{b^2}{4a^2} - \frac{c}{a} \Rightarrow \left[x + \frac{b}{2a}\right]^2 = \frac{b^2 - 4ac}{4a^2}.$$

Then taking the square root of both sides of this equation yields

$$\sqrt{\left[x + \frac{b}{2a}\right]^2} = \left|x + \frac{b}{2a}\right| = \frac{\sqrt{b^2 - 4ac}}{2a} \Rightarrow x + \frac{b}{2a} = \pm\frac{\sqrt{b^2 - 4ac}}{2a}$$

which when solved results in the quadratic formula for $x = x_{1,2} = (-b \pm \sqrt{b^2 - 4ac})/(2a)$.

After deriving that formula, Limin applied it to the same quadratic equation for $\Lambda_c = \lambda_c^2/4 > 0$ examined above: Namely, $\Lambda_c^2 - m\Lambda_c - 2m = 0$ where $m \geq 1$. Making the identifications that $x = \Lambda_c$, $a = 1$, $b = -m$, and $c = -2m$, he then found the roots

$$\Lambda_c = x_{1,2} = \frac{m \pm \sqrt{m^2 + 8m}}{2}.$$

Limin explained that the "−" root $x_2 < 0$ because $\sqrt{m^2 + 8m} > m$ for $m \geq 1$ and hence had to be neglected. Thus only the "+" root $x_1 > 0$, being consistent with $\Lambda_c > 0$, was valid and yielded

$$\Lambda_c = \frac{\lambda_c^2}{4} = \frac{m + \sqrt{m^2 + 8m}}{2}.$$

In the event, this completed Limin's oral presentation even though we had prepared one more application of the quadratic formula. He had involved the audience in the teaching process by using a method of delivery whereby questions were posed at appropriate intervals. For instance, in the derivation of the quadratic formula, Limin had asked what was the value of $\sqrt{\alpha^2}$. Both members of the committee from the English Department had answered α. He replied with, "Really? Do you think that $\sqrt{(-5)^2} = -5$?" They then changed their answer to $\pm\alpha$. Limin responded by saying, "It's a single valued function." They were stumped. So Limin asked me to answer the question and I did so by offering the proper definition of the absolute value function

$$\sqrt{\alpha^2} = |\alpha| = \left\{ \begin{array}{ll} \alpha & \text{for } \alpha \geq 0 \\ -\alpha & \text{for } \alpha < 0 \end{array} \right. .$$

This scenario was repeated for each question and by the time Limin finished his first application, they stopped the exam having decided he was qualified to teach college algebra courses after all. That was a shame since our prospective second application required using the quadratic formula to deduce the inverse of the hyperbolic sine function, denoted by argsinh, as follows:

$$y = \sinh(v) = \frac{e^v - e^{-v}}{2}.$$

Multiplying this equation by $2e^v$, results in the quadratic for e^v:

$$(e^v)^2 - 2y(e^v) - 1 = 0.$$

Making the identifications that $x = e^v$, $a = 1$, $b = -2y$, and $c = -1$, yields the roots

$$e^v = x_{1,2}(y) = \frac{2y \pm \sqrt{4y^2 + 4}}{2} = y \pm \sqrt{y^2 + 1}.$$

Since $v = 0$ corresponds to $y = \sinh(0) = 0$ and $e^0 = 1$, the x_2 root must be neglected because $x_2(0) = -1$, while the x_1 root is retained because $x_1(0) = 1$, its correct value. Thus

$$e^v = y + \sqrt{y^2 + 1} > 0 \Rightarrow v = \ln\left(y + \sqrt{y^2 + 1}\right) = \text{argsinh}(y).$$

If Limin's actual presentation was too much for them, this one would have been even worse!

That incident is reminiscent of C.P. Snow's *Two Cultures* ([232]) describing the dichotomy between advocates of the sciences and the humanities. In his famous 1959 Cambridge University lecture with this title, he related that his assertion The Second Law of Thermodynamics (entropy increases in closed systems) should be just as much common knowledge as Shakespeare's works was both coolly and negatively received by the humanities types. I believe that Limin, although displaying complete innocence in using a Socratic style of exposition during his presentation, knew exactly what would transpire should he do so and hence, did it on purpose to affect this end. I have always thought that his behavior in this instance showed a great deal of cleverness.

So Limin taught college algebra during his last year of graduate work at WSU and it paid off handsomely for him. While I was in the Mathematics doctoral program at RPI, they had a policy that a student was required to teach at least one year in their Calculus sequence in order to obtain a Ph.D. This requirement was instituted by the Chair George Handelman, who felt such experience to be necessary to prepare students for potential faculty positions at colleges or universities. Limin's desire to teach at the university level was motivated by exactly the same logic. During his last semester I had him present his thesis results at Arizona State University's Dynamic Days which were organized to allow students to showcase their dissertation research for prospective employers. Although his presentation was well received, it did not generate any job offers of postdoctoral or faculty positions at universities or four-year colleges.

Finally, Limin applied for an open Assistant Professorship at Columbia Basin College (CBC) in Pasco, Washington, which has a two-year program in Mathematics leading to an Associates Degree. His visit to CBC went excellently, as did the telephone interview with their Dean of Applied Sciences, when the latter called to have me elaborate upon the items included in my recommendation letter. He was particularly concerned about Limin's ability to teach students at CBC and I assured him the level of college algebra instruction at WSU was no more difficult than that of a community college class. Hence they hired Limin for this position and he performed extremely well in that capacity, ultimately being promoted to Associate and eventually to Full Professor. Many community colleges are reluctant to hire Ph.D. recipients feeling that those individuals will view such a position as underemployment and not make a long-term investment in their institution, *i.e.*, there is always the fear that they will leave for more prestigious jobs requiring a Ph.D., should positions of this sort become available to them. In his case, such an admonishment could not have been more wrong and fortunately, CBC was willing to take the chance of hiring him. Although he is the only one of my Ph.D. students employed at this level, his commitment to community college undergraduate education has worked out to the mutual benefit of both Limin and CBC.

As far as remuneration is concerned, I believe that people often sell community colleges too short in this regard. Let me cite Limin Zhang's case as an object lesson when comparing the salary structure of research universities versus community colleges. Bear in mind that I worked at WSU for 45 years, the last 35 of which were as a Full Professor. When I retired in 2015 my salary, which started at $11,200, had finally reached the $90,000 level. Limin Zhang's Columbia Basin College's salary for 2015 was $94,876. As noted before, it has paid off handsomely for him over the years and proved *The Vicar of Wakefield* adage that "handsome is that handsome does."

.

7

Chemical Turing Patterns and Diffusive Instabilities: Hexagonal Planform Nonlinear Stability Analysis

The first time I became acquainted with the term Turing instabilities was in the introduction and formulation section of my *Royal Society* paper Wollkind and Segel ([285]). If that seems like an incongruous statement, let me explain this assertion more fully. My NSF postdoctoral research appointment with Harry Frisch described in Chapter 3 only paid me for 11 months during its first year, so I was given the month of August of 1969 off for the purpose of finally writing a paper based on that thesis problem completed under the direction of Lee Segel discussed in Chapter 2. The division of labor for this effort was as follows: I was to draft the formulation part of Section 1 and Sections 2-7 of it describing the linear and two-dimensional nonlinear stability analyses of our alloy solidification problem, while Segel concentrated on the introduction part of Section 1 and its final Section 8 conclusions. Here is the relevant paragraph of the Section 1 introduction:

> Our analysis also provides a concrete discussion of nonlinear effects in a symmetry breaking instability occurring in a dissipative system. Turing ([256]), Gmitro and Scriven ([66]), and Prigogine and Nicolis ([178]) have stressed the importance of such instabilities, particularly in chemical and biological contexts. The metallurgical problem investigated here, with the convective terms omitted as relatively unimportant, serves as a good model for pattern-formation instabilities both because it is driven by a 'surface engine' and because diffusive transport is more important than convective transport. Consequently the study reported here may be of interest to some who have no direct concern with the metallurgical problem.

Let us examine this topic in some detail. Almost seventy years ago, Turing ([256]) proposed the chemical basis of morphogenesis in a landmark paper with that title. In particular, he postulated the existence of chemical morphogens which formed the basis of embryo-morphogenesis through the development of prepatterns. Specifically, he investigated the possibility of an instability occurring in purely dissipative systems involving chemical reactions far from equilibrium and the transport process of diffusion but no hydrodynamic motion. When restricted to two chemical species, an activator and an inhibitor, the existence of such instabilities requires an autocatalytic or positive feedback reaction for the activator and a diffusive advantage for the inhibitor as necessary conditions. Then, an initially homogeneous state, which would be stable in the absence of diffusion, can be destabilized resulting in a re-equilibrated inhomogeneous symmetry breaking pattern. That concept along with the one of positional information has made Turing theory a fundamental paradigm for explaining developmental biological processes ranging from embryology to limb formation and coat patterning (Murray, [154]). In addition, this Turing diffusive instability mechanism was proposed as a means for better understanding ecological pattern formation by introducing spatial effects into the dynamical systems describing predator-prey and consumer-resource

DOI: 10.1201/9781003195603-7

interactions (Okubo and Levin, [158]). That introduction produced ecological partial differential interaction-dispersion equations very similar in nature to the chemical reaction-diffusion equations just described. At first, the need for the activator species to diffuse significantly less rapidly than the inhibitor posed a major obstacle for designing an experiment which exhibited chemical Turing instability patterns, since in aqueous media nearly all simple ions and molecules have diffusion coefficients within a factor of two of 1.5×10^{-5} cm^2/sec. So it was felt that if one wanted to compare theoretically predicted diffusive instabilities to observational data this would have to be done in an ecological context.

Toward that end we considered the nondimensional interaction-dispersion equations given by

$$\frac{\partial \mathcal{H}}{\partial \tau} = \mathcal{H}\mathcal{F}(\mathcal{H}, \mathcal{P}) + \mu \nabla_2^2 \mathcal{H},$$

$$\frac{\partial \mathcal{P}}{\partial \tau} = \mathcal{P}\mathcal{G}(\mathcal{H}, \mathcal{P}) + \nabla_2^2 \mathcal{P} - \chi \boldsymbol{\nabla}_2 \cdot \left(\frac{\mathcal{P}}{\mathcal{H}} \boldsymbol{\nabla}_2 \mathcal{H} \right);$$

for our predator-prey mite ecosystems of Chapter 4, where $\boldsymbol{\nabla}_2 \equiv (\partial/\partial x, \partial/\partial y)$, $\nabla_2^2 \equiv \boldsymbol{\nabla}_2 \cdot \boldsymbol{\nabla}_2$, $\mu = D_1/D_2$ with $D_{1,2} \equiv$ prey, predator dispersal or motility coefficients, and $\chi \equiv$ predator aggregation or preytaxis coefficient. Taking the May mite model interaction terms for $\mathcal{F}$ and $\mathcal{G}$, John Collings and I as part of his postdoctoral research, in collaboration with my masters student Maria Barba, performed both a linear and one-dimensional nonlinear stability analysis of the community equilibrium point of that system over the temperature range $T \in (0, T_1) \cup (T_2, T_D)$ where $T_D = 26.31$. These represent those temperatures for which this equilibrium is linearly stable for the May mite model and the activator prey interaction exhibits a positive feedback or a so-called Alee effect characterized by

$$\frac{\partial \mathcal{F}}{\partial \mathcal{H}}(\mathcal{H}_e, \mathcal{H}_e) = \theta_c > 0 \text{ or equivalently, } \varphi > \Delta,$$

using the notation of Chapter 4, as is required for such ecological Turing diffusive instabilities. In addition, since the McDaniel mite moves so much more slowly than its inhibitory predator the other necessary condition $\mu < 1$ for such instabilities is also fulfilled. The results of these stability analyses were published in the *Journal of Mathematical Biology* as Wollkind *et al.* ([281]), which was mentioned in Chapter 4 as well.

Next, I chose an extension of the nonlinear stability analysis of this model to two-dimensions by considering a hexagonal planform for my then current Ph.D. student Laura Stephenson in much the same manner that Rukmini Sriranganathan, David Oulton, and I had extended Wollkind and Segel ([285]) to three-dimensions for the alloy solidification problem described in Chapter 2. Before we actually started that analysis, I accepted an invitation from one of Lee's Weizmann Institute Ph.D. students, Leah Keshet of the University of British Columbia, to give an invited address entitled "Modeling the dynamics and pattern formation of a temperature sensitive predator-prey mite system on fruit trees" at the 1992 Theoretical Biology and Biomathematics Gordon Conference held in Tilton, NH. During this conference, I had the occasion to speak with Qi Ouyang, a postdoctoral colleague of Harry Swinney at the University of Texas in Austin, who gave a poster presentation exhibiting photographs of the chemical Turing patterns they had recently produced that were depicted in Fig. 1.2. Once I returned from this conference, I told Laura we were switching gears and would start work on that problem instead. The reasons for this were two fold: First, there were now very good pattern formational data on that problem with which to compare our eventual theoretical predictions, and second, this was an extremely important scientific topic being an experimental confirmation of Turing's ([256]) theoretical predictions after nearly forty years of continuous effort in trying to do so.

At last, Ouyang and Swinney ([163]) had managed to overcome the difficulty that in aqueous solutions chemical species had similar diffusion coefficients by conducting their experiments involving the iodide-chlorite chemical reaction-diffusion activator-inhibitor system in a polyacrylamide gel reactor with a starch indicator which, besides preventing convection, resulted in a marked reduction of the effective iodide/chlorite diffusion coefficient ratio. Turing patterns consisting of parallel stripes and hexagonal arrays of spots or honeycombs (see Fig. 1.2) occupying a single thin layer appeared as the system's control parameters of reservoir pool species concentrations consisting of chlorine dioxide, iodine, and malonic acid were tuned (see Fig. 7.1). The mechanism suggested for this reduction by Lengyel and Epstein ([115]) was that the starch indicator reversibly formed an immobile complex with the activator iodide species and the iodine rapidly enough to allow, in essence, a circumvention of the differential diffusivity requirement. Thus, Turing instabilities could be generated over a parameter range where ordinarily the system would have exhibited oscillatory behavior in the absence of starch. We next quantified this reduction mechanism proposed by Lengyel and Epstein ([115]) and applied it to the CDIMA model system. In doing so, we Pulled our first Rabbit Out of a Hat, as will be demonstrated below.

We accomplished this by the introduction of a Turing pattern indicator, such as starch, which reversibly forms an immobile complex with the activator species, making it possible to produce those patterns in a system that would otherwise not exhibit them. Consider the following chemical reactions:

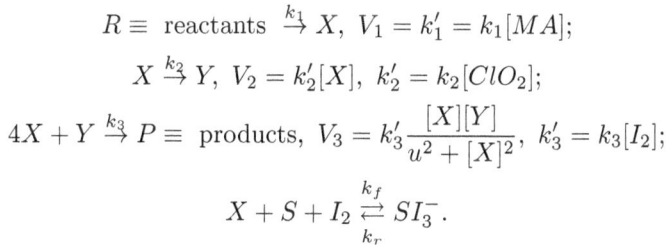

$$R \equiv \text{ reactants } \xrightarrow{k_1} X, \ V_1 = k_1' = k_1[MA];$$

$$X \xrightarrow{k_2} Y, \ V_2 = k_2'[X], \ k_2' = k_2[ClO_2];$$

$$4X + Y \xrightarrow{k_3} P \equiv \text{ products, } V_3 = k_3' \frac{[X][Y]}{u^2 + [X]^2}, \ k_3' = k_3[I_2];$$

$$X + S + I_2 \underset{k_r}{\overset{k_f}{\rightleftarrows}} SI_3^-.$$

Here the chlorine dioxide (ClO_2), iodine (I_2), malonic acid (MA), and pattern indicator (S) concentrations were assumed to remain constant, where in the above reaction scheme the V_i's represent velocities and the bracketed characters, concentrations, while $X \equiv$ iodide (I^-) and $Y \equiv$ chlorite (ClO_2^-) ions, the concentrations of which are our dynamical variables, may be regarded as functions of space and time denoted by $s \equiv (s_1, s_2)$ and τ, respectively. Further, the latter intermediate species have self-diffusion coefficients D_1 and D_2, taken to be constant as is the case for the reaction rates k_1, k_2, k_3, k_r, and k_f, while u represents a uniform shaping concentration selected historically to provide agreement with experimental data, all of which will be assigned later. Then introducing the following dimensionless variables and parameters

$$t = k\tau, \ r = \frac{s}{\sqrt{D_2/k}}, \ x = \frac{[X]}{u}, \ y = \frac{k_3'[Y]}{k_2' u^2};$$

$$\alpha = \frac{k_1'}{5k_2' u}, \ \beta = \frac{k_3'}{k_2' u}, \ \mu = \frac{D_1}{D_2}, \ K - \frac{k_f}{k_r}[S][I_2], \ k = \frac{k_2'}{1+K};$$

and employing the law of mass action and Fick's second law (Lin and Segel, [119]) with this

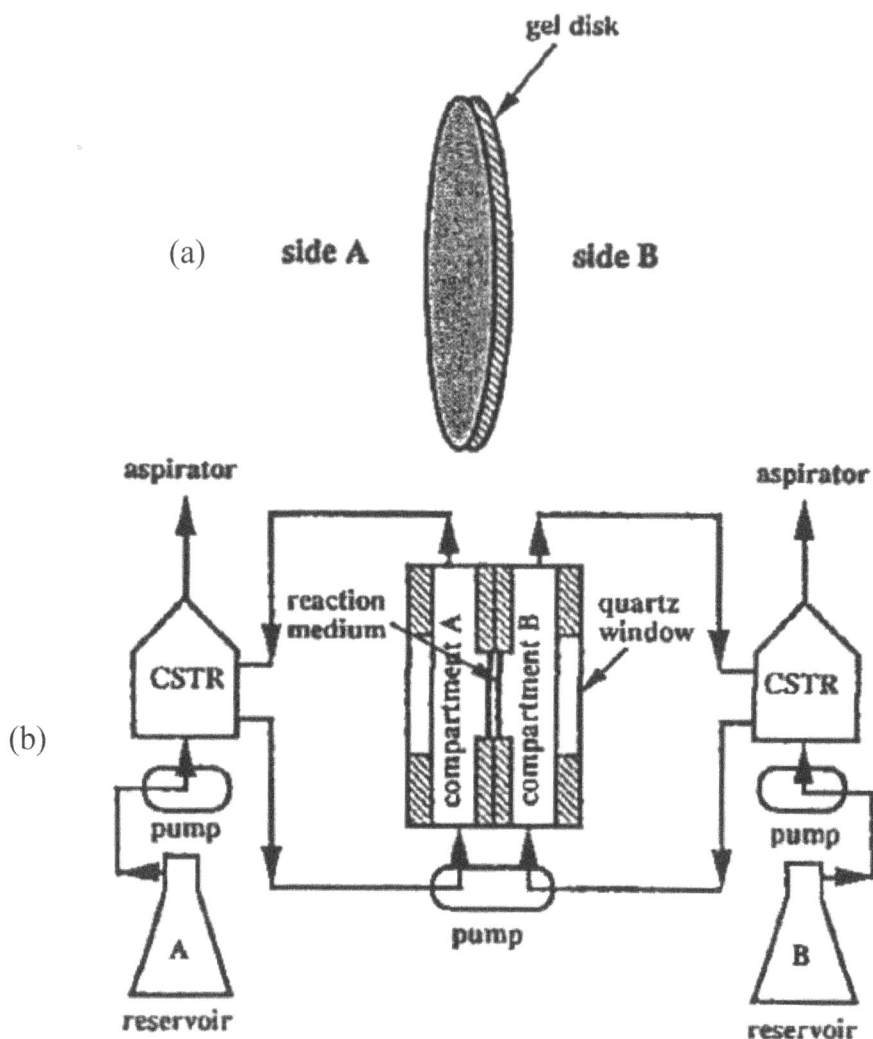

FIGURE 7.1
Schematic diagrams from Ouyang and Swinney ([164]) for the CDIMA gel reactor: (a) the reaction medium, a thin gel disk; (b) the reaction system for a so-called two-sided fed reactor. Here, ClO_2 is contained in reservoir A; $MA \equiv CH_2(COOH)_2$, in reservoir B; and I_2, in both reservoirs; while CSTR is an acronym for *C*onstantly *S*tirred *T*ank Reactor.

scaling, we deduced the nondimensional governing activator-inhibitor/immobilizer reaction-diffusion system defined on an unbounded spatial domain (the r_1-r_2 plane)

$$\frac{\partial x}{\partial t} = F(x,y;\alpha) + \frac{\mu}{1+K}\nabla_2^2 x, \quad \frac{\partial y}{\partial t} = (1+K)\beta G(x,y) + \nabla_2^2 y, \quad \nabla_2^2 \equiv \sum_{i=1}^{2}\frac{\partial^2}{\partial r_i^2},$$

where

$$F(x,y;\alpha) = 5\alpha - x - \frac{4xy}{1+x^2}, \quad G(x,y) = x - \frac{xy}{1+x^2},$$

with the equilibrium point

$$F(x_0,y_0;\alpha) = G(x_0,y_0) = 0 \Rightarrow x_0 = x_0(\alpha) = \alpha, \; y_0 = y_0(\alpha) = 1+\alpha^2.$$

In the above, we implicitly made use of the facts that the tri-iodide complex (SI_3^-) did not diffuse and satisfied the chemical quasi-equilibrium relation

$$x^* = x \text{ where } x^* = \frac{[SI_3^-]}{Ku}$$

which required $\varepsilon = k/k_r \ll 1$ as a necessary condition (see Chapter 10). In addition, we also employed the so-called quasi-two-dimensional approximation that allowed us to consider the axial coordinate z, scaled with the height of the gel disk in which the chemical reactions occurred, as a parameter rather than an independent variable and to introduce the pool species concentration gradient relations for $0 < z < 1$ given by

$$[ClO_2] = [ClO_2]_0(1-z), \; [MA] = [MA]_0 z, \; [I_2] = [I_2]_0;$$

analogous to the laboratory reservoir configurations of Ouyang and Swinney ([163]), where the observed patterns formed perpendicular to these concentration gradients. Further, given that those experimental patterns typically had a characteristic wavelength to gel disk diameter ratio of 0.01, and consequently, the disk boundary did not significantly influence those patterns (Graham *et al.*, [73]), it seemed reasonable to consider our activator-inhibitor/immobilizer equations on an unbounded spatial domain as a first approximation.

The equilibrium point $x = x_0(\alpha) = \alpha$, $y = y_0(\alpha) = 1+\alpha^2$ to our model system represented a uniform steady-state spatially homogeneous exact solution to these governing equations. It was the stability of this state to one-dimensional perturbations with which we were first concerned and hence sought a solution of those equations of the form

$$x(\mathbf{r},t) \sim x_0(\alpha) + A_1(t)\cos(q_c r_1) + A_1^2(t)[x_{20} + x_{22}\cos(2q_c r_1)]$$
$$+ A_1^3(t)[x_{31}\cos(q_c r_1) + x_{33}\cos(3q_c r_1)]$$

with an analogous expansion for $y(\mathbf{r},t)$, where the amplitude function $A_1(t)$ as usual satisfied

$$\frac{dA_1}{dt} \sim \sigma_0(\beta;\alpha,\mu,K)A_1 - a_1(\alpha;\mu,K)A_1^3.$$

Here $q_c = q_c(\alpha;\mu,K)$ was the critical wavenumber of linear stability theory, while σ_0 denoted the growth rate associated with that most dangerous mode and a_1, the corresponding Landau constant. We found that

$$q_c^2(\alpha;\mu,K) = \frac{5(1+K)(3\alpha^2-5)}{\mu(1+\alpha^2)[5+2\alpha(10)^{1/2}/(1+\alpha^2)^{1/2}]}$$

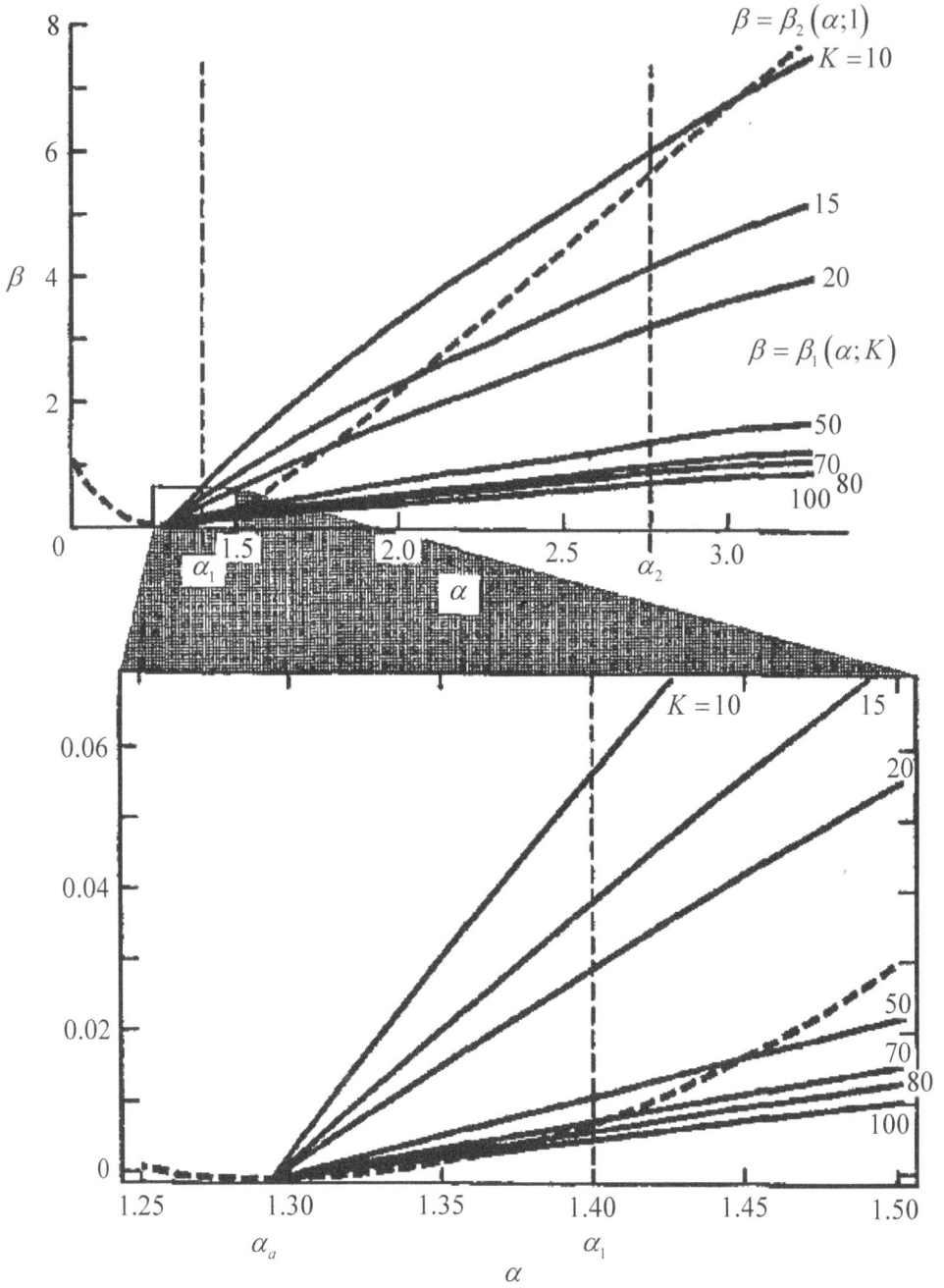

FIGURE 7.2

One-dimensional stability diagram in the α-β plane for the CDIMA/starch model system when $\mu = 1$ and $K = 10$, 15, 20, 50, 70, 80, and 100. Here $\beta_1(\alpha_a; K) = 0$.

and diffusive instabilities ($\sigma_0 > 0$) occurred whenever (see Fig. 7.2)

$$0 < \beta_1(\alpha; K) < \beta < \beta_2(\alpha; \mu)$$

where

$$\beta_1(\alpha; K) = \frac{3\alpha - 5/\alpha}{1 + K}$$

and

$$\beta_2(\alpha; \mu) = \frac{(3\alpha^2 - 5)^2}{\mu\alpha[13\alpha^2 + 5 + 4\alpha(10)^{1/2}(1 + \alpha^2)^{1/2}]},$$

which re-equilibrated ($a_1 > 0$) to form a striped pattern of characteristic dimensional wavelength (see Fig 7.3a)

$$\lambda_c^*(\alpha) = \frac{2\pi}{q_c^*(\alpha)}$$

such that

$$q_c^*(\alpha) = \left(\frac{k}{D_2}\right)^{1/2} q_c(\alpha; \mu, K) = \left(\frac{k_2'}{D_1}\right)^{1/2} q_c(\alpha; 1, 0)$$

provided (see Fig. 7.3b)

$$\alpha_1 < \alpha < \alpha_2$$

where

$$\alpha_1 = 1.40 \text{ and } \alpha_2 = 2.77.$$

Finally, we took

$$D_1 = D_2 = \chi^2 D_x \Rightarrow \mu = 1$$

with

$$\chi = 0.40 \text{ and } D_x = 7 \times 10^{-6} \frac{\text{cm}^2}{\text{sec}}.$$

This constitutive relation reflects the fact that a fully hydrolyzed saturated gel will result in an ionic diffusion coefficient which has been uniformly reduced from its common aqueous solution value D_x (in this case associated with the temperature $T = 280$ K), the amount of that reduction as measured by χ^2 being dependent on the characteristic pore diameter of the gel itself (Ouyang et al., [162]).

In this context, we note that when Pearson ([168]) modeled CDIMA/indicator experiments in gels by employing a similar activator-inhibitor/immobilizer reaction-diffusion system to that catalogued above, but with u assigned the value of 0 M (where M $\equiv$ moles or gram molecular mass) in his basic dimensional model and then took $\chi = 1$, he predicted a $\lambda_c^* = 0.40$ mm instead of the observed value of $\lambda_c^* = 0.17$ mm, an overprediction which would be adjusted correctly upon adoption of our $\chi = 0.40$ instead. Indeed, this was our rationale for adopting that constitutive relation which was the second Rabbit Pulled Out of a Hat.

Previous to our adoption of this constitutive relation, it was customary for modelers, such as Pearson, to assume that the diffusion coefficients of the small molecular ionic reactants in the gel were the same as their aqueous diffusion values. In other words, they diffused as freely in the gel as they did in water, although the larger starch indicator molecules were considered to be essentially immobile in that medium. This assumption was usually defended by the assertion that the gel had an average pore size of 10 nm (where nm $\equiv 10^{-9}$ m) and

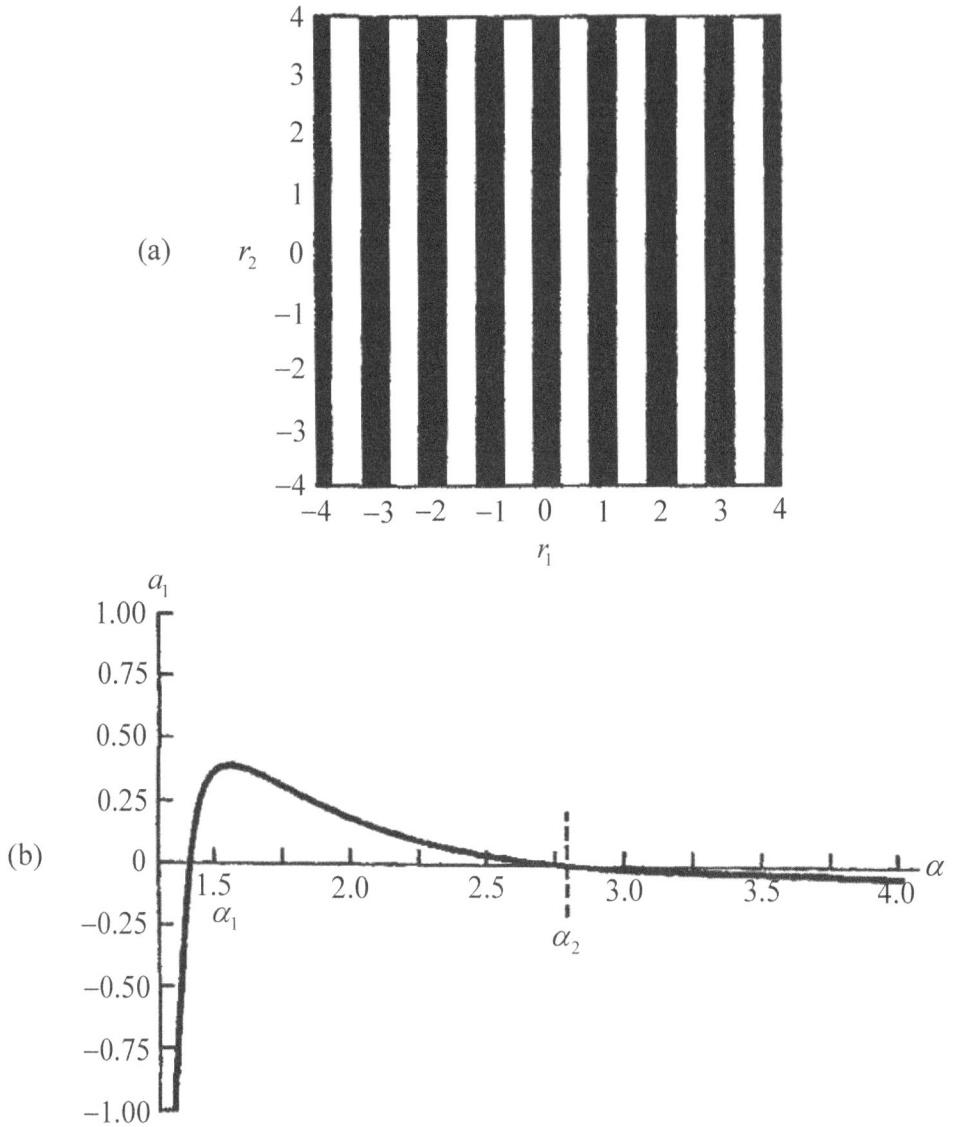

FIGURE 7.3
(a) Contour plot of stripes (critical point II of the hexagonal planform stability analysis).
(b) Plot of a_1 versus α for the CDIMA/starch model system when $\mu = 1$ and $K = 100$.

at least 90% void space (Ouyang and Swinney, [162]). This did not seem intuitively correct to me especially given the discrepancy between theory and experiment described above.

At the time, Yin Luo, a former postdoctoral associate of Irving Epstein at Brandeis, had accepted a similar position with my colleague Ed Pate at WSU. She said that the gel acted like the gelatin dessert Jello and slowed down both the activator and inhibitor ions by exactly the same amount. In our initial publication, Stephenson and Wollkind ([239]) on one-dimensional chemical Turing patterns, we referenced that conversation as Luo, Y: Private communication (1994) to justify our adoption of this constitutive relation.

Subsequently, I had the occasion to discuss this matter with Harry Swinney himself at the SIAM Conference on the Applications of Dynamical Systems held in Snowbird, Utah, during October of 1992. Upon his return from that conference, he put Qui Ouyang to work on this problem, which resulted in their publication of Ouyang *et al.* ([162]), which was an experimental verification of my conjecture about ionic diffusion in gels that confirmed this constitutive relation and has been referenced in all my later papers on chemical Turing patterns as a justification for its adoption. What has always seemed strange to me was that they did not update their review paper Ouyang and Swinney ([164]) on Turing patterns to reflect this new result (perhaps they never had the chance) and in that review restated their old assertion cited above.

Every once in a while, a theoretical prediction that has not yet been confirmed experimentally leads experimentalists to design a particular experiment to test this prediction and Ouyang *et al.* ([162]) is an example of just such an occurrence. The more common occurrence is that a theoretical modeling prediction confirms an existing experimental result and this confirmation serves as a partial validation of that particular model for the phenomenon under examination. This was the case for our predicted dimensional wavelength formula $\lambda_c^*(\alpha) = 2\pi/q_c^*(\alpha)$, which is independent of the complexification strength K since $q_c^*(\alpha)$ does not depend on the latter quantity, as demonstrated above. We note that in the experiments of Ouyang and Swinney ([164]) and Lengyel and Epstein ([115]), the observed dimensional pattern wavelength λ_c^* was indeed independent of K in agreement with our theoretical prediction.

Let us recapitulate the behavior of our amplitude equation for the conditions summarized above: Given these conditions, the amplitude function $A(t)$ undergoes a standard supercritical pitchfork bifurcation (Walgraef, [267]) at the Turing boundary $\beta = \beta_2$ when $\alpha_1 < \alpha < \alpha_2$, from which we may conclude that:

1. For $\beta > \beta_2$ and $\alpha_1 < \alpha < \alpha_2$, the undisturbed state $A = 0$ is stable, yielding a uniform homogeneous pattern $x(\boldsymbol{r}, t) \to x_0(\alpha) = \alpha$ as $t \to \infty$.

2. For $\beta_1 < \beta < \beta_2$ and $\alpha_1 < \alpha < \alpha_2$, $A = A_e = \sqrt{\sigma/a_1}$ is stable, yielding a one-dimensional periodic Turing pattern consisting of stationary parallel stripes

$$x(\boldsymbol{r}, t) \to x_e(r_1) \sim x_0(\alpha) + A_e \cos\left(\frac{2\pi r_1}{\lambda_c}\right) \text{ as } t \to \infty$$

of characteristic wavelength

$$\lambda_c = \frac{2\pi}{q_c},$$

as depicted in Fig. 7.3a.

Here, the spatial variable is measured in units of λ_c, while elevations appear dark and depressions, light in accordance with the chemical quasi-equilibrium relation and the observed character of the experimental Turing patterns. Note that when compared to the photographic images of Fig. 1.2 dark corresponds to the blue portions and light to the yellow ones. The system undergoes a Hopf bifurcation at the boundary $\beta = \beta_1$ and hence there is oscillatory behavior for $0 < \beta < \beta_1$, while when $\alpha_a = (5/3)^{1/2} < \alpha < \alpha_1$ or $\alpha > \alpha_2$, the bifurcation is subcritical (the implication of such subcriticality is the subject of the nonlinear stability analysis of that model equation introduced in Chapter 15). These one-dimensional nonlinear stability results are summarized by means of the regions in Fig. 7.2, where the vertical dotted lines denote the loci $\alpha = \alpha_{1,2}$ of Fig 7.3b; the dashed curve, the Turing boundary; and the solid ones, the Hopf boundaries for various values of K. Observe from this figure, that there exists a critical complexification strength $K_c = 10.54$, such that Turing patterns of this sort can only occur for $K > K_c$ or defining an effective diffusion coefficient ratio suggested by the form of our reaction-diffusion equations

$$\mu_K = \frac{\mu}{1 + K}$$

for

$$\mu_K < \mu_{K_c} = \frac{\mu}{1 + K_c} = 0.087\mu$$

or when, in this case, $\mu = 1$, for $\mu_K < 0.087$. Hence, Turing patterns can occur even if $\mu \geq 1$, provided the complexification strength is strong enough where, in the absence of the starch indicator, the system would exhibit the oscillatory behavior of traveling waves. Indeed, according to Yin Luo, Lengyel and Epstein ([115]) were trying to produce scroll waves of that type when they accidentally discovered Turing patterns due to this effect. Yin said she actually observed such Turing patterns earlier when trying to produce scroll waves but Epstein dismissed them as an experimental error before realizing their significance.

In order to investigate the possibility of occurrence for our CDIMA/indicator model system of those hexagonal patterns, observed by Ouyang and Swinney ([163]), depicted in Fig. 1.2, we sought weakly nonlinear two-dimensional solutions of these equations which to lowest order satisfied

$$x(\boldsymbol{r}, t) - x_0(\alpha) \sim A_1(t) \cos[q_c r_1 + \varphi_1(t)] + A_2(t) \cos\left[q_c \frac{r_1 - \sqrt{3} r_2}{2} - \varphi_2(t)\right]$$

$$+ A_3(t) \cos\left[q_c \frac{r_1 + \sqrt{3} r_2}{2} - \varphi_3(t)\right] = f(r_1, r_2, t)$$

where, for $(j, k, \ell) \equiv$ even permutation of (1,2,3),

$$\frac{dA_j}{dt} \sim \sigma A_j - 4a_0 A_k A_\ell \cos(\varphi_j + \varphi_k + \varphi_\ell) - A_j[a_1 A_j^2 + 2a_2(A_k^2 + A_\ell^2)],$$

$$A_j \frac{d\varphi_j}{dt} \sim 4a_0 A_k A_\ell \sin(\varphi_j + \varphi_k + \varphi_\ell),$$

with an analogous expansion for $y(\boldsymbol{r}, t)$. Recall, that this is the same hexagonal planform solution as described in Chapter 2 for alloy solidification. Segel who first employed a solution of this sort when investigating Bénard convection cells said one should do such an analysis once in one's research career because it was good for the soul, but implied that once was enough. As will become obvious here and in later chapters, I have not strictly abided by this admonishment. Given that the methodology to be employed in all these analyses is the same as that introduced in Chapter 2, we need only catalogue the relevant results in what follows, using both the notation and tabular entries defined in that chapter.

Indeed, I have become one of the principal practitioners of these weakly nonlinear stability analyses to examine pattern formational aspects for models of phenomena. That seems to have become a lost art. With the advent of numerical techniques, most researchers have used a linear stability analysis to determine the parameter range over which instabilities occur for their models and then employ various simulation techniques to generate the pattern formation behavior in that range. In fact, Segel had trained Leah Keshet to approach modeling problems in this manner so that, when she saw me employing it for the first time, I had to explain to her the methodology he taught me to use. That is, weakly nonlinear stability theory which, while incorporating the nonlinearities of a particular model system, basically pivots its perturbation procedure about the critical point of linear stability theory. As noted in the caption of Fig. 2.8, the advantage of such an approach, over strictly numerical procedures, is that it allows one to deduce quantitative relationships between system parameters and stable patterns which are valuable for comparison with experimental or observational evidence and difficult to accomplish using simulation alone. Researchers who prefer these numerical techniques to weakly nonlinear ones claim that the latter are only valid in the neighborhood of the marginal stability curve in this parameter space and that the simulation methods have global validity. It has been my experience, as shall be seen in later chapters, that it is often possible to extrapolate these weakly nonlinear stability results to regions of parameter space far removed from the marginal stability curve and that these predictions agree very well with numerical results for particular choices of parameters in those regions. Realize also that these numerical simulations must be performed for each different set of parameter values the pattern formation behavior of which one wishes to determine. This is in the spirit of a formula being worth a thousand pictures.

Returning to our amplitude-phase equations, as in Chapters 2 and 3, these have potentially stable equivalence classes of critical points that can be represented by $\varphi_1 = \varphi_2 = \varphi_3 = 0$ with I: $A_1 = A_2 = A_3 = 0$; II: $A_1^2 = \sigma/a_1$, $A_2 = A_3 = 0$; and III$^\pm$: $A_1 = A_2 = A_3 = A_0^\pm$ where

$$A_0^\pm = \frac{-2a_0 \pm [4a_0^2 + (a_1 + 4a_2)\sigma]^{1/2}}{a_1 + 4a_2}.$$

Let us review our results of those chapters. The stability of these critical points can be posed in terms of σ. Thus, critical point I, the undisturbed state, is stable in this sense for $\sigma < 0$, while the stability behavior of critical points II and III$^\pm$, which depends upon the signs of a_0 and $2a_2 - a_1$, as well were catalogued in Table 2.1. Then, as before, critical points I and II represent the uniform or homogeneous state and the stripes or bands depicted in Fig. 7.3a, respectively.

We next offer a similar morphological interpretation of critical points III$^\pm$ relative to the Turing patterns under investigation. Continuing the review of our previous results, note that to lowest order the equilibrium iodide concentration associated with these critical points satisfies

$$x(\boldsymbol{r}, t) \to x_0(\alpha) + A_0^\pm f_0(r_1, r_2) \text{ as } t \to \infty$$

where

$$f_0(r_1, r_2) = \cos\left(\frac{2\pi r_1}{\lambda_c}\right) + 2\cos\left(\frac{\pi r_1}{\lambda_c}\right)\cos\left(\frac{\sqrt{3}\pi r_2}{\lambda_c}\right)$$

such that $A_0^+ > 0$ and $A_0^- < 0$. We plot the function $f_0(r_1, r_2)$ in Fig. 7.4, where $r_{1,2}$ are

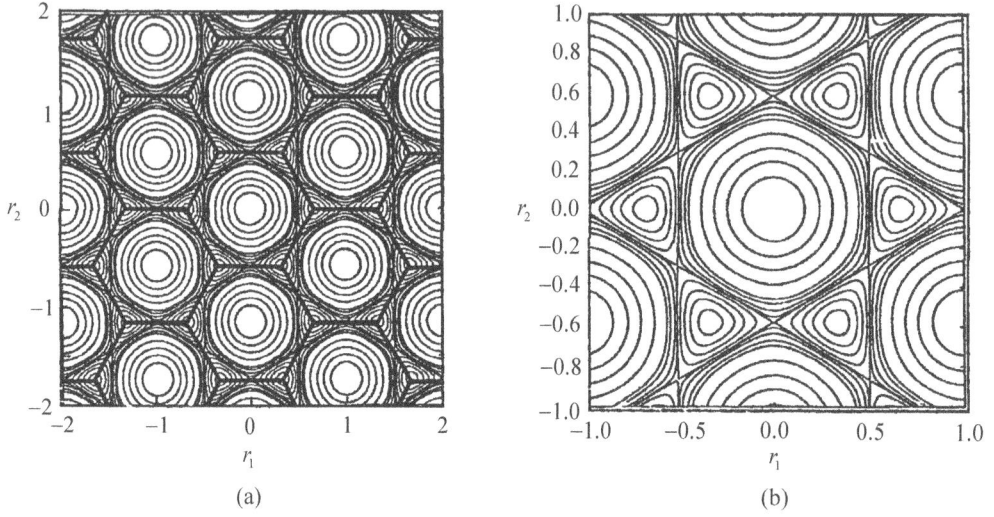

(a)   (b)

FIGURE 7.4

Contour plots of $f_0(r_1, r_2)$. In this context, the axes are again being measured in units of λ_c. Here, (a) exhibits the relevant hexagonal symmetry while (b), an enlargement of its central region, emphasizes the level curve behavior of an individual cell, each of which has an elevation ranging from 3 at its center to $-3/2$ at its vertices. The third, fourth, and fifth rings from the center are at $3/4$, zero, and $-3/5$ elevation, respectively, while the boundary of each cell is of variable depth with its high points at elevation -1 occurring at the midpoint of each edge (see Fig. 2.10).

again being measured in units of λ_c. From the contour plot of Fig. 7.4a we see that these equilibrium iodide concentrations possess hexagonal symmetry. In particular, focusing our attention upon the single hexagonal cell enlargement of Fig. 7.4b, we observe that each individual f_0 cell has an elevated nearly circular central region with a maximum elevation of three at its center which is bounded by a level curve of zero elevation, depicted as the fourth such curve in that figure counting outward from this point (see Fig. 2.10). The peripheral portion of each cell exterior to that central region is depressed with the hexagonal cellular boundary being of variable depth which ranges from $-3/2$ at its vertices to -1 at the midpoint of its edges. Since $A_0^+ > 0$ for $a_0 \leq 0$ and $A_0^- < 0$ for $a_0 \geq 0$ (see Table 2.1), we concluded that the contour plots of $III^\pm$ had circular elevations for critical point III^+ when stable and circular depressions for III^-, with both the diameter of these regular hexagons and distances between centers equal to $2\lambda_c/\sqrt{3}$. Those contour plots are represented in Fig. 7.5, where part (a) is for III^+ and part (b) for III^-. As in the analogous plot for stripes of Fig. 7.3a, elevations appear dark and depressions light in this figure. Hence, recalling, that the Turing patterns under consideration are classified by their light regions (see Fig. 1.2), we identified the hexagonal arrays of honeycombs or nets with critical point III^+ and of spots or dots with critical point III^-, respectively. Further, given the hexagonal close-packed nature of these arrays associated with $III^\pm$, we shall also refer to them collectively as hexagons in what follows.

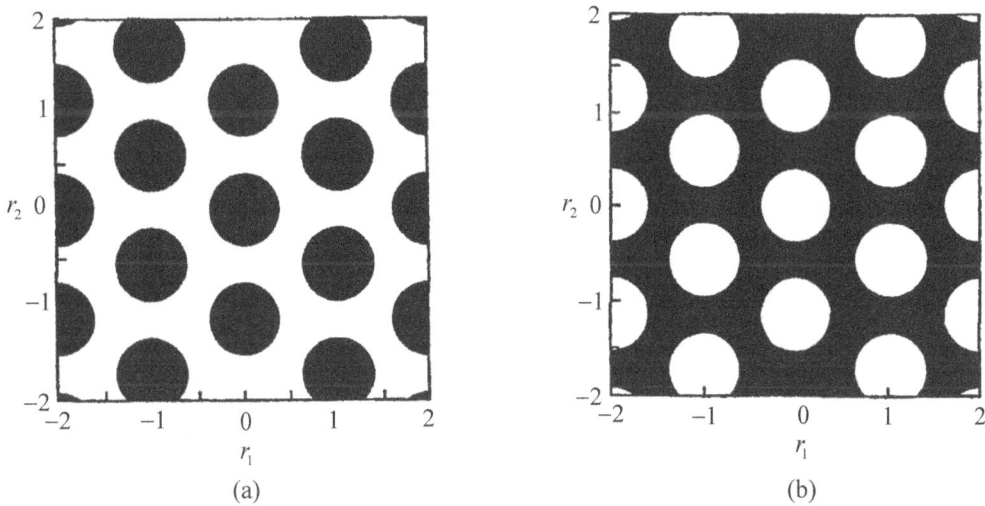

(a)

(b)

FIGURE 7.5
Contour plots for the critical points (a) III^+ and (b) III^-.

Having summarized these morphological identifications and their stability properties catalogued in Table 2.1, we determined the formulae for the other two hexagonal planform Landau constants which had the same functional dependence as $a_1 = a_1(\alpha; \mu, K)$: Namely,

$$a_0 = a_0(\alpha; \mu, K) \text{ and } a_2 = a_2(\alpha; \mu, K);$$

by employing Fredholm solvability conditions (sketched in Chapters 2 and 3) in the identical manner that the expression for a_1 was obtained and examined the signs of a_0, $2a_2 - a_1$, and

$a_1 + 4a_2$ for $\alpha_1 < \alpha < \alpha_2$ after, as in Fig. 7.3b, assigning μ and K the typical values

$$\mu = 1 \text{ and } K = 100.$$

From these results we observed that, besides the zeroes α_1 and α_2 of a_1, depicted in this figure, there exist the following other significant values of α

$$\alpha_3 = 1.44, \ \alpha_5 = 1.48, \ \alpha_c = 1.88, \ \alpha_6 = 2.36, \ \alpha_4 = 2.53$$

such that

$$a_1 + 4a_2 = 0 \text{ for } \alpha = \alpha_3 \text{ or } \alpha_4, \ a_1 + 4a_2 > 0 \text{ for } \alpha_3 < \alpha < \alpha_4;$$
$$2a_2 - a_1 = 0 \text{ for } \alpha = \alpha_5 \text{ or } \alpha_6, \ 2a_2 - a_1 > 0 \text{ for } \alpha_5 < \alpha < \alpha_6,$$
$$2a_2 - a_1 < 0 \text{ for } \alpha < \alpha_6 \text{ or } \alpha > \alpha_6;$$
$$a_0 = 0 \text{ for } \alpha = \alpha_c, \ a_0 > 0 \text{ for } \alpha > \alpha_c, \ a_0 < 0 \text{ for } \alpha < \alpha_c.$$

We note that this behavior is independent of the choice for μ and K, as was also true for a_1.

Finally, upon determining the functions

$$\sigma_1 = \sigma_1(\alpha; 1, 100), \ \sigma_2 = \sigma_2(\alpha; 1, 100),$$

obtained from the definitions of $\sigma_{1,2}$ in terms of $a_{0,1,2}$ (defined in Chapter 2) for $\mu = 1$ and $K = 100$, we produced the loci

$$\beta = \beta_{\sigma_i}(\alpha) = \beta_c[\alpha; 1, 100, \sigma_i(\alpha; 1, 100)] \text{ for } i = 1 \text{ and } 2,$$

where, as in Chapter 2,
$$\beta = \beta_c(\alpha; \mu, K, s)$$

represents the generalized marginal stability curve corresponding to $\sigma = s$ in the α-β plane (see Chapter 12 for a detailed development of that procedure) and thus as to be expected

$$\beta_c(\alpha; \mu, K, 0) = \beta_2(\alpha; \mu).$$

Since we had now evaluated all the quantities required for the identification of the stable Turing patterns of Table 2.1, the regions corresponding to these patterns could next be represented graphically in the α-β plane of Fig. 7.6, where the loci $\beta = \beta_{\sigma_i}(\alpha)$, defined above, have been denoted by $\sigma = \sigma_i$, $i = 1$ and 2, in that figure.

Then, from Fig 7.6, we can observe that for $a_1 + 4a_2 > 0$ all (when $2a_2 - a_1 > 0$) or part (when $2a_2 - a_1 < 0$) of the region $(\sigma, a_1 > 0)$, where the one-dimensional analysis predicted stable striped or banded Turing patterns, is further divided into two subregions characterized by stable hexagonal patterns consisting of either spots (when $a_0 > 0$) or nets (when $a_0 < 0$), respectively. In the overlap regions satisfying

$$\beta_1(\alpha; 100) < \beta_{\sigma_2}(\alpha) < \beta < \beta_{\sigma_1}(\alpha) < \beta_2(\alpha; 1)$$

where bands and nets $(\alpha_c^- < \alpha < \alpha_c)$ or bands and spots $(\alpha_c < \alpha < \alpha_c^+)$ are predicted, initial conditions determine which stable Turing pattern of each pair will be selected. Here, $\alpha_c^\pm$ are defined implicitly by

$$\beta_{\sigma_1}(\alpha_c^\pm) = \beta_1(\alpha_c^\pm; 100)$$

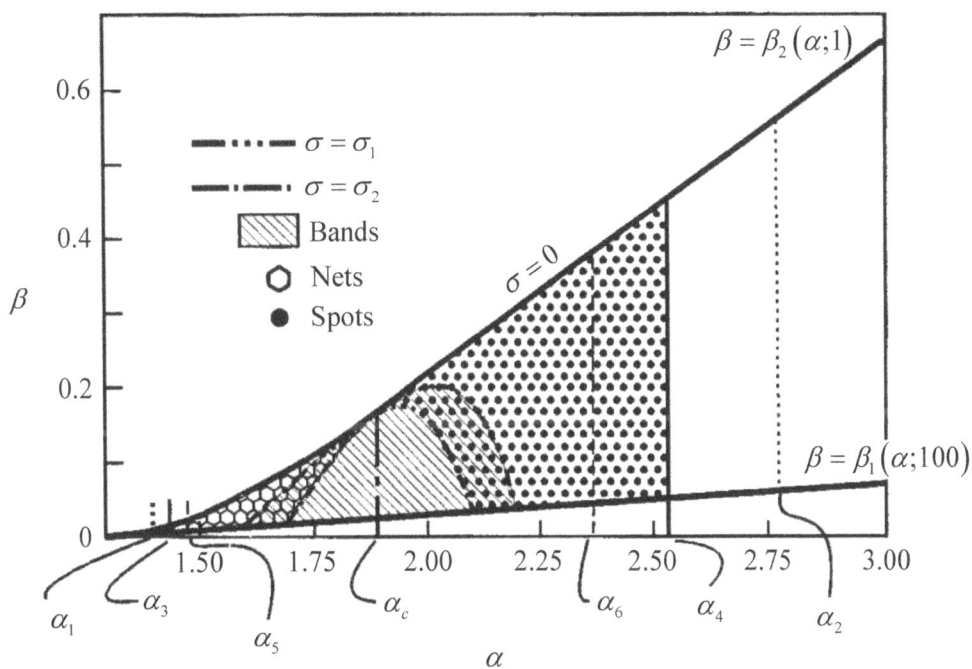

FIGURE 7.6
Hexagonal planform two-dimensional stability diagram in the α-β plane for the CDIMA/starch model system with $\mu = 1$ and $K = 100$ denoting the predicted Turing patterns summarized in Table 2.1 in conjunction with those identifications of this chapter.

which, from Fig. 7.6, implies that

$$\alpha_c^- = 1.58 \text{ and } \alpha_c^+ = 2.19.$$

There also exists a region of bistability corresponding to $\sigma_{-1} < \sigma < 0$, the uniform state being stable for $\sigma < 0$ and hexagons for $\sigma_{-1} < \sigma < \sigma_2$. Given that $\sigma_{-1} = -4a_0^2/(a_1 + 4a_2) < 0$ for $a_1 + 4a_2 > 0$ and $a_0 \neq 0$, the hexagons persisting in that overlap region would be subcritical in nature. As can be seen, however, from Fig. 7.7

$$|a_0| \ll a_1 + 4a_2$$

in this parameter range and thus the loci $\sigma = \sigma_{-1}$ and $\sigma = 0$ are virtually indistinguishable over that range. Hence, unlike the type between hexagons and bands, this bistability is beyond experimental resolution and thus only the locus $\sigma = 0$ appears explicitly in Fig. 7.6. Further, as noted in Chapter 2, to justify the truncation procedure inherent to our hexagonal planform asymptotic expansion, it is necessary that the Landau coefficients of the amplitude-phase equations satisfy the size constraints

$$\frac{|a_0|}{(a_1 + 4a_2)^2} \ll 1.$$

Observing that the inequality condition depicted in Fig. 7.7 also guarantees the satisfaction of this constraint, we can conclude that such a truncation procedure is valid for our hexagonal planform weakly nonlinear stability analysis of the CDIMA/indicator model system.

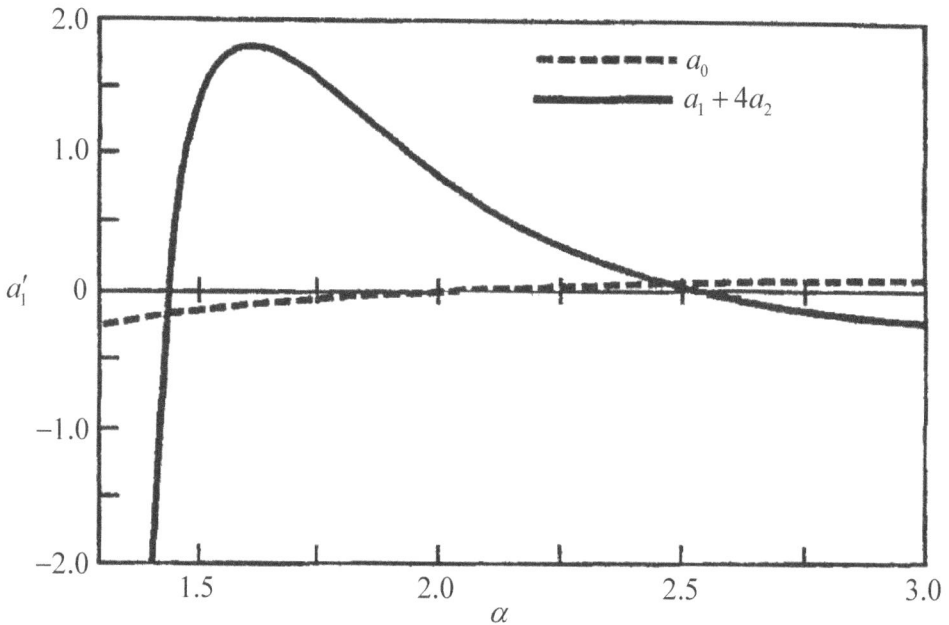

FIGURE 7.7
Plots of a_0 and $a_1 + 4a_2$ versus α for $\mu = 1$ and $K = 100$.

We were now ready to compare these theoretical predictions, summarized in Fig. 7.6, with relevant experimental observation. We first considered those experiments which involved the banded and hexagonal spot or net patterns that emerged upon increase of the $[MA]_0$ reservoir concentration. To do so, we examined the possible succession of Turing patterns predicted when a member of the one-parameter horizontal family of lines $\beta \equiv \beta_0$ is traversed in the direction of increasing α. In particular, to produce a theoretical prediction consistent with these experimental observations, it was necessary for us to assume the constant u-condition $u = u_0$, which yields the transit curve $\alpha = m[MA]_0$ and $\beta = \beta_0$ where

$$m = \frac{k_1 z}{5k_2[ClO_2]_0 u_0(1-z)} \quad \text{and} \quad \beta_0 = \frac{k_3[I_2]_0}{k_2[ClO_2]_0 u_0(1-z)}.$$

For fixed values of the other parameters, including z, this horizontal line was traversed in the direction of increasing α as $[MA]_0$ increased. Adjusting these parameters appropriately, we concluded from an examination of Fig. 7.6, that such a transit line was capable of generating all those Turing pattern sequences catalogued in Table 7.1 as $[MA]_0$ increased.

TABLE 7.1
Predicted Turing pattern sequence versus β_0. Here, the notation A/B indicates the bistability of structures A and B while $\beta = 0.18, 0.20$ are its maximum values on $\sigma = \sigma_{2,1}$.

β_0	Predicted Sequence
(0.05,0.16)	I, III$^+$, III$^+$/II, II, II/III$^-$, III$^-$
0.16	I, II, II/III$^-$, III$^-$
(0.16,0.18)	I, III$^-$, III$^-$/II, II, II/III$^-$, III$^-$
(0.18,0.20)	I, III$^-$, III$^-$/II, III$^-$
(0.20,0.44)	I, III$^-$

Note that the first row of Table 7.1 corresponds to the complete succession of Turing patterns consisting of the uniform state, nets, nets/bands, bands, bands/spots, and spots observed experimentally by Ouyang and Swinney ([164]) as $[MA]_0$ was increased. In addition to pure bands and hexagons, Ouyang and Swinney ([163]) also observed distinct stationary mixed states which appeared as if they were overlaps of hexagons and bands, consistent with our predicted intervals of bistability. Specifically this complete sequence depends on β_0 being in the range

$$0.05 = \beta_2(\alpha_3; 1) < \beta_0 < \beta_2(\alpha_c; 1) = 0.16.$$

Further, if β was decreased along the vertical lines $\alpha = \alpha_0^{\pm}$ in Fig. 7.6 where $\alpha_{cr}^- < \alpha_0^- < \alpha_c$ and $\alpha_c < \alpha_0^+ < \alpha_{cr}^+$, with $\alpha_{cr}^{\pm}$ defined implicitly by $\beta_{\sigma_2}(\alpha_{cr}^{\pm}) = \beta_1(\alpha_{cr}^{\pm}; 100)$, the Turing pattern behavior is reminiscent of the morphological solidification sequences predicted along the loci $U = U_0^{\pm}$ of Fig. 7.8, which reprises the results of Chapter 2. Upon examination of Fig. 7.6, we see that $\alpha_{cr}^- = 1.68$ and $\alpha_{cr}^+ = 2.09$. Hence, specifically when $\alpha_0^+ = 2$, we predicted the Turing pattern sequence of spatial homogeneity, spots, spots/bands, and bands as β was decreased. For fixed values of the other parameters such a transition should occur as the $[I_2]_0$ concentration decreases, which is consistent with Ouyang and Swinney's ([163]) experimental observations. More generally, the transit vertical lines $\alpha \equiv \alpha_0$ are capable of generating all those Turing pattern sequences catalogued in Table 7.2.

The morphological stability investigation of Wollkind *et al.* ([295]) was an extension of Segel's ([211]) six-disturbance analysis of the Rayleigh-Bénard problem of buoyancy-driven convection. In particular, it extended Segel's analysis to include the possibility of $2a_2 - a_1 < 0$, given that this quantity was identically positive for the Rayleigh-Bénard problem, as indicated in Chapter 3. Recall from Chapter 2 that the equivalence class of critical points

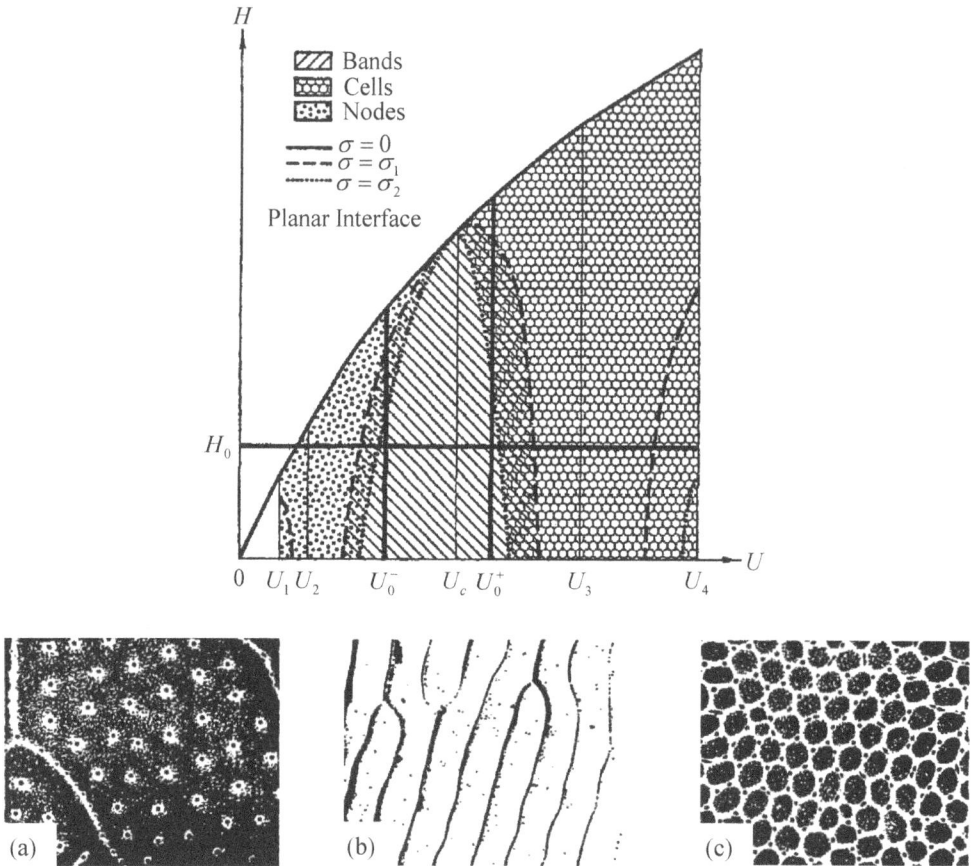

FIGURE 7.8

A plot in the nondimensional U-H plane depicting the predicted three-dimensional stability results of Chapter 2. Here, the photographs labeled (a), (b), and (c), correspond to nodes, bands, and cells, respectively. Compare with Figs. 2.2 and 2.8.

TABLE 7.2

Predicted Turing pattern sequence versus α_0. Here, the notation A/B again indicates the bistability of structures A and B.

α_0	Predicted Sequence
(1.44,1.58)	I, III$^+$
(1.58,1.68)	I, III$^+$, III$^+$/II
(1.68,1.88)	I, III$^+$, III$^+$/II, II
1.88	I, II
(1.88,2.09)	I, III$^-$, III$^-$/II, II
(2.09,2.19)	I, III$^-$, III$^-$/II
(2.19,2.53)	I, III$^-$

designated as II in all our hexagonal planform nonlinear stability analyses actually contains the three solutions

$$A_i^2 = \frac{\sigma}{a_1}, \; A_j = A_k = 0, \; (i,j,k) = \text{ even permutation of } (1,2,3).$$

Hence, the region of Fig. 7.6 identified with bands is itself a locus of multiple stable states. These represent a family of bands or stripes aligned parallel to the r_2-axis, as depicted in Fig. 7.3a, plus two similar families of bands making angles of $\pm 60°$ to them, for which stable co-existence with a member of the original family or one another is impossible. Then, as initial conditions varied from point-to-point in the planar patterned region of the gel, we concluded, as in Chapter 2, that such families of stable bands could give rise to polygonal arcs, the boundaries of which would appear quite random in orientation. Indeed, a number of Turing patterns classified as stripes by Ouyang and Swinney ([163, 164]) have the appearance of such curved elongated cells in the relevant photographic reproductions contained therein. That conclusion was another Rabbit being Pulled Out of a Hat for this problem.

When our graphic artist Karen Parvin drew the contour plots that were to make up Fig. 7.5, she was supposed to use the zero-elevation fourth curve from the center of the hexagonal cell of Fig. 7.4 to denote the demarcation between its dark and light regions, but mistakenly used the fifth such curve instead producing those contours depicted in part (b) of both Figs. 7.9 and 7.10. After having her correct that error by producing the plots appearing in Fig. 7.5, we had Karen make two more contour plots using the third curve from the center as the demarcation between the dark and light regions, depicted in part (a) of both Figs. 7.9 and 7.10 because that mistake gave us an idea. This idea concerned itself with a topic only handled implicitly in the then current pattern formation literature (Walgraef, [267]): Namely, the observational implications of various methodologies for selecting the dark and light regions in contour plots of Turing patterns.

To explain that topic we need to recapitulate the observations of Ouyang and Swinney ([163]) described earlier. They performed a series of experiments involving the CDIMA/starch reaction occurring in a polyacrylamide gel reactor observing one-dimensional stripes and two-dimensional hexagonal arrays of nets or spots forming from an initially uniform state as a control parameter was varied. These Turing patterns as depicted in Fig. 1.2 appear as a yellow (light) oxidized (low $[SI_3^-]$) region on a blue (dark) reduced (high $[SI_3^-]$) background, generated by the color change of the starch indicator during the redox reaction. In Figs. 7.3a and 7.5 the threshold starch tri-iodide concentration (recall $x^* = x$) for that color change was implicitly chosen to coincide with the homogeneous state value of α. Hence, all spatial regions characterized by $x \geq \alpha$ would appear dark and

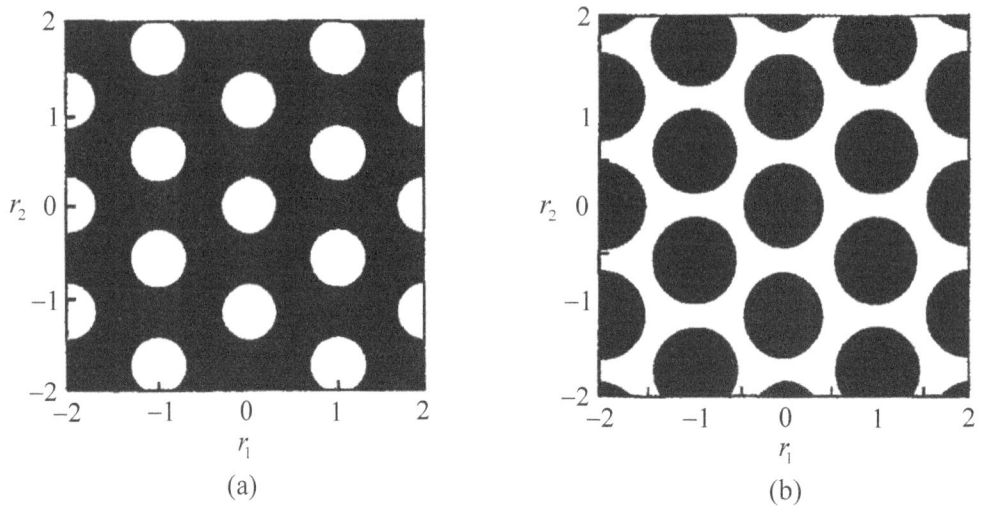

(a)

(b)

FIGURE 7.9
Contour plots for the critical points (a) III$^-$ and (b) III$^+$ based on a lower threshold protocol which represent (a) spots and (b) honeycombs.

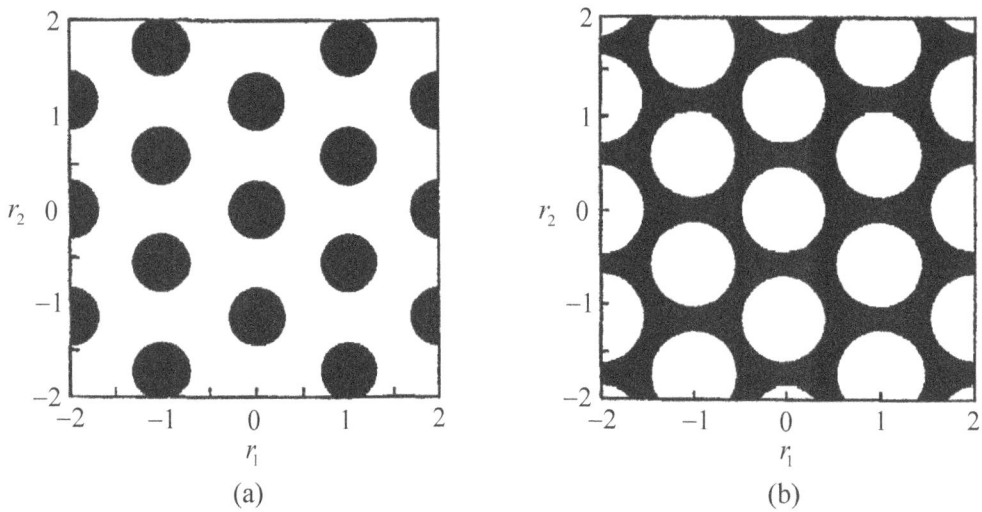

(a)

(b)

FIGURE 7.10
Contour plots for the critical points (a) III$^+$ and (b) III$^-$ based on a higher threshold protocol which represent (a) "spots" and (b) "honeycombs."

$x < \alpha$, light. Thus, given this morphological interpretation, the homogeneous state $x = \alpha$ would appear uniformly dark. The adoption of such a protocol has also been standard operating procedure for representing patterns obtained by numerical simulations of CDIMA model reaction-diffusion systems (Lengyl and Epstein, [115]). Interchanging these colors will result in Turing patterns that preserve stripes, while switching the two hexagon types and is a distinguishing feature of all protocols based on a zero-deviation threshold (see Figs. 7.3a and 7.5). In fact, any critical transition threshold x_c which satisfies $x_c \leq \alpha$ will guarantee that the homogeneous state appears uniformly dark.

In this context, when viewing the photographs depicted in Fig. 1.2 at his Gordon conference poster presentation, my first question to Qi Ouyang was, what is the color of the homogeneous state which had not been exhibited. His answer of 'blue' told me that the yellow regions of the exhibited photographs represented the Turing patterns for his experiments. This being the case, it is instructive to examine what effect the adoption of a different protocol based on another member of that class would have on the appearance of our Turing patterns. Should we choose a lower threshold protocol satisfying $x_c < \alpha$, then the dark regions would predominate, as depicted in Fig. 7.9. Extrapolating from this occurrence for hexagonal arrays, we can infer that such a lower threshold protocol would trigger striped patterns consisting of narrower light stripes alternating with wider dark interstripes (see Fig. 7.11 and Chapter 12). Having investigated these lower threshold protocols, there is also some merit in examining the morphological consequences of the adoption of a higher threshold one satisfying $x_c > \alpha$ instead. For this situation, the homogeneous state $x = \alpha$ would appear uniformly light. Hence, under these circumstances, it seems reasonable to view Turing patterns as being generated by the formation of dark regions against a light background. Given such a morphological interpretation III$^+$ would represent "spots" and III$^-$, "honeycombs," with light regions now predominating as depicted in Fig. 7.10.

It is of interest to note that all of those patterns described in the previous paragraph were actually observed by Ouyang and Swinney ([164]). That the adoption of such protocols to replicate these sort of patterns was not as well-known a theoretical pattern generation mechanism as one might have suspected can be attested to by Hans Meinhardt's ([141]) commentary provided for Kondo and Asai's ([100]) *Nature* letter that investigated the formation of stripes in marine angelfish employing an activator-inhibitor reaction-diffusion model, which incorporated the kinetics of Turing ([256]). Analogous to our contour plots, high and low concentrations of the activator species were represented by dark and light regions, respectively, in their computer simulated patterns. When commenting on this work in conjunction with the actual appearance of the adult angelfish *Pomacanthus imperator* (see Fig. 7.11), whose photograph was featured on the cover of the August 31, 1995, issue in question with the caption "Turing patterns come to life," Meinhardt, perhaps the foremost specialist in biological pattern formation at the time, stated that the light (yellow) stripes on the fish still required further explanation since those shown by Kondo and Asai were very narrow with respect to the dark (blue) spaces in between, while all the models of which he was aware could only produce stripes and interstripes of the same width.

Since $x_0 = \alpha = m[MA]_0$ for fixed values of the other parameters included in m, our threshold for the color change described above has been based on a sliding scale. Let us define that threshold by $x_c = \alpha_{crit} = m[MA]_{0_{crit}}$ where $[MA]_{0_{crit}}$ is the value of $[MA]_0$ corresponding to the starch tri-iodide concentration at which the Turing pattern indicator color change occurs. Hence, we can conclude that the generated Turing patterns will be of zero-deviation, lower, or higher threshold type depending on whether $[MA]_0$ is equal to, greater than, or less than this $[MA]_{0_{crit}}$, respectively. In order to demonstrate this result most easily, first select

FIGURE 7.11
Albert Kok's Wikimedia commons image of the Imperial or Emperor angelfish (*Pomacan-thus imperator*).

$\alpha_{crit} \le \alpha_1$. Then, all of the $\alpha \in (\alpha_1, \alpha_2)$ in the patterned interval of Fig. 7.6 would satisfy $\alpha > \alpha_{crit}$ or $[MA]_0 > [MA]_{0_{crit}}$ and we would have lower threshold patterns for which the dark regions predominate such as those in Fig. 7.9. Next, select $\alpha_{crit} \ge \alpha_2$ instead. Then all of those same $\alpha \in (\alpha_1, \alpha_2)$ would satisfy $\alpha < \alpha_{crit}$ or $[MA]_0 < [MA]_{0_{crit}}$ and we would have higher threshold patterns for which the light regions predominate such as those in Fig. 7.10. I consider our coming up with these higher and lower threshold protocol methods of pattern formation a classic example of how Pulling Rabbits Out of Hats can sometimes occur purely by accident.

This completed Laura Stephenson's thesis work and hence, the dissemination phase could begin. In point of fact, once the one-dimensional analysis was finished, we decided to publish it as part of a paper for a reaction-diffusion model with general reaction functions $F(x, y; \alpha)$ and $G(x, y)$, the results of which were then applied to our CDIMA system, as well as Prigogine's Brussellator reaction and its Schnackenberg simplification (see Wollkind and Dichone, [284]). Given the amount of attention the Turing pattern experiments were drawing, I felt there was some merit in publishing this work as soon as possible and since Turing's original paper was biological in nature chose to submit it to the *Journal of Mathematical Biology*. I drew Segel and my colleague, Ed Pate, as referees, who both accepted our paper, which appeared as Stephenson and Wollkind ([239]), although Lee suggested that the hexagonal planform analysis which was previewed in it should be submitted to an applied mathematics modeling journal rather than a biological one. I chose the *SIAM Journal on Applied Mathematics* for that purpose, figuring it was time to give them another chance after the aerosol convection debacle related in the last chapter.

Having made this choice, I called Greg Kriegsmann of the New Jersey Institute of Technology, the *SIAP* editor, told him I had a paper getting ready for submittal, and asked if they published color figures, since it was my intention to produce the contour plots of Figs. 7.3a, 7.5, 7.9, and 7.10 in blue and yellow. Greg said it was up to the referees to decide whether or not the use of color enhanced the papers they were reviewing and I should submit the manuscript with those figures in color, leaving that decision to them. Since this seemed fair enough to me, I did exactly what he requested. The referees agreed with my choice of the color contour plots, but suggested a few minor revisions, chief among them being a discussion of Rovinsky and Menzinger's ([193]) recent results as applied to our problem. Those authors considered the interaction of the Turing and Hopf bifurcations in the model used by Lengyl and Epstein ([114]), to which our system reduced for $K = 0$ and $0 < \mu < 1$, by performing a weakly nonlinear stability analysis about the degenerate point where those bifurcations occur simultaneously. Upon examination of Fig. 7.6, it can be noted that this degenerate point (α_K, β_K), where the Turing and Hopf boundaries intersect or $\beta_K = \beta_1(\alpha_K; \mu) = \beta_2(\alpha_K; K)$, is given by (1.36,0.004) for $K = 100$, $\mu = 1$, and thus lies in the subcritical bifurcation region relevant to our Turing instabilities since $\alpha_K < \alpha_1$. Hence, this so-called codimension two degenerate bifurcation point lies outside our parameter range of interest. Therefore, Rovinsky and Menzinger's ([193]) predicted spatiotemporal patterns, occurring in the neighborhood of such a point when that bifurcation is supercritical, have no bearing on the scenario considered in our paper. We then modified our original submission to take the referees' suggestions into account, including the observations just pointed out and resubmitted this revised manuscript to *SIAP*, which accepted that revision for publication. Imagine my surprise when, a few weeks later, I got a phone call from Kriegsmann informing me that since the *SIAP* managing editor had decided to overrule the referees' decision and not allow any color figures for our paper, he was requesting that we replace them with black and white contour plots instead, given that, in the opinion of his managing editor, our color versions did not have the suitable contrast to generate what are called 'half-tone figures' from them, which appear dark and light in print. Needless to say, I was very unhappy about the situation and would not have submitted our paper to *SIAP* had Greg told me that this was a possibility from the outset, but having little alternative asked Karen to produce the desired black and white contour plots, which were then sent to *SIAP* as replacements for our original color ones. Although the whole process of *SIAP*'s receipt of the original manuscript to their publication of its final version took three years, certain things involved in that process tend to make more of an impression than others. Indeed, going through this publication Wollkind and Stephenson ([296]) for the first time, I did find a particular occurrence, to be described below, highly ironic in light of all the trouble we went through to replace the original color figures. Realize that, unlike today's publication process where everything is done electronically, in those days, journals set type from the manuscript and incorporated the figures from their originals which were submitted separately. Upon careful inspection of the black and white contour plots in Wollkind and Stephenson ([296]), corresponding to our Figs. 7.3a, 7.5,7.9, and 7.10, which have been reproduced from them, I noticed that the labels for the r_1 and r_2 axes of their Fig. 7.10b differed from those for their Figs. 7.3a; 7.5a,b; 7.9a,b; and 7.10a, in that they were in bold-face. The import of this is that only the original color contour plots had bold-faced labels for their axes while the replacement ones did not. Hence, their Fig. 7.10b was a half-tone reproduction of the original color contour plot that the *SIAP* managing editor said was impossible for them to generate!

In addition I presented our results at the following 2 conferences, 2 colloquia, and 2 workshops:

- The AMS (American Mathematical Society) 899th Meeting special session at the University of Central Florida and the SIAM 45th Anniversary Meeting minisymposium at Stanford University;

- A Department of Mathematics colloquium at Harvey Mudd College and for a Distinguished Lecture Series of the Institute of Applied Mathematics at the University of British Columbia;

- The Pacific Northwest Workshop on Mathematical Biology at Washington State University and The IMA workshop on Pattern Formation and Morphogenesis at the University of Minnesota.

The latter presentation is of particular note. James Murray of the University of Washington, Philip Maini of Oxford University, and Hans Othmer of the University of Minnesota organized this workshop, held in September of 1998 as part of the 1998-1999 IMA special emphasis year on Mathematics in Biology, with the Fall being devoted to Theoretical Problems in Developmental Biology and Immunology, and asked me to present a talk entitled Chemical Turing patterns: A paradigm for morphogenesis. I found this invitation to be somewhat ironic given the original admonishment by Segel, who was visiting the IMA during the 1998-1999 academic year as part of their program, that our hexagonal pattern formation analysis did not belong in a mathematical biology journal, although he seemed to have changed his mind during the intervening three years.

This workshop acquainted me with dryland vegetative pattern formation, which was to play an important role in my later modeling endeavors (see Chapters 11, 12, and 14), through a talk featuring tiger bush in sub-Saharian Africa by Rene Lefever, a colleague of Prigogine at the University of Brussels. Also, after my talk, there was an extended discussion concerning the anomalous black-eyed hexagonal patterns that were observed during the CDIMA/indicator experiments performed by Gunaratne *et al.* ([75]) and reported in the chemical Turing pattern review of Ouyang and Swinney ([164]). These experiments employed a polyvinyl alcohol gel disk, which served as its own Turing pattern indicator by turning dark red in its reduced state, but remaining clear in its oxidized state. For those experiments, these authors discovered a complex stationary periodic black-eye array which they felt was unexpected from the general theory of pattern formation (see Fig. 7.12a). Such structures were only obtained in experiments involving polyvinyl alcohol gel disks of high concentration. This pattern formed from a normal hexagonal one, which was the initial instability to the uniform state, when $[MA]_0$ increased slowly. They described it as consisting of two hexagonal lattices: One of white spots and the other of smaller black spots located at the center of each white spot and at the center of the dark region in each equilateral triangle with three neighboring white spots as its vertices (see Fig. 7.12b). Then, upon further increase of $[MA]_0$, this pattern disappeared and was replaced by stripes. Specifically, the transition from the clear uniform state to the normal hexagonal spot pattern and from the latter to the black-eye array, which occurred at $[MA]_0 = 7$ and 8 mM, respectively, were nonhysteretic, while that from the black-eye pattern to stripes was hysteretic with the hysteresis occurring over the 11.6-13.8 mM interval (see Fig. 7.12c). This discussion got me thinking of trying to replicate these experimental results theoretically. I thought that would involve the superposition of both hexagonal-type patterns, but since they did not co-exist as bistable states for any parameter range (see Fig. 7.6), it seemed implausible this mechanism could be used as a black-eye explanation.

On the last day of the workshop, Hans Othmer, who along with Philip Maini had organized the sessions, asked me, while we both were sharing an elevator, just how thin was the layer where the patterns formed during the CDIMA experiments modeled in my talk. This question was related to the validity of the quasi-two-dimensional approximation adopted in the proposed mathematical models for these experiments. The justification for that assumption was the occurrence of those Turing patterns in an extremely thin layer relative to the thickness of the reactor's gel disk. His question, however, got me to realize that, in spite of this assumption, the thickness of the pattern-forming layer was not infinitesimal, but rather finite. Suddenly, I had the mechanism to superpose both the hexagonal patterns in order to create the black-eye arrays. To reproduce the black-eye hexagonal array sequence described above, it was necessary to relax the infinitesimal thickness constraint, which we implicitly assumed when taking $z = z_0$ during all the interpretations of our quasi-two-dimensional model results presented previously in Wollkind and Stephenson ([296]). This epiphany-like realization represented the fifth and most important Rabbit Pulled Out of a Hat. Toward that end, we introduced a thin layer of thickness Δz located in the interval $z \in (z_1, z_2)$, where $z_{2,1} = z_0 \pm \Delta z/2$. This replaced our transit curve $\beta \equiv \beta_0$ of Table 7.1 particularized to $z = z_0$ by a band of width

$$\Delta\beta \cong \frac{\beta_0}{1 - z_0}\Delta z$$

centered about β_0, such that for a fixed value of $[MA]_0$ the locus of interest in the α-β plane was a line segment through (α_0, β_0) joining the end points $(\alpha_0 \pm \Delta\alpha/2, \beta_0 \pm \Delta\beta/2)$ for $z \in (z_1, z_2)$, where $\alpha_0 = m_0[MA]_0$ and

$$\Delta\alpha \cong \frac{\alpha_0}{z_0(1 - z_0)}\Delta z,$$

with m_0 being the m, as previously defined, particularized to $z = z_0$. Here, both $\Delta\alpha$ and $\Delta\beta$ were determined using the linear interpolation for $f(z)$ where $\Delta f \cong f'(z_0)\Delta z$, which approximates a curve by its tangent line if $|\Delta z| \ll 1$ with

$$\alpha(z) = \frac{k_1[MA]_0}{5k_2[ClO_2]_0 u_0}\frac{z}{1 - z}, \quad \beta(z) = \frac{k_3[I_2]_0}{k_2[ClO_2]_0 u_0}\frac{1}{1 - z};$$

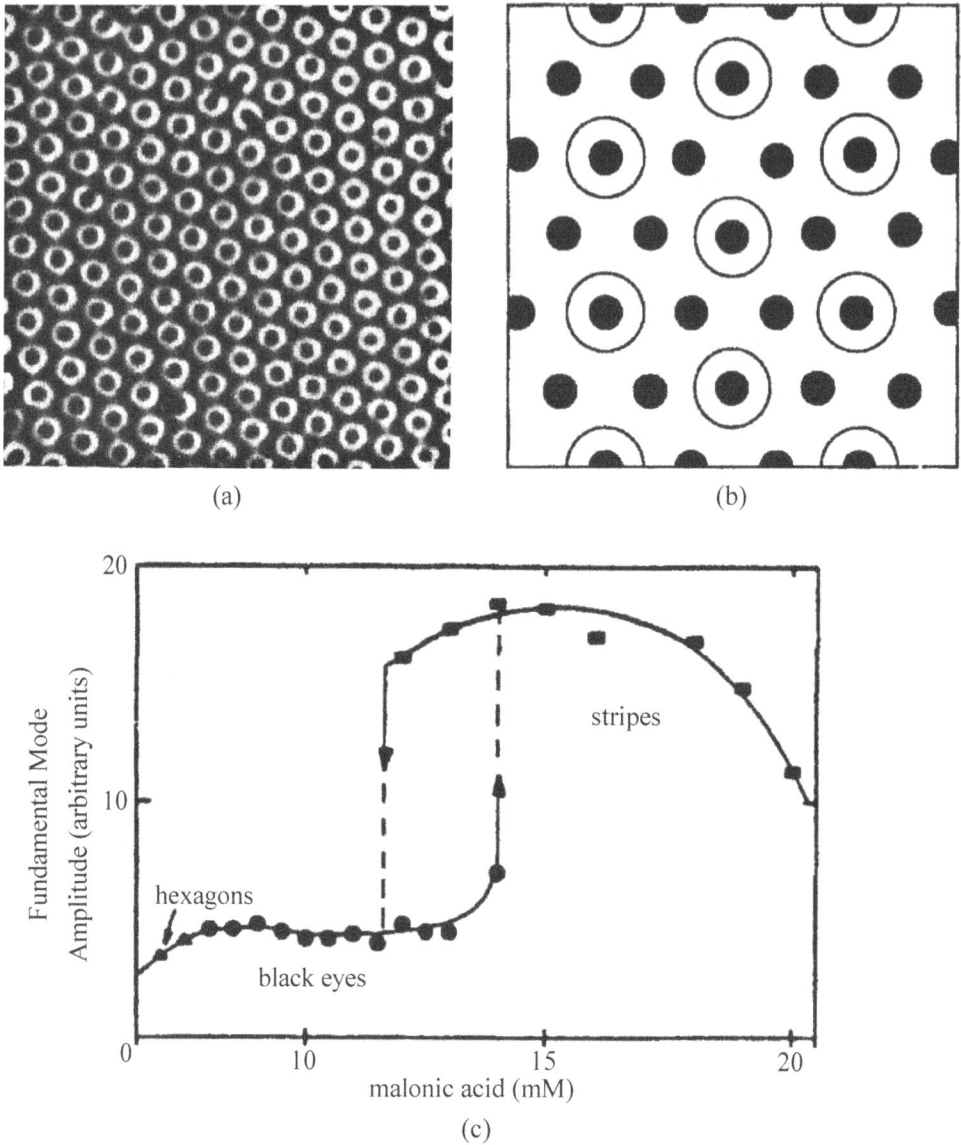

(a)

(b)

(c)

FIGURE 7.12

(a) Photographic reproduction from Ouyang and Swinney ([164]) of the black-eye hexagonal patterns for the CDIMA/indicator system in a polyvinyl alcohol gel which serves as its own pattern indicator. (b) Schematic diagram of the two hexagonal lattices that are described in the text as composing this pattern. (c) Reproduction from Gunaratne *et al.* ([75]) of the fundamental modes versus $[MA]_0$ plot for these black-eye experiments.

where $\alpha_0 = \alpha(z_0)$, $\beta_0 = \beta(z_0)$, and

$$\left(\frac{z}{1-z}\right)' = \left(\frac{1}{1-z}\right)' = \frac{1}{(1-z)^2}.$$

Hence,

$$\alpha'(z_0) = \frac{k_1[MA]_0}{5k_2[ClO_2]_0 u_0} \frac{1}{(1-z_0)^2} = \frac{\alpha_0}{z_0(1-z_0)};$$
$$\beta'(z_0) = \frac{k_3[I_2]_0}{k_2[ClO_2]_0 u_0} \frac{1}{(1-z_0)^2} = \frac{\beta_0}{1-z_0}.$$

Finally, considering a band of this sort, flanking $\beta_c = \beta_2(\alpha_c; 1) = 0.16$ in Fig. 7.6, and superposing those patterns predicted in the top and bottom portions of the layer as $[MA]_0$ is increased, we were able to obtain the observed sequence under investigation.

Let us describe that procedure in detail. The optical transmission technique inherent to these experimental observations permitted us for comparison purposes to superpose the predicted Turing patterns for the top $z \cong z_2$ and bottom $z \cong z_1$ surfaces of the thin layer. Upon examination of Fig. 7.6, we concluded that the patterns at the top and bottom were represented by the predicted sequences listed in the third and first rows of Table 7.1, respectively, with the Turing patterns for $z \cong z_2$ emerging from the uniform state before those for $z \cong z_1$, as $[MA]_0$ increased, given that the slope of the line defined above generating this band is positive. Moreover, we truncated these sequences by limiting the $[MA]_0$ range so that each terminated in the region where stripes were the only stable pattern and further assumed that the critical $[MA]_0$, value at which the polyvinyl alcohol color change occurred, corresponded to a point in this region as well. Hence the III$^\pm$ hexagonal states in these sequences are of the higher threshold variety as depicted in Fig. 7.10, while the uniform state I appears clear. In what follows, we shall use the notation $\{\ldots, \ldots\}$ to represent the layer patterns where its first and second entries are for those on the upper $z \cong z_2$ and lower $z \cong z_1$ surfaces, respectively. The superposition of these patterns in those sequences that can coexist for a particular value of $[MA]_0$ in the allowable range results in the following superposed combinations: $\{I, I\}$, $\{III^-, I\}$, $\{III^-, III^+\}$, and $\{II, II\}$ as $[MA]_0$ increases, which are identifiable with a homogeneous, "honeycomb," black-eye, and striped pattern, respectively.

Justification of these identifications required us to re-examine the resolution of our hexagonal structures of Fig. 7.10. Only regions of relatively high tri-iodide concentration appear dark in those patterns. From the definition of III$^-$, we deduced that the highest such concentrations for the "honeycomb" of Fig. 7.10b were located in the circular regions about the vertices of the hexagons of Fig. 7.4a or equivalently inscribed within the equilateral triangles depicted in Fig. 7.4b. Should the lighting conditions be sensitive enough for their resolution, those regions would appear as black dots standing out against the dark background of the "honeycomb" patterns. When superposed with the black "spots" of III$^+$ in Fig. 7.10a, this configuration yields the black-eyed pattern described by Gunaratne et al. ([75]) and Ouyang and Swinney ([164]) depicted in Fig. 7.12a,b, where the black hexagonal lattice comes from a combination of Figs. 7.10a,b and the white one from Fig. 7.10b. When superposed with the uniform state of I, it yields the normal "honeycomb" hexagonal pattern since the latter situation, unlike the former, fails to provide sufficient illumination at the key vertices to make them visible (note, in this context, that the corresponding vertices of III$^+$ are brighter than their light background). This circumstance is analogous in leopard or jaguar patterning to the black spots on the dark coat of its melanistic panther color phase only being visible in direct sunlight, but not in the shade (see Fig. 7.13). A resolution of this sort for the "honeycomb" and "spot" patterns is consistent with the appearance in Fig. 7.8 of the cells and nodes, respectively, of the metallurgical problem of Chapter 2 to which they are now correlated. As one last detail, we observed that the interplay of the bi-stability regions of the pattern sequences of the third and first rows of Table 7.1 characteristic of the top and bottom of the layer, respectively, gives rise to the hysteretic behavior between the black-eyed and striped patterns reported by Ouyang and Swinney ([164]), as depicted in Fig. 7.12c.

FIGURE 7.13
Photographic reproduction from Meyer ([144]) of the melanistic phase of the jaguar (*Panthera onca*).

To make these qualitative comparisons between theoretical predictions and experimental observations more quantitative in nature, we next assigned the model parameters their typical values of

$$k_1 = 9.0 \times 10^{-4}/\text{sec}, \ k_2 = 1.2 \times 10^3/(\text{M sec}), \ k_3 = 1.5 \times 10^{-4}/\text{sec},$$

appropriate for $T = 280$ K; and

$$[I_2]_0 = 7.2 \times 10^{-4} \text{ M}, \ [ClO_2]_0 = 1.6 \times 10^{-3} \text{ M}, \ u_0 = 3.5 \times 10^{-6} \text{ M}, \ z_0 = 0.9;$$

the latter condition being consistent with the location of the chemical front near the malonic acid reservoir boundary of $z = 1$ during these experiments. We then found that the intersection point (α_0, β_0) between the transit curve $\beta \equiv \beta_0$ and the Turing boundary $\beta = \beta_2$ satisfied

$$\alpha_0 = \alpha_c, \ \beta_0 = \beta_c,$$

when
$$[MA]_0 = 7 \text{ mM},$$

for the scenario proposed above, in agreement with the polyvinyl alcohol gel sequence depicted in Fig. 7.12c. The dimensional critical wavelength of the resulting pattern corresponded to
$$\lambda_c^* = 0.20 \text{ mm},$$

employing our formula for that quantity deduced earlier, in accordance with the experimental measurements of Gunaratne et al. ([75]). In doing so, we pulled our final Rabbit Out of a Hat.

Gunaratne et al. ([75]) offered an alternate explanation for their periodic black-eye array. They associated the black dots at the hexagonal vertices with the black "spot" pattern and postulated such a black hexagonal lattice was a spatial harmonic of the primary white spotted one, the former being generated as a secondary mode by the resonant interaction of the basic modes of the latter. Employing that hypothesis however, Gunaratne et al. ([75]) were unable to explain why this secondary mode did not grow continuously beyond the onset of the primary instability, while stating that they did not understand this difference between their theory and experiment. Gunaratne et al. ([75]) then suggested that either there might not be sufficient sensitivity to detect this harmonic closer to the onset of normal hexagons or perhaps the secondary modes were not so-called "slaved" to the primary ones, in the sense of bifurcation theory (Walgraef, [267]). Given that our explanation for the occurrence of black-eyed patterns alleviated this deficiency between theory and experiment, we considered ours to be an improvement on theirs.

This completed my and Laura's analyses of the chemical Turing instability problem and all that remained was the dissemination of our final results in the usual way. Philip Maini and Hans Othmer published a Springer volume entitled *Mathematical Models for Biological Pattern Formation* that was composed of refereed chapters of all their workshops' invited speakers' talks. We took this opportunity to summarize those results, which appeared as Wollkind and Stevenson ([277]), having the title of our presentation, but concluding with the addition of the black-eye pattern explanation just described.

Three years later, I was asked to participate in a mini-symposium on Turing patterns in biology at the 2004 SIAM summer meeting and presented these results in my invited address. I have always been very appreciative of this invitation for the following two reasons: It allowed me to include our black-eyed pattern formation explanation in an oral presentation and to see Lee A. Segel once more before his untimely death in February of 2005 from pancreatic cancer. That meeting was held in the Portland, Oregon conference center and, since Lee attended it, we were able to spend a good deal of time together. In a conversation with some mutual colleagues, including Leah Keshet, he said that good doctoral students always thought they could improve upon their thesis advisers' results, "...isn't that right, David," and poked me in the ribs. Further, Segel added that, although I had made great strides in comprehensive applied mathematical modeling in the C. C. Lin sense of comparing theoretical predictions with observational or experimental data of the phenomena involved, this effort would not be truly comprehensive in the traditional sense of the totality of phenomena investigated until that number exceeded the four areas of materials science, fluid mechanics, predator-prey ecological interactions, and chemical Turing patterns covered to date by my research publications. I asked him what would he say when that number equaled five and Segel replied, "We'll see." Unfortunately, we never did.

As is obvious from the content of this book, Segel has played a very influential role in my professional career and I have missed him immensely since his death, especially in regard to those discussions we periodically had about our latest research results. It is a sobering thought to realize that by the time Segel was my present age he had been dead for five years (with apologies to the folk singer musician comedian Tom Lehrer, who was one of Lee's undergraduate classmates and originally made the same statement about Mozart using two years, instead of five). His continuing influence on my work however can be attested to by the number of references to him that are included in the rest of this book, mostly involving research finished after Segel died. Indeed, when faced with a decision on the choice of ways to make a particular point in writing about these results, I still ask myself how Lee would have done it. Also, my constant goal remains to produce research results that would have made him proud to be my thesis adviser.

This chapter about my collaboration with Laura Stephenson could not be considered complete unless I included the following statement in reference to her amazing ability to perform very complicated calculations without making any errors. Simply put, I never have found a single mistake in all of Laura's computations required for the nonlinear stability analyses of the chemical Turing pattern formation investigations of the reaction-diffusion model systems that constituted our research. Neither I nor any of my other Ph. D. students can make that claim. As a check on my students' work, we often performed these calculations in parallel. Any discrepancy between Laura's calculations and mine were invariably due to my error and not hers. She was a researcher who did not make any mistakes in her expansions and whose preliminary calculations were always her final ones. Those of my graduate students who have performed similar hexagonal planform nonlinear stability analyses on their reaction-diffusion or related interaction-dispersion model systems all used Laura's calculations as a template for their work. No greater testimonial to the accuracy of her computations can possibly be made. Laura Stephenson is indeed the singular exception which serves to prove the age old rule "that to err is human."

8

Evolution Equation Phenomenon I: Lubrication Theory of Liquids: Hexagonal Planform Nonlinear Stability Analysis

All the topics I selected for my Ph.D. students after Laura Stephenson were based upon articles appearing in the journal *Science*, published weekly by AAAS (American Association for the Advancement of Science). These pattern formation analyses are covered in this chapter and Chapters 9-14. The four problems contained in this and Chapters 9-11 could be reduced to the analysis of a spatio-temporal nonlinear partial differential model evolution equation, a different one for each problem, and hence the titles of these chapters. In particular, the two represented in this and the next chapter are the Lennard-Jones lubrication and damped Kuramoto-Sivashinsky equations (see Cross and Hohenberg, [42]). The lubrication theory of liquids and ion-sputtering of solids (see Chapter 9) were the thesis topics for Mei (Emily) Tian and Adoon Pansuwan, my WSU and Mahidol University in Bangkok, Thailand, Ph.D. students, respectively, arranged in the order of the publication dates for the papers from their dissertations. These phenomena have been grouped together in Chapters 8 and 9 because both of them give rise to very similar thin-film pattern formation on the nm scale, with the relevant contour plots appearing in Chapter 9.

I first met Mei Tian when she enrolled in MATH 570, the Fall semester of my academic year-long continuum mechanics sequence. As for all my graduate courses, there were no in-class examinations, with the students' grades being determined solely by their scores on original problem sets, closely related to material presented in class, and a take-home final examination. After the first graded problem set had been returned, Mei disputed every lost point. Her basic contention was that she was right and I, wrong. This performance was repeated for each problem set and the final exam. Hence, I was not surprised when Mei, in spite of having received an "A" grade in MATH 570, chose not to enroll in its sequel MATH 571. What did surprise me was that, after passing her doctoral qualifying examination, she asked if I would be willing to supervise her Ph.D. dissertation. I was. We decided to consider a problem just featured as a perspective on thin liquid film pattern formation entitled "The artistic side of intermolecular forces" by Günter Reiter ([185]) describing the article "Spinodal dewetting in liquid crystal and liquid metal films" by Stephan Herminghaus *et al.* ([79]), both of which appeared in the volume 282 number 5390 issue of *Science*. Reiter ([185]) began his perspective by stating:

Liquid films thinner than about 1 μm are often unstable and develop various morphologies (see Fig. 8.1). The stability of such films may be essential, for both numerous applications and various scientific experiments, and so this behavior can be rather annoying. As reported on page 916 of this issue Herminghaus *et al.* ([79]) have recently addressed the question of why such films are unstable. They have presented some convincing results that indicated the relevance of intermolecular forces for what they called "spinodal dewetting." It is similar to the process of spinodal decomposition, in which

DOI: 10.1201/9781003195603-8

a binary mixture becomes unstable and moves along a cusplike curve (spinode) toward phase separation.

Reiter ([185]) concluded by stating that Herminghaus *et al.*'s ([79]) major achievement was their demonstration that the same coupling mechanism between van der Waals long-range attractive and Born short-range repulsive forces could account for the different patterns produced in a given experiment and their subsequent suggestion that the exact form of the associated intermolecular potential could be inferred from knowledge of this interfacial morphology together with layer thickness as an inverse problem. He then declared that unfortunately it was still unclear why almost the identical patterns can be found in both polystyrene and gold films.

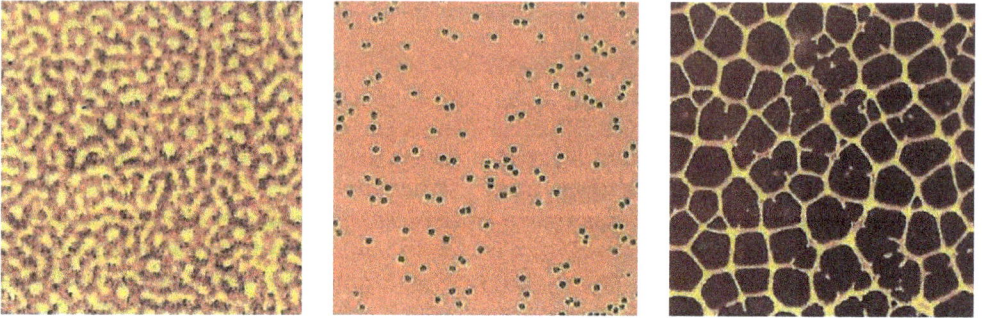

FIGURE 8.1
Three different morphologies exhibited by a thin polystyrene film on a silicon substrate appearing in Reiter ([185]) depicting from left to right an undulation pattern and the early and later stages of cellular pattern formation. Here, the photographs on the far left and right represent the predicted morphological phase separation equilibrium states associated with the hexagonal planform critical points II and III$^-$, respectively, while critical point III$^+$ corresponds to a close-packed configuration of nanodroplets (not pictured here), all separated by flat ultra-thin films.

The primary purpose of our research was to answer questions of this sort. As stated above, liquid layers thinner than about 1 μm are often unstable and spontaneously form various stationary equilibrium morphologies including surface ridges, which exhibit clearly defined critical wavelengths and hexagonal cellular patterns or uniform distributions of nanodroplets separated by very thin films, as depicted in the photographic reproductions of experimental results (see Fig. 8.1 where lighter colors represent thicker sections of the sample). These experiments involved thin liquid polymer, crystal, or metallic films coating a solid substrate open to the ambient air. Hence, we considered a thin liquid layer of mean thickness h_0 bounded below by a planar solid surface located at $z = 0$ and from above by an interface satisfying $z = h(x, y, \tau)$ which separates it from a passive gas (see Fig. 8.2). The liquid layer is thin enough so that intermolecular forces must be taken into account and variations in its density may be neglected, but thick enough so that a continuum mechanical approach will still be valid. Then in the long-wavelength limit the Navier-Stokes and continuity partial differential equations for this liquid layer are given by

$$(p + \rho_0 g z + \phi)_z = P_z = 0, \quad \mu u_{zz} = P_x, \quad \mu v_{zz} = P_y, \quad u_x + v_y + w_z = 0;$$

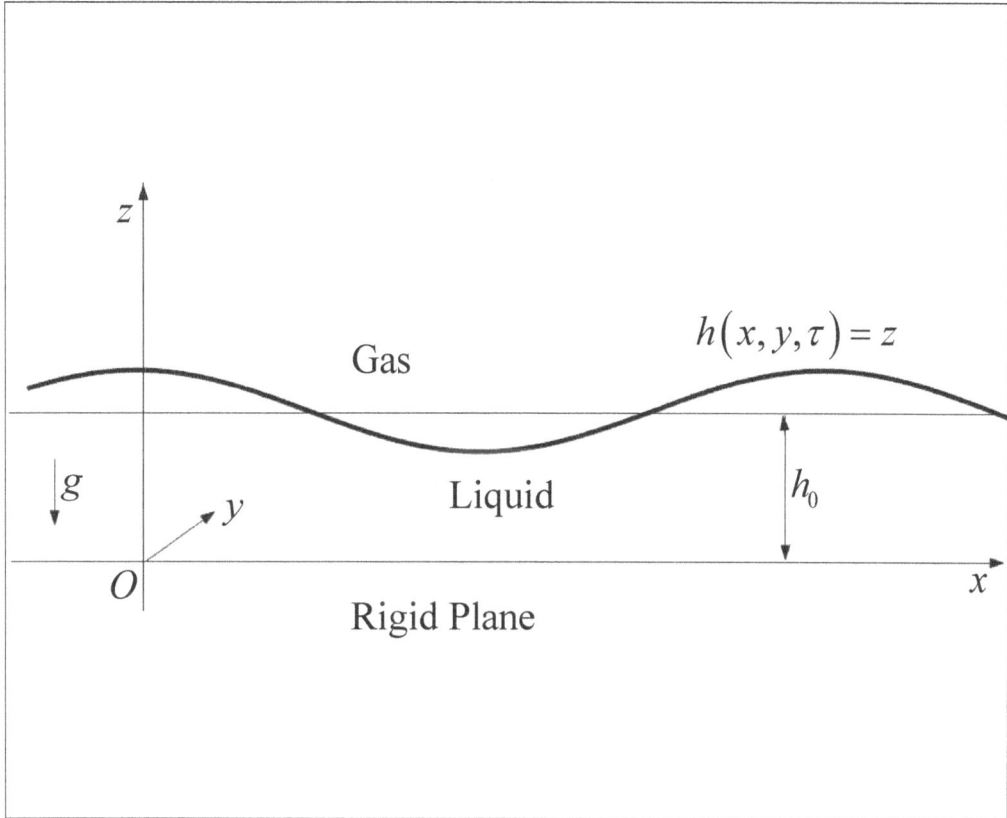

FIGURE 8.2
The physical configuration showing the dimensional Cartesian coordinate system.

with the no-slip and the no-penetration boundary conditions on the solid surface

$$u = v = w = 0 \text{ at } z = 0;$$

and the balance of the tangential and normal components of momentum and the kinematic boundary conditions at the interface

$$\mu u_z = \gamma_x, \ \mu v_z = \gamma_y, \ p_0 - p = \gamma(h_{xx} + h_{yy}), h_\tau + u h_x + v h_y = w \text{ at } z = h.$$

Here, τ is time and the Cartesian coordinate system (x, y, z), with corresponding velocity components (u, v, w), has been employed, while μ and ρ_0 are the constant shear viscosity and density of the Newtonian fluid layer. In addition, g is the acceleration due to gravity, while p and p_0 refer to the pressure in the liquid layer and passive gas, respectively. Further, $\phi = \phi(h)$ and $\gamma = \gamma(h)$ are the intermolecular body force potential per unit volume and interfacial surface tension coefficient per unit length, both of which depend on the layer thickness h, the latter dependence being due to a nonzero $\gamma'(h)$ caused by the Marangoni effect (see Chapter 5). The long-wavelength asymptotic procedure entails introducing a positive nondimensional scaling parameter k such that $w = O(k); z, u, v, h = O(1); x, y, \tau, P, \gamma' = O(1/k);$ and $\gamma = O(1/k^3)$. Then, taking the limit as $k \to 0$ in the full system of partial differential equations (PDEs) and boundary conditions (BCs), one retains only those terms appearing above. Solving these PDEs and all the BCs, except for the last two interfacial ones, we obtained the velocity components

$$\mu u = P_x \left(\frac{z^2}{2} - hz \right) + \gamma_x z, \ \mu v = P_y \left(\frac{z^2}{2} - hz \right) + \gamma_y z,$$

$$\mu w = -(P_{xx} + P_{yy}) \frac{z^3}{6} + [(P_x h)_x + (P_y h)_y - (\gamma_{xx} + \gamma_{yy})] \frac{z^2}{2};$$

which upon substitution into the kinematic boundary condition at the interface yielded

$$3\mu h_\tau - (h^3 P_x)_x - (h^3 P_y)_y + 3 \frac{(h^2 \gamma_x)_x + (h^2 \gamma_y)_y}{2} = 0 \text{ at } z = h.$$

Using the definition for P in conjunction with the interfacial normal momentum balance boundary condition, we converted this into the following PDE only involving $h = h(x, y, \tau)$:

$$3\mu h_\tau + \boldsymbol{\nabla} \cdot [h^3 \boldsymbol{\nabla} \{\gamma(h) \nabla^2 h\}] - \boldsymbol{\nabla} \cdot [h^3 \{\phi'(h) + \rho_0 g\} \boldsymbol{\nabla} h] + 3 \boldsymbol{\nabla} \cdot \frac{[h^2 \gamma'(h) \boldsymbol{\nabla} h]}{2} = 0$$

where $\boldsymbol{\nabla} \equiv (\partial/\partial x, \partial/\partial y)$ and $\nabla^2 \equiv \boldsymbol{\nabla} \cdot \boldsymbol{\nabla}$. We completed the formulation of our problem by choosing the Lennard-Jones potential and the Langmuir surface tension functions (Mitlin, [146])

$$\phi(h) = \frac{a}{h^3} - \frac{b}{h^9}; \ a, b > 0; \ \gamma(h) = \Gamma(T) = \gamma_0 - \gamma_1(T - T_0) \text{ at } z = h; \ \gamma_{0,1} > 0;$$

where the terms proportional to a and b in $\phi(h)$ were due to the long-range van der Waals attractive and extremely short-range Born repulsive forces, respectively; and consistent with the long-wavelength approximation the temperature T in $\Gamma(T)$ satisfied

$$T_{zz} = 0 \text{ for } 0 < z < h; \ T = T_0 \text{ at } z = 0; \ T_z = -m_0 \leq 0 \text{ at } z = h.$$

Here, we assumed that the planar solid surface was a perfect conductor maintained at a fixed temperature $T_0 = O(1)$, while the interface was a poor conductor with an imposed heat flux $m_0 = O(1)$ to the environment (Davis, [46]). Defining h_c by $\phi'(h_c) = 0$ and solving for T, we found that

$$b = \frac{ah_c^6}{3} \text{ and } T = T_0 - m_0 z,$$

respectively, which upon substitution yielded

$$\phi(h) = \left(\frac{a}{h_c^3}\right)\left[\left(\frac{h_c}{h}\right)^3 - \frac{(h_c/h)^9}{3}\right] \text{ and } \gamma(h) = \gamma_0 + \gamma_1 m_0 h,$$

where $a = O(1/k)$ in $\phi(h)$ is related to the Hamaker van der Waals force constant, while $\gamma_0 = O(1/k^3)$, $\gamma_1 = O(1/k)$ in $\gamma(h)$ are the capillarity and thermal Marangoni coefficients.

Then incorporating these functions into our nonlinear PDE for h and nondimensionalizing the resulting equation by introducing the scale factors $\rho_0 h_c^2/\mu$ for time and h_c for both length and layer thickness, we obtained

$$h_\tau + S\boldsymbol{\nabla} \cdot [h^3 \boldsymbol{\nabla}(\nabla^2 h)] + A\boldsymbol{\nabla} \cdot [(h^{-1} - h^{-7})\boldsymbol{\nabla} h] - G\boldsymbol{\nabla} \cdot (h^3 \boldsymbol{\nabla} h) + M\boldsymbol{\nabla} \cdot \frac{(h^2 \boldsymbol{\nabla} h)}{2} = 0$$

where

$$S = \frac{\gamma_0 \rho_0 h_c}{3\mu^2}, \ A = \frac{a\rho_0}{\mu^2 h_c}, \ G = \frac{g\rho_0^2 h_c^3}{3\mu^2}, \ M = \frac{\gamma_1 m_0 \rho_0 h_c^2}{\mu^2},$$

and a second term proportional to M, $\boldsymbol{\nabla} \cdot [h^3 \boldsymbol{\nabla}(h\nabla^2 h)]$, had been neglected since it was negligible with respect to the one retained by virtue of the long-wavelength approximation. Finally, we introduced the rescaled variables (Williams and Davis, [274])

$$t = \frac{A^2 \tau}{S}, \ \boldsymbol{r} \equiv (r_1, r_2) = \left(\frac{A}{S}\right)^{1/2}(x, y), \ H(\boldsymbol{r}, t) = h(x, y, \tau),$$

which transformed our nonlinear PDE into

$$H_t + \boldsymbol{\nabla}_2 \cdot [H^3 \boldsymbol{\nabla}_2(\nabla_2^2 H)] + \boldsymbol{\nabla}_2 \cdot [(H^{-1} - H^{-7})\boldsymbol{\nabla}_2 H]$$
$$- \varepsilon \boldsymbol{\nabla}_2 \cdot (H^3 \boldsymbol{\nabla}_2 H) + \beta \boldsymbol{\nabla}_2 \cdot \frac{(H^2 \boldsymbol{\nabla}_2 H)}{2} = 0$$

where

$$\varepsilon = \frac{G}{A} = \frac{g\rho_0 h_c^4}{3a} > 0, \ \beta = \frac{M}{2A} = \frac{\gamma_1 m_0 h_c^3}{2a} \geq 0,$$
$$\boldsymbol{\nabla}_2 \equiv \left(\frac{\partial}{\partial r_1}, \frac{\partial}{\partial r_2}\right), \ \nabla_2^2 \equiv \boldsymbol{\nabla}_2 \cdot \boldsymbol{\nabla}_2.$$

This was the spatio-temporal model evolution equation we wished to analyze for the type of interfacial phase-separation morphologies catalogued earlier. Although bifurcation analyses relevant to the linear problem and numerical simulations relevant to the nonlinear one were either conducted on, or reviewed for, evolution equations of that sort in the references already cited, their primary emphasis was on the related phenomenon of thin liquid film rupture and none of them included all of the features contained in this equation.

At the same time we started our research, Oron and Bankoff ([159]) performed both a weakly nonlinear stability analysis and a numerical simulation on a one-dimensional version of an evolution equation similar to ours, but with $\varepsilon = 0$, the thermal Marangoni effect replaced by one appropriate for more general heat transfer, and the repulsive force term in the intermolecular potential taken proportional to h^{-4} rather than h^{-9}. In particular, they felt that such a term was superior to one proportional to h^{-9} for representing actual substrates which are often coated or microscopically rough since the latter Lennard-Jones Born potential had been deduced for an idealized clean smooth surface. We shall defer a discussion of this matter until the comparisons included in the last part of the chapter.

To that date, no two-dimensional weakly nonlinear stability analysis had been performed on any of these thin liquid film lubrication-type evolution equations. To alleviate this deficiency we wished to perform a hexagonal-planform weakly nonlinear stability analysis on our evolution equation. Toward this end we noted that there existed a planar interface solution

$$H_0(\alpha) = \alpha = \frac{h_0}{h_c} > 0$$

of our equation which satisfied the implicit far-field boundary condition

$$H \text{ remains bounded as } r_1^2 + r_2^2 \to \infty$$

and represented a layer of uniform depth. It was the weakly nonlinear stability of this solution to longitudinal-planform one-dimensional and hexagonal-planform two-dimensional perturbations with which we were concerned, the former being a special case of the latter. As in the previous chapter, we first performed a one-dimensional analysis of this state by considering a solution to our evolution equation of the form

$$H(\boldsymbol{r}, t) \sim \alpha + A_1(t)\cos(qr_1) + A_1^2(t)[H_{20} + H_{22}\cos(2qr_1)]$$
$$+ A_1^3(t)[H_{31}\cos(qr_1) + H_{33}\cos(3qr_1)]$$

where, as usual, the amplitude function $A_1(t)$ satisfied the Landau equation

$$\frac{dA_1}{dt} \sim \sigma A_1 - a_1 A_1^3$$

and $q = 2\pi/\lambda$, λ being the wavelength of that class of spatially periodic perturbations under investigation (see below for a justification of this truncation procedure).

Then our first-order problem yielded the secular equation

$$\sigma = (\beta\alpha^2 - \varepsilon\alpha^3 + \alpha^{-1} - \alpha^{-7})q^2 - \alpha^3 q^4 = \sigma_0(q^2; \beta, \alpha, \varepsilon)$$

from which we deduced the stability criterion

$$\beta \le \beta_0(\alpha; \varepsilon) = \varepsilon\alpha - \alpha^{-3} + \alpha^{-9};$$

or, when that was violated, the instability criterion

$$\beta > \beta_0(\alpha; \varepsilon).$$

Under this condition, after Sekimura *et al.* ([220]), who used an identical approach on a secular equation of the same form, we selected a fixed value of β_c satisfying

$$\beta > \beta_c \ge 0$$

such that, for $\beta_c = 0$ and $q^2 = q_c^2 = Q_c(\alpha; \varepsilon) = \alpha^{-4} - \alpha^{-10} - \varepsilon > 0$,

$$\sigma = \alpha^2 Q_c(\alpha; \varepsilon)\beta = \alpha^3 \frac{Q_c^2(\alpha; \varepsilon)}{4} = \delta^2 > 0.$$

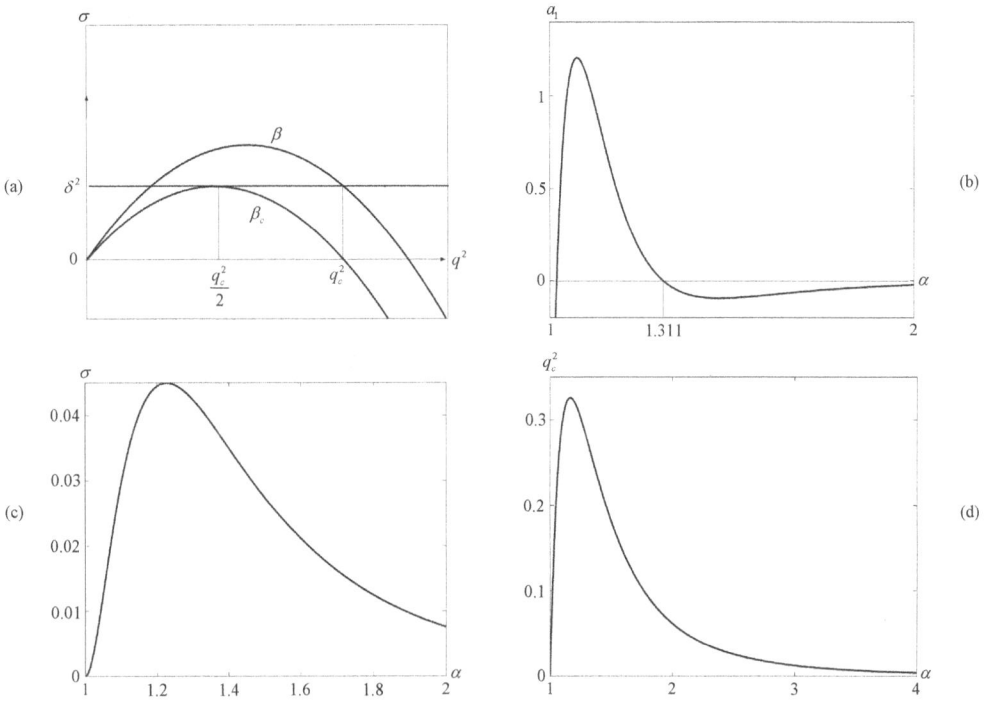

FIGURE 8.3
(a) Schematic plots of $\sigma = \sigma_0(q^2; \beta, \varepsilon)$ and $\sigma = \sigma_0(q^2; \beta_c, \varepsilon)$ where $\beta > \beta_c \geq 0$ is such that $\sigma_0(q_c^2; \beta, \varepsilon) = \sigma_0(q_c^2/2; \beta_c, \varepsilon) = \delta^2 \ll 1$. Plots of (b) a_1, (c) $\sigma = \alpha^3 Q_c(\alpha; \beta_c, \varepsilon)/4 = \delta^2$, and (d) $q_c^2 = Q_c(\alpha; \beta_c, \varepsilon)$ versus α for $\beta_c = 0$ and $\varepsilon = 10^{-11}$.

That scenario is depicted in Fig. 8.3a. We now offer a more detailed explanation of this approach to our secular equation given that its application represented the first Rabbit Pulled Out of a Hat. It involves selecting a critical value $\beta_c \geq 0$ of the bifurcation parameter so that the associated parabolic secular equation $\sigma = \sigma_0(q^2; \beta_c, \alpha, \varepsilon)$ has its maximum that occurs at $q^2 = q_c^2/2$ satisfying $\sigma_0(q_c^2/2; \beta_c, \alpha, \varepsilon) = \delta^2 \ll 1$, where the critical wavenumber squared $q_c^2 > 0$ is defined by $\sigma_0(q_c^2; \beta_c, \alpha, \varepsilon) = 0$. Then, the bifurcation parameter β is perturbed about this critical value in such a way that the resulting secular equation satisfies $\sigma_0(q_c^2; \beta, \alpha, \varepsilon) = \delta^2$. We showed that the constraint of $q_c^2 > 0$ or equivalently the instability criterion particularized to the isothermal case of $\beta_c = 0$ treated by Mitlin ([146]), occurred whenever the nondimensional mean layer thickness α satisfied a spinoidal-decomposition (Cahn and Hilliard, [23])-like condition $\alpha_1 < \alpha < \alpha_2$ where $\alpha_1 \cong 1 + \varepsilon/6$, $\alpha_2 \cong \varepsilon^{-1/4}$ should $\varepsilon \ll 1$, which is typically the case. For example, given the representative values (Kheshgi and Scriven, [96]; Mitlin, [146])

$$\rho_0 g = 10^4 \frac{\text{kg}}{\text{m}^2 \text{ sec}^2}, \quad 3a = 10^{-21} \frac{\text{kg m}^2}{\text{sec}^2}, \quad h_c = 1 \text{ nm},$$

we found $\varepsilon = 10^{-11}$ and $\alpha_2 \cong 562$. Observe that $q_c^2 > 0$ for all $\alpha > 1$ if $\varepsilon = 0$, which was the reason why Mitlin ([146]) retained that gravity effect in the first place. Then, as usual, in our nonlinear expansion we took $q \equiv q_c$, the square of which is plotted in Fig. 8.3d with $\varepsilon = 10^{-11}$ for the isothermal situation of $\beta_c = 0$ indicated above. Under these conditions and suppressing all other dependencies except the bifurcation parameter for ease of exposition, $\sigma = \sigma_0(\beta) = \delta^2$ which is plotted in Fig. 8.3c. Thus, in the limit as $\beta \to \beta_c = 0$, or equivalently as $\delta \to 0$, we could conclude that $\sigma \to 0$ as is required for the application of weakly nonlinear stability theory to compute the Landau coefficients by employing the Fredholm alternative solvability conditions described in Chapters 2 and 3. Since the expressions are simpler, more detail will be provided in this and the next three chapters when evaluating the hexagonal-planform Landau coefficients than was done in Chapters 2 and 7 where that procedure was only indicated symbolically.

Continuing with our summary of the results of this one-dimensional expansion, the second-order problems could be solved in a straight-forward manner to obtain

$$H_{20} = 0, \quad H_{22} = \frac{5\alpha^{-9} - 2\alpha^{-3} + \beta}{\alpha^2(6q_c^2 - \beta/\alpha)}.$$

Although there were also two third-order problems, as usual, it was permissible to concentrate on the one proportional to $\cos(q_c r_1)$, which contained the Landau coefficient a_1 for the Fredholm-type method of solvability mentioned above. That problem could be represented symbolically as

$$a_1 - 2\sigma_0(\beta)H_{31}(\beta) = (21\alpha^2 q_c^2 + \alpha^{-2} - 7\alpha^{-8} + 3\varepsilon\alpha^2 - 2\beta\alpha)\frac{H_{22}(\beta)}{2}$$

$$+ q_c^2 \frac{3\alpha q_c^2 + 28\alpha^{-9} - \alpha^{-3} + 3\varepsilon\alpha - \beta}{4}.$$

Here, we again have only explicitly denoted the β-dependence of the quantities in question for ease of exposition. Now taking the limit as $\beta \to \beta_c = 0$, or equivalently as $\delta \to 0$, employing our formulae for q_c^2 and H_{22}, and assuming the requisite continuity at $\beta = \beta_c = 0$, we obtained

$$a_1(\alpha; \varepsilon) = \frac{1}{6}(11\alpha^{-4} - 14\alpha^{-10} - 9\varepsilon)(5\alpha^{-9} - 2\alpha^{-3})$$

$$+ \frac{1}{4}(2\alpha^{-3} + 25\alpha^{-9})(\alpha^{-4} - \alpha^{-10} - \varepsilon),$$

which is plotted for $\varepsilon = 10^{-11}$ in Fig. 8.3b. As noted in earlier chapters, a main feature of weakly nonlinear stability theory is a phenomenological interpretation of the problem under examination based on the long-time behavior of the truncated amplitude equation and that behavior, as catalogued in Fig. 2.9, depended upon the signs of σ_0 and a_1. We had already determined the sign of σ_0 as a function of $0 < \alpha < \alpha_2$ which was the same as that for q_c^2 or

$$\sigma_0 > 0 \text{ for } \alpha_1 < \alpha < \alpha_2, \ \sigma_0 < 0 \text{ for } 0 < \alpha < \alpha_1.$$

It only remained to examine our formula for a_1 to determine its sign as a similar function of α. Upon such an examination, we found that for $\varepsilon = 10^{-11}$ (see Fig. 8.3b)

$$a_1 > 0 \text{ for } \alpha_3 = 1.019 < \alpha < \alpha_4 = 1.311,$$
$$a_1 < 0 \text{ for } 0 < \alpha < \alpha_3 \text{ or } \alpha_4 < \alpha < \alpha_2;$$

and hence tabulated the phenomenological behavior of our problem in Table 8.1.

TABLE 8.1
Predicted morphological behavior from Fig. 2.9 denoting the relevant α-intervals for $\varepsilon = 10^{-11}$ corresponding to the three possible cases. Observe that $\sigma_0 < 0$, $a_1 > 0$ cannot occur.

α-interval	σ_0	a_1	Fig. 2.9 case	Interfacial morphology range
$(0, \alpha_1 \cong 1)$	$-$	$-$	(iv)	Metastable rupture range
$(\alpha_1, \alpha_3 = 1.019)$	$+$	$-$	(iii)	Rupture range
$(\alpha_3, \alpha_4 = 1.311)$	$+$	$+$	(i)	Pattern formation range
$(\alpha_3, \alpha_2 \cong 562)$	$+$	$-$	(iii)	Rupture range

Having identified those α-intervals in this table corresponding to the cases catalogued in Fig. 2.9, we next provided an explanation for the phenomenological interpretation of each of these cases included in that table:

(i) For $\alpha_1 < \alpha_3 < \alpha < \alpha_4 < \alpha_2$, it follows that $\sigma_0, a_1 > 0$. Thus, we concluded that, in this parameter range, our solution represented a periodic one-dimensional re-equilibrated pattern consisting of stationary parallel liquid ridges (see Fig. 9.3 for the corresponding contour plot) separated by very thin films and having a characteristic wavelength of

$$\lambda_c = \frac{2\pi}{Q_c^{1/2}(\alpha; \varepsilon)},$$

in qualitative agreement with Oron and Bankoff's ([159]) supercritical prediction. For this problem, with $\varepsilon = 10^{-11}$, we can then deduce from Fig. 8.3b,c that

$$0 < \sigma_0 = \delta^2 \le 0.045, \ 0 < a_1 \le 1.20;$$

and hence,

$$A_e^2 = \frac{\sigma_0}{a_1} \cong \delta^2,$$

which implies that

$$\lim_{t \to \infty} H(\boldsymbol{r}, t) \sim \alpha + \delta \cos\left(\frac{2\pi r_1}{\lambda_c}\right) \text{ as } \delta \to 0.$$

This is a necessary requirement for the justification of the truncation procedure inherent to the asymptotic representation of our solution with $\lambda = \lambda_c$ and $\sigma = \sigma_0$. That truncation can be most easily accomplished by rewriting the full amplitude equation associated with it in the form

$$\frac{1}{a_1}\frac{dA_1}{dt} = \frac{\sigma_0}{a_1}A_1 - A_1^3 + O(A_1^5);$$

introducing the rescaled variables

$$\eta = \sigma_0 t, \ \mathcal{A}(\eta) = \frac{A_1(t)}{\delta}$$

where both $\mathcal{A}, d\mathcal{A}/\eta = O(1)$ as $\delta \to 0$; employing $\sigma_0/a_1 \cong \delta^2$; and cancelling the resulting common δ^3 factor, which transforms this equation into

$$\frac{d\mathcal{A}}{d\eta} = \mathcal{A} - \mathcal{A}^3 + O(\delta^2).$$

Then, upon neglecting terms of $O(\delta^2)$, we obtain the truncated amplitude equation. Similarly, the truncation inherent to our expansion expression results upon neglecting terms of $O(\delta^4)$, while retaining those through $O(\delta^3)$ for its full representation. In this context, we can deduce from Fig. 8.3d that

$$0 < q_c^2 < 0.325 \text{ or } 0 < q_c < 0.570$$

and hence conclude $q_c = O(1)$ for the α-range in question, which is a consequence of the proper selection of scale factors and introduction of nondimensional variables.

(iii) For $\alpha_1 < \alpha < \alpha_3$ or $\alpha_4 < \alpha < \alpha_2$, our unstable solution resulted in a dewetting-type rupture of the thin liquid film, since $\sigma_0 > 0$, $a_1 < 0$ as indicated by Oron and Bankoff ([159]).

(iv) For $0 < \alpha < \alpha_1$, since $\sigma_0, a_1 < 0$, there is a subcritical instability or a metastable state. We noted that a numerical simulation of an analogous lubrication equation in the absence of short-range intermolecular Born repulsive forces and the Marangoni effect by Khesghi and Scriven ([96]) resulted in rupture. Since our a_1 reduces to the asymptotic representation $-19\alpha^{-7}/6$ as $\varepsilon \to 0$ under these conditions, we also identified such subcritical behavior with thin liquid film rupture as did Oron and Bankoff ([159]). Given this behavior, we concluded that the inclusion of Born repulsive forces was necessary for the occurrence of pattern formation.

(ii) Since the conditions $0 < \alpha < \alpha_1$ and $\alpha_3 < \alpha < \alpha_4$ corresponding to $\sigma_0 < 0$ and $a_1 > 0$, respectively, are mutually exclusive given that $\alpha_1 < \alpha_3$, this case could never occur as indicated in the caption of Table 8.1. Hence, the planar interface solution was always unstable for our thin liquid film problem.

We complete the chronology of the longitudinal-planform analysis of our thin liquid film Lennard-Jones model lubrication equation by offering the following two commentaries on it, in the spirit of the semi-autobiographical nature of this account. The first concerns the scenario depicted in Fig. 8.3a. As pointed out earlier, this represented a Rabbit Pulled Out of a Hat because, without the introduction of the Marangoni effect involving β and the selection of a β_c, the weakly nonlinear stability analysis could not have been performed. This particular Rabbit however was due to the realization that the Sekimura *et al.* ([220]) approach needed to be employed for our problem because of the form of its secular equation.

Their paper entitled "Pattern formation of scale cells in Lepidoptera by differential origin-dependent cell adhesion," the thesis topic of co-author Mei Zhu, now at Pacific Lutheran University, had just been published in the *Bulletin of Mathematical Biology* and made me realize that this methodology was required for performing weakly nonlinear stability analyses on any problem with a secular equation of the same form. This is reminiscent of the scene from David Lean's 1962 Academy Award winning epic movie *Lawrence of Arabia*, in which Jack Hawkins' General Edmund Allenby responds to Peter O'Toole's T. E. Lawrence's comment of, "You're a clever man, sir," by saying, "No, but I know a good thing when I see one, that's fair surely." In the event, the genesis of Pulling this Rabbit Out of a Hat was "knowing a good thing when you see one."

The second concerns the argument I had with Mei Tian over the sign of our longitudinal Landau constant a_1. After we had developed its formula, but before producing Fig. 8.3b, examining its behavior as a function of α, she told me it was obvious to her that $a_1 < 0$ identically. I did a "back of the envelope" calculation on that formula with $\varepsilon \cong 0$ due to its small size obtaining

$$a_1(\alpha) \cong \frac{1}{6}(11\alpha^{-4} - 14\alpha^{-10} - 9\varepsilon)(5\alpha^{-9} - 2\alpha^{-3}) + \frac{1}{4}(2\alpha^{-3} + 25\alpha^{-9})(\alpha^{-4} - \alpha^{-10}).$$

Then, since the second factor in the first term would be zero for $\alpha = \alpha_0 = (5/2)^{1/6} = 1.165$, and the second term would be positive for this value given that $\alpha_0^{-4} > \alpha_0^{-10}$, $a_1(\alpha_0) > 0$, which implied a_1, a continuous function of α, had to be positive in an interval containing α_0. I showed these calculations to Mei and said, "This is the one time you don't want to be right, because if I am wrong you won't have a thesis problem!" Incidentally, note that $(\alpha_3 + \alpha_4)/2 = \alpha_0$ exactly!

Wishing to refine the pattern formation predictions of our one-dimensional longitudinal planform analysis of this model lubrication evolution equation for $\alpha_3 < \alpha < \alpha_4$, we next sought a two-dimensional hexagonal planform solution of it which to lowest order satisfied

$$H(\boldsymbol{r}, t) \sim \alpha + f(r_1, r_2, t)$$

where $f(r_1, r_2, t)$ was that function involving the amplitude and phase functions $A_{1,2,3}(t)$ and $\varphi_{1,2,3}(t)$, respectively, defined in the previous chapter. In doing so, we closely followed the approach of Wollkind and Stephenson ([296]) for the chemical Turing formation analyses presented therein and hence it is our intention in this chapter merely to sketch that methodology with a focus on the relevant stability results. Recall, that the weakly nonlinear stability behavior of the amplitude-phase equations satisfied by $A_{1,2,3}(t)$ and $\varphi_{1,2,3}(t)$ depended only on the values of its growth rate σ and Landau coefficients $a_{0,1,2}$. Since

$$\sigma = \sigma_0 = \delta^2 \text{ and } a_1 = a_1(\alpha; \varepsilon)$$

were already determined by our longitudinal planform analysis with $\beta_c = 0$, as plotted for $\varepsilon = 10^{-11}$ in Fig. 8.3b,c, we only needed to evaluate the remaining two Landau coefficients a_0 and a_2. As in Chapter 2 and Wollkind and Stephenson ([296]), we determined the solvability conditions for these Landau coefficients by introducing the transformation

$$A_{2,3}(t) = \frac{B_1(t)}{2}, \quad \varphi_{1,2,3}(t) \equiv 0;$$

and seeking an expansion for $H(\boldsymbol{r}, t)$ involving the appropriate terms through third-order of the form

$$H_{mjN\ell} A_1^m(t) B_1^j(t) \cos\left(\frac{Nq_c r_1}{2}\right) \cos\left(\frac{\ell\sqrt{3}q_c r_2}{2}\right);$$

with

$$H_{0000} = \alpha, \quad H_{1020} = H_{0101} = 1, \text{ and } H_{m0N0} = H_{mn} \text{ for } N = 2n.$$

Proceeding in the same manner as we did to evaluate the Landau coefficient a_1, the Fredholm-type solvability conditions for H_{0220} and H_{1220}, respectively, then yielded

$$a_0 = \frac{1}{4}(\alpha^{-4} - \alpha^{-10} - \varepsilon)(2\alpha^{-2} - 5\alpha^{-8})$$

and

$$a_2 = \frac{1}{8}(13\alpha^{-4} - 19\alpha^{-10} - 9\varepsilon)(5\alpha^{-9} - 2\alpha^{-3}) + \frac{1}{4}(2\alpha^{-3} + 25\alpha^{-9})(\alpha^{-4} - \alpha^{-10} - \varepsilon).$$

Having determined these Landau coefficients, we returned to the full hexagonal planform amplitude-phase equations the critical points of which were catalogued and orbital stability behavior summarized in Chapter 2, employing the quantities σ_{-1}, σ_1, and σ_2, defined therein.

We next offered the following morphological interpretation of these potentially stable critical points relative to the thin liquid film patterns under investigation: I and II represented the planar films of uniform depth and the surface ridges, respectively, described in our longitudinal planform analysis while III$^-$ and III$^+$ could be identified with hexagonal network-like structures of regular cells and close-packed arrays of nanodroplets (see Fig. 9.4 for the corresponding contour plots), respectively, both patterns of which were separated by relatively flat ultra-thin films. We began summarizing the orbital stability properties of these critical points by noting that since critical point I was unstable for the longitudinal planform analysis, it was also unstable for the hexagonal planform analysis as well, that one-dimensional perturbation expansion being a special case of this two-dimensional one. The stability behavior of critical points II and III$^\pm$ which depended on the signs of a_0 and $2a_2 - a_1$ had been catalogued in Table 2.1, under the assumptions that a_1, $a_1 + a_2$, $a_1 + 4a_2 > 0$, this table being reproduced below for convenience of reference.

TABLE 8.2
Orbital stability behavior for critical points II and III$^\pm$ with a_0 and $2a_2 - a_1$. This is a reproduction of Table 2.1.

a_0	$2a_2 - a_1$	Stable Structures
$+$	$-, 0$	III$^-$ for $\sigma > \sigma_{-1}$
$+$	$+$	III$^-$ for $\sigma_{-1} < \sigma < \sigma_2$, II for $\sigma > \sigma_1$
0	$-$	III$^\pm$ for $\sigma > 0$
0	$+$	II for $\sigma > 0$
$-$	$+$	III$^+$ for $\sigma_{-1} < \sigma < \sigma_2$, II for $\sigma > \sigma_1$
$-$	$-, 0$	III$^+$ for $\sigma > \sigma_{-1}$

Recall, in summarizing the orbital stability behavior of the critical points of those equations appearing in Table 8.2, it was necessary to employ the quantities

$$\sigma_{-1} = -\frac{4a_0^2}{a_1 + 4a_2}, \quad \sigma_1 = 16a_1\frac{a_0^2}{(2a_2 - a_1)^2}, \quad \sigma_2 = \frac{32(a_1 + a_2)a_0^2}{(2a_2 - a_1)^2}.$$

Having reviewed these stability criteria and made those morphological interpretations, we returned to our Landau coefficient formulae for $\beta_c = 0$ and examined the signs of a_0, $2a_2 - a_1$,

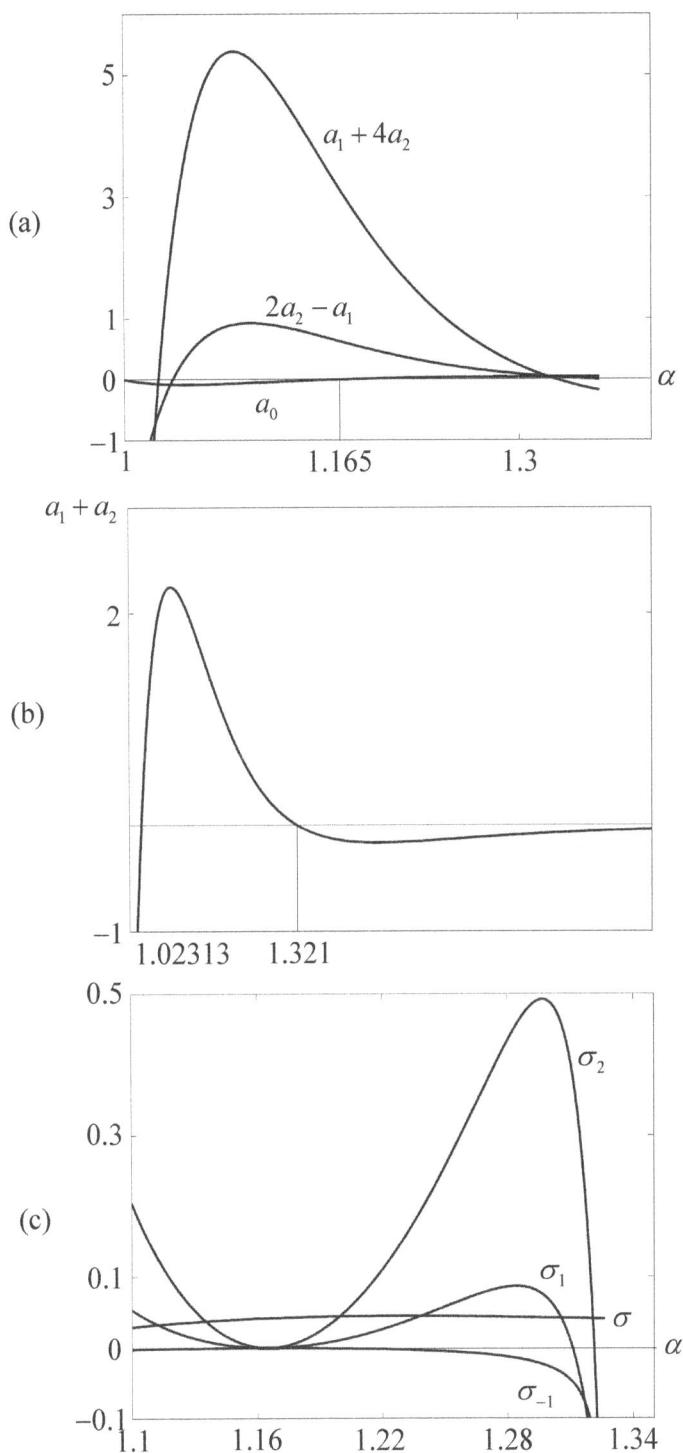

FIGURE 8.4
Plots of (a) $a_1 + 4a_2$, $2a_2 - a_1$, a_0; (b) $a_1 + a_2$; and (c) $\sigma_{-1,1,2}$, as well as σ reproduced from Fig. 8.3c versus α for $\beta_c = 0$ and $\varepsilon = 10^{-11}$.

$a_1 + 4a_2$, and $a_1 + a_2$ versus α in Fig. 8.4a,b with $\varepsilon = 10^{-11}$. Then, from the results of this examination, we observed that besides $\alpha_{3,4}$, the zeroes of a_1 identified earlier, there existed the following other significant values of α:

$$\alpha_3 < \alpha_5 < \alpha_6 < \alpha_7 < \alpha_c < \alpha_4 < \alpha_8 < \alpha_9$$

such that

$$a_1 + a_2 = 0 \text{ for } \alpha = \alpha_5 \text{ or } \alpha_8, \ a_1 + a_2 > 0 \text{ for } \alpha_5 < \alpha < \alpha_8;$$
$$a_1 + 4a_2 = 0 \text{ for } \alpha = \alpha_6 \text{ or } \alpha_9, \ a_1 + 4a_2 > 0 \text{ for } \alpha_6 < \alpha < \alpha_9;$$
$$a_0 = 0 \text{ for } \alpha = \alpha_c, \ a_0 < 0 \text{ for } \alpha < \alpha_c, \ a_0 > 0 \text{ for } \alpha > \alpha_c;$$
$$2a_2 - a_1 = 0 \text{ for } \alpha = \alpha_7, \ 2a_2 - a_1 < 0 \text{ for } \alpha < \alpha_7, \ 2a_2 - a_1 > 0 \text{ for } \alpha > \alpha_7;$$

where these values of α were given by

$$\alpha_5 = 1.023, \ \alpha_6 = 1.026, \ \alpha_7 = 1.037, \ \alpha_c = 1.165, \ \alpha_8 = 1.321, \ \alpha_9 = 1.328.$$

Note, that $\alpha_c = \alpha_0 = (5/2)^{1/6}$, as defined in conjunction with the commentary on the sign of a_1. Given that the stability behavior outlined in Table 8.2 depended not only on the signs of a_0 and the various combinations of Landau constants appearing in Fig. 8.4a,b but also on the relative sizes of σ and $\sigma_{-1,1,2}$, we next plotted the growth rate σ of Fig. 8.3c and those quantities $\sigma_{-1,1,2}$ versus α in Fig. 8.4c for $\beta_c = 0$ and $\varepsilon = 10^{-11}$. Our attention here has been focused on the special case of $\beta_c = 0$, since the numerical simulations and experimental observations with which we wished to compare these analytical theoretical predictions were for such isothermal situations. In spite of this fact, it was still necessary for us to have included the Marangoni effect involving β in order subsequently to employ the Sekimura *et al.* ([220]) approach of analyzing our secular equation. We were finally ready to make theoretical pattern formation predictions based upon the existence and morphological stability behavior of our critical points summarized in Table 8.2. Recalling the definition of these critical points with $\varphi_{1,2,3}(t) \equiv 0$, in terms of $A_{1,2,3}(t)$: Namely,

$$\text{II: } A_1^2 = \frac{\sigma}{a_1}, \ A_2 = A_3 = 0;$$

$$\text{III}^\pm: A_1 = A_2 = A_3 = A_0^\pm = \frac{-2a_0 \pm [4a_0^2 + (a_1 + 4a_2)\sigma]^{1/2}}{a_1 + 4a_2};$$

and their orbital stability criteria outlined in this table, we saw that they had to satisfy certain inequality constraints as necessary conditions to yield stable patterns given by

$$\text{II: } a_1, \ 2a_2 - a_1 > 0;$$
$$\text{III}^+: a_1 + 4a_2, \ a_1 + a_2 > 0 \text{ and } a_0 < 0;$$
$$\text{III}^-: a_1 + 4a_2, \ a_1 + a_2, \ a_0 > 0.$$

In the above, we implicitly made use of the facts that $\sigma, \sigma_{1,2} > 0$, and $\sigma_{-1} < 0$. Hence, from Figs. 8.3b and 8.4a,b, we deduced the following α-intervals of interest relevant to II and III$^\pm$:

$$\text{II: } (1.037, 1.311); \ \text{III}^+: (1.026, 1.165); \ \text{III}^-: (1.165, 1.321).$$

It alone remained for us to refine these predictions by employing the stability criteria of Table 8.2, in conjunction with the plots of Fig. 8.4c, comparing the sizes of σ and $\sigma_{-1,1,2}$. Toward this end, we generated Fig. 8.5, the six parts of which comprised both an extension and enlargement of Fig. 8.4c, emphasizing these α-subintervals that exhibited qualitatively different morphological stability behavior. We examined the parts of this figure on a case by case basis as follows:

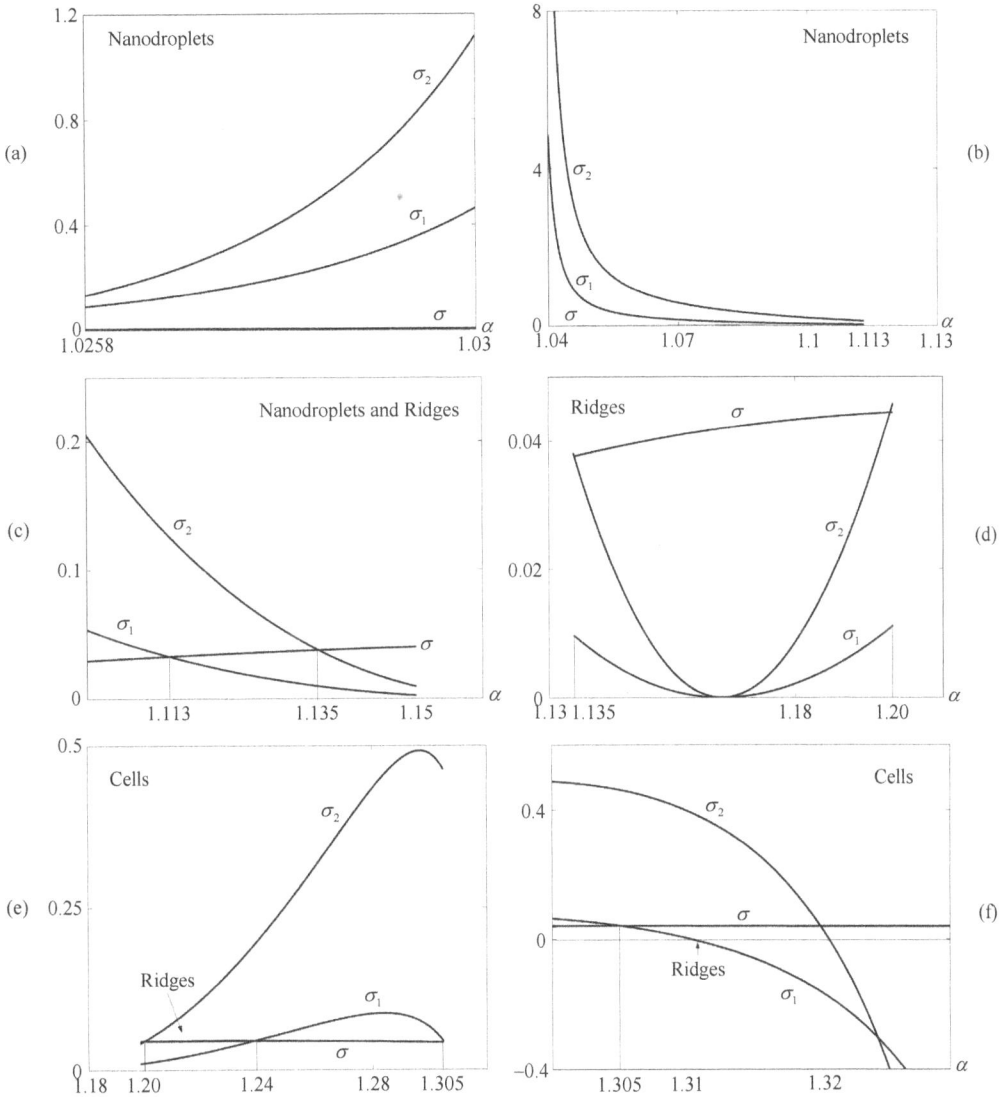

FIGURE 8.5
An extension for (a) $\alpha \in (1.026, 1.037)$, (b) $\alpha \in (1.037, 1.113)$, and enlargement for (c) $\alpha \in (1.113, 1.135)$, (d) $\alpha \in (1.135, 1.20)$, (e) $\alpha \in (1.20, 1.305)$, (f) $\alpha \in (1.305, 1.321)$ of the σ, σ_1, and σ_2 plots of Fig. 8.4c indicating the morphological predictions catalogued in Table 8.3.

(a) $\alpha \in (1.026, 1.037]$: Then $a_0 < 0$ and $2a_2 - a_1 \leq 0$. Hence, since $\sigma > 0 > \sigma_{-1}$ over the whole range of Fig. 8.5 (see Fig. 8.4c), we concluded from the last row of Table 8.2 that nanodroplets were the only stable pattern. Note, that $\sigma_{1,2}$ play no substantive role in this regime and have only been plotted in Fig. 8.5a for the sake of completeness.

(b) $\alpha \in (1.037, 1.113]$: Then, $a_0 < 0$ and $2a_2 - a_1 > 0$. Hence, since $0 < \sigma \leq \sigma_1 < \sigma_2$, we concluded from the fifth row of Table 8.2 that nanodroplets were the only stable pattern.

(c) $\alpha \in (1.113, 1.135)$: Then, $a_0 < 0$ and $2a_2 - a_1 > 0$. Hence, since $0 < \sigma_1 < \sigma < \sigma_2$, we concluded from the fifth row of Table 8.2 that there was bistability between nanodroplets and ridges.

(d) $\alpha \in [1.135, 1.20]$: We further partitioned this case into the three subcases:

     (i) $\alpha \in [1.135, 1.165)$: Then, $a_0 < 0$ and $2a_2 - a_1 > 0$. Hence, since $0 < \sigma_1 < \sigma_2 \leq \sigma$, we concluded from the fifth row of Table 8.2 that ridges were the only stable pattern.

     (ii) $\alpha = 1.165$: Then, $a_0 = 0$ and $2a_2 - a_1 > 0$. Hence, since $\sigma > 0 = \sigma_1 = \sigma_2$, we concluded from the fourth row of Table 8.2 that ridges were the only stable pattern.

     (iii) $\alpha \in (1.165, 1.20]$: Then, $a_0 > 0$ and $2a_2 - a_1 > 0$. Hence, since $0 < \sigma_1 < \sigma_2 \leq \sigma$, we concluded from the second row of Table 8.2 that ridges were the only stable pattern.

Taken together, these subcases yielded the result that ridges were the only stable pattern in this whole subinterval. Note that for the rest of the subintervals to be catalogued below, $a_0 > 0$ and $2a_2 - a_1 > 0$, as in subcase (iii) of (d), and thus we again employed the second row of Table 8.2.

(e) $\alpha \in (1.20, 1.305]$: We further partitioned this case into the two subcases:

     (i) $\alpha \in (1.20, 1.24)$: Hence, since $0 < \sigma_1 < \sigma < \sigma_2$, we concluded there was bistability between ridges and cells.

     (ii) $\alpha \in [1.24, 1.305]$: Hence, since $0 < \sigma \leq \sigma_1 < \sigma_2$, we concluded cells were the only stable pattern.

(f) $\alpha \in (1.305, 1.321)$: We further partitioned this case into the three subcases:

     (i) $\alpha \in (1.305, 1.311)$: Hence, since $0 < \sigma_1 < \sigma < \sigma_2$, we concluded that there was bistability between cells and ridges.

     (ii) $\alpha \in [1.311, 1.32)$: Then $a_1 \leq 0$. Hence, since $\sigma_1 \leq 0 < \sigma < \sigma_2$, we concluded that cells were the only stable pattern because ridges did not exist.

     (iii) $\alpha \in [1.32, 1.321)$: Then, $a_1 < 0$. Hence, since $\sigma_1 < 0 < \sigma_2 < \sigma$, we concluded that there were no stable patterns because ridges did not exist and cells were unstable.

We summarized these morphological stability predictions by means of Table 8.3.

Here, unlike the hexagonal planform stability analyses of Chapters 2 and 7, we compared the relative sizes of σ and $\sigma_{1,2}$ directly in Fig. 8.5, rather than employing the indirect generalized marginal stability curve method of obtaining those loci appearing in Figs. 2.8, 7.6, and 7.8, given the very different nature of our secular equation. Again, as in those figures, observe from Fig. 8.4a that

$$|a_0| \ll a_1 + 4a_2$$

in the parameter range of interest and thus, σ_{-1}, although negative is virtually indistinguishable from σ (see Fig. 8.4c). Further, as mentioned for all these hexagonal planform

TABLE 8.3

Predicted morphological stability behavior from Fig. 8.5 denoting the relevant α-intervals for $\varepsilon = 10^{-11}$ corresponding to the possible stable patterns identified in that figure.

α-range	Fig 8.5 case(s)	Stable pattern
(1.026,1.113)	(a) & (b)	Nanodroplets
(1.113,1.135)	(c)	Nanodroplets & Ridges
[1.135, 1.20]	(d)	Ridges
(1.20,1.24)	(e):(i)	Ridges & Cells
[1.24, 1.305]	(e):(ii)	Cells
(1.305,1.311)	(f):(i)	Cells & Ridges
[1.311, 1.32)	(f):(ii)	Cells

analyses to justify the truncation procedure inherent to the asymptotic representations introduced from the outset, it is necessary that the Landau constants satisfy the additional size constraint

$$\frac{|a_0|}{(a_1 + 4a_2)^2} \ll 1.$$

Noting that the satisfaction of this constraint is also guaranteed by the previous inequality, we concluded, as in Chapter 7, that such a truncation procedure was valid for our hexagonal planform weakly nonlinear stability analysis of the thin liquid film spatio-temporal model evolution equation under investigation.

To facilitate comparison of our morphological stability predictions with the results of earlier pattern formation studies involving thin liquid films, we next described both those theoretical and experimental outcomes in detail. The three photographs of different morphologies exhibited by unstable thin polymer films, appearing in Reiter ([185]) and reproduced in Fig. 8.1, depict an undulation pattern and the early and late stages of cellular pattern formation in which the elevated rims of the initial circular cylindrical depressions coalesced to form network-like regular hexagonal cells of nearly identical size. In previous work, Reiter ([184]) characterized that pattern as a polygonal cellular structure, where the most often found polygon was a hexagon with its vertices joining three edges and these hexagons were of uniform size by virtue of the coalescence process of the elevated rims of the initial depressions. The morphologies exhibited by such structures are precisely the topographies expected for type III$^-$ patterns (in this context, note that the highest points of these networks occur at those vertices); hence, our identification of the latter with cells.

Let us turn to the other morphology observed by Reiter ([185]), the undulation pattern. We have identified such structures with surface ridges characteristic of type II patterns. This identification is consistent with Oron and Bankoff's ([159]) theoretical prediction for a one-dimensional weakly nonlinear stability analysis of their evolution equation. There also exist numerical simulations of one-dimensional versions of such equations which produced stationary solutions by Sharma and Jameel ([221]), Mitlin and Petviashvilli ([147]), and Oron and Bankoff ([159]). Restricting their analyses to a single disturbance wavelength, they found a stationary arch-type solution for certain layer thicknesses. Mitlin and Petviashvilli ([147]) and Oron and Bankoff ([159]) interpreted this solution as a one-dimensional spatially periodic stable liquid ridge-type pattern while Sharma and Jameel ([221]) made the interpretation that it represented nanodroplets separated by relatively flat ultra-thin films. In deciding between these interpretations, it is instructive to point out Reiter's ([185]) having reported that none of his experimental states represented an equilibrium situation of isolated droplets. Although we felt that isolated nanodroplets did not represent the best identification of stationary one-dimensional periodic patterns, a uniform close-packed distribution of such droplets referred to as morphological phase separation by Sharma and Jameel ([221]) and Jameel and Sharma ([86]) was consistent with the topography of type III^+ patterns for thin liquid films. Hence, we have identified the III^+ critical point with morphological phase separation of this sort.

Returning to the parallel surface ridges predicted by one-dimensional nonlinear stability analyses and numerical simulations, Khanna and Sharma ([95]) point out that such regular arrangements of cylindrical hills and valleys represent a pattern never witnessed in thin liquid film experiments. Instead, interconnected bicontinuous patterns, as depicted in the left-hand photograph of Fig. 8.1, are observed. Recall, as related in Chapters 2 and 7, the equivalence class of critical points designated by II actually contains the three solutions

$$A_i^2 = \frac{\sigma}{a_1}, \ A_j = A_k = 0, \ (i, j, k) = \text{ even permutation of } (1, 2, 3).$$

Thus, each of the α-intervals of Table 8.3 identified with ridges is itself a locus of multiple stable states. Then, as initial conditions vary from point to point on the interfacial surface, such families of ridges can give rise to polygonal arcs the boundaries of which would appear quite random in orientation (see Chapter 7). Indeed, the interconnected bicontinuous pattern classified as an undulation by Reiter ([185]) has the appearance of such curved elongated ridges in the relevant photographic reproduction at the far left in Fig. 8.1. Hence, we identified our type II critical point with such interconnected bicontinuous patterns.

We were now ready to compare our two-dimensional morphological stability predictions with earlier isothermal numerical simulations and experimental observations. Sharma and Khanna ([222]) performed two-dimensional numerical simulations on an equation analogous to ours for $\varepsilon = \beta = 0$, but having its repulsive force term taken proportional to $\exp(-h/\ell_0)$ with the correlation length $\ell_0 = 2.5$ nm. They found two completely different morphological patterns by which pseudo-dewetting, *i.e.*, elevations separated by ultra-thin flat films - could occur based upon the mean layer thickness. For relatively thick films, the rims of the circular depressions which formed first from an initial bicontinuous pattern developed into a polygonal structure by repeated coalescence. In contrast to this scenario, for relatively thin films the bicontinuous pattern fragmented directly to produce a uniform array of microdroplets. For a range of intermediate thickness, the bicontinuous structure composed of long hills and valleys resolved itself into a mixture with these other two patterns, the type and proportion of which depended, in our notation, on the relative distance $h_0 - h_c$. These simulation results were qualitatively consistent with our theoretical analytical predictions.

Xie *et al.* ([302]) investigated true dewetting of polystyrene films on a silicon substrate at a fixed annealing temperature $T = T_0 = 388$ K as a function of mean layer thickness in the range $h_0 \in [4.5 \text{ nm}, 35 \text{ nm}]$. For $h_0 = 7.5$ nm they observed a bicontinuous surface pattern, while for $h_0 = 12.5$ nm they saw a bistable state between configurations resembling Voronoi tessellation patterns and these surface undulations. For even thinner films, such as $h_0 = 4.5$ nm, the initial bicontinuous pattern broke up into small uniformly distributed droplets that subsequently coarsened to form large isolated drops by coalescence and hence dewetted the surface (see Fig. 8.6a). They contrasted this behavior to the dewetting of much thicker films, such as $h_0 = 35$ nm in which rupture occurred by the formation of circular holes (see Fig. 8.6b).

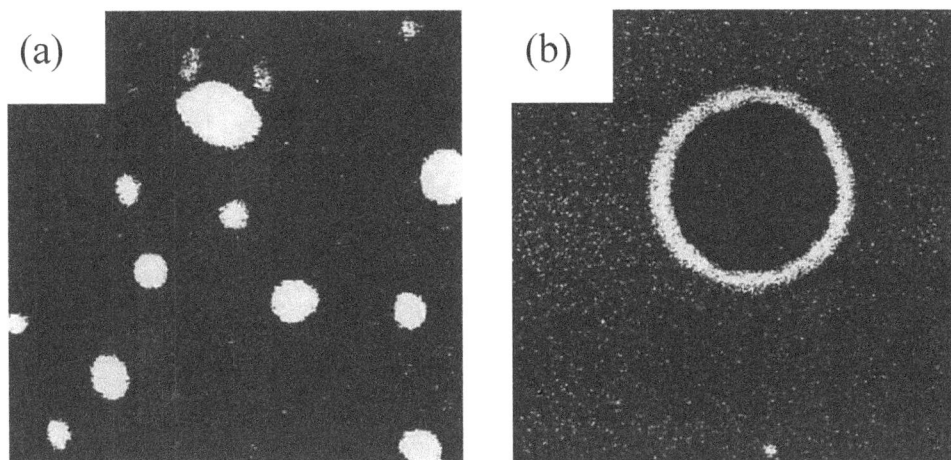

FIGURE 8.6
Images of dewetting of polystyrene films on a silicone substrate at various layer depths h_0 (Xie *et al.*, [303]): (a) Isolated droplets at $h_0 = 4.5$ nm; and (b) Circular holes at $h_0 = 35$ nm.

From the two-dimensional morphological stability predictions of Table 8.3, in conjunction with the one-dimensional results of Table 8.1, we drew the following general conclusions: Although the planar interface solution was never stable, supercritical equilibrium patterns occurred for an interval of mean layer thickness $\alpha \in (1.026, 1.32)$, with subcritical rupture occurring outside this interval. These pseudo-dewetting patterns consisted of bicontinuous surface ridges and hexagonal network-like cells or close-packed configurations of nanodroplets separated by relatively flat ultra-thin films. In particular, those morphological phase separation patterns were crucially dependent upon the inclusion of the short-range intermolecular repulsive force with cells tending to be stable for the thicker layers; nanodroplets, for the thinner ones; and ridges, for layers of intermediate thickness. These theoretical predictions were in qualitative accord with both the relevant isothermal experimental evidence and numerical simulations of similar model equations cited above, as well as being consistent with dewetting-type rupture occurring for such situations by hole formation in relatively thick layers but by drop formation in thinner ones. As in all hexagonal planform nonlinear stability analyses for the parameter range of interest, here $A_0^+ > 0$ and $A_0^- < 0$, with stability occurring when $a_0 A_0^{\pm} < 0$. Thus, III$^+$ could only be stable when $a_0 < 0$ and III$^-$, when $a_0 > 0$. Hence, the sign of a_0 determined which of these hexagonal patterns would occur. As I pointed out to Mei, once our analysis was completed, we should have been able to anticipate from the outset that $a_0 < 0$ for $\alpha < \alpha_c$ and $a_0 > 0$ for $\alpha > \alpha_c$ where $a_0 = 0$ at $\alpha = \alpha_c$, since the reverse transition in sign about that zero would have predicted a morphological sequence with increasing layer depth, which made no sense in the context of this dewetting rupture behavior. Again, we had just Pulled a Rabbit Out of a Hat!

In order to make these qualitative comparisons of those theoretical analytical predictions with experimental evidence and numerical simulations more quantitative in nature, we next examined the effect on our results of taking a different choice for h_c, this critical thickness of the Lennard-Jones potential being a measure of the coupling between the long-range van der Waals attractive and short-range Born repulsive intermolecular forces. Recall, h_0 was related to α by $h_0 = \alpha h_c$ and $\alpha = \alpha_c$ corresponded to the crossover value of the mean layer thickness at which hexagonal patterns switch character. Further note that when $\beta_c = 0$, h_c only influences our problem by its presence in the very small dimensionless parameter ε, which may be rewritten as

$$\varepsilon = \left(\frac{h_c}{h_g}\right)^4 \text{ where } h_g = \left(\frac{3a}{\rho_0 g}\right)^{1/4}.$$

We observed in this context that the results of our one-dimensional longitudinal planform analysis were extremely robust with respect to changes in ε. In particular, α_1, α_3, and α_4 were virtually invariant as ε ranged over a fairly wide interval flanking $\varepsilon = 0$. Then, although α_2 varies drastically with ε since $\alpha_2 \sim \varepsilon^{-1/4}$ as $\varepsilon \to 0$, note that the asymptotic representation for the corresponding dimensional thickness

$$h_2 = \alpha_2 h_c \cong \varepsilon^{-1/4} h_c = h_g$$

is actually independent of h_c. Similarly, the results of our two-dimensional hexagonal planform analysis were also robust with respect to changes in ε. Hence, to determine the isothermal predictions of our problem for different h_c, we only needed to employ the significant values of α for $h_c = 1$ nm from Table 8.3 in conjunction with the relationship $h_0 = \alpha h_c$. For example, by taking

$$h_c = 7.725 \text{ nm},$$

we deduced a crossover layer depth related to $\alpha_c = 1.165$ of

$$h_{cr} = \alpha_c h_c = 9 \text{ nm}$$

and a patterned layer interval of $h_0 \in (7.926 \text{ nm}, 10.197 \text{ nm})$, corresponding to the related interval $\alpha \in (1.026, 1.32)$ of Table 8.3, in quantitative agreement with the observations of Xie *et al.* ([303]) and the simulations of Sharma and Khanna ([222]), while $\varepsilon = 3.56 \times 10^{-8}$ and the upper instability bound was retained at

$$h_2 = \alpha_2 h_c \cong h_g = 562 \text{ nm}.$$

This implies that the planar interface solution would be linearly stable when $h_0 > h_g$ consistent with such equilibrium behavior for Marangoni convection in relatively thin layers (see Chapter 5) which was the reason Mitlin ([146]) retained a gravitational effect in his isothermal model. Given that the existence of an $\alpha = \alpha_c$ at which $a_0 = 0$ played such a fundamentally important role in our morphological stability predictions, we felt there was some merit in examining this condition more thoroughly. We observed that for this value of α, which was independent of ε, $a_1 = a_2 > 0$ and hence $2a_2 - a_1 = a_1 > 0$. Therefore, in spite of the possibility of bistability existing between the two types of hexagonal patterns when $a_0 = 0$ and $2a_2 - a_1 < 0$ (see the third row of Table 8.2 which actually only refers to neutral stability in this instance) that particular possibility was precluded for our specific model system. Thus, the vanishing of the quadratic terms in the amplitude-phase equations, guaranteed from Table 8.2, that ridges alone but never hexagonal patterns could be stable under this condition. In order to examine these ridges in more detail, we next developed an expression for their characteristic dimensional wavelength λ_c^*, which, from the length scale factors introduced earlier, was related to λ_c by

$$\lambda_c^* = h_c \left(\frac{S}{A}\right)^{1/2} \lambda_c = h_c^2 \left(\frac{\gamma_0}{3a}\right)^{1/2} \frac{2\pi}{q_c}.$$

From the definition of q_c

$$\frac{1}{q_c} \sim (\alpha^{-4} - \alpha^{-10})^{-1/2} \text{ as } \varepsilon \to 0$$

while

$$(\alpha^{-4} - \alpha^{-10})^{-1/2} \sim \alpha^2 \text{ when } \alpha \gg 1$$

since the left- and right-hand sides of the above relation with $\alpha = 2$ yielded the values 4.03 and 4, respectively. Then, substituting this one-term asymptotic representation into λ_c^*, we obtained

$$\lambda_c^* \sim 2\pi \left(\frac{\gamma_0}{3a}\right)^{1/2} (h_c\alpha)^2 = 2\pi \left(\frac{\gamma_0}{3a}\right)^{1/2} h_0^2,$$

which, upon taking the typical surface-tension value (Sharma and Ruckenstein, [223])

$$\gamma_0 = 50\frac{\text{dynes}}{\text{cm}} = 50 \times 10^{-3}\frac{\text{kg}}{\text{sec}^2},$$

in conjunction with the previous assignment for $3a$, yielded

$$\lambda_c^* \sim 2\pi \left(\frac{50}{10^{-18} \text{ m}^2}\right)^{1/2} h_0^2 = 10\pi\sqrt{2}\frac{h_0^2}{\text{nm}}.$$

We noted that this asymptotic expression for λ_c^* was independent of h_c and consistent with the results of Xie et al. ([303]) and Sharma and Khanna ([223]), both of whom found that λ_c^* for their ridges were proportional to h_0^2, the mean layer thickness squared, as well as with that of Bishof et al. ([15]), whose thin liquid metal film ridge patterns also exhibited the same relationship. This latter behavior is depicted in Fig. 8.7 for those experiments involving thin liquid gold (Au) films. All of these correlations represented our last Rabbit Pulled Out of a Hat for this problem.

We concluded with a final comparison between our one-dimensional longitudinal planform results and those obtained if the short-range repulsive force term used by Oron and Bankoff ([159]) was employed. Recall that the latter authors felt their repulsive intermolecular force potential term proportional to h^{-4} was superior to the analogous Lennard-Jones potential term proportional to h^{-9}. We wished to explore the matter further by comparing the morphological stability predictions obtained by this choice for that repulsive force potential term with ours.

Toward that end we first considered a intermolecular force potential related to the Oron-Bankoff ([159]) choice for this term

$$\phi(h) = \left(\frac{a}{h_c^3}\right)\left[\left(\frac{h_c}{h}\right)^3 - \frac{(h_c/h)^4}{3}\right]$$

which produced a spatio-temporal evolution partial differential model equation that differed from our final version only in the replacement of its H^{-7} term by H^{-2}. Next, we performed the same one-dimensional longitudinal planform weakly nonlinear stability analysis on this equation as had been performed on our equation and obtained a growth rate still satisfying

$$\sigma = \frac{\alpha^3 Q_c(\alpha; \varepsilon)}{4}$$

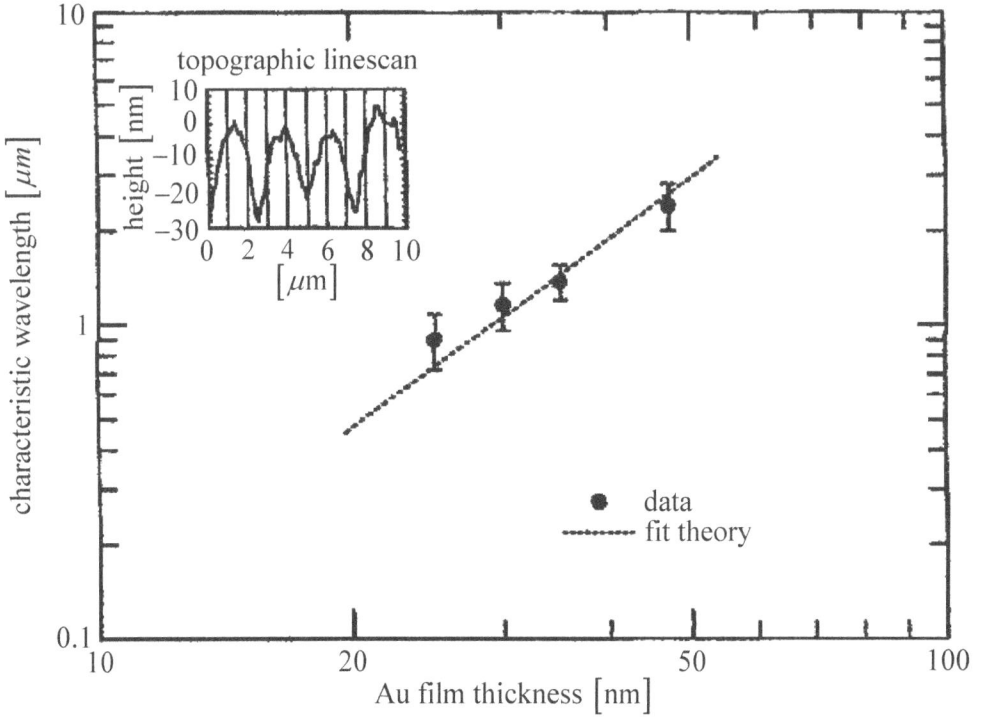

FIGURE 8.7
Log-log plot of the characteristic dimensional wavelength of ridges λ_c^* formed on the surface of a thin liquid gold (Au) film versus its mean thickness h_0 (Bishof *et al.*, [15]). Here, the theoretical dotted line is based upon a second-order power law relating λ_c^* to h_0, while the inset shows a topological linescan of the surface modulation of these ridges at a thickness of 47 nm. Note that if $\lambda_c^* = c_0 h_0^2$, where λ_c^* and h_0 are numerical values measured in μm's and nm's, respectively, then taking a logarithm of base 10, denoted by log, of both sides of this equation yields $\log(\lambda_c^*) = \log(c_0) + 2\log(h_0)$, which is a line of slope 2 in the $\log(h_0)$ - $\log(\lambda_c^*)$ plane. *Reproduced by permission of the American Physical Society which holds the copyright.*

but now with

$$Q_c(\alpha; \varepsilon) = \alpha^{-4} - \alpha^{-5} - \varepsilon > 0 \text{ for } \alpha \in (\alpha_1, \alpha_2)$$

where $\alpha_1 \sim 1 + \varepsilon$, $\alpha_2 \sim \varepsilon^{-1/4}$ as $\varepsilon \to 0$

and a Landau constant formula

$$a_1 = \frac{1}{24}(22\alpha^{-4} - 23\alpha^{-5} - 18\varepsilon)(5\alpha^{-4} - 4\alpha^{-3}) + \frac{1}{2}\alpha^{-3}(\alpha^{-4} - \alpha^{-5} - \varepsilon).$$

Then analogous to our results for what was denoted by case (i) earlier, we deduced that

$$a_1 > 0 \text{ for } \alpha \in (\alpha_3, \alpha_4) \text{ where } \alpha_3 = 1.027, \; \alpha_4 = 1.470 \text{ when } \varepsilon = 10^{-11}$$

and found in this α-interval that

$$0 < \sigma_0 \leq 0.0036, \; 0 < a_1 \leq 0.037.$$

Given the relative sizes of the upper bounds of these quantities, in comparison to those obtained for the corresponding ones from the similar analysis of our equation, we observed that the truncation procedure inherent to such a development would not be as simple to implement in this instance. Hence, we felt that the form of the Lennard-Jones potential repulsive force term was more appropriate for the purpose of weakly nonlinear stability theory, due to its convenience of application than the one chosen by Oron and Bankoff ([159]). As pointed out by Thess and Orszag ([246]), weakly nonlinear stability analyses of this type do not necessarily apply to all pattern-formation systems but only hold under certain restrictive conditions that need to be established on a case by case basis.

We closed by offering an additional rationale for our research. The study of long-wavelength hydrodynamic instabilities generated by intermolecular forces in thin films has a wide variety of applications, ranging from industrial processes to biological phenomena, all of which involved a liquid layer coating a solid substrate. The development of various models of thin liquid films to be analyzed for their stability, such as our prototype spatio-temporal evolution partial differential model equation, has been motivated by the desire to provide a better understanding of these diverse applications. Recall, Reiter ([185]) stated that Herminghaus et al.'s ([79]) major achievement was their demonstration that the same coupling mechanism between van der Waals attractive and Born repulsive forces could account for the different patterns produced in a given experiment and their subsequent suggestion that the exact form of the associated intermolecular potential could be inferred from knowledge of interfacial morphology, together with layer thickness as an inverse problem, which was precisely what our selection of h_c provided. Further, recall he then declared that unfortunately it was still unclear why almost the identical patterns can be found in both polystyrene and gold films. The primary reason for this thin liquid film research was to answer questions of that sort by explaining more completely the pattern formational behavior of our Lennard-Jones lubrication model evolution equation.

Let us next discus how the results from Mei's research were disseminated in the two conference presentations and related journal publications listed below:

I organized a mini-symposium entitled "Mathematical Modeling of Biological Pattern Formation" for the 2001 Annual SIAM meeting held in San Diego. Such SIAM mini-symposia consist of four speakers, each giving 30 minute presentations. Typically the organizer gives the first or last talk and invites the other three participants comprising the mini-symposium, as described for the 1985 SIAM meeting in Chapter 4. This one had the following Summary:

Many modeling techniques originally developed in the physical sciences have been adapted so they can be used to describe biological phenomena. The four speakers provide examples of such adaptations drawn from the areas of microbiology, developmental biology, biomechanics, and biofluids. The dynamical and pattern formational behavior of the model systems involved are deduced by a variety of analytical and numerical methods. Then these theoretical predictions are compared to experimental or observational data and placed in the context of some recent pattern formation studies. Applied mathematicians, biologists, chemists, computer scientists, engineers, and physicists should find it of interest.

Its four participants with their affiliations and talk titles were given in order of presentation by:

- Herbert Levine, Univ. of Cal. at San Diego, The Physics of Collective Micro-Organism Patterns;

- Mei Zhu, Pacific Lutheran Univ., Pattern Formation of Scale Cells in Lepidoptera Wings by Adhesion;

- Robert Dillon, Wash. State Univ., A Microscale Model of Flagellar and Ciliary Motion;

- David Wollkind, Wash. State Univ., Nonlinear Stability Analyses of Pattern Formation in Thin Liquid Films.

Recall that Herbert Levine was a speaker at the Los Alamos special session described in Chapter 2. He had become a faculty member at UC San Diego and is now at Rice University. His research career parallels mine, in that Herb started in the areas of materials science and fluid mechanics, before modeling biological phenomena as well. Hence, he was a perfect speaker for this mini-symposium. So were Mei Zhu and Robert Dillon, since Mei complemented my talk by presenting the secular equation method of analysis she helped develop for Sekimura *et al.* ([220]) and Bob, a colleague of mine in the WSU Department of Mathematics, had been both a Ph.D. student and postdoctoral scholar of Hans Othmer, while the latter was at the University of Utah.

Specifically, my talk included an extension of the exposition presented in this chapter, given that its one-dimensional longitudinal planform analysis considered the effect of a nonzero β_c, which also appeared in Mei's thesis. In particular, we found that such a β_c preserved the qualitative results of our $\beta_c = 0$ isothermal analysis, while yielding the quantitative changes when $\varepsilon = 10^{-11}$ that $\alpha_3 = 0.93$, $\alpha_4 = 1.27$; and $\alpha_3 = 0.80$, $\alpha_4 = 1.16$; should $\beta_c = 1$ and 7, these β_c corresponding to $\gamma_1 m_0 h_c = 2/3$ and $14/3$ dynes/cm, respectively, which are typical nonisothermal values for that parameter as employed by Sharma and Ruckenstein ([223]). We published a paper (Tian and Wollkind, [248]) based on that talk in the journal *Interfaces and Free Boundaries*. Incidentally, although the name on her thesis is "Mei Tian," by the time that this paper appeared she was using the name "Emily Mei Tian" instead. Hence, the reference to this paper is attributed to E.M. Tian and D.J. Wollkind.

Andrea Bertozzi, then at Duke University and now at UCLA, organized a mini-symposium on thin liquid film rupture and pattern formation at that conference which was held in the same room as mine, but a day earlier. After she had trouble turning down the lights in the room and I adjusted them for her, Andrea said to me, "I am so happy that SIAM decided to send someone to this room to control the lighting"; to which I replied that my name was David Wollkind and SIAM hadn't sent me. Rather, I was running my own mini-symposium on mathematical modeling of biological pattern formation in the same room the next day and toward that end had previously checked out the lighting controls. I then said that my mini-symposium talk was on nonlinear stability analyses of pattern formation in thin liquid biofilms and might be of interest to her mini-symposium participants and attendees. One of those participants was Andrew Bernoff of Harvey Mudd College whose thin liquid film model lubrication equation, employing a polar coordinate system (r, θ) where $x = r\cos(\theta)$ and $y = r\sin(\theta)$, predicted stable equilibrium solutions which consisted of depressed circles of constant radius. When comparing that result with actual experimental observation, Andy displayed the middle photograph of Fig. 8.1. I first asked him whether this was from Günter Reiter's ([185]) *Science* article entitled "The artistic side of intermolecular forces" and when, he answered in the affirmative, told him that the yellow circles in his slide were elevated instead of being depressed and further did not represent an equilibrium state, but only the early stages of cellular pattern formation. Andy responded with, "But that's not what my model predicts," to which I replied, "Exactly."

The next year, the Department of Mathematics at WSU organized a conference in honor of John Cannon's 65th birthday, in conjunction with the Pacific Northwest PDE seminar series. The main topics of this conference were inverse, ill-posed, control, and free boundary problems, as well as other nonlinear PDE problems arising from scientific and industrial applications. One of its plenary speakers was Ralph Showalter from UT Austin, whom we shall meet again in Chapter 17. I was also invited to be a plenary speaker at this conference and presented the same talk as my 2001 Annual SIAM meeting mini-symposium contribution given that it covered all the topics delineated above, except for ill-posed problems. The collection of papers presented at this PDE conference including Showalter ([227]) were published as a special issue of the journal *Dynamics of Continuous, Discrete & Impulsive Systems, Series A: Mathematical Analysis*. Since we did not wish to duplicate Tian and Wollkind ([248]), our paper Tian and Wollkind ([249]) published in that issue was restricted to the isothermal case, along the lines of the development presented in this chapter, but without its figures and concluded with the following Acknowledgement: This paper is a particularization to the isothermal case of a problem treated more generally in Tian and Wollkind ([248]). We have chosen to use that isothermal model as this chapter's template for ease of exposition.

Further, when Mei interviewed for an Assistant Professorship in the Department of Mathematics and Statistics at Wright State University in Dayton, Ohio, she gave a colloquium based on our slide show from these conferences and was offered the job immediately since her presentation served as a perfect prototype of comprehensive applied mathematical modeling in the sciences. Finally, we also performed another extension of the one-dimensional longitudinal planform analysis of our model equation: Namely, one which extended that analysis to fifth-order, motivated by my admonishment to Mei of not wanting the Landau constant a_1 to be identically negative. Carrying the one-dimensional analysis to fifth-order, an extension requiring an enormous amount of extra calculation, was the only thing possible in this event. Although it was not actually essential for our problem, we did sc anyhow to examine a broader patterned range in α by considering a solution of our nondimensionalized model lubrication equation of the form

$$\begin{aligned}
H(\boldsymbol{r}, t) \sim \alpha &+ A\cos(qr) + A^2[H_{20} + H_{22}\cos(2qr)] \\
&+ A^3[H_{31}\cos(qr) + H_{33}\cos(3qr)] \\
&+ A^4[H_{40} + H_{42}\cos(2qr) + H_{44}\cos(4qr)] \\
&+ A^5[H_{51}\cos(qr) + H_{53}\cos(3qr) + H_{55}\cos(5qr)]
\end{aligned}$$

where now the amplitude function $A = A(t)$ satisfied $dA/dt \sim \sigma A - a_1 A^3 - a_3 A^5$ and in the above expansion the subscripts have been dropped from q_c, r_1, and $A_1 = A_1(t)$. Here, the relevant coefficients through third-order have the same values as determined previously. We then solved for the remaining coefficients using the usual procedure and obtained the fifth-order Landau constant formula $a_3 = a_3(\alpha; \varepsilon)$ when $\beta_c = 0$, which is plotted in Fig. 8.8 for $\varepsilon = 10^{-11}$. Note from this figure, that $a_3 > 0$ in the α-intervals $(0, 1.0491)$ and $(1.0491, 2)$. In particular, this implies that $a_3 > 0$ in the rupture subintervals $(\alpha_1, 1.019)$ and $(1.311, 2)$ of Table 8.1. Recall that in these subintervals the other coefficients of the amplitude equation satisfy $\sigma > 0$ and $a_1 < 0$. We shall show in Chapter 15 that for these signs of its coefficients, the fifth-order Landau equation admits a stable critical point satisfying $2a_3 A_e^2 = (a_1^2 + 4\sigma a_3)^{1/2} - a_1$ and conjecture in Chapters 11, 13, and 14 that such solutions correspond to localized structures which can be related to dewetting-type rupture occurring in those subintervals by isolated large drop or hole formation, respectively, as described in conjunction with the experimental observations of Xie *et al.* ([303]). In the absence of this analysis, we used the results of Oulton and Wollkind ([160]) to deduce the stability behavior of the fifth-order Landau equation for $a_3, \sigma > 0$ and $a_1 < 0$.

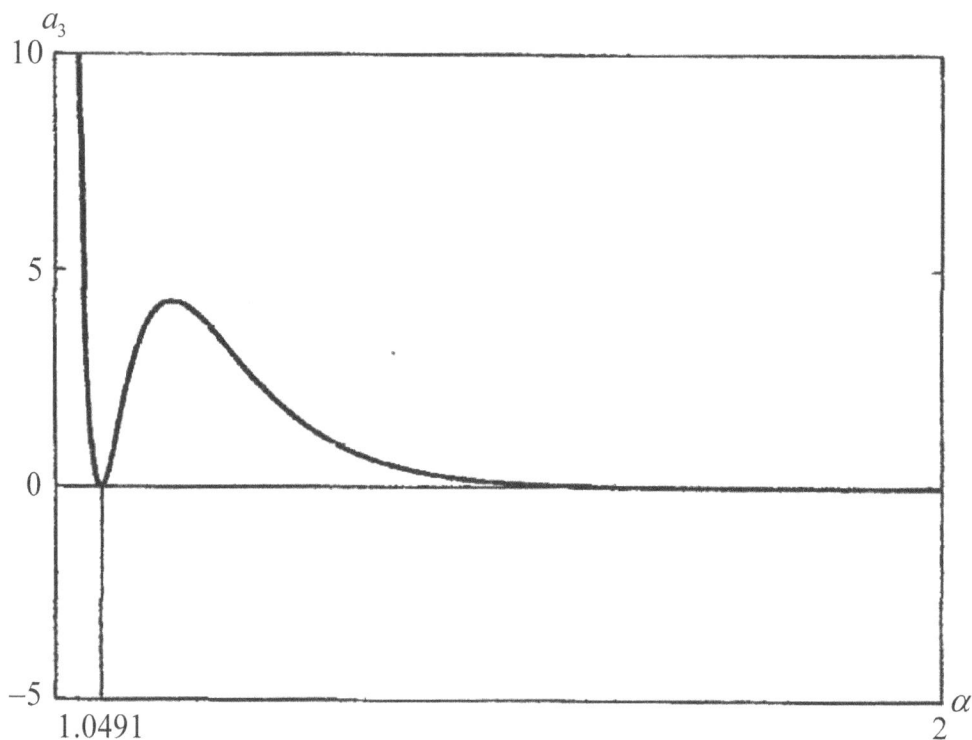

FIGURE 8.8
Plot of the fifth-order Landau constant a_3 versus α for $\beta_c = 0$ and $\varepsilon = 10^{-11}$.

9

Evolution Equation Phenomenon II: Ion-Sputtering of Solids: Hexagonal and Rhombic Planform Nonlinear Stability Analyses

While I was working with Mei Tian, Professor Yongwimon Lenbury, the Chair of the Mathematics Department at Mahidol University in Bangkok, Thailand, met with me and my chair, Alan Genz, to arrange a Memorandum of Agreement (MOA) with WSU, whereby doctoral graduate students in her department who had completed all of their requirements except for the thesis could spend up to a year in Pullman doing their dissertation research under the direction of one of our faculty members. These students would have the visa status of Visiting Scholar and be financed by a Thailand educational scholarship. Although not enrolled at WSU, they were still provided office space and attended classes that might help them to perform this thesis research. After finishing that research, each submitted a paper, which could be co-authored, to a reputable scientific journal for publication. Mahidol required proof of acceptance of these papers before students were allowed to defend their theses. Eventually, I supervised three such students, the first one of which was Adoon Pansewan, followed by Nichaphat Boonkorkuea and Inthira Chaiya. Adoon was accompanied by Chontita Rattanakul, one of Professor Lenbury's recent Ph.D. recipients, who spent four months doing postdoctoral research with me. She was also supported by a Thailand grant and, speaking perfect English, helped him adjust to life at WSU. Both began by attending a graduate mathematical modeling class I taught in the summer of 2002.

For Adoon's thesis topic, I selected the subject of Stefan Facsko *et al.*'s ([58]) report entitled "Formation of ordered nanoscale semiconductor dots by ion sputtering" that had appeared in the volume 285 number 5433 issue of *Science*. This report opened with the following paragraph:

> To date, two approaches for the fabrication of semiconductor quantum dots have been pursued. In the top-down approach, lithographic methods are used for direct patterning of quantum dots, whereas the bottom-up approach relies on self-organized processes. In contrast to serial electron-beam lithography, self-organization phenomena open the way for the formation of a regular array of quantum dots on large areas in a single technological step. Self-organized semiconductor quantum dots have been produced by the Stranski-Krastanow growth mode in molecular beam epitaxy and metal-organic vapor phase epitaxy in which coherent island formation occurs during the growth of lattice-mismatched semiconductors (Springholz *et al.*, [235]). Here, we present a controlled and cost-effective method for the production of well-ordered quantum dots by ion-bombardment of semiconductor surfaces that is based on a self-organized mechanism induced by ion-sputtering of solid surfaces, where the formation kinetics is determined by etching instead of growth.

We wished to model this phenomenon by investigating spontaneous pattern formation on semiconductor or metallic solid surfaces during ion-sputtered erosion at normal incidence (see Fig. 9.1) and began by examining all the references included in Facsko *et al.* ([58]).

DOI: 10.1201/9781003195603-9

From those references we found that this erosion process could be represented by a damped Kuramoto-Sivashinsky spatio-temporal nonlinear partial differential model evolution equation describing the deviation of a solid surface from its mean planar position. This governing evolution equation was defined on an unbounded two-dimensional spatial domain and included the deterministic effects of ion bombardment, surface diffusion, and ion sputtering. Let us explain that process more fully.

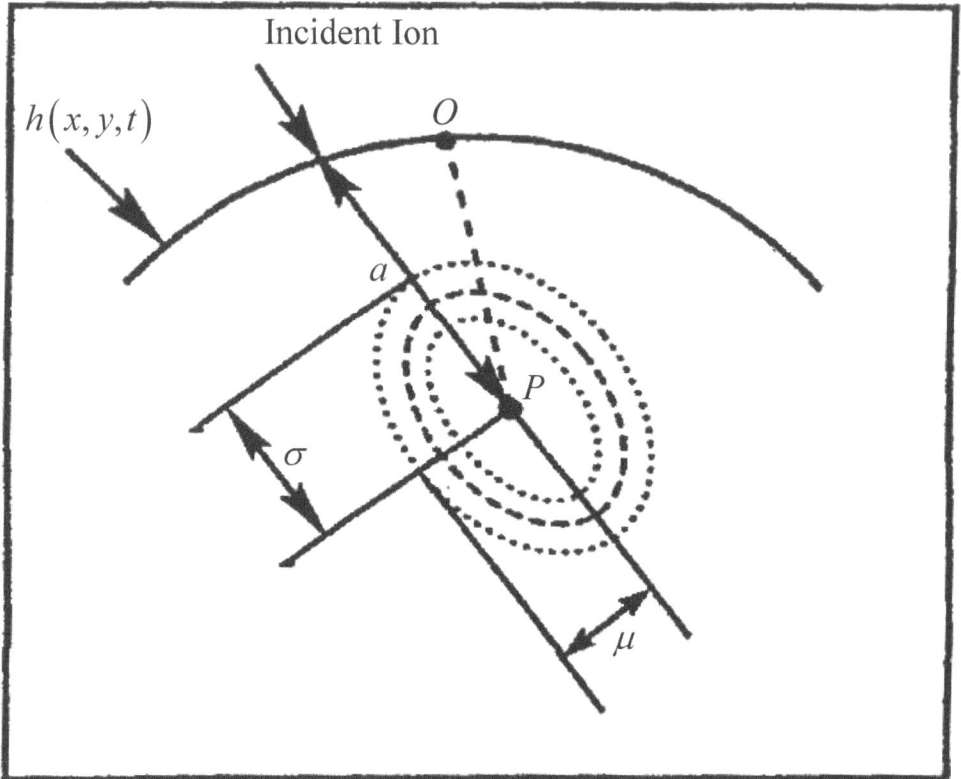

FIGURE 9.1
Schematic diagram from Makeev *et al.* ([133]) of the normal incidence ion-bombardment of a solid surface . Here, $h(x, y, t)$ represents the nondimensional deviation of the solid surface from its mean planar position and a is the ion penetration depth, while σ and μ are the ion-beam parallel and perpendicular lengths, respectively, that characterize the shape of the energy collision cascade. The dashed line from the points P to O depicts the trajectory of the sputtering yield material.

Sputtering, the removal of material from the surface of semiconductor or metallic solids through ion bombardment, is an important thin film processing technique (reviewed by Makeev *et al.*, [133]). The erosion rate for such surfaces can be characterized by the ion flux, defined as the number of particles arriving per unit area per unit time, and the sputtering yield, defined as the amount of material leaving the surface per unit incident particle. In the sputtering process, the incoming ions penetrate the solid and transfer their kinetic energy to the substrate material by inducing a collision cascade that allows some of the latter to gain sufficient energy to be removed from the surface or sputtered. One might suspect that erosion of this sort would tend to erase every possible interfacial feature and produce only a uniformly smooth morphology; however, under certain circumstances, periodic patterns are actually etched on the surface. That spontaneous self-organization is generated by the interplay of this ion-sputtering roughening with the smoothing mechanism of surface diffusion. In particular, these patterns were predicted to consist of coherent ripples (Park $et\ al.$, [167]) when that ion bombardment was at off-normal incidence and periodic arrays of quantum dots or holes (Kahng $et\ al.$, [91]) when at normal incidence. Specifically, crystalline nanoscale cone-shaped quantum dots (islands) arranged in regular hexagonal distributions or relatively uniform arrays of holes (vacancies) have been produced experimentally on gallium antimonide or platinum surfaces bombarded with argon ions at normal incidence by Facsko $et\ al.$ ([58]) and Michely and Comsa ([145])), respectively.

In order to model this process, we considered a thin solid film of dimensional thickness $H(\boldsymbol{r},\tau)$ undergoing normal-incidence ion-bombardment induced erosion where $\boldsymbol{r}=(r_1,r_2)$ represents a transverse laboratory Cartesian coordinate system and τ is time. A number of theoretical models had been developed in the references cited above to study the evolution of periodic roughening-type instabilities from coherent ripples to hexagonal arrays of quantum dots or holes in such instances by numerical simulations. These models were derived from conservation of mass at the interfacial solid surface and had as their point of departure the general continuity equation

$$H_\tau + K\nabla_2^4 H + \nu_1 H_{r_1 r_1} + \nu_2 H_{r_2 r_2} + D_1 H_{r_1 r_1 r_1 r_1} + D_2 H_{r_2 r_2 r_2 r_2}$$
$$+ D_{12} H_{r_1 r_1 r_2 r_2} + J_0 Y_0(H)$$
$$= \frac{1}{2}(\lambda_1 H_{r_1}^2 + \lambda_2 H_{r_2}^2) + H_{r_1}(\gamma_1 + \xi_1 H_{r_1 r_1} + \xi_2 H_{r_2 r_2}) + \Omega_1 H_{r_1 r_1 r_1} + \Omega_2 H_{r_1 r_2 r_2} + \eta.$$

Here, K denotes the thermal surface diffusion coefficient; J_0, the deterministic component of the ion flux; $Y_0(H)$, the sputtering yield; η, a noise term resulting from the stochastic component of the ion flux; $\boldsymbol{\nabla}_2 \equiv (\partial/\partial r_1, \partial/\partial r_2)$; $\nabla_2^2 \equiv \boldsymbol{\nabla}_2 \cdot \boldsymbol{\nabla}_2$; and $\nabla_2^4 \equiv (\nabla_2^2)^2$; while for the special case of normal incidence ion bombardment

$$\gamma_1 = \xi_1 = \xi_2 = \Omega_1 = \Omega_2 = 0, \ \nu_1 = \nu_2 = \nu,$$
$$D_1 = D_2 = \frac{D_{12}}{2} = D_0, \ \lambda_1 = \lambda_2 = \lambda_0;$$

which reduced this equation to its isotropic form

$$H_\tau + \nu\nabla_2^2 H + D\nabla_2^4 H + J_0 Y_0(H) = \frac{\lambda_0}{2}|\boldsymbol{\nabla}_2 H|^2 + \eta$$

where the effective surface diffusion coefficient $D = K + D_0$ and $|\boldsymbol{\nabla}_2 H|^2 = H_{r_1}^2 + H_{r_2}^2$.

Then introducing $\ell_0 = (2D/\nu)^{1/2}$ and $\tau_0 = 8D/\nu^2$ as scale factors for length and time, respectively; defining the dimensionless variables

$$(x,y) = \frac{(r_1,r_2)}{\ell_0}, \ t = \frac{\tau}{\tau_0}, \ h = \frac{H - H_0 + w_n\tau}{\ell_0},$$

where

$$w_n = J_0 h_0 \ell_0^2 \equiv \text{normal velocity of erosion};$$

adopting the sputtering yield relation, consistent with that of Makeev and Barabási ([132]),

$$Y_0(H) = h_0 \ell_0^2 \left[1 + \left(\frac{h_0}{\ell_0} \right) \sinh(h) \right]$$

$$\Rightarrow J_0 Y_0(H) = w_n \left[1 + \left(\frac{h_0}{\ell_0} \right) \sinh(h) \right] = w_0(h);$$

and, after Kahng *et al.* ([91]), taking

$$\eta \equiv 0;$$

that isotropic form was converted into the following damped Kuramoto-Sivashinsky equation

$$h_t + 4\nabla^2 h + 2\nabla^4 h + 2\beta \sinh(h) = \alpha |\nabla h|^2,$$

where

$$\alpha = \frac{\tau_0 \lambda_0}{2\ell_0} \in \mathbb{R}, \ 2\beta = \tau_0 J_0 h_0^2 \geq 0.$$

Here, $h = h(x, y, t)$ is the nondimensional deviation of the interface from its mean planar position $H_0 - w_n \tau_0 t$; h_0 represents the maximal interfacial deviation from that mean position; $\nabla \equiv (\partial/\partial x, \partial/\partial y)$; and $\nabla^2 \equiv \nabla \cdot \nabla$. Further, we viewed α and β as the nondimensional versions of λ_0, the tilt-dependent coefficient of the erosion rate, and DJ_0, the product of the effective surface diffusion coefficient times the deterministic ion-bombardment flux, respectively. Hence, β, which served as the bifurcation parameter for our analyses, is a measure of that damping effect to this pattern formation process caused by the interaction between these latter two smoothing mechanisms. Finally, ν is the absolute value of the coefficient of negative capillarity. This is the spatio-temporal model evolution equation we wished to analyze for the interfacial morphologies catalogued earlier. Although Kahng *et al.* ([91]) had conducted a two-dimensional numerical simulation on this equation for its undamped state of $\beta = 0$, no weakly nonlinear stability analyses had been performed on either that or our fully damped version with $\beta > 0$. To alleviate this deficiency we wished to perform longitudinal planform one-dimensional and both hexagonal and rhombic (see below) planform two-dimensional weakly nonlinear stability analyses of the planar interface solution $h \equiv 0$ for that equation. Toward this end, we noted that such a solution, representing a planar layer of uniform dimensional thickness $H = H_0 - w_n \tau$, implicitly satisfied the far-field boundary condition

$$h \text{ remains bounded as } x^2 + y^2 \to \infty.$$

As in the previous two chapters, we first performed a one-dimensional longitudinal planform analysis of this trivial state by considering the following solution to our evolution equation

$$h(x, y, t) \sim A_1(t) \cos(qx) + A_1^2(t)[h_{20} + h_{22}\cos(2qx)]$$
$$+ A_1^3(t)[h_{31}\cos(qx) + h_{33}\cos(3qx)]$$

where the amplitude function $A_1(t)$ satisfied the Landau equation

$$\frac{dA_1}{dt} \sim \sigma A_1 - a_1 A_1^3$$

and $q = 2\pi/\lambda$, λ being the wavelength of that class of spatially periodic perturbations under investigation. Then noting, that to this order (see Chapter 5)

$$\sinh(h) = \frac{e^h - e^{-h}}{2} \sim h + \frac{h^3}{6},$$

our linear problem yielded the secular equation

$$\sigma = 2[\beta_0(q^2) - \beta] \text{ with } \beta_0(q^2) = q^2(2 - q^2).$$

Thus, the parabola $\beta = \beta_0(q^2)$ served as its marginal stability curve in the q^2-β plane of Fig. 9.2. As can be seen from that figure, the maximum value of this parabola occurs at its vertex (q_c^2, β_c) where $q_c^2 = 1$ and $\beta_c = \beta_0(q_c^2) = 1$. Hence, for $\beta > \beta_c = 1$ there exists no q^2 associated with growing modes, while for $0 \le \beta < \beta_c$ there exists a band of such wavenumbers squared centered about $q^2 = q_c^2$. Therefore, the trivial solution is linearly stable for $\beta > 1$, neutrally stable for $\beta = 1$, and unstable for $0 \le \beta < 1$ (analogous to Fig. 2.7). Given this linear stability behavior we, as usual, equated the q and σ in our expansion to $q = q_c = 1$ and $\sigma = \sigma_0(\beta) = 2(1 - \beta)$.

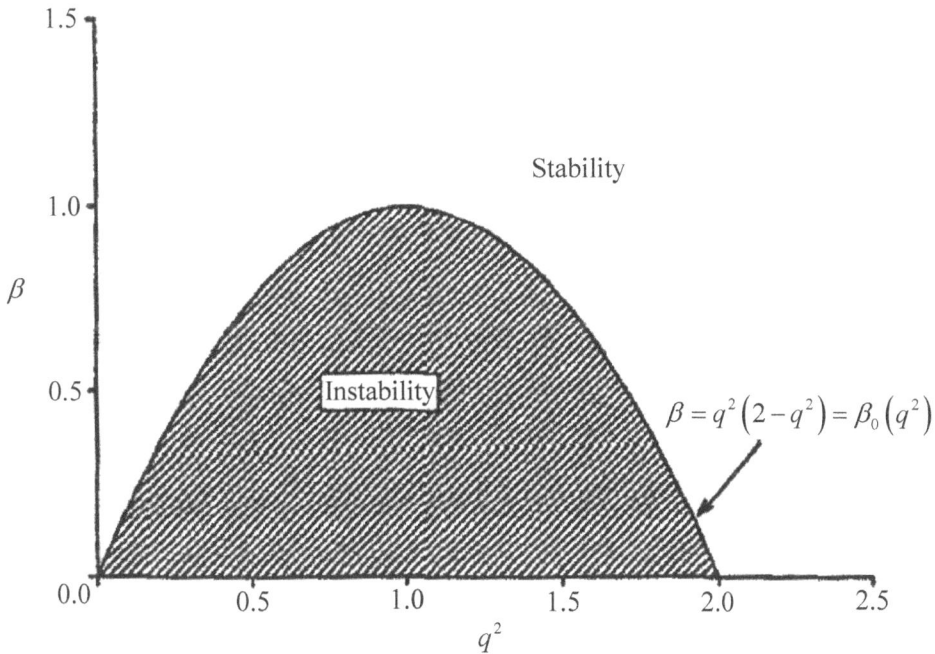

FIGURE 9.2
The marginal stability curve $\beta = \beta_0(q^2)$ plotted in the q^2-β plane denoting linear stability behavior.

Continuing with our description of the results of this one-dimensional expansion, the second-order problems could be solved in a straight-forward manner to yield

$$h_{20}(\beta; \alpha) = \alpha[4(2 - \beta)], \quad h_{22}(\beta; \alpha) = -\alpha[4(10 - \beta)].$$

Although there were also two third-order problems, we needed only consider the one proportional to $\cos(x)$ containing the Landau constant a_1

$$a_1 - 2\sigma_0(\beta)h_{31}(\beta; \alpha) = \frac{\beta}{4} - 2\alpha h_{22}(\beta; \alpha)$$

for our Fredholm method of solvability. Then, employing our previous results, taking the limit as $\beta \to \beta_c = 1$ of this relation, and assuming the requisite continuity of h_{31} at $\beta = \beta_c$, we obtained the solvability condition

$$a_1 = a_1(\alpha) = \frac{1}{4} + \frac{\alpha^2}{18} > 0,$$

which then yielded the solution

$$h_{31}(\beta; \alpha) = \frac{1}{16} + \frac{\alpha^2}{72(10 - \beta)}.$$

Although this procedure did not actually require it, for the sake of completeness we also solved the other third-order problem in a straight-forward manner to obtain

$$h_{33}(\beta; \alpha) = \frac{6\alpha^2 - \beta(10 - \beta)}{48(33 - \beta)(10 - \beta)}.$$

As in the previous chapter, we now made a phenomenological interpretation of the problem under examination based upon the signs of the growth rate σ_0 and the Landau constant a_1 in the truncated Landau equation. Since $a_1 > 0$, the four possible cases catalogued in Fig. 2.9 were reduced to the following two:

(i) σ_0, $a_1 > 0$: There exists a stable equilibrium solution $A_e^2 = \sigma_0/a_1$. Since $\sigma_0 > 0$ ($0 \leq \beta < 1$), linear theory would predict instability whereas our nonlinear analysis showed the existence of this finite amplitude supercritically stable equilibrium state. Specifically, that stable equilibrium point corresponded to a steady-state re-equilibrated spatially nonuniform pattern given by

$$\lim_{t \to \infty} h(x, y, t) = h_e(x, y) = \delta \cos(x) + O(\delta^2), \quad -\infty < x < \infty;$$

where

$$\delta = A_e + h_{31}(1; \alpha)A_e^3 > 0 \text{ with } h_{31}(1; \alpha) = \frac{1}{16} + \frac{\alpha^2}{648},$$

which represented a periodic one-dimensional interfacial morphology consisting of coherent stationary parallel ripples having a characteristic wavelength of

$$\lambda_c = \frac{2\pi}{q_c} = 2\pi \text{ and } \lambda_c^* = \ell_0 \lambda_c = 2\pi \ell_0 = 2\pi \left(\frac{2D}{\nu}\right)^{1/2}$$

in dimensionless and dimensional variables, respectively, the latter quantity being generated by the competition between surface tension (ν) and effective surface diffusion (D). We depicted this deviation function in the contour plot of Fig. 9.3, which is the figure referenced in the previous chapter in relation to its thin liquid film ridges (an interchange of the terms ripples and ridges would seem more logical but we have adopted the standard terminology employed for them).

(ii) $\sigma_0 < 0$, $a_1 > 0$: The planar or smooth interface solution $h \equiv 0$ was stable to both infinitesimal and one-dimensional finite amplitude disturbances. Since $\sigma_0 < 0$ ($\beta > 1$) linear theory predicts stability of the undisturbed state $A_1 \equiv 0$ and our nonlinear effects enhanced this stabilizing behavior. Unlike the planar interface thin liquid layers of the previous chapter this smooth interfacial morphology was not identically unstable for our ion-sputtered solid surface.

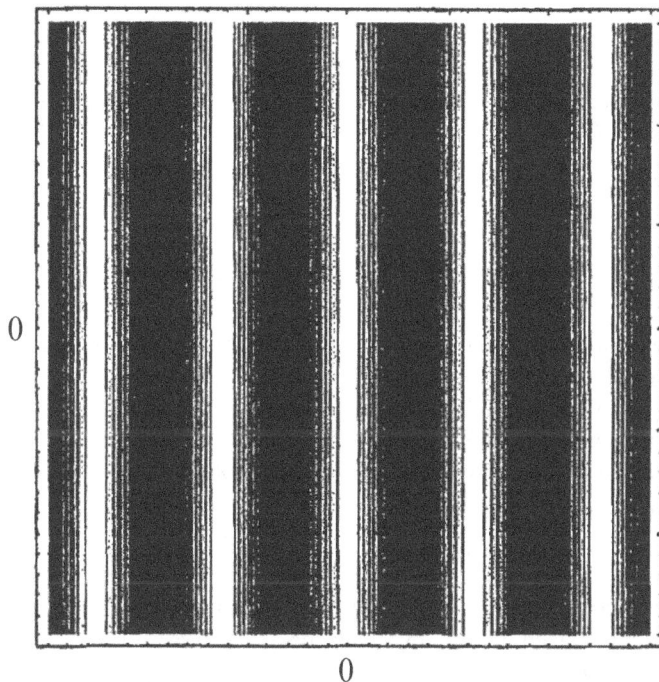

FIGURE 9.3
Density plot for the ripples of this chapter and ridges of Chapter 8 (critical point II of the hexagonal planform stability analysis) where the spatial variables are measured in units of λ_c. Here, elevations appear light and depressions, dark, in accordance with experimental observation. The associated contour plot for this pattern appears in Fig. 10.5.

In order to refine these one-dimensional predictions, we investigated two-dimensional ion-sputtered erosion patterns by seeking a hexagonal planform solution of our damped Kuramoto-Sivashinsky equation of the same form as that described in Chapter 8, but for the zero deviation and with the spatial variables (x, y) replacing (r_1, r_2). Proceeding in the identical manner as in that chapter, by employing the notation h_{ijNk} for the coefficient of each higher-order term in the relevant h-expansion of the form $A_1^i(t)B_1^j(t)\cos(Nq_c x/2)\cos(k\sqrt{3}q_c y/2)$, we found that

$$h_{i0N0} = h_{in} \text{ for } N = 2n, \ \sigma = \sigma_0(\beta), \ a_1 = a_1(\alpha),$$

as defined earlier, while, in particular, the other two Landau constants satisfied

$$a_0 - \sigma h_{0220} = -\frac{\alpha}{8};$$

$$a_2 - 2\sigma h_{1220} + (8a_0 + \alpha)h_{0220} = \frac{\beta}{4} - 3\alpha h_{0202} \text{ where } h_{0202} = \frac{-\alpha}{16(5-\beta)}.$$

Finally, taking the limit of these relations as $\beta \to 1$ and employing our previous results, we obtained the solvability conditions

$$a_0 = a_0(\alpha) = \frac{-\alpha}{8}, \ a_2 = a_2(\alpha) = \frac{1}{4} + \frac{3\alpha^2}{64} > 0.$$

Note, from our relation for a_2, that its solvability condition does not contain the component $\lim_{\beta \to 1} h_{0220}$ since the coefficient of the latter quantity, namely $8a_0 + \alpha$, is identically equal to zero by virtue of the value for a_0. As pointed out by Wollkind *et al.* ([291]) such independence can be expected as a general rule, although in this specific instance we may also conclude by solving our relation for a_0 that $h_{0220} \equiv 0$. Finally, observe that, for our values of $a_1(\alpha)$ and $a_2(\alpha)$,

$$2a_2(\alpha) - a_1(\alpha) = \frac{1}{4} + \frac{11\alpha^2}{288} > 0.$$

Having determined the formulae for the growth rate and Landau coefficients of the hexagonal planform amplitude-phase equations involving the amplitude and phase functions $A_{1,2,3}$ and $\varphi_{1,2,3}$, respectively, we catalogued the orbital stability behavior of its critical points for the special case of $2a_2 - a_1 > 0$ summarized in Chapter 3. As in that chapter, the potentially stable critical points of these equations can be catalogued by considering the following member of each relevant equivalent class corresponding to $A_1 = A_0$, $A_2 = A_3 = B_0$, $\varphi_1 = \varphi_2 = \varphi_3 = 0$ where

$$\text{I: } A_0 = B_0 = 0; \ \text{II: } A_0^2 = \frac{\sigma_0}{a_1}, \ B_0 = 0;$$

$$\text{III}^\pm: A_0 = B_0 = A_0^\pm = \frac{-2a_0 \pm [4a_0^2 + (a_1 + 4a_2)\sigma_0]^{1/2}}{a_1 + 4a_2}.$$

Recall that the orbital stability criteria for these critical points can be posed in terms of σ. Thus, critical point I was stable for $\sigma < 0$, while the stability properties of II and III$^\pm$, which depended only on the sign of a_0, were catalogued in Table 3.2 under the assumption that $2a_2 - a_1 > 0$, this table being reproduced below for convenience of reference.

The quantities σ_{-1}, σ_1, and σ_2 appearing in that table, originally defined in Chapters 2 and 3, were defined again in Chapter 8, while over the parameter range in question, $A_0^+ > 0$ and $A_0^- < 0$.

TABLE 9.1
Orbital stability behavior of critical points II and III$^{\pm}$ with a_0 when $2a_2 - a_1 > 0$. This is
a reproduction of Table 3.2.

a_0	Stable structures
$+$	III$^-$ for $\sigma_{-1} < \sigma < \sigma_2$, II for $\sigma > \sigma_1$
0	II for $\sigma > 0$
$-$	III$^+$ for $\sigma_{-1} < \sigma < \sigma_2$, II for $\sigma > \sigma_1$

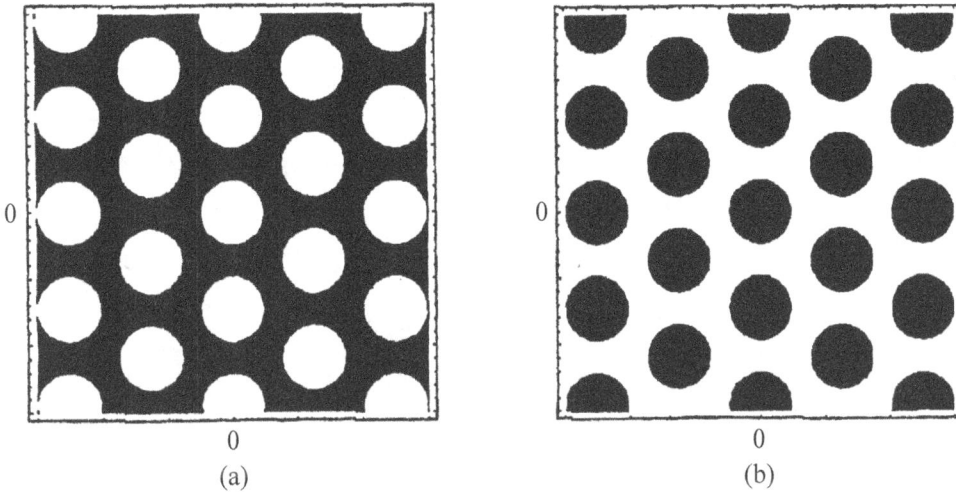

(a) (b)

FIGURE 9.4
Contour plots for critical points (a) III$^+$ and (b) III$^-$, the (a) quantum dots and (b) holes
of this chapter or the (a) nanodroplets and (b) cells of Chapter 8. Here, the spatial variables
are again being measured in λ_c with elevations appearing light and depressions, dark, as in
Fig. 9.3.

We next offered the following morphological interpretation of these potentially stable critical
points relative to the ion-sputtered erosion patterns under investigation: I and II represented
the planar or smooth interface and the coherent ripples, respectively, described in our longitudinal planform analysis, while III$^+$ and III$^-$ could be identified with the hexagonal
close-packed arrays of quantum dots and holes, respectively. Fig. 9.4 depicts their corresponding contour plots, which is the figure referenced in the previous chapter in relation
to thin liquid film arrays of nanodroplets and cells. In this figure, elevations appear uniformly light and depressions, uniformly dark, in accordance with experimental observation
and numerical simulation of the ion-sputtered erosion and thin liquid film morphological
separation patterns to which we ultimately compared them. In deciding between Chapters 8
and 9 where to include the contour plots of Fig. 9.3 and Fig. 9.4, we were influenced by
these figures original appearance being in the paper based upon Adoon's thesis.

In order to compare our theoretical predictions with experimental observation and numerical
simulation, we represented the stability results of Table 9.1 graphically in the α-β plane of
Fig. 9.5. Note, in this context, from the formula for $a_0(\alpha)$ that there exists an α_c such that

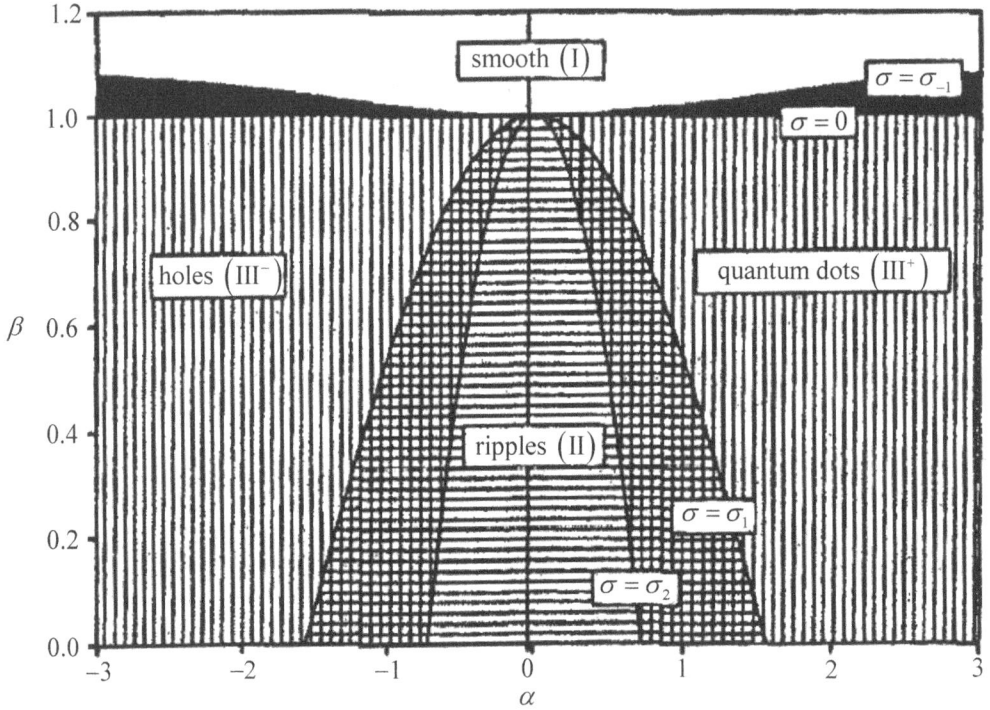

FIGURE 9.5
Stability diagram in the α-β plane for the damped Kuramoto-Sivashinski equation denoting the ion-sputtered erosion patterns predicted in Table 9.1. Note, that unlike our previous hexagonal planform nonlinear stability analyses, $\sigma = 0$ and $\sigma = \sigma_{-1}$ do not virtually coincide.

(see Fig. 9.6)

$$a_0 = 0 \text{ for } \alpha = \alpha_c, \; a_0 > 0 \text{ for } \alpha < \alpha_c, \; a_0 < 0 \text{ for } \alpha > \alpha_c,$$

which is given by

$$\alpha_c = 0.$$

Then, it was necessary to generate the loci $\beta = \beta_j(\alpha)$ associated with $\sigma = \sigma_j$ for $j = -1, 1,$ and 2, respectively. Employing our previous formulae, we solved $\sigma_0(\beta) = \sigma_j[a_0(\alpha), a_1(\alpha), a_2(\alpha)]$ for β when $j = -1, 1,$ and 2, and obtained

$$\beta = \beta_{-1}(\alpha) = \frac{79\alpha^2 + 360}{70\alpha^2 + 360} \text{ for } \sigma = \sigma_{-1},$$

$$\beta = \beta_1(\alpha) = \frac{-455\alpha^4 - 1,008\alpha^2 + 5,184}{121\alpha^4 + 1,584\alpha^2 + 5,184} \text{ for } \sigma = \sigma_1,$$

$$\beta = \beta_2(\alpha) = \frac{-2,003\alpha^4 - 8,784\alpha^2 + 5,184}{121\alpha^4 + 1,584\alpha^2 + 5,184} \text{ for } \sigma = \sigma_2,$$

respectively. Since all the quantities required for the identification of the ion-sputtered erosion patterns predicted in Table 9.1 have been evaluated, we could represent graphically the regions corresponding to these patterns in the α-β plane of Fig. 9.5, where the loci just determined have been denoted by $\sigma = \sigma_j$ with $j = -1, 1,$ and 2, in that figure, while $\sigma = \sigma_0(\beta) = 0$ or $\beta = 1$ has been denoted by $\sigma = 0$. Then, from Fig. 9.5, we observed that part (when $0 < \sigma < \sigma_2$) of the region ($0 \le \beta < 1$), where the one-dimensional analysis predicted coherent ripple patterns, was further divided into subregions characterized by hexagonal patterns consisting of either quantum dots (when $\alpha > \alpha_c = 0$) or holes (when $\alpha < \alpha_c = 0$), respectively. In the overlap regions satisfying $\sigma_1 < \sigma < \sigma_2$ or

$$0 < \beta_2(\alpha) < \beta < \beta_1(\alpha) < 1,$$

where ripples and quantum dots ($0 = \alpha_c < \alpha < \alpha_1^+$) or ripples and holes ($\alpha_1^- < \alpha < \alpha_c = 0$) were predicted, the initial conditions determined which stable equilibrium pattern of each pair would be selected. Here, $\alpha_1^\pm$ were defined implicitly by

$$\beta_1(\alpha_1^\pm) = 0,$$

which implied that

$$\alpha_1^\pm = \pm 1.56.$$

There also existed a region of bistability, denoted by uniform shading in Fig. 9.5, corresponding to $\sigma_{-1} < \sigma < 0$ or

$$1 < \beta < \beta_{-1}(\alpha) < \frac{79}{70},$$

the smooth interface state being stable for $\sigma < 0$ or $\beta > 1$ and hexagons, for $\sigma_{-1} < \sigma < \sigma_2$. Given that $\sigma_{-1} < 0$ for $a_0 \ne 0$ or $\alpha \ne \alpha_c = 0$, the hexagons persisting in this overlap region would be subcritical in nature. Finally, to justify the truncation procedure inherent to the asymptotic representation of our hexagonal planform amplitude-phase equations, it was necessary that its coefficients satisfy the size constraint

$$\frac{|a_0|}{(a_1 + 4a_2)^2} \ll 1.$$

As can be seen from Fig. 9.6,

$$|a_0| \ll (a_1 + 4a_2)^2$$

in the parameter range of interest and hence, we concluded that such a truncation procedure was valid for our hexagonal planform weakly nonlinear stability analysis of the damped Kuramoto-Sivashinsky equation. Further, given the hexagonal close-packed nature of the arrays associated with the critical points $III^{\pm}$, we have referred to them collectively as hexagons above. Finally, our rationale for including greater detail in this nonlinear stability exposition than those contained in previous chapters, was that the damped Kuramoto-Sivashinsky model evolution equation lent itself more readily to such a development due to the simplicity of its results.

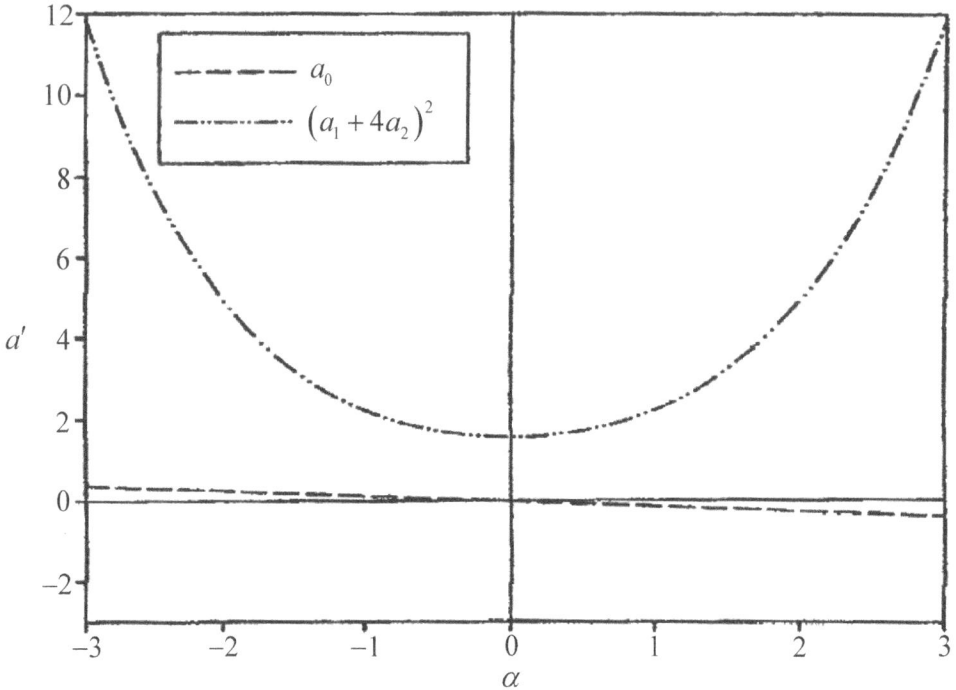

FIGURE 9.6
Plots of a_0 and $(a_1 + 4a_2)^2$ versus α.

To facilitate comparison of our morphological stability predictions with the results of the pattern formation studies involving normal-incidence ion-sputtered erosion of semiconductor or metallic surfaces referenced in the papers cited earlier, we described those experimental outcomes and numerical simulations in detail. Facsko *et al.* ([58]) produced a regular array of cone-shaped crystalline nanoscale quantum dots arranged in a nearly perfect hexagonal lattice on the solid surface of the semi-conductor gallium antimonide with Miller indices of (100). Recall that Miller indices, a notational system for planes in a crystalline lattice, had been introduced in Chapter 2, in conjunction with its Fig. 2.12. Unlike epitaxy methods, where such self-organized semiconductor patterns are generated through deposition, the formation process in this instance was erosion, induced by surface bombardment with argon ions at normal incidence. Thus, these highly-ordered close-packed nanostructures that they observed were actually argon ion-sputtered erosion patterns, where this self-organization was determined through the etching of the gallium antimonide surfaces instead of their growth (Kadar *et al.*, [90]).

To investigate the origin and dynamics of such quantum dot formation under normal-incidence ion sputtering, Kahng *et al.* ([91]) numerically integrated a continuum evolution equation for the dimensional deviation of the interfacial surface from its mean planar position $H_0 - w_n\tau$ that is derivable from the isotropic continuity equation upon adoption of the constant sputtering yield

$$J_0 Y_0(H) \equiv w_0(0) = w_n,$$

instead of our $\sinh(h)$-dependent one. These numerical simulations imposed periodic boundary conditions and, to improve the uniformity of the patterns generated, were carried out in the absence of noise ($\eta \equiv 0$), using instead a random initial surface configuration. Observe that such an approach differed markedly from the one employed for all the other simulations to be referenced, in which $\eta(r_1, r_2, \tau)$ was taken to be an uncorrelated white noise term with zero mean. In this context, the two-dimensional problem treated by Kahng *et al.* ([91]) was equivalent to the simulation of the undamped or $\beta = 0$ version of our damped Kuramoto-Sivashinsky equation. Specifically, they performed their numerical simulations for the idealized parameter values

$$\nu = 0.6169 \times 10^{-14} \frac{cm^2}{sec}, \quad D = 2 \times 10^{-28} \frac{cm^4}{sec}, \quad \lambda_0 = \pm 1 \frac{nm}{sec}$$

(note that since these authors assumed an implicit nondimensionalization by only assigning the numerical values of $\nu = 0.6169$, $D = 2$, and $\lambda_0 = \pm 1$, we took the liberty of supplying their appropriate magnitude and dimension upon comparison with the relevant experimental data to be introduced below), obtaining a hexagonal lattice of closely packed islands or quantum dots when $\lambda_0 = 1$ nm/sec and a similar lattice of depressions or holes when $\lambda_0 = -1$ nm/sec. Hence, Kahng *et al.* ([91]) concluded that their predicted morphologies for $\lambda_0 > 0$ were reminiscent of the hexagonal semiconductor patterns reported by Facsko *et al.* ([58]), while those for $\lambda_0 < 0$ resembled the relatively uniform distribution of vacancies or holes which resulted from argon ion-sputtering of platinum under normal incidence (Michely and Comsa, [145]). Further, upon examination of the experimental reference Rusponi *et al.* ([198]) one sees that the normal-incidence neon ion-sputtered erosion patterns occurring on their silver surfaces consisted of checkerboards of square pyramid-like mounds and pits or a ripple structure of elongated mounds and channels being induced on that metallic surface. Finally, in Rusponi *et al.* ([196, 197]) which dealt with argon ion-sputtering of silver and copper, respectively, these authors also reported coherent ripple formation after bombardment at normal incidence. Such a transition occurred from a smooth morphology as substrate temperature decreased through a critical value (Rusponi *et al.*, [196]). Hence, the admonishment of Kahng *et al.* ([91]) that ripple formation of this sort could only occur when the ion sputtering was at off-normal or so-called grazing incidence (Park *et al.*, [167]) was obviously not valid for metallic crystalline materials.

We were now ready to compare our two-dimensional ion-sputtered erosion morphological predictions with the normal-incidence experimental observations and numerical simulations just described. The same order was used in making these comparisons, as employed above, to describe those results. Thus, we began with the quantum dot semiconductor experiments of Facsko *et al.* ([58]) that originally motivated our research. To compare our theoretical predictions with their results, we first had to decide what values to assign the material and experimental parameters appearing in our model equation since most of the relevant quantities were unreported. Given that Facsko *et al.* ([58]) also produced the same quantum dot formation on germanium surfaces bombarded with argon ions as they did on gallium anti-monide ones, we used parameter values as measured by Chason *et al.* ([31]) who investigated temperature-dependent erosion during xenon ion sputtering of the group IV semiconductor germanium. Toward that end, we took

$$\nu = 2 \times 10^{-15} \frac{\text{cm}^2}{\text{sec}}, \ D = 0.8 \times 10^{-27} \frac{\text{cm}^4}{\text{sec}}, \ h_0 = 0.1 \text{ nm},$$

where these were appropriate values for normal-incidence ion sputtering at a substrate temperature $T_S = 423$ K. Then employing these values in the scale factors for length and time, we found that

$$\ell_0 = 8.94 \text{ nm}, \ \tau_0 = 1.60 \times 10^3 \text{ sec}.$$

Thus, from the definitions of β and w_n, we saw that $\beta = \beta_c = 1$ corresponded to

$$J_0 = 1.25 \times 10^{13} \text{ cm}^{-2} \text{ sec}^{-1}, \ w_n = 1.00 \frac{\text{nm}}{\text{sec}},$$

which were typical values of those quantities for the experiments of Chason *et al.* ([31]) and Facsko *et al.* ([31]), respectively. In order to determine the proper value to select for the parameter λ_0 which is the tilt dependent coefficient for the erosion rate, we used the stochastic simulation results of Cuerno *et al.* ([44]). These authors devised a discrete lattice dynamics model to mimic the effects of normal-incidence ion sputtering on solid surfaces by introducing two rules of interfacial motion, one to account for erosion and the other, for surface diffusion. They computed the mean equilibrium erosion velocity of the interface $v(m)$ as a function of the average interfacial tilt m and obtained the parabolic curve

$$v(m) = -w_n + \left(\frac{\lambda_0}{2} \right) m^2.$$

Cuerno *et al.* ([44]) generated a set of twenty-one data points $\{(m_i, v_i)\}_{i=1}^{21}$ by taking the spatial average of a noisy undamped Kuramoto-Sivashinsky equation and fit that parabolic curve to these generated data points with a high degree of accuracy as depicted in Fig. 9.7. Comparing those parameter identifications with the above value for w_n, we deduced that to two significant figures

$$\frac{\lambda_0}{2} = 0.01 \frac{\text{nm}}{\text{sec}}$$

was the proper choice of this coefficient for our purposes. Finally, for these selected values of τ_0, ℓ_0, and $\lambda_0/2$, we found from the definition of α that

$$\alpha = 1.79.$$

Upon examination of our hexagonal planform morphological stability results contained in Fig. 9.5, we concluded that such a positive value of $\alpha > \alpha_1^+ = 1.56$ was compatible with a prediction of quantum dot formation when $\beta = \beta_c = 1$. Further, for the selected value of ℓ_0, we obtained from the definition of the pattern dimensional wavelengh λ_c^* that

$$\lambda_c^* = 56.2 \text{ nm}.$$

Since Facsko *et al.* ([58]) stated that the diameter d_c^* of these quantum dots was closely approximated by λ_c^*, this compares quite favorably with their measured value of $d_c^* = 50$ nm (see Fig. 9.8 and Fig. 9.9). Hence, our theoretical predictions were in both very good qualitative and quantitative agreement with their quantum dot experimental observations. Given all the procedural steps involved, we definitely considered such a close correlation as

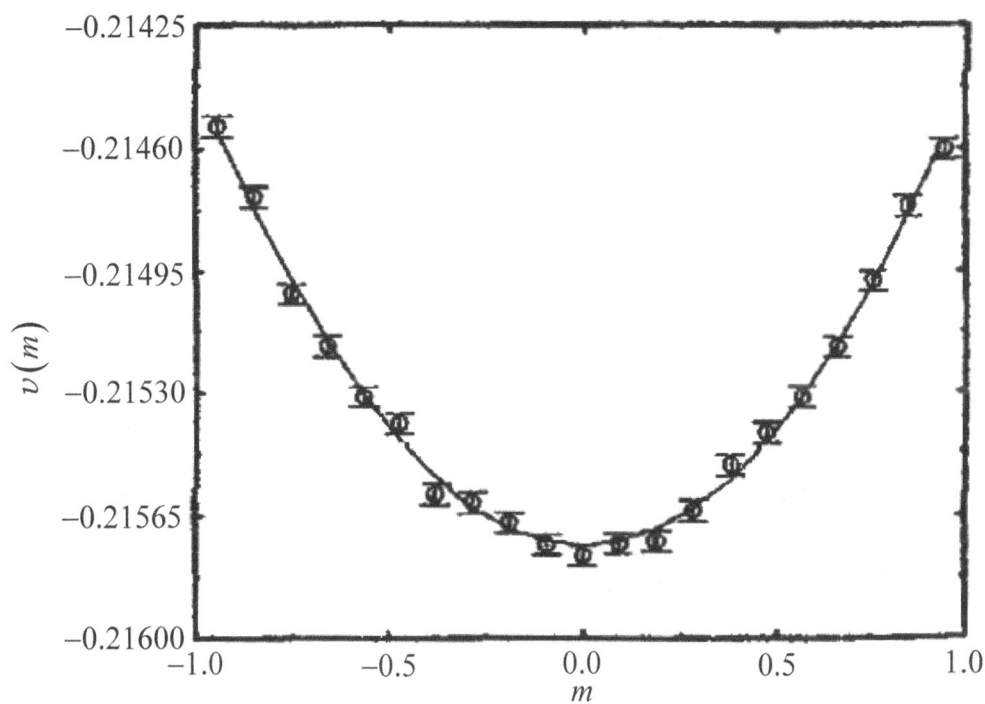

FIGURE 9.7
Parabolic fit of Cuerno *et al.* ([44]) to their generated data points in the *m*-*v* plane.

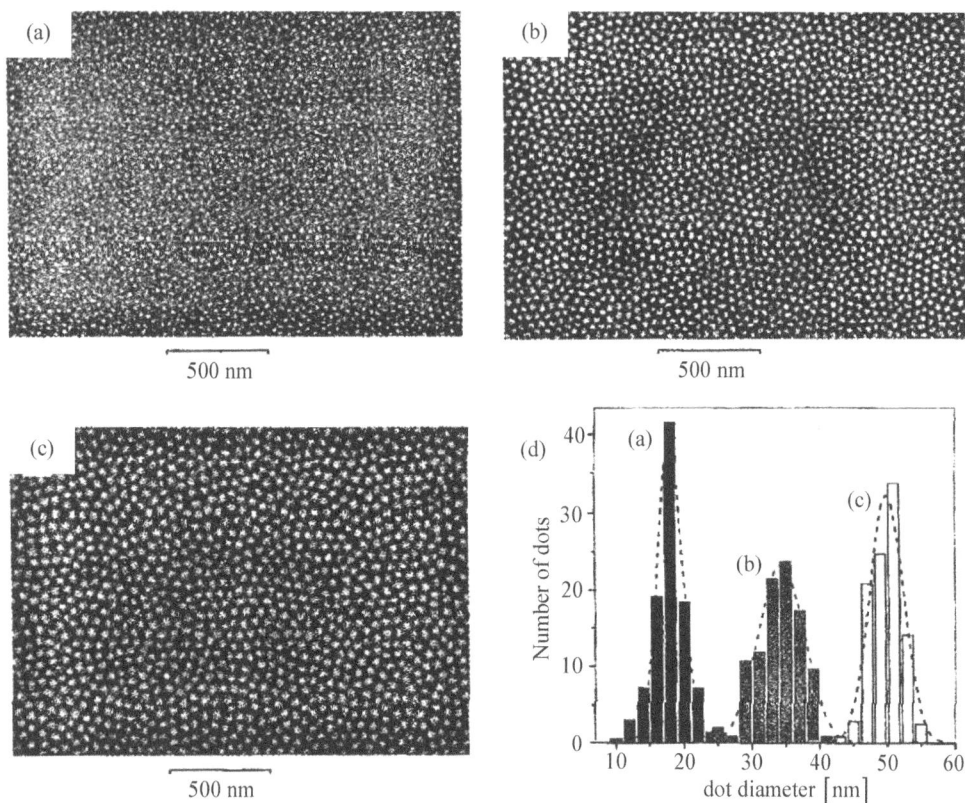

FIGURE 9.8

Scanning electron microscope images of the quantum dots for the increasing ion flux exposure times of (a), (b), and (c) observed by Facsko *et al.* ([58]) in their argon ion-sputtering of the semiconductor gallium antimonide at normal incidence and the corresponding size distribution of its dot diameters as depicted in (d). These times are 40, 200, and 400 sec for (a), (b), and (c), respectively. For longer exposure times the pattern did not change and its hexagonal order was fully maintained. Hence (c) represented the final stage of the dot formation.

Quantum Dots

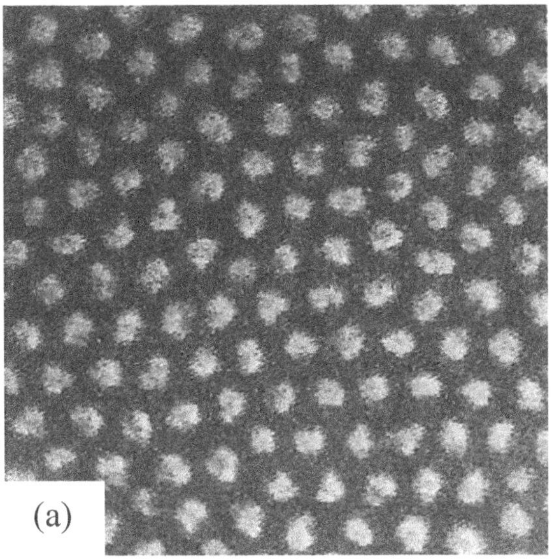

(a)

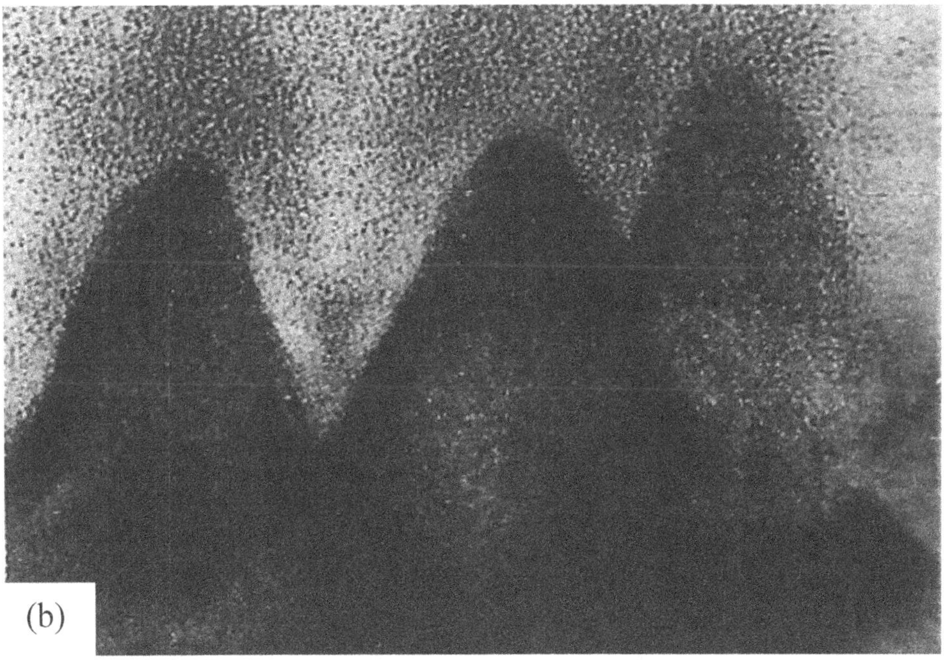

(b)

Miller indices of (100)

FIGURE 9.9

(a) An extract of the quantum dot image of Fig. 9.8c and (b) a transmission electron microscope cross-sectional image of its (100) surface nanostructure. Facsko *et al.* ([58]) stated that the spatial period or wavelength λ_c^* of the hexagonal pattern was approximately equal to the diameter d_c^* of the cone-shaped quantum dots.

Pulling a Rabbit Out of a Hat.

Before continuing with these comparisons, it is necessary to summarize Chontita Rattanakul's postdoctoral research results. Given that Rusponi *et al.* ([198]) produced a checkerboard pattern etched on their silver surface after normal incidence neon ion sputtering, she and I performed a rhombic planform weakly nonlinear stability analysis on our damped Kuramoto-Sivashinsky model equation. That is, we considered a solution of that equation to lowest order of the form

$$h(x, y, t) \sim A_1(t)\cos(x) + B_1(t)\cos(z) \text{ with } z = x\cos(\varphi) + y\sin(\varphi)$$

where

$$\frac{dA_1}{dt} \sim \sigma\, A_1 - A_1(a_1 A_1^2 + b_1 B_1^2) = F(A_1, B_1),$$

$$\frac{dB_1}{dt} \sim \sigma\, B_1 - B_1(b_1 A_1^2 + a_1 B_1^2) = G(A_1, B_1),$$

and $\varphi \equiv$ the rhombic angle, while each higher-order term in that expansion was of the form

$$h_{ijnk} A_1^i(t) B_1^j(t) \cos(nx + kz).$$

Then substituting this expansion into our damped Kuramoto-Sivashinsky model equation, we obtained a sequence of problems each of which was proportional to one of these terms. Solving those problems, we found that

$$h_{i0n0} = h_{0i0n} = h_{in}, \ \sigma = \sigma_0(\beta) = 2(\beta - 1), \ a_1 = a_1(\alpha) = \frac{1}{4} + \frac{\alpha^2}{18},$$

as defined earlier, while

$$h_{111(\pm 1)} = \frac{\alpha}{2}\frac{\pm \cos(\varphi)}{2 - \beta + 4\cos(\varphi)[\cos(\varphi) \pm 1]}$$

and b_1, in particular, satisfied

$$b_1 - 2\sigma h_{2101} = \frac{\beta}{2} - \alpha[1 + \cos(\varphi)]h_{1111} + \alpha[\cos(\varphi) - 1]h_{111(-1)}.$$

Now taking the limit of this relation as $\beta \to 1$ and making use of our previous results, we obtained the solvability condition

$$b_1 = b_1(\alpha, \varphi) = \frac{1}{2} + \alpha^2 \frac{\cos^2(\varphi)[4\cos^2(\varphi) - 3]}{[4\cos^2(\varphi) - 1]^2}.$$

Having developed these formulae for its growth rate and Landau constants, we turned our attention to the rhombic planform amplitude equations which possessed the following equivalence classes of critical points (A_0, C_0) such that $F(A_0, C_0) = G(A_0, C_0) = 0$ with $A_0, C_0 \geq 0$ given by

$$\text{I: } A_0 = C_0 = 0; \ \text{II: } A_0^2 = \frac{\sigma}{a_1}, \ C_0 = 0; \ \text{V: } A_0 = C_0 \text{ with } A_0^2 = \frac{\sigma}{a_1 + b_1}.$$

Here $a_1 > 0$ and we assumed $a_1 + b_1 > 0$, as well. Hence, critical points II and V would only occur provided $\sigma > 0$ or $0 \leq \beta < 1$.

We first investigated the stability of these critical points by seeking a solution of our rhombic planform amplitude equations of the form

$$A_1(t) = A_0 + \varepsilon_1 \mathcal{A}(t) + O(\varepsilon_1^2), \ \ B_1(t) = C_0 + \varepsilon_1 \mathcal{B}(t) + O(\varepsilon_1^2) \text{ with } |\varepsilon_1| \ll 1,$$

and, employing the methodology introduced in Chapter 4, showed that the perturbation quantities $\mathcal{A}(t)$, $\mathcal{B}(t)$ satisfied the linear homogeneous ordinary differential equation system

$$\frac{d\mathcal{A}}{dt} = c_{11}\mathcal{A} + c_{12}\mathcal{B}, \ \ \frac{d\mathcal{B}}{dt} = c_{21}\mathcal{A} + c_{22}\mathcal{B},$$

where

$$c_{11} = \frac{\partial F}{\partial A_1}(A_0, C_0) = \sigma - 3a_1 A_0^2 - b_1 C_0^2,$$

$$c_{22} = \frac{\partial G}{\partial B_1}(A_0, C_0) = \sigma - 3a_1 C_0^2 - b_1 A_0^2,$$

$$c_{12} = \frac{\partial F}{\partial B_1}(A_0, C_0) = c_{21} = \frac{\partial G}{\partial A_1}(A_0, C_0) = -2b_1 A_0 C_0.$$

Then letting $[\mathcal{A}, \mathcal{B}](t) = [\mathcal{C}_1, \mathcal{C}_2]e^{pt}$ where $\mathcal{C}_1^2 + \mathcal{C}_2^2 \neq 0$, we obtained the following linear homogeneous system for the constants $\mathcal{C}_1$ and $\mathcal{C}_2$:

$$(p - c_{11})\mathcal{C}_1 - c_{12}\mathcal{C}_2 = 0, \ \ -c_{21}\mathcal{C}_1 + (p - c_{22})\mathcal{C}_2 = 0;$$

which, upon imposition of the Pat Munroe algorithm to guarantee the nontriviality property for these constants, yielded the following quadratic in p:

$$(p - c_{11})(p - c_{22}) - c_{12}^2 = 0.$$

Next, particularizing this quadratic to the specific (A_0, C_0) values of the critical points and noting that under these conditions it had the associated roots $p_1 = c_{11} + c_{12}$ and $p_2 = c_{22} - c_{12}$, since either $c_{12} = 0$ for I and II or $c_{11} = c_{22}$ for V, we concluded that

$$\text{I: } p_{1,2} = \sigma; \ \text{II: } p_1 = -2\sigma, \ p_2 = \frac{(a_1 - b_1)\sigma}{a_1}; \ \text{V: } p_1 = -2\sigma, \ p_2 = \frac{2(b_1 - a_1)\sigma}{a_1 + b_1}.$$

Finally, requiring $p_{1,2} < 0$, we deduced the stability criteria:

$$\text{I is stable for } \sigma < 0;$$
$$\text{II is stable for } \sigma > 0, \ b_1 > a_1;$$
$$\text{V is stable for } \sigma > 0, \ a_1 > b_1.$$

Observe that since those criteria are mutually exclusive there can never be pair-wise bistability between these critical points. Also note that $\sigma > 0$ is both an existence condition and a stability criterion for critical points II and V. Again, given the simplicity of this stability analysis, we have included more detail than retained heretofore in corresponding hexagonal planform ones.

Before examining the implications of those stability criteria, we made a morphological interpretation of the potentially stable critical points relative to the ion-sputtered erosion patterns under investigation. Then to lowest order the interfacial deviation associated with these critical points was given by

$$\lim_{t\to\infty} h(x, y, t) = he(x, y) \sim A_0 \cos(x) + C_0 \cos(z) \text{ where } z = x\cos(\varphi) + y\sin(\varphi).$$

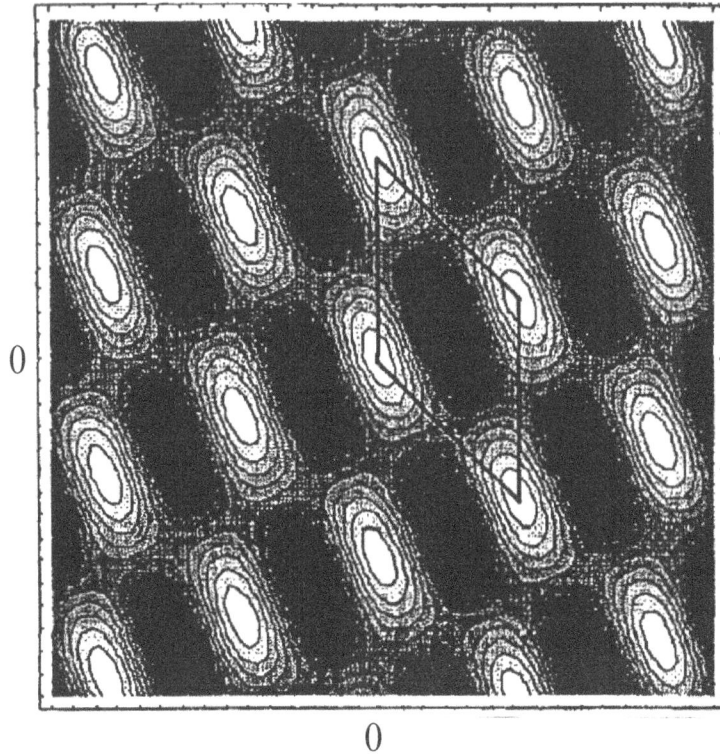

FIGURE 9.10
Contour plot of critical point V with $\varphi = 59.5°$ (or equivalentlty 1.038). This is a representative stable rhombic pattern for $\alpha = 0.50$ (see Table 9.2). Here, the quadrilateral formed
by the solid lines depicts the rhombic symmetry of the rectangular pattern.

Thus critical points I and II corresponded to the smooth planar surface and coherent ripples, respectively, already discussed in the longitudinal and hexagonal planform analyses. For critical point V, $0 < \varphi \leq \pi/2$ and, in order to make an analogous interpretation of that critical point, we considered this deviation function with $A_0 = C_0$ and sequentially took $\varphi = 59.5\pi/180$ and $\pi/2$, the contour plots of which appear in Fig. 9.10 and Fig. 9.11, respectively. From the checkerboard structure of this array for $\varphi = \pi/2$ (or equivalently 90°) in Fig. 9.11, it was clear that this state should be identified with an ion-sputtered erosion pattern of square planform. Similarly from Fig. 9.10, we concluded that for $\varphi = 59.5\pi/180 = 1.038$ (or equivalently 59.5°) it could be identified with a rhombic array of rectangles as indicated in this figure.

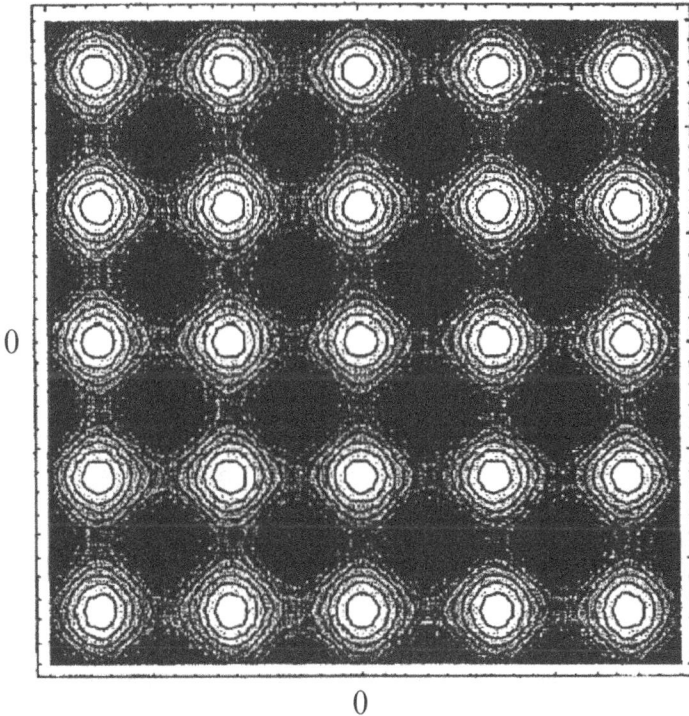

FIGURE 9.11
Contour plot of critical point V with $\varphi = \pi/2$. This is a checkerboard square pattern.

We next examined the existence and stability of critical points II and V. Toward that end we investigated the signs of $a_1 + b_1$ and $a_1 - b_1$ for $\alpha \in \mathbb{R}$ and $0 < \varphi \leq \pi/2$. To illustrate this procedure, consider the stability of the square planform obtained by setting $\varphi = \pi/2$. Given that

$$0 < b_1\left(\alpha, \frac{\pi}{2}\right) = \frac{1}{2} < \frac{1}{4} + \frac{\alpha^2}{18} = a_1(\alpha) \text{ whenever } \alpha^2 > \frac{9}{2},$$

we concluded that square rhombic patterns would be stable versus ripples for

$$|\alpha| > \frac{3\sqrt{2}}{2} \text{ provided } \sigma > 0 \text{ or } 0 \leq \beta < 1.$$

Then we investigated this behavior for fixed values of α and since our Landau constants were symmetric in that parameter only needed to consider $\alpha \geq 0$. Observe in this context that

$$b_1(0, \varphi) = \frac{1}{2} > \frac{1}{4} = a_1(0),$$

and thus, for $\alpha = 0$ and $0 \leq \beta < 1$, ripples were stable versus rhombic patterns. For $\alpha > 0$, we found there existed two φ-intervals (φ_m, φ_M) and $(\varphi_\ell, \varphi_r)$, flanking $\varphi = \pi/3$ (see Table 9.2), in which both $a_1(\alpha) \pm b_1(\alpha, \varphi) > 0$, or equivalently, $-1 < b_1(\alpha, \varphi)/a_1(\alpha) < 1$, where

$$0 < \varphi_m(\alpha) < \varphi_M(\alpha) < \pi/3 < \varphi_\ell(\alpha) < \varphi_r(\alpha) \leq \pi/2,$$

and thus, rhombic patterns of these characteristic angles were stable versus ripples. Note that, although this limit exists, such an occurrence implies

$$\lim_{\alpha \to \alpha_c} b_1(\alpha, \varphi) \neq b_1(\alpha_c, \varphi)$$

or $b_1(\alpha, \varphi)$ has a jump discontinuity at $\alpha = \alpha_c = 0$ where $a_0(\alpha_c) = 0$, the conclusive demonstration of which for general rhombic-planform nonlinear stability analyses is being deferred until Chapter 13 when we treat the mussel-bed Turing pattern formation problem of Cangelosi *et al.* ([25]), as well as relevant upper and lower threshold rhombic planform patterns.

TABLE 9.2
The φ-range for stable rhombic patterns versus α.

α	φ_m	φ_M	φ_ℓ	φ_r
0.05	1.03	1.04	1.06	1.07
0.10	1.00	1.02	1.06	1.09
0.50	0.85	0.93	1.15	1.22
1.00	0.75	0.87	1.21	1.35
2.00	0.67	0.81	1.23	1.40
3.00	0.48	0.68	1.38	1.57

Having catalogued those rhombic planform predictions, we returned to the numerical simulation results of Kahng *et al.* ([91]). Employing their idealized parameter values introduced above, we found that

$$\ell_0 = 2.55 \text{ nm}, \quad \tau_0 = 42.04 \text{ sec}, \quad \alpha = \pm 8.26, \quad \lambda_c^* = 16 \text{ nm}.$$

Given that the problem analyzed by Kahng *et al.* ([91]) corresponded to $\beta = 0$, we restricted our attention in this instance to the α-axis of Fig. 9.5. In that context, observe this figure has only been drawn for $|\alpha| \leq 3$. Since the deleted intervals $|\alpha| > 3$ possessed the same qualitative bifurcation properties as those associated with $1.56 = |\alpha_1^{\pm}| \leq |\alpha| \leq 3$ in Fig. 9.5, we could extrapolate the morphological stability behavior of the former intervals from that of the latter. We then saw that the value of $\alpha = 8.26$ was compatible with a hexagonal lattice of quantum dots, while $\alpha = -8.26$ was compatible with a similar pattern involving holes. Both of these theoretical predictions were, therefore, in agreement with the $\lambda_0 = \pm 1$ nm/sec numerical simulation results of Kahng *et al.* ([91]) described earlier. As to their general conclusion that $\lambda_0 > 0$ would give rise to hexagonal lattices of quantum dots and $\lambda_0 < 0$, to similar lattices of holes, such an assertion is compatible with our morphological stability predictions of Fig. 9.5 when $|\alpha| > 1.56$. For $0.73 < |\alpha| < 1.56$ where

$$\beta_2(\alpha_2^{\pm}) = 0 \to \alpha_2^{\pm} = \pm 0.73,$$

our hexagonal planform morphological stability results of Fig. 9.5 predict bistability between the relevant hexagonal pattern and ripples, while for $0 < |\alpha| < 0.73$ only ripples are predicted.

To understand exactly what was transpiring for this case, we had to examine our rhombic planform results in conjunction with these predictions. From those results, we concluded that for $\alpha \neq 0$, any such ripples were unstable with respect to rhombic patterns having characteristic angles $\varphi \in (\varphi_m, \varphi_M)$ or $\varphi \in (\varphi_\ell, \varphi_r)$. In particular, for the α-range over which our hexagonal planform analysis predicts ripples as the only stable pattern, rhombic arrays of this sort are difficult to distinguish from hexagonal lattices since the allowable intervals of their characteristic angles closely flank $\pi/3 \cong 1.047$ (see Table 9.2, Fig. 9.10, and Fig. 9.12). Then, substituting this type of rhombic pattern as the stable morphology in those α-intervals, hexagonal or nearly hexagonal arrays could be anticipated for all values of $\alpha \neq 0$ when $\beta = 0$. This substitution procedure most certainly represented the Pulling of a Rabbit Out of a Hat. We also noted that, for $0 < \beta < 1$, the same argument as just employed was equally valid given the behavior of the intercepts of horizontal lines of constant β with the curves $\beta = \beta_{1,2}(\alpha)$ in the α-β plane of Fig. 9.5. Further, we observed that, for $1 \leq \beta < 79/70$, there existed no rhombic arrays and thus, it was unnecessary to take them into account during our interpretation of the experiments of Facsko *et al.* ([58]). Therefore, in light of these theoretical, experimental, and numerical outcomes, we conjectured that $\lambda_0 > 0$ was to be expected for ion-sputtering of semiconductor materials.

Next, we considered the morphological stability predictions of our model relevant to the experimental observations of Michely and Comsa ([145]) and Rusponi *et al.* ([196, 197, 198]) involving the ion-sputtering of metals. Clearly, the lattice of vacancies produced by Michely and Comsa ([145]) at a substrate temperature T_S from 550 K to 625 K on a normal-incidence argon-ion sputtered platinum surface with Miller indices of (111), depicted in Fig. 9.13c,d, when compared with the hexagonal or nearly hexagonal arrays of holes predicted in our stability diagram of Fig. 9.5, along with the rhombic substitution procedure, required the parameters $\alpha < 0$ (or equivalently $\lambda_0 < 0$) and $0 \leq \beta < 1$. They referred to such depressions as vacancy islands to distinguish them from normal islands or cones (quantum dots). Our rhombic planform stability analysis for $\varphi = \pi/2$ predicted that checkerboard arrays of pyramids or pits required $0 \leq \beta < 1$ and $|\alpha| > 3\sqrt{2}/2$. Now, making the companion conjecture for ion-sputtering of metallic crystalline materials, that λ_0 was expected to be nonpositive or $\lambda_0 \leq 0$, this latter requirement reduced to $\alpha < -3\sqrt{2}/2$, consistent with the square checkerboard of pits observed by Rusponi *et al.* ([198]) on a neon ion-sputtering silver surface with Miller indices of (001) at normal incidence, depicted in Fig. 9.13a.

Finally, from Figure 9.5 for our hexagonal planform morphological stability analysis, in conjunction with the results of our rhombic planform morphological stability analysis, we can conclude that the coherent ripples observed by Rusponi *et al.* ([196, 197, 198]), which occurred under argon or neon ion-sputtering of silver or copper erosion surfaces with Miller indices of (110) at normal incidence (see, for example, Fig. 9.13b), required $\alpha = \alpha_c = 0$, or equivalently $\lambda_0 = 0$, since such patterns are only stable for both planforms at the critical value of that parameter and when $0 \leq \beta < \beta_c = 1$. For $\beta > \beta_c = 1$ and $\alpha = 0$, the planar or smooth interface solution is the only stable state. Hence, there is a so-called exchange of stabilities for $\alpha = 0$ between these two states at $\beta = \beta_c = 1$.

We used this result to show that the predictions of our model were compatible with the transition between those states Rusponi *et al.* ([196]) produced experimentally upon decreasing

regular hexagonal array   rhombic array

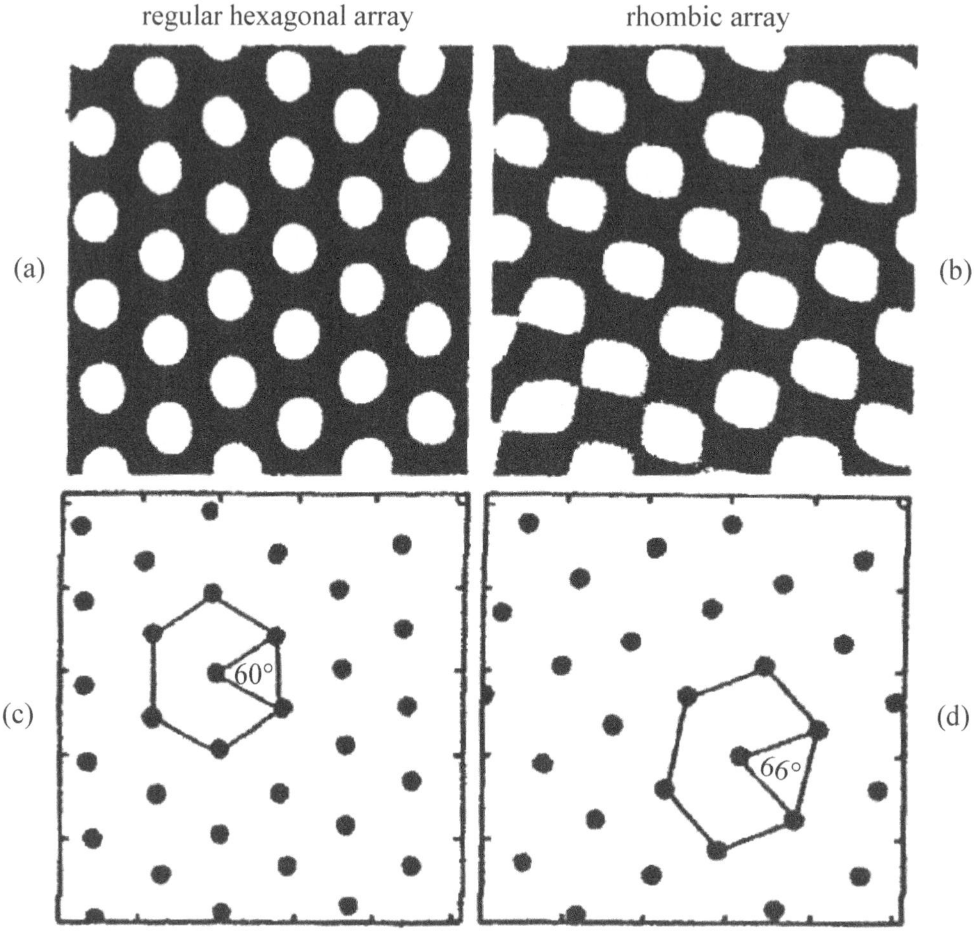

FIGURE 9.12
Contour & lattice plots (a) & (c) of a regular hexagonal array and (b) & (d) of a rhombic array of characteristic angle $\varphi = 66°$ or 1.150. Note the similarities in those arrays for rhombic angles close to $\varphi = 60°$. These represent chemical Turing patterns (see Fig. 1.2) from Gunaratne *et al.* ([75]).

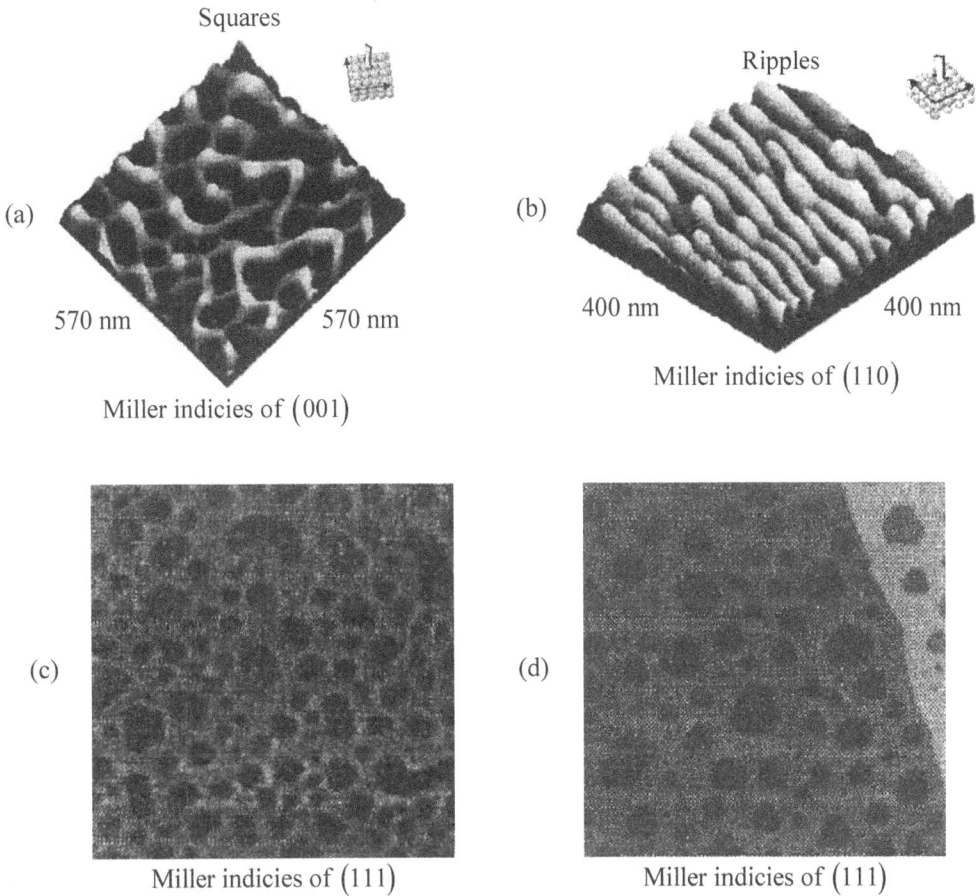

Squares

(a)

570 nm   570 nm

Miller indicies of (001)

Ripples

(b)

400 nm   400 nm

Miller indicies of (110)

(c)

Miller indicies of (111)

(d)

Miller indicies of (111)

FIGURE 9.13

(a) & (b): Three-dimensional images of the surface morphologies resulting after the normal-incidence neon-ion sputtering of silver by Rusponi *et al.* ([198]) for (a) the (001) & (b) the (110) surfaces, respectively, where the former produced a checkerboard pattern of pits & the latter, a coherent ripple structure (note $\lambda_c^* = 55$ nm!). (c) & (d): Scanning tunneling microscopy topographical images of the (111) surface morphology after the normal incidence argon ion-sputtering of platinum by Michely and Comsa ([145]) at (c) $T_S = 550$ K & (d) $T_S = 625$ K, showing the distribution of vacancies or holes produced which appear dark in these images.

substrate temperature. First recalling, from the definitions of β and τ_0, that

$$\beta = \frac{4J_0 h_0^2 D}{\nu^2}$$

and then noting that the thermal diffusion coefficient K satisfied a relationship of the form (Makeev and Barabási, [131])

$$K = K_0 \exp\left(-\frac{T_0}{T_S}\right) = K(T_S),$$

where K_0 and T_0 were positive characteristic values, we derived the following substrate temperature dependence of the effective surface diffusion coefficient

$$D = K(T_S) + D_0 = D(T_S).$$

Thus, since the other quantities appearing in our formula for β were virtually invariant over the substrate temperature range of interest, incorporation of this function into that formula yielded

$$\beta = \frac{4J_0 h_0^2 D(T_S)}{\nu^2} = \beta(T_S).$$

Hence, because $K(T_S)$ increased exponentially with substrate temperature given that

$$K'(T_S) = \left(\frac{T_0}{T_S^2}\right) K(T_S) > 0,$$

we deduced there existed a critical value of this temperature T_c, defined implicitly by

$$\beta(T_c) = \beta_c = 1$$

such that

$$\beta < \beta_c = 1 \text{ for } T_S < T_c, \ \beta > \beta_c = 1 \text{ for } T_S > T_c,$$

which served as a point of transition between the smooth and rippled morphologies along the vertical line $\alpha = 0$ of Fig. 9.5 when substrate temperature was decreased through it, in accordance with the experimental evidence of Rusponi *et al.* ([196]) depicted in Fig. 9.14.

Having demonstrated that, given the proper identification of its parameter values, our theoretical model predictions agreed very well with relevant experimental observations and numerical simulations, we discussed a general symmetry property of the damped Kuramoto-Sivashinsky governing equation responsible for this correlation. As pointed out by Kahng *et al.* ([91]), for their undamped version of that equation the morphological reversal which occurred upon the change in sign of α could be anticipated from its qualitative behavior. Specifically, this equation was invariant under the simultaneous transformation $h \to -h$ and $\alpha \to -\alpha$ by virtue of both $\sinh(-h) = -\sinh(h)$ and $|\nabla(-h)|^2 = |\nabla h|^2$ which indicated that the change in sign of α did not affect its interfacial dynamics, but rather turned quantum dot patterns into mirror imaged ones involving holes. This also indicated that such patterns were intrinsically nonlinear. Had linear terms alone been responsible for their formation then the surface morphology would not have depended on the sign of α (Kahng *et al.*, [91]). This being the case, there was some merit for us to examine further the two nonlinear effects contained in that equation.

Ripples

Smooth

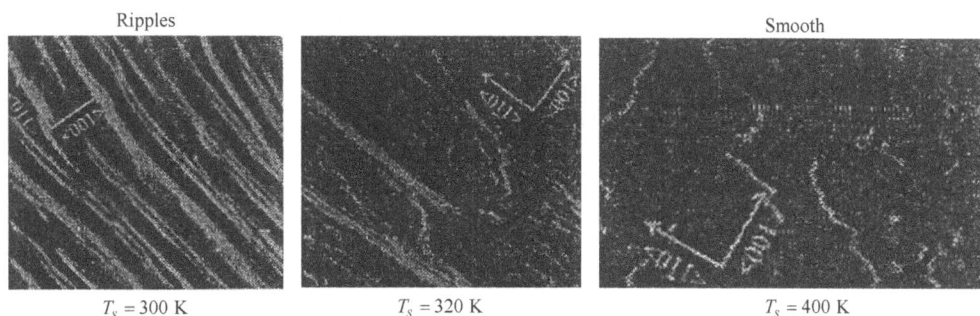

$T_s = 300$ K $T_s = 320$ K $T_s = 400$ K

FIGURE 9.14
Scanning tunneling microscopy images of the (110) surface morphologies resulting after the normal-incidence argon ion-sputtering of silver by Rusponi *et al.* ([196]) at the three different substrate temperatures from right to left of $T_S = 400$ K, 320 K, and 300 K, showing the transition from a smooth surface to coherent ripples as that temperature is decreased.

We first considered our sputtering yield constitutive relation with $H = h_0 + H_0 - w_n \tau$ or equivalently $h = h_0/\ell_0$ in its normalized form

$$\frac{Y_0(h_0 + H_0 - w_n \tau)}{h_0 \ell_0^2} = 1 + \frac{h_0}{\ell_0}\ \sinh\left(\frac{h_0}{\ell_0}\right).$$

Then employing the series for sinh through third-order, we obtained the representation

$$1 + \frac{h_0}{\ell_0}\ \sinh\left(\frac{h_0}{\ell_0}\right) \sim 1 + \left(\frac{h_0}{\ell_0}\right)^2 + \frac{\left(\frac{h_0}{\ell_0}\right)^4}{6},$$

which is consistent with the asymptotic result of Makeev and Barabási ([132])

$$C_1 + C_2 h_0^2 + O(h_0^4) \text{ where } C_1 = 1 \text{ and } C_2 > 0,$$

for that normalized yield. Hence, we initially introduced a general secondary yield term in our constitutive relation of the form

$$r(h) = h + \eta_0 h^3 + O(h^5)$$

rather than the particular function $\sinh(h)$ appearing there. Since that function only contained odd powers of h and hence $r(-h) = -r(h)$, our damped Kuramoto-Sivashinsky equation would still exhibit the symmetry property discussed above. Then our hexagonal and rhombic planform weakly nonlinear stability analyses resulted in the third-order Landau constants

$$a_1 = \frac{3\eta_0}{2} + \frac{\alpha^2}{18}, a_2 = \frac{3\eta_0}{2} + \frac{3\alpha^2}{64},$$
$$b_1 = 3\eta_0 + \alpha^2 \frac{\cos^2(\varphi)[4\cos^2(\varphi) - 3]}{[4\cos^2(\varphi) - 1]^2},$$

which, of course, reduced to their previously obtained formulae for $\eta_0 = 1/6$. Given that the transition between smooth and rippled morphologies with substrate temperature observed by Rusponi *et al.* ([196]) was nonhysteretical, we moreover assumed $\eta_0 > 0$ in order to eliminate the possibility of such metastability by guaranteeing

$$b_1 - a_1 = a_1 = a_2 = \frac{3\eta_0}{2} > 0 \text{ for } \alpha = 0.$$

Finally, since Makeev and Barabási ([132]) did not find any oscillatory behavior for their normalized yield function, we adopted the specific odd nonperiodic secondary source term

$$r(h) = \sinh(h) \Rightarrow \eta_0 = \frac{1}{6},$$

this selection having been made for the sake of definiteness, as well as being motivated by the form of that interfacial model equation analyzed by Wollkind and Vislocky ([298]), which was also the rationale for our choice of scale factors when nondimensionalizing the damped Kuramoto-Sivashinsky equation. We close this particular discussion by noting that the factor appearing before our sputtering yield constitutive relation and the dimension of specific values of J_0 should more properly have been $h_0 \ell_0^2/$ion and ion cm^{-2}sec^{-1}, respectively, but since those quantities only appear as their product $w_n = J_0 h_0 \ell_0^2$, in which case the ion designation cancels out, we suppressed it for ease of exposition.

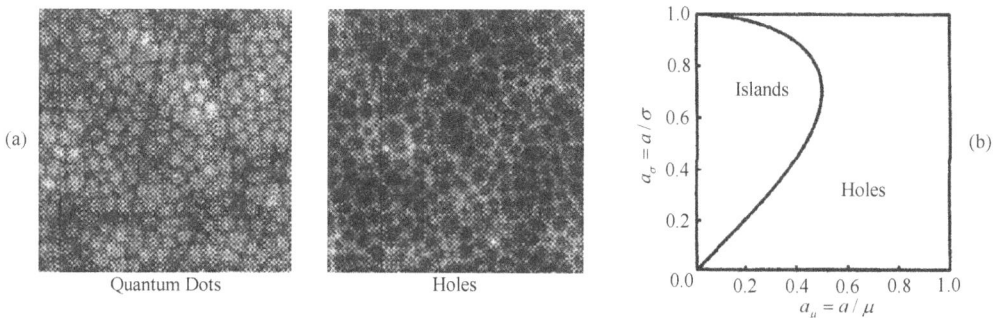

FIGURE 9.15
(a) Surface morphologies predicted by Kahng *et al.* ([91]) from their simulations for the undamped ($\beta = 0$) Kuramoto-Sivashinsky equation with $\nu = 0.6169$, $D = 2$, and $\lambda_0 = \pm 1$ where $\lambda_0 = 1$ produced quantum dots and $\lambda_0 = -1$, holes. (b) Phase diagram in the a_μ-a_σ plane generated by Kahng *et al.* ([91]) showing the parameter regimes corresponding to island or quantum dot (for $\lambda_0 > 0$) and hole or vacancy (for $\lambda_0 < 0$) formation where $a_\mu = a/\mu$ and $a_\sigma = a/\sigma$, a being the ion penetration depth, while σ and μ are the ion-beam parallel and perpendicular lengths, respectively, that characterized the shape of the energy collision cascade of Fig. 9.1. Note from the latter figure that, although depicted here, $0 \leq a_\sigma \leq 1$ cannot occur. *Reproduced by permission of the American Institute of Physics which holds the copyright.*

In light of the experimental results just summarized, the suggested dependence of λ_0 on crystallographic orientation was reminiscent of the role played by surface free energy when comparing observed versus predicted interfacial morphology in Chapter 2 during dilute binary alloy solidification involving possible symmetries about the normal to the interface for different face centered cubic crystalline lattices, as depicted in Fig 2.12. Indeed, Kahng *et al.* ([91]) made such an attempt by proposing that λ_0 depended instead on those experimental quantities defined in Fig. 9.1 and offering the following representation of it suitable for the isotropic situation of normal-incidence ion-bombardment

$$\lambda_0 = F_0 \frac{a_\sigma^2 - a_\sigma^4 - a_\mu^2}{2a_\mu^2}$$

where $a_\sigma = a/\sigma$ and $a_\mu = a/\mu$. Here, the factor $F_0 > 0$ was a measure of the flux dependence of this coefficient; and a, the ion-penetration depth; while σ and μ were the ion-beam parallel and perpendicular lengths, respectively, which characterized the shape of the energy collision cascade. Before examining this representation, we described in more detail Kahng *et als.*'s ([91]) numerical simulation results (see Fig. 9.15a). These were plotted on 256×256 square grids for $\lambda_0 = \pm 1$, in their idealized parameter values listed earlier, with $\lambda_0 = 1$ producing quantum dots (called islands) and $\lambda_0 = -1$, holes. Since for those parameter values $\lambda_c^* = 16$, they found approximately $256/16 = 16$ quantum dots or holes on each side of those grids. Then, returning to this representation and restricting ourselves to the symmetric case first treated by Cuerno and Barabási ([44]) of

$$\mu = \sigma \text{ or equivalently } a_\mu = a_\sigma,$$

we found that λ_0 was reduced to

$$\lambda_0 = -F_0 \frac{a_\sigma^2}{2} < 0.$$

Even for the general case, from the way they were defined in Fig. 9.1, it seemed to us that

$$a > \sigma \text{ or equivalently } a_\sigma > 1 \Rightarrow a_\sigma^2 > 1;$$

and hence, since under this condition

$$a_\sigma^2 - a_\sigma^4 = a_\sigma^2(1 - a_\sigma^2) < 0,$$

λ_0 would still satisfy

$$\lambda_0 = F_0 \frac{a_\sigma^2(1 - a_\sigma^2) - a_\mu^2}{2a_\mu^2} < 0$$

identically. In spite of that fact, Kahng *et al.* ([91]) depicted a region in their a_μ-a_σ phase space which was identified with $\lambda_0 \geq 0$, but only occurred if $0 \leq a_\sigma \leq 1$ (see Fig. 9.15b). Thus λ_0, as represented by their formula, seemed to preclude any such occurrence and therefore could neither predict quantum dot ($\lambda_0 > 0$) nor ripple ($\lambda_0 = 0$) formation, but only holes ($\lambda_0 < 0$) for normal-incidence ion-bombardment should that condition prove to be true and not our misinterpretation.

In order to decide between these possibilities, I sent an email to the senior author of Kahng *et al.* ([91]), Albert-László Barabási of the University of Notre Dame, asking him about whether our interpretation was correct and if not, why not. I received an automatic email response saying my original message had been classified as junk and hence was unread by the recipient. I then sent him the same message by regular mail including copies of Fig. 9.1 and Fig. 9.15 and marked the mailing envelope "Confidential" to prevent his secretary from discarding it out of hand (since Barabási was then the Emil T. Hofman Professor of Physics at Notre Dame, I assumed a secretary would handle his mail). Never receiving a reply to that mail, I next sent the same message to the first author of the paper in question, Professor Byungnam Kahng at Konkuk University in Seoul, Korea. He replied that our interpretation was correct and did not know how their mistake could be fixed, but would keep trying to correct it. I then sent my complete correspondence with Kahng to Barabási and asked him if he would care to contribute anything else to that communication. This was 17 years ago and I have not received a follow up message from either one of them since that time. This being the case in the paper from Adoon's thesis, Pansuwan *et al.* ([166]), I closed the paragraph examining Kahng *et al.*'s ([91]) λ_0-representation with the following statement:

> One would suspect given these deficiencies that some inadequacies or inaccuracies exist in its derivation and thus, an effort should be made to re-examine the procedure used to produce the relevant formulae. In this context, we note that separate formulae were originally developed for λ_1 and λ_2 which reduced to their isotropic common value λ_0 upon setting the impact angle $\theta = 0$, as required for normal incidence ion sputtering (reviewed by Makeev *et al.*, [133]).

Kahng *et al.* ([91]) claimed that their results were a theoretical confirmation of Facsko *et al.*'s ([58]) or Rusponi *et al.*'s ([197, 198]) experimental observations of remarkably well-ordered arrays of quantum dots or relatively uniform distributions of holes on the surface of gallium antimonide or copper and silver, respectively, during normal-incidence ion-sputtering erosion, depending upon whether the tilt-dependent coefficient of the erosion rate $\lambda_0 > 0$ or $\lambda_0 < 0$. As we have just shown, $\lambda_0 > 0$ was not possible for their model, while from our previous exposition it is clear that Rusponi *et al.* ([196, 197, 198]) dealt with either ripple or square array formation and only Michely and Comsa ([145]) described relatively uniform distributions of holes, but on the surface of platinum rather than on that of either copper or silver. Further, Kahng *et al.* ([91]) also claimed that ripples cannot occur during normal-incidence ion sputtering, but only when that bombardment is at off-normal or grazing incidence. Of course, as noted earlier, Rusponi *et al.* ([196, 197, 198]) observed ripple formation at normal incidence. Indeed, Rusponi *et al.* ([196]) even went so far as to state, "We want to emphasize that the ripple structure appears on the silver (110) surface at *normal incidence*..." where that italics is included in their sentence. This fact was especially inconvenient for Kahng *et al.* ([91]), since in a previous Kahng-Barabási simulation paper, Park *et al.* ([167]), they had only been able to produce ripples from their undamped Kuramoto-Sivashinsky equation for the nonisotropic choice of parameter values $\nu_1 = 0.6169$, $\nu_2 = 0.01$, $\lambda_1 = 1$, and $\lambda_2 = -4$, while maintaining $D = 2$, which implied that the ion bombardment was at off-normal incidence. Observe that should ν_2 and λ_2 have been taken equal to ν_1 and λ_1, respectively, then this selection would have been reduced to that isotropic case of $\nu = 0.6169$ and $\lambda_0 = 1$ treated by Kahng *et al.* ([91]) for normal-incidence ion bombardment, which produced their quantum dot formation. Realize that, unlike for our damped Kuramoto-Sivashinsky equation, Kahng *et al.* ([91]) could not produce any patterns if they took $\lambda_0 = 0$, since there then would be no nonlinearities in their undamped Kuramoto-Sivashinsky equation which already had $\beta = 0$. Recall, as pointed out at the end of our introductory Chapter 1, that pattern formation of this sort can only occur if the model under examination is nonlinear. Thus, our model is unified, in the sense that with $\lambda_0 = 0$ or equivalently $\alpha = 0$ and $0 < \beta < 1$ it can also account for the formation of ripples at normal incidence, which Kahng *et al.* ([91]) could not. Hence, the introduction of a secondary yield term in our constitutive relation which provided that damping represented a Rabbit Pulled Out of a Hat.

Before leaving this topic, we wished to make two additional comments related to these nonlinearities. The first dealt with the extrapolation of our weakly nonlinear stability predictions to a parameter range where finite amplitude solutions can be expected. Strictly speaking, our theoretical predictions were only valid in a restricted neighborhood of the marginal stability curve $\beta = 1$. Since our predictions were in agreement with the $\beta = 0$ numerical simulations of Kahng *et al.* ([91]), for which the growth rate took on its largest possible value $\sigma_0 = 2$, however, we had some confidence that an extrapolation of this sort was permissible. The second dealt with the deterministic form of our model system. As mentioned earlier, most authors who modeled ion-sputtered erosion have employed a noisy Kuramoto-Sivashinsky equation to account for the fact that such ion-bombardment is intrinsically stochastic in nature. Although the basic framework of our damped Kuramoto-Sivashinsky equation is deterministic, by selecting λ_0 as we did, such stochasticity was implicitly built into our model given the way Cuerno *et al.* ([44]) determined that coefficient originally (see Fig. 9.7). This is in the spirit of the hybrid predator-prey mite model developed in Chapter 4, which had an analytical ordinary differential equations framework with coefficients determined by a discrete-time simulation procedure (see Fig. 4.4).

This completed the work on Adoon's thesis. I knew from the outset that its dissemination phase might be more difficult than usual because of Mahidol University's requirement of a paper based on this dissertation be accepted for publication by a reputable journal before he could defend it. Unless we were willing to wait several years for his defense to occur, that required the paper in question to be written fairly quickly and submitted to a journal known to have a reasonably short turnaround time between submittal and acceptance. Given my experience with *Mathematical and Computer Modelling* (described in Chapter 6), I selected that journal for our prospective paper and began to write it immediately after Adoon's return to Thailand, exactly one year from his date of arrival. Under normal circumstances this would have been a collaborative effort, but due to the time constraint, I essentially wrote our paper by myself but, as was the case for all of my Mahidol Ph.D. students, listed Adoon's name as the first author with Chontita Rattanakul and Yongwimon Lenbury as his Mahidol co-authors and Lynn Harrison, Indika Rajapakse, and Kevin Cooper as the other WSU ones. Lynn's contribution was that her Master's thesis involved a longitudinal planform weakly nonlinear stability analysis of a damped Kuramoto-Sivashinsky equation with our $\sinh(h)$ replaced by $\sin(h)$ and had been used by both Adoon and Chontita as templates for their analyses, while Indika and Kevin played a significant role in the preparation of the figures for our paper to be related below, as had Chontita, who, along with Yongwimon and Adoon, actually wrote a first draft of that paper which I ultimately rewrote extensively. Once this manuscript was finished and all its figures generated, I submitted it to *Mathl. Comput. Modelling* in December of 2003, receiving the reviews back in time for the revised paper to be accepted by July of 2004. Those requested revisions, which have been incorporated into the exposition of our research presented in this chapter, included, among other things, the extrapolation of the stability properties to the deleted intervals $|\alpha| > 3$ in Fig. 9.5, that I would normally judge obvious enough to be unnecessary. From long experience, however, I have found it much easier to comply with such requests and keep the reviewers happy than to try disputing them. Now armed with that acceptance letter of our paper from *Mathl. Comput. Modelling*, Adoon was able to defend his thesis successfully and be granted a Ph.D. in Applied Mathematics from Mahidol University. Incidentally, that thesis was Adoon's version of the analyses of the ion-sputtered erosion pattern formation problem and thus did not have to be identical to our paper. This, unfortunately, was not the end of the story about that paper Pansuwan *et al.* ([166]). This paper now went into its production stage, at which point the copy editor of *Mathl. Comput. Modelling* requested we provide electronic files of its manuscript and figures. Although only five years after Wollkind and Stephenson ([296]) was published, the production procedure, described in Chapter 7, in reference to that paper had already completely changed. After complying with this request, we were informed by the copy editor that the software for our files of Fig. 9.2 and Fig. 9.5 was not suitable for what the journal required to include those figures in the printed paper and unless substitute compatible files for them were provided, our paper would not be published by *Mathl. Comput. Modelling*. This is when Indika and Kevin entered the picture (I am using that term in both its colloquial and literal sense). Indeed, the software Chontita employed to generate those figures was nonstandard and did not allow for their normal reproduction so Kevin, who ran our departmental computer center, and Indika, who was a Ph.D. student of my colleague Valipuram Manoranjan, helped us produce versions of them that could be reproduced by *Mathl. Comput. Modelling.* Although many authors only credit such an effort by an acknowledgement, it has always been my policy to give co-authorships to anybody without whose contribution a paper would not be publishable. Hence our paper finally appeared in the volume 41, numbers 8-9, April-May, 2005, issue of the Elsevier journal *Mathl. Comput. Modelling* on pages 939-964.

We concluded the dissemination of our results by presenting them at the following three venues:

1. Mechanical and Materials Engineering Symposium, WSU on October 13, 2005;

2. Showcase 2007, WSU on March 23, 2007; and

3. Wright State University Colloquium, Dayton, Ohio, on September 14, 2007.

Each of these was entitled "Nonlinear stability analyses of pattern formation on solid surfaces during normal-incidence ion-sputtered erosion," the first and third of which were invited slide shows, while the second was a contributed poster presentation.

In the context of Mahidol University's requirement of journal acceptance of a paper based on the dissertation before a thesis defense would be allowed, most American universities feel that such a policy results in the premature publication of research results at the expense of its completeness should more time have been taken until submission. I believe that the timing of the publication of Pansuwan *et al.* ([166]) represents an object lesson for this admonishment. Simultaneously to our performance of that research, Stefan Facsko and his co-workers devised a damped Kuramoto-Sivashinsky equation, the simulation of which could be used to model their original normal-incidence argon ion-sputtering gallium antimonide semiconductor quantum dot experiments. Employing our notation this equation was of the form

$$h_t + 4\nabla^2 h + 2\nabla^4 h + 2\beta h = \alpha|\boldsymbol{\nabla} h|^2 + \eta.$$

Hence, it differed from ours by the presence of the white noise term η and the retention of only the linear part h of our secondary sputtering yield nonlinear function $\sinh(h)$. Their paper Facsko *et al.* ([57]), received by the journal *Physical Review B* on November 24, 2003, introduced this linear damping term to provide additional dissipation for the purpose of suppressing spatio-temporal chaos and interpreted it as the continuum effect of the mechanism of redeposition of the sputtered material on the substrate surface. They asserted that in the case of a corrugated morphology, a considerable amount of the sputtered particles hits the surface and is redeposited resulting in a net exchange of material from higher to lower lying regions. This mechanism, which tends to decrease both h_0, the maximum deviation of the interface from its mean planar position, and w_n, the normal velocity of erosion, was first described by Michely and Comsa ([145]), but had been ignored heretofore in modeling endeavors. Observe that the h_0-dependence of our formula $w_n = J_0 h_0 \ell_0^2$ for the normal velocity of erosion is consistent with this interpretation and represents another Rabbit Pulled Out of a Hat. Note that Facsko *et al.* ([57]) took $2\lambda_0 = 0.01$, which is the same order of magnitude as our particular choice for this parameter. Mention of their redeposition mechanism was included in both the two slide-show presentations catalogued above but not in our paper given the simultaneous dates of submission.

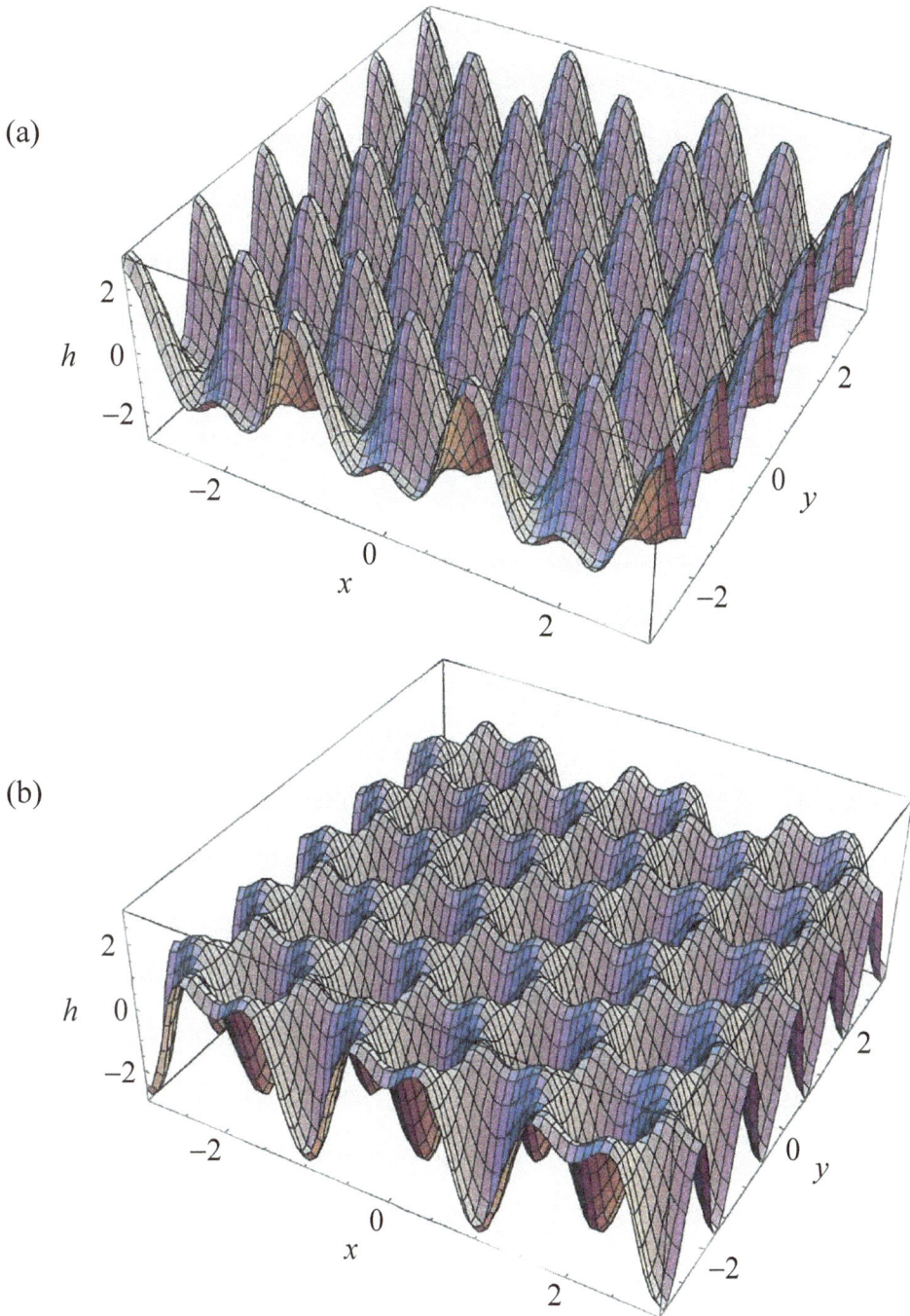

FIGURE 9.16

Surface plots $h = h^{\pm}(x, y)$ corresponding to critical points (a) III$^+$ and (b) III$^-$ where $h^{\pm}(x, y) = \pm[2\cos(\pi x/\lambda_c)\cos(\sqrt{3}\pi y/\lambda_c) + \cos(2\pi x/\lambda_c)]$ and these nondimensional spatial variables have been measures in λ_c. Note the resemblence that (a) bears to the experimental quantum dots pictured in Fig. 9.9b and (b), to the cells pictured in Fig. 8.1. Compare with the contour plots of Fig. 9.4.

Finally, these slide shows also included the three-dimensional surface plots contained in Fig. 9.16, Fig. 9.17, and Fig. 9.18 for the critical points III$^\pm$, II, and V with $\varphi = \pi/2$, respectively. Here, Fig. 9.16 came from Adoon's thesis, while Fig. 9.17 and Fig. 9.18 came from Laura's. All of our previous critical point representations were two-dimensional contour-plot ones, but due to the three-dimensional nature of the photographic images of the ion-sputtered erosion patterns, surface plots are more appropriate for comparison purposes (see the captions of those figures). That Fig. 9.17 and Fig. 9.18 came from Laura's thesis indicates the 2D-papers from it contained the first rhombic planform weakly nonlinear stability analysis I ever performed, which being unnecessary for the purposes of the exposition of Chapter 7 had not been included there.

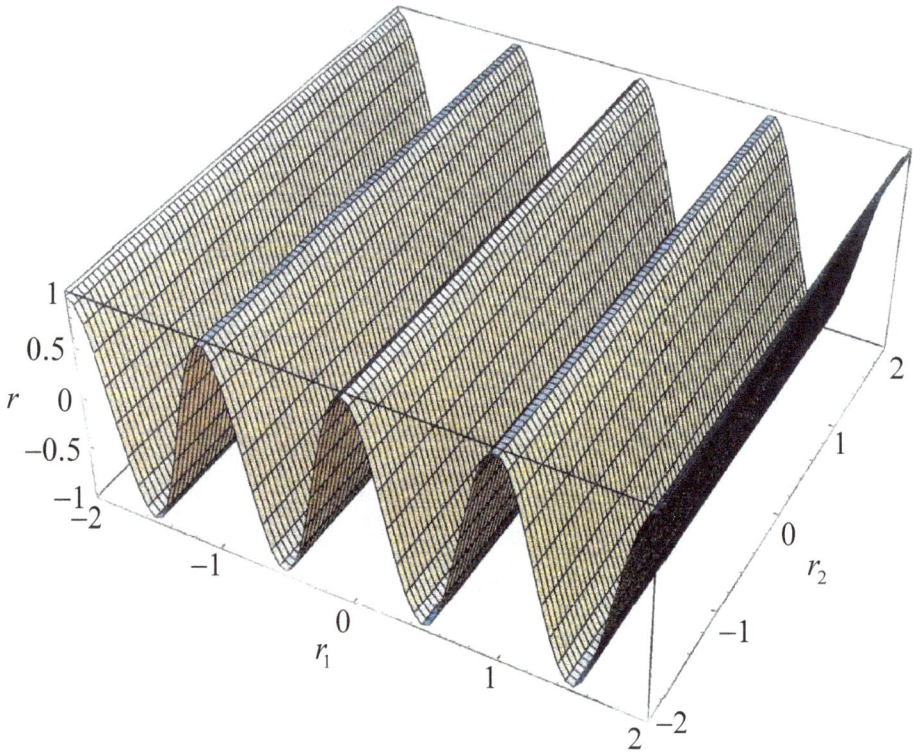

FIGURE 9.17
Surface plot $r = H(r_1, r_2) = \cos(2\pi r_1/\lambda_c^*)$ corresponding to critical point II or critical point V with $\varphi = 0$ where these dimensional spatial variables have been measured in λ_c^*. Note the resemblance this bears to the experimental ripples pictured in Fig. 9.13b. Compare with the contour plot of Fig. 9.3.

I assigned the rhombic planform weakly nonlinear stability analysis of our damped Kuramoto-Sivashinsky equation to Chontita because she was going to be working with me for only four months and we needed a problem that could be completed in that time. I had no idea of the importance of having those results, in conjunction with the corresponding hexagonal planform ones, would play in our interpretation of the ion-sputtering solid surface morphologies because in Laura's case, after Kuske and Matkowsky ([105]), we basically kept these results separate from each other. Given this occurrence however, I had both

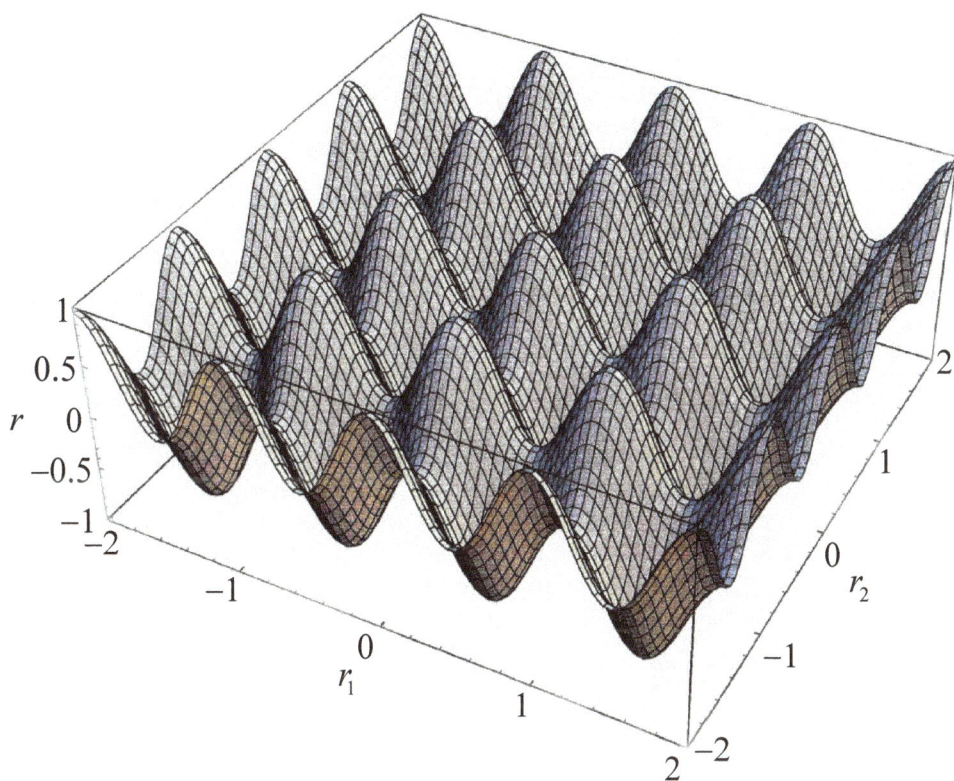

FIGURE 9.18

Surface plot $r = H(r_1, r_2) = \cos(2\pi r_1/\lambda_c^*) + \cos(2\pi r_2/\lambda_c^*)$ corresponding to critical point V with $\varphi = \pi/2$ where these dimensional spatial variables have been measured in λ_c^*. Note the resemblance this bears to the experimental checkerboard pattern pictured in Fig. 9.13a. Compare with the contour plot of Fig. 9.11.

analyses performed on subsequent pattern formation problems selected for my Ph.D. students. Specifically for the problems of Chapters 10-13, the results of the rhombic planform analysis only mediated the hexagonal planform pattern formation predictions (as in this chapter), but for the problem of Chapter 14 they actually generated the final pattern formation predictions, as will be seen in the next five chapters where the concept of upper and lower threshold rhombic patterns is introduced. Since no rhombic type patterns had been observed in a hydrodynamic setting, neither Segel in his analyses of the Bénard problem, nor Mei and I in our thin liquid film lubrication problem of Chapter 8, considered them. The same reasoning also applied to Rukmini Sriranganathan's thesis described in Chapter 2.

Critics of research universities often assert their faculty members believe classroom teaching to be incompatible with the mission of these institutions. I have always, on the other hand, tried to dovetail my teaching and research. In that spirit, Adoon and I devised the following problem set for my graduate modeling course, which he and Chontita had attended the previous summer:

Consider an interfacial deviation $h(x, y, t)$ of an originally planar surface where $t \equiv$ time and $(x, y) \equiv$ a transverse laboratory Cartesian coordinate system, all of which are dimensionless. Upon seeking a real hexagonal planform solution for $h(x, y, t)$ of the usual form, assume it is found that the coefficients of the amplitude-phase equations satisfy

$$\sigma = 2(1 - \beta), a_0 = -\alpha, \ a_1 = a_2 = 4(1 + \alpha^2),$$

for $\beta > 0$ and $\alpha \in \mathbb{R}$, an experimentally controllable and a material parameter, respectively. Here, the equivalence classes of potentially stable critical points for these equations have the following morphological identifications: I represents a smooth surface; II, parallel ripples; and III$^{\pm}$, hexagonal arrays of elevated dots or circular holes, respectively. The orbital stability criteria for those critical points can be posed in terms of σ. Thus, critical point I is stable in this sense for $\sigma < 0$, while the orbital stability behavior of II and III$^{\pm}$ which depends only on the sign of a_0 (since $2a_2 - a_1 = a_1 > 0$) and on the quantities σ_j for $j = -1$, 1, and 2 defined by

$$\sigma_{-1} = \frac{-4a_0^2}{a_1 + 4a_2}, \ \sigma_1 = 16\frac{a_1 a_0^2}{(2a_2 - a_1)^2}, \ \sigma_2 = 32\frac{(a_1 + a_2)a_0^2}{(2a_2 - a_1)^2},$$

has been summarized in Table 9.1. Construct a morphological stability diagram analogous to Fig. 9.5 in the physically relevant portion of the α-β plane consistent with these predictions identifying those regions corresponding to stable smooth surfaces, dots, ripples, and holes.

This problem, based on Adoon's thesis results, yielded a morphological stability diagram qualitatively identical in form to Fig. 9.5, as was our intention.

We close this chapter by cataloguing the explicit functional representations of the relevant curves appearing in that diagram and listing its quantitative properties in comparison to those of Fig. 9.5 as follows:

$$\beta = \beta_{-1}(\alpha) = \frac{11\alpha^2 + 10}{10(\alpha^2 + 1)} \text{ for } \sigma = \sigma_{-1},$$

$$\beta = \beta_1(\alpha) = \frac{-\alpha^2 + 1}{\alpha^2 + 1} \text{ for } \sigma = \sigma_1,$$

$$\beta = \beta_2(\alpha) = \frac{-7\alpha^2 + 1}{\alpha^2 + 1} \text{ for } \sigma = \sigma_2.$$

Here, as in Fig. 9.5,

$$a_0 = 0 \text{ for } \alpha = \alpha_c = 0, \ a_0 > 0 \text{ for } \alpha < \alpha_c, \ a_0 < 0 \text{ for } \alpha > \alpha_c;$$

and

$$\sigma = 0 \text{ for } \beta = \beta_c = 1, \ \sigma > 0 \text{ for } 0 \leq \beta < \beta_c, \ \sigma < 0 \text{ for } \beta > \beta_c;$$

while

$$\beta_{-1}(\alpha) \to \frac{11}{10} \text{ as } \alpha \to \pm\infty;$$
$$\beta_1(\alpha_1^\pm) = 0$$

implies that

$$\alpha_1^\pm = \pm 1;$$

and

$$\beta_2(\alpha_2^\pm) = 0$$

implies that

$$\alpha_2^\pm = \pm\frac{1}{\sqrt{7}} = \pm 0.38.$$

We regarded the similarity between the morphological stability diagram for this problem and Fig. 9.5 to be a Rabbit Pulled Out of a Hat. Given the simplicity of the calculations involved for these idealized growth rate and Landau constant functions, it has proven to be useful in illustrating morphological stability results of hexagonal planform weakly nonlinear stability analyses. Due to this simplicity, it was also perfect for the written-portion of a three-question Ph.D. preliminary examination, each question of which is expected to take an hour, and had been employed by me in that capacity on a number of occasions. Further, ending his stay at WSU by helping me devise this problem set for the same graduate course with which he started it, Adoon Pansuwan had, in essence, brought that Ph.D. experience full circle. Finally, upon leaving he presented me with a new Jean jacket to replace my worn out one that I had used all the winter months during his stay!

10

Evolution Equation Phenomenon III: Nonlinear Optical Pattern Formation: Hexagonal and Rhombic Planform Nonlinear Stability Analyses

The two problems which I selected for my next three Ph.D. students involved hexagonal and rhombic planform stability analyses of evolution equations related to nonlinear optical and vegetative pattern formation. The models represented for these phenomena, in this and the next chapter, could be reduced to the appropriate spatio-temporal modified Swift-Hohenberg and fourth-order logistic nonlinear partial differential evolution equations (see Cross and Hohenberg, [42]). Having split up the hexagonal and rhombic planform nonlinear stability analyses for the ion-sputtered solid surface erosion problem described in the previous chapter between Adoon and Chontita, I decided to do the same thing for these two problems. In particular, Dean Edmeade and Francisco Alvarado, from WSU, analyzed the nonlinear optical pattern formation problem and Francisco and Nichaphat Boonkorkuea, from Mahidol University, the nonlinear vegetative one, with Francisco performing the rhombic planform analyses on both problems.

For Dean's thesis topic I selected a subject highlighted in the 'This Week in Science' entry entitled "Complex Nonlinear Laser Patterns" of the volume 285 number 5425 issue of *Science*, which referenced the report by Scheuer and Orenstein ([202]) on pages 230-233 of that issue about the generation of laser patterns in nonlinear semiconductor microcavities. In the first paragraph of that report it was stated:

> The generation of global field patterns was explored recently in optical cavities incorporating a nonlinear medium (Firth and Scroggie, [60]). These studies focused mainly on externally driven cavities and exhibited the formation of ordered patterns of bright or dark spots with hexagonal symmetry. Other patterns such as rolls (repeated elongated field distributions) had been observed as well (Firth and Scroggie,[60]).

Always being on the lookout for phenomena that exhibited both types of hexagonal patterns (spots and honeycombs) and stripes (see Fig. 1.2), I immediately made a copy of the Firth and Scroggie ([60]) paper in Europhysics Letters entitled "Spontaneous pattern formation in an absorptive system," which described the transition from a spotted to a striped to a honeycombed pattern induced by the injection of a laser pump field into an atomic sodium vapor optical ring cavity.

Note that the issue of *Science* in question predates the one featuring the ion-sputtered erosion pattern of solids by eight weeks. When Yongwimon Lenbury was arranging for Adoon and Chontita to come to Pullman and asked me on what problem they would be working, I sent her a copy of Firth and Scroggie ([60]). By the time Adoon actually started work on his thesis, I had decided to switch him to the Facsko *et al.* ([58]) problem instead, as described in the last chapter, while having Dean work on this nonlinear optical ring-cavity model driven by a gas laser. Let us explain that model in more detail.

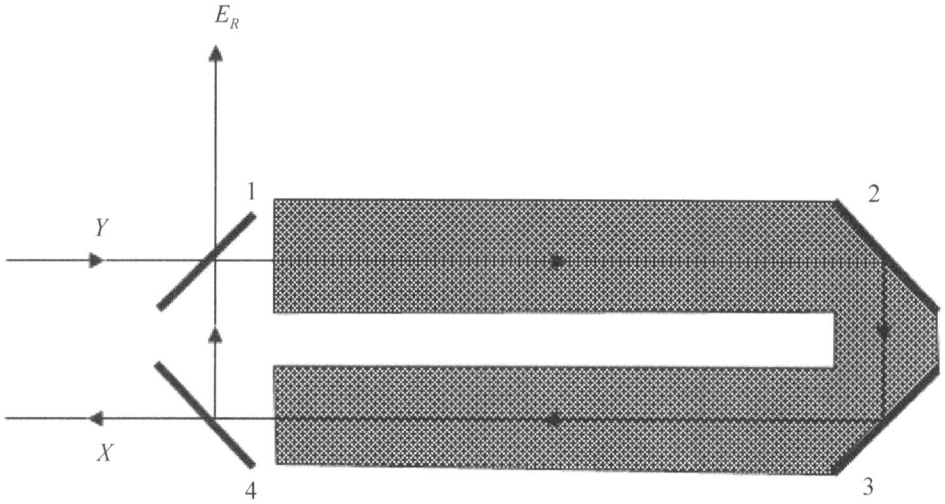

FIGURE 10.1
Schematic diagram of a ring cavity filled with atomic sodium vapor (shaded region) where Y and X are related to the envelopes of the injected pump and internal cavity fields, respectively, while E_R bears the same relationship to the corresponding reflected field. Here, mirrors 1 and 4 have very small transmissitivity coefficients while mirrors 2 and 3 are 100% reflective.

Consider the injection of a laser pump field into a ring cavity containing a purely absorptive two-level atomic sodium vapor medium (see Fig. 10.1). Pattern formation in this phenomenon is modeled by coupling Maxwell's equation for the intracavity field with the Bloch equations for the atomic variables of polarization and population difference. That system may be reduced to a single nonlinear modified Swift-Hohenberg model equation describing the real part of the deviation of the intracavity field envelope function from its uniform plane-wave solution on an unbounded two-dimensional transverse spatial domain provided the decay rate for the field is negligible, when compared with the dephasing rate for the polarization and the decay rate for the population difference. Diffraction of radiation can induce transverse patterns consisting of stripes and hexagonal arrays of bright spots or honeycombs, as well as square patterns (Scroggie and Firth, [209]), in an initially uniform plane-wave configuration. We began with a sketch of the reduction procedure required to obtain the governing modified Swift-Hohenberg model equation.

The nondimensional complex system of coupled Maxwell-Bloch equations for this phenomenon was given by (Firth and Scroggie, [60])

$$\frac{\partial X}{\partial t} = -(1 + i\theta)X + Y - \beta P + i\chi \nabla_2^2 X,$$

$$\varepsilon_1 \frac{\partial P}{\partial t} = fX - (1 + i\Delta)P,$$

$$\varepsilon_2 \frac{\partial f}{\partial t} = 1 - f - \frac{X^*P + XP^*}{2}.$$

Here, $t \equiv$ time, $(x, y) \equiv$ transverse coordinates, $\nabla_2 \equiv (\partial/\partial x, \partial/\partial y)$, and $\nabla_2^2 \equiv \nabla_2 \cdot \nabla_2$.

Further, X and Y were related to the internal cavity and injected pump fields; P and f, to the atomic polarization and population difference between the upper and lower energy levels; while θ and Δ, were the cavity mistuning and atomic detuning parameters; β and χ, the coefficients of absorption and diffraction; and $\varepsilon_{1,2} = r/r_{1,2}$, where r was the dimensional decay rate associated with X and $r_{1,2}$ bore a similar relationship to P and f, being the dephasing and decay rates for the polarization and population difference, respectively. In particular, $\theta = (\Omega_c - \Omega_0)/r$ and $\Delta = (\Omega_a - \Omega_0)/r$, where Ω_c was the frequency of the longitudinal cavity mode nearest to the resonant mode Ω_0 and Ω_a was the atomic transition frequency. Finally, the oscillation $e^{-i\Omega_0 t/r}$ had been factored out of the cavity, pump, and polarization fields, where $i = \sqrt{-1}$. Hence, the dynamical variables X and P represented the complex valued envelopes of these fields, while f, measured in occupation probabilities, satisfied $0 < f \leq 1$ and Y, the pump field envelope function, was assumed to be a real positive constant. In this context, the asterisked quantities denote complex conjugates in which the imaginary part of the quantity changes sign. In addition, it had been assumed that the dependent variables were initially independent of the cavity longitudinal spatial dimension. Then, as demonstrated directly by Lugiato and Oldano ([129]), these envelope functions retained that independence for all later time and thus needed only to be defined on the cavity two-dimensional transverse spatial domain.

Should $\varepsilon_{1,2} \ll 1$, as is typically the case (Moloney and Newell, [148]), one could employ a steady-state assumption on the two Bloch equations to obtain the quasi-equilibrium conditions satisfied implicitly by the atomic variables

$$P = \frac{fX}{1 + i\Delta};$$

$$f = 1 - \frac{X^*P + XP^*}{2} = 1 - \mathrm{Re}(X^*P)$$

since

$$(XP^*)^* = X^*P^{**} = X^*P \text{ and } z + z^* = 2\mathrm{Re}(z).$$

These equations could be solved simultaneously to yield explicit representations for the atomic variables as follows: Upon substituting P into the equation for f and taking the real part of the resulting product of it times X^*, we found that

$$f = 1 - \frac{f|X|^2}{1 + \Delta^2} \text{ where } |X|^2 = XX^* = [\mathrm{Re}(X)]^2 + [\mathrm{Im}(X)]^2$$

$$\Rightarrow f = \frac{1 + \Delta^2}{1 + \Delta^2 + |X|^2};$$

which, in conjunction with that equation for P, yielded

$$P = \frac{(1 - i\Delta)X}{1 + \Delta^2 + |X|^2} \text{ since } 1 + \Delta^2 = (1 + i\Delta)(1 - i\Delta).$$

Hence, substitution of this expression for P into the Maxwell equation reduced the latter to a single nonlinear evolution equation for the intracavity field envelope function X:

$$\frac{\partial X}{\partial t} = -X\left[1 + i\theta + \frac{\beta(1 - i\Delta)}{1 + \Delta^2 + |X|^2}\right] + Y + i\chi\nabla_2^2 X.$$

Given that the pump field envelope function Y was a real positive constant, there existed a stationary uniform equilibrium solution $X \equiv X_e$ to this reduced equation satisfying

$$Y = X_e\left[1 + \frac{\beta}{D} + \left(\theta - \frac{\beta\Delta}{D}\right)i\right] \text{ with } D = 1 + \Delta^2 + \alpha \text{ for } \alpha = |X_e|^2$$

or

$$Y^2 = YY^* = \alpha \left[\left(1 + \frac{\beta}{D} \right)^2 + \left(\theta - \frac{\beta\Delta}{D} \right)^2 \right],$$

which was single valued in Y provided $0 < \beta < \beta_{crit}$, where β_{crit} was defined implicitly by (Mandel *et al.*, [135]; Wollkind and Dichone, [284])

$$27\beta_{crit}(1 + \Delta^2)(1 + \theta^2) = (\beta_{crit} - 2 + 2\theta\Delta)^3.$$

Observe from Fig. 10.2 that for $\beta < \beta_{crit}$, $dY^2/d\alpha > 0$, while for $\beta = \beta_{crit}$ there then existed an α-value where $dY^2/d\alpha = 0$ and for $\beta > \beta_{crit}$, an α-interval where $dY^2/d\alpha < 0$. This equilibrium solution X_e satisfied the transverse far-field boundary condition

$$|X|^2 \text{ remains bounded as } x^2 + y^2 \to \infty$$

and represented an initial uniform plane-wave configuration. It was the stability of this solution to various periodic transverse-type disturbances with which we were concerned in what follows.

Toward that end, we considered a solution to this reduced equation of the form

$$X = X_e(1 + A)$$

and, retaining terms through third-order in A, obtained

$$\frac{\partial A}{\partial t} \equiv A_t \sim \sum_{n=1}^{3} \sum_{\ell=0}^{n} \frac{1}{(n-\ell)!\ell!} \frac{\partial^n F(0,0)}{\partial A^{n-\ell} \partial A^{*\ell}} A^{n-\ell} A^{*\ell} + i\chi \nabla_2^2 A$$

where

$$F(A, A^*) = -(1 + A) \left[1 + i\theta + \frac{\beta(1 - i\Delta)}{1 + \Delta^2 + \alpha(1 + A)(1 + A^*)} \right].$$

Finally, after Firth and Scroggie ([60]), we concentrated on the special resonant excitation case of $\Delta = 0$, given that this assumption allowed us to simplify our computations enormously, while still preserving the salient pattern formation features of the more general case of $\Delta \neq 0$ (see the observation at the end of the chapter). Under this condition, the function $F(A, A^*)$ was reduced to

$$G(A, A^*) = -(1 + A) \left[1 + i\theta + \frac{\beta}{1 + \alpha(1 + A)(1 + A^*)} \right].$$

Let us define its various partial derivatives evaluated at $A = A^* = 0$:

$$\gamma_{n\ell} \equiv \frac{1}{(n-\ell)!\ell!} \frac{\partial^n G(0,0)}{\partial A^{n-\ell} \partial A^{*\ell}}$$

which were given by

$$\gamma_{10} = -\left(1 + \frac{\beta}{D^2} + i\theta \right), \ \gamma_{11} = \frac{\alpha\beta}{D^2} \text{ with } D = 1 + \alpha;$$

$$\gamma_{20} = \frac{\alpha\beta}{D^3}, \ \gamma_{21} = \frac{2\alpha\beta}{D^3}, \ \gamma_{22} = \frac{-\beta\alpha^2}{D^3};$$

$$\gamma_{30} = -\frac{\beta\alpha^2}{D^4}, \ \gamma_{31} = \frac{\alpha\beta(1 - 2\alpha)}{D^4}, \ \gamma_{32} = -\frac{3\beta\alpha^2}{D^4}, \ \gamma_{33} = \frac{\beta\alpha^3}{D^4}.$$

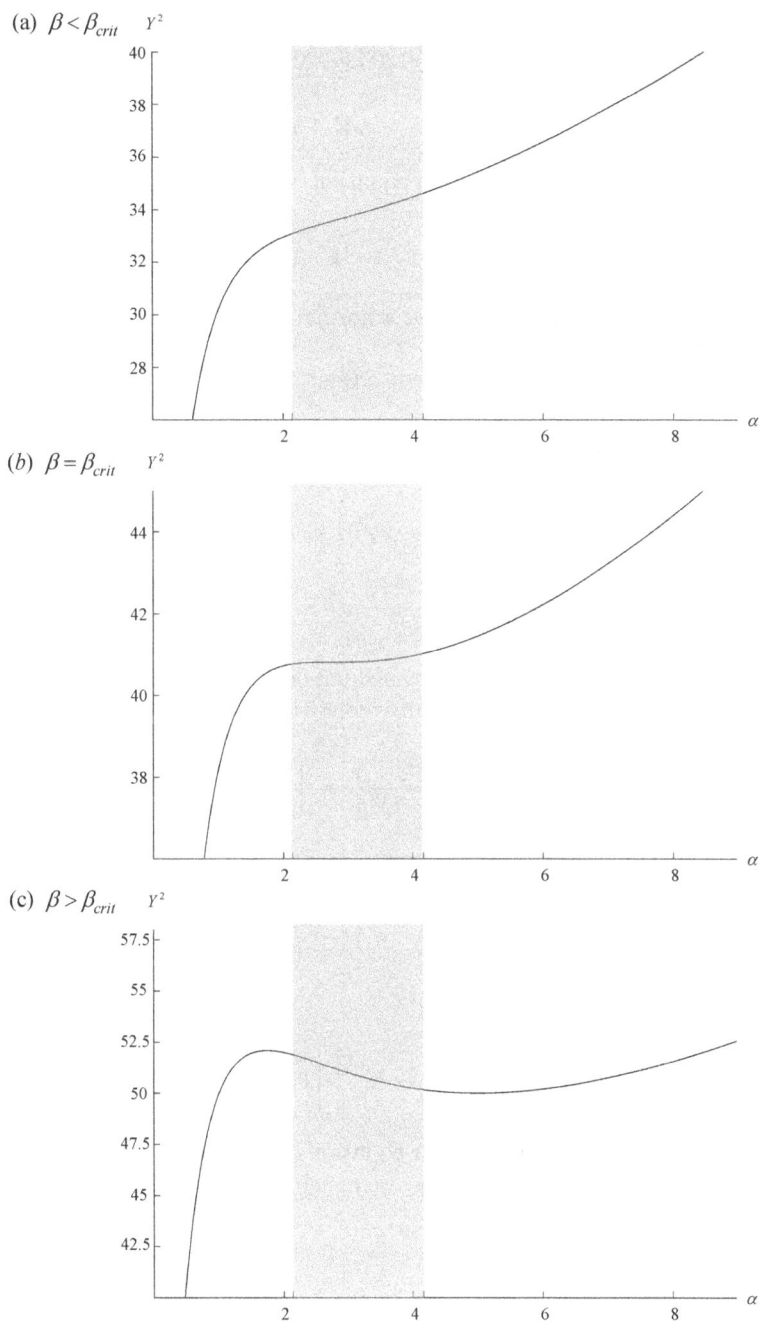

FIGURE 10.2

Plots of Y^2 versus α for $\Delta = 0$, $\theta = -1$, and β equals (a) $8.8 < \beta_{crit} = 10.2$, (b) β_{crit}, and (c) $12 > \beta_{crit}$. Here, the shaded interval corresponds to the patterned region.

In particular, we first examined the associated linear stability problem by considering

$$A(x,y,t) = \varepsilon_1 \mathcal{A}_1(x,y,t) + O(\varepsilon_1^2) \text{ where } |\varepsilon_1| \ll 1$$

and, retaining only first-order terms in the expansion introduced above, found that $\mathcal{A}_1$ satisfied

$$\frac{\partial \mathcal{A}_1}{\partial t} = \gamma_{10}\mathcal{A}_1 + \gamma_{11}\mathcal{A}_1^* + i\chi\nabla_2^2\mathcal{A}_1.$$

Then, taking the complex conjugate of this equation, we obtained

$$\frac{\partial \mathcal{A}_1^*}{\partial t} = \gamma_{10}^*\mathcal{A}_1^* + \gamma_{11}\mathcal{A}_1 - i\chi\nabla_2^2\mathcal{A}_1^*.$$

After Lugiato and Lefever ([128]), we sought a normal-mode solution of these perturbation quantities of the form

$$[\mathcal{A}_1, \mathcal{A}_1^*](x,t) = [k_1, k_2]e^{\sigma t}\cos(qx),$$

where $|k_1|^2 + |k_2|^2 \neq 0$ and $\lambda = 2\pi/q \equiv$ wavelength of the periodic disturbance under investigation, which yielded the linear homogeneous system of equations for k_1 and k_2:

$$\left[\sigma + 1 + \frac{\beta}{D^2} + i(\theta + \chi q^2)\right]k_1 - \frac{\alpha\beta}{D^2}k_2 = 0,$$

$$\frac{-\alpha\beta}{D^2}k_1 + \left[\sigma + 1 + \frac{\beta}{D^2} - i(\theta + \chi q^2)\right]k_2 = 0.$$

Imposing the Pat Munroe algorithm of the vanishing of the determinant of the matrix of its coefficients required to satisfy the nontriviality condition, resulted in the secular equation

$$\sigma^2 + 2\left[1 + \frac{\beta}{(\alpha+1)^2}\right]\sigma + \left[1 + \frac{\beta}{(\alpha+1)^2}\right]^2 - \left[\frac{\alpha\beta}{(\alpha+1)^2}\right]^2 + (\theta + \chi q^2)^2 = 0.$$

Finally, by employing the difference of squares

$$\left[1 + \frac{\beta}{(\alpha+1)^2}\right]^2 - \left[\frac{\alpha\beta}{(\alpha+1)^2}\right]^2 = \left[1 + \frac{\beta(1-\alpha)}{(\alpha+1)^2}\right]\left[1 + \frac{\beta}{\alpha+1}\right],$$

this secular equation became

$$\sigma^2 + 2\left[1 + \frac{\beta}{(\alpha+1)^2}\right]\sigma + \left[1 + \frac{\beta(1-\alpha)}{(\alpha+1)^2}\right]\left[1 + \frac{\beta}{\alpha+1}\right] + (\theta + \chi q^2)^2 = 0.$$

The marginal stability curve for that secular equation on which $\sigma = 0$ had its lowest threshold in (α, β) space when $\theta + \chi q^2 = 0$. Hence, we concluded that the critical wavenumber q_c satisfied

$$q_c^2 = -\frac{\theta}{\chi} > 0$$

and, for $q^2 \equiv q_c^2$, this secular equation had the corresponding roots

$$\sigma_R(\alpha, \beta) = -1 + \frac{\beta(\alpha-1)}{(\alpha+1)^2}, \quad \sigma_I(\alpha, \beta) = -1 - \frac{\beta}{\alpha+1},$$

where σ_R, the growth rate of the most dangerous mode, yielded the marginal stability curve

$$\beta = \beta_0(\alpha) = \frac{(\alpha+1)^2}{\alpha-1} \text{ for } \alpha > 1,$$

plotted in Fig. 10.3, while σ_I was strongly stabilizing. In this context, we noted that under these conditions a similar linear stability analysis of the uniform plane wave solution to the original system of governing equations would yield a fifth-order secular equation (fifth order since P as well as X were complex and f, real) having the following asymptotic behavior for $\varepsilon_2 = O(\varepsilon_1)$:

$$\sigma^{(1,2)} = O(1), \ \sigma^{(3)} \sim \frac{-1}{\varepsilon_1}, \ \sigma^{(4,5)} \sim \frac{\Sigma^\pm}{\varepsilon_1} \text{ where } \Sigma^\pm = O(1) \text{ as } \varepsilon_1 \to 0$$

such that to lowest order $\Sigma^\pm$ satisfied

$$r_0\Sigma^2 + (r_0 + 1)\Sigma + 1 + \alpha = 0 \text{ where } r_0 = \frac{\varepsilon_2}{\varepsilon_1} = \frac{r_1}{r_2} = O(1),$$

which, since all its coefficients are positive, implied that

$$\text{Re}(\Sigma^\pm) < 0 \text{ as } \varepsilon_1 \to 0.$$

Specifically, for $\varepsilon_1 = \varepsilon_2$ or $r_0 = 1$, these reduced to

$$\Sigma^\pm \sim -1 \pm i\alpha^{1/2} \text{ as } \varepsilon_1 \to 0.$$

Further, we made the observation that $\sigma^{(1,2)} \sim \sigma_{R,I}$ as $\varepsilon_1 \to 0$, this correspondence being consistent with our steady-state assumption. Thus, with no loss of generality for the nonlinear optical instabilities under examination, we could consider our basic system in the limit as $\varepsilon_1 \to 0$ should that quantity be very small, since then the neglected roots were highly stabilizing; hence justifying this steady-state assumption for that basic system. Such an argument was implicit to the quasi-equilibrium assumption employed in the chemical Turing pattern model of Chapter 7, as well.

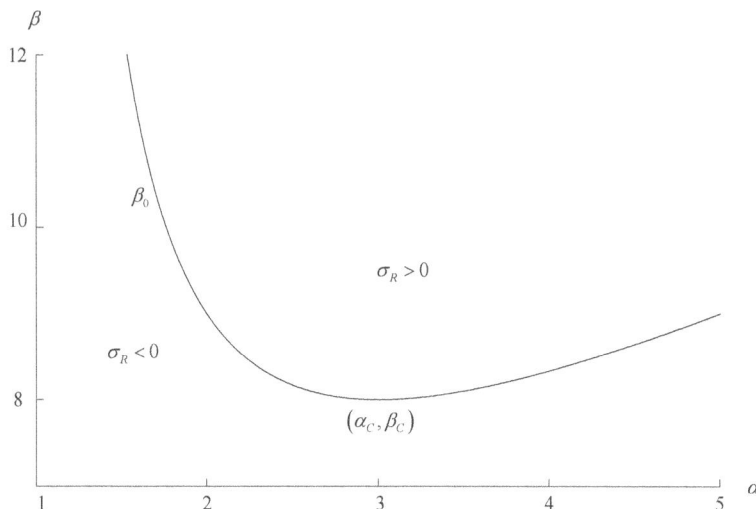

FIGURE 10.3
Plot in the α-β plane of the marginal stability curve $\beta = \beta_0(\alpha)$ with minimum point $(\alpha_c, \beta_c) = (3, 8)$ on which the growth rate $\sigma_R(\alpha, \beta_0) = 0$.

Substituting $A = R + iI$ into our third-order partial differential equation for A with $\Delta = 0$, taking

$$\chi q_c^2 = -\theta = 1 \text{ or } \theta = -1$$

after Firth and Scroggie ([60]), and separating the resultant equation into its real and imaginary parts, we derived partial differential equations for R and I which had growth rates σ_R and σ_I, respectively. Making use of these facts to deduce an asymptotic quasi-equilibrium condition for I from the latter and employing this to eliminate it from the former, we obtained the modified Swift-Hohenberg equation for R appearing in Firth and Scroggie ([60])

$$R_t \sim \sigma_R(\alpha, \beta)R - \omega_0(\alpha, \beta)R^2 - \omega_1(\alpha, \beta)R^3 + \left[\frac{\chi^2}{\sigma_I(\alpha, \beta)}\right](\nabla_2^2 + q_c^2)^2 R,$$

where

$$\omega_0(\alpha, \beta) = \frac{\beta\alpha(\alpha - 3)}{(\alpha + 1)^3}, \quad \omega_1(\alpha, \beta) = \frac{\beta\alpha[8\alpha - (\alpha + 1)^2]}{(\alpha + 1)^4}.$$

Since this served as our evolution equation, we described that procedure in some detail. First, we considered the PDE in question for A when $\Delta = 0$:

$$A_t \sim \gamma_{10}A + \gamma_{11}A^* + \gamma_{20}A^2 + \gamma_{21}AA^* + \gamma_{22}A^{*2}$$
$$+ \gamma_{30}A^3 + \gamma_{31}A^2A^* + \gamma_{32}AA^{*2} + \gamma_{33}A^{*3} + i\chi\nabla_2^2 A.$$

Then, taking $A = R + iI$ with $\theta = -\chi q_c^2 = -1$ and representing $\gamma_{10} = \gamma_{10}^{(r)} + i\chi q_c^2$ where $\gamma_{10}^{(r)} = -(1 + \beta/D^2)$ in this equation, we obtained

$$R_t + iI_t \sim \gamma_{10}^{(r)}(R + iI) + \gamma_{11}(R - iI) + \gamma_{20}(R^2 + 2iRI - I^2) + \gamma_{21}(R^2 + I^2)$$
$$+ \gamma_{22}(R^2 - 2iRI - I^2) + \gamma_{30}(R^3 + 3iR^2I - 3RI^2 - iI^3)$$
$$+ \gamma_{31}(R^3 + iR^2I + RI^2 + iI^3) + \gamma_{32}(R^3 - iR^2I + RI^2 - iI^3)$$
$$+ \gamma_{33}(R^3 - 3iR^2I - 3RI^2 + iI^3) + i\chi(\nabla_2^2 + q_c^2)(R + iI).$$

Next, separating the real and imaginary parts of that equation and retaining terms through third-order on the right-hand side of the former and through second-order on the corresponding side of the latter, while noting in this context that I actually represented a second-order effect since the operator $\chi(\nabla_2^2 + q_c^2)$ represented a first-order one (see below), we deduced that

$$R_t \sim [\gamma_{10}^{(r)} + \gamma_{11}]R + (\gamma_{20} + \gamma_{21} + \gamma_{22})R^2$$
$$+ (\gamma_{30} + \gamma_{31} + \gamma_{32} + \gamma_{33})R^3 - \chi(\nabla_2^2 + q_c^2)I,$$

$$I_t \sim [\gamma_{10}^{(r)} - \gamma_{11}]I + \chi(\nabla_2^2 + q_c^2)R.$$

Upon simplifying the coefficients in these equations we made the following identifications:

$$\gamma_{10}^{(r)} + \gamma_{11} = -1 + \frac{\beta(\alpha - 1)}{(\alpha + 1)^2} = \sigma_R(\alpha, \beta),$$

$$\gamma_{10}^{(r)} - \gamma_{11} = -1 - \frac{\beta}{\alpha + 1} = \sigma_I(\alpha, \beta),$$

$$\gamma_{20} + \gamma_{21} + \gamma_{22} = \frac{\beta\alpha(3 - \alpha)}{(\alpha + 1)^3} = -\omega_0(\alpha, \beta),$$

$$\gamma_{30} + \gamma_{31} + \gamma_{32} + \gamma_{33} = \frac{\beta\alpha[(\alpha + 1)^2 - 8\alpha]}{(\alpha + 1)^4} = -\omega_1(\alpha, \beta).$$

That the growth rates for R and I then turned out to be the two roots of our secular equation, respectively, which was the reason they had been denoted previously with those subscripts, definitely represented a Rabbit Pulled Out of a Hat.

Finally, employing these identifications in those equations, assuming in addition that the one for I satisfied the quasi-equilibrium condition

$$I \sim - \left[\frac{\chi}{\sigma_I(\alpha, \beta)} \right] (\nabla_2^2 + q_c^2)R,$$

and using this asymptotic relation to eliminate it from the one for R, we rederived Firth and Scroggie's ([60]) modified Swift-Hohenberg equation

$$R_t \sim \sigma_R(\alpha, \beta)R - \omega_0(\alpha, \beta)R^2 - \omega_1(\alpha, \beta)R^3 + \left[\frac{\chi^2}{\sigma_I(\alpha, \beta)} \right] (\nabla_2^2 + q_c^2)^2 R.$$

It has always been my method of operation when investigating a new phenomenon never to accept anything appearing in previous modeling papers until I had re-examined the relevant developments, which often resulted in the incorporation of corrections or modifications, but in this case upon using Firth and Scroggie's ([60]) very clever procedure merely confirmed their result.

We wished to analyze that modified Swift-Hohenberg evolution equation for the type of non-linear optical patterns mentioned earlier. We noted that this evolution equation admitted the trivial $R \equiv 0$ state, which corresponded to our stationary uniform plane-wave equilibrium solution $X \equiv X_e$, and hence implicitly satisfied the companion far-field boundary condition

$$R \text{ remains bounded as } x^2 + y^2 \to \infty.$$

In our development, we had paralleled and synthesized the approaches of both Lugiato and Oldano ([129]), who performed a linear stability analysis on the original system, and Firth and Scroggie ([60]), who numerically integrated the reduced equation for X, while using their modified Swift-Hohenberg equation as a guide to select the appropriate parameter range favoring pattern formation. We further noted that these numerical simulations were, of necessity, performed on a square grid with periodic boundary conditions. Given that the characteristic wavelength of the nonlinear optical patterns to be investigated was very small with respect to the territorial length scale of the cavity cross-sectional area, it again seemed reasonable as a first approximation for us to consider our evolution equation on an unbounded transverse spatial domain, since under this condition the actual cross-sectional boundary did not significantly influence the pattern (Graham et al., [73]). In this context, the boundedness property of the transverse far-field condition physically represented the analytical analogue of the periodic boundary conditions adopted for numerical simulation.

It was the weakly nonlinear stability of the plane wave solution of our evolution equation to two-dimensional hexagonal planform perturbations with which Dean and I were concerned. Although Firth and Scroggie ([60]) had stated their intention to publish such an analysis on this equation in the future, from the negative results of our literature search on the subject, they seem never to have actually done so, fortunately for us. As a necessary prelude to this investigation, we first performed a one-dimensional longitudinal planform weakly nonlinear stability analysis of its zero state, along the lines of that developed in Chapter 9, by seeking a solution to our evolution equation through third-order terms of the form

$$R(x, t) \sim A_1(t) \cos(q_c x) + A_1^2[R_{20} + R_{22} \cos(2q_c x)]$$
$$+ A_1^3(t)[R_{31} \cos(q_c x) + R_{33} \cos(3q_c x)],$$

where the amplitude function $A_1(t)$ satisfied the Landau equation

$$\frac{dA_1(t)}{dt} \sim \sigma A_1(t) - a_1 A_1^3(t)$$

and q_c was the critical wavenumber of linear stability theory defined above. Substituting this solution into the modified Swift-Hohenberg equation we obtained a sequence of problems, one for each pair of n and m values that corresponded to a term of the form $A_1^n(t)\cos(mq_c x)$ included explicitly in that expansion.

Then, the linear stability problem for $n = m = 1$ not surprisingly yielded the secular equation

$$\sigma = \sigma_R(\alpha, \beta) = \frac{\beta}{\beta_0(\alpha)} - 1,$$

where $\beta_0(\alpha)$ is as defined earlier. Thus, the locus $\beta = \beta_0(\alpha)$ of Fig. 10.3 served as its marginal stability curve in the α-β plane with a minimum point at (α_c, β_c) where

$$\alpha_c = 3, \quad \beta_c = 8.$$

Hence,

$$\sigma_R < 0 \text{ for } 0 < \beta < \beta_0(\alpha), \ \sigma_R = 0 \text{ for } \beta = \beta_0(\alpha), \ \sigma_R > 0 \text{ for } \beta > \beta_0(\alpha) \geq \beta_c.$$

Therefore, $R \equiv 0$ was linearly stable for $0 < \beta < \beta_0(\alpha)$, neutrally stable for $\beta = \beta_0(\alpha)$, and unstable for $\beta > \beta_0(\alpha) \geq \beta_c = 8$.

Continuing our description of the results of this one-dimensional expansion procedure, the second-order problems corresponding to $n = 2$ and $m = 0$ or 2 could be solved in a straightforward manner to yield

$$R_{20} = \frac{\omega_0}{2}\left(\frac{1}{\sigma_I} - \sigma_R\right), \ R_{22} = \frac{\omega_0}{2}\left(\frac{9}{\sigma_I} - \sigma_R\right).$$

Although there were also two third-order problems, it was permissible for us to concentrate our attention exclusively on the one corresponding to $n = 3$ and $m = 1$, which contained the Landau coefficient a_1 for the Fredholm alternative solvability method to be employed

$$a_1 - 2\sigma_R R_{31} = \omega_0(2R_{20} + R_{22}) + \frac{3\omega_1}{4}.$$

Now, taking the limit of this equation as $\beta \to \beta_0(\alpha)$, making use of our previous results, and assuming the requisite continuity at $\beta = \beta_0(\alpha)$, we obtained the solvability condition

$$a_1 = \frac{\alpha\beta_0(\alpha)}{(\alpha+1)^4}\left\{\frac{-19}{9}\left[\frac{\alpha(\alpha-3)}{\alpha-1}\right]^2 + \frac{3}{4}[8\alpha - (\alpha+1)^2]\right\} = a_1(\alpha).$$

As demonstrated in Chapters 7 and 8, the stability behavior of the Landau equation truncated through terms of third-order and thus the pattern formation aspect of our model system was crucially dependent upon the sign of a_1. Hence, in order to determine this behavior, we needed to examine the formula for a_1. Toward that end, we plotted $a_1(\alpha)$ versus α in Fig. 10.4. From this figure, we observed that a_1 had two zeroes at $\alpha = \alpha_{1,2}$ such that

$$a_1 < 0 \text{ for } 1 < \alpha < \alpha_1 \text{ or } \alpha > \alpha_2, \ a_1 > 0 \text{ for } \alpha_1 < \alpha < \alpha_2;$$

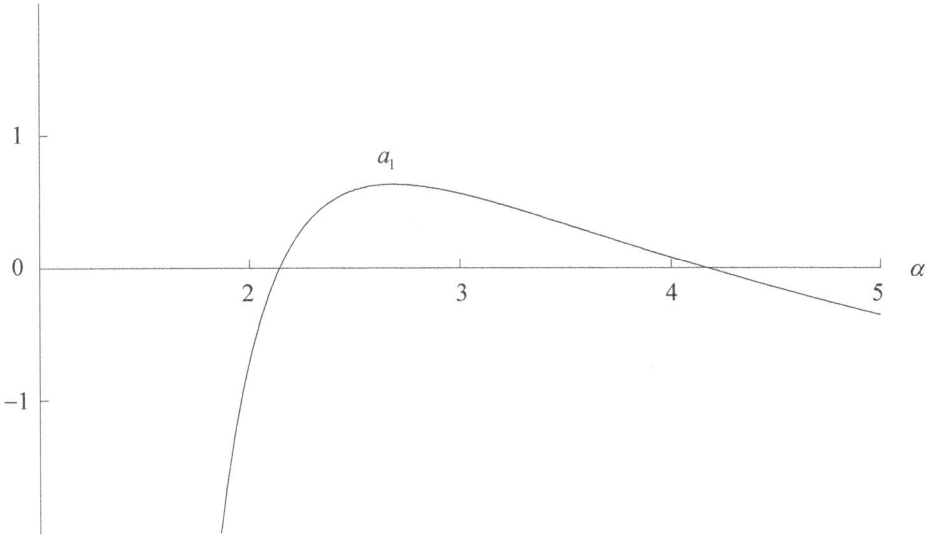

FIGURE 10.4
Plot of $a_1 = a_1(\alpha)$ versus α.

where

$$\alpha_1 = 2.143 \text{ and } \alpha_2 = 4.167.$$

Again, given these conditions for $\sigma_R(\alpha, \beta)$ and $a_1(\alpha)$, we noted that the Landau amplitude function $A_1(t)$ underwent a pitchfork bifurcation at $\beta = \beta_0(\alpha)$ when $\alpha_1 < \alpha < \alpha_2$ from which we concluded, as in Chapter 9, that

1. For $0 < \beta < \beta_0(\alpha)$ and $\alpha_1 < \alpha < \alpha_2$, the undisturbed state $A_1 = 0$ was stable since $\sigma_R < 0$ and $a_1 > 0$, yielding a uniform plane-wave configuration since $R(x, t) \to 0$ as $t \to \infty$.

2. For $\beta > \beta_0(\alpha)$ and $\alpha_1 < \alpha < \alpha_2$, $A_1 = A_e = \sqrt{\sigma_R/a_1}$ was stable, since $\sigma_R, a_1 > 0$ yielding a periodic one-dimensional pattern consisting of stationary parallel stripes

$$R(x, t) \to R_e(x) \sim A_e \cos\left(\frac{2\pi x}{\lambda_c}\right) \text{ as } t \to \infty$$

with characteristic wavelength, exactly fitting the cavity, of

$$\lambda_c = \frac{2\pi}{q_c} = 2\pi \chi^{1/2}.$$

These supercritical stripes are represented in the contour plot of Fig. 10.5 where, after Firth and Scroggie ([60]), regions of higher intensity ($R > 0$ in this case) appear bright and those of lower intensity ($R < 0$) dark. Here, the axes are being measured in units of λ_c. When $\alpha < \alpha_1$ or $\alpha > \alpha_2$, that bifurcation was subcritical. The consequences of such subcritical behavior will be discussed in Chapter 15. In what follows, we shall be concentrating on the behavior of our modified Swift-Hohenberg evolution equation in its supercritical regime where $\alpha_1 < \alpha < \alpha_2$.

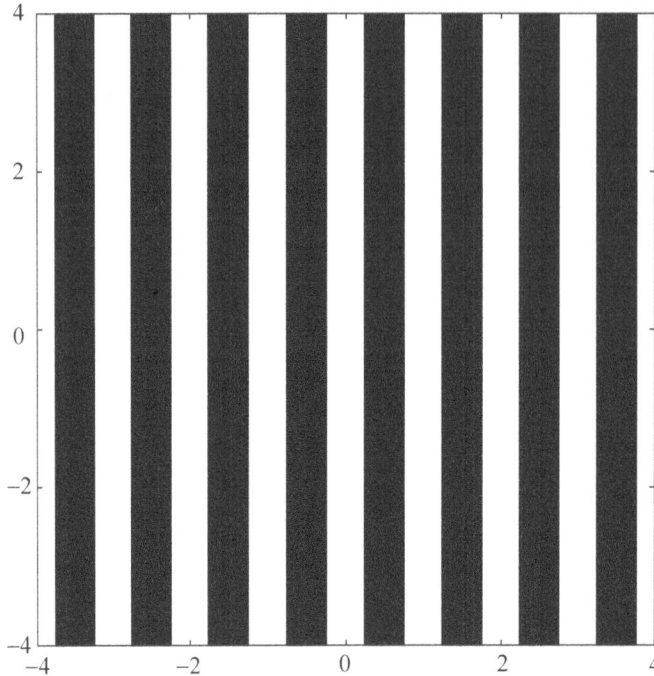

FIGURE 10.5

Contour plot in the x-y plane for supercritical stripes (critical point II of the hexagonal and rhombic planform stability analyses) where the spatial variables are measured in units of λ_c. Here, elevations appear light and depressions, dark, in accordance with experimental observation. This transition occurs at $R = 0$, which is a protocol that will be labeled zero threshold for our rhombic planform stability analysis. We shall show from this analysis that striped patterns are only stable where a protocol of the zero threshold-type holds. Hence, this contour plot is the proper representation for critical point II. The associated density plot for this pattern appears in Fig. 9.3.

In order to investigate the possibility of occurrence for our modified Swift-Hohenberg equation of the type of two-dimensional optical patterns mentioned earlier, we considered a hexagonal planform solution of it that to lowest order satisfied

$$R(x,y,t) \sim A_1(t)\cos[q_c x + \varphi_1(t)] + A_2(t)\cos\left[\frac{q_c(x - \sqrt{3}y)}{2} - \varphi_2(t)\right]$$
$$+ A_3(t)\cos\left[\frac{q_c(x + \sqrt{3}y)}{2} - \varphi_3(t)\right]$$

where, for $(j, k, \ell) \equiv$ even permutation of (1,2,3),

$$\frac{dA_j}{dt} \sim \sigma A_j - 4a_0 A_k A_\ell \cos(\varphi_j + \varphi_k + \varphi_\ell) - A_j[a_1 A_j^2 + 2a_2(A_k^2 + A_\ell^2)],$$

$$A_j \frac{d\varphi_j}{dt} \sim 4a_0 A_k A_\ell \sin(\varphi_j + \varphi_k + \varphi_\ell).$$

Here, we have represented this hexagonal expansion in its real form. Given that many non-linear stability analyses employ its equivalent complex form, we examine that equivalence for the sake of completeness. Consider the corresponding complex expansion (Kuske and Matkowsky, [105])

$$R(x, y, t) \sim R_1 \mathcal{A}_1(t) e^{i \boldsymbol{Q}_1 \cdot \boldsymbol{r}} + R_1^* \mathcal{A}_1^*(t) e^{-i \boldsymbol{Q}_1 \cdot \boldsymbol{r}} + R_2 \mathcal{A}_2(t) e^{i \boldsymbol{Q}_2 \cdot \boldsymbol{r}} + R_2^* \mathcal{A}_2^*(t) e^{-i \boldsymbol{Q}_2 \cdot \boldsymbol{r}}$$
$$+ R_3 \mathcal{A}_3(t) e^{i \boldsymbol{Q}_3 \cdot \boldsymbol{r}} + R_3^* \mathcal{A}_3^*(t) e^{-i \boldsymbol{Q}_3 \cdot \boldsymbol{r}}$$

where

$$\boldsymbol{r} = x\boldsymbol{e}_1 + y\boldsymbol{e}_2; \quad \boldsymbol{Q}_j = q_c[\cos(\theta_j)\boldsymbol{e}_1 + \sin(\theta_j)\boldsymbol{e}_2], \quad j = 1, 2, 3;$$

with

$$\boldsymbol{e}_1 = (1, 0), \quad \boldsymbol{e}_2 = (0, 1); \quad \theta_1 = 0, \quad \theta_2 = \frac{2\pi}{3}, \quad \theta_3 = \frac{4\pi}{3};$$

while, for $(j, k, \ell) \equiv$ even permutation of (1,2,3),

$$\frac{d\mathcal{A}_j}{dt} \sim \sigma \mathcal{A}_j - 4a_0 \mathcal{A}_k^* \mathcal{A}_\ell^* - \mathcal{A}_j[a_1 |\mathcal{A}_j|^2 + 2a_2(|\mathcal{A}_k|^2 + |\mathcal{A}_\ell|^2)].$$

Now, noting that

$$\boldsymbol{Q}_1 \cdot \boldsymbol{r} = q_c x, \quad \boldsymbol{Q}_2 \cdot \boldsymbol{r} = \frac{q_c(-x + \sqrt{3}y)}{2}, \quad \boldsymbol{Q}_3 \cdot \boldsymbol{r} = \frac{-q_c(x + \sqrt{3}y)}{2};$$

writing each complex amplitude $\mathcal{A}$ in its generic polar form

$$\mathcal{A}(t) = A(t)e^{i\varphi(t)};$$

taking all the

$$R_j = \frac{1}{2};$$

and recalling that

$$\cos(\varphi) = \frac{e^{i\varphi} + e^{-i\varphi}}{2};$$

this complex representation reduces to our real one. Further, substituting these polar functions into the complex amplitude equations, computing the indicated derivatives, noting that

$$\mathcal{A}^*(t) = A(t)e^{-i\varphi(t)},$$

and multiplying the result by $e^{-i\varphi_j(t)}$, we obtain

$$\frac{dA_j}{dt} + iA_j \frac{d\varphi_j}{dt} \sim \sigma A_j - 4a_0 A_k A_\ell e^{-i(\varphi_j + \varphi_k + \varphi_\ell)} - A_j[a_1 A_j^2 + 2a_2(A_k^2 + A_\ell^2)],$$

the real and imaginary parts of which yield our amplitude-phase equations, respectively, thus completing this equivalency demonstration.

The nonlinear stability behavior of these amplitude-phase equations depended, as usual, only on the values of their growth rate and Landau coefficients. We determined this growth rate and these Landau coefficients by proceeding in the identical manner to that employed in the previous two chapters. Then, denoting the coefficient of each higher-order term in the relevant R-expansion of the form $A_1^n(t)B_1^j(t)\cos(Mq_cx/2)\cos(k\sqrt{3}q_cy/2)$ by R_{njMk}, we found that

$$R_{n0M0} = R_{nm} \text{ for } M = 2m, \ \sigma = \sigma_R(\alpha, \beta), \ a_1 = a_1(\alpha),$$

as defined earlier, and

$$R_{1131} = \frac{\omega_0}{4/\sigma_I - \sigma_R};$$

while, in particular, the other two Landau coefficients satisfied

$$4a_0 - \sigma_R R_{1111} = \omega_0,$$

$$a_2 - \sigma_R R_{2111} + (\omega_0 - 4a_0)\frac{R_{1111}}{2} = \omega_0\left(R_{2000} + \frac{R_{1131}}{2}\right) + \frac{3\omega_1}{4}.$$

Finally, taking the limit of these equations as $\beta \to \beta_0(\alpha)$ and employing our previous results, we obtained the solvability conditions

$$a_0 = \frac{\beta_0(\alpha)\alpha(\alpha - 3)}{4(\alpha + 1^3)} = \frac{\alpha(\alpha - 3)}{4(\alpha^2 - 1)} = a_0(\alpha),$$

$$a_2 = \frac{\alpha\beta_0(\alpha)}{4(\alpha + 1)^4}\left\{-5\left[\frac{\alpha(\alpha - 3)}{\alpha - 1}\right]^2 + 3[8\alpha - (\alpha + 1)^2]\right\} = a_2(\alpha).$$

Note that the expression $a_2(\alpha)$ does not explicitly contain the component $\lim_{\beta \to \beta_0(\alpha)} R_{1111}/2$ since the coefficient of the latter quantity, namely, $\lim_{\beta \to \beta_0(\alpha)}(\omega_0 - 4a_0)$ is identically equal to zero by virtue of the formula for $a_0(\alpha)$. Although, as pointed out by Wollkind *et al.* ([291]) and demonstrated in Chapter 9, such independence can be expected in single equation models, thus eliminating the necessity of determining R_{1111}, in this instance, we can conclude, for the sake of completeness, that

$$R_{1111} = -\frac{\beta_0(\alpha)\alpha(\alpha - 3)}{(\alpha + 1)^3} = -4a_0(\alpha).$$

Having determined formulae for their growth rate and Landau coefficients, we now returned to the six-disturbance hexagonal-planform amplitude phase-equations. Given that the techniques for analysis of those equations were introduced in Chapters 2 and 3 and reviewed in Chapters 7-9, we shall employ them whenever necessary in the remainder of this book by simply referencing the relevant parts of those chapters without reproducing definitions or orbital stability tables.

We made the following nonlinear optical morphological interpretations of the equivalence classes of potentially stable critical points for these equations catalogued in those chapters: I and II, represented the uniform plane-wave configuration and the striped pattern plotted in Fig. 10.5, respectively, as described in our longitudinal planform stability analysis, while the hexagonal arrays of III$^+$ and III$^-$, depicted in parts (a) and (b) of Fig. 10.6, represented bright spotted and honeycombed patterns, respectively.

Having summarized those morphological identifications, we returned to our expressions for the Landau coefficients. First, we examined the signs of a_0, $2a_2 - a_1$, $a_1 + 4a_2$, and $a_1 + a_2$ as functions of α by plotting these quantities versus α in Fig. 10.7. From the results of

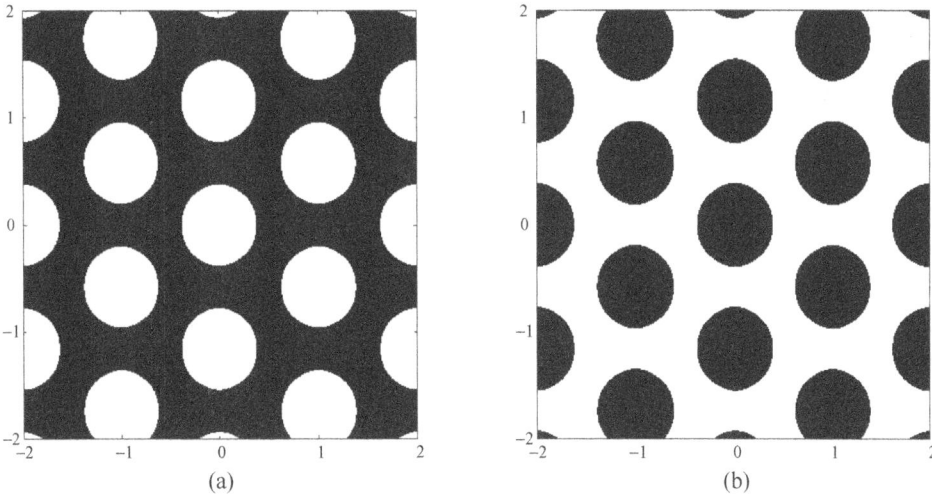

(a)   (b)

FIGURE 10.6
Contour plots for critical points (a) III$^+$ and (b) III$^-$, the (a) spots and (b) honeycombs of this chapter. Here, the spatial variables are again being measured in λ_c with regions of high intensity ($R > 0$) appearing light and low intensity ($R < 0$), dark, as in Fig. 10.5.

this examination, we observed that besides α_c, α_1, and α_2 defined earlier, there existed the following other significant values of α which are tabulated in Table 10.1:

$$\alpha_7 < \alpha_5 < \alpha_3 < \alpha_1 < \alpha_c < \alpha_2 < \alpha_4 < \alpha_6 < \alpha_8$$

such that

$$a_1 + a_2 = 0 \text{ for } \alpha = \alpha_3 \text{ or } \alpha_4, \ a_1 + a_2 > 0 \text{ for } \alpha_3 < \alpha < \alpha_4;$$
$$a_1 + 4a_2 = 0 \text{ for } \alpha = \alpha_5 \text{ or } \alpha_6, \ a_1 + 4a_2 > 0 \text{ for } \alpha_5 < \alpha < \alpha_6;$$
$$2a_2 - a_1 = 0 \text{ for } \alpha = \alpha_7 \text{ or } \alpha_8, \ 2a_2 - a_1 > 0 \text{ for } \alpha_7 < \alpha < \alpha_8;$$
$$a_0 = 0 \text{ for } \alpha = \alpha_c, \ a_0 < 0 \text{ for } 1 < \alpha < \alpha_c, \ a_0 > 0 \text{ for } \alpha > \alpha_c.$$

TABLE 10.1
The significant α-values for hexagonal planform optical pattern formation.

α_7	α_5	α_3	α_1	α_c	α_2	α_4	α_6	α_8
1.646	2.024	2.075	2.143	3	4.167	4.290	4.383	5.107

We next restricted our attention to those $\alpha \in (\alpha_3, \alpha_4)$ and observed that then

$$2a_2 - a_1 > 0.$$

Thus, under this additional assumption the relevant entries of Table 9.1 which only involved the sign of a_0 may be employed to determine the orbital stability of our critical points. In order to compare these theoretical predictions with numerical simulations and relevant experimental evidence, we represented the stability results of that table graphically in the

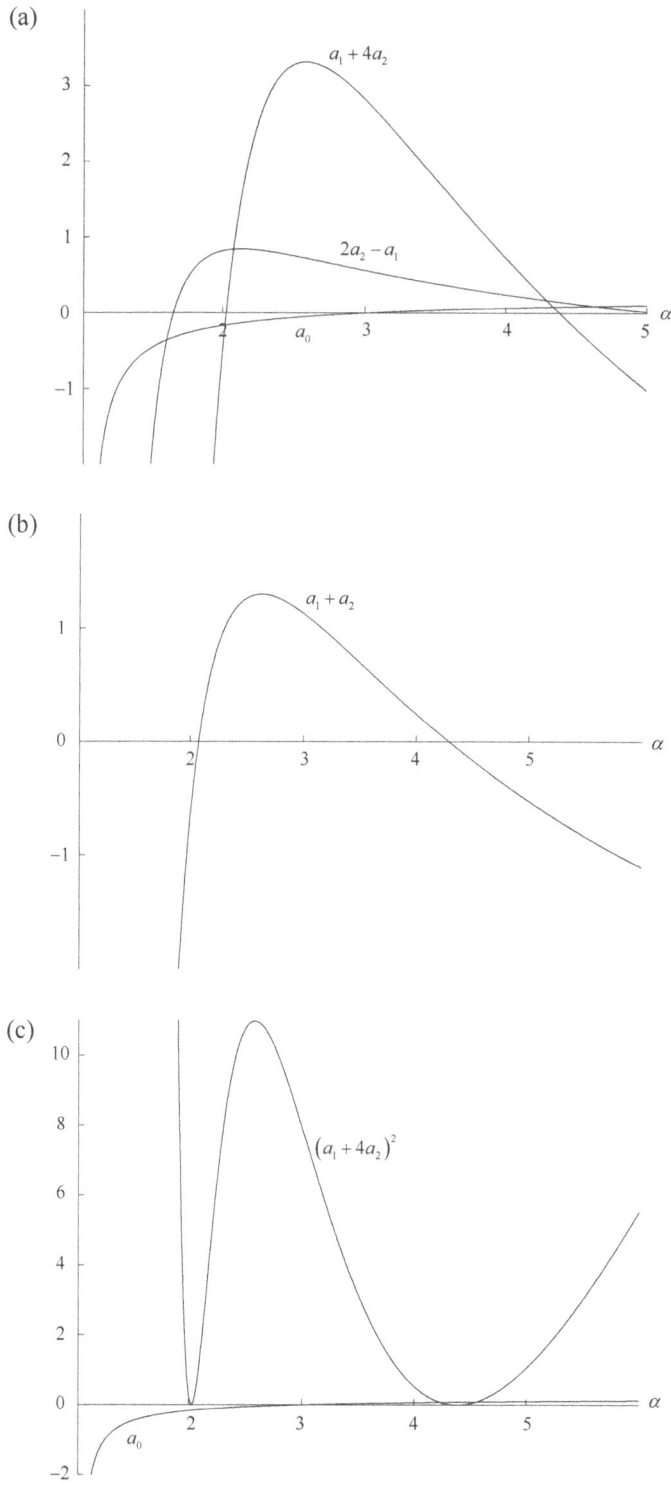

FIGURE 10.7
Plots of (a) a_0, $2a_2 - a_1$, and $a_1 + 4a_2$; (b) $a_1 + a_2$; (c) a_0 and $(a_1 + 4a_2)^2$; versus α.

α-β plane. To do so, it was necessary as usual for us to generate the loci associated with $\sigma = \sigma_j$ for $j = -1, 1$, and 2, respectively. Solving

$$\sigma_R(\alpha, \beta) = \sigma_j[a_0(\alpha), a_1(\alpha), a_2(\alpha)], \quad \text{with } j = -1, 1, \text{ and } 2,$$

for β we obtained

$$\beta = \beta_0(\alpha)\{1 + \sigma_j[a_0(\alpha), a_1(\alpha), a_2(\alpha)]\} = \beta_j(\alpha) \text{ with } j = -1, 1, \text{ and } 2,$$

respectively. Since all the quantities required for the transverse optical patterns of Table 9.1 had been evaluated, we represented graphically the regions corresponding to these patterns in the α-β plane of Fig. 10.8, where the marginal stability curve and these loci have been denoted by β_j with $j = -1, 0, 1$, and 2, in that figure, and identified this correspondence in Table 10.2. From the latter table, we saw that stable spots could only occur for $\alpha < \alpha_c = 3$ and stable honeycombs for $\alpha > \alpha_c = 3$.

In this context, we observed from Fig. 10.8a that all of the loci intersected each other at the point $(\alpha_c, \beta_c) = (3, 8)$ and noted that in the left half of that figure for $\alpha < \alpha_c$ (see the top half of Table 10.2 and Fig. 10.8b) the α-component of the intersection points of β_1 and β_2 with β_0 correspond to α_1 and α_3, respectively, while in its right half for $\alpha > \alpha_c$ (see the bottom half of Table 10.2 and Fig. 10.8c) the α-components of these intersection points correspond to α_2 and α_4. Although the stability behavior summarized in Table 9.1 was strictly for $a_1 > 0$ or $\alpha \in (\alpha_1, \alpha_2)$, we extended these results to the case of $a_1 + a_2 > 0$ or $\alpha \in (\alpha_3, \alpha_4)$ in Table 10.2. This extrapolation was accomplished by realizing that in the relevant α-intervals of $(\alpha_3, \alpha_1]$ and $[\alpha_2, \alpha_4)$ only $\text{III}^\pm$ structures existed since $a_1 < 0$ and these hexagonal structures were stable for

$$\beta_{-1}(\alpha) < \beta < \beta_2(\alpha).$$

TABLE 10.2
Pattern formation predictions for Fig. 10.8.

α-interval	β-range	Stable pattern
(α_3, α_c)	$\beta < \beta_{-1}$	Plane wave
	$\beta_{-1} < \beta < \beta_0$	Plane wave and spots
$(\alpha_3, \alpha_1]$	$\beta_0 < \beta < \beta_2$	Spots
(α_1, α_c)	$\beta_0 < \beta < \beta_1$	Spots
	$\beta_1 < \beta < \beta_2$	Spots and stripes
	$\beta > \beta_2$	Stripes
α_c	$\beta > \beta_c$	Stripes
(α_c, α_2)	$\beta_0 < \beta < \beta_1$	Honeycombs
	$\beta_1 < \beta < \beta_2$	Honeycombs and stripes
	$\beta > \beta_2$	Stripes
$[\alpha_2, \alpha_4)$	$\beta_0 < \beta < \beta_2$	Honeycombs
(α_c, α_4)	$\beta < \beta_{-1}$	Plane wave
	$\beta_{-1} < \beta < \beta_0$	Plane wave and honeycombs

Note that in those α-intervals there are no stable patterns for $\beta > \beta_2$ and metastability between the plane wave and subcritical striped states for $\beta < \beta_{-1}$. Again, considering $\alpha \in (\alpha_1, \alpha_2)$, we observed from Figs. 10.8, in conjunction with Table 10.2, that part $(\beta_0 < \beta < \beta_2)$ of the region $(\beta > \beta_0)$, where the longitudinal planform analysis predicted

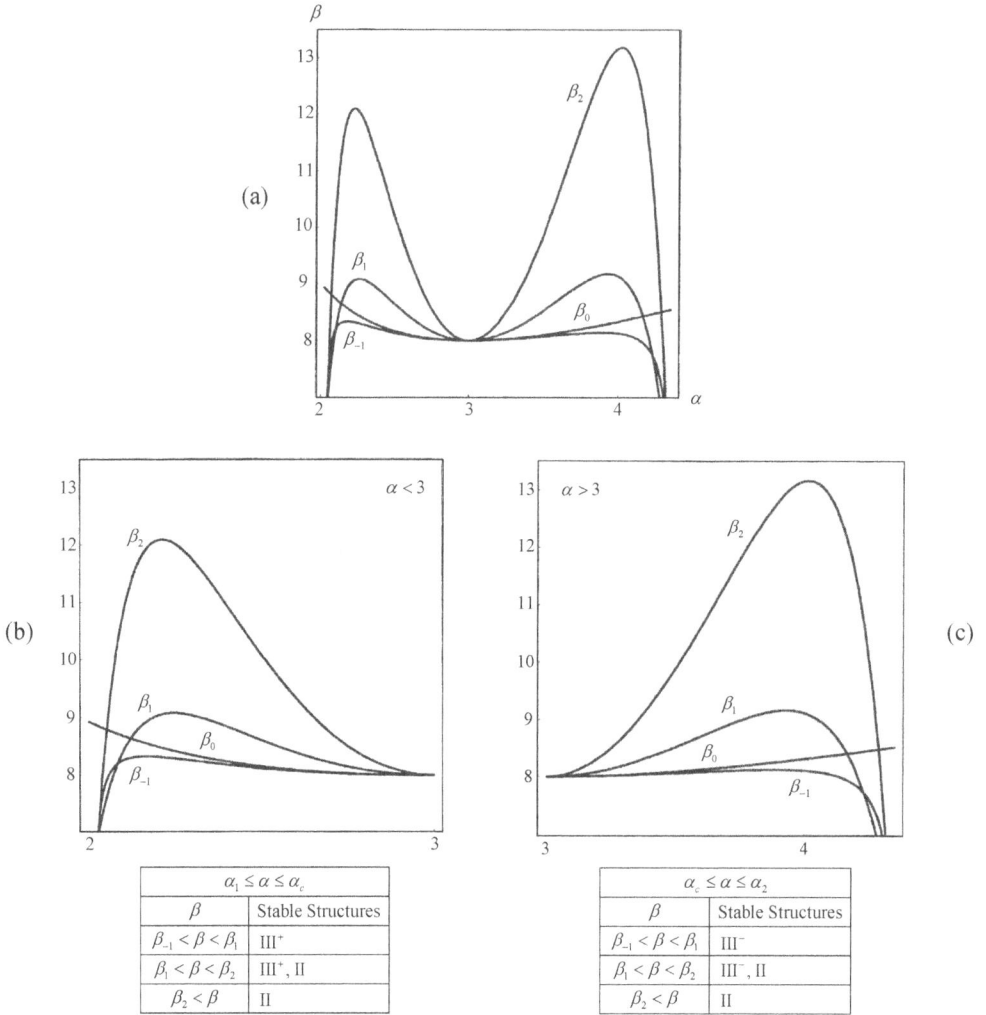

$\alpha_1 \le \alpha \le \alpha_c$	
β	Stable Structures
$\beta_{-1} < \beta < \beta_1$	III$^+$
$\beta_1 < \beta < \beta_2$	III$^+$, II
$\beta_2 < \beta$	II

$\alpha_c \le \alpha \le \alpha_2$	
β	Stable Structures
$\beta_{-1} < \beta < \beta_1$	III$^-$
$\beta_1 < \beta < \beta_2$	III$^-$, II
$\beta_2 < \beta$	II

FIGURE 10.8

(a) Stability diagram in the α-β plane for the modified Swift-Hohenberg equation with the predicted nonlinear optical patterns summarized in Table 10.2 denoted by the enlargements for (b) $\alpha_1 < \alpha < \alpha_c$ and (c) $\alpha_c < \alpha < \alpha_2$. Here, $\beta_j = \beta_0(1 + \sigma_j)$ for $j = -1, 0, 1, 2$.

supercritical striped patterns, was further divided into two sub-regions characterized by hexagonal patterns consisting of either spots (when $\alpha < \alpha_c$) or honeycombs (when $\alpha > \alpha_c$), respectively. In the overlap regions satisfying $\sigma_1 < \sigma < \sigma_2$ or $\beta_1(\alpha) < \beta < \beta_2(\alpha)$, where stripes and spots ($\alpha_1 < \alpha < \alpha_c = 3$) or stripes and honeycombs ($3 = \alpha_c < \alpha < \alpha_2$) were predicted, initial conditions determine which stable equilibrium pattern of each pair would be selected. Returning to the extended interval (α_3, α_4), there also existed a region of bistability corresponding to $\beta_{-1}(\alpha) < \beta < \beta_0(\alpha)$ or $\sigma_{-1} < \sigma < 0$, the plane wave being stable for $\sigma < 0$ and hexagons for $\sigma_{-1} < \sigma < \sigma_2$. Given that $\sigma_{-1} < 0$ for $a_0 \neq 0$ or $\alpha \neq \alpha_c = 3$, the hexagons predicted in this overlap region would be subcritical in nature. Finally, to justify the truncation procedure inherent to our asymptotic representation, it was necessary that the Landau coefficients of these amplitude-phase equations satisfied the size constraint $|a_0|/(a_1 + 4a_2)^2 \ll 1$ (Wollkind et al., [291]), as demonstrated in Fig. 10.8c. Hence, we concluded that this truncation procedure was valid for our modified Swift-Hohenberg evolution equation.

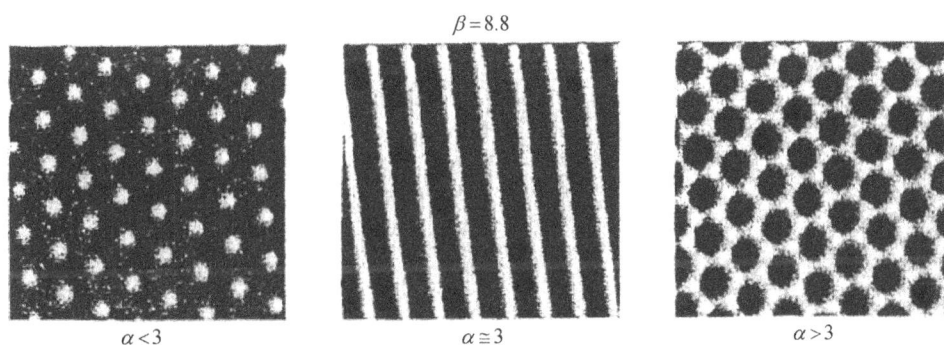

$\beta = 8.8$

$\alpha < 3$ $\alpha \cong 3$ $\alpha > 3$

FIGURE 10.9
Adapted from plots relevant to the spot-stripe-honeycomb transition as α is increased for the simulations of Firth and Scroggie ([60]) with $\Delta = 0$, $\theta = -1$, $\chi = 1$, and $\beta = 8.8$ by Wollkind and Dichone ([284]).

We were finally ready to compare these theoretical predictions with the relevant numerical simulations of Firth and Scroggie ([60]). These authors numerically integrated the reduced equation for X on a square grid using periodic boundary conditions with $\Delta = 0$, $\theta = -1$, and χ normalized to unity. Implicitly exploiting our relationship for a_0 and assuming that α_c could be determined from the expression for ω_0, they concluded that $\alpha_c = 3$ and constructed contour plots of $\text{Re}(X)$ in the x-y plane by selecting a value of β greater than $\beta_c = 8$ and increasing α across the instability region. In this event, Firth and Scroggie ([60]) observed a transition from stable III$^+$-to II-to III$^-$-type patterns for $\alpha < 3$, $\alpha \cong 3$, and $\alpha > 3$, respectively. To compare that simulation result with our theoretical predictions, we selected the typical fixed value of $\beta = 8.8$ employed by Firth and Scroggie ([60]) in the three parts of the contour plots depicting their results reproduced in Fig. 10.9. Upon examination of Fig. 10.8 and Table 10.2, we deduced the predicted morphological sequence given in Table 10.3 as α was increased along that horizontal transit line.

In order to interpret these results with respect to Firth and Scroggie's ([60]) simulations depicted in Fig. 10.9, we assumed the maximal possible domain for hexagons when considering

TABLE 10.3
Morphological stability predictions versus α for $\beta = 8.8$.

α-interval	Stable pattern
(2.08,2.14)	Spots
(2.14,2.16)	Spots and stripes
(2.16,2.45)	Spots
(2.45,2.73)	Spots and stripes
(2.73,3.30)	Stripes
(3.30,3.62)	Honeycombs and stripes
(3.62,4.10)	Honeycombs
(4.10,4.17)	Honeycombs and stripes
(4.17,4.28)	Honeycombs

regions of bistability in Table 10.3. This assumption was consistent with the fact that those authors were unable to observe the stable coexistence of stripes and hexagonal patterns. Making such an interpretation for $\beta = 8.8$, yielded the predicted sequence of morphologies

$$\text{III}^+ \text{ for } 2.08 < \alpha < 2.73, \quad \text{II for } 2.73 < \alpha < 3.30, \quad \text{III}^- \text{ for } 3.30 < \alpha < 4.28,$$

in qualitative agreement with the results of Firth and Scroggie ([60]). This completed Dean's thesis work and he successfully defended that dissertation during August of 2004, becoming WSU's Department of Mathematics first African-American Ph.D. recipient in the process.

Then, in order to refine these predictions, as well as to explain the square patterns Scroggie and Firth ([209]) had produced by performing numerical simulations on their related optical model for a sodium vapor cell with a feedback mirror where pattern formation was induced by external pumping (Ackemann *et al.*, [2]; Ackemann and Lange, [1]), Francisco Alvarado, who had just that month chosen to do his Ph.D. dissertation research under my direction, and I began performing a rhombic planform weakly nonlinear stability analysis on the modified Swift-Hohenberg evolution equation. That is, we considered a solution of this equation to lowest order of the form

$$R(x,y,t) \sim A_1(t)\cos(q_c x) + B_1(t)\cos(q_c z) \text{ with } z = x\cos(\psi) + y\sin(\psi)$$

where

$$\frac{dA_1}{dt} \sim \sigma A_1 - A_1(a_1 A_1^2 + b_1 B_1^2), \quad \frac{dB_1}{dt} \sim \sigma B_1 - B_1(b_1 A_1^2 + a_1 B_1^2),$$

and $\psi \equiv$ the rhombic angle (not φ, as in Chapter 9, since it has already been used for the angle in the complex polar representation), while each higher-order term in that expansion was of the form

$$R_{njmk} A_1^n(t) B_1^j(t) \cos(q_c[mx + kz]).$$

Then substituting this expansion into our modified Swift-Hohenberg model equation, we obtained a sequence of problems each of which was proportional to one of these terms. Solving those problems, we found that

$$R_{n0m0} = R_{0n0m} = R_{nm}, \sigma = \sigma_R(\alpha,\beta), \ a_1 = a_1(\alpha),$$

as defined earlier, while

$$R_{111(\pm 1)} = \frac{\omega_0}{[1 \pm 2\cos(\psi)]^2/\sigma_I - \sigma_R}$$

and b_1, in particular, satisfied

$$b_1 - 2\sigma_R R_{2101} = \frac{3\omega_1}{2} + \omega_0[2R_{2000} + R_{1111} + R_{111(-1)}].$$

Now, taking the limit of this relation as $\beta \to \beta_0(\alpha)$ and making use of our previous results, we obtained the solvability condition for the other rhombic planform Landau coefficient

$$b_1 = \frac{\alpha\beta_0(\alpha)}{(\alpha+1)^4}\left\{-2\left[\frac{\alpha(\alpha-3)}{\alpha-1}\right]^2 \frac{3+16\cos^4(\psi)}{[1-4\cos^2(\psi)]^2} + \frac{3}{2}[8\alpha-(\alpha+1)^2]\right\}$$
$$= b_1(\alpha,\psi).$$

Having developed these formulae for its growth rate and Landau coefficients, we turned our attention to the rhombic planform amplitude equations. Recall from Chapter 9, that this system possessed the following equivalence classes of critical points (A_0, C_0):

$$\text{I: } A_0 = C_0 = 0; \quad \text{II: } A_0^2 = \frac{\sigma}{a_1}, \ C_0 = 0; \quad \text{V: } A_0 = C_0 \text{ with } A_0^2 = \frac{\sigma}{a_1+b_1};$$

with the stability criteria:

$$\text{I is stable for } \sigma < 0;$$
$$\text{II is stable for } \sigma > 0, \ b_1 > a_1;$$
$$\text{V is stable for } \sigma > 0, \ a_1 > b_1.$$

Before examining the implications of those stability criteria, we made a morphological interpretation of the potentially stable rhombic planform critical points relative to the nonlinear optical patterns under investigation. Then, to lowest order, the equilibrium intensity function associated with these critical points was given by

$$\lim_{t\to\infty} R(x,y,t) = R_e(x,y) \sim A_0\cos(q_c x) + C_0\cos(q_c z)$$

$$\text{where } z = x\cos(\psi) + y\sin(\psi).$$

Thus, critical points I and II corresponded to the plane wave configuration and supercritical striped patterns, respectively, already discussed in the longitudinal and hexagonal planform analyses. To make an analogous interpretation of critical point V, we considered this intensity function with $A_0 = C_0 > 0$ and introduced the concept of higher, zero, and lower threshold patterns based upon the mean level of intensity. To do so, we first examined how, what in fluid mechanics are called, the mean motion terms (Segel, [212]) from our nonlinear stability analyses had altered that intensity level. These homogeneous higher-order terms in $R_e(x,y)$ for critical point V are given by

$$R_{2000}A_0^2 + R_{0200}C_0^2 \text{ with } A_0 = C_0 \text{ or } 2R_{20}C_0^2 \text{ since } R_{2000} = R_{0200} = R_{20}.$$

Then, adding these terms to the original zero mean level of intensity and employing our previous results, we found that this mean level to second-order now satisfied

$$R_m = 2R_{20}C_0^2 = \frac{\omega_0 C_0^2}{1/\sigma_I - \sigma_R}.$$

We next adopted the protocol that in our contour plots of critical point V the elevations, which satisfy $R > R_m$, would appear light, and depressions, which satisfy $R < R_m$, dark, where $R = R_m$ represented the threshold value at which this transition occured. When $R_{20} was greater than, equal to, or less than zero we labeled such patterns as being of higher, zero, or lower threshold, respectively. Hence, to determine the proper threshold type, we had to examine the sign of R_{20}. Since this quantity satisfied $R_{20} = \omega_0/[2(1/\sigma_I - \sigma_R)]$ and recalling that for stable rhombic patterns $\sigma = \sigma_R(\alpha, \beta) > 0$, while $\sigma_I < 0$, we concluded that this behavior of R_{20} depended on the sign of $-\omega_0$, which is equivalent to that of $-a_0$. Therefore, our stable rhombic patterns will be of higher, zero, or lower threshold-type depending upon whether $\alpha < \alpha_c = 3$, $\alpha = 3$, or $\alpha > 3$, respectively. To make a physical interpretation of these results, we noted that to lowest order the equilibrium intensity pattern associated with critical point V satisfied

$$ R_e(x, y) \sim C_0 \left[\cos\left(\frac{2\pi x}{\lambda_c}\right) + \cos\left(\frac{2\pi z}{\lambda_c}\right) \right] = C_0 g(x, z) $$

$$ \text{for } z = x\cos(\psi) + y\sin(\psi). $$

The three parts of Fig. 10.10 are threshold contour plots of $g(x, z)$ for the typical rhombic angle of Fig. 9.12 or $\psi = 1.150$ (66°) with the threshold values of 1, 0, and -1, respectively. Here, the spatial variables are being measured in units of λ_c with regions exceeding that threshold in each part appearing light and regions less than it, dark. Given their appearance in Fig. 10.10, we identified these higher, zero, and lower threshold-type rhombic arrays with itensity patterns of pseudo spots, rectangles, and pseudo honeycombs (nets), respectively, denoting them by V$^+$, V^0, and V$^-$ in what follows. We repeated this process and obtained the threshold contour plots of $g(x, z)$ for $\psi = \pi/2$ (90°) or $z = y$, which appear in Fig. 10.11. From the checkerboard structure of these arrays it was clear they should be identified with optical patterns of square planform.

$$ \psi = 1.150 \ \left(66° \right) $$

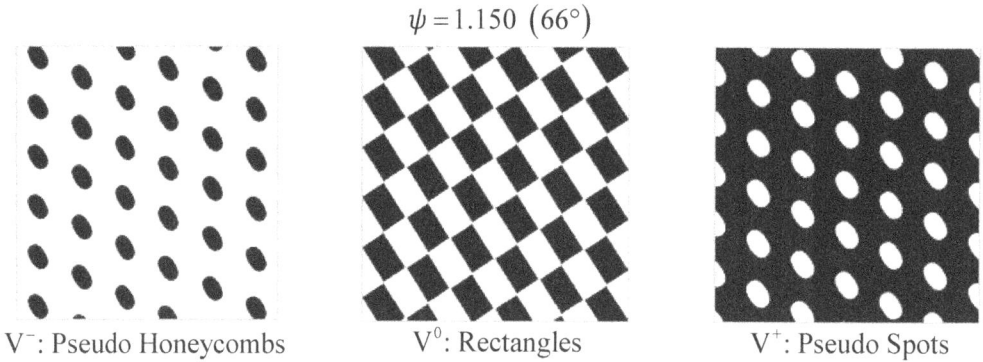

V$^-$: Pseudo Honeycombs   V^0: Rectangles   V$^+$: Pseudo Spots

FIGURE 10.10
Rhombic patterns in the x-y plane relevant to $g(x, z)$ for $\psi = 1.150$ (66°) with threshold values from left to right of -1, 0, and 1, which are denoted by V$^-$, V^0, and V$^+$, respectively, and represent arrays of pseudo honeycombs, rectangles, and pseudo spots. Here, the spatial variables are being measured in units of λ_c with regions below that threshold in each part appearing dark and regions above it, light. Note that $|g(x, z)| \leq 2$ and $\psi = 1.150$ is the typical rhombic angle depicted in the plot of Fig. 9.12. Observe that for this angle the V$^\pm$ patterns approximate the close-packed structure of hexagonal arrays.

The same morphological interpretation was implicitly employed in the Chapter 9 argument about the difficulty in distinguishing rhombic arrays of angles close to $\pi/3$ from hexagonal

$$\psi = \pi/2 \; (90°)$$

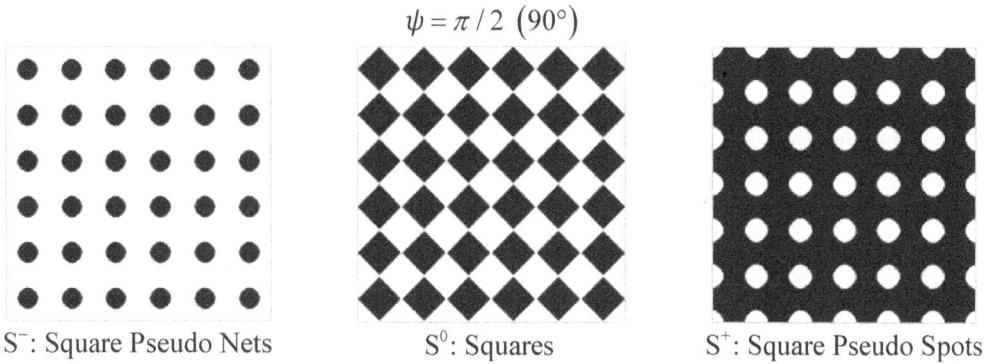

S⁻: Square Pseudo Nets   S⁰: Squares   S⁺: Square Pseudo Spots

FIGURE 10.11
Square patterns relevant to $g(x,y)$ for $\psi = \pi/2$ (90°) with threshold values from left to right of -1, 0, and 1, which are denoted by S⁻, S⁰, and S⁺, respectively, and represent arrays of square pseudo nets, squares, and square pseudo spots, respectively. Here, the spatial variables are again being measured in units of λ_c with regions below that threshold in each part appearing dark and regions above it, light. In particular, contrast the appearance of these patterns with the corresponding ones from Fig. 10.10. Sekimura *et al.* ([220]) referred to S⁺ collectively as "square spots" in their lepidoptera wing pattern formation square-type planform nonlinear stability analysis to distinguish them from the hexagonal arrays corresponding to III⁺ of what were referred to as spots and "spots," respectively, for the chemical Turing patterns described in Chapter 7.

ones, which we next describe explicitly for those ion-sputtered erosion patterns by means of the following interlude: In this instance, the lower, zero, and higher threshold rhombic patterns are based on the mean position of the interface. Adding the homogeneous higher-order terms in the equilibrium deviation function to the original dimensional mean position of the interface and employing our previous results of Chapter 9, we find that this mean position to second-order satisfies

$$H_m = H_0 - w_n\tau + 2h_{20}\ell_0 C_0^2.$$

Adopting the protocol that in our contour plots of critical point V the elevations, which satisfy $H > H_m$, would appear light, and depressions, which satisfy $H < H_m$, dark, then $H = H_m$ or $h = 2h_{20}C_0^2$ represents the threshold value at which this transition occurs. When h_{20} is less than, equal to, or greater than zero, we label such patterns as being of lower, zero, or higher threshold, respectively. Hence, to determine the proper threshold type, we must examine the sign of h_{20}. Since this quantity was given by $h_{20} = \alpha/[4(2-\beta)]$, where $\alpha = \tau_0\lambda_0/(2\ell_0)$ and recalling that for stable rhombic patterns $\sigma = \sigma_0(\beta) = 2(1-\beta) > 0$, we can conclude that this behavior of h_{20} depends on the sign of λ_0, which, being the tilt dependent coefficient of the erosion rate, is a real-valued parameter with the dimension of velocity. Therefore, our stable rhombic patterns would be of lower, zero, or higher threshold-type depending upon whether $\lambda_0 < 0$, $\lambda_0 = 0$, or $\lambda_0 > 0$, respectively. Denoting the higher and lower threshold patterns by V⁺, from a similar investigation to that completed, we could make the morphological interpretation that V⁺ represent pseudo island and pseudo hole ion-sputtered erosion patterns, respectively, consistent with our corresponding interpretation of III⁺ as island and hole hexagonal arrays. In this context, observe that, with no loss of generality, all the contour plots of the rhombic arrays depicted for illustrative purposes by the figures of Chapter 9 were of the V$^+$-type. Indeed, this mechanism for determining low, zero, and high threshold rhombic patterns, definitely represents the Pulling of a Rabbit Out of a Hat, the genesis for which will be explained in our final Chapter 18. Here, we have taken the historical liberty of employing this mechanism rather than the pattern persistence argument actually used by both Chontita and Francisco which, although yielding the same result seems, in retrospect, more speculative and hence much less convincing to us now.

Having made these interpretations, we now return to an examination of the existence and stability of the nonlinear optical rhombic planform critical points II and V. Toward that end, we determined the signs of $a_1 \pm b_1$ for $\alpha_1 < \alpha < \alpha_2$ and $0 < \psi \leq \pi/2$. To illustrate this procedure, we defined the Landau coefficient ratio $\gamma_1(\alpha, \psi) = b_1(\alpha, \psi)/a_1(\alpha)$ and considered both the existence and stability of the square patterns depicted in Fig. 10.11 by plotting that quantity in Fig. 10.12 versus α for $\psi = \pi/2$. Here, there existed two α-intervals of stable square rhombic patterns where $a_1(\alpha) \pm b_1(\alpha, \pi/2) > 0$ or, equivalently, $-1 < \gamma_1(\alpha, \pi/2) < 1$ given by

$$\alpha \in (2.215, 2.315) \text{ or } \alpha \in (3.876, 4.042),$$

provided, in addition, that $\sigma_R > 0$ or $\beta > \beta_0(\alpha)$. These intervals have been indicated by shading in Fig. 10.12.

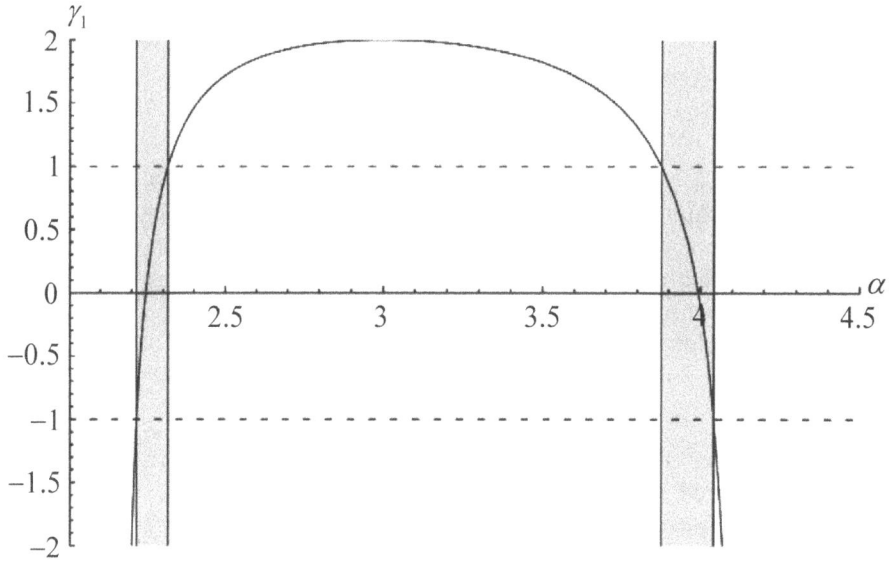

FIGURE 10.12
Plot of γ_1 versus α for $\psi = \pi/2$.

Then we investigated this behavior for fixed values of α. Observe in this context that

$$b_1(\alpha_c, \psi) = \frac{3\omega_1(\alpha_c, \beta_c)}{2} = 2a_1(\alpha_c) > a_1(0) \text{ or } \gamma_1(\alpha_c, \psi) = 2$$

and thus, for $\alpha = \alpha_c$ and $\beta > \beta_0(\alpha)$, stripes are stable versus rhombic patterns. We found that for $\alpha \neq \alpha_c$ there existed two ψ-intervals (ψ_m, ψ_ℓ) and (ψ_r, ψ_M), flanking $\psi = \pi/3$, in which both $a_1(\alpha) \pm b_1(\alpha, \varphi) > 0$, or equivalently, $-1 < \gamma_1(\alpha, \varphi) = b_1(\alpha, \varphi)/a_1(\alpha) < 1$, where

$$0 < \psi_m(\alpha) < \psi_\ell(\alpha) < \frac{\pi}{3} < \psi_r(\alpha) < \psi_M(\alpha) \le \frac{\pi}{2},$$

and thus, rhombic patterns of these characteristic angles are stable versus stripes. These values are tabulated in Table 10.4 for some α's flanking $\alpha = \alpha_c = 3$. Note, although this limit exists, such an occurrence again implies, as was pointed out in Chapter 9, that

$$\lim_{\alpha \to \alpha_c} b_1(\alpha, \varphi) \ne b_1(\alpha_c, \varphi)$$

or $b_1(\alpha, \varphi)$ has a jump discontinuity at $\alpha = \alpha_c$ where $a_0(\alpha_c) = 0$.

TABLE 10.4
The ψ-range for stable rhombic patterns versus α.

α	ψ_m	ψ_ℓ	ψ_r	ψ_M
2.90	0.995	1.017	1.076	1.072
3.05	1.022	1.033	1.062	1.127
3.10	0.977	1.018	1.075	1.096
3.20	0.945	0.989	1.103	1.142

We then reconsidered our prediction relevant to the simulations of Firth and Scroggie ([60]) depicted in Fig. 10.9. In order to improve the quantitative accuracy of that prediction, it was necessary for us to adopt the approach employed in Chapter 9 for comparing the theoretical ion-sputtered erosion pattern predictions with the simulation results of Kahng *et al.* ([91]). Recall, from our hexagonal planform analysis of the modified Swift-Hohenberg model equation, that we predicted striped patterns associated with critical point II occuring for $2.73 < \alpha < 3.30$ when $\beta = 8.8$. To understand exactly what was transpiring here in that interval, where only stripes were predicted, we had to examine our rhombic planform results in conjunction with this prediction. From those results, we concluded that for any $\alpha \ne \alpha_c = 3$ all such stripes were unstable with respect to rhombic patterns having characteristic angle of either $\psi \in (\psi_m, \psi_\ell)$ or $\psi \in (\psi_r, \psi_M)$. In particular, for the α-interval over which our hexagonal planform analysis predicted stripes as the only stable pattern, rhombic arrays of this sort are difficult to distinguish from hexagonal lattices since their allowable characteristic angles closely flank $\pi/3 = 1.047$ (see Table 10.4 and Fig. 10.10). Hence, substituting this type of rhombic pattern as the stable morphology for the intervals $2.73 < \alpha < 3$ and $3 < \alpha < 3.30$, then hexagonal or nearly hexagonal arrays can be anticipated for all $\alpha \ne 3$ over the instability region of Table 10.3 when $\beta = 8.8$. Note that V^+ pseudo spots are the predicted pattern for $2.73 < \alpha < 3$, while V^- pseudo honeycombs are the predicted pattern for $3 < \alpha < 3.30$. Finally, from the hexagonal planform results of Fig. 10.8, in conjunction with these rhombic planform morphological stability predictions, we can conclude that the occurrence of stripes requires $\alpha = \alpha_c = 3$ since such patterns are only stable for both planforms at this critical value of α and when $\beta > \beta_c = 8$. Then, employing these conclusions in the α-interval $2.73 < \alpha < 3.30$, our predicted theoretical morphological sequence reduced to

Spots for $2.08 < \alpha < 3$, Stripes for $\alpha = 3$, Honeycombs for $3 < \alpha < 4.28$;

in quantitative agreement with Firth and Scroggie's ([60]) simulation results of Fig 10.9. This theoretical prediction was consistent with the aforementioned fact that those authors were unable to observe the stable coexistence of stripes and hexagons.

Lugiato and Oldano ([129]) formulated their ring cavity laser model of Fig. 10.1 to demonstrate that a nonlinear optical system could produce stationary patterns analogous to the chemical reaction-diffusion Turing structures considered in Chapter 7, but with diffraction in the optical system taking the place of diffusion in the chemical one. Since the experimental motivation for the numerical simulations Firth and Scroggie ([59]) performed on this ring cavity model was the sort of transition between the two types of hexagonal patterns that Ackemann *et al.* ([2]) observed in a sodium vapor cell with a feedback mirror, where optical pattern formation was induced by external pumping, we next compared our theoretical predictions with the latter authors' primary experimental results. Figure 10.13, adapted from those results, represents a plot of a steady-state plane-wave configuration measure $\alpha/(1 + \alpha)$ versus the rate of external pumping Y. In particular, for the portion of this curve having positive slope they found spots occurred at the left-hand end of the patterned interval and honeycombs at the right-hand end, while for the portion having negative slope the locations of occurrence of these two types of hexagonal patterns were interchanged.

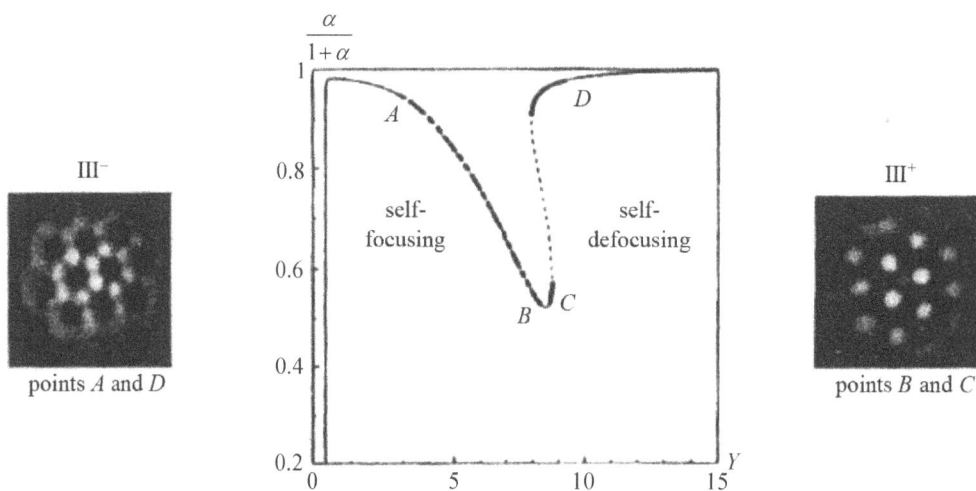

FIGURE 10.13
Adapted from a plot relevant to the transition between spots (III$^+$) and honeycombs (III$^-$) as Y is increased for the experiments of Ackemann *et al.* ([2]) in nonlinear Kerr optical media by Wollkind and Dichone ([284]).

To compare these experimental results with our theoretical predictions, we first re-examined Fig. 10.2, which is a plot of our relationship between α and Y^2 with $\Delta = 0$ and $\theta = -1$ for various fixed values of β. Specifically, in this instance, β_{crit} which was defined implicitly, satisfied $\beta_{crit} = 10.2$ and the three parts of that figure have been plotted for (a) $\beta = 8.8$, (b) $\beta = 10.2$, and (c) $\beta = 12$, respectively. Observe from these plots that

$$\frac{dY^2}{d\alpha} > 0 \text{ for } \beta = 8.8, \quad \frac{dY^2}{d\alpha} \geq 0 \text{ for } \beta = 10.2, \quad \frac{dY^2}{d\alpha} < 0 \text{ for } \beta = 12,$$

over the patterned region which has been denoted by shading in Fig. 10.2. Next, we determined the morphological sequences analogous to that deduced by Dean for $\beta = 8.8$ when

β was assigned the other two values employed here instead. From Fig. 10.8 and Table 10.2, we then obtained the predicted morphological sequences:

$$\text{III}^+ \text{ for } 2.1 < \alpha < 2.5, \text{ II for } 2.5 < \alpha < 3.55,$$
$$\text{III}^- \text{ for } 3.55 < \alpha < 4.25; \text{ when } \beta = 10.2;$$

and

$$\text{III}^+ \text{ for } 2.2 < \alpha < 2.3, \text{ II for } 2.3 < \alpha < 3.8,$$
$$\text{III}^- \text{ for } 3.8 < \alpha < 4.20; \text{ when } \beta = 12.$$

Combining these results, while identifying the sign of the slope of the curve in Fig. 10.13 with that of the reciprocal of $dY^2/d\alpha$ in Fig. 10.2b,c, we deduced that for $\beta = 10.2$ this slope was positive, spots occurred at the left-hand end of the patterned interval and honeycombs at its right-hand end, as the external pump rate Y increased, and for $\beta = 12$ the slope was negative and that behavior was reversed. Further, we could again anticipate that, except for $\alpha = 3$, the α-intervals over which stripes are listed as stable in those sequences actually gave rise to the allowable rhombic patterns predicted by our analysis. These rhombic patterns have a much greater characteristic angle range than did those associated with the sequence when $\beta = 8.8$. Then, for α in this interval, those patterns may be expected to switch from one to another as initial conditions vary over the points in the transverse domain. Since such point-to-point pattern switching is a property of optical turbulence, we identified this interval with that state, which has been denoted by the dashed curves appearing in Fig. 10.13 between the hexagonal arrays occurring at its left- and right-hand ends. This situation is highly reminiscent of the development from a hexagonal state of what Ouyang and Swinney ([162]) called chemical turbulence in their Turing pattern formation study. Hence, in spite of the fact that the optical device employed by Ackemann et al. ([2]) differed from a ring cavity configuration, we demonstrated the potential of our theoretical predictions, for the proper choice of parameter values, to provide a reasonable correlation with their experimental results. Indeed, we selected these β-values in order to improve both the qualitative and quantitative accuracy of that correlation since those selections produced patterned Y-interval lengths, as well as a point of vertical tangency for $\beta = 10.2$ corresponding to $\alpha = 2.73$ where $dY^2/d\alpha = 0$, in agreement with the plot in Fig. 10.13. This whole interpretation of those experimental results of Ackemann et al. ([2]) summarized by that figure with respect to our theoretical predictions most definitely represented the ultimate in Pulling a Rabbit Out of a Hat!

We continued this discussion by considering the other commonly occurring supercritical optical pattern for our problem, namely, squares. To facilitate that discussion, we employed Fig. 10.14, which is a composite of Figs. 10.8a and 10.12, in that it not only replicates Fig. 10.8a but also includes the shaded α-intervals from Fig. 10.12 where square patterns are stable. Examining these intervals, we saw that each contained a point at which $\beta = \beta_2$ took on its maximum. Since stable squares occurred for $\beta > \beta_0$ and stable hexagons for $\beta_{-1} < \beta < \beta_2$, this tended to maximize the extent of the overlap region $\beta_0 < \beta < \beta_2$, where optical bistability existed between those two patterns. We then compared these predictions with the experimental observations described by Ackemann and Lange ([1]) and the numerical simulations performed by Scroggie and Firth ([209]) on their model for this experiment, the latter consisting of the same optical configuration employed by Ackemann et al. ([2]), but with pattern formation being induced by a driving field made up of two orthogonally polarized components instead. Ackemann and Lange ([1]) reported that, when the input

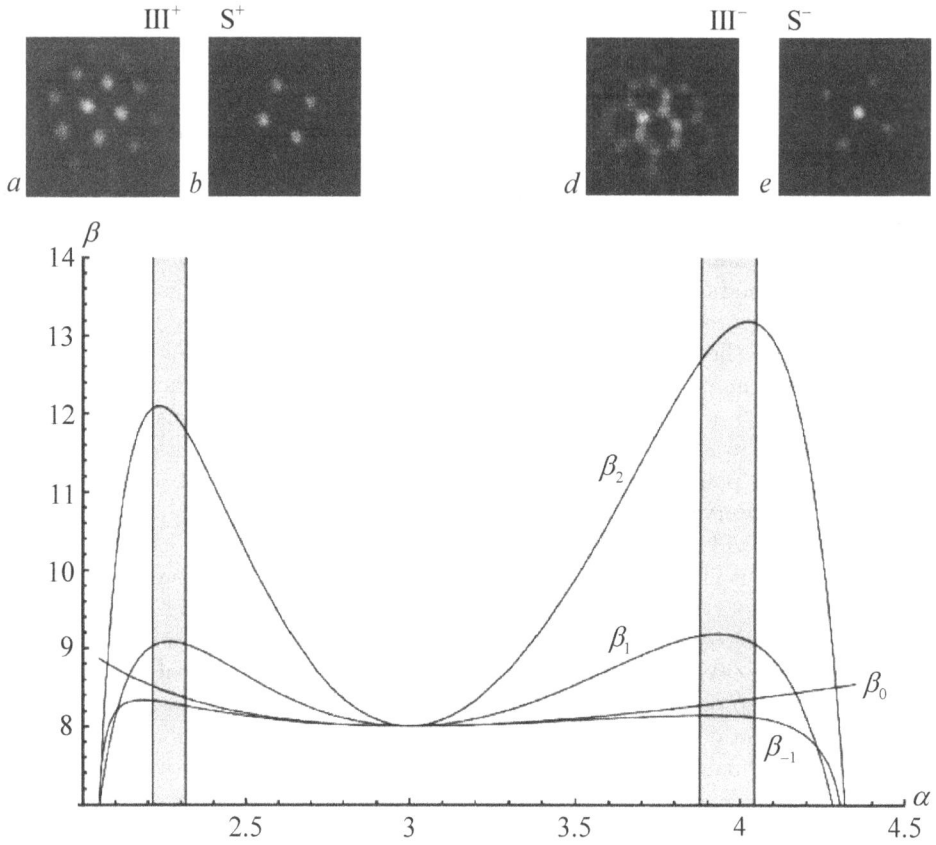

FIGURE 10.14

A composite of Figs. 10.8 and 10.12. Here the hexagonal patterns $III^{\pm}$, represented by the photographic insets labeled a and d, respectively, occur for $\beta_{-1} < \beta < \beta_2$ in the shaded intervals, while the square patterns $S^{\pm}$, represented by those labeled b and e, respectively, occur for $\beta > \beta_0$ in these same intervals where the "+" patterns lie to the left and the "−" ones to the right, consistent with our predictions of Figs. 10.8b,c and the threshold images of Fig. 10.11.

driving field was plane polarized, squares were observed of both $S^{\pm}$-types while, when that polarization was elliptical, a transition occurred to hexagonal patterns with III^{+} spots for positive ellipticity and III^{-} honeycombs, for negative. We reproduced their plane polarized results by taking $\beta > \beta_2$ and their elliptical ones by taking $\beta_0 < \beta < \beta_2$ in the shaded intervals of Fig. 10.14, with the left-hand one corresponding to positive ellipticity and the right-hand one to negative. Scroggie and Firth ([209]) performed their numerical simulations by using the results of a de facto weakly nonlinear stability analysis to aid them in determining the parameter range favoring pattern formation. They reported explicit expressions for a_1, b_1, and a_0 but, since these formulae depended on a q_c^2 that was defined by an implicit transcendental relation, were unable to obtain closed-form conditions in their parameter space under which different patterns could be found and hence did not bother to report a_2 due to its complexity. For the case of plane polarization, Scroggie and Firth ([209]) plotted a_1 and $b_1(\psi)$ versus ψ with appropriate values chosen for their other parameter values. From this plot, they deduced that $a_1 > b_1(\psi) > 0$ whenever ψ was in the narrow intervals $\psi_m < \psi < \psi_\ell$ or $\psi_r < \psi < \psi_M$ where $48° \in (\psi_m, \psi_\ell)$ and $90° \in (\psi_r, \psi_M)$. Then, Scroggie and Firth ([209]) ran simulations for these parameter values and found that initial conditions determined which of those two types of rhombic patterns would be observed. The same thing is true for our problem with $\psi = \pi/6$ and $\psi = \pi/2$, since both can exist and be stable in the shaded intervals of Fig. 10.14 when $\beta > \beta_0$, given that $b_1(\alpha, \pi/6) = b_1(\alpha, \pi/2)$. For the case of elliptical polarization with the model parameter values selected appropriately, their numerical simulations produced either type of hexagonal or square pattern. Hence, Scroggie and Firth ([209]) demonstrated the existence of some particular sets of parameter values that allowed them to generate simulated patterns consistent with the experimental morphologies reported by Ackemann and Lange ([1]) for the same situations and hence, with our corresponding results summarized in Fig. 10.14, as well. This correlation can be considered, that, of Pulling another Rabbit Out of the same Hat!!

So far, we had been concentrating on the supercritical behavior of our model system and thus left for last a discussion of its subcritical behavior. We began by considering the onset of subcritical hexagons. From Fig. 10.8 and Table 10.2, for a fixed value of $\alpha = \alpha_0$, such that $\alpha_0 \in (\alpha_3, \alpha_c) \cup (\alpha_c, \alpha_4)$ and $\beta < \beta_1$, we obtained the morphological sequence

$$\text{I for } \beta < \beta_{-1}, \text{ I/III for } \beta_{-1} < \beta < \beta_0, \text{ III for } \beta_0 < \beta < \beta_1,$$

where III represents III^{+} or III^{-}, depending on whether $\alpha_0 \in (\alpha_3, \alpha_c)$ or $\alpha_0 \in (\alpha_c, \alpha_4)$, respectively, and here we are again, as in Table 7.1, using the notation B/C to indicate the bistability between patterns B and C. Employing a morphological persistence argument, we concluded that the transition from a plane-wave configuration to hexagons would occur supercritically at $\beta = \beta_0$ for increasing β, but that the reverse transition from hexagons to a plane-wave configuration would occur subcritically at $\beta = \beta_{-1}$ for decreasing β, thus resulting in a region of hysteresis. Although Firth and Scroggie ([60]) saw the formation of III^{+} or III^{-} patterns close to threshold when varying β, while holding α fixed, they did not mention the occurrence of such hysteresis possibly because of the difficulty in distinguishing numerically between β_{-1} and β_0 for this instance (see Fig. 10.8). There does, however, exist a general theoretical result relevant to our prediction. Aranson et al. ([7]) analyzed a modified Swift–Hohenberg equation of the same form as ours but with the idealized parameter values of

$$\sigma_R = -1, \ \omega_0 = -\beta, \ \omega_1 = 1, \ \sigma_I = \chi^2$$

and numerically found a subcritical bifurcation that gave rise to localized solutions in two spatial dimensions over a range of real β. Here, β is the modification parameter which when zero reduces it to an ordinary Swift-Hohenberg equation in normal form.

We finally turned to the possibility of occurrence of one-dimensional stable stripes even when $a_1 < 0$. Firth and Scroggie ([60]) asserted that the sign of ω_1 determined whether such one-dimensional bifurcations occurred supercritically ($a_1 > 0$) or subcritically ($a_1 < 0$) for their system with $\Delta = 0$ and $\theta = -1$. From our supercriticality condition of $\alpha_1 < \alpha < \alpha_2$ where $\alpha_1 = 2.143$ and $\alpha_2 = 4.167$, we see that this assertion is different than ours since $\omega_1 > 0$ whenever

$$\alpha^- < \alpha < \alpha^+ \text{ where } \alpha^\pm = 3 \pm 2\sqrt{2} \text{ or } \alpha^- \cong 0.172 \text{ and } \alpha^+ \cong 5.828.$$

Firth and Scroggie ([60]) then performed a numerical simulation of that system for $\alpha = 5$ and $\beta = 10$, obtaining stable stripes. That choice of parameters yields a $\sigma > 0$ since $\beta_0(5) = 9$ and an $a_1 < 0$ since $\alpha_2 = 4.167 < 5$. This is precisely the unstable case for the amplitude equation truncated through terms of third order, to be treated in Chapter 15, that can give rise to a stable re-equilibrated solution should a term of the proper sign be retained at fifth-order. For our model, such a solution would represent a striped pattern. This behavior is reminiscent of that occurring for the analysis and simulation performed by Geddes *et al.* ([65]) on a nonlinear optical model system involving a Kerr medium, where pattern formation was driven by a pump field. Although squares did not saturate theoretically at third-order, these authors nonetheless generated stable square patterns by numerical simulation and suggested such saturation would occur theoretically at quintic-order (see Chapter 14).

We closed with a few additional observations about this problem. Firth and Scroggie ([60]) were concerned with the spontaneous formation of stationary patterns relevant to optical bistability rather than the occurrence of pulsed, oscillatory, or chaotic temporal or spatiotemporal patterns, as is often the major focus for problems in nonlinear optics (Moloney and Newell, [148]). For their $\Delta = 0$ analysis, they pointed out that, since their modified Swift-Hohenberg equation was based on a perturbation expansion, any results from it would only be strictly valid in the vicinity of the marginal curve $\beta = \beta_0(\alpha)$. Thus, although one could not then formally predict a spot-stripe-honeycomb transition at $\alpha = \alpha_c = 3$, when α was increased for constant $\beta > \beta_c$, Firth and Scroggie ([60]) stated that this was never-the-less essentially the observed behavior from their simulations. Besides these $\Delta = 0$ simulations already described in this chapter, they also numerically integrated their basic system with $\Delta \neq 0$ and obtained results broadly similar to those with $\Delta = 0$. For example, upon increasing α with β held constant, Firth and Scroggie ([60]) saw the spot-stripe-honeycomb transition now occurring at a minimum threshold $\alpha = \alpha_c(\Delta)$ where $\alpha_c(0) = \alpha_c$. That being the case, the range of validity of our results was much wider than could be expected otherwise, an extrapolation implicitly exploited earlier.

In conclusion, our theoretical predictions, when compared with relevant numerical simulations and experimental evidence from existing nonlinear optical pattern formation studies, provided consistency in the former case and very good agreement when parameter values were chosen appropriately in the latter one. Hence, this laser-injected atomic sodium optical ring cavity problem, involving a single evolution equation for the real part of the intracavity field, was compatible with our long-range aim of employing the simplest reasonable natural science models that preserve the essential features of pattern formation and are still consistent with observation.

Again, once this research had been completed, it was time as usual to disseminate those results. Since that research actually was a composite of Dean's and Francisco's dissertations, which followed each other sequentially, this dissemination had to be delayed until both were finished. We eventually published those results in the *IMA (Institute of Mathematics and* its Applications) *Journal of Applied Mathematics* in an article entitled "Non-linear stabil-ity analyses of optical pattern formation in an atomic sodium vapour ring cavity" and gave poster presentations of them at a WSU Showcase and an American Institute of Physics Meeting, all during the year of 2008.

We have deferred until now a description of how Dean and Francisco became my Ph.D. students. Dean started by working with a geometer in the department who sent him to the library to find himself a thesis problem. Although my two WSU Ph.D. students from outside the Mathematics Department, Jesse Logan (see Chapter 4) from Entomolgy and Iwan Alexander (see Chapter 5) from Geology, actually brought their own thesis problems with them, I have provided the problems for all of my other Ph.D. students. Dean in par-ticular was very unimpressed with this mode of choosing his dissertation topic. Taking my graduate modeling class at the time, he came to me and asked if I had a thesis problem for him. That was when I decided to have Dean do the hexagonal planform nonlinear optical pattern formation analysis, as related earlier. Upon completion of his Ph.D., Dean decided to return to his high school in the Bronx suburb of New York City to teach mathematics. Francisco started by working with one of our other applied mathematical modelers. On leave from the faculty of the Guadalajara campus of the Technological University of Monterrey in Mexico and with a wife running a restaurant in its home city, he wanted to finish his Ph.D. thesis as soon as possible and return home. When Francisco's original advisor told him that would take at least two more years, he came to me after discussing the matter with some other doctoral graduate students and asked if I had a Ph.D. problem that could be completed in at most a year's time. Given my experience with Chontita, I knew that a rhombic planform analysis would be the perfect vehicle for this sort of thesis problem and suggested he work with me on such an analysis of the modified Swift-Hohenberg model evolution equation for nonlinear optical pattern formation. I told him that my intention upon the completion of that thesis was to publish a joint paper with both him and Dean as co-authors, which indeed came to pass. It has always been my policy to tell Ph.D. students in advance exactly what their dissertations would include, unlike many advisors who leave this as an open-ended decision they will determine at some later date.

Indeed, we finished his thesis research with plenty of time to spare, after which he took the preliminary doctoral examination. Unfortunately, given what I referred to as the J. Iwan D. Alexander graduate school rule in Chapter 5, Francisco had to wait 90 more days before he could defend that dissertation. In order to occupy his time, while waiting those three extra months, he asked me to give him some more researech to do. Since I had already made arrangements with Yongwimon Lenbury to work with her Mahidol University student Nichaphat Boonkorkuea, who would arrive that fall, on a evolution equation for nonlinear vegetative pattern formation in an arid environment (see Chapter 11), which required both hexagonal and rhombic planform analyses to be completed in a year's time, it seemed obvious to me that, given Francisco's Ph.D. experience, the rhombic planform analysis of this problem would be tailor-made for him. So I had Francisco perform it, with Nichaphat performing the hexagonal planform analysis, as well as checking his work for any errors, just as Francisco had checked Dean's work in a similar manner. Although by then he had returned to Guadalajara, the writing of the *IMA Journal of Applied Mathematics* paper (Wollkind *et al.*, [279]) was a joint effort between the two of us, with Francisco generating all twenty-three of the required figures that ultimately appeared in it. During the review process we were asked to replace our plots of Fig. 10.8 by ones explicitly identifying the morphological stability regions in them and Francisco produced those as well (see Fig. 10.15). Just to prove Segel's old adage about typographocal errors, this paper had a glaring one in the third basic equation of its model system (unnoticed until the preparation of Chapter 17 of Wollkind

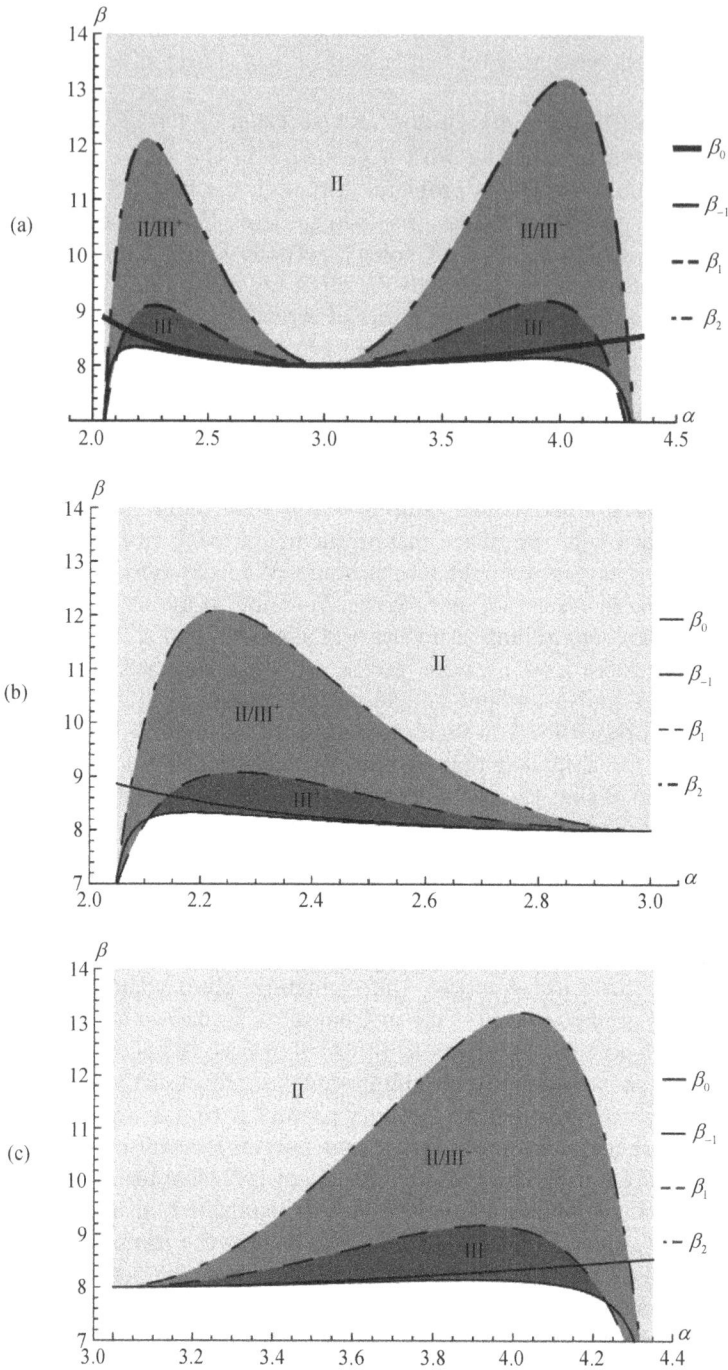

FIGURE 10.15
Plots equivalent to those of Fig. 10.8 but with its morphological stability regions identified explicitly.

and Dichone, [284]) which appears as

$$\varepsilon_2 \frac{\partial f}{\partial t} = f - \frac{X^* P + X P^*}{2},$$

instead of its proper form where the "f" term is replaced by "$1 - f$," an obvious mistake since, representing a population difference *decay* rate, its "f" term must be preceded by a *minus* sign.

11

Evolution Equation Phenomenon IV: Nonlinear Vegetative Pattern Formation: Hexagonal and Rhombic Planform Nonlinear Stability Analyses

This phenomenon served as the thesis topic for three of my last four Ph.D. students: Namely, Nichaphat Boonkorkuea and Inthira Chaiya (see Chapter 14) from Mahidol University and my co-author Bonni Dichone (nee Kealy) from WSU. Of those four, only Richard Cangelosi from WSU was exempt and his mussel bed Turing pattern formation problem was so closely related to Bonni's that they have been grouped together as diffusive versus differential flow instabilities in Chapters 12 and 13, respectively. Nichiphat, Bonni, and Inthira each analyzed a different model for nonlinear vegetative pattern formation in an arid environment. Although I originally became aware of this phenomenon when Rene Lefever presented a paper on one-dimensional tiger bush patterns at the IMA workshop in 1998, as related in Chapter 7, the review article by Max Rietkerk, Steffan Dekker, Peter de Ruiter, and Johan van de Koppel ([187]) entitled "Self-organized patchiness and catastrophic shifts in ecosystems" that appeared in the volume 305 number 5692 issue of *Science*, on pages 1926-1929, was what really got me interested in it. This review contained a bifurcation diagram of equilibrium dryland vegetative density versus resource input which showed transitions from a homogeneous to various self-organized patchy states that included spotted, bicontinuous, and gapped patterns. Always on the lookout for such arrays (again see Fig. 1.2), this occurrence piqued my interest and I decided to use my methods of theoretical analysis on some of the reviewed nonlinear models of dryland vegetative pattern formation to compare predictions from the former with numerical simulations from the latter.

When Yongwimon Lenbury arranged to have Nichaphat Boonkorkuea spend a year doing her Ph.D. dissertation research with me, I picked out for that thesis problem the spatio-temporal model evolution equation proposed by Lefever *et al.* ([110]) and Lejeune *et al.* ([113]) to investigate nonlinear vegetative patterns in an arid isotropic environment. They modeled pattern formation in this instance by a spatio-temporal partial differential evolution equation with a logistic source term defined on an unbounded flat domain, describing the total plant biomass per unit area divided by its carrying capacity that, under a weak gradient-low density truncation, reduced to fourth-order in its spatial variables. Then, for appropriate conditions, patterns could be generated in an initially homogeneous environment by the balance between the effects of short-range facilitation and long-range competition. As indicated at the end of the last chapter, we wished to perform both hexagonal and rhombic planform weakly nonlinear stability analyses on that truncated fourth-order partial differential logistic evolution equation and then compare these theoretical predictions with existing numerical simulations and relevant observational evidence. In particular, this specific nondimensional propagator-inhibitor logistic equation was of the form

$$\frac{\partial \rho}{\partial t} = f(\rho) + \frac{1}{2}(\beta - \rho)\nabla^2 \rho - \frac{1}{8}\rho\nabla^4 \rho$$

DOI: 10.1201/9781003195603-11

where

$$f(\rho) = (1 - \mu)\rho + (\Lambda - 1)\rho^2 - \rho^3 \text{ and } \beta = L^2.$$

Here, $\rho = \rho(x, y, t) \equiv$ dimensionless total plant biomass density with $(x, y) \equiv$ a transverse Cartesian coordinate system and $t \equiv$ time; $\boldsymbol{\nabla} \equiv (\partial/\partial x, \partial/\partial y)$; $\nabla^2 = \boldsymbol{\nabla} \cdot \boldsymbol{\nabla}$; and $\nabla^4 \equiv (\nabla^2)^2$; while L and Λ were the fascilitation-to-competition range and interaction ratios, respectively; and μ was the mortality-to-growth rate ratio which served as a measure of the environment's aridity. We noted that in this equation time and space had been scaled with the inverse of the vegetative growth rate, which corresponds roughly to the period for the plants to achieve adult size, and with the interplant competition length, which is approximated by the radius of the superficial root system of the dominant plants, respectively. Observe that this development is consistent with the scaling laws

$$\nabla^2 \sim O(\varepsilon^{1/2}); \; \rho, \; \Lambda - 1 \sim O(\varepsilon); \; \beta - \rho \sim O(\varepsilon^{3/2}); \; \frac{\partial}{\partial t}, \; 1 - \mu \sim O(\varepsilon^2);$$

for which each term in that equation is of $O(\varepsilon^3)$ where the parameter ε can be identified with the mean biomass density α (defined below), that is relatively small for a dryland (arid or semiarid) environment. These scaling laws differed from those adopted by Lefever *et al.* ([110]) and that discrepancy will be discussed in more detail at the end of this chapter.

We first sought a uniform stationary solution, $\rho \equiv \rho_0$, of this equation satisfying $f(\rho_0) = 0$ and found that either

$$\rho_0 = 0 \text{ or } \rho_0 = \rho_0^{\pm} = \frac{1}{2}[\Lambda - 1 \pm \sqrt{(\Lambda - 1)^2 + 4(1 - \mu)}];$$

where $\rho_0 = 0$, which always existed, represented bare ground, while $\rho_0^{\pm}$, when real and positive, represented spatially homogeneous plant distributions. If $0 \leq \Lambda, \mu \leq 1$, then $\rho_0^+ \geq 0$, $\rho_0^- \leq 0$ and hence, only ρ_0^+ was biologically meaningful, while if $\Lambda > 1$ then $\rho_0^+ > 0$ for $0 \leq \mu \leq \mu^*$ and $\rho_0^- > 0$ for $1 \leq \mu \leq \mu^*$, where $\mu^* = 1 + (\Lambda - 1)^2/4$. We next performed linear stability analyses of these distributions to homogeneous disturbances by considering solutions of the form

$$\rho(x, y, t) = \rho_0 + \varepsilon_1 \rho_1(t) + O(\varepsilon_1^2) \text{ with } f(\rho_0) = 0 \text{ and } |\varepsilon_1| \ll 1,$$

which yielded the perturbation ordinary differential equation

$$\frac{d\rho_1}{dt}(t) = f'(\rho_0)\rho_1(t) \text{ where } f'(\rho_0) = 1 - \mu + 2(\Lambda - 1)\rho_0 - 3\rho_0^2.$$

Since $f'(0) = 1 - \mu$, $f'(\rho_0^{\pm}) = \rho_0^{\pm}(\Lambda - 1 - 2\rho_0^{\pm})$; and there was linear stability for that equation when $f'(\rho_0) < 0$ and instability, when $f'(\rho_0) > 0$; we concluded that $\rho_0 = 0$ was linearly stable for $\mu > 1$ and unstable for $0 \leq \mu < 1$; $\rho_0 = \rho_0^+$ was linearly stable everywhere it existed; and $\rho_0 = \rho_0^-$, linearly unstable everywhere it existed. We were interested in determining conditions under which a nonzero homogeneous distribution of vegetation that was linearly stable in the absence of spatial effects could become unstable to heterogeneous perturbations. Since this was reminiscent of the diffusive instabilities that occurred during chemical Turing pattern formation (see Chapter 7), we referred to them as ecological Turing instability patterns. Thus, only the solution $\rho_0^+ \equiv \alpha$ needed to be considered. We then restricted our attention to the parameter range $0 < \Lambda \leq 1$ and $0 \leq \mu < 1$ to eliminate the possibility of bistability between that homogeneous distribution and the zero state (Lejeune *et al.*, [113]); adopted the transverse far-field boundary condition that ρ remained bounded as $x^2 + y^2 \to \infty$, implicitly satisfied by $\rho \equiv \alpha$; and deduced, from $f(\alpha) = 0$ and $\alpha > 0$, the relationship $\mu = 1 - (1 - \Lambda)\alpha - \alpha^2$, which for the special representative case of $\Lambda = 1$ reduced to $\mu = 1 - \alpha^2$, relating the parameter μ to α.

As a necessary prelude to the two-dimensional hexagonal and rhombic planform nonlinear stability analyses of this homogeneous solution to our propagator-inhibitor logistic equation, we performed an analogous one-dimensional longitudinal planform analysis by considering

$$\rho(x, y, t) \sim \alpha + A_1(t) \cos(qx) + A_1^2(t)[\rho_{20} + \rho_{22} \cos(2qx)]$$
$$+ A_1^3(t)[\rho_{31} \cos(qx) + \rho_{33} \cos(3qx)]$$

where the amplitude function $A_1(t)$ satisfied the Landau equation

$$\frac{dA_1(t)}{dt} \sim \sigma A_1(t) - a_1 A_1^3(t)$$

and $q = 2\pi/\lambda$, λ being the wavelength of the class of spatially periodic perturbations under investigation. Substituting that solution into our evolution equation, we obtained a sequence of problems, one for each pair of m and n values which corresponded to a term of the form $A_1^n(t) \cos(mqx)$ appearing in that expansion. Then, the $n = m = 0$ problem was satisfied identically by virtue of $f(\alpha) = 0$, while the linear problem for $n = m = 1$ yielded the secular equation (here, the Λ-dependence of various quantities will be suppressed for ease of exposition)

$$\sigma = \alpha(\Lambda - 1 - 2\alpha) + \frac{(\alpha - \beta)q^2}{2} - \frac{\alpha q^4}{8},$$

which is a parabola in the q^2-σ plane. If $\beta \geq \alpha$, then $\sigma < 0$ for our parameter range of interest, while if $\beta < \alpha$ then this parabola had a maximum value at its vertex (q_c^2, σ_c) with

$$q_c^2(\alpha, \beta) = 2\left(1 - \frac{\beta}{\alpha}\right), \quad \sigma_c(\alpha, \beta) = \frac{2D(\alpha, \beta)}{\alpha}$$

where the discriminant of this quadratic in q^2 is given by

$$D(\alpha, \beta) = \frac{(\alpha - \beta)^2}{4} - \frac{\alpha^2(2\alpha + 1 - \Lambda)}{2}.$$

If, in addition $D(\alpha, \beta) > 0$, then $0 < \sigma < \sigma_c(\alpha, \beta)$ for $q_1^2 < q^2 < q_2^2$ where

$$q_{1,2}^2(\alpha, \beta) = q_c^2(\alpha, \beta) \pm \frac{4D^{1/2}(\alpha, \beta)}{\alpha}.$$

This scenario is plotted in Fig. 11.1 for the special case of $\Lambda = 1$ when $\alpha = 0.2$ and $\beta = 0.02$. Hence, we equated the q and σ in our expansion to $q \equiv q_c(\alpha, \beta)$ and $\sigma = \sigma_c(\alpha, \beta)$. Then,

$$\sigma_c < 0 \text{ for } \beta > \beta_c, \ \sigma_c = 0 \text{ for } \beta = \beta_c, \ \sigma_c > 0 \text{ for } \beta < \beta_c,$$

where $\beta_c = \beta_c(\alpha)$ is defined implicitly by $D(\alpha, \beta_c) = 0$ or explicitly by

$$\beta_c(\alpha) = \alpha[1 - \sqrt{2(2\alpha + 1 - \Lambda)}] \text{ for } 0 < \alpha < \frac{2\Lambda - 1}{4}.$$

Therefore, the homogeneous solution was linearly stable for $\beta > \beta_c(\alpha)$, neutrally stable for $\beta = \beta_c(\alpha)$, and unstable for $0 < \beta < \beta_c(\alpha)$. Thus, $\beta = \beta_c(\alpha)$ served as its marginal sability curve in the α-β plane. This situation is plotted in Fig. 11.2 for the special case of $\Lambda = 1$. In that instance, $\beta_c(\alpha) = \alpha(1 - 2\alpha^{1/2})$ for $0 < \alpha < 1/4$, where $\beta_c(0) = \beta_c(1/4) = 0$ while, since $\beta_c'(\alpha) = 1 - 3\alpha^{1/2}$, $\beta_c'(\alpha_0) = 0$ implies $\alpha_0 = 1/9 = 0.111$ and $\beta_c(\alpha_0) = 1/27 = 0.037$.

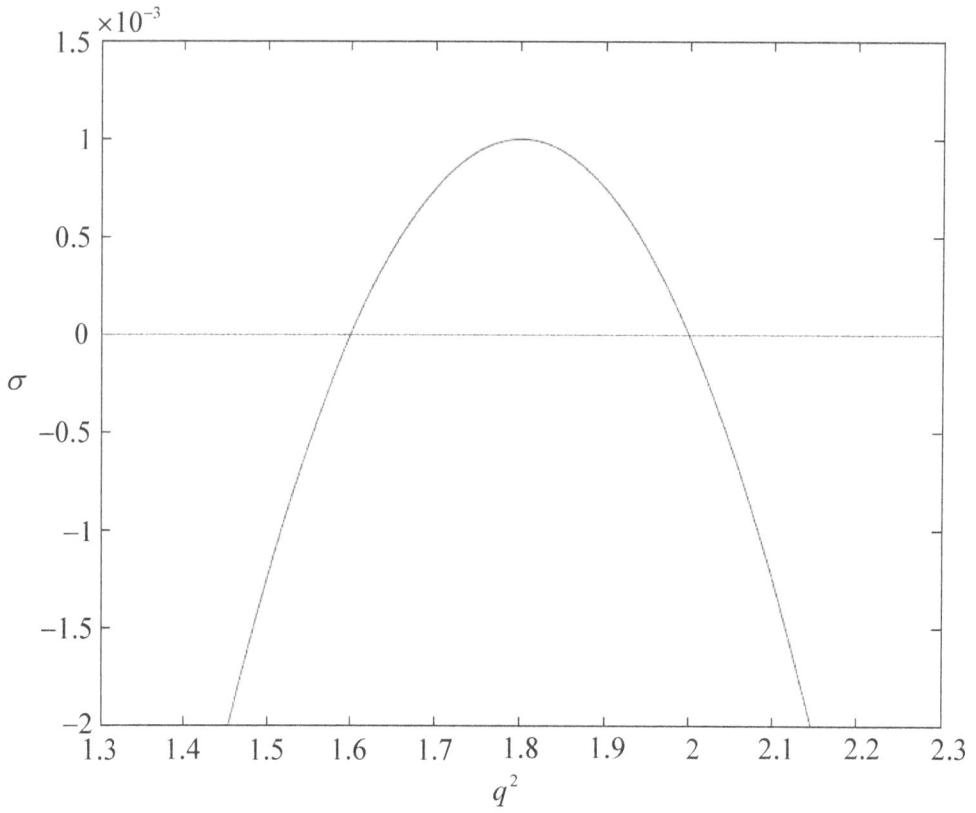

FIGURE 11.1
Plot of the parabolic secular equation in the q^2-σ plane for $\Lambda = 1$, $\alpha = 0.2$, and $\beta = 0.02$ with vertex (q_c^2, σ_c).

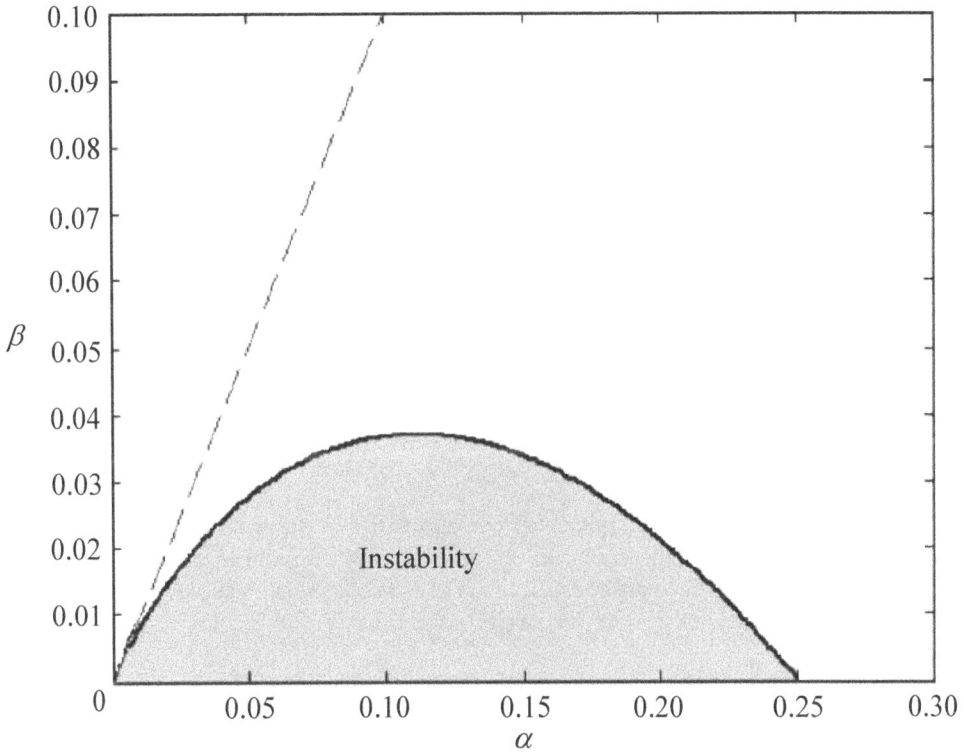

FIGURE 11.2
Plot of the marginal stability curve $\beta = \beta_c(\alpha)$ in the α-β plane for $\Lambda = 1$ on which the maximum growth rate $\sigma_c(\alpha, \beta_c) = 0$. That locus is denoted by the solid curve while the dashed line denotes $\beta = \alpha$.

Here, we restricted our analysis to the critical wavenumber alone. In the discussions at the end of this chapter we present the results of those side-band instability analyses, originally introduced contemporaneously by Segel ([213]) and Newell and Whitehead ([156]), that examines the consequences of investigating other wavenumbers in the side-band centered about that critical wavenumber as well (see Chapters 2 and 12).

Continuing our description of the results of this one-dimensional expansion procedue, the second-order problems corresponding to $n = 2$ and $m = 0$ or 2 could be solved in a straightforward manner to yield

$$\rho_{20} = \frac{(\Lambda - 1)/2 - 3\alpha/2 + q_c^2/4 - q_c^4/16}{\sigma_c + (\alpha - \beta)q_c^2/2 - \alpha q_c^4/8}$$

and

$$\rho_{22} = \frac{(\Lambda - 1)/2 - 3\alpha/2 + q_c^2/4 - q_c^4/16}{\sigma_c + 3(\beta - \alpha)q_c^2/2 + 15\alpha q_c^4/8}.$$

Although there were also two third-order problems, it was permissible, as usual, to concentrate our attention exclusively on the one corresponding to $n = 3$ and $m = 1$ containing the Landau coefficient a_1 for the Fredholm alternative method of solvability we used. That problem could be represented by $a_1 - 2\sigma_c(\beta)\rho_{31}(\beta) = r_{31}(\beta)$ where

$$r_{31}(\beta) = \frac{3}{4} + (3\alpha + 1 - \Lambda)[2\rho_{20}(\beta) + \rho_{22}(\beta)] - \frac{[\rho_{20}(\beta) + 5\rho_{22}(\beta)]q_c^2(\beta)}{2}$$
$$+ \frac{[\rho_{20}(\beta) + 17\rho_{22}(\beta)/2]q_c^2(\beta)}{8}.$$

Here, we only denoted the β-dependence of the quantities in question for ease of exposition. Then, taking the limit of this equation as $\beta \to \beta_c(\alpha)$, employing our previous results, and assuming the requisite continuity at $\beta = \beta_c$, we obtained the solvability condition represented symbolically as $a_1 = r_{31}(\beta_c)$ or given explicitly by

$$a_1 = \frac{1}{36\alpha(1 + 2\alpha - \Lambda)}[-866\alpha^2 + \alpha(-798 + 706\Lambda) - 192$$
$$+ 299\sqrt{2\alpha^2(1 + 2\alpha - \Lambda)} + 388\Lambda - 146\Lambda^2 - 119(\Lambda - 1)\sqrt{2(1 + 2\alpha - \Lambda)}]$$
$$= a_1(\alpha).$$

As demonstrated in Chapters 7, 8, and 10, the stability behavior of the Landau equation truncated through terms of third-order and thus, the pattern formation aspect of our model system was crucially dependent upon the sign of a_1. Hence, in order to determine that behavior, we examined this formula for a_1. Toward that end, we plotted $a_1(\alpha)$ versus α in Fig. 11.3 with $\Lambda = 1$. From this figure, we observed that a_1 had two zeroes at $\alpha = \alpha_{1,2}$ such that

$$a_1 < 0 \text{ for } 0 < \alpha < \alpha_1 \text{ or } \alpha > \alpha_2, \ a_1 > 0 \text{ for } \alpha_1 < \alpha < \alpha_2;$$

where

$$\alpha_1 = 0.0535 \text{ and } \alpha_2 = 0.2108.$$

Again, given these conditions for $\sigma_c(\alpha, \beta)$ and $a_1(\alpha)$, we noted that the Landau amplitude function $A_1(t)$ underwent a pitchfork bifurcation at $\beta = \beta_c(\alpha)$ when $\alpha_1 < \alpha < \alpha_2$, from which we concluded that

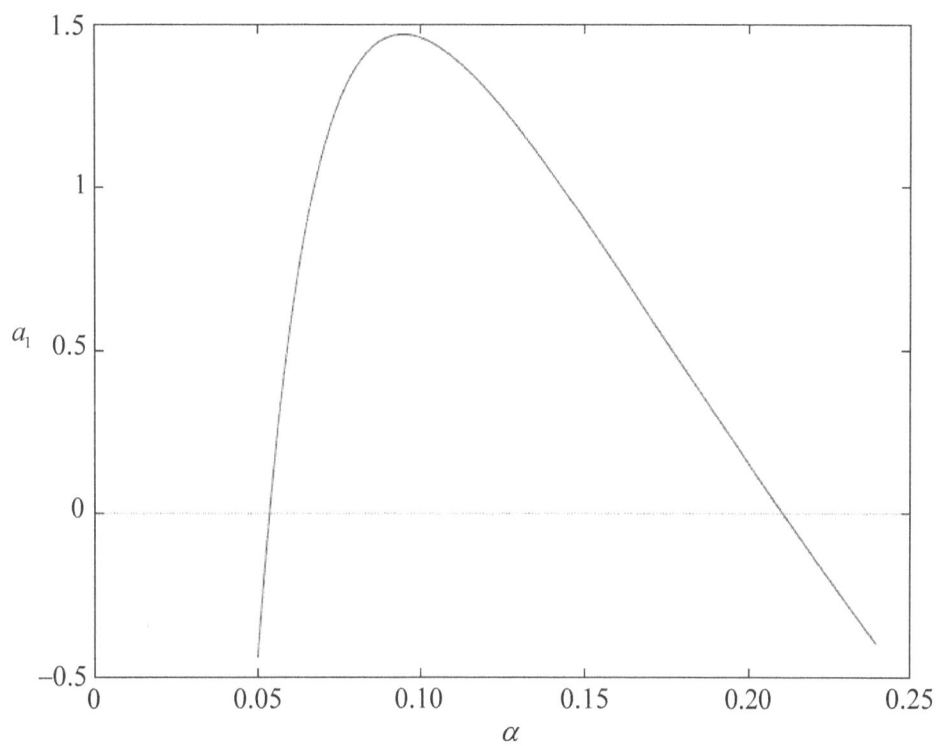

FIGURE 11.3
Plot of $a_1 = a_1(\alpha)$ versus α for $\Lambda = 1$.

(i) For $\beta > \beta_c(\alpha)$ and $\alpha_1 < \alpha < \alpha_2$, the undisturbed state $A_1 = 0$ was stable since $\sigma_c < 0$ and $a_1 > 0$ yielding a uniform homogeneous distribution since $\rho(x, y, t) \to \alpha$ as $t \to \infty$.

(ii) For $0 < \beta < \beta_c(\alpha)$ and $\alpha_1 < \alpha < \alpha_2$, $A_1 = A_e = \sqrt{\sigma_c/a_1}$ was stable since σ_c, $a_1 > 0$ yielding a periodic one-dimensional pattern consisting of stationary parallel stripes

$$\rho(x, y, t) \to \rho_e(x) \sim \alpha + A_e \cos\left(\frac{2\pi x}{\lambda_c}\right) \text{ as } t \to \infty$$

with characteristic wavelength of

$$\lambda_c = \frac{2\pi}{q_c},$$

which is plotted in the x-ρ plane of Fig. 11.4 for $\Lambda = 1$, $\alpha = 0.12$, and $\beta = 0.02$. These supercritical stripes are represented in the contour plot of Fig. 11.5 where, after Lejeune *et al.* ([113]), regions of higher density ($\rho > \alpha$, in this case) appear dark (green) and those of lower density ($\rho < \alpha$), light (tan). Here, the axes are being measured in units of λ_c. When $\alpha < \alpha_1$ or $\alpha > \alpha_2$, that bifurcation was subcritical. Recall that the consequences of such subcritical behavior will be discussed in Chapter 15.

In what follows, we again concentrated on the behavior of our propagator-inhibitor logistic equation in its supercritical regime where $\alpha_1 < \alpha < \alpha_2$.

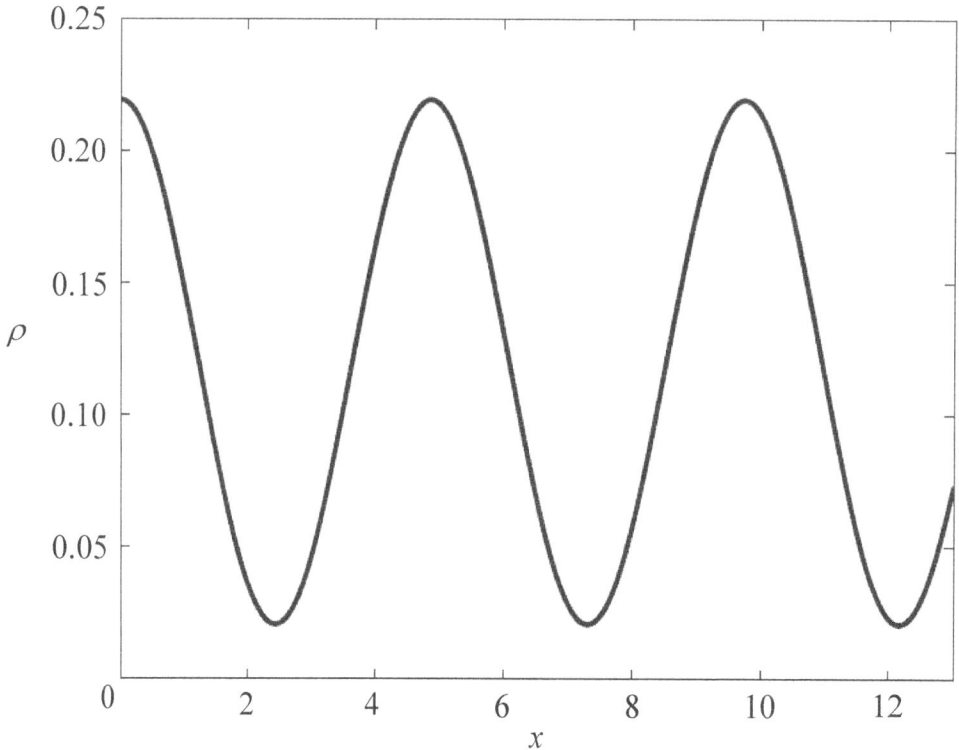

FIGURE 11.4
Plot of the equilibrium density $\rho_e(x)$ versus x for $\Lambda = 1$, $\alpha = 0.12$, and $\beta = 0.02$.

In order to investigate the possibility of occurrence for this evolution equation of the type of two-dimensional vegetative patterns described in Rietkerk *et al.* ([187]), we considered a hexagonal planform solution of it that to lowest order satisfied

$$\rho(x, y, t) \sim \alpha + A_1(t) \cos[q_c x + \varphi_1(t)] + A_2(t) \cos\left[\frac{q_c(x - \sqrt{3}y)}{2} - \varphi_2(t)\right]$$

$$+ A_3(t) \cos\left[\frac{q_c(x + \sqrt{3}y)}{2} - \varphi_3(t)\right]$$

where, for $(j, k, \ell) \equiv$ even permutation of (1,2,3),

$$\frac{dA_j}{dt} \sim \sigma A_j - 4a_0 A_k A_\ell \cos(\varphi_j + \varphi_k + \varphi_\ell) - A_j[a_1 A_j^2 + 2a_2(A_k^2 + A_\ell^2)]$$

$$A_j \frac{d\varphi_j}{dt} \sim 4a_0 A_k A_\ell \sin(\varphi_j + \varphi_k + \varphi_\ell).$$

The nonlinear stability behavior of these amplitude-phase equations depended, as usual, only on the values of their growth rate and Landau coefficients. We determined this growth rate and these Landau coefficients by proceeding in the identical manner to that employed in the previous three chapters. Then, denoting the coefficient of each higher-order term in the relevant ρ-expansion of the form $A_1^n(t) B_1^j(t) \cos(Mq_c x/2) \cos(k\sqrt{3}q_c y/2)$ by ρ_{njMk}, we found that

$$\rho_{n0M0} = \rho_{nm} \text{ for } M = 2m, \ \sigma = \sigma_c(\alpha, \beta), \ a_1 = a_1(\alpha),$$

as defined earlier, and

$$\rho_{1131} = \frac{\Lambda - 1 - 3\alpha + q_c^2/2 - q_c^4/8}{\sigma_c + (\beta - \alpha)q_c^2 + \alpha q_c^4};$$

while, in particular, the other two Landau coefficients satisfied

$$4a_0 - \sigma_c \rho_{1111} = 1 - \Lambda + 3\alpha - \frac{q_c^2}{2} + \frac{q_c^4}{8} = r_{1111}(\alpha, \beta),$$

$$a_2 - \sigma_c \rho_{2111} + \left(4a_0 + \Lambda - 1 - 3\alpha + \frac{q_c^2}{2} - \frac{q_c^4}{8}\right) \frac{\rho_{1111}}{2}$$

$$= \left(1 - \Lambda - 3\alpha - q_c^2 + \frac{5q_c^2}{8}\right) \frac{\rho_{1131}}{2} + \left(1 - \Lambda + 3\alpha - \frac{q_c^2}{4} + \frac{q_c^4}{16}\right) \rho_{2000} + \frac{3}{4}.$$

Finally, taking the limit of these equations as $\beta \to \beta_c(\alpha)$ and employing our previous results, we obtained the solvability conditions

$$a_0 = \frac{r_{1111}[(\alpha, \beta_c(\alpha))]}{4} = \frac{5\alpha + 2(1 - \Lambda) - \sqrt{2(1 + 2\alpha - \Lambda)}}{4} = a_0(\alpha),$$

$$a_2 = \frac{1}{8\alpha(1 + 2\alpha - \Lambda)}[-133\alpha^2 + 16\alpha(-8 + 7\Lambda) - 32 + 49\sqrt{2\alpha^2(1 + 2\alpha - \Lambda)}$$

$$+ 56\Lambda - 24\Lambda^2 - 20(\Lambda - 1)\sqrt{2(1 + 2\alpha - \Lambda)}]$$

$$= a_2(\alpha).$$

Note that the expression $a_2(\alpha)$ does not explicitly contain the component $\lim_{\beta \to \beta_c(\alpha)} \rho_{1111}/2$ since the coefficient of the latter quantity, namely, $\lim_{\beta \to \beta_c(\alpha)} (4a_0 + \Lambda - 1 - 3\alpha + q_c^2/2 - q_c^4/8)$

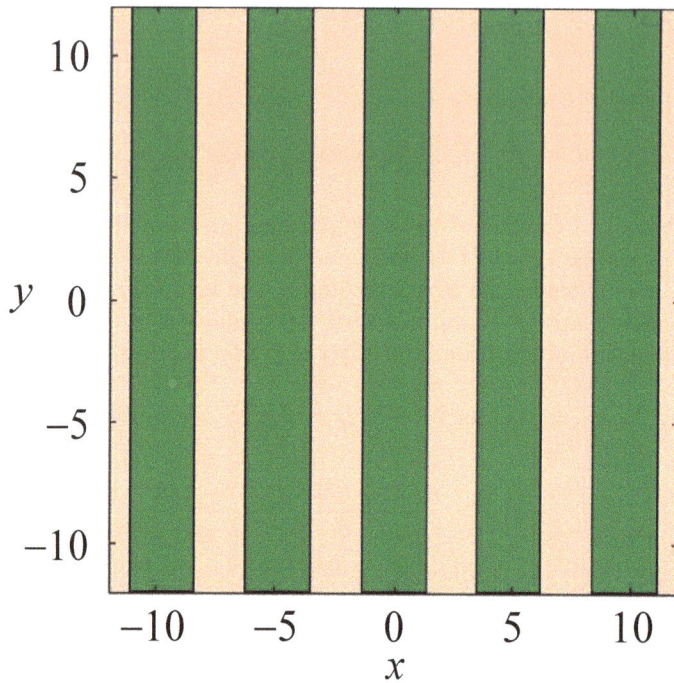

FIGURE 11.5

Contour plot in the x-y plane for the supercritical stripes (critical point II of the hexagonal and rhombic planform stability analyses) of Fig. 11.4 where the spatial variables are measured in units of λ_c. Here high densities ($\rho_e > \alpha$) appear dark (green) and low ones ($\rho_e > \alpha$), light (tan), in accordance with the numerical simulations of Lejeune *et al.* ([113]). This transition occurs at $\rho_e = \alpha$, which is a protocol that will be labeled zero threshold for our rhombic planform stability analysis. We shall show from this analysis that striped patterns are only stable where a protocol of the zero threshold-type holds. Hence this contour plot is the proper representation for critical point II.

is identically equal to zero by virtue of the formula for $a_0(\alpha)$. As pointed out by Wollkind *et al.* ([291]) and demonstrated in Chapters 9 and 10, such independence can be expected in single equation models, thus eliminating the necessity of determining ρ_{1111}, often called a free-mode.

Having determined formulae for the six-disturbance hexagonal-planform amplitude phase-equations' growth rate and Landau coefficients, we made the following nonlinear vegetative morphological interpretations of the equivalence classes of potentially stable critical points for these equations: I and II, represented the uniform homogeneous and the striped pattern plotted in Fig. 11.5, respectively, as described in our longitudinal planform stability analysis, while the hexagonal arrays of III^+ and III^-, depicted in Figs. 11.6 and 11.7, represented spotted and gapped patterns, respectively.

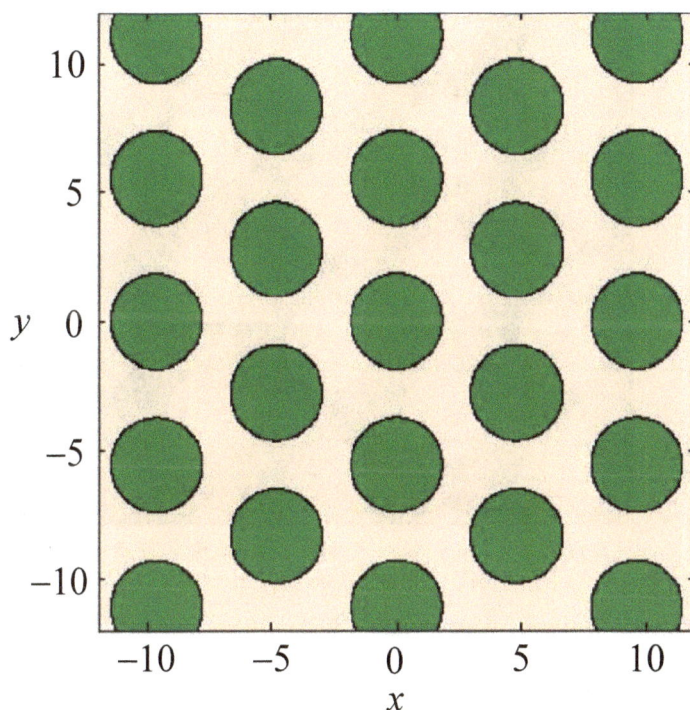

FIGURE 11.6
Contour plot for critical point III^+, representing vegetative spots, in the x-y plane with $\Lambda = 1$, $\alpha = 0.1$, and $\beta = 0.015$. Here, the spatial variables are again being measured in λ_c with regions of high density ($\rho > \alpha$) appearing dark (green) and low density ($\rho < \alpha$), light (tan), as in Fig. 11.5.

Having summarized those morphological identifications, we returned to our expressions for the Landau coefficients. First, we examined the signs of $a_1 + 4a_2$, $a_1 + a_2$, a_0, and $2a_2 - a_1$, by plotting those quantities versus α in Figs. 11.8 and 11.9 for $\Lambda = 1$. From the results of this examination, we observed that besides α_1 and α_2, defined earlier, there existed the

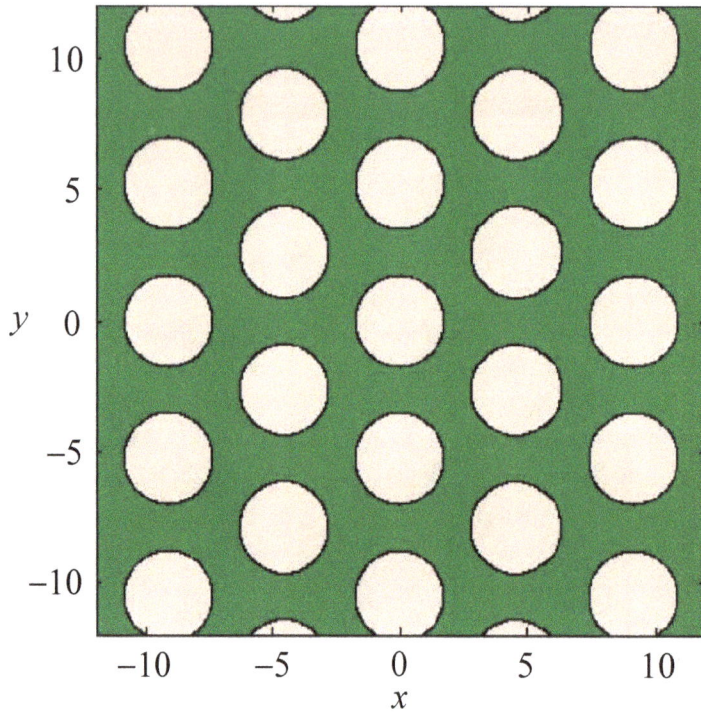

FIGURE 11.7

Contour plot for critical point III$^-$, representing vegetative gaps, in the x-y plane with $\Lambda = 1$, $\alpha = 0.2$, and $\beta = 0.01$. Here, the spatial variables are again being measured in λ_c with regions of high density ($\rho > \alpha$) appearing dark (green) and low density ($\rho < \alpha$), light (tan), as in Fig. 11.6.

following other significant values of α, compiled in Table 11.1 for $\Lambda = 0.825$ and 1:

$$\alpha_1 < \alpha_3 < \alpha_5 < \alpha_7 < \alpha_c < \alpha_2 < \alpha_4 < \alpha_6 < \alpha_8$$

such that

$$a_1 + a_2 = 0 \text{ for } \alpha = \alpha_3 \text{ or } \alpha_4, \ a_1 + a_2 > 0 \text{ for } \alpha_3 < \alpha < \alpha_4;$$
$$a_1 + 4a_2 = 0 \text{ for } \alpha = \alpha_5 \text{ or } \alpha_6, \ a_1 + 4a_2 > 0 \text{ for } \alpha_5 < \alpha < \alpha_6;$$
$$2a_2 - a_1 = 0 \text{ for } \alpha = \alpha_7 \text{ or } \alpha_8, \ 2a_2 - a_1 > 0 \text{ for } \alpha_7 < \alpha < \alpha_8,$$
$$2a_2 - a_1 < 0 \text{ for } \alpha < \alpha_7 \text{ or } \alpha > \alpha_8;$$
$$a_0 = 0 \text{ for } \alpha = \alpha_c, \ a_0 < 0 \text{ for } 0 < \alpha < \alpha_c, \ a_0 > 0 \text{ for } \alpha > \alpha_c.$$

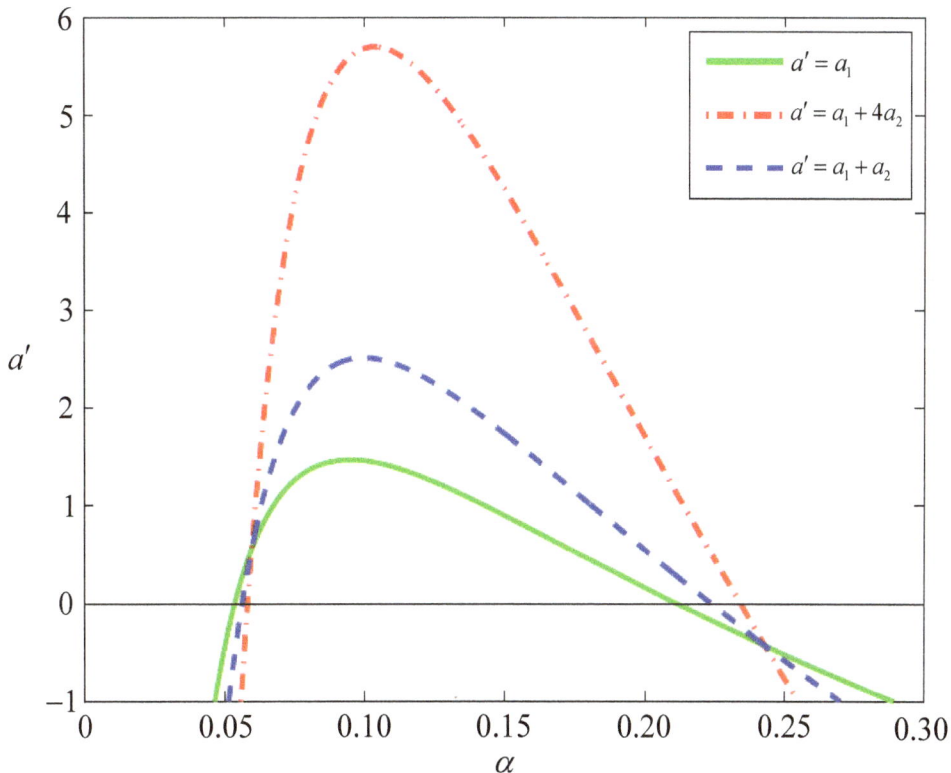

FIGURE 11.8
Plots of $a_1 + 4a_2$ and $a_1 + a_2$ versus α with $\Lambda = 1$ where the plot of a_1 of Fig. 11.3 is presented for the purpose of comparison.

Thus, given this behavior of $2a_2 - a_1$, the relevant entries of Table 8.2, which involved the signs of both that quantity and a_0, must be employed to determine the orbital stability of our critical points. To compare our theoretical predictions with numerical simulations and observational evidence, we needed to represent the results of this table graphically in the α-β plane. To do so, it was necessary for us to generalize the approach used to produce the marginal stability $\sigma_c = 0$ curve $\beta = \beta_c(\alpha)$, in order to generate the analogous loci associated

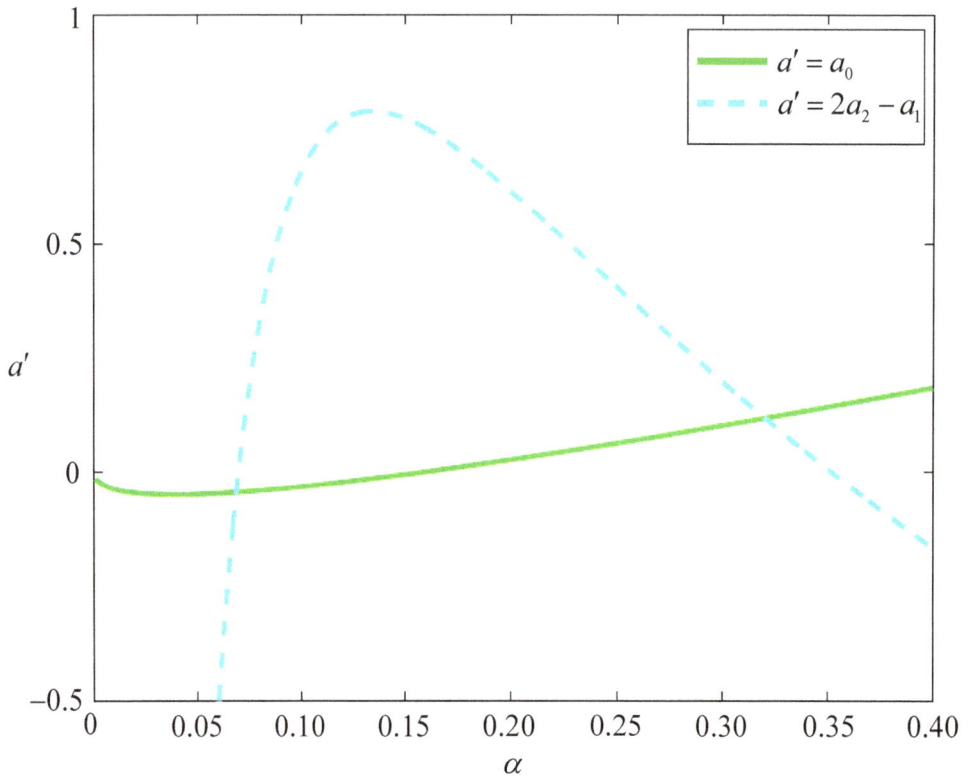

FIGURE 11.9
Plots of a_0 and $2a_2 - a_1$ versus α with $\Lambda = 1$.

TABLE 11.1
The significant α-values for hexagonal planform vegetative pattern formation.

Λ	α_1	α_3	α_5	α_7	α_c	α_2	α_4	α_6	α_8
0.825	0.0064	0.0090	0.0110	0.0237	0.1059	0.1430	0.1521	0.1600	0.2522
1	0.0535	0.0562	0.0581	0.0701	0.1600	0.2108	0.2232	0.2337	0.3518

with $\sigma = \sigma_j$ for $j = -1, 1$, and 2, respectively. That is, we solved

$$\sigma_c(\alpha, \beta) = \frac{2D(\alpha, \beta)}{\alpha} = \sigma_j[a_0(\alpha), a_1(\alpha), a_2(\alpha)] \text{ with } j = -1, 1, \text{ and } 2,$$

for β and obtained

$$\beta = \alpha - \sqrt{2\alpha^2(2\alpha + 1 - \Lambda) + 2\alpha\sigma_j[a_0(\alpha), a_1(\alpha), a_2(\alpha)]} = \beta_j(\alpha)$$

with $j = -1, 1$, and 2,

respectively. Note that, for $\sigma_c = 0$, $\beta = \beta_j(\alpha)$ reduces to our marginal stability curve $\beta = \beta_c(\alpha)$.

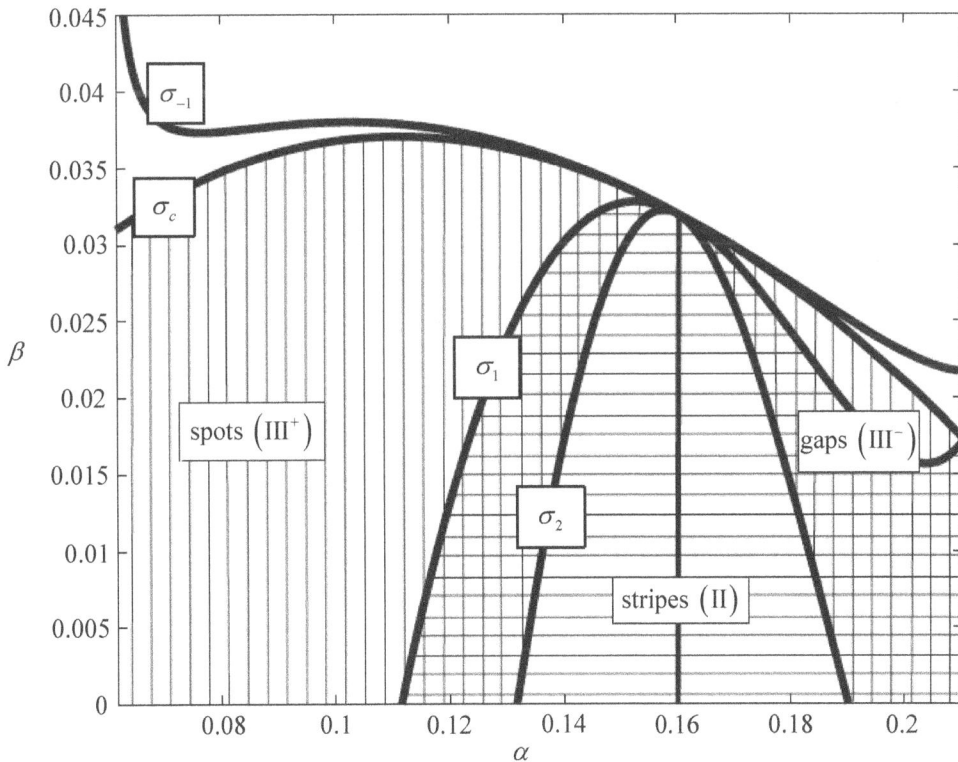

FIGURE 11.10
Stability diagram in the α-β plane for the propagator-inhibitor logistic equation with $\Lambda = 1$ identifying the predicted nonlinear vegetative patterns summarized in Table 8.2.

Since all the quantities required for the identification of the vegetative patterns of Table 8.2 had now been evaluated, we represented graphically the regions corresponding to those patterns in the α-β plane of Fig. 11.10 with $\Lambda = 1$ and $0.06 \leq \alpha \leq 0.2108$, where the loci $\beta = \beta_c(\alpha)$ and $\beta_j(\alpha)$ have been denoted by σ_c and σ_j with $j = -1$, 1, and 2, respectively. Note that all these curves intersect at $\alpha = \alpha_c = 0.16$ where $a_0 = 0$. In this context, from Fig. 11.10 we observed that

$$\beta_1(\alpha_1^-) = \beta_2(\alpha_2^\pm) = 0$$

where

$$\alpha_1^- = 0.1115, \quad \alpha_2^- = 0.1317, \quad \alpha_2^+ = 0.0.1902.$$

Finally, as usual, to justify the truncation procedure inherent to the asymptotic representation of our expansions, it was necessary that its Landau coefficients satisfied the size constraint

$$|a_0| \ll (a_1 + 4a_2)^2,$$

which, as can be seen from Fig. 11.11, was valid in the parameter range of interest. Hence we concluded that such a truncation procedure was justified for our hexagonal planform nonlinear stability analysis of the spatio-temporal propagator-inhibitor logistic evolution equation and thus, the morphological vegetative pattern predictions from Fig. 11.10 could be regarded as conclusive.

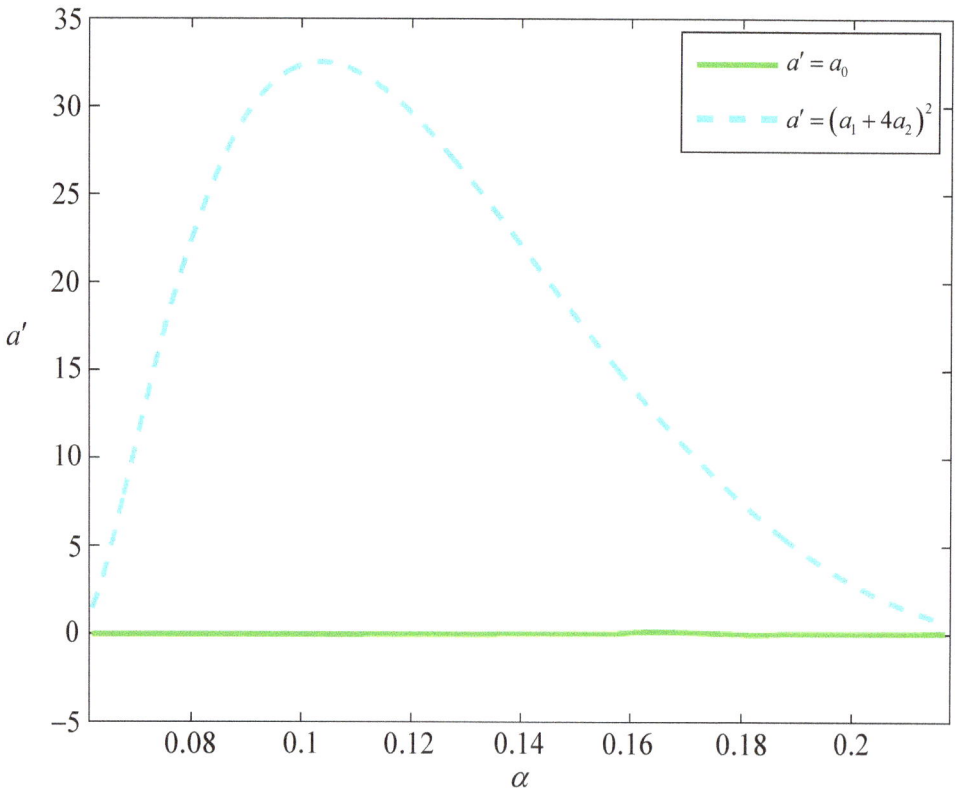

FIGURE 11.11
Plots of $(a_1 + 4a_2)^2$ and a_0 of Figs. 11.8 and 11.9 versus α.

Before comparing these theoretical hexagonal planform predictions developed in Nichaphat's thesis with relevant numerical simulations and observational evidence, it was necessary for us to summarize the rhombic planform results produced by her and Francisco on the propagation-inhibition model evolution equation. In that context, we sought a solution of this equation given by

$$\rho(x,y,t) \sim \alpha + R(x,z,t) \text{ with } z = x\cos(\varphi) + y\sin(\varphi),$$

where $R(x,z,t)$ had the form of the deviation function employed in the rhombic planform analysis of the nonlinear optical pattern formation problem of Chapter 10 but with its rhombic angle of ψ replaced by φ consistent with the analogous rhombic analysis of Chapter 9, and found that

$$R_{n0m0} = R_{0n0m} = \rho_{nm}, \ \sigma = \sigma_c(\alpha,\beta), \ a_1 = a_1(\alpha),$$

as defined earlier, while

$$R_{111(\pm1)} = \frac{\Lambda - 1 - 3\alpha + q_c^2/2 + q_c^4/8}{\sigma_c + (\beta-\alpha)[1/2 \pm \cos(\varphi)]q_c^2 + [3/8 \pm \cos(\varphi) + \cos^2(\varphi)]\alpha q_c^4}$$

and b_1, in particular, satisfied

$$b_1(\alpha,\varphi) - 2\sigma_c(\alpha,\beta)R_{2101}(\alpha,\beta,\varphi) = r_{2101}(\alpha,\beta,\varphi)$$

where

$$r_{2101}(\alpha,\beta,\varphi) = \frac{3}{2} + (1 - \Lambda + 3\alpha)[2R_{2000} + R_{1111} + R_{111(-1)}]$$
$$- \frac{\{R_{2000} + [3/2 + \cos(\varphi)]R_{11111} + [3/2 - \cos(\varphi)]R_{111(-1)}\}q_c^2}{2}$$
$$+ \frac{\{R_{2000} + [5/4 + 4\cos(\varphi) + 2\cos^2(\varphi)]R_{1111}\}q_c^4}{8}$$
$$+ \frac{[5/4 - 4\cos(\varphi) + 2\cos^2(\varphi)]R_{111(-1)}q_c^4}{8}.$$

Taking the limit of this equation as $\beta \to \beta_c(\alpha)$, we obtained the solvability condition that

$$b_1(\alpha,\varphi) = r_{2101}[\alpha,\beta_c(\alpha),\varphi].$$

Proceeding in the same manner as in the previous two chapters, we found that

$$b_1(\alpha_c,\varphi) = \frac{3}{2} = 2a_1(\alpha_c)$$

where α_c was defined implicitly by $a_0(\alpha_c) = 0$ and had the explicit formulation given by

$$\alpha_c = \alpha_c(\Lambda) = \frac{2(5\Lambda - 4) + \sqrt{2(7 - 5\Lambda)}}{25} \text{ for } 0.5 < \Lambda < 1.4;$$

note, in this context, that $\alpha_c(1) = 0.16$ and $\alpha_c(0.825) = 0.1059$ in accordance with Table 11.1; hence, we concluded that for $\alpha = \alpha_c(\Lambda)$ and $0 < \beta < \beta_c(\alpha_c)$ stripes were stable versus rhombic patterns; while for $\alpha \neq \alpha_c$, there were two φ-intervals, $(\varphi_m, \varphi_\ell)$ and (φ_r, φ_M), that flanked $\varphi = \pi/3$, where rhombic patterns were stable versus stripes when $0 < \beta < \beta_c(\alpha)$ satisfying

$$0 < \varphi_m(\alpha) < \varphi_\ell(\alpha) < \frac{\pi}{3} < \varphi_r(\alpha) < \varphi_M(\alpha) < \frac{\pi}{2}.$$

TABLE 11.2
The φ-range for stable rhombic patterns versus α for $\Lambda = 1$.

α	φ_m	φ_ℓ	φ_r	φ_M
0.12	0.8123	0.9025	1.2319	1.4645
0.13	0.8720	0.9396	1.1829	1.3225
0.15	0.9881	1.0106	1.0932	1.1335
0.18	0.8465	0.9355	1.1327	1.1761
0.19	0.7161	0.8536	1.1903	1.2530
0.20	0.5358	0.7173	1.2760	1.3511

These angles are tabulated in Table 11.2 for the Λ-value employed in Fig. 11.10 or for $\Lambda = 1$.

We were now ready to compare our hexagonal and these rhombic planform predictions with the relevant numerical simulations and observational evidence mentioned above. We began, naturally, by comparing our predictions with the numerical simulations performed for the propagation-inhibition logistic evolution equation by Lejeune and his co-workers: Namely, those contained in Lejeune and Tlidi ([111]), Couteron and Lejeune ([38]), and Lejeune *et al.* ([112, 113]). In particular, Lejeune *et al.* ([113]) numerically integrated this equation on a square grid with periodic boundary conditions for $\Lambda = 0.825$ and $L = 0.1$ after constructing the corresponding bifurcation diagram of associated vegetative states as a function of μ to select appropriate values of this aridity control parameter favoring hexagonal pattern formation. These authors reported the following sequence of spatially periodic vegetation states obtained from that bifurcation diagram as aridity was increased: First, a pattern of spots of sparser vegetation or gaps appeared forming a hexagonal lattice, which then transformed into an alternation of stripes of sparser and thicker vegetation, and finally into a hexagonal pattern of vegetation spots separated by bare ground. Recalling that $\beta = L^2$ and α is inversely related to $\mu = 1 - (1 - \Lambda)\alpha - \alpha^2$, we saw that the sequence of morphologies

$$\text{III}^+, \ \text{III}^+/\text{II}, \ \text{I}, \ \text{II}/\text{III}^-, \ \text{III}^-$$

obtained from Fig. 11.10, when the horizontal line

$$\beta \equiv \beta_0 \text{ where } 0.015 < \beta_0 < \beta_c(\alpha_c) = \beta_c(0.16) = 0.032$$

was traversed in the direction of increasing α through the patterned region, was in qualitative agreement with the sequence just catalogued. Next, Lejeune *et al.* ([113]) performed the numerical simulations for the selected values of Λ and L as reported above and generated vegetation patterns consisting of hexagonal gaps, alternating stripes, or hexagonal spots when μ was assigned the values of 0.965, 0.970, or 0.980, sequentially (see Fig. 11.12). Although $\Lambda = 1$, chosen for ease of exposition, is a special case, since such a choice eliminates the quadratic term of $f(\rho)$, it is a representative value of that parameter for our purposes. The qualitative behavior of our stability analyses is the same for the other values of $0 < \Lambda < 1$, as it was for $\Lambda = 1$. This is demonstrated conclusively relevant to the simulation results of Lejeune *et al.* ([113]) by Table 11.1, which lists the significant α values for $\Lambda = 0.825$, which was the value employed during that simulation, as well as those for $\Lambda = 1$.

To compare our theoretical predictions with these simulation results, we first constructed a stability diagram analogous to Fig. 11.10 but for $\Lambda = 0.825$ rather than $\Lambda = 1$. Figure 11.13 represents such a diagram plotted in the α-β plane. The critical value of density α_c has

FIGURE 11.12
Simulation results for the propagation-inhibitor logistic equation obtained by Lejeune *et al.*
([113]) with $\Lambda = 0.825$ and $L = 0.1$ ($\beta = L^2 = 0.01$) when $\mu = 0.965$, 0.970, and 0.980,
from left to right. Here, black corresponds to the highest phytomass density and thus the
patterns identified by H_π, S, H_0, represent our critical points III$^-$, II, III$^+$, respectively.

been denoted by the vertical line $\alpha = \alpha_c$ in both Figs. 11.10 and 11.13 and separates the
two types of vegetative hexagonal patterns from each other, since spots can only occur for
$\alpha < \alpha_c$, while gaps can only occur for $\alpha > \alpha_c$. Then, using the relation between μ and α with
$\Lambda = 0.825$, it was possible for us to convert Fig. 11.13 into Fig. 11.14, which represented
the corresponding stability diagram plotted in the μ-β plane. Again, note that the vertical
line in this figure is given by $\mu = \mu_c = 0.9702$, where the latter quantity corresponds
to the $\alpha_c(0.825) = 0.1059$. Finally, to compare the simulation results just described with
our theoretical predictions, we selected $\beta = L^2 = 0.01$ consistent with the value of $L = 0.1$
employed by Lejeune *et al.* ([113]). Upon examination of Fig. 11.14, we deduced the predicted
morphological sequence of vegetative patterns summarized in Table 11.3 as the horizontal
line $\beta = 0.01$ was traversed in the direction of increasing μ. Observe from Table 11.3 that
the complete patterned region occurs for $\mu \in (\mu_c^-, \mu_c^+)$ where

$$\mu_c^- = 0.9623, \ \mu_c^+ = 0.9951$$

were the two critical points of bifurcation as identified by Lejeune *et al.* ([113]), while the
three μ values of 0.965, 0.970, or 0.980, chosen by the latter authors for their simulation
purposes, lay within the subintervals for which only gaps, only stripes, or only spots were
stable, respectively.

Hence our theoretical predictions were in quantitative agreement with these simulation results. Further, note from Figs. 11.13 and 11.14 that the term "Homogeneous" appearing in
the stable pattern column of the first two entries of Table 11.3 refers to uniform vegetative
distributions of relatively high density, while the same term, appearing in the corresponding column of its last two entries, refers to similar distributions of relatively low density.
The results of our weakly nonlinear analyses were only strictly valid in the vicinity of the
marginal stability curve. Often, however, in situations of this sort, it is possible to extend
these results to regions of the relevant parameter space substantially removed from the
marginal curve. In order to determine the validity of such an extrapolation, it is standard
operating procedure to perform numerical simulations of the original governing partial differential, equation(s) in those extended regions and compare these simulation results with
those theoretical predictions. This is precisely what was done by Firth and Scroggie ([60]),

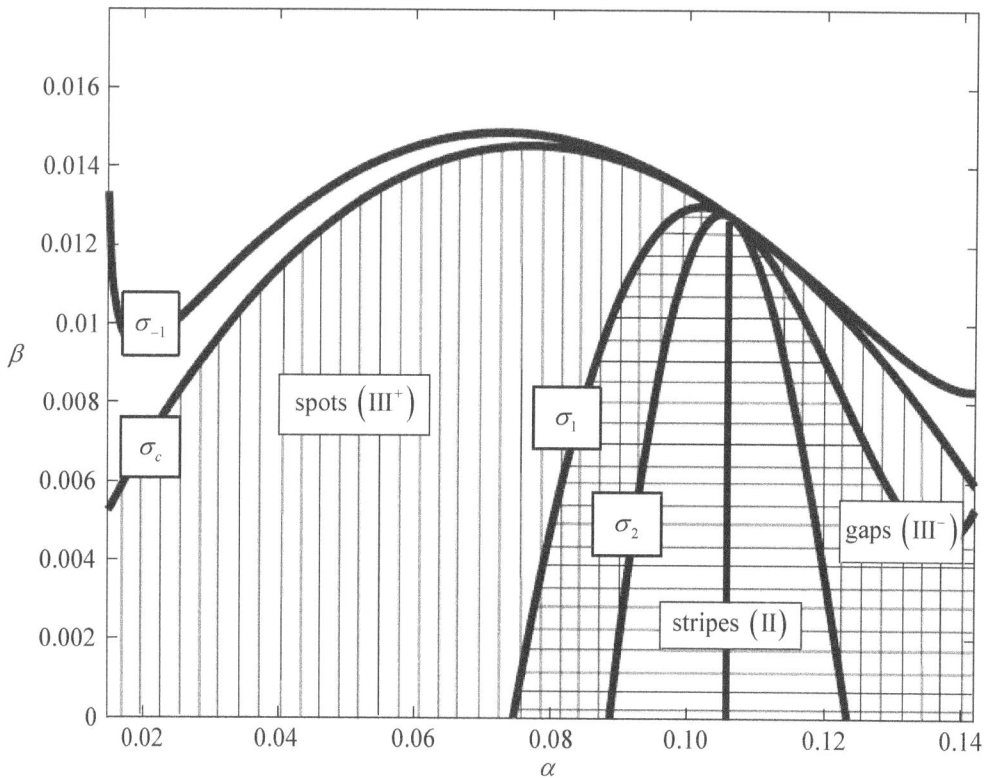

FIGURE 11.13
Stability diagram in the α-β plane for the propagator-inhibitor logistic equation analogous to Fig. 11.10 but with $\Lambda = 0.825$ instead.

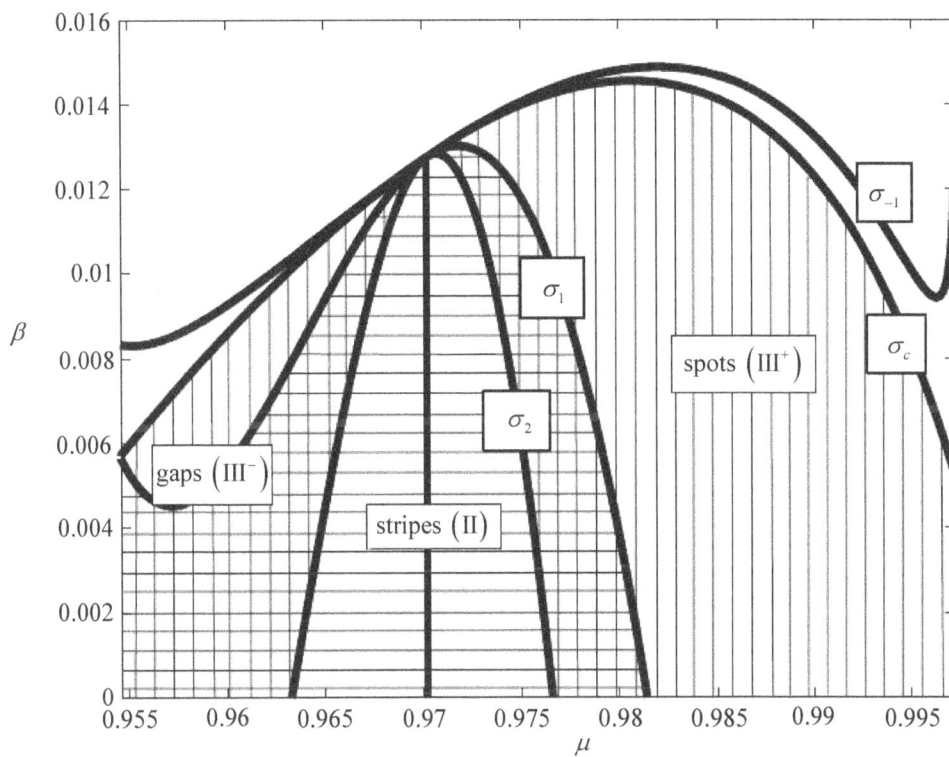

FIGURE 11.14
Stability diagram in the μ-β plane corresponding to Fig. 11.13.

TABLE 11.3

Table 11.3: Morphological stability predictions versus μ for $\beta = 0.01$ and $\Lambda = 0.825$.

μ-interval	Stable pattern
(0.9547,0.9623)	Homogeneous
(0.9623,0.9630)	Homogeneous and gaps
(0.9630,0.9657)	Gaps
(0.9657,0.9676)	Gaps and stripes
(0.9676,0.9735)	Stripes
(0.9735,0.9767)	Stripes and spots
(0.9767,0.9932)	Spots
(0.9932,0.9951)	Spots and homogeneous
(0.9951,0.9969)	Homogeneous

as described in Chapter 10 and more recently by Golovin *et al.* ([69]) in their chemical Turing pattern formation problem of the Brusselator model with superdiffusion. In the latter instance, their theoretical predictions compared very well with their corresponding numerical simulations. Indeed, Lejeune *et al.* ([113]) also compared their arid environmental vegetative pattern formation theoretical predictions and numerical simulations by reporting the maximum and minimum phytomass density values for those simulated patterns and the corresponding values for the analytical bifurcation diagram that was a plot of the stationary states of their truncated amplitude equations versus μ. We reproduced those results in Table 11.4, which also contains the difference between these values not provided explicitly by Lejeune *et al.* ([113]).

TABLE 11.4

Comparison of analytical and numerical ρ-results for $\beta = 0.01$ and $\Lambda = 0.825$ from Lejeune *et al.* ([113]). Here Difference = Maximum − Minimum for each column.

		Minimum		Maximum		Difference	
μ	Pattern	Analytical	Numerical	Analytical	Numerical	Analytical	Numerical
0.965	Gaps	0.051	0.057	0.152	0.149	0.101	0.092
0.970	Stripes	0.050	0.048	0.163	0.165	0.113	0.117
0.980	Spots	0.031	0.024	0.192	0.196	0.161	0.172

Considering the amplitude of that difference between these maximum and minimum values, our table shows this relative deviation between the analytical and numerical values to be less than 10% in each of those cases, as pointed out by Lejeune *et al.* ([113]). These cases, summarized earlier upon the introduction of Table 11.3, were for $\Lambda = 0.825$ and $\beta = 0.01$ with the aridity parameter $\mu = 0.965$, 0.970, and 0.980 corresponding to gaps, stripes, and spots, respectively, all of which, as can be seen from Fig. 11.14, lie relatively deep within the patterned region.

So far, we had only been concerned with the bifurcation behavior of our evolution equation when $0 < \Lambda \leq 1$ and $\beta_{crit} < \beta < \beta_c$ where $\beta_c = \beta_c(\alpha_c)$ and β_{crit} corresponded to the local minimum on $\beta = \beta_1(\alpha)$ for $\alpha > \alpha_c$, which in Fig. 11.10 were given by $\beta_{crit} = 0.015$ and $\beta_c = 0.032$. We next examined what transpired if that figure were considered in conjunction with the transit line

$$\beta \equiv \beta_0 \text{ where } 0 < \beta_0 < \beta_{crit} = 0.015.$$

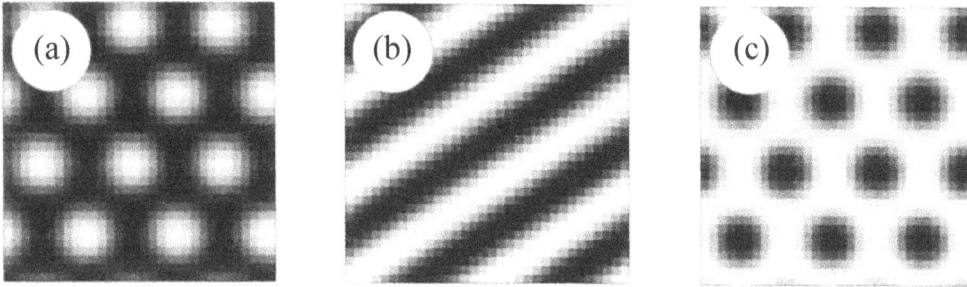

FIGURE 11.15
Simulation results for the propagator-inhibitor logistic equation obtained by Lejeune and
Tlidi ([111]) with $\Lambda = 1.2$ and $L = 0.2$ ($\beta = L^2 = 0.04$) when (a) $\mu = 0.98$, (b) $\mu = 1.00$,
and (c) $\mu = 1.01$. Here, black again corresponded to the highest phytomass density and
hence these patterns could be identified with our critical points III$^-$ for (a), II for (b), and
III$^+$ for (c).

Then, we obtained the sequence of morphologies

$$\text{III}^+, \text{ III}^+/\text{II}, \text{ II}, \text{ II}/\text{III}^-$$

as α was increased along that horizontal transit line. The simulation results of Lejeune and
Tildi ([111]) and Couteron and Lejeune ([38]) were in a parameter range that yielded bifurcation behavior of this generic sort. In particular, these authors numerically integrated the
propagation-inhibition logistic equation on a square grid with periodic boundary conditions
for $\Lambda = 1.2$ and $L = 0.2$ obtaining a hexagonal pattern of gaps for $\mu = 0.98$, alternating
stripes for $\mu = 1.00$, and a hexagonal pattern of spots for $\mu = 1.01$ (see Fig. 11.15).

To compare our theoretical predictions with these simulation results, we again constructed
stability diagrams analogous to Figs. 11.13 and 11.14, but for $\Lambda = 1.2$ in both the α-
β plane of Fig. 11.16 and the μ-β plane of Fig. 11.17. In doing so, we first pointed out
that this required a relaxation of our original constraint of $0 \leq \Lambda$, $\mu \leq 1$ and second, that
neither Lejeune and Tlidi ([111]), nor Couteron and Lejeune ([38]), constructed a bifurcation
diagram relevant to their two-dimensional simulations, although Lejeune *et al.* ([112]) did
generate a so-called one-dimensional bifurcation diagram for this case (see below). Observe
that Figs. 11.13 and 11.14, generated for $\Lambda = 0.825$, are in qualitative agreement with
Figs. 11.16 and 11.17, respectively, generated for $\Lambda = 1.2$. We also note that the vertical
lines in those figures are given by

$$\alpha = \alpha_c(1.2) = 0.2166, \ \mu = \mu_c = 0.9964,$$

respectively. Next, we selected $\beta = L^2 = 0.04$, consistent with the value of $L = 0.2$ employed
by Lejeune and Tlidi ([111]) and Couteron and Lejeune ([38]). Then, upon examination of
Fig. 11.17, we deduced the predicted morphological sequence of vegetative patterns summarized in Table 11.5 as the horizontal transit line $\beta = 0.04$ was traversed in the direction
of increasing μ.

Observe from Table 11.5, that this morphological sequence is both in qualitative agreement
with our predicted sequence and quantitatively compatible with the simulation results of

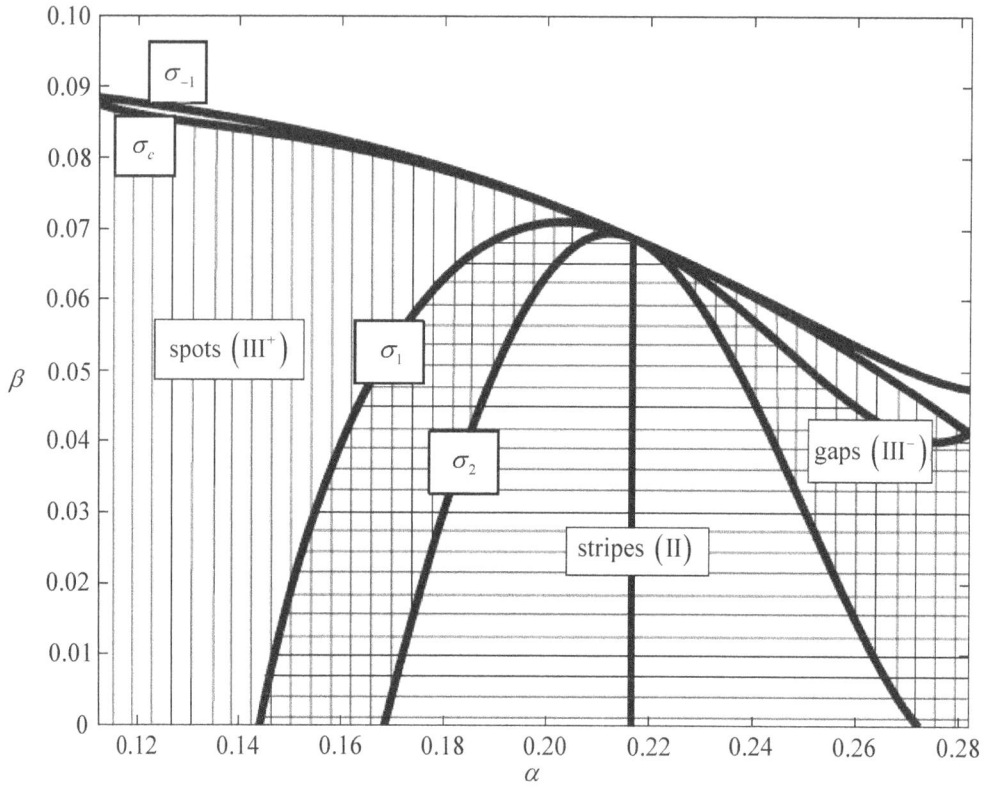

FIGURE 11.16
Stability diagram in the α-β plane for the propagator-inhibitor logistic equation with $\Lambda = 1.2$.

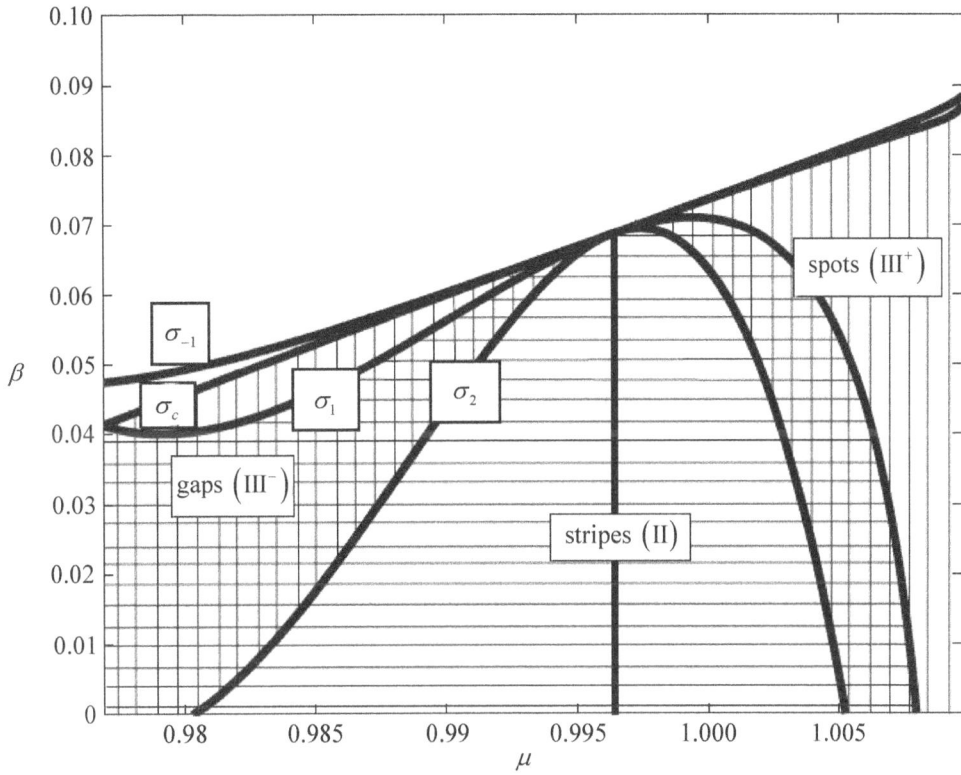

FIGURE 11.17
Stability diagram in the μ-β plane corresponding to Fig. 11.16. Note that the values of μ in this figure lie within the interval $0 \leq \mu \leq \mu^* = 1.01$ where $\rho_0^+ = \alpha > 0$ (see Fig. 11.16).

TABLE 11.5
Morphological stability predictions versus μ for $\beta = 0.04$ and $\Lambda = 1.2$.

μ interval	Stable pattern
(0.9769,0.9891)	Gaps and stripes
(0.9891,1.0029)	Stripes
(1.0029,1.0064)	Stripes and spots
(1.0064,1.0098)	Spots

Fig. 11.15 once the μ-values of that table have been rounded off to the two places of accuracy reported by Lejeune and Tlidi ([111]) and Couteron and Lejeune ([38]). Upon examination of Fig. 11.17, we could see that the parameter values of Table 11.5 lie even deeper within the patterned region of this figure than the parameter values of Table 11.4 did in Fig. 11.15. In this context, Lejeune *et al.* ([113]) reported their simulation μ values to three places of accuracy. We noted that for all the simulation results of Lejeune and his co-workers described here which generated striped patterns

$$\mu = \mu_c$$

to the relevant degree of accuracy (displayed for the purpose of emphasis).

In order for us to demonstrate that this occurrence was not coincidental, we needed to reconsider our rhombic planform results in conjunction with the morphological predictions summarized in Table 8.2. Toward this end, we began by pointing out that, although these rhombic results had been obtained for the special case of $\Lambda = 1$, they were qualitatively valid for any other value of Λ satisfying the condition $\Lambda - 1 \sim O(\varepsilon)$, *e.g.*, specifically for $\Lambda = 0.825$ or 1.2. Consistent with our previous definitions when $\beta_0 = 0$, we implicitly redefined generalized $\alpha_2^\pm$ by

$$\beta_2(\alpha_2^\pm) = \beta_0 \text{ where } 0 \le \beta_0 < \beta_c = \beta_c(\alpha_c)$$

and observed from Tables 11.3, 11.4, and 11.5 and Figs. 11.10, 11.13, and 11.16 that our hexagonal planform analysis predicted stripes as the only stable pattern for

$$\alpha_2^- < \alpha < \alpha_2^+.$$

From the results of our rhombic planform analysis, we concluded that for any

$$\alpha \in (\alpha_2^-, \alpha_c) \cup (\alpha_c, \alpha_2^+)$$

all such stripes were unstable with respect to rhombic patterns having characteristic angle

$$\varphi \in (\varphi_m, \varphi_\ell) \text{ or } \varphi \in (\varphi_r, \varphi_M).$$

We next examined the morphology of such rhombic patterns. Using the same approach as developed in Chapter 10, we considered the higher-order homogeneous terms in the equilibrium vegetative density deviation function $R_e(x, y)$ for critical point V to be given by $2\rho_{20}C_0^2$. Then, adding these terms to the original zero mean level of density and employing our previous results, we found that this mean level to second-order now satisfied

$$R_m = 2\rho_{20}C_0^2 = -r_{1111}(\alpha, \beta)\frac{C_0^2}{\sigma_c + (\alpha - \beta)q_c^2/2 - \alpha q_c^4/8}.$$

Substituting $q_c^2 = 2(\alpha - \beta)/\alpha$ into the denominator of that expression and simplifying, yielded

$$R_m = -r_{1111}(\alpha, \beta)\frac{C_0^2}{\sigma_c + (\alpha - \beta)^2/(2\alpha)}.$$

We adopted the protocol that in our contour plots of this critical point the elevations, which satisfy $R > R_m$, would appear dark, and depressions, which satisfy $R < R_m$, light, where $R = R_m$ represented the threshold value at which this transition occured. When R_m was greater than, equal to, or less than zero we labeled such patterns as being of higher, zero, or lower threshold, respectively. Hence, to determine the proper threshold type, we had to examine the sign of R_m. Recalling that for stable rhombic patterns $\sigma = \sigma_c(\alpha, \beta) > 0$, we concluded that this behavior of R_m depended on the sign of $-r_{1111}(\alpha, \beta)$, which is equivalent to that of $-a_0$ near the marginal stability curve $\beta = \beta_c(\alpha)$. Therefore, our stable rhombic patterns would be of higher, zero, or lower threshold-type depending upon whether $\alpha < \alpha_c$, $\alpha = \alpha_c$, or $\alpha > \alpha_c$, respectively. To make a physical interpretation of these results we noted that to lowest order the equilibrium density deviation associated with this critical point satisfied

$$R_e(x, y) \sim C_0 \left[\cos\left(\frac{2\pi x}{\lambda_c}\right) + \cos\left(\frac{2\pi z}{\lambda_c}\right) \right] = C_0 g(x, z) \text{ where } C_0 > 0$$

$$\text{for } z = x \cos(\varphi) + y \sin(\varphi).$$

The three parts of Fig. 11.18 are threshold contour plots of $g(x, z)$ for the typical rhombic angle of Fig. 10.10 or $\varphi = 1.150 \ (66°)$ with the threshold values of 1, 0, and -1, respectively. Here, the spatial variables are being measured in units of λ_c with regions exceeding that threshold in each part appearing dark and regions less than it, light. Given their appearance in Fig. 11.18 we identified these higher, zero, and lower threshold-type rhombic arrays with density patterns of pseudo spots, rectangles, and pseudo gaps (nets), respectively. In particular, for α in the relevant intervals $\alpha \in (\alpha_2^-, \alpha_c) \cup (\alpha_c, \alpha_2^+)$, rhombic arrays of this sort are difficult to distinguish from hexagonal distributions since the allowable bands of their characteristic angles closely flank $\pi/3 \cong 1.047$ (see Table 11.2 and Fig. 11.18). Hence, if we substituted these type of rhombic patterns as the stable morphology for the intervals

$$\alpha_2^- < \alpha < \alpha_c \text{ and } \alpha_c < \alpha < \alpha_2^+$$

(that is pseudo spots for the former and pseudo gaps for the latter), then hexagonal or nearly hexagonal arrays could be anticipated for all $\alpha \neq \alpha_c$ over the instability region. Finally, from Figs. 11.10 and 11.13-11.16, in conjunction with the morphological stability results of the rhombic planform analysis, we concluded that the occurrence of stripes required $\alpha = \alpha_c$, or equivalently $\mu = \mu_c$, since such patterns were only stable for both planforms at this critical value of α or μ and when $\beta < \beta_c = \beta_c(\alpha_c)$ in agreement with the simulation results of Lejeune and his co-workers noted earlier. We considered this agreement a series of Rabbits Pulled Out of Hats!

Here, where our analysis predicted a stable parallel striped pattern as described above, the equivalence class for the critical point designated as II in the hexagonal planform analysis actually contained the three solutions

$$A_i^2 = \frac{\sigma}{a_1}, \ A_j = A_k = 0, \ (i, j, k) = \text{ even permutation of } (1, 2, 3);$$

As noted in previous chapters, these represented a family of stripes aligned parallel to the y axis as per our original identification plus two similar families of stripes making angles of $\pm 60°$ with them, for which stable coexistence with a member of either the original family or one another was impossible. Then for $\alpha = \alpha_c$ (or equivalently $\mu = \mu_c$) and $\beta < \beta_c$ as initial conditions vary from point to point over a flat environment, such families of stripes could give rise to polygonal arcs the boundaries of which would appear quite random in

$$\varphi = 1.150 \ \left(66^\circ\right)$$

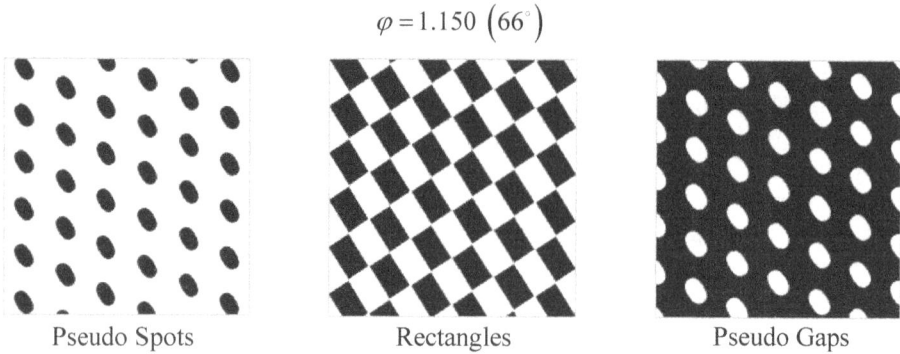

| Pseudo Spots | Rectangles | Pseudo Gaps |

FIGURE 11.18
Rhombic patterns in the x-y plane relevant to $g(x,z)$ for $\varphi = 1.150$ (66°) with threshold
values from left to right of 1, 0, and −1 and represent arrays of pseudo spots, rectangles,
and pseudo gaps, respectively. Here, the spatial variables are being measured in units of λ_c
with regions below that threshold in each part appearing light and regions above it, dark.
Note that $|g(x,z)| \leq 2$ and $\varphi = 1.150$ is the typical rhombic angle depicted in the plot of
Fig. 10.10. Observe that for this angle, the pseudo spot and gap patterns approximate the
close-packed structure of hexagonal arrays.

orientation. Indeed, some of the simulation results classified as striped patterns by Lejeune and Tlidi ([111]) and Couteron and Lejeune ([38]) had the appearance of such curved
elongated stripes or interconnected bicontinuous patterns (see Chapter 8) in the relevant
reproductions contained therein.

So far, we had limited our discussion to analyses for which the wavenumber was restricted
to the critical wavenumber alone. In order to investigate the consequences of considering
other wavenumbers in the instability side band centered about this critical wavenumber, it
would be necessary to convert the Landau-type amplitude ordinary differential equations
in time to Ginzburg-Landau type partial differential equations by adding the appropriate
spatial derivative terms to them. As reviewed in detail by Wollkind *et al.* ([291]), such an
analysis always yields two additional stripe-type instabilities besides those discussed above:
Namely, zig-zag and cross-band relevant to the interaction of oblique and perpendicular
modes, respectively (see Chapter 2). Specifically, it should be pointed out that the stripes
simulated by Lejeune *et al.* ([113]) for $\beta = 0.01$, $\Lambda = 0.825$, and $\mu = 0.97$ were of the zig-zag
variety, as can be seen from Fig. 11.12.

Up until now, we have been concerned with $a_1 > 0$ type behavior for the simulation results
of Lejeune and his co-workers when $\Lambda = 1.2$ and $L = 0.2$. Lejeune *et al.* ([112]) constructed a
one-dimensional bifurcation diagram corresponding to this problem for $\Lambda = 1.2$ and $L = 0.2$.
They found isolated vegetation patches as a simulation outcome occurring outside a patterned aridity interval when $\mu > \mu^* = 1.01$ and interpreted such localized structures to be
a spatial compromise between the periodic patchy vegetative and bare stable states. We
plotted a_1 versus α in Fig. 11.19 for $\Lambda = 1.2$ relevant to this situation and noted that
$\alpha = 0.1$ was equivalent to $\mu = 1.01$. Observe that Fig. 11.3 for $\Lambda = 1$ is in qualitative agree-

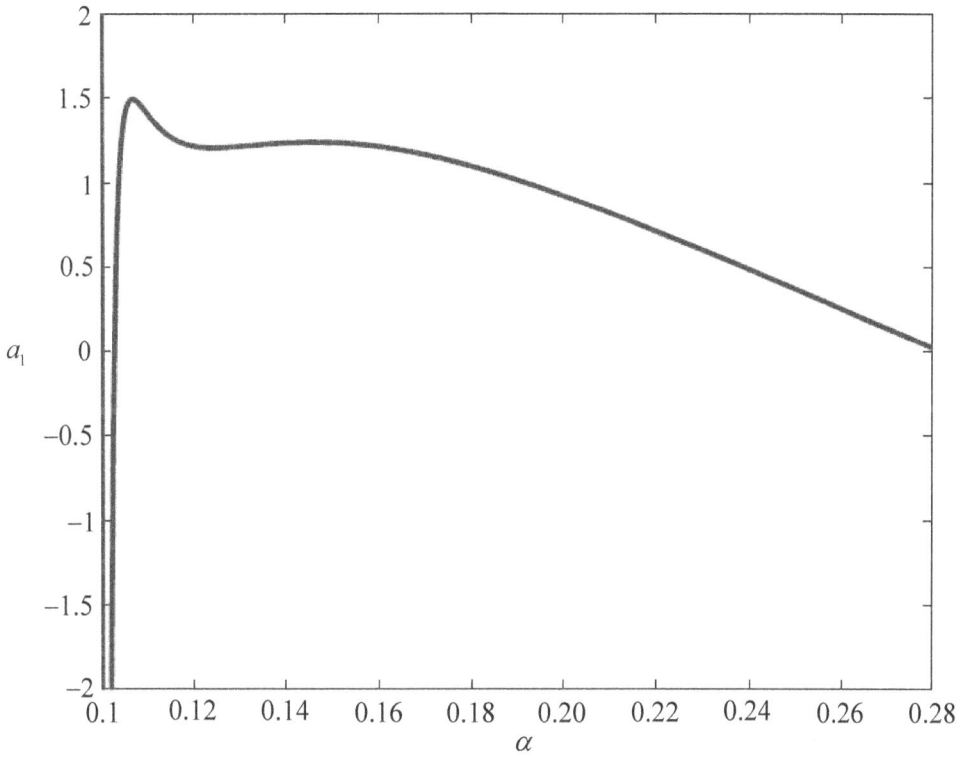

FIGURE 11.19
Plot of $a_1 = a_1(\alpha)$ versus α for $\Lambda = 1.2$.

ment with Fig. 11.19 for $\Lambda = 1.2$. Here, the latter case yielded the values $\alpha_1 = 0.1028$ and $\alpha_2 = 0.2821$. Note from Fig. 11.18 that, for the μ range included in Table 11.5, the transit curve $\beta = 0.04$ lies totally within the region where σ, $a_1 > 0$. Should μ be extended just beyond the bounds of that table, however, $a_1 < 0$, although $\sigma > 0$ (see Figs. 11.17-11.19). This behavior is highly reminiscent of the morphological phase separation instabilities ($\sigma > 0$) analyzed in Chapter 8. Recall that, outside the patterned thickness interval dewetting-type, rupture occurred where $a_1 < 0$ by isolated drop formation in relatively thin layers and isolated hole formation in thicker ones. Analogous behavior was discovered by Golovin *et al.* ([69]) in their examination of chemical Turing pattern formation for the Brusselator problem with superdiffusion. Given the similarity of behavior among all these phenomena, we conjectured that outside the bounds of Table 11.5 where $\sigma > 0$ and $a_1 < 0$ there would be desertification characterized by the formation of isolated vegetation patches for $\mu > 1.0098$, consistent with the findings of Lejeune *et al.* ([112]), and by the formation of isolated patches of bare ground for $\mu < 0.9769$ (see Chapter 15).

FIGURE 11.20
Patterns of the perennial grass, *Paspalum vaginatum*, observed by Meron *et al.* ([143]) in the Northern Negev desert in Israel (200 mm mean annual rainfall): (a) a labyrinth-like pattern and closeups showing (b) spots, (c) stripes, and (d) gaps. Here, the typical distance between spots, stripes, or gaps was about 0.1 m.

It remained for us to compare our theoretical predictions with observational evidence. Striking periodic vegetation patterns covering widespread areas of arid or semi-arid regions of Africa, Australia, the Americas, Asia, and the Near East became noticeable through aerial photography nearly 80 years ago (reviewed by Meron, [142]). In this instance, an arid or semi-arid region refers to an environment that is characterized by an extended dry season where yearly potential evaporation exceeds yearly rainfall and water availability is a limiting factor of plant growth (Lejeune *et al.*, [113]). The reported patterns (see Fig. 11.20) consisted of gaps (pearled or spotted bush), labyrinths, stripes (tiger bush), and spots (leopard bush). These included bushy vegetation punctuated by bare spots in Niger, labyrinths of perennial grasses in Israel, striped patterns of bushy vegetation in Niger, and spots of trees or shrubs in Ivory Coast and French Guiana (Rietkerk *et al.*, [187]). From our one-dimensional morphological interpretation of the equivalence class for critical point II, it followed that the patterns usually described by ecologists as tiger bush, lanes, groves, and bands should be associated with what we had termed parallel stripes while those described as labyrinths, mazes, and arcs should be associated with what we had termed bicontinuous patterns. Similarly, leopard or pearled bush should be associated with critical points III$^+$ or III$^-$, respectively.

We wished to re-examine further these hexagonal patterns in order to determine sufficient conditions under which they could give rise to regions characterized by bare ground in agreement with relevant observational evidence. Toward that end, we returned to our hexagonal symmetry function f_0 of Figs. 2.10 and 7.4. Focusing our attention upon the single hexagonal cell depicted in these figures, we recalled that each such individual f_0 cell had an elevated central region with a maximum elevation of 3 at its center which was bounded by a level curve of zero elevation appearing as the nearly circular loci found in Figs. 2.10 and 7.4. The peripheral portion of each cell exterior to that central region was depressed, with the hexagonal cellular boundary of variable depth ranging from $-3/2$ at its vertices to -1 at the midpoint of its edges. Specifically, at the center of these cells, the density function characteristic of III$^-$ had a value

$$\rho \sim \alpha_0^- + 3A_e^- \text{ where } A_e^- < 0,$$

while on its boundaries the density function characteristic of III$^+$ ranged from

$$\alpha_0^+ - \frac{3A_e^+}{2} \text{ to } \alpha_0^+ - A_e^+ \text{ where } A_e^+ > 0.$$

Even though (see Figs. 11.10, 11.13, and 11.16)

$$\alpha_0^+ < \alpha_c < \alpha_0^-,$$

we saw that there was about the same tendency for these density values to be equal to, or less than, zero, and hence represent bare ground in agreement with the actual appearance of leopard or pearled bush, respectively.

Finally, we investigated the predicted wavelength of these vegetation patterns. From the definition of λ_c we deduced that

$$\lambda_c = \sqrt{2}\pi \left(1 - \frac{\beta}{\alpha}\right)^{1/2}$$

and observed that, for $\beta \equiv \beta_0$, λ_c was a decreasing function of α. Then, evaluating this function on the marginal stability curve $\beta = \beta_c(\alpha)$, we obtained

$$\lambda_c = \sqrt{2}\pi[2(2\alpha + 1 - \Lambda)]^{1/4},$$

which, for the special case of $\Lambda = 1$ reduced to $\lambda_c = \pi/\alpha^{1/4}$, these critical wavelengths being generated by the interplay between short-range facilitative and long-range competitive plant interactions. Specifically, for $\alpha = \alpha_0^{\pm}$, we saw that $\lambda_c^+ > \lambda_c^-$ and hence concluded that the vegetative distributions in leopard bush had a tendency to be more widely spaced than the bare patches which regularly punctuate the vegetation cover in pearled bush. Further, the characteristic wavelength of tiger bush which corresponds to $\alpha = \alpha_c = 0.16$ for this case satisfied $\lambda_c = 5$ and was intermediate in value between the distances $\lambda_c^{\pm}$. After Lejeune and Tlidi ([111]), employing a length scale of 5 m in this situation, yielded the associated dimensional wavelength $\lambda_c^* = 25$ m, whereas, after Meron *et al.* ([143]), using one of 2 cm (0.02 m), yielded $\lambda_c^* = 0.1$ m, both consistent with field observations (see Fig. 11.20).

We conclude this discussion by offering some additional comments about our analyses in relation to those of Lejeune and his co-workers. As indicated at the start of this chapter, Lefever *et al.* ([110]) introduced scaling laws at variance to those we employed. Specifically, they took

$$\frac{\partial}{\partial t} \sim O(\varepsilon^{3/2}), \ L \sim O(\varepsilon^{1/2}),$$

while retaining our other scales. Then the terms

$$\frac{\partial \rho}{\partial t}, \ (L^2 - \rho)\nabla^2 \rho \sim O(\varepsilon^{5/2}),$$

while all the rest of the terms in the governing evolution equation remained of $O(\varepsilon^3)$. Within the framework of our nonlinear stability analyses, we examined the validity of our scaling laws where they differed from theirs: Namely

$$\frac{\partial}{\partial t} \sim O(\varepsilon^2), \ L^2 - \rho \sim O(\varepsilon^{3/2}).$$

The first condition was consistent with the asymptotic scalings employed for all methods of weakly nonlinear stability theory (Matkowsky, [137]) while using the marginal stability curve, since our analysis was expected to be asymptotically valid as $\beta \to \beta_c(\alpha)$, we found that

$$L^2 - \rho \sim \beta - \alpha \sim -\alpha\sqrt{2(2\alpha + 1 - \Lambda)} \sim O(\varepsilon^{3/2}),$$

which was consistent with the second condition. Observe, as noted earlier, that with our scaling laws all the terms retained in the governing evolution equation were now of $O(\varepsilon^3)$. In their seminal work Lefever and Lejeune ([109]) actually retained more terms of different order during their weak gradient-low density truncation procedure, which was the reason we did not attempt to compare our theoretical predictions with the simulation results contained therein. Also, Lejeune and Tlidi ([111]) claimed that they never observed rhombic patterns in their numerical simulations because such patterns were unstable with respect to stripes and hexagons. Our rhombic planform results showed this assertion to be in error with respect to stripes, except when $\alpha = \alpha_c$. Their simulation results were presented for both small-scale domains as in Fig. 11.15 and large-scale ones. For the latter extended domains, the striped patterns in the small windows were converted to zig-zag patterns and the hexagonal patterns became much less regular. All of these developments were consistent with our earlier discussion concerning potential bistability of various types of stripes for $\alpha = \alpha_c$ and between rhombic and hexagonal patterns for $\alpha \neq \alpha_c$. In point of fact, Golovin *et al.* ([70]) found bistability between squares and hexagons in certain regions of their gravity number-capillary number parameter space for a Bénard-Marangoni surface-tension driven convection problem with poorly conducting boundaries.

That completed Nichiphat's research and she returned to Thailand to defend her thesis once a paper from this dissertation had been accepted for publication. Again, as in Adoon's case, I wrote that paper with her, Yongwimon Lenbury, and Francisco as my co-authors. In addition, I chose to submit it to the *Journal of Biological Dynamics*. Since Nichiphat and her advisor were in a hurry to get this paper published, Yongwimon submitted it while I was still proofreading their LaTeX version of my manuscript, actually making her the initial corresponding author of record. Hence, the reviews for our paper were sent to Yongwimon who forwarded them to me by email. She considered these reviews to be favorable and felt we could easily complete their suggested revisions, the major one of which was contained in Lejeune's report, who identified himself by concluding it with the statement, "I am Olivier Lejeune." Before explaining why I felt this suggested modification was unnecessary, let me first put our contributions in the context of those of Lejeune and his co-workers. We, in essence, had performed systematic longitudinal, hexagonal, and rhombic planform nonlinear stability analyses on the isotropic flat dryland vegetative pattern formation model evolution equation they had investigated and then compared our theoretical results with their numerical simulations. The basic difference between us and them in reporting hexagonal planform analytical results was that whereas Lejeune and his co-workers employed standard one-parameter bifurcation diagrams in the μ-equilibrium density plane (see Fig. 2.13) for fixed values of L, we used our more comprehensive two-parameter stability diagram in the μ-β plane with $\beta = L^2$ of the same form as that developed for this purpose throughout the book. What Lejeune wanted us to do was perform a numerical analysis on our hexagonal-planform amplitude-phase equations instead of employing the stability criteria for its critical points. Since he and his co-workers had already performed numerical analyses on the governing partial differential evolution equation, such an analysis on these truncated ordinary differential equations would be tangential to the main goal of the paper which was to compare the results of our weakly nonlinear stability analyses with those numerical simulations. Indeed, as I pointed out to the handling editor Mark Lewis, a leading mathematical ecologist from the University of Alberta, Lejeune *et al.* ([113]) had not performed such a numerical analysis on their complex amplitude equations when comparing the analytical and simulation results contained in Table 11.4, added to the paper as our only response to that suggestion, so why should we need to do so? My argument, which Bonni considered Pulling a Rabbit Out of a Hat, must have convinced Lewis since the editorial board of the *Journal of Biological Dynamics* accepted our revision, which included all the other very minor revisions suggested by the reviewers, for publication. In spite of Yongwimon's opinion to the contrary, Lejeune's suggested modification would have been difficult to implement in a timely manner. Thus, most fortuitously knowing the identity of this reviewer helped me to refute his argument. This is only one of the many reasons that I have always thought referees should be required to identify themselves to authors during the review process. Be that as it may, except for my colleagues, only Robert May, as related in Chapter 4, and Olivier Lejeune, in this instance, of all my supposedly anonymous reviewers have ever revealed their identity to me during this process. Finally, given that I was the co-author of record who actually communicated these responses to the editorial board, the *J. Biol. Dyn.* eventually listed me as the corresponding author instead of Yongwimon when our paper (Boonkorkuea *et al.*, [17]) appeared in print.

We completed the dissemination of this research with the following three talks in 2009, all entitled "Nonlinear stability analyses of vegetative pattern formation in an arid environment": At WSU Showcase 2009, in Pullman, WA, on March 29; at the Pacific Northwest Conference on Comprehensive Mathematical Modeling in the Natural and Engineering Sciences in Pullman, WA, on June 6; and at the Technological University of Monterey in Guadalajara, Mexico, on September 14; where the second was a keynote address at this NSF-sponsored conference presented in the spirit of a testimonial to Lee A. Segel, organized by me and my colleague Robert Dillon, while the third was an invited lecture arranged by Francisco Alvarado as part of his university's prestigious Academic Leaders Program. I was very impressed with everything about Francisco's university but most especially by the fact that there were 500 attendees at my talk!

12

Diffusive Versus Differential Flow Instabilities I: Dryland Turing Pattern Formation: Hexagonal and Square Planform Nonlinear Stability Analyses

As mentioned in Chapter 1, I first met Bonni Dichone (then, Kealy) in the Fall Semester of 2007 when she entered our doctoral program and enrolled in my two-semester graduate-level continuum mechanics course. In January of 2008, Bonni asked me if I would be her Ph.D. adviser and we have been collaborating on a variety of comprehensive applied mathematical modeling papers and books ever since. Besides the continuum mechanics sequence, she took all three of my other graduate modeling and applied analysis courses, as well as my undergraduate class in differential equations at WSU. In the context of the subject matter of this chapter, Bonni did her Ph.D. dissertation with me on vegetative pattern formation in arid flat environments and served as my postdoctoral associate for the next two years extending that research. Indeed, she began working on her Ph.D. thesis immediately upon becoming my advisee, before passing the doctoral qualifying examination, which was very unusual for most WSU Mathematics graduate students.

One of the systems reviewed by Rietkerk *et al.* ([187]) was the differential flow instability model from Christopher Klausmeier's ([98]) report published in the volume 284 number 5421 issue of *Science*. Here, the positive feedback control between an activator consumer (plants) and an inhibitory limiting resource (surface water) was considered the driving force underlying pattern formation in conjunction with the advective effect of the surface water, while the diffusion of the latter was neglected even though it exceeded the dispersal of the plants which was included. This seminal model was unable to predict pattern formation in flat environments for which there could be no downhill surface water flow. The problem I selected for Bonni's thesis introduced a two-component interaction-diffusion plant-surface water system based on Klausmeier's ([98]) differential flow instability model but with the advective term of his surface water equation replaced by a diffusive one instead, to explain more fully the occurrence of tiger bush (or banded thicket) patterns observed in arid flat environments such as those on dryland plateaus reported by Couteron *et al.* ([39]). This was in the spirit of Kondo and Miura's ([101]) review entitled "Reaction-Diffusion Model as a Framework for Understanding Biological Pattern Formation."

Toward that end, we considered a coupled partial differential interaction-diffusion equation model for $N(X, Y, \tau) \equiv$ plant biomass density and $W(X, Y, \tau) \equiv$ surface water content, where $(X, Y) \equiv$ a two-dimensional spatial coordinate system and $\tau \equiv$ time of the form

$$\frac{\partial N}{\partial \tau} = F(N, W) + D_1 \nabla_2^2 N, \quad \frac{\partial W}{\partial \tau} = G(N, W) + D_2 \nabla_2^2 W; \quad \nabla_2^2 \equiv \frac{\partial^2}{\partial X^2} + \frac{\partial^2}{\partial Y^2};$$

defined on an unbounded flat domain with its interaction terms given by (Klausmeier, [98])

$$F(N, W) = RJWN^2 - MN, \quad G(N, W) = A - LW - RWN^2.$$

DOI: 10.1201/9781003195603-12

Here, water was being supplied uniformly due to precipitation at rate A and lost due to evaporation at rate LW; plants took up water at rate $Rf(W)g(N)N$, where $f(W)$ represented the functional response of plants to water, while $g(N)$ described how plants increased water infiltration and, for simplicity, it was assumed that these functions were linear thus satisfying $f(W) = W$, $g(N) = N$; J was the yield of plant biomass per unit water consumed; plant biomass was being lost due to mortality and maintenance at rate MN; and finally, D_1 and D_2 were the dispersal and diffusion coefficients of the plants and water, respectively.

Specifically, Klausmeier's ([98]) interaction-dispersion-advection model system for vegetation and surface water contained the slope-gradient advection term

$$V \frac{\partial W}{\partial X}$$

where V was the downhill water flow speed in the negative X-direction in its surface water equation, rather than our diffusive one. Numerical integration of that system yielded no periodic vegetative patterns for flat ground and migrating stripes when the downhill flow of surface water was allowed for hillsides. Sherratt ([225]) and Sherratt and Lord ([226]) presented detailed theoretical linear stability and nonlinear bifurcation analyses, respectively, of pattern formation in a one-dimensional version of the Klausmeier ([98]) model, derived formulae for the wavelength and migration speed of the predicted patterns, and systematically investigated how these depended on model parameters. We have extended that model system in the case of flat ground for which $V = 0$, by replacing the advection term in the surface water equation with

$$D_2 \nabla_2^2 W.$$

In doing so, we have reversed the procedure employed by Rietkerk *et al.* ([186]) in extending their partial differential interaction-diffusion equation system for plant density, soil water, and surface water to the case of sloping ground. Our main purpose in introducing this modification, as stated above, was to devise an extension of Klausmeier's ([98]) model which, while retaining its interaction terms, permitted the occurrence of striped vegetative patterns for flat ground in accordance with the observations of Couteron *et al.* ([39]).

We first sought uniform stationary solutions of our system given by

$$N \equiv N_0, \ W \equiv W_0$$

satisfying

$$F(N_0, W_0) = G(N_0, W_0) = 0$$

and found the following equilibrium points:

$$N_0 = 0, \ W_0 = \frac{A}{L};$$

$$N_0 = N_e^{\pm} = \frac{AJ}{2M} \pm \left[\left(\frac{AJ}{2M} \right)^2 - \frac{L}{R} \right]^{1/2} \quad \text{for} \ \left(\frac{AJ}{2M} \right)^2 \geq \frac{L}{R},$$

$$W_0 = W_e^{\pm} = \frac{M}{RJN_e^{\pm}}.$$

Here the equilibrium point with $N_0 = 0$, which corresponded to bare ground or no vegetation, always existed while, when $N_e^{\pm}$ were real and positive, the other two equilibrium points corresponded to a situation of spatially homogeneous vegetative distributions.

We next examined the linear stability behavior of these equilibrium points and, as usual for prospective Turing instabilities, were interested in those states that were stable in the absence of diffusive effects but could become unstable once those effects were taken into account.

We examined this behavior for our equilibrium points sequentially. To examine the linear stability of the bare ground equilibrium point, we considered a normal mode solution of our system of the form

$$N(X, Y, \tau) = 0 + \varepsilon_1 N_{11} \cos(QX)e^{\Sigma \tau} + O(\varepsilon_1^2),$$

$$W(X, Y, \tau) = \frac{A}{L} + \varepsilon_1 W_{11} \cos(QX)e^{\Sigma \tau} + O(\varepsilon_1^2);$$

where

$$|\varepsilon_1| \ll 1; \quad N_{11}^2 + W_{11}^2 \neq 0; \quad Q \geq 0.$$

Here, Q and Σ represented the wavenumber and growth rate of the perturbation quantities, respectively. Upon substituting this solution into our basic equations, neglecting terms of $O(\varepsilon_1^2)$, and canceling the resulting common factor, we obtained an uncoupled system which implied that

$$\Sigma = \Sigma_1 = -(D_1 Q^2 + M) < 0 \text{ or } \Sigma = \Sigma_2 = -(D_2 Q^2 + L) < 0.$$

Thus, we concluded that this equilibrium point was linearly stable for all parameter values and need not be considered in any investigation of Turing-type pattern formation for our model. For ease of exposition, we then denoted the components of the first community equilibrium point (N_e^+, W_e^+) by $N_e = N_e^+$, $W_e = W_e^+$; and introducing the nondimensional variables and parameters

$$(x, y) = \left(\frac{L}{D_2}\right)^{1/2} (X, Y), \ t = L\tau, \ n = \frac{N}{N_e}, \ w = \frac{W}{W_e};$$

$$a = \frac{AR^{1/2}}{L^{3/2}}, \ \alpha = \frac{M}{L}, \ \mu = \frac{D_1}{D_2},$$

$$\beta = \left(\frac{R}{L}\right) N_e^2 = (v + \sqrt{v^2 - 1})^2 \text{ for } v = \frac{a}{2\alpha} \geq 1;$$

transformed our basic system into

$$\frac{\partial n}{\partial t} = \alpha(wn^2 - n) + \mu \nabla^2 n, \ \frac{\partial w}{\partial t} = 1 - w + \beta(1 - wn^2) + \nabla^2 w;$$

where $\nabla^2 \equiv \partial^2/\partial x^2 + \partial^2/\partial y^2$. Observe that this community equilibrium point corresponds to $n = w \equiv 1$ in our dimensionless variables. Hence, to examine its linear stability, we considered a solution to the nondimensional system of the form

$$n(x, t) = 1 + \varepsilon_1 n_{11} \cos(qx)e^{\sigma t} + O(\varepsilon_1^2),$$

$$w(x, t) = 1 + \varepsilon_1 w_{11} \cos(qx)e^{\sigma t} + O(\varepsilon_1^2);$$

where $q \geq 0$ and σ represented the dimensionless wave number and growth rate of these perturbation quantities (i.e., $|\varepsilon_1| \ll 1$), while the constants n_{11} and w_{11} satisfied the nontriviality condition $n_{11}^2 + w_{11}^2 \neq 0$; and found in the usual way from the Pat Munroe algorithm that the vanishing of the determinant of its coefficients yielded the quadratic secular equation or dispersion relation given by

$$\mathcal{D}(\sigma; q^2) = \sigma^2 + [(1 + \mu)q^2 + \beta + 1 - \alpha]\sigma + \mu q^4 + [(\beta + 1)\mu - \alpha]q^2 + \alpha(\beta - 1) = 0.$$

Recalling that a quadratic secular equation had roots with negative real parts provided its coefficients were positive, to guarantee the onset of a diffusive instability ($q^2 > 0$) from a uniform steady state that was stable to linear homogeneous perturbations ($q^2 = 0$), we required

$$\beta + 1 - \alpha > 0, \ \alpha(\beta - 1) > 0 \text{ for } q^2 = 0;$$

$$\beta < \beta_0(q^2; \alpha, \mu) = \frac{(q^2 + 1)(\alpha - \mu q^2)}{\mu q^2 + \alpha} \text{ for } q^2 > 0.$$

Since, from its definition

$$\beta \geq 1 \text{ for } a \geq 2\alpha,$$

we noted that the conditions for $q^2 = 0$ were satisfied identically, provided α lay in the range

$$0 < \alpha < 2 \Rightarrow \beta + 1 - \alpha \geq 2 - \alpha > 0.$$

For fixed values of α and μ, the curve $\beta = \beta_0(q^2; \alpha, \mu)$ in the $q^2 \geq 0$ and $\beta \geq 1$ portion of the q^2-β plane was marginal, in that it separated the linearly stable region where $\beta > \beta_0(q^2; \alpha, \mu)$ from the unstable region where $1 < \beta < \beta_0(q^2; \alpha, \mu)$. This marginal stability curve, the linearly stable region, and the stable region could be characterized by $\sigma_c = 0$, $\text{Re}(\sigma_c) < 0$, and $\sigma_c > 0$, respectively, where σ_c represented that root of our secular equation having the largest real part. That curve had a maximum at the point (q_c^2, β_c) with components (derived later in a more general context)

$$q_c^2(\alpha; \mu) = \frac{-\alpha + \sqrt{2\alpha(\alpha - \mu)}}{\mu},$$

$$\beta_c(\alpha; \mu) = \frac{(\alpha + \mu)^2}{\mu[3\alpha - \mu + 2\sqrt{2\alpha(\alpha - \mu)}]} \text{ for } \alpha > 2\mu;$$

since $q_c^2 = 0$ and $\beta_c = 1$, should $\alpha = 2\mu$. That curve is plotted in Fig. 12.1. Note, when $\beta > \beta_c$, there existed no squared wavenumbers q^2 corresponding to growing disturbances, while when $1 < \beta < \beta_c$ there existed an interval of such squared wavenumbers centered about q_c^2 for which $\sigma_c > 0$. In particular, taking $q^2 \equiv q_c^2$, as we did in our nonlinear stability analyses, it was possible for fixed α and μ to represent $\sigma_c = \sigma_c(\beta)$ such that for β sufficiently close to β_c, σ_c was still real when $\beta > \beta_c$. Then $\sigma_c(\beta)$ satisfied the conditions $\sigma_c(\beta) < 0$ for $\beta > \beta_c$, $\sigma_c(\beta_c) = 0$, and $\sigma_c(\beta) > 0$ for $1 < \beta < \beta_c$ (see Fig. 12.1). Thus, the locus $\beta = \beta_c(\alpha; \mu)$ was a marginal stability curve in the α-β plane with μ fixed. We plot that locus in Fig. 12.2 for $\alpha > 2\mu$ and $\beta > 1$ with $\mu = 0.001$ corresponding to the typical diffusion coefficient values as reported by Rietkerk *et al.* ([186])

$$D_1 = 0.1\frac{\text{m}^2}{\text{d}}, \ D_2 = 100\frac{\text{m}^2}{\text{d}}.$$

For comparison purposes of our eventual results with those of Klausmeier ([98]) and Sherratt ([225]), we wished to plot this Turing bifurcation curve of Fig. 12.2 as a surface in α-μ-a space. This was something that I started thinking about during Bonni's preliminary doctoral oral exam, being prompted by a Robert Dillon question. Given that its eventual determination represents just the sort of Rabbit Pulled Out of a Hat, as described in Chapter 1 that now never seems to appear in research disseminations, which was a deficiency we tried to alleviate by attempting to include it without success in the publication Kealy and Wollkind ([92]) from Bonni's thesis, I am going to give a detailed historical account of the complete evolution of its derivation in what follows.

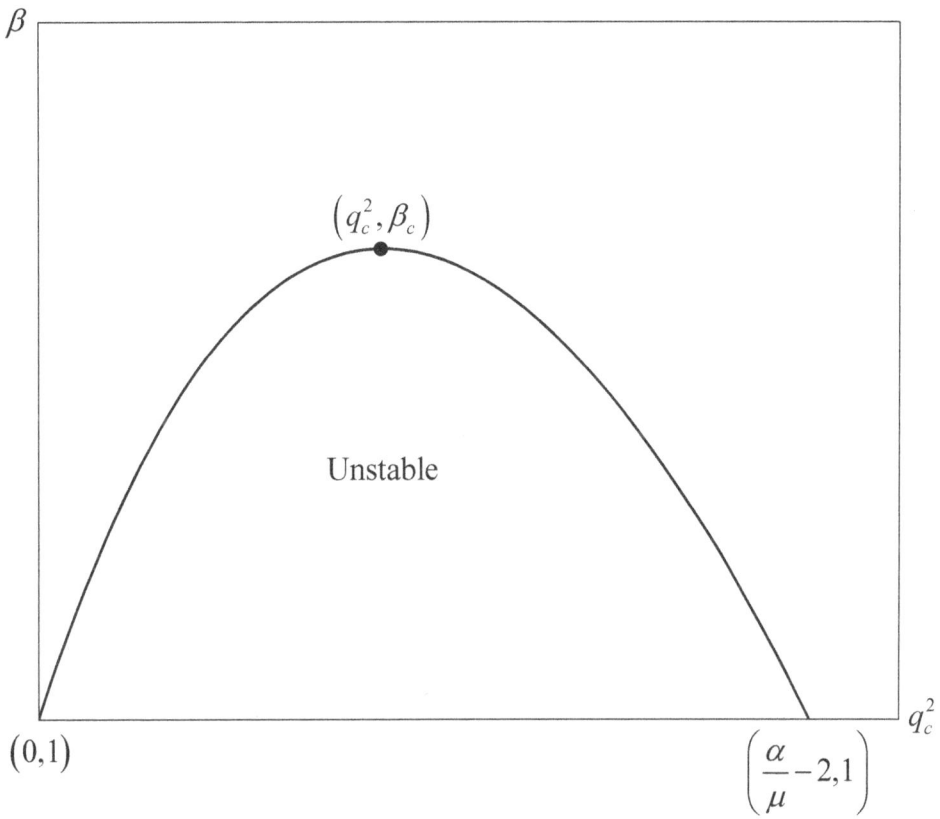

FIGURE 12.1
Plot of $\beta = \beta_0(q^2; \alpha, \mu)$ in the q^2-β plane for $\beta > 1$ and $a > 2\mu$.

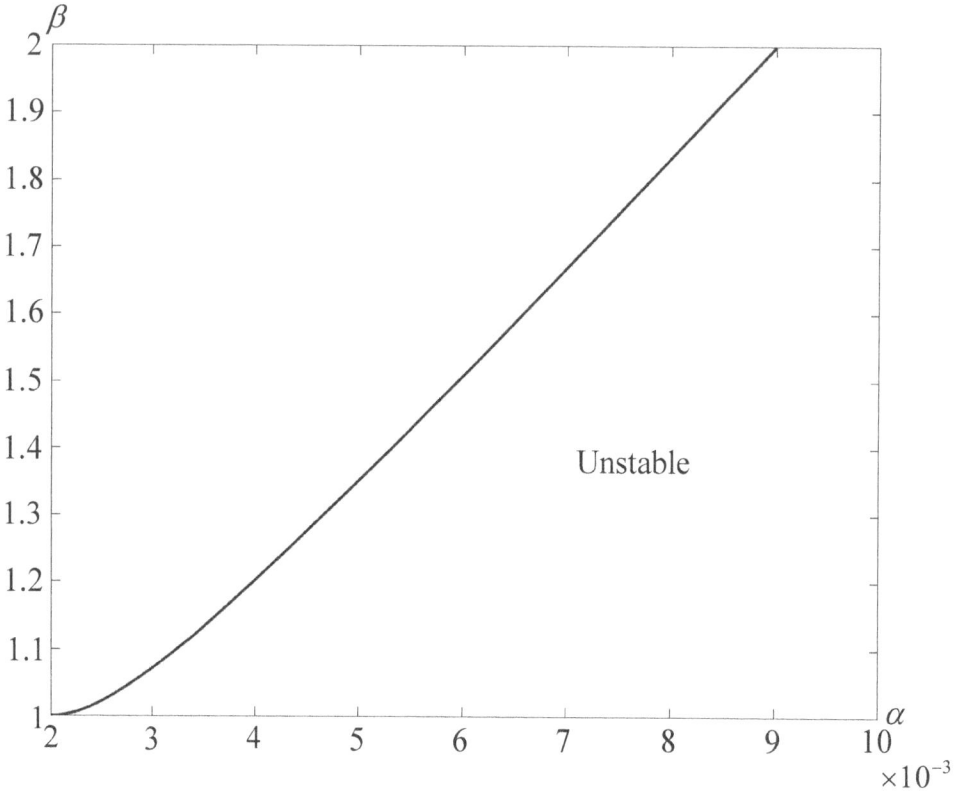

FIGURE 12.2
Plot of the marginal stability curve $\beta = \beta_c(\alpha; \mu)$ in the α-β plane for $\beta > 1$, $a > 2\mu$, and $\mu = 0.001$.

Noting that the expression $v + \sqrt{v^2 - 1}$ where $v = a/(2\alpha)$ appeared in the definition of β and recalling that the same expression appeared in the inverse hyperbolic cosine function which can be demonstrated by the interlude:

$$v = \cosh(z) = \frac{e^z + e^{-z}}{2} \geq 1 \Rightarrow (e^z)^2 - 2v(e^z) + 1 = 0$$

$$\Rightarrow e^z = v + \sqrt{v^2 - 1} \Rightarrow z = \ln(v + \sqrt{v^2 - 1}) = \mathrm{argcosh}(\nu);$$

where argcosh denotes the inverse hyperbolic cosine function; allowed us to proceed to solve the relation

$$\beta = (v + \sqrt{v^2 - 1})^2$$

to obtain a as a function of β:

$$\ln(\beta) = 2\ln(v + \sqrt{v^2 - 1})$$

$$\Rightarrow \left(\frac{1}{2}\right)\ln(\beta) = \ln(\beta^{1/2}) = \ln(v + \sqrt{v^2 - 1}) = \mathrm{argcosh}(v)$$

$$\Rightarrow v = \frac{a}{2\alpha} = \cosh(\ln(\beta^{1/2})) = \frac{e^{\ln(\beta^{1/2})} + e^{-\ln(\beta^{1/2})}}{2}$$

$$\Rightarrow a = \alpha(\beta^{1/2} + \beta^{-1/2}).$$

This most definitely represented the Pulling of the "a" Rabbit Out of the "argcosh" Hat. Once we had obtained a as a function of β by using the indirect method just developed, it occurred to us that this function could be obtained directly instead. That is, from the definition for β,

$$\beta^{1/2} = v + \sqrt{v^2 - 1} \Rightarrow \beta^{-1/2} = \frac{1}{v + \sqrt{v^2 - 1}} = v - \sqrt{v^2 - 1}$$

since $(v + \sqrt{v^2 - 1})(v - \sqrt{v^2 - 1}) = 1$. Then

$$\beta^{1/2} + \beta^{-1/2} = (v + \sqrt{v^2 - 1}) + (v - \sqrt{v^2 - 1}) = 2v = \frac{a}{\alpha}$$

$$\Rightarrow a = \alpha(\beta^{1/2} + \beta^{-1/2}).$$

This is an example of a method of which one thinks of immediately that takes a longer time to produce the desired result than a more elegant alternative method that takes a much longer time to deduce. Then, the best strategy to achieve that result in the shortest time is to employ the first rather than the second method, even though the former is slower than the latter once both have been deduced. Hence, making use of this function yields the Turing instability surface

$$a = a_c(\alpha; \mu) = \alpha[\beta_c^{1/2}(\alpha; \mu) + \beta_c^{-1/2}(\alpha; \mu)];$$

and the plane $a = 2\alpha$ plotted in Fig. 12.3, corresponding to $\beta = \beta_c(\alpha; \mu)$ and 1, respectively. Finally, we observed that an analogous examination of the linear stability of the remaining community equilibrium point would yield the same nondimensional system, except that now $\beta = (v - \sqrt{v^2 - 1})^2 < 1$ for $v > 1$ and hence, violating one of the $q^2 = 0$ conditions, it also need not be considered in any investigation of Turing-type pattern formation for our model.

In order to account for the nonlinear terms in this ecological Turing pattern formation plant-surface water nondimensional system, we performed both longitudinal and hexagonal

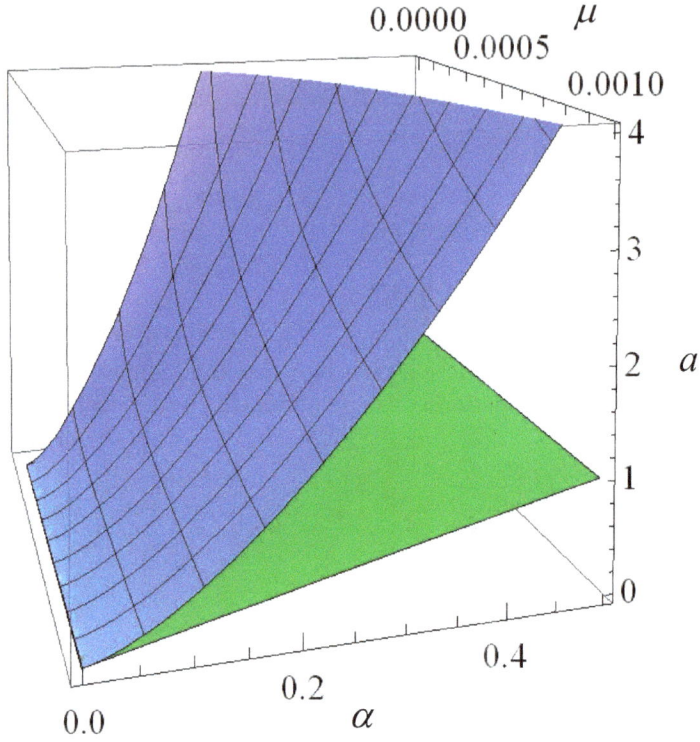

FIGURE 12.3
Three-dimensional plots of the Turing instability surface $a = a_c(\alpha; \mu)$ in blue and the planar surface $a = 2\alpha$ in green for $2\mu < \alpha < 0.5$ and $0.0001 < \mu < 0.0010$. The cross sections with the $\mu = 0.0010$ plane correspond to $\beta = \beta_c(\alpha; \mu)$ in Fig. 12.2 and $\beta = 1$, respectively.

planform stability analyses of its first community equilibrium point in the usual way, by considering β as a bifurcation parameter and taking the limit as $\beta \to \beta_c(\alpha; \mu)$. Having developed analyses of this sort from first principles in the previous chapters concerning the evolution equations introduced in Chapters 8-11, we summarize Bonni's nonlinear stability results graphically by plotting the relevant Landau coefficients, represented symbolically as functions of α for a fixed value of μ. In particular, for our longitudinal planform analysis, we found that the Landau amplitude equation for $A_1 = A_1(t)$

$$\frac{dA_1}{dt} \sim \sigma A_1 - a_1 A_1^3$$

had coefficients

$$\sigma = \sigma_c(\beta; \alpha, \mu), \; a_1 = a_1(\alpha; \mu);$$

where $a_1(\alpha; \mu)$ is plotted versus α in Fig. 12.4 for $\alpha > 2\mu$, with the typical value of $\mu = 0.001$ employed in Fig. 12.2. That Landau coefficient had asymptotic representation

$$a_1(\alpha; \mu) \sim m_0 \alpha + \frac{80 m_0 \mu^2}{2\mu - \alpha} \text{ as } \mu \to 0$$

with

$$m_0 = \frac{10\sqrt{2} - 7}{36} \cong 0.2.$$

Note, Fig. 12.4 consists of two parts: A left-hand panel for which a_1 is plotted for the same α-domain as in Fig. 12.3: Namely, $2\mu < \alpha < 0.5$; and a right-hand one, which is an enlargement of the former plot restricted to the lower end of this domain. From Fig. 12.4 we observed that a_1 had a zero at $\alpha = \alpha_0(\mu)$ and satisfied

$$a_1 < 0 \text{ for } 2\mu < \alpha < \alpha_0(\mu), \; a_1 > 0 \text{ for } \alpha > \alpha_0(\mu);$$

where, specifically,

$$\alpha_0(0.001) = 0.0101$$

or, more generally,

$$\alpha_0(\mu) \cong 10\mu.$$

Here a_1 had a linear asymptote of the form $m_0\alpha$, which virtually coincided with it for $\alpha \geq \alpha_0(\mu)$ and its asymptotic representation had a zero at $\alpha = 10\mu$ by virtue of

$$\frac{80\mu^2}{10\mu - 2\mu} = \frac{80\mu^2}{8\mu} = 10\mu.$$

Since the ecologically meaningful values $\alpha_{tree} = 0.045$ and $\alpha_{grass} = 0.45$ (Klausmeier, [98]) lay in the α-interval of supercriticality for the typical ratios of the coefficient of plant dispersion to that of surface water diffusion $0.0001 < \mu < 0.001$ (Rietkerk et al., [186]), we treated a_1 as being identically positive during the rest of Bonni's analyses.

Again, given this condition, the amplitude function $A_1(t)$ underwent a standard supercritical pitchfork bifurcation at $\beta = \beta_c$ from which we concluded:

(i) For $\beta > \beta_c$, the undisturbed state $A_1 = 0$ was stable, yielding a uniform homogeneous vegetative pattern $n(x,t) \sim 1$.

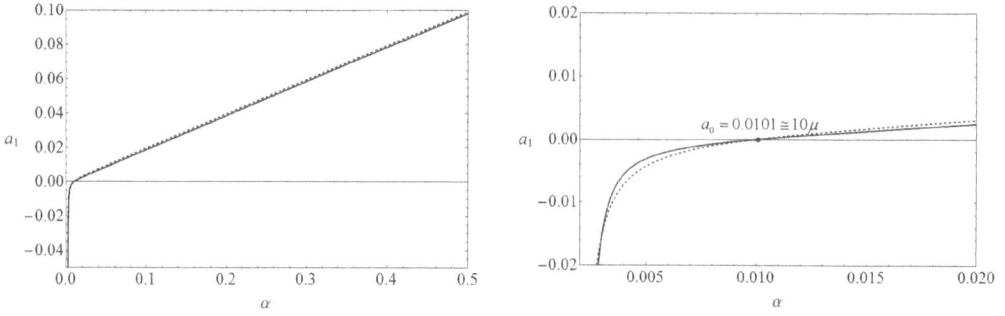

FIGURE 12.4
Plots of the Landau coefficient $a_1(\alpha; \mu)$ versus α for $\alpha > 2\mu$ with $\mu = 0.001$ (solid curve) and its asymptotic approximation (dashed curve) deduced by the method of matched asymptotes (see Chapter 4). The right-hand panel is an enlargement about its zero $\alpha_0 = 0.0101$.

(ii) For $1 < \beta < \beta_c$, the re-equilibrated state $A_1 = A_e = (\sigma_c/a_1)^{1/2}$ was stable, yielding a periodic one-dimensional vegetative pattern consisting of stationary parallel stripes

$$n(x, t) \sim n_e(x) = 1 + A_e \cos\left(\frac{2\pi x}{\lambda_c}\right)$$

of characteristic wavelength

$$\lambda_c = \frac{2\pi}{q_c} \quad \text{and} \quad \lambda_c^* = \left(\frac{D_2}{L}\right)^{1/2} \lambda_c,$$

in dimensionless and dimensional variables, respectively.

We synthesized the pattern formation results of our linear and longitudinal planform nonlinear stability analyses in the α-a plane of Fig. 12.5 for $\mu = 0.001$. Here, we plotted simultaneously both the Turing boundary curve $a = a_c(\alpha; \mu)$ and the straight line $a = 2\alpha$ of Fig. 12.3 for $\alpha_0 < \alpha < 0.5$ and $\mu = 0.001$, with the three regions $0 < a < 2\alpha$, $2\alpha < a < a_c(\alpha; \mu)$, and $a > a_c(\alpha; \mu)$ corresponding to bare ground, stationary striped vegetative patterns, and non zero homogeneous distributions of vegetation, respectively, being identified in that parameter space. As indicated by Sherratt ([225]) in his analogous Fig. 12.6, when the rainfall is too low for stripe formation, vegetation is absent while when the rainfall is too high for stripes, there is a non zero homogeneous distribution of vegetation. Hence, for a given amount of plant loss, stripes can only occur in an intermediate rainfall interval bounded by those lower and upper threshold values.

Wishing to refine these one-dimensional pattern formation predictions summarized in Fig. 12.5 so that they could be compared with the other types of dryland vegetative patterns described in the previous chapter, we next performed the same two-dimensional hexagonal planform analysis of the community equilibrium solution to our system as that employed therein by considering perturbation expansions consisting to lowest order of three modes, of which the wavenumber vectors were 120° apart in x-y plane and the long-time asymptotic behavior was governed by a set of six ordinary differential equations for its three amplitudes and phases containing the growth rate σ and the three Landau coefficients $a_{0,1,2}$. Given that σ and a_1 had been determined by the longitudinal planform analysis, only a_0 and a_2 still

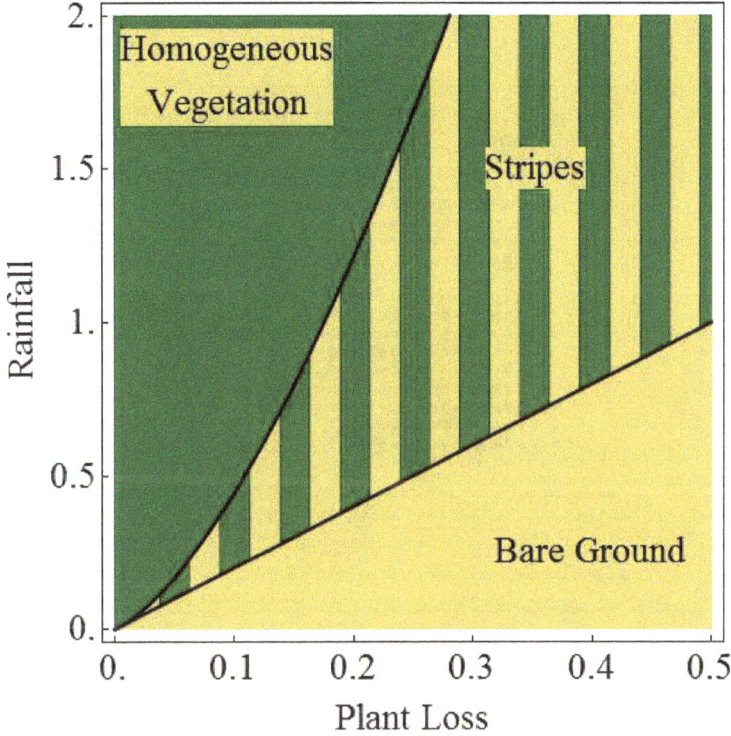

FIGURE 12.5
Stability diagram in the α-a plane for the one-dimensional linear and nonlinear analyses of our interaction-diffusion plant-suface water model system with $\alpha > \alpha_0(\mu) \cong 10\mu$ and $\mu = 0.001$ denoting the predicted vegetative patterns. The curves depicted in this figure are cross sections of the $\mu = 0.0010$ plane with the surfaces in Fig. 12.3. Hence, the upper one is the Turing boundary $a = a_c(\alpha; \mu)$ with $\mu = 0.001$ and the lower one, the straight line $a = 2\alpha$. Here, α and a are measures of the plant loss and rainfall, respectively, accounting for their axis-labels.

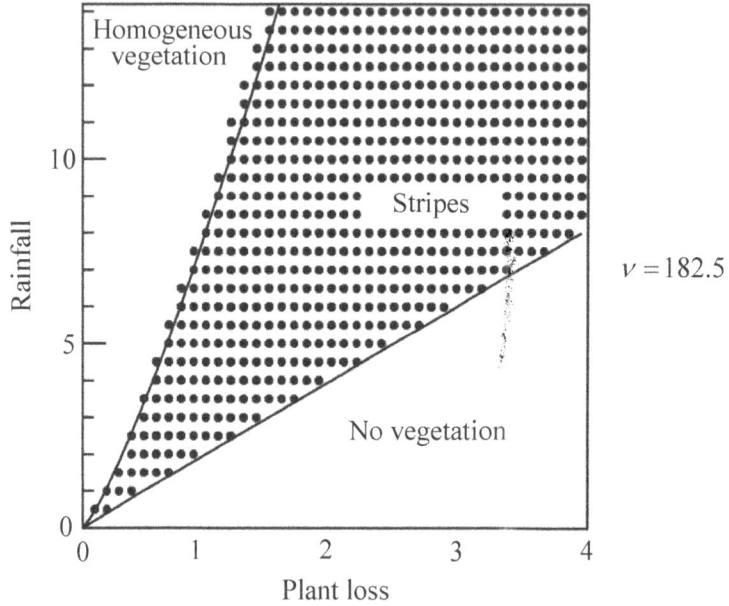

FIGURE 12.6
An illustration from Sherratt ([225]) in the same plane as Fig. 12.5 comparing his theoretical prediction (the region between the two solid curves) and numerical calculations (the filled circles) for stripe formation in the Klausmeier ([98]) model. Here $\nu = V/(D_1 L)^{1/2}$ is a nondimensional flow velocity which measures the strength of this downhill advective effect.

needed to be evaluated. Proceeding in the identical manner as the one used to determine a_1 in that analysis, we then imposed the relevant Fredholm-type solvability conditions to yield formulae for those remaining two Landau coefficients represented symbolically by

$$a_0 = a_0(\alpha; \mu), \ a_2 = a_2(\alpha; \mu);$$

catalogued the relevant equivalence classes of critical points for the amplitude-phase equations; summarized their orbital stability behavior; and identified the potentially stable ones with various two-dimensional vegetative pattern, obtaining the following critical point identifications: I, homogeneous distributions; II, stripes; III^+, spots; and III^-, gaps. Note, in this context, that I and II represented the same identifications as depicted in Fig. 12.5. The contour plots relevant to critical points $\text{III}^\pm$ are depicted in Fig. 12.7, where part (a) is for III^+ and part (b), for III^-. Again, the spatial variables are measured in units of λ_c while dark and light regions correspond to high and low densities, respectively, in accordance with the aerial photograph appearing in Fig. 12.8. These identifications and figures, not surprisingly, were identical to those in Chapter 11 and Fig. 12.7 has been included to demonstrate the differences that inevitably occur when two separate groups try to create the same thing since Nichaphat generated all the figures of Chapter 11, while Bonni and her colleague Richard Cangelosi produced the corresponding ones of this chapter.

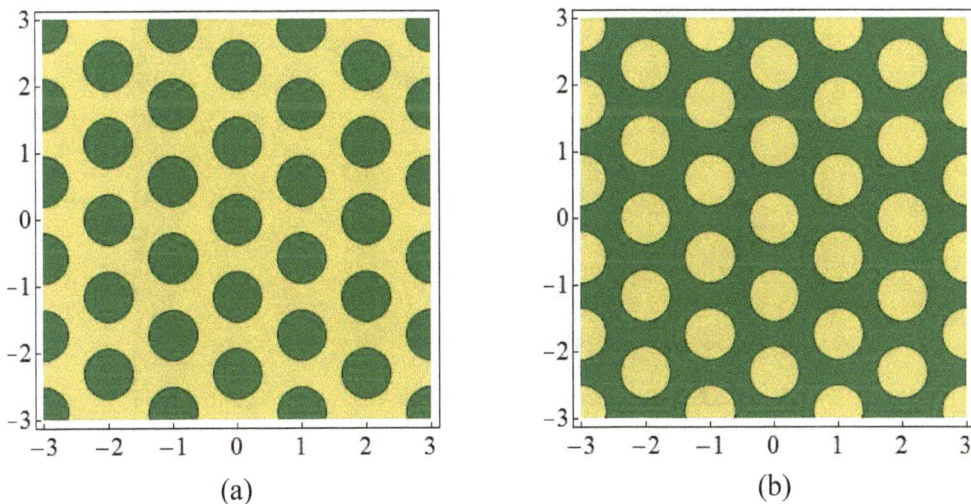

FIGURE 12.7
Contour plots of the vegetative (a) spots of critical point III^+ and (b) gaps of III^-.

We then plotted $a_1 + 4a_2$, $a_1 + a_2$, $2a_2 - a_1$, a_2, and a_0 versus α for $\alpha > \alpha_0(\mu) \cong 10\mu$ with $\mu = 0.001$, which is reproduced in Fig. 12.9 for $\alpha_0 < \alpha < 0.5$, and found that a_0 was identically positive, while all the other quantities in that figure could be treated as positive as well, having zeroes less than α_0 given approximately in μ-units by 7.5, 8.4, 4.6, and 7, respectively. Thus, employing the first row of Table 9.1 since a_0, $2a_2 - a_1 > 0$, we determined, as usual, that our critical point I was stable for $\sigma < 0$; II, for $\sigma > \sigma_1$; and III^-, for $\sigma_{-1} < \sigma < \sigma_2$, while these conditions precluded the stability of critical point III^+ and hence our model did not predict the occurrence of any vegetative spotted patterning of this type. In order to represent these predictions graphically in the α-a plane, it was necessary

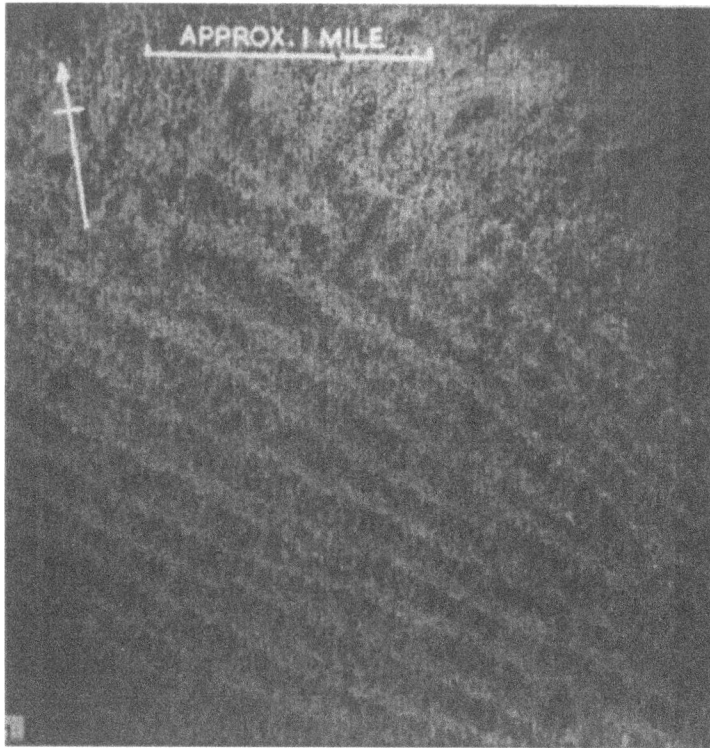

FIGURE 12.8
An aerial photograph of tiger bush patterns consisting of Acacia trees on a plateau in the Go-Gub area of Somaliland from Lefever and Lejeune ([109]). The bands of dense vegetation (dark) and their separating lanes (bright) are approximately 100 m and 50 m wide, respectively.

for us to genealize the procedure that produced our marginal stability curve. Toward that end, we solved our secular equation with $\sigma = \sigma_c$ to obtain

$$\beta_{\sigma_c}(q^2; \alpha, \mu) = \frac{-\mu q^4 + [\alpha - \mu - \sigma_c(\mu+1)]q^2 - \sigma_c^2 + (\alpha-1)\sigma_c + \alpha}{\mu q^2 + \alpha + \sigma_c}.$$

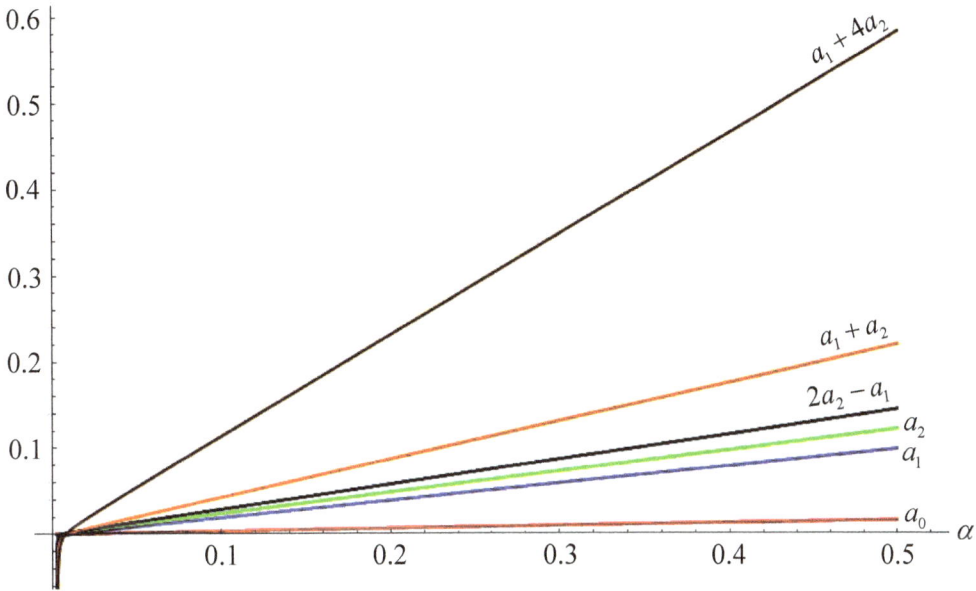

FIGURE 12.9
Plots of $a_1 + 4a_2$, $a_1 + a_2$, $2a_2 - a_1$, and a_0 versus α for $\alpha > \alpha_0(\mu) \cong 10\mu$ with $\mu = 0.001$ where the plots of a_1 and a_2 are presented for comparison purposes.

To find the critical point of this function, we next considered $d\beta_{\sigma_c}(q_c^2; \alpha, \mu)/dq^2 = 0$ or

$$[-2\mu q_c^2 + \alpha - \mu - \sigma_c(\mu+1)](\mu q_c^2 + \alpha + \sigma_c)$$
$$= \mu\{-\mu q_c^4 + [\alpha - \mu - \sigma_c(\mu+1)]q_c^2 - \sigma_c^2 + (\alpha-1)\sigma_c + \alpha\},$$

which implied that

$$(\mu q_c^2)^2 + 2(\sigma_c + \alpha)(\mu q_c^2) + \sigma_c^2 + 2\alpha\mu\sigma_c + \alpha(2\mu - \alpha) = 0$$

or solving

$$\mu q_c^2 = -\sigma_c - \alpha + \sqrt{2\alpha[\alpha - \mu + (1-\mu)\sigma_c]}.$$

Hence,

$$\beta_c = \beta_{\sigma_c}(q_c^2; \alpha, \mu) = \frac{-2\mu q_c^2 + \alpha - \mu - \sigma_c(\mu+1)}{\mu}$$

which upon substituting for μq_c^2 yielded

$$\beta_c = \frac{3\alpha - \mu + (1-\mu)\sigma_c - 2\sqrt{2\alpha[\alpha - \mu + (1-\mu)\sigma_c]}}{\mu}.$$

Then defining

$$f^\pm(\alpha, \mu, \sigma_c) = 3\alpha - \mu + (1 - \mu)\sigma_c \pm 2\sqrt{2\alpha[\alpha - \mu + (1 - \mu)\sigma_c]},$$

multiplying the numerator and the denominator of this fraction by $f^+(\alpha, \mu, \sigma_c)$ to rationalize it, and noting from the difference of squares that

$$
\begin{aligned}
f^+(\alpha, \mu, \sigma_c)f^-(\alpha, \mu, \sigma_c) &= [3\alpha - \mu + (1 - \mu)\sigma_c]^2 - 8\alpha[\alpha - \mu + (1 - \mu)\sigma_c] \\
&= (3\alpha - \mu)^2 + 2(3\alpha - \mu)(1 - \mu)\sigma_c + (1 - \mu)^2\sigma_c^2 \\
&\quad - 8\alpha(\alpha - \mu) - 8\alpha(1 - \mu)\sigma_c \\
&= 9\alpha^2 - 6\alpha\mu + \mu^2 - 8\alpha^2 + 8\alpha\mu \\
&\quad + [2(3\alpha - \mu) - 8\alpha](1 - \mu)\sigma_c + (1 - \mu)^2\sigma_c^2 \\
&= \alpha^2 + 2\alpha\mu + \mu^2 + (6\alpha - 2\mu - 8\alpha)(1 - \mu)\sigma_c \\
&\quad + (1 - \mu)^2\sigma_c^2 \\
&= (\alpha + \mu)^2 - 2(\alpha + \mu)(1 - \mu)\sigma_c + (1 - \mu)^2\sigma_c^2 \\
&= [\alpha + \mu - (1 - \mu)\sigma_c]^2;
\end{aligned}
$$

we obtained

$$\beta_{\sigma_c}(\alpha; \mu) = \frac{[\alpha + \mu - (1 - \mu)\sigma_c]^2}{\mu[3\alpha - \mu + (1 - \mu)\sigma_c + 2\sqrt{2\alpha\{\alpha - \mu + (1 - \mu)\sigma_c\}}]};$$

and observed that for $\sigma_c = 0$, the critical wavenumber defined above reduced to

$$q_c^2(\alpha; \mu) = \frac{-\alpha + \sqrt{2\alpha(\alpha - \mu)}}{\mu}$$

and

$$\beta_{\sigma_c=0}(\alpha; \mu) = \frac{(\alpha + \mu)^2}{\mu[3\alpha - \mu + 2\sqrt{2\alpha(\alpha - \mu)}]} = \beta_c(\alpha; \mu),$$

consistent with the components of the maximum point of the marginal stability curve.

In order to generate the loci in the α-a plane associated with $\sigma_c = \sigma_j$ for $j = -1$, 1, and 2, we defined

$$\beta = \beta_j(\alpha; \mu) = \beta_{\sigma_j}(\alpha; \mu) \text{ where } \sigma_j = \sigma_j[a_0(\alpha; \mu), a_1(\alpha; \mu), a_2(\alpha; \mu)]$$

and again, as in Fig. 12.5, the analogous generalized marginal curves related to them by

$$a = a_{\sigma_j}(\alpha; \mu) = \alpha[\beta_j^{1/2}(\alpha; \mu) + \beta_j^{-1/2}(\alpha; \mu)],$$

which are plotted versus α in Figs. 12.10 and 12.11 for $\mu = 0.001$.

Since all the quantities required for the identification of the predicted vegetative patterns in the first row of Table 9.1 had been evaluated, we could identify the graphical regions corresponding to these patterns in the α-a plane of Fig. 12.11 containing the loci $a_{\sigma_j}(\alpha; \mu)$ for $j = -1$, 1, and 2 with $\mu = 0.001$ by plotting the line $a = 2\alpha$ as well and cataloguing this correspondence in Table 12.1 that also includes a similar interpretation relative to Fig. 12.10. Upon examination of Fig. 12.11 in conjunction with Table 12.1, we saw that when compared to our one-dimensional pattern formation predictions summarized in Fig. 12.5, the principal refinement provided by our two-dimensional analysis was a region flanking the Turing boundary $a = a_c$ in which hexagonal close-packed arrays of vegetative gaps could occur. Characterized by $a_{\sigma_2} < a < a_{\sigma_{-1}}$, this region was further divided into the following three subregions:

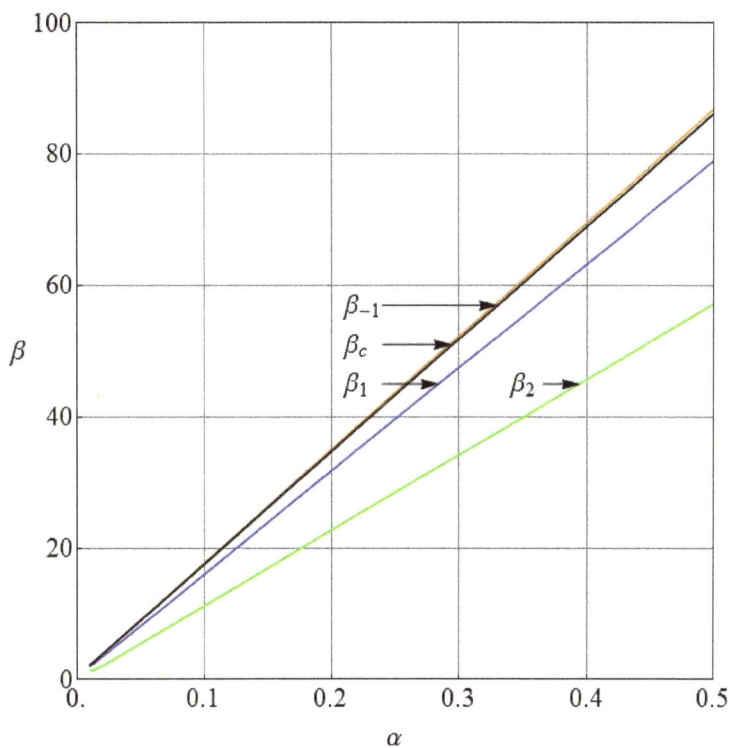

FIGURE 12.10
Plots of the marginal curves β_j for $j = -1, 1, 2$ and β_c versus α with $\mu = 0.001$.

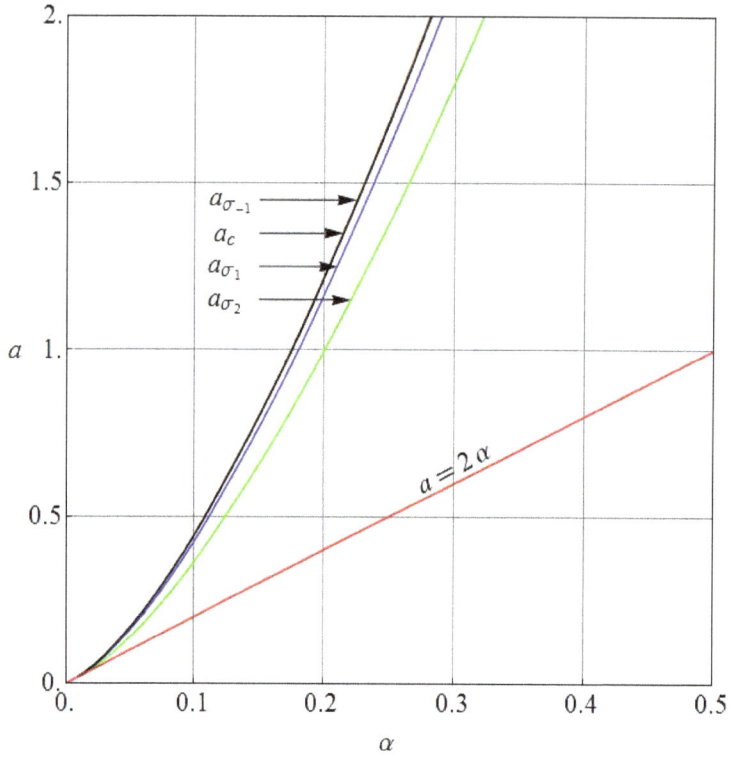

FIGURE 12.11

Plots of the marginal curves $a_{\sigma_j}(\alpha; \mu)$ for $j = -1, 1, 2$, as well as a_c and the straight line $a = 2\alpha$ of Fig. 12.5 versus α with $\mu = 0.001$.

TABLE 12.1

Morphological stability predictions for Figs. 12.10 and 12.11.

β Range	α Range	Stable pattern
$\beta > \beta_{-1}$	$a > a_{\sigma_{-1}}$	Homogeneous
$\beta_c < \beta < \beta_{-1}$	$a_c < a < a_{\sigma_{-1}}$	Homogeneous and gaps
$\beta_1 < \beta < \beta_c$	$a_{\sigma_1} < a < a_c$	Gaps
$\beta_2 < \beta < \beta_1$	$a_{\sigma_2} < a < a_{\sigma_1}$	Gaps and stripes
$1 < \beta < \beta_2$	$2\alpha < a < a_{\sigma_2}$	Stripes
	$0 < a < 2\alpha$	Bare ground

(i) $a_{\sigma_2} < a < a_{\sigma_1}$: Here, since $\sigma_1 < \sigma_c < \sigma_2$, there was bistability between gaps and stripes where as usual initial conditions determined which one of this pair of predicted stable equilibrium patterns would be selected.

(ii) $a_{\sigma_1} < a < a_c$: Here, since $0 < \sigma_c < \sigma_1 < \sigma_2$, gaps were the only stable equilibrium pattern.

(iii) $a_c < a < a_{\sigma_{-1}}$: Here, since $\sigma_{-1} < \sigma_c < 0$, there was bistability between gaps and homogeneously distributed vegetation patterns where the hexagonal gap vegetative patterns persisting in this subregion would be subcritical in nature.

Finally, to justify the truncation procedure inherent to the asymptotic representation of the hexagonal planform expansions, it was necessary that our Landau coefficients as usual satisfied the size constraint

$$a_0 \ll (a_1 + 4a_2)^2$$

which we concluded from Fig. 12.9 was valid for our parameter range of interest and hence the vegetative pattern formation predictions of Table 12.1 could be regarded as conclusive.

We were now ready to compare these theoretical predictions for our diffusive instability model with both the existing numerical simulations of Klausmeier ([98]) and the bifurcation results of Sherratt ([225]) and Sherratt and Lord ([226]) on their corresponding differential flow instability model. Given the inherent propensity of our model system for generating stripes, we began by comparing the pattern formation results summarized in Fig. 12.5 with those of Sherratt ([225]) and Klausmeier ([98]). Recall that the interaction-dispersion-advection model system which these authors analyzed differed from the interaction-diffusion model merely by the replacement of our surface water diffusion effect with a downhill slope gradient one. Sherratt ([225]) found the strength of this advective effect to be measured by the dimensionless flow velocity

$$\nu = \frac{V}{(D_1 L)^{1/2}}$$

in that the uphill migrating stripes predicted by their analyses would only occur provided the threshold condition

$$\nu > \nu_c \cong \frac{2\sqrt{2}a^2}{\alpha^{5/2}}$$

was satisfied over the relevant parameter range of Klausmeier ([98]). Sherratt ([225]) summarized his pattern formation results in Fig. 12.6. That figure, which was plotted in the α-a plane for the typical parameter value $\nu = 182.5$ (Klausmeier, [98]), bears a striking qualitative resemblance to our Fig. 12.5 once his regime of migrating stripes is associated with ours of stationary ones. Although his stripe upper stability boundary had to be calculated numerically, Sherratt ([225]) offered the following asymptotic represenation for that curve derived from his simplified approximation for ν_c

$$a \sim 8^{-1/4}\nu^{1/2}\alpha^{5/2} \text{ as } \nu \to \infty.$$

The lower stripe stability boundary was the same as our correspoding one in Fig. 12.5: $a = 2\alpha$. Examining the asymptotic behavior of our Turing instability boundary function in light of

$$q_c^2 \sim \frac{(2\sqrt{2}-1)\alpha}{\mu} \text{ and } \beta_c \sim \frac{(3-2\sqrt{2})\alpha}{\mu} \text{ as } \mu \to 0,$$

we deduced that

$$a = a_c(\alpha;\mu) \sim (3+2\sqrt{2})^{-1/2}\mu^{-1/2}\alpha^{3/2} \text{ as } \mu \to 0,$$

a type of power law very similar in form to Sherratt's ([225]) analogous representation should the obvious identification $\mu \equiv 1/\nu$ be made. Sherratt ([225]) then investigated how both the wavelength and migration speed of his migrating stripes varied with the Klausmeier ([98]) model parameters over his patterned region. In particular, representing this dimensionles wavelength to leading order by

$$\Lambda_c \sim \nu^{1/2}\Lambda_0(\alpha, a),$$

where $\Lambda_0(\alpha, a)$ was a prescribed function independent of ν, he plotted that quantity versus a, with $\alpha = 0.45$, in part (a) and versus α, with $a = 1$, in part (b) of Fig. 12.12, for $\nu = 182.5$.

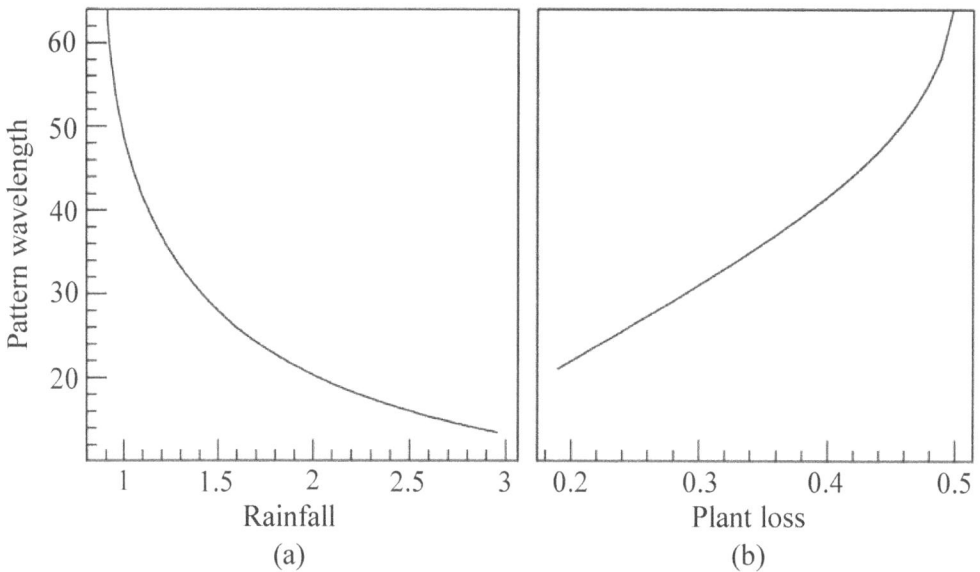

(a)   (b)

FIGURE 12.12
An illustration from Sherratt ([225]) of the stripe wavelength Λ_c versus (a) rainfall a with plant loss $\alpha = 0.45$ and (b) versus α with $a = 1$, both for $\nu = 182.5$ in the Klausmeier ([98]) model over the patterned region of Fig. 12.6.

385

To compare the α and a variations of our dimensionless wavelength for stripes with these results, we needed to reformulate the wavenumber expression for q_c^2 so that it was only a function of β. There were two different ways of doing this: A direct and an indirect method. The first, which consisted of considering the marginal stability curve $\beta = \beta_c(\alpha;\mu)$, solving for

$$\frac{\alpha}{\mu} = 3\beta - 1 + 2\sqrt{2\beta(\beta-1)},$$

and substituting that result into the original expression for q_c^2 to obtain

$$q_c^2 = \beta - 1 + \sqrt{2\beta(\beta-1)} = Q_c^2(\beta),$$

required a significant amount of calculation and cleverness to yield this simplification. The other, which was much more straight-forward to implement, only entailed reformulating our diffusive instability inequality in terms of μ

$$\mu < \mu_0(q^2;\alpha,\beta) = \frac{\alpha(q^2+1-\beta)}{q^2(q^2+\beta+1)} \text{ for } q^2 \geq \beta - 1 \geq 0,$$

and proceeding exactly as before to determine that this marginal stability curve $\mu = \mu_0(q^2;\alpha,\beta)$ had a maximum at its critical point (q_c^2,μ_c) where

$$q_c^2 = Q_c^2(\beta), \ \mu_c = \mu_0(q_c^2;\alpha,\beta) = \frac{\alpha}{2Q_c^2(\beta)+\beta+1}.$$

This approach also had the added advantage of demonstrating the necessary condition for a diffusive instability

$$\mu \leq \mu_c < \frac{\alpha}{\beta+1} < 1$$

by virtue of the parameter inequality satisfied when $q^2 = 0$. Now employing this wavenumber formula and making use of the definition of β, we derived the requisite relationship

$$\lambda_c = \lambda_0(\alpha,a) = \frac{2\pi}{Q_c(\beta)} \text{ with } \beta = (v+\sqrt{v^2-1})^2 \text{ for } v = \frac{a}{2\alpha}.$$

Further, noting that Q_c^2 satisfied the quadratic

$$(Q_c^2)^2 + 2(1-\beta)(Q_c^2) + 1 - \beta^2 = 0,$$

from which it was calculated, we observed that β satisfied the companion quadratic

$$\beta^2 + 2Q_c^2\beta - (Q_c^2+1)^2 = 0 \Rightarrow \beta = -Q_c^2 + \sqrt{Q_c^4 + (Q_c^2+1)^2} = z_c(Q_c).$$

This allowed us to invert our functional relationship for λ_c by implying that

$$a = s(\lambda_c)\alpha \text{ where } s(\lambda_c) = z_c^{1/2}(Q_c) + z_c^{-1/2}(Q_c) \text{ with } Q_c = \frac{2\pi}{\lambda_c}.$$

We felt that the derivation of these formulae represented Pulling two Rabbits Out of a Hat!

Plotting $\lambda_c = \lambda_0(\alpha,a)$ versus a for $\alpha = 0.45$, and versus α for $a = 1$, in the left- and right-hand panels of Fig. 12.13, respectively, we saw that the behavior of our wavelength with variations of either a or α was exactly the same as Sherratt ([225]) portrayed in Fig. 12.12. We observed in this context that although his Λ_c was dependent upon v, our λ_c was independent of μ, and noted that the spatial variables of the Klausmeier ([98]) nondimensional model system were scaled by the dispersion length of the plants $(D_1/L)^{1/2}$, while our spatial variables had been scaled by the diffusion length of the surface water $(D_2/L)^{1/2}$. Hence, introducing the dimensional wavelength appropriate to the Klausmeier ([98]) model system, we obtained that

$$\Lambda_c^* = \left(\frac{D_1}{L}\right)^{1/2} \Lambda_c \sim \left(\frac{D_1}{L}\right)^{1/2} \nu^{1/2} \Lambda_0 = \nu^{1/2} \Lambda_0^*,$$

while our formulation, in conjunction with the definition of μ, implied that the corresponding dimensional wavelength

$$\lambda_c^* = \left(\frac{D_2}{L}\right)^{1/2} \lambda_c = \frac{(D_1/L)^{1/2}}{\mu^{1/2}} \lambda_0 = \frac{1}{\mu^{1/2}} \lambda_0^*,$$

where Λ_0^* and λ_0^* had been scaled with the plants' dispersion length. Thus, using the same scaling for these reference wavelengths, yielded the expected $1/\mu^{1/2}$ dependence of λ_c^* when compared with the dependence of Λ_c^* on $\nu^{1/2}$ given the identification of $\mu \equiv 1/\nu$. Both Bonni and I considered our resolution of this apparent anomaly to be another Rabbit Pulled Out of a Hat.

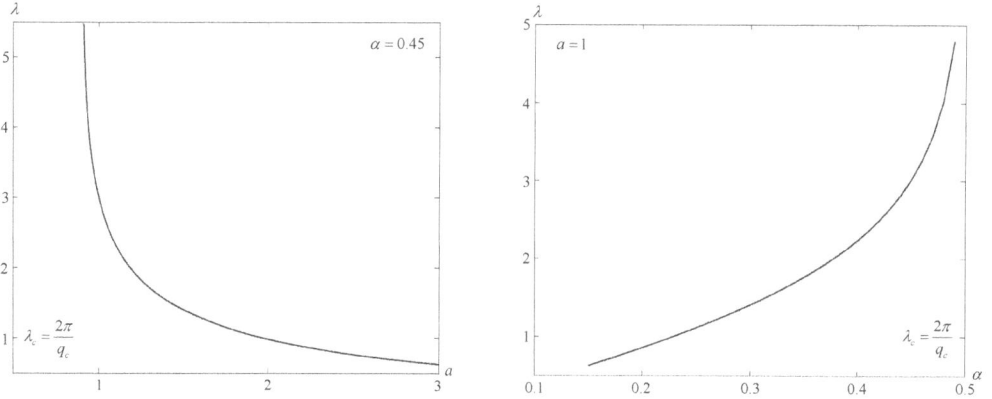

FIGURE 12.13
An illustration of the variation of our pattern wavelength λ_c with (left-hand panel) rainfall a when $\alpha = 0.45$ and (right-hand panel) plant loss α when $a = 1$ over the patterned region of Fig. 12.5. Observe that λ_c increases as a decreases for fixed α or as α increases for fixed a.

Klausmeier's ([98]) numerical results were reported in a plot identical to Sherratt's ([225]) of Fig. 12.6 but that included contours of various constant wavelength Λ_c^* in the migrating stripe patterned region as well. After Klausmeier ([98]), we reproduced our Fig. 12.5 but used the relationship $a = s(\lambda_c)\alpha$ to plot lines corresponding to various constant wavelengths λ_c in the stationary stripe patterned region of the α-a plane of Fig. 12.14 and observed that, along the vertical line $\alpha = 0.45$ and the horizontal line $a = 1$, the wavelength predictions summarized in this figure reduced to the behavior depicted in the left- and right-hand panels of Fig. 12.13, respectively. In that context, we also noted that the line $a = 2\alpha$ corresponded to $\lambda_c \to \infty$.

It remained for us to compare our theoretical predictions with observational evidence. We began by summarizing the description of striped vegetative tiger bush patterns contained in

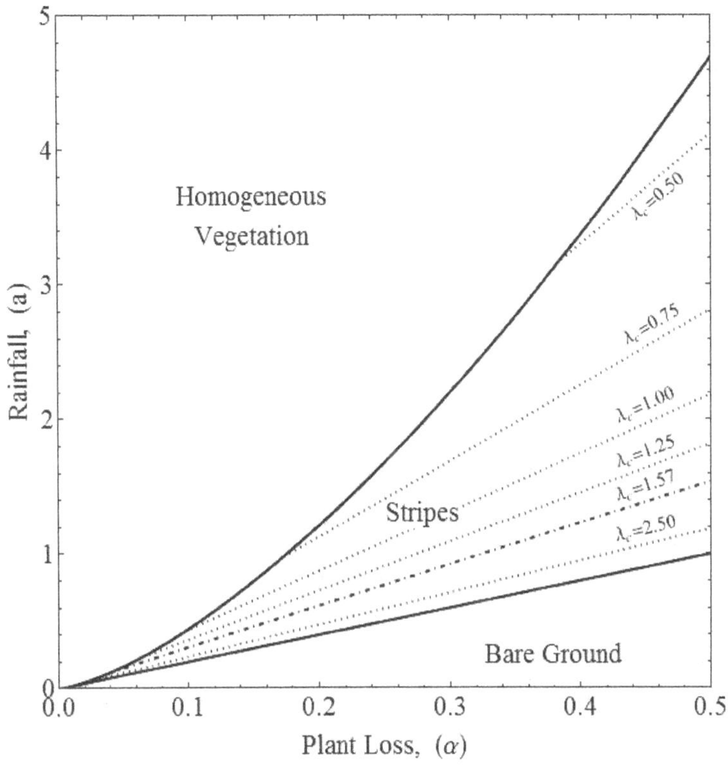

FIGURE 12.14
A reproduction of the α-a plane of Fig. 12.5 denoting the lines of various constant wavelength λ_c in the striped vegetation patterned region as determined by the relationship $a = s(\lambda_c)\alpha$. Observe that the line corresponding to $\lambda_c = 1.57$ has been designated specially.

the seminal study of such patterns by Lefever and Lejeune ([109]). First reported in British Somaliland by Macfaydyen ([130]), that striped appearance which resembles the coat patterning of a tiger accounts for its name in French of *la brousse tigerée* or in English "of the tiger bush" (Clos-Arceduc, [34]). These vegetation stripes consisted entirely of grass, of grass and scrub, or of trees and bushes. It had been found occurring on soils ranging from sand to silt to clay. In the case that the ground surface slopes, the vegetation stripes tended to form orthogonal to this ground slope and an upslope migration of these stripes had been observed for grasses. A similar upslope migration occured for such patterns consisting entirely of trees and bushes, but on a much slower time scale. When the ground surface is practically flat these vegetation band patterns are static and oriented with respect to neighboring declivities or, in their absence, with respect to the system borders. Note that the observation of tiger bush wavelength increasing as average plant density decreases (Lefever and Lejeune, [109]) is consistent with our inverse relationship between λ_c and β implicit to $\lambda_c = \lambda_0(\alpha, a)$. Couteron *et al.* ([39]) catalogued those differences between these types of tiger bush (banded thicket) patterns just described that occur at two sites in West Africa having soil composed of a sand-clay mixture: One found on a gentle slope in northern Burkina Faso; and the other, on a virtually flat plateau in southwest Niger. It was clear that our analysis provided good qualitative agreement with tiger bush patterns occurring in arid flat environments, while those of Klausmeier ([98]) and Sherratt ([225]) provided such an agreement with patterns of this type occurring in similar, but gently sloping environments. Hence, we identified our parallel stationary diffusive instability stripes with these tiger bush patterns.

In order to demonstrate that our model also provided good quantitative agreement when compared with observed tiger bush patterning, we considered the scenario depicted in the aerial photograph which comprises Fig. 12.8. This photograph shows stripes of *Acacia bussei* trees in the Go-Gub area of Somaliland. The bands of dense vegetation that appear dark in this photograph are approximately 100 m wide, while the width of the separating lanes that appear bright is about 50 m. Thus, we wished to deduce biologically meaningful values for the parameters of our model that predicted the dimensional wavelength $\lambda_c^* = 150$ m. Assigning the evaporation rate L the value employed by Klausmeier ([98]) $L = 4/\text{yr}$ and introducing it, as well as our previous D_2-value, we obtained $\lambda_c^* = 95.5$ m $\approx \lambda_c = 150$ m which implied that $\lambda_c = 1.57$. Then from $a = s(\lambda_c)\alpha$ and the fact that $s(1.57) = 3.08$, we deduced that the $\lambda_c = 1.57$ contour of Fig. 2.14 satisfied the linear relation $a = 3.08\alpha$ which has been designated specially in this figure. Restricting our attention to Fig. 12.13, we saw that $\alpha = 0.45$, $a = 1.386$ and $\alpha = 0.325$, $a = 1$ were in this relation. Klausmeier ([98]) stated that plausible values for these parameters were $\alpha_{tree} = 0.045$ and $a_{tree} \in [0.077, 0.230]$. We observed that when $\alpha = 0.045$ our relation yielded an intermediate value of $a = 0.1386$ in this interval. Note also from Fig. 12.5, that for $\alpha = 0.045$ and $\mu = 0.001$, $a = 0.1386$ lies at the upper end of the striped patterned interval $0.09 < a < 0.1442$ although our asymptotic representation for the Turing boundary curve would incorrectly reduce the right-hand end point of this interval to 0.1250, showing the advantage of obtaining the actual stability boundary rather than an approximation to it. Furthermore, in order to partition this pattern so that its stripes and interstripes are in the required 2 to 1 width ratio, we need only pose the proper threshold value for vegetation, as postulated in previous chapters. We illustrate this procedure in Fig. 12.15, which is a plot of $n_e(x)$ in item (ii) of our longitudinal planform results. Note from this figure, that if we assume a low threshold value of $n_c = 1 - A_e/2$ between dense ($n > n_c$) and sparse ($n < n_c$) vegetation, then the pattern will have the proper partition (see below for a description of this procedure). Therefore, the vegetative pattern formation predictions of our analysis were in both good qualitative and quantitative agreement with the occurrence of these tiger bush stripes consisting of acacia trees.

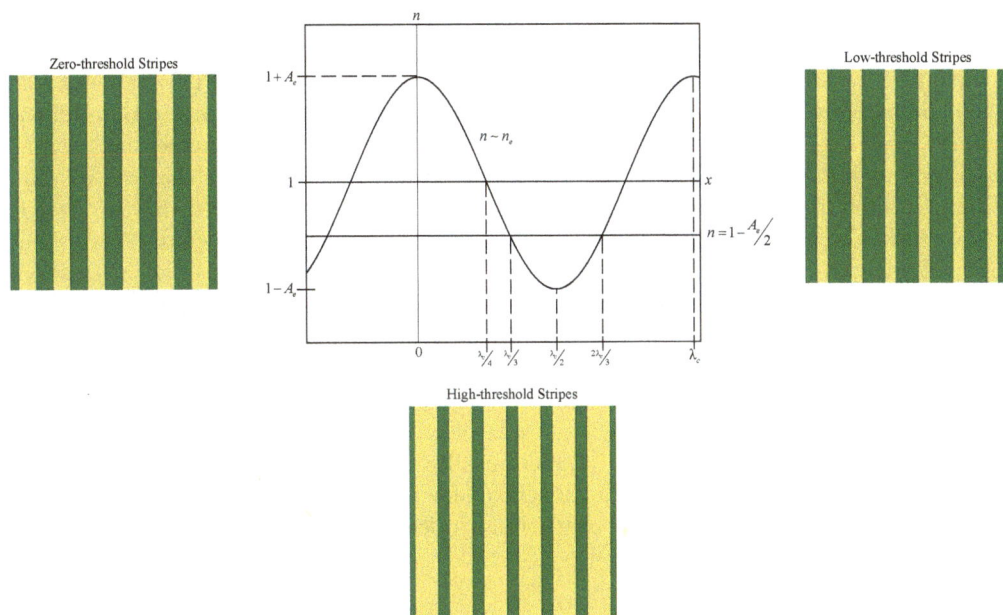

FIGURE 12.15
A schematic plot of the equilibrium plant density n_e versus x. Here a low-threshold value of $n_c = 1 - A_e/2$ yields a 2 to 1 width ratio between stripes ($n_e > n_c$) and interstripes ($n_e < n_c$) where the former has width $2\lambda_c/3$ and the latter, $\lambda_c/3$ (see the right-hand inset), while the zero-threshold value of $n_c = 1$ yields a 1 to 1 width ratio between them (see the left-hand inset) and the high-threshold value of $n_c = 1 + A_e/2$, a 1 to 2 width ratio (see the bottom inset).

We next offered an ecological interpretation of the close-packed vegetative distribution of gaps predicted as well by our hexagonal planform analysis. As pointed out in the previous chapter, it has been standard operating procedure to associate this stable critical point III$^-$ with pearled or spotted bush patterns consisting of bushy vegetation punctuated by bare spots, occurring in Niger as reported by Rietkerk *et al.* ([187]). Given its honeycomb appearance, it can also be associated with those nets or web-like patterns of dense shrubs and trees surrounding sparsely covered areas occurring in western Siberia reported by these same authors. Then from Fig. 12.11, we concluded that as plant loss α was increased or water input a was decreased our model predicted a transition from homogeneous vegetative distributions to stationary patterns of gaps, to gaps or stripes, to stripes, and finally to bare ground, with the wavelength increasing in the diffusive instability patterned region.

We closed by comparing our Turing pattern formation results with those of Sherratt and Lord ([226]). These authors presented a detailed study of the patterned solutions of the full nonlinear Klausmeier ([98]) model system using numerical bifurcation analyses. In the parameter space, where Sherratt's ([225]) linear analysis predicted instabilities, they found periodic traveling wave solutions corresponding to migrating vegetation stripes. Sherratt and Lord's ([226]) main result was that stable patterned solutions of this type also exist in part of the region where Sherratt ([225]) predicted only bare ground would occur. This result was consistent with one of Wang *et al.*'s ([268]) statements about how differential-flow instability pattern formation for a generic interaction-dispersion-advection model system compares with the differential-diffusivity patterns produced by its associated interaction-diffusion model system: Namely, that the potential outcome of a differential-flow instability is periodic traveling waves having a wider domain of pattern occurrence in parameter space when compared with the stationary Turing patterns of the associated interaction-diffusion system. Wang *et al.* ([268]), who studied pattern formation in young mussel beds employing an interaction-dispersion-advection mussel-algae system, also stated that the relevance of interaction-diffusion models to biological systems had been criticized due to their extreme sensitivity to parameter values as a consequence of the parameter range, over which such models gave rise to diffusive instabilities being often really very small. Clearly, our interaction-diffusion model did not suffer from this particular deficiency. Indeed, that model in some sense had more biological relevance than did its associated flow-instability version. This is because of the fact that our analysis required the diffusivity ratio μ to be relatively small, whereas the analyses of Sherratt ([225]) and Sherratt and Lord ([226]) required the dimensionless flow-rate parameter ν to be relatively large. Although the former condition was valid for our plant-surface water interaction-diffusion system and the smaller the better, the latter condition was somewhat inconvenient for the Klausmeier ([98]) model because, should ν be too large, surface water would tend to form gullies instead of flowing downhill as a sheet.

Finally, we examined the parameter range over which our results were valid. Recall that to guarantee the linear stability of our community equilibrium point to homogeneous perturbations, we required $\alpha < 2$. Sherratt ([225]) and Sherratt and Lord ([226]) assumed this condition to be satisfied identically for semi-arid environments. When violated, Klausmeier ([98]) pointed out that his nonspatial model, obtained by setting spatial derivatives to zero (note this is equivalent to our nonspatial model as well), could give rise to limit cycle behavior wherein plants and surface water oscillate about the community equilibrium point or yield excitable behavior associated with the trivial equilibrium point. The ecologically realistic parameter values of α estimated by Klausmeier ([98]) were far from those that lead to limit cycles or excitability since both of these behaviors can only occur provided plant biomass and surface water are on similar time scales, while, in fact, the latter changes on a much faster time scale than does the former.

We ended by noting that, contemporaneously to Sherratt ([225]), Ursino ([257]) performed a linear stability analysis on a one-dimensional version of a unifed Klausmeier model system containing both the advective and diffusive effects in her surface water (which she referred to as soil moisture) equation. Having introduced the same nondimensional variables and parameters as employed by Klausmeier ([98]), the advective and diffusive terms in that equation had coeffcients ν and $1/\mu$, respectively, which she defined in terms of soil characteristics. All of her results were represented graphically and Ursino ([257]) obtained them by directly applying Rovinsky and Menzinger's ([194]) linear stability criterion, deduced for their general chemical reaction-diffusion-advection model system, to her problem. She pointed out that when $1/\mu = 0$ her model reduced to Klausmeier's ([98]), while, unlike the latter, this diffusive term in her water equation allowed an instability to appear also on flat ground for which $\nu = 0$. Hence, these two limiting cases yielded linear stability predictions essentially equivalent to those of Sherratt ([225]) and ours, respectively. Ursino ([257]) observed that, although her linear stability analysis determined the critical conditions for the onset of instability to very small initial disturbances, it said nothing about the resulting equilibrium patterns caused by the consideration of finite amplitude disturbances. Hence, our study of pattern formation in this related interaction-dispersion-diffusion system by employing the methods of weakly nonlinear stability theory extended her unified Klausmeier model's results to such perturbations for the case of $\nu = 0$.

This completed Bonni's thesis research and she successfully defended that dissertation in December of 2011. Then, as usual, the dissemenation phase began. Actually, we had presented a talk on the longitudinal planform nonlinear stability analysis for this problem at the ICIAM (International Conference on Industrial and Applied Mathematics) held in Vancouver, B.C., Canada, the summer before and submitted a paper relevant to that talk to the *Bulletin of Mathematical Biology*. There were two referee reports. The first reviewer was upset because we had presented two different ways of obtaining a number of results and demanded that only one be included in a revised version of the manuscript, while suggesting which one should be retained in each instance. There went my attempt at explaining our deduction of the relation $a = a_c(\alpha; \mu)$ since this reviewer preferred the second rather than the first derivation presented in the original manuscript and reproduced in this chapter as well. The second reviewer, who identified himself as a biologist, objected to the fact that we had only included a one-dimensional analysis of our problem and suggested that it should be extended to a two-dimensional analysis. Since the hexagonal planform nonlinear stability analysis, that I intended to include in a follow-up paper, had already been completed this was a fairly easy suggestion with which to comply and we submitted the requested revision in two weeks. Of course our revision was accepted once we added a further explanation for the form of the hexagonal planform expansion requested by the second reviewer who, not realizing it had already been done, was duly impressed by our having completed that two-dimensional analysis in such a short time and said that the authors should be commended for this effort which went above and beyond the normal call of duty! Hence, Kealy and Wollkind ([92]) appeared on pages 803-833 of volume 74 of the *Bull. Math. Biol.*

With Kealy and Wollkind ([92]) in press, we submitted a paper to the JMM (Joint Mathematics Meetings) to be held in Boston at the start of January in 2012. The way this meeting works is that authors of submitted papers are asked to suggest various invited special sessions for which that contribution might be appropriate. Then the organizers of these special sessions select those papers that they wish to include in their sessions. In our case, we were invited to present this paper at five such sessions. Thus, we prepared five different versions of our talk, each one of which was created specifically for a particular session, and Bonni, who attended this meeting, presented all of them. That caught the attention of a science representative from the weekly British news magazine *The Economist*, who wanted to feature our research in the latest issue of his magazine. Thus, the work of Kealy and Wollkind ([92]) on vegetative pattern formation in arid flat environments was featured in an article entitled "Mathematical Ecology: Spot Check" that appeared in the Science and Technology section on page 77 of the 14-20 January 2012 issue of *The Economist*, having Mitt Romney on its cover. Ironically enough, vegetative spots were not a pattern predicted by our model. Although we were allowed to edit the article for accuracy, its title was chosen by a copy editor without any input from us.

We co-organized a minisymposium on diffusion-driven pattern formation in ecological, developmental biological, and behavioral phenomena at the 2012 SIAM Conference on Nonlinear Waves and Coherent Structures held on the University of Washington's Seattle campus later that summer. Bonni presented one of our JMM talks entitled "Vegetative pattern formation model systems: Comparison of Turing diffusive and differential flow instabilities" in the opening talk at this session. During that presentation, Alan Newell of 'Newell and Whitehead' fame, who is an expert on nonlinear optical pattern formation (see Chapter 10), inquired about the possibility of occurrence of square patterns for our model system. In order to answer this question conclusively, it was necessary for us to extend the weakly nonlinear stability analyses of Kealy and Wollkind ([92]) by performing a two-dimensional square planform analysis of its community equilibrium point. As mentioned earlier in this chapter, Bonni, after obtaining her Ph.D., worked with me for two more years extending her research on this problem, having been granted a teaching post-doctoral appointment by WSU. The first thing we did was perform the extension just described. After Kuske and Matkowsky ([105]), who examined roll, square, and hexagonal flame instability patterns for their combustion model, we sought a square planform solution of our system to lowest order of the form

$$n(x, y, t) \sim 1 + A_1(t) \cos\left(\frac{2\pi x}{\lambda_c}\right) + B_1(t) \cos\left(\frac{2\pi y}{\lambda_c}\right)$$

with an analogous expansion for $w(x, y, t)$, where, just as in a rhombic planform analysis since this was equivalent to one with $\varphi \equiv \pi/2$, the amplitudes $A_1(t)$ and $B_1(t)$ satsfied the equations

$$\frac{dA_1}{dt} \sim \sigma A_1 - A_1(a_1 A_1^2 + b_1 B_1^2),$$
$$\frac{dB_1}{dt} \sim \sigma B_1 - B_1(b_1 A_1^2 + a_1 B_1^2).$$

Proceeding in the same manner as in previous chapters, we found that

$$\sigma = \sigma_c(\beta; \alpha, \mu), \ a_1 = a_1(\alpha; \mu);$$

as defined in the longitudinal planform analysis, while imposition of the usual Fredholm alternative yielded the solvability condition

$$b_1 = b_1(\alpha; \mu),$$

which is plotted in Fig. 12.16 versus α with $\mu = 0.001$. As in our rhombic planform analyses, the equivalence classes of critical points (A_0, C_0) for the amplitude equations are given by:

$$\text{I: } A_0 = C_0 = 0; \ \text{II: } A_0^2 = \frac{\sigma}{a_1}, \ C_0 = 0; \ \text{V: } A_0 = C_0 \text{ with } A_0^2 = \frac{\sigma}{a_1 + b_1};$$

with the orbital stability criteria:

$$\text{I is stable for } \sigma < 0; \ \text{II is stable for } \sigma > 0, \ b_1 > a_1;$$
$$\text{V is stable for } \sigma > 0, \ a_1 > b_1.$$

Here when stable I, II, and V represent homogeneous, striped, and square patterns, respectively. Observe from Fig. 12.16 that since $b_1 > a_1$ for $\alpha > \alpha_0(\mu)$ with $\mu = 0.001$ square patterns are never stable versus stripes and hence, in answer to Alan Newell's question, they can not occur, which I had conjectured at the time, its confirmation being another Rabbit Pulled Out of a Hat!

We next re-examined the morphological predictions of Fig. 12.11 in more detail. Given that $\alpha_{tree} = 0.045$, we considered an enlargement of that figure by concentrating on its region of $[0.01, 0.05] \times [0, 0.15]$ in the α-a plane which is plotted in Fig. 12.17. Observe from this figure that the plot of $a = a_{\sigma_{-1}}$ seems to be visibly coincident with the Turing boundary $a = a_c$. Note in this context that for the parameter value $\alpha = 0.045$ relevant to tiger bush (Klausmeier, [98])

$$a_{\sigma_{-1}} = 0.1448, \ a_c = 0.1442, \ a_{\sigma_1} = 0.1386, \ a_{\sigma_2} = 0.1198, \ a = 2\alpha = 0.09.$$

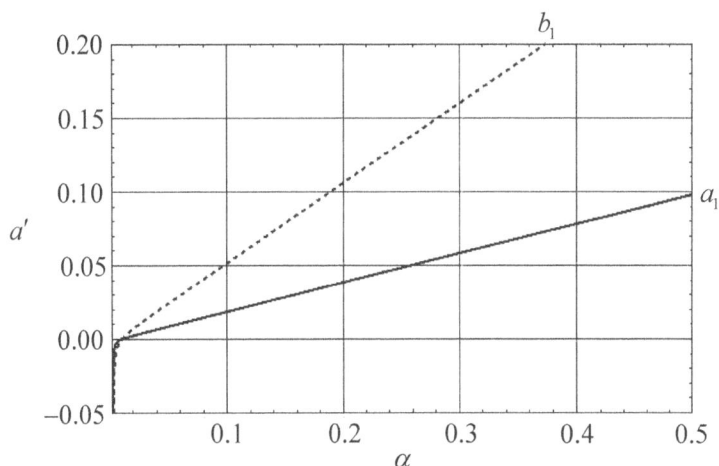

FIGURE 12.16
Plots of $b_1(\alpha;\mu)$ and $a_1(\alpha;\mu)$ represented by the dashed and solid curves, respectively, versus α with $\mu = 0.001$.

The locus $\alpha = 0.045$ is designated by the vertical dotted line appearing in Fig. 12.17. Since the behavior displayed here was typical of that occurring for all α in the biologically relevant range depicted in this figure and such small differences cannot be distinguished ecologically, we took

$$a_{\sigma_{-1}} \cong a_c.$$

Under this simplification, the rainfall column of the morphological stability predictions in Table 12.1 reduced to that of Table 12.2.

TABLE 12.2
Simplified morphological stability predictions along a rainfall gradient.

α Range	Stable pattern
$a > a_{\sigma_{-1}} \cong a_c$	Homogeneous
$a_{\sigma_1} < a < a_c$	Gaps
$a_{\sigma_2} < a \leq a_{\sigma_1}$	Gaps and stripes
$2\alpha < a \leq a_{\sigma_2}$	Stripes
$0 < a < 2\alpha$	Bare ground

Observe that the three types of contour plots for stripes depicted in the insets to Fig. 12.15 correspond to, what in previous chapters have been referred to as, low-, zero-, and high-threshold patterns. Note that all this chapter's other contour plots implicitly employed a zero threshold. Those plots, as well as the zero-threshold stripes depicted in Fig. 12.15, used the dimensional vegetative solution value of N_e where $\beta = RN_e^2/L$ as the threshold to trigger the color change from light to dark. Thus all spatial regions characterized by $N = N_e n \geq N_e$ or $n \geq 1$ appeared dark and those characterized by $N < N_e$, light, where again dark regions corresponded to high plant biomass density and light ones to low plant biomass density or bare ground. Then, defining $N_c = [L\beta_2(\alpha;\mu)/R]^{1/2}$, we adopted the

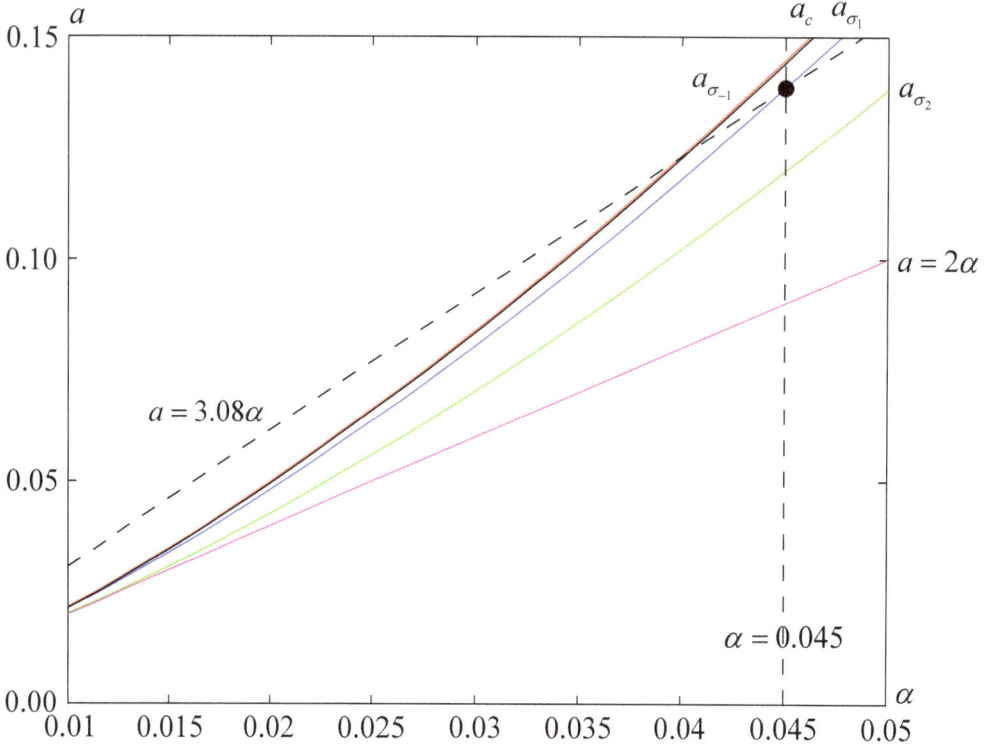

FIGURE 12.17

An enlargement of Fig. 12.11 with the region relevant to $\alpha = 0.045$ emphasized. The following identifications by color have been employed: $a = a_{\sigma_{-1}}$ (red), $a = a_c$ (black), $a = a_{\sigma_1}$ (blue), and $a = a_{\sigma_2}$ (green) with $a = 2\alpha$ (magenta). Note that $a_{\sigma_{-1}}$ and a_c are visibly coincident. Hence, for ecologically meaningful values, we have taken $a_{\sigma_{-1}} = a_c$. Here $\bullet \equiv (0.045, 0.1386)$. Observe that the curve $a = a_{\sigma_1}$, the straight line $a = 3.08\alpha$, and the vertical line $\alpha = 0.045$ all intersect at this point.

protocol that N_c represented this threshold instead. Thus, where $\beta > \beta_2$, or equivalently $a > a_{\sigma_2}$ (see Figs. 12.10 and 12.11, as well as Table 12.1), $N_e > N_c$ and low threshold patterns would occur while, where $\beta < \beta_2$ or $a < a_{\sigma_2}$, $N_e < N_c$, and high threshold patterns would occur. The insets depicted in Fig. 12.15 were generated in the following way: The stable equilibrium stripes in nondimensional variables satisfy to lowest order

$$n_e(x) = 1 + A_e g(x) \text{ where } A_e > 0 \text{ and } g(x) = \cos\left(\frac{2\pi x}{\lambda_c}\right).$$

Using the threshold values of $1/2, 0$, and $-1/2$ for $g(x)$ yielded the striped patterns identified as high-, zero-, and low-threshold stripes in that figure which had corresponding thresholds n_c of $1 + A_e/2, 1$, and $1 - A_e/2$, respectively, partitioning these patterns into stripe to interstripe width ratios of 1 to 2, 1 to 1, and 2 to 1 as discussed earlier. Those insets depict representative contour plots of these three types of striped patterns. Observe, that having adopted this protocol, the stripes occurring for $a_{\sigma_2} < a \leq a_{\sigma_1}$ in Table 12.2 would be of low-threshold type; those for $a = a_{\sigma_2}$, zero-threshold type; and those for $2\alpha < a < a_{\sigma_2}$, high-threshold type.

Before continuing the examination of Fig. 12.17 in more detail, we discussed our results in the context of those statements made by van der Stelt *et al.* ([261]) soon after Kealy and Wollkind ([92]) appeared. These authors performed a nonlinear stability analysis in the limit of large advection on a one-dimensional version of what they termed a generalized Klausmeier-Gray-Scott model which was a rescaled version of Ursino's ([257]) when restricted to Fickian diffusion. Further, van der Stelt *et al.* ([261]) stated that the nonlinear stability results of Morgan *et al.* ([149]) for their form of this model in the absence of advection were strongly related to the corresponding ones of Kealy and Wollkind ([92]). To test the validity of these assertions, we first needed to consider the Gray-Scott nondimensionalized reaction-diffusion model system for the chemical species $U = U(X, Y, \tau)$ and $V = V(X, Y, \tau)$ where $(X, Y) \equiv$ a two-dimensional coordinate system and $\tau \equiv$ time, given by (Pearson, [169])

$$\frac{\partial U}{\partial \tau} = F(1 - U) - UV^2 + D_U \Delta_2 U, \quad \frac{\partial V}{\partial \tau} = UV^2 - (F + k)V + D_V \Delta_2 V;$$

defined on an unbounded planar domain. Here, $\Delta_2 \equiv \partial^2/\partial X^2 + \partial^2/\partial Y^2$ and $D_{U,V}$ were the species diffusion coefficients, while F and k represented flow and reaction rates, respectively. Van der Stelt *et al.* ([261]) formulated their generalized Klausmeier-Gray-Scott model from this traditional one by adding an advection term of the form $C \, \partial U/\partial X$ to the right-hand side of the U-equation and replacing $F + k$ in the V-equation with a positive constant E independent of F. Observe that when $C = 0$ this reduced to the generalized Klausmeier-Gray-Scott model system analyzed by Morgan *et al.* ([149]). Now introducing the rescaled variables and parameters

$$(x, y) = \frac{(X, Y)}{d}, \quad t = F\tau; \quad n = \frac{V}{V_e}, \quad w = \frac{U}{U_e};$$

$$d = \sqrt{\frac{D_U}{F}}, \quad V_e = \sqrt{\beta F}, \quad U_e = \frac{E}{V_e} = \frac{1}{1 + \beta}, \quad \alpha = \frac{E}{F}, \quad \mu = \frac{D_V}{D_U};$$

where $4E^2\beta = (\sqrt{F} + \sqrt{F - 4E^2})^2$ for $F \geq 4E^2$; that system was transformed into our interaction-diffusion model one. Then, by virtue of the transformation just demonstrated, a one-dimensional version of this model was isomorphic to the Kealy-Wollkind interaction-diffusion system with $\nabla^2 \equiv \partial^2/\partial x^2$. That being the case, it was extremely fortuitous that referee number two for Kealy and Wollkind ([92]) forced us to add the hexagonal planform analysis to our original longitudinal planform one since the latter merely replicated the work of Morgan *et al.* ([149]) already in print.

So far, we have limited our discussion to analyses for which the wavenumber was restricted to the critical wavenumber of linear stability theory alone. In order to investigate the consequence of considering other wavenumbers in the instability sideband centered about this critical wavenumber, we would need to convert our Landau-type amplitude equations in time to Ginzburg-Landau partial differential equations by adding the appropriate spatial derivative terms to them (see Chapters 2 and 11). That was precisely what Morgan *et al.* ([149]) did in their analysis. In particular, for $\mu = 0.01$ ($D_U = 1$, $D_V = 0.01$) and $\alpha = 0.96$ ($F = 0.09$, $E = 0.086$), they showed that stationary spatially periodic solutions occurred in a subinterval of that instability interval (the so-called Busse bubble of the Eckhaus sideband). Given the isomorphism just described, this result may be directly applied to our problem. Then, as reviewed in detail by Wollkind *et al.* ([291]), a two-dimensional analysis would yield two additional instabilities besides these parallel modes: Namely, zig-zag and cross-band relevant to the interaction of oblique and perpendicular modes, respectively. All of these states, as well as the two other parallel families at $\pm 60°$ with the original one (see Chapter 11), collectively could produce quite complicated labyrinthine patterns (see Figs. 12.18 and 12.19) in the striped regimes of Table 12.2. Indeed, stationary irregular mosaic vegetative patterns have also been observed in arid flat environments (Klausmeier, [98]) and an analogous interaction, to that just described for labyrinths, of the III$^-$ critical point with these II states, in the gap overlap region with stripes of that table could easily produce mosaics of this sort (again see Figs. 12.18 and 12.19). Chen and Ward ([33]) noted that the parameter values $\mu = 0.5$ ($D_V = 10^{-5}$, $D_U = 2D_V$) and $\alpha = (F + k)/F = 3.95$ ($F = 0.02$, $k = 0.059$) for the Gray-Scott system relevant to modeling its specific chemical reaction did not produce Turing patterns. This raised the question of what parameter ranges the results of analyses of the chemical Gray-Scott model were valid for our ecological system. Recall, we required $\alpha < 2$ and Sherratt ([225]) and Sherratt and Lord ([226]) assumed this condition was satisfied identically for semi-arid or arid environments. It was certainly satisfied by the ecologically meaningful α and μ ranges of Klausmeier ([98]) and Rietkerk *et al.* ([186]) given by $10\mu \cong \alpha_0(\mu) < \alpha < 0.5$ and $0.0001 \leq \mu \leq 0.001$, depicted in Fig. 12.3. Note that the condition $F \geq 4E^2$ for Turing pattern formation in the rescaled generalized Klausmeier-Gray-Scott system treated by Morgan *et al.* ([149]) is equivalent to the condition $a \geq 2\alpha$ for our system and its relation of $\beta = V_e^2/F$, equivalent to ours of $\beta = RN_e^2/L$. Observe that the Gray-Scott system having the different parameter restriction $F \geq 4(F + k)^2$ for Turing patterns than these models could behave in quite a different manner. Note that for the parameter values employed by Chen and Ward ([33]) this inequality was violated. Thus, only some of its chemical reaction results would actually be applicable to our ecological interaction.

Returning to our examination of Fig. 12.17, we observed its most significant feature to be that the intersection point of the $\lambda_c = 1.57$ linear locus $a = 3.08\alpha$ with $\alpha = 0.045$ was precisely on the curve $a = a_{\sigma_1}$ and this a-value of 0.1386 served as the upper bound of allowable values for striped patterns to occur in the $a_{\sigma_2} < a \leq a_{\sigma_1}$ region of Table 12.2 where the stripes in this interval were of what have been termed low-threshold type consistent with the appearance of Acacia tree tiger bush. We considered that result as definitely a Rabbit Pulled Out of a Hat!

Von Hardenberg *et al.* ([266]) proposed this Turing diffusive instability mechanism as a means for better understanding the occurrence of self-organized vegetative patterns in arid

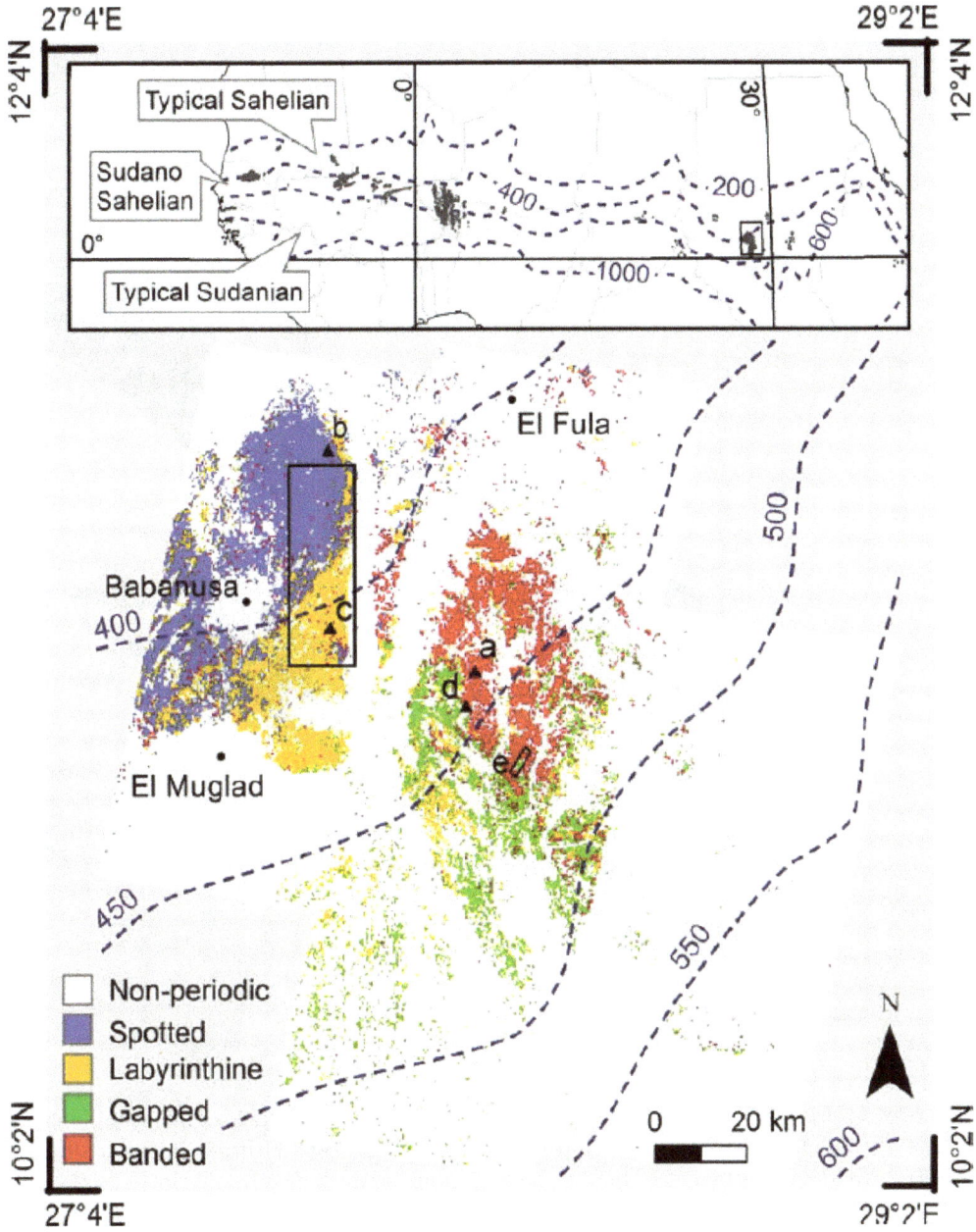

FIGURE 12.18

Map of the location of vegetation morphologies in Sudan reported by Deblauwe *et al.* ([48]). Here, the dashed blue curves represent isohyets or points having the same amount of rainfall measured in mm per year. Observe that there is a transition from gapped (site point d) to striped (site point a) vegetation as rainfall decreases while the overlap (site point e) between these patterns is depicted in Fig. 12.19. Note that Olivier Lejeune is a co-author of this paper.

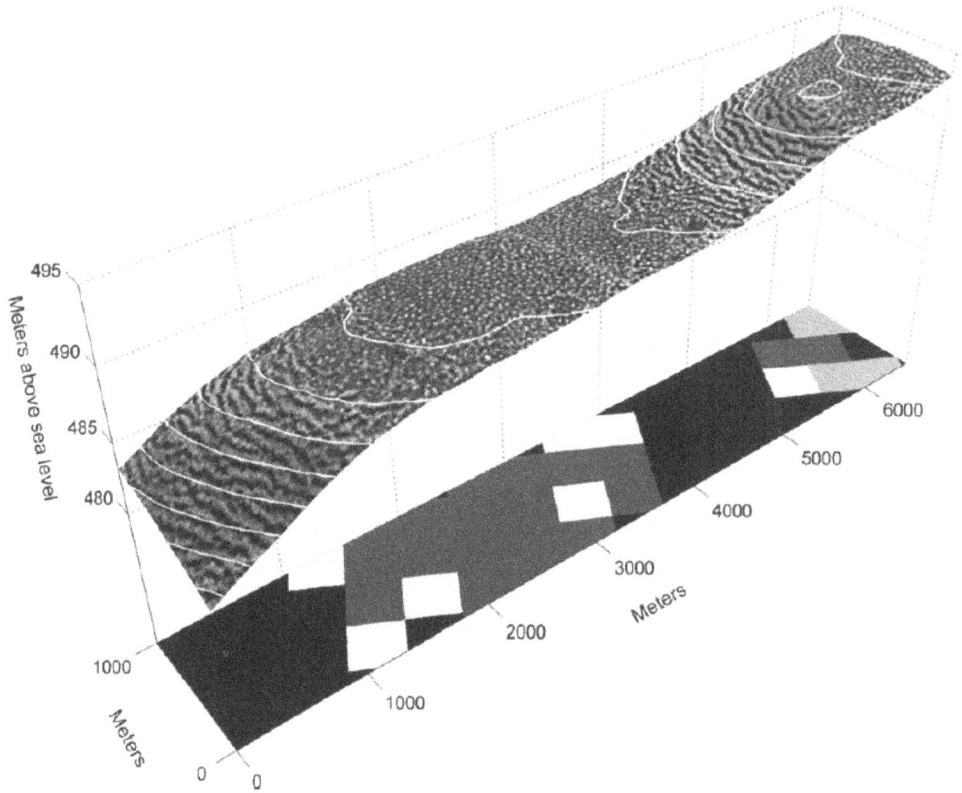

FIGURE 12.19
Profile of gapped and striped vegetative patterns from Deblauwe *et al.* ([48]) occuring at the site point e in Fig. 12.18. Here, the numbers indicate altitude in meters and the patterns are identified by color: bands (black), gapped (dark gray), labyrinths (light gray), and nonperiodic (white). Observe the elevated central plateau region of virtually constant altitude above sea level.

or semiarid environments. Here again, the positive feedback control between an activator consumer (plants) and an inhibitory limiting resource (groundwater) was considered the driving force underlying pattern formation in conjunction with the differential diffusivity existing between the two. We now, after von Hardenberg *et al.* ([266]), offered an aridity classification scheme along a rainfall gradient in Table 12.3 based upon the results of Table 12.2 particularized to the specific values of $\alpha = 0.045$ and $\mu = 0.001$.

TABLE 12.3
Aridity classification scheme along a rainfall gradient for $\alpha = 0.045$ and $\mu = 0.001$.

Aridity Classification	a Range	Stable patterns
Dry-subhumid	$a > 0.1442$	Homogeneous
Semiarid	$0.1198 < a < 0.1442$	Gaps or Stripes
Arid	$0.0900 < a < 0.1198$	Stripes
Hyperarid	$0 < a < 0.0900$	Bare ground

We ended by repeating the von Hardenberg *et al.* ([266]) assertion that the importance of model systems, such as ours, was their predicted sequence of stable states along a rainfall gradient can be used to motivate the aridity classification scheme offered in Table 12.3 that is characterized by three rainfall thresholds $2\alpha < a_{\sigma_2} < a_c$ which, when particularized to $\alpha = \alpha_{tree} = 0.045$ for Acacia trees and $\mu = 0.001$, become $p_0 = 0.0900 < p_1 = 0.1198 < p_2 = 0.1442$. Here, we employed the notation of von Hardenberg *et al.* ([266]) for these three rainfall thresholds where p refers to precipitation and in Table 12.3 introduced the following possible aridity classes based upon the inherent vegetative states of our system:

- *Dry-subhumid* $(a > p_2)$: The only vegetative state the system supports corresponds to a uniform homogeneous distribution.

- *Semiarid* $(p_1 < a < p_2)$: The only vegetative states the system supports correspond to gaps or stripes of low-threshold type. Note that in this region vegetation predominates.

- *Arid* $(p_0 < a < p_1)$: The only vegetative state the system supports corresponds to stripes of high-threshold type. Note that in this region bare ground predominates.

- *Hyperarid* $(0 < a < p_0)$: The only possible stable state the system supports is bare ground.

As pointed out by von Hardenberg *et al.* ([266]), the advantage of the proposed aridity classification scheme pertains to the information it contains about dynamical aspects of drylands. Recalling that the bare ground state always exists and is stable, any region whose aridity class implies coexistence of this state with stable vegetative patterns of high-threshold type, as the one labeled arid in Table 12.3 does, is vulnerable to desertification. At the same time, this region lends itself to recovery operations by human intervention, such as crust disturbance for soil, seed augmentation for plants, or irrigation for surface water. Meron *et al.* ([143]) provided a mechanistic explanation for vegetative patterning along such a precipitation gradient. These authors asserted that the vegetative patterns result from instabilities induced by positive feedback between plants and water. Specifically, vegetation in areas where the plant biomass is slightly greater than that in an adjacent area will be more successful in absorbing water. Thus, the biomass differential between the two regions will increase even more, further expanding the difference in their abilities to absorb water. This, in turn, will increase the biomass differential. That cycle will continue until a steady state is achieved, in which the adjacent region dries out completely. Note that a process of this sort occurs in all directions for bare gaps but only in two directions for bare interstripes.

This completed Bonni's postdoctoral research extension of her thesis problem and hence the dissemination phase began. Given that we were both also working with Rick Cangelosi on his mussel bed pattern formation problem, which had priority since it served as his dissertation, a paper from this extension had to be deferred until that research was completed. When the relevant manuscript from this extension research was finally prepared and submitted, I had more trouble getting it published than with any other paper since the aerosol convection debacle recounted in Chapter 6. First, we submitted it to the *Journal of Theoretical Biology*, the editor of which claimed her journal no longer considered mathematical ecology papers in spite of the fact that Sherratt has been publishing such research there with no problem. Then, I submitted it along with a paper from Inthira Chaiya's thesis problem (see Chapter 14) to the *Journal of Mathematical Biology*, where we had just published the mussel bed pattern formation research (see Chapter 13). They made us rewrite both manuscripts before sending them out for review, saying our mode of presentation was too similar to that of the mussel bed pattern formation paper. Then the referees rejected both papers for, in my opinion, spurious or specious reasons. In the case of the paper under discussion, one reviewer's main objection was that its major innovation of low- and high-threshold patterns had already been developed in our mussel bed pattern formation paper, while the other reviewer asserted that being an extension of Kealy and Wollkind ([92]) it should instead be submitted to the same journal where the latter had been published! The details of the reviews for the paper related to Inthira's thesis problem will be discussed in Chapter 14, the subject of which is her vegetative pattern formation model system.

Just after we had rewritten our manuscripts as related above, I received an email from the *American Journal of Plant Sciences* describing their special issue "Researches on Vegetation," to be published later that summer and asking me to consider it as a place to submit any future papers of mine similar in content to that of Kealy and Wollkind ([92]). We then prepared our manuscripts for both papers in the format required by this journal and submitted them to be considered for publication in that special issue. This allowed us to include an additional ecological interpretation of the bistability between vegetative distributions of gaps and stripes occurring in the semiarid region for the paper from Bonni's postdoctoral research. In that context, Deblauwe *et al.* ([48]) reported a plateau in Sudan where only gapped and one-dimensional isotropic banded vegetative patterns occurred, with a transition from the former to the latter as rainfall decreased (see Figs. 12.18 and 12.19). An occurrence of this sort was consistent with our model's morphological predictions summarized in Table 12.2. We noted that such bistability was accompanied by an interval of hysteresis in which gapped patterns would persist as rainfall decreased until $a < a_{\sigma_2}$ where its transition to stripes would occur, while upon an increase of rainfall, stripes would persist until $a > a_{\sigma_1} > a_{\sigma_2}$, where its transition to gaps would occur. Given the publication dates of the respective papers, Bonni and I were unaware of this work of Deblauwe and his co-workers cited above when preparing to submit Kealy and Wollkind ([92]) and only realized its existence at this time. Since our model did not predict spotted patterns, it was of importance, in the spirit of comprehensive applied mathematical modeling, to offer an actual example of a field observation where gaps and stripes alone occurred and we considered such an occurrence to be another Rabbit Pulled Out of a Hat!

The third time being the charm, the *American Journal of Plant Sciences* accepted this paper from Bonni's postdoctoral research, as well as the one from Inthira's thesis after some minor revisions which were not much trouble to implement, with the exception of their copy editor's request of its figures be of higher resolution, which required Richard Cangelosi, who had generated the figures for both papers, several iterations before producing ones to her satisfaction. Hence, they appeared back to back in the volume 6 *Amer. J. Plant Sci.*'s special issue 8 "Researches on Vegetation" as Kealy-Dichone *et al.* ([93]) pp. 1256-1277 and Chaiya *et al.* ([28]) pp. 1278-1300.

We finished the dissemination of Bonni's postdoctoral research by presenting various aspects of it at five different meetings between 2012 and 2015. I gave three of these presentations:

- A Vegetative Pattern Formation Aridity Classification Scheme along a Rainfall Gradient: An Example of Desertification Control, WSU Showcase 2012, Pullman, WA, March 30, 2012;

- A Model for Soil-Plant-Surface Water Relationships in Arid Flat Environments, AAAS Pacific Division, Annual Meeting, Las Vegas, NV, June 17, 2013, and PNW MAA, Annual Meeting, Missoula, MT, June 28, 2014;

while Bonni gave the other two:

- The Sustainability of Vegetation Pattern Formation along a Rainfall Gradient in an Arid Flat Environment, Society of Mathematical Biology Annual Meeting, Tempe, AZ, June 11, 2013;

- A Model for Soil-Plant-Surface Water Relationships in Arid Flat Environments, Centeniel MAA Meeting MathFest, Washington, D.C., August 7, 2015.

Bonni also presented one more talk on this subject at Gonzaga University in Spokane,WA, while successfully interviewing for an Assistant Professorship in their Mathematics Department. Given that a major complaint of Gonzaga graduates during exit interviews was the lack of applied modeling and analysis courses offered by the Mathematics Department, they hired her to rectify that deficiency and have not lived to regret it. Indeed, Bonni now starting her ninth year, has done such a good job of establishing an applied mathematics program and supervising modeling projects for senior theses that she was promoted early to Associate Professor in the Fall of 2018 and tenured the next year. During this time, our collaboration has continued, as documented by that research described in the rest of this book.

13

Diffusive Versus Differential Flow Instabilities II: Mussel Bed Turing Pattern Formation: Hexagonal and Rhombic Planform Nonlinear Stability Analyses

I first met Richard Cangelosi when his classmate Bonni convinced him to enroll in my graduate courses after one year in our program and he soon, like her, became my Ph.D. advisee. Rick turned into a strong advocate of comprehensive applied mathematical modeling and, being a self-starter, did yeoman work in finding papers related to his thesis topic. The problem I selected for this thesis concerned mussel bed pattern formation and was closely related to Bonni's, in that, it involved a diffusive versus a differential flow instability. As advertised, the genesis for this problem was another *Science* report: Namely, Johan van de Koppel, Joanna Gascoigue, Guy Theraulaz, Max Rietkerk, Wolf Mooij, and Peter Herman ([259]) entitled, "Experimental evidence for spatial self-organization and its emergent effects in mussel bed ecosystems," that appeared in the volume 322 issue 5902 of *Science* on pages 739-741. That paper referenced a theoretical model analysis of the ecosytem relevant to these experiments by van de Koppel *et al.* ([260]).

In particular, van de Koppel *et al.* ([259]) and Liu *et al.* ([121]) devised field experiments consisting of young beds of mussels on a homogeneous substrate, covered by a relatively quiescent layer of marine water, containing algae as a food source and observed a variety of clumped or gapped type patterns. Van de Koppel *et al.* ([260]) tested in the laboratory the hypothesis that the observed patterns were self-organized. These laboratory experiments, which consisted of young beds of mussels initially distributed uniformly on a homogeneous substrate covered by a static marine layer containing algae, replicated those patterns observed in the field. They formulated a coupled one-dimensional partial differential equation two-component model system for mussel and algae densities, including the motility effect of the mussel and the advective effect of the tide on the algae but neglecting the lateral dispersal effect for the latter. Van de Koppel *et al.* ([260]) provided an explanation for mussel bed pattern formation based upon differential flow instabilities occurring in their model by virtue of the interaction between short-range facilitation with neighbors and long-range competition for algae. Wang *et al.* ([268]) conducted a systematic analysis of the linear stability results for that model and compared them with nonlinear results obtained numerically, focusing upon their sensitivity to model parameters and assessing their influence on the speed, amplitude, and wavelength of the migrating banded patterns. Liu *et al.* ([122]) extended that simulation to two spatial dimensions. As in the Klausmeier ([98]) model system, these analyses demonstrated that ecological patterns consisting of migrating mussel bands could only occur in the van de Koppel *et al.* ([260]) model system, provided its tidal advective strength exceeded a certain critical level and hence this model was unable to predict periodic pattern formation for the carefully controlled experiments of van de Koppel *et al.* ([259]) and Liu *et al.* ([121]), in which there was virtually no tidal advection. We wished to employ a two-dimensional extension of the van de Koppel *et al.* ([260]) model that replaced this tidal advective effect with the lateral dispersal of algae instead, in order to resolve that discrepancy by explaining these stationary experimental patterns more fully. This was again in the spirit of the review of Kondo and Miura ([101]) entitled, "Reaction-Diffusion Model as a Framework for Understanding Biological Pattern Formation."

Although, at first glance, that model might seem to be an exploitation system because of the predation involved, it was actually, like Bonni's, a consumer-resource model since predator-prey systems require an activator prey species and an inhibitory predaceous one that this mussel-algae interaction did not satisfy because, here, the mussel was the activator and the algae, the inhibitor. Our system was an extension of a pair of PDEs employed by Wang *et al.* ([268]) to model field observations of young mussel beds, which had been modified so that it could be applied to the laboratory experiments devised by van de Koppel *et al.* ([259]). Specifically, we considered a coupled partial differential interaction-advection-diffusion equation model defined on an unbounded domain for $M(X, Y, \tau) \equiv$ mussel biomass density on the sediment and $A(X, Y, \tau) \equiv$ algae concentration in the lower water layer overlying the mussel bed, where $(X, Y) \equiv$ transverse two-dimensional coordinate system and $\tau \equiv$ time, of the form (van de Koppel *et al.*, [260]; Wang *et al.*, [268]; Liu *et al.*, [122])

$$\frac{\partial M}{\partial \tau} = ecAM - d_M \frac{k_M}{k_M + M} M + D_M \Delta_2 M,$$

$$\frac{\partial A}{\partial \tau} = (A_{up} - A)\rho - \frac{c}{H} AM - V \frac{\partial A}{\partial X} + D_A \Delta_2 A;$$

where $\Delta_2 \equiv \partial^2/\partial X^2 + \partial^2/\partial Y^2$; $e \equiv$ conversion constant relating ingested algae to mussel biomass production; $c \equiv$ consumption constant; $d_M \equiv$ maximal per capita mussel mortality rate; $k_M \equiv$ value of M at which mortality is half maximal; $A_{up} \equiv$ uniform concentration of algae in the upper reservoir water layer; $\rho \equiv$ rate of exchange between the lower and upper water layers; $H \equiv$ height of the lower water layer; $V \equiv$ speed of the tidal current assumed to be acting in the positive X-direction; and D_M and $D_A \equiv$ motility and dispersal coefficients of the mussel and algae, respectively (see Fig. 13.1).

The diffusion of algae consisted of two components: A lateral dispersal one D_A caused by their swimming mechanism of flagellar motion (see Fig. 13.2) and a passive vertical one, ρ, caused by the photo-gyrogravitactic driven exchange between the upper reservoir and lower benthic water layers (Williams and Bees, [273]). In addition, mussels are filter feeders, drawing seawater containing algae across their gills into their mouths through an inlet siphon and then discharging the filtered seawater back into that benthic boundary layer through an outlet siphon (Wang *et al.*, [268]). Further, d_M represented mussel loss due to predation, dislodgement, and maintenance, while D_M was a measure of mussel motility caused by their pedal locomotion mechanism and its concomitant byssal thread deployment (see Fig. 13.3). Young mussels use such threads as hiking ropes to pull themselves along. We completed the mathematical formulation of this problem by introducing the nondimensional variables and parameters

$$(x, y) = (X, Y)\sqrt{\frac{\omega}{D_A}} \text{ for } \omega = \frac{ck_M}{H}, \ t = d_M \tau, \ m = \frac{M}{k_M}, \ a = \frac{A}{A_{up}};$$

$$r = \frac{ecA_{up}}{d_M}, \ \alpha = \frac{\rho}{\omega}, \ \gamma = \frac{d_M}{\omega}, \ \nu = \frac{V}{\sqrt{\omega D_A}}, \ \mu = \frac{D_M}{D_A \gamma};$$

which transformed our system into

$$\frac{\partial m}{\partial t} = rma - \frac{m}{1+m} + \mu \Delta m, \ \gamma \frac{\partial a}{\partial t} = \alpha(1-a) - ma - \nu \frac{\partial a}{\partial x} + \Delta a;$$

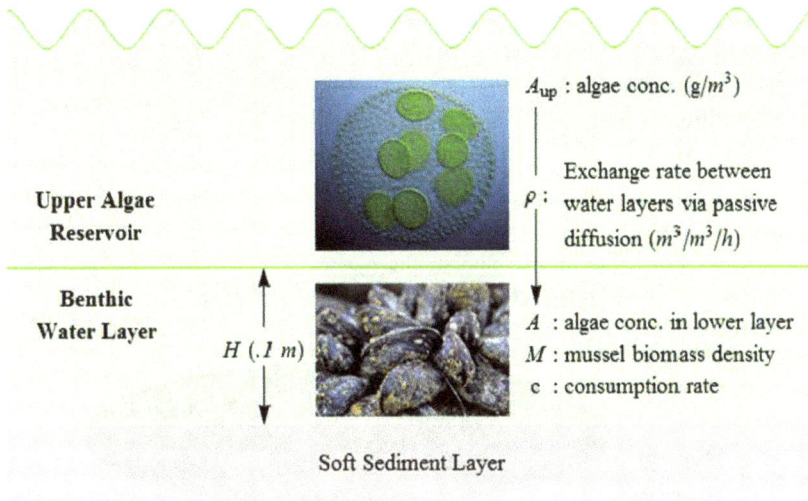

FIGURE 13.1

Schematic diagram of the mussel-algae interaction-diffusion model. Concentration of algae in the upper water layer A_{up} is taken to be constant. Algae passively diffuse into the lower benthic layer at a rate ρ and become available to the mussel for consumption at rate c.

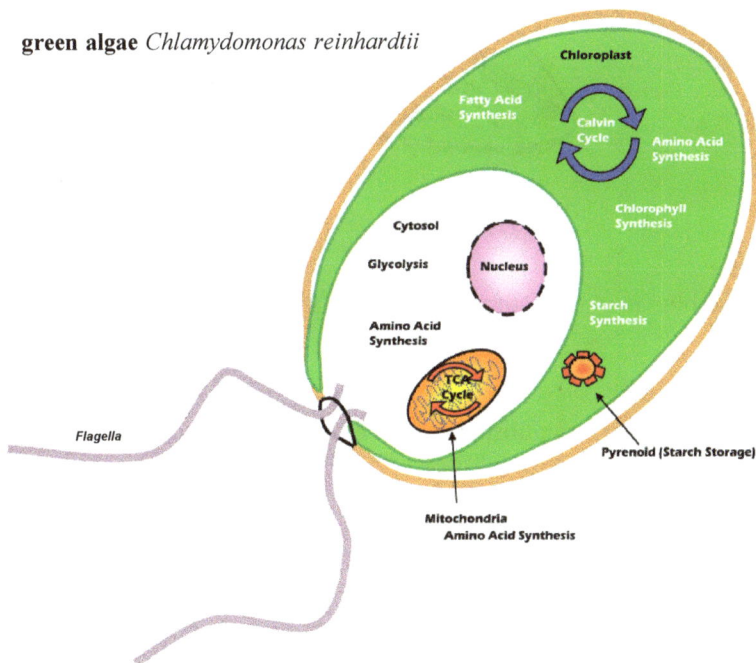

FIGURE 13.2

Depiction of flagellar motion for the green algae *Chlamydomonas reinhardtii* from Boyle and Morgan ([19]).

where $\Delta \equiv \partial^2/\partial x^2 + \partial^2/\partial y^2$ and m, a remained bounded as $x^2 + y^2 \to \infty$. Since our primary concern was with the field experiments of van de Koppel *et al.* ([259]) and Liu *et al.* ([121]), in which tidal advection was minimized as much as possible and with the former's laboratory experiments in which it was eliminated entirely, we next set $\nu \equiv 0$ and adopted that interaction-diffusion mussel-algae system to model them. Hence, this system bore the same relationship to Liu *et al.*'s ([122]), as that devised by Kealy and Wollkind ([92]) did to Klausmeier's ([98]).

blue mussel *Mytilus edulis*

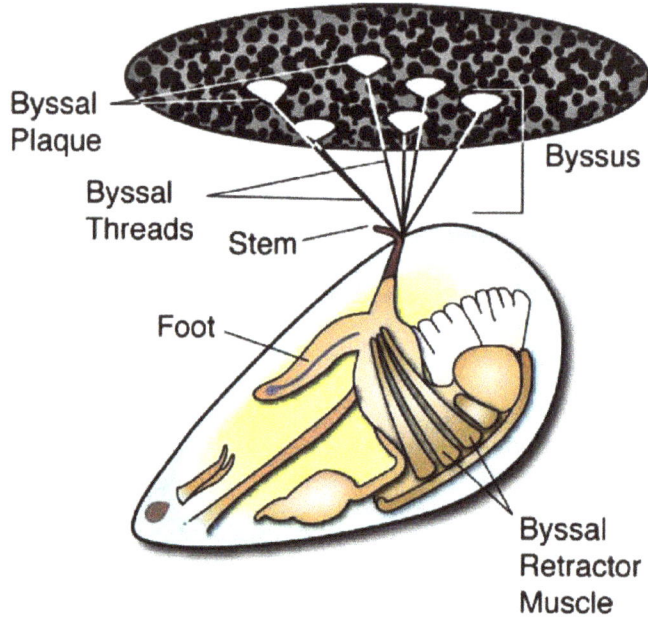

FIGURE 13.3
Anatomy of the blue mussel *Mytilus edulis* and its byssus structures from Silverman and Roberto ([229]).

Seeking uniform stationary equilibrium points of our system given by $(m, a) \equiv (m_e, a_e)$, we found the following two points (see Fig. 13.4):

$$m_e = 0, \ a_e = 1;$$

$$m_e = \alpha \frac{r-1}{1-\alpha r}, \ a_e = \frac{1-\alpha r}{r(1-\alpha)};$$

and noted that the first equilibrium point corresponded to bare sediment with no mussel biomass which existed for all parameter values, while the second corresponded to a nonzero homogeneous distribution of mussel that only existed for $1 < r < \alpha^{-1}$ and $0 < \alpha < 1$. Since, as in the last chapter, staightforward calculations showed the bare sediment equilibrium point to be linearly stable for $0 < r < 1$, it was the stability of the community equilibrium point with which we were concerned. Hence, we performed longitudinal, hexagonal, and rhombic planform nonlinear stability analyses on that equilibrium point. Here, we shall only summarize the results of these analyses for this problem given that their particular details are analogous to those described in the previous chapters.

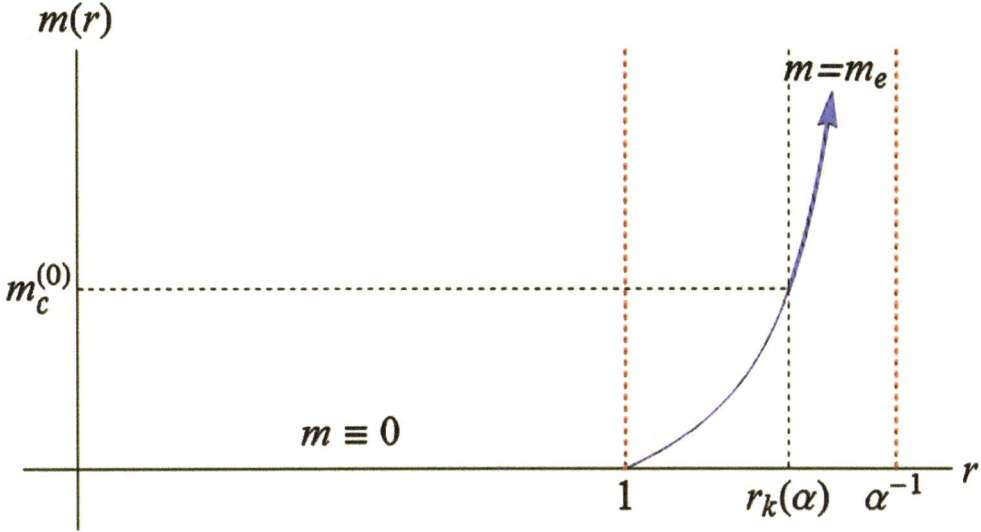

FIGURE 13.4
Plot of the mussel component of the equilibrium points versus the parameter $0 < r < \alpha^{-1}$ for a fixed value of $0 < \alpha < 1$. Here, $r_k(\alpha)$ and $m_c^{(0)} = 1.5$ are as to be defined later.

The results of our longitudinal planform linear stability analysis were as follows: To guarantee the onset of an ecological Turing diffusive instability from a uniform steady state that was stable to linear homogeneous perturbations, we required:

$$\frac{a_e \gamma m_e}{\alpha(1+m_e)^2} < 1, \quad c_0 = \frac{\alpha(r-1)(1-r\alpha)}{(1-\alpha)} > 0 \text{ for } q^2 = 0;$$

and

$$\mu < \mu_0(q^2; r, \alpha) = \frac{m_e}{(1+m_e)^2} \frac{q^2 - (1-\alpha)}{q^2(q^2 + \alpha/a_e)} \text{ for } q^2 \neq 0;$$

where, as usual, $q = 2\pi/\lambda$, λ being the wavelength of the class of spatially periodic perturbations under investigation. We observed that for the typical parameter values of van de Koppel et al. ([260])

$$A_{up} = 1 \text{ gm m}^{-3}, \ H = 0.1 \text{ m}, \ \rho = 100 \text{ h}^{-1}, \ e = 0.21,$$
$$c = 0.1 \text{ m}^3 \text{ gm}^{-1} \text{ h}^{-1}, \ d_M = 0.02 \text{ h}^{-1}, \ k_M = 150 \text{ gm m}^{-2};$$

our relevant nondimensional quantities were given by

$$r = 1.050, \ \gamma = \frac{4}{3} \times 10^{-4}, \ \alpha = \frac{2}{3};$$

and hence $m_e = 0.111$, $a_e = 0.857$. Clearly, these values satisfied both the $q^2 = 0$ inequality constraints. Indeed, for any r and α in the parameter range for the existence of the community equilibrium point, the second inequality was satisfied identically, while should γ be sufficiently small so that, as here, $\gamma \ll 1$ then the first inequality was also satisfied identically as well.

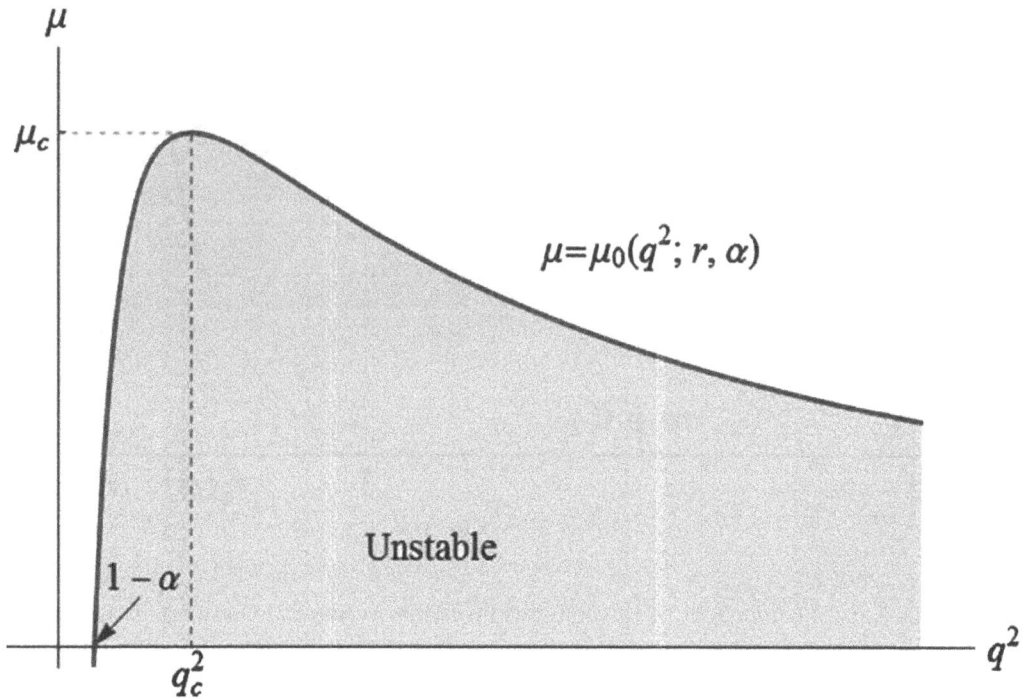

FIGURE 13.5
Plot of $\mu = \mu_0(q^2; r, \alpha)$ in the q^2-μ plane for $1 < r < \alpha^{-1}$ and $0 < \alpha < 1$ with $\alpha = 2/3$ and $r = 1.3$.

For fixed values of r and α, the curve $\mu = \mu_0(q^2; r, \alpha)$ in the first quadrant of the q^2-μ plane was marginal in the sense that it separated the linearly stable region where $\mu > \mu_0$ from the unstable region where $0 < \mu < \mu_0$. The marginal stability curve, the linearly stable region, and the unstable region could be characterized by $\sigma_0 = 0$, $\text{Re}(\sigma_0) < 0$, and $\sigma_0 > 0$, respectively, where σ_0 represented that growth rate having the greatest real part. This marginal stability curve had a maximum at the point (q_c^2, μ_c) where

$$q_c^2 = (1 - \alpha)[1 + (1 - \alpha r)^{-1/2}], \quad \mu_c = \frac{m_e}{(1 + m_e^2)^2} \frac{1}{2q_c^2 + \alpha/a_e}.$$

Hence, when $\mu > \mu_c$, there existed no squared wavenumbers q^2 corresponding to growing disturbances, while when $0 < \mu < \mu_c$ there existed an interval of such q^2 centered about q_c^2 for which $\sigma_0 > 0$ (see Fig. 13.5). In particular, taking $q^2 \equiv q_c^2$, as we did in our weakly nonlinear stability analyses, it was possible, assuming r and α fixed, to represent $\sigma_0 = \sigma_c(\mu)$ such that, for μ sufficiently close to μ_c, σ_0 would still be real when $\mu > \mu_c$. Then $\sigma_c(\mu)$ satisfied the conditions $\sigma_c(\mu) < 0$ for $\mu > \mu_c$, $\sigma_c(\mu_c) = 0$, and $\sigma_c(\mu) > 0$ for $0 < \mu < \mu_c$. Hence, the locus $\mu = \mu_c(r; \alpha)$ with μ_c, defined as above, was a marginal stability curve in the r-μ plane for $0 < \alpha < 1$ fixed. That locus is plotted in Fig. 13.6 with $\alpha = 2/3$ for $1 < r < \alpha^{-1} = 1.5$. Defining $\mu = \mu_0/\gamma$, where $\mu_0 = D_M/D_A$, we observed from our previous results, in conjunction with Fig. 13.6 that a requirement for the occurrence of an ecological Turing diffusive instability in this instance was the satisfaction of the inequality condition

$$0 < \mu_0 = \gamma\mu < \gamma\mu_c < \frac{a_e \gamma m_e}{\alpha(1 + m_e)^2} < 1.$$

(a)

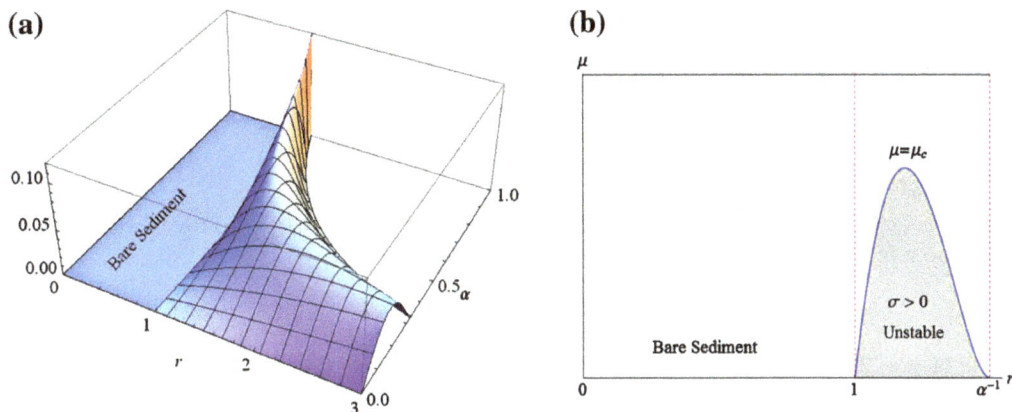

(b)

FIGURE 13.6
(a) Three-dimensiomnal plot of the marginal stability surface $\mu = \mu_c(r, \alpha)$ for $0 < r < 3$ and $0 < \alpha < 1$. Here, $0 < r \leq 1$ corresponds to bare sediment. (b) Marginal stability curve in the r-μ plane given by the cross-section of the plane $\alpha = 2/3$ with the surface of part (a) for $1 < r < \alpha^{-1} = 1.5$.

In this context, we had determined the critical conditions for the onset of diffusive instabilities that could occur provided $0 < r < \alpha^{-1}$ and $0 < \alpha < 1$. To predict the long-time behavior and spatial pattern of such growing disturbances, it was necessary for us to take the nonlinear terms of our governing equations into account. The main nonlinear result of our longitudinal planform analysis was the determination, using the usual Fredholm alternative solvability condition, of its Landau coefficient, represented here symbolically by $a_1 = a_1(r; \alpha)$, which is plotted versus r for the fixed value of $\alpha = 2/3 \in [0.5, 1)$ in Fig. 13.7. From this figure, we observed that $a_1(r; \alpha)$ had two zeroes at $r = r_{1,2}(\alpha)$ such that $a_1 > 0$ for $r_1 < r < r_2$ and $a_1 < 0$ for $r < r_1$ or $r > r_2$; where, in particular, $r_1(2/3) = 1.065$ and $r_2(2/3) = 1.462$. The values of $r_{1,2}(\alpha)$ for some other typical values of α in that interval are tabulated in Table 13.1.

TABLE 13.1
The zeroes $r_{1,2}(\alpha)$ of $a_1(r; \alpha)$ for some typical values of $\alpha \in [0.5, 1)$.

α	0.5	0.6	0.7	0.8	0.9
r_1	1.048	1.065	1.062	1.047	1.026
r_2	1.931	1.618	1.395	1.229	1.101
α^{-1}	2.000	1.667	1.429	1.250	1.111

Next, we performed a hexagonal planform nonlinear stability analysis on our model system and determined its additional Landau coefficients $a_0 = a_0(r; \alpha)$ and $a_2 = a_2(r; \alpha)$. Then, we examined the signs of a_0, $a_1 + 4a_2$, and $2a_2 - a_1$ versus r in Fig. 13.8 for that representative

(a)   **(b)**

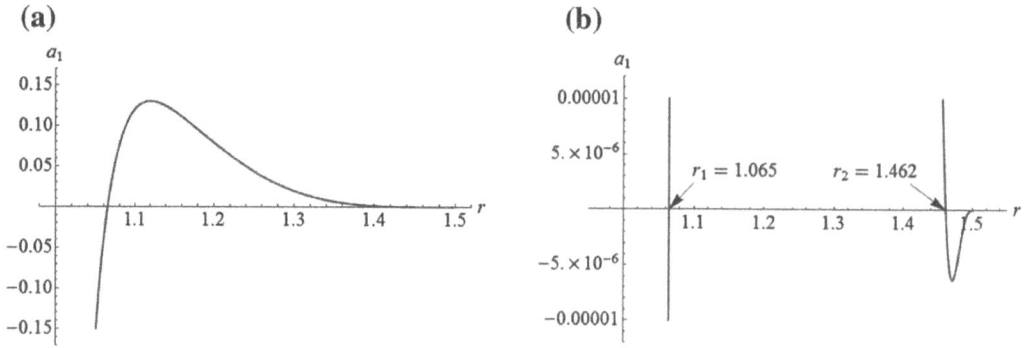

FIGURE 13.7
(a) Plot of $a_1(r; \alpha)$ versus r for $\alpha = 2/3$. (b) Re-scaled plot of part (a) to reveal the behavior near the r-axis. Here, $\alpha^{-1} = 1.5$ in this instance.

value of $\alpha = 2/3 \in [0.5, 1)$ employed previously. From this figure, we observed that, besides $r_{1,2}$, there existed the following other significant values of r:

$$1 < r_1 < r_3 < r_5 < r_0 < r_6 < r_4 < r_2 < \alpha^{-1}$$

such that

$$a_0 = 0 \text{ for } r = r_0, \; a_0 > 0 \text{ for } r > r_0, \; a_0 < 0 \text{ for } r < r_0;$$
$$a_1 + 4a_2 = 0 \text{ for } r = r_3 \text{ or } r_4, \; a_1 + 4a_2 > 0 \text{ for } r_3 < r < r_4;$$
$$2a_2 - a_1 = 0 \text{ for } r = r_5 \text{ or } r_6, \; 2a_2 - a_1 > 0 \text{ for } r_5 < r < r_6,$$
$$2a_2 - a_1 < 0 \text{ for } r < r_5 \text{ or } r > r_6.$$

These values are compiled in Table 13.2 for typical $\alpha \in [0.5, 1)$.

TABLE 13.2
The other significant r-values for some typical $\alpha \in [0.5, 1)$.

α	r_3	r_5	r_0	r_6	r_4
0.5	1.173	1.273	1.592	1.753	1.856
0.6	1.126	1.182	1.390	1.497	1.568
2/3	1.100	1.136	1.289	1.370	1.424
0.7	1.088	1.117	1.247	1.315	1.362
0.8	1.056	1.068	1.141	1.181	1.209
0.9	1.028	1.030	1.061	1.078	1.092

Finally, we offered the following morphological mussel bed pattern identifications of the potentially stable critical points of the hexagonal planform amplitude-phase equations: I and II represented the nonzero uniformly homogeneous and banded states, respectively, while III$^+$ could be associated with hexagonal close-packed arrays of clumps and III$^-$, with similar arrays of gaps where the periodic mussel bed patterns were of critical wavelength $\lambda_c = 2\pi/q_c$ and $\lambda_c^* = \sqrt{D_A/\omega}\lambda_c$ in dimensionless and dimensional variables, respectively. Contour plots relevant to these patterns are depicted in Fig. 13.9 where the spatial variables are being measured in units of λ_c. Here, regions of high mussel density $m > m_e$ appear blue

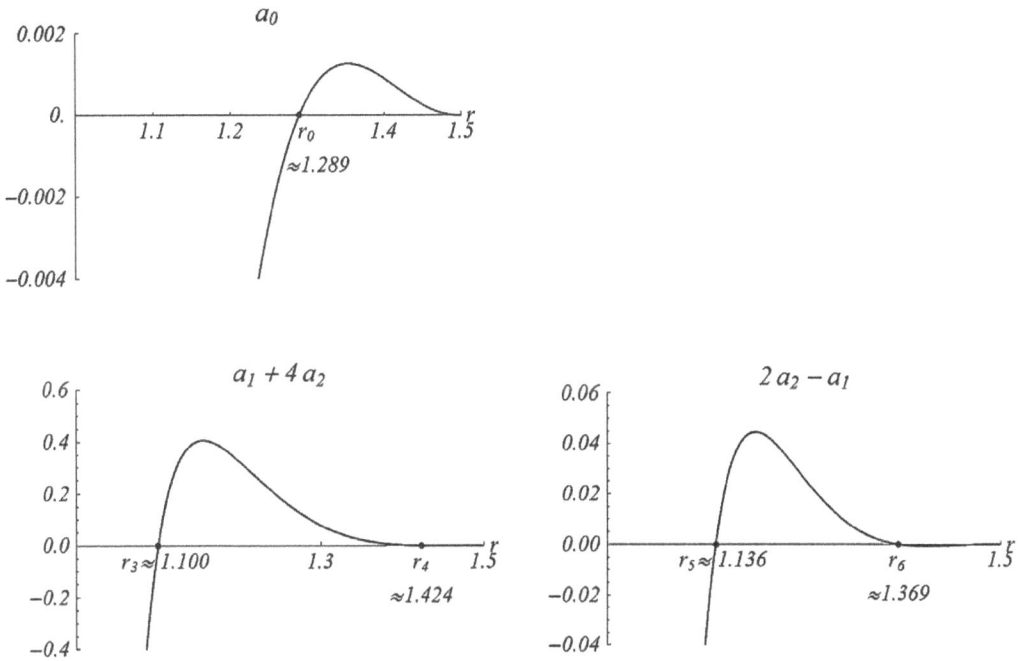

FIGURE 13.8
Plots of the hexagonal planform Landau coefficients with $r \in [1, \alpha^{-1})$ for $\alpha = 2/3$. Top: $a_0(r; \alpha)$ where r_0 represents its root. Bottom left: $(a_1 + 4a_2)(r; \alpha)$ where r_3 and r_4 represent its roots. Bottom right: $(2a_2 - a_1)(r; \alpha)$ where r_5 and r_6 represent its roots.

and those of low density $m < m_e$, yellow, where m_e is the community equilibrium point's mussel density component.

III$^+$: Mussel Clumps   II: Mussel Bands   III$^-$: Mussel Gaps

FIGURE 13.9
Contour plots of the potentially stable hexagonal planform periodic mussel bed patterns associated with critical points II and III$^\pm$.

In order, ultimately, to compare our hexagonal-planform predictions with experimental observation, we needed to represent the stability results of Table 8.2, applied to this problem, graphically in the r-μ plane. To do so, it was first necessary for us to generalize the approach which produced the marginal stability $\sigma_c = 0$ curve $\mu = \mu_c(r;\alpha)$, in order to generate the analogous loci associated with $\sigma_c = \sigma_j$ for $j = -1, 1$, and 2, respectively. After Edelstein-Keshet ([54]) and HilleRisLambers *et al.* ([80]), we employed a quasi-steady-state assumption on our secular equation to simplify this procedure. That is, we set $\gamma = 0$ in this equation and hence reduced that quadratic in σ to a linear equation. Here, we made use of the fact that this parameter had a typically small value and that this assumption did not alter σ_c significantly, but merely neglected the other root of the original quadratic. That root being negative and very large in absolute value had a highly stabilizing influence on our problem and hence played no role in the diffusive instabilities under examination. Proceeding in this fashion, we Pulled a Rabbit Out of a Hat and found that this generalized marginal curve in the r-μ plane could be represented by

$$\mu = \mu_{\sigma_c}(r;\alpha) = \frac{(1+m_e)^2 - \sigma_c}{2q_k^2(\sigma_c, r;\alpha) + \alpha/a_e}$$

where

$$q_k^2(\sigma, r;\alpha) = f(\sigma, r;\alpha) + \sqrt{f^2(\sigma, r;\alpha) + \frac{\alpha f(\sigma, r;\alpha)}{a_e}}$$

with

$$f(\sigma, r;\alpha) = \frac{c_0 + \alpha\sigma/a_e}{\frac{m_e}{(1+m_e)^2} - \sigma};$$

which, as to be expected, satisfied the reduction condition $\mu = \mu_{\sigma_c=0}(r;\alpha) = \mu_c(r;\alpha)$ since $q_k^2(0, r;\alpha) = q_c^2$ by virtue of $f(0, r;\alpha) = (1+m_e)^2 c_0/m_e = 1 - \alpha$. Then, we incorporated the Landau coefficients into the expressions for σ_j and obtained

$$\sigma_j = \sigma_j[a_0(r;\alpha), a_1(r;\alpha), a_2(r;\alpha)] = \sigma_j(r;\alpha),$$

which upon substitution for the σ_c in our generalized marginal curve, yielded the desired loci in the r-μ plane $\mu = \mu_{\sigma_j(r;\alpha)}(r;\alpha) = \mu_j(r;\alpha)$ with $j = -1, 1$, and 2. We plotted these loci, as well as the marginal stability curve $\mu = \mu_c(r;\alpha)$ in the r-μ plane of Fig. 13.10 with $\alpha = 2/3$ and, since all the quantities required for the identification of the hexagonal planform periodic mussel bed patterns of Fig. 13.9 predicted in Table 8.2 had now been evaluated, represented graphically the regions corresponding to those patterns in that plane.

Further, to justify the truncation procedure inherent to the hexagonal planform asymptotic representation, it was necessary that its Landau coefficients, as usual, satisfied the size constraint $|a_0| \ll (a_1 + 4a_2)^2$, which, as can be seen from Fig. 13.8, was valid for the parameter range of interest. Hence, we concluded that such a truncation procedure was justified for our hexagonal planform nonlinear stability analysis.

Lastly, we performed a rhombic planform nonlinear stability analysis on our model system and determined its other Landau coefficient $b_1 = b_1(r, \phi; \alpha)$ where $\phi \equiv$ rhombic angle. Defining

$$\eta(r, \phi; \alpha) = \frac{b_1(r, \phi; \alpha)}{a_1(r; \alpha)},$$

for $r_1 < r < r_2$, we catalogued the stability criteria for the rhombic planform critical points:

I: $\mu > \mu_c$; II: $0 < \mu < \mu_c$ and $\eta > 1$; V: $0 < \mu < \mu_c$ and $-1 < \eta < 1$.

Then, we observed that $\eta(r_0, \phi; 2/3) = 2$, except at $\phi = \pi/3$ where $\eta = 1.24105$ from Fig. 13.11, which is a plot of η versus ϕ for $r = r_0 = 1.2894436353646677$. Hence, the function $\eta(r_0, \phi; 2/3)$ had a jump discontinuity at this point. Since $\eta > 1$ for all values of ϕ, we concluded from our rhombic planform stability criteria that bands were the only stable pattern when $r = r_0$ where $a_0(r_0; \alpha) = 0$. Such behavior depended upon our taking $r = r_0$ precisely, which was our rationale for representing those values in Table 13.3 to sixteen significant figures.

TABLE 13.3
Values of $r_0(\alpha)$ defined implicitly by $a_0(r_0; \alpha) = 0$ for some typical values of α.

α	r_0
0.5	1.5921041084455716
0.6	1.3897661053573491
2/3	1.2894436353646677
0.7	1.2467252890661253
0.8	1.1410766780540762
0.9	1.0608935136041813

For $r \neq r_0$, there existed intervals of rhombic angles ϕ for which $-1 < \eta < 1$ as depicted in Fig. 13.12. Hence, for those r-values rhombic patterns of these characteristic angles as compiled in Table 13.4 were stable versus bands.

In order to demonstrate why there were no central intervals for the first and last two entries of this table, we examined Fig. 13.12 in more detail. Observe from this three-dimensional plot that the planes for those values of r do not intersect the central portion of the η-surface for that figure, while the planes for the middle two values of this table, as well as that for $r = 1.187$ of Fig. 13.13 do. Restricting ourselves to the interval of interest, $0 \leq \phi \leq \pi/2$ in this figure, we see that there are two ϕ-intervals of stable rhombic patterns flanking

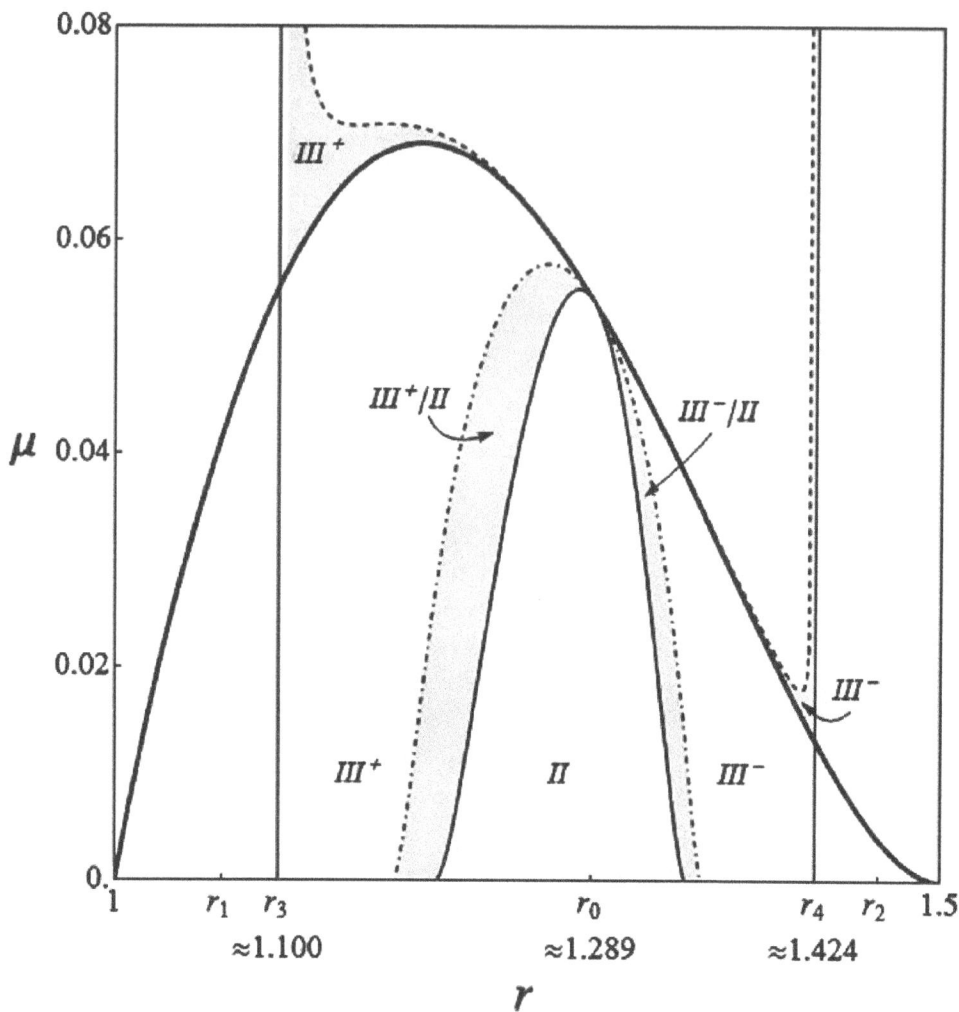

FIGURE 13.10

Stability diagram in the r-μ plane for the hexagonal planform analysis of our two-dimensional interaction–diffusion model system with $r_3 < r < r_4$ and $\alpha = 2/3$, denoting the predicted periodic mussel bed patterns summarized in Table 8.2. Here, the heavy black curve represents $\mu = \mu_c(r; \alpha)$; the dashed curve, $\mu = \mu_{-1}(r; \alpha)$; the dashed-dotted curve, $\mu = \mu_1(r; \alpha)$; and the solid curve, $\mu = \mu_2(r; \alpha)$; while the notation A/B indicates a region of bistability between the A and B patterns. Note that III^+ patterns can only occur for $r < r_0$; and III^- ones, for $r > r_0$.

415

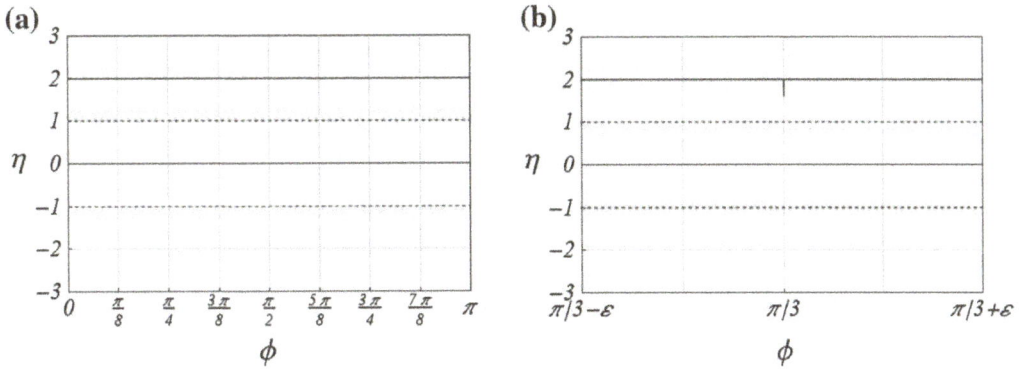

FIGURE 13.11
Plots of η versus (a) $\phi \in (0, \pi)$ and (b) $\phi \in (\pi/3 - \varepsilon, \pi/3 + \varepsilon)$ where $\varepsilon = 10^{-4}$ for $r = 1.2894436353646677$ and $\alpha = 2/3$.

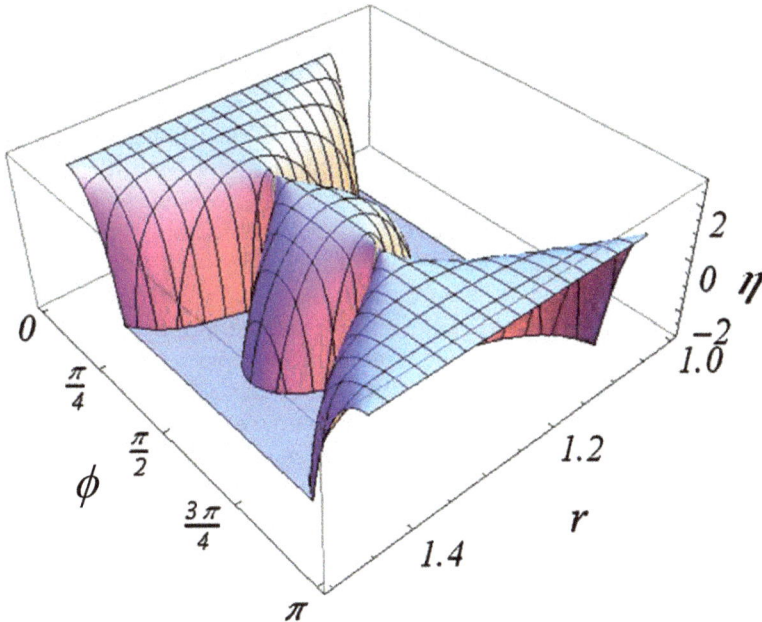

FIGURE 13.12
Three-dimensional surface plot of η versus r and ϕ for $\alpha = 2/3$.

TABLE 13.4
Intervals of stable rhombic angles for various values of r with $\alpha = 2/3$.

r	Left-hand Interval	Central Interval
1.065	0.0244-0.0423 (1.40°-2.4°)	None
1.100	0.3295-0.4964 (18.9°-28.4°)	None
1.200	0.7002-0.8314 (40.1°-47.6°)	1.3081-$\pi/2$ (74.9°-90°)
1.300	1.0079-1.0231 (57.7°-58.6°)	1.0762-1.1015 (61.7°-63.1°)
1.400	0.6130-0.7581 (35.1°-43.4°)	None
1.462	0.0028-0.0049 (0.2°-0.3°)	None

$\phi = \pi/3$, both lying between $\eta = -1$ and $\eta = 1$ which have been designated by horizontal lines. That figure has been plotted for the extended interval $\pi/2 \leq \phi \leq \pi$ in order to demonstrate graphically the symmetry about $\phi = \pi/2$ characteristic of rhombic patterns.

To make a morphological interpretation of these stable rhombic patterns, we considered the plots contained in Fig. 13.14 for $\alpha = 2/3$ and $\phi = 1$ or 57.3°. The top one is an η versus r plot where the r-intervals for stable rhombic patterns of this characteristic angle flanking $r = r_0$ have been designated by shading. The bottom ones are threshold contour plots of rhombic patterns of angle $\phi = 1$ from left to right with threshold values of 1, 0, and -1, corresponding to $r < r_0$, $r = r_0$, and $r > r_0$, respectively. Here, the spatial variables were measured in units of λ_c and regions exceeding that threshold in each part appear blue, while those below it appear yellow. Hence, the three parts of the bottom contour plot in this figure corresponded to what we have termed higher, zero, and lower threshold patterns, respectively. Traditionally, most pattern formation analyses of this type have used the homogeneous solution value of m_e as the threshold to trigger the color change from yellow to blue as in Fig. 13.9. Thus, all spatial regions characterized by $m \geq m_e$ appear blue and those characterized by $m < m_e$ appear yellow. Here, blue regions correspond to high mussel biomass densities and yellow ones to low densities or bare sediment. This is equivalent to our zero threshold case of $V^{(0)}$ in Fig. 13.14. Explicitly denoting the mussel component of the community equilibrium point by $m_e = m_e(r; \alpha)$, we defined $m_c = m_e(r_0; \alpha)$ and adopted the protocol that m_c represented this critical threshold instead. Then observing from Fig. 13.4 that $m_e(r; \alpha)$ is an increasing function of r for a fixed value of α, we concluded that, in the left-hand shaded interval of the top plot of Fig. 13.14 where $r < r_0$ or $m_e < m_c$, the higher threshold rhombic patterns of the bottom contour plot V^+ would occur, while in the right-hand shaded interval of the top plot of that figure where $r > r_0$ or $m_e > m_c$, the lower threshold rhombic patterns of the bottom contour plot V^- would occur. Given their appearance and occurrence, in comparison with those of $III^\pm$, we labeled mussel bed patterning corresponding to these higher and lower threshold stable rhombic patterns as arrays of pseudo clumps and pseudo gaps, respectively. In Fig. 13.15, we plotted similar stable $V^\pm$ patterns for both $\phi = \pi/4$ and $\pi/2$, that again would occur for $r < r_0$ and $r > r_0$, respectively.

Before incorporating these rhombic planform results into our hexagonal planform ones that were summarized in Fig. 13.10, we modified our procedure for choosing the morphological threshold $r_0(\alpha)$ and $m_c = m_e(r_0; \alpha) = m_c(\alpha)$ relevant to the rhombic patterns just introduced so that the new threshold $m_k(\alpha) \equiv m_c^{(0)}$ would be independent of α. Toward that end, we deduced a close approximation to $r_0(\alpha)$ denoted by $r_k(\alpha)$ defined implicitly such

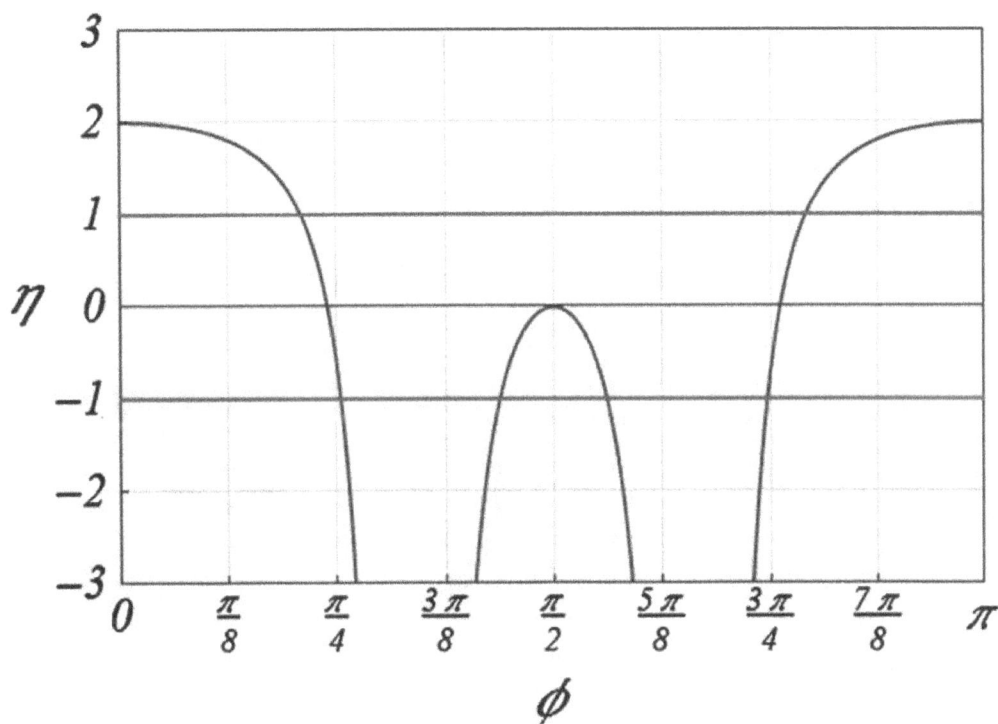

FIGURE 13.13
Plot of η versus ϕ for $\alpha = 2/3$ and $r = 1.187$. This is a cross-section of the η-surface of Fig. 13.12 with the plane $r = 1.187$. Here, observe $\eta(r, \phi; \alpha) = \eta(r, \pi - \phi; \alpha)$.

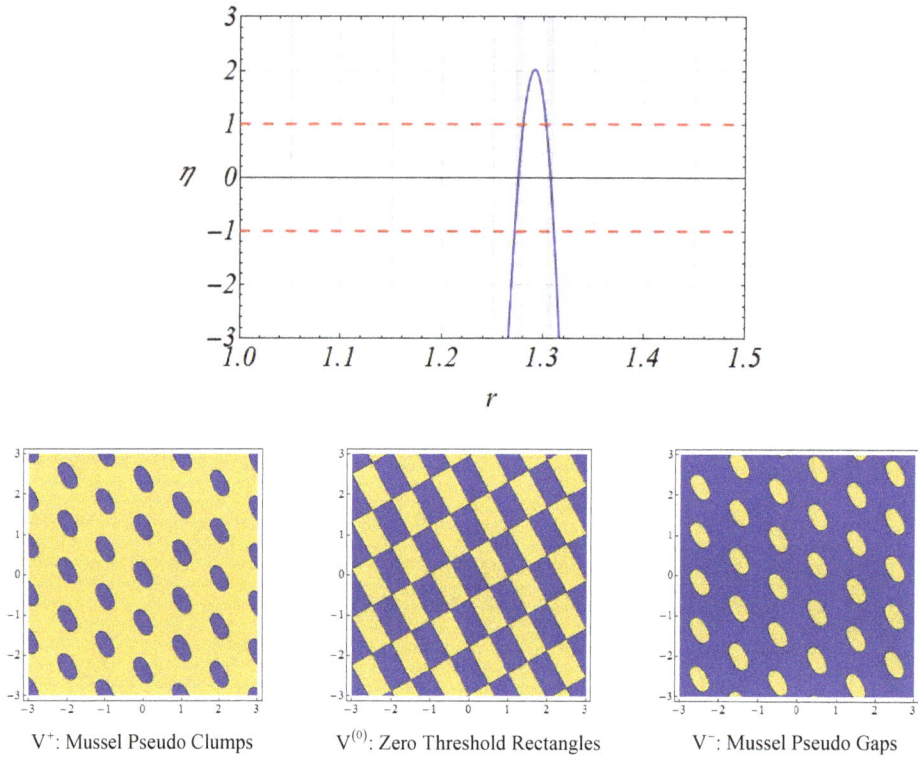

FIGURE 13.14
Top: Plot of η versus r for $\alpha = 2/3$ and $\phi = 1$. Bottom: Contour plots of stable rhombic patterns of that characteristic angle from left to right for $r < r_0$, $r = r_0$, and $r > r_0$. Observe that for this angle, the $V^{\pm}$ patterns approximate the close-packed structure of the hexagonal arrays $III^{\pm}$, which is why they have been labeled as pseudo clumps and pseudo gaps.

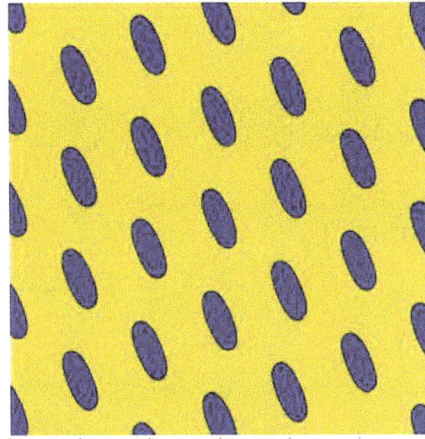

$$\phi = \frac{\pi}{2} \qquad \text{V}^+: \text{Mussel Pseudo Clumps} \qquad \phi = \frac{\pi}{4}$$

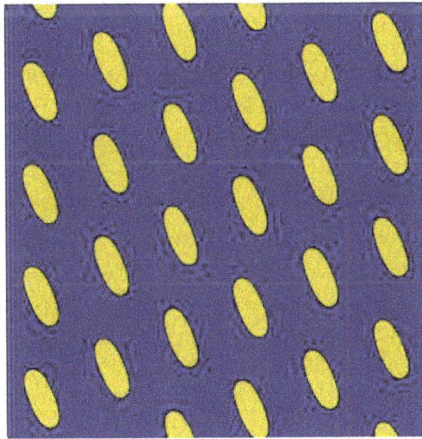
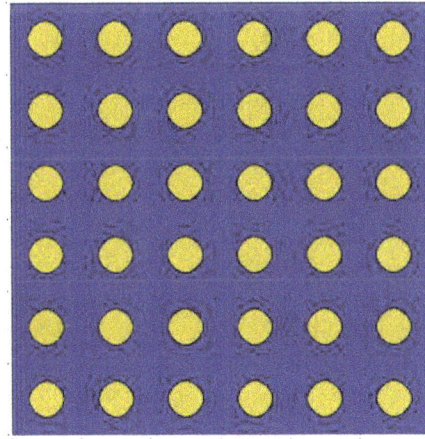

$$\phi = \frac{\pi}{4} \qquad \text{V}^-: \text{Mussel Pseudo Gaps} \qquad \phi = \frac{\pi}{2}$$

FIGURE 13.15
Contour plots of stable rhombic patterns of characteristic angle $\phi = \pi/4$ and $\pi/2$.

that

$$m_e(r_k; \alpha) = \alpha \frac{r_k - 1}{1 - \alpha r_k} = m_k(\alpha) \equiv m_c^{(0)},$$

where $m_c^{(0)}$ represented a constant independent of α. Hence, we introduced a function $r_k(\alpha)$ for which $m_c^{(0)} = 1.5$ and used these threshold values of r and mussel density in what followed (see Fig. 13.4). That function was given explicitly by

$$r_k(\alpha) = \frac{1}{\alpha} \frac{m_c^{(0)} + \alpha}{m_c^{(0)} + 1} = \frac{1.5 + \alpha}{2.5\alpha},$$

which is compared with $r_0(\alpha)$ in Table 13.5.

TABLE 13.5
Values of $r_k(\alpha)$ for $m_c^{(0)} = 1.5$ in comparison with $r_0(\alpha)$.

α	r_0	r_k	$\frac{2(r_k - r_0)}{r_k + r_0}$	m_k
0.5	1.592	1.600	0.005	1.5
0.6	1.390	1.400	0.007	1.5
2/3	1.289	1.300	0.008	1.5
0.7	1.247	1.257	0.008	1.5
0.8	1.141	1.150	0.008	1.5
0.9	1.061	1.067	0.006	1.5

We felt that so close a correlation as exhibited in this table to be a Rabbit Pulled Out of a Hat! Hence, we considered $m_c^{(0)} = 1.5$ as the critical mussel density threshold such that blue regions corresponded to $m > 1.5$ and yellow ones to $m < 1.5$. Then, when $m_e < 1.5$ or equivalently, $r < r_k$, high threshold patterns in which yellow regions predominated would be generated, while when $m_e > 1.5$, or equivalently, $r > r_k$, low threshold ones in which blue regions would predominate occurred. This corresponded in dimensional variables to $M_c^{(0)} = 1.5 k_M = 0.0225$ gm/cm^2, a biologically reasonable threshold value. We next synthesized our rhombic and hexagonal planform morphological stability results for $\alpha = 2/3$ in the r-μ plane of Fig. 13.16 and identified regions corresponding to the predicted mussel bed patterning. In that figure, after Geddes et al. ([65]), we superimposed upon the hexagonal pattern identifications included in Fig. 13.10 the rhombic pattern predictions for $r_1 < r < r_2$ cataloged in Table 13.4. Here, bands only occurred for $r = r_0 \cong r_k$ since critical point II is only stable for both planforms along that vertical line from $0 < \mu < \mu_c$. Thus, in Fig. 13.16 for $r \neq r_0$, we have replaced all the regions of banded mussel pattern (II) predictions appearing in Fig. 13.10 with rhombic mussel pattern (V) ones instead. These have been identified in Fig. 13.16 as rhombic arrays of pseudo clumps (V$^+$) for $r_1 < r < r_k$ and of pseudo gaps (V$^-$) for $r_k < r < r_2$ where $0 < \mu < \mu_c$. Further, note that the $\mu > \mu_c$ region of that figure which can be identified with homogeneous mussel patterns ranges from a relatively sparse distribution ($m_e < 1.5$) for $r_1 < r < r_k$ to a relatively dense one ($m_e > 1.5$) for $r_k < r < r_2$ (see Fig. 13.4) and hence have been designated by I$^\mp$, respectively.

So far, we had been concerned with the morphological stability behavior of our model system for $a_1 > 0$. We then considered its behavior for $a_1 < 0$. Recall from Chapter 11 that a comparison of simulation (Lejeune et al., [112]) and weakly nonlinear stability (Boonkorkuea et al., [17]) results for a particular evolution equation describing vegetative pattern formation in arid isotropic environments, led to the conjecture that when $a_1 < 0$ localized

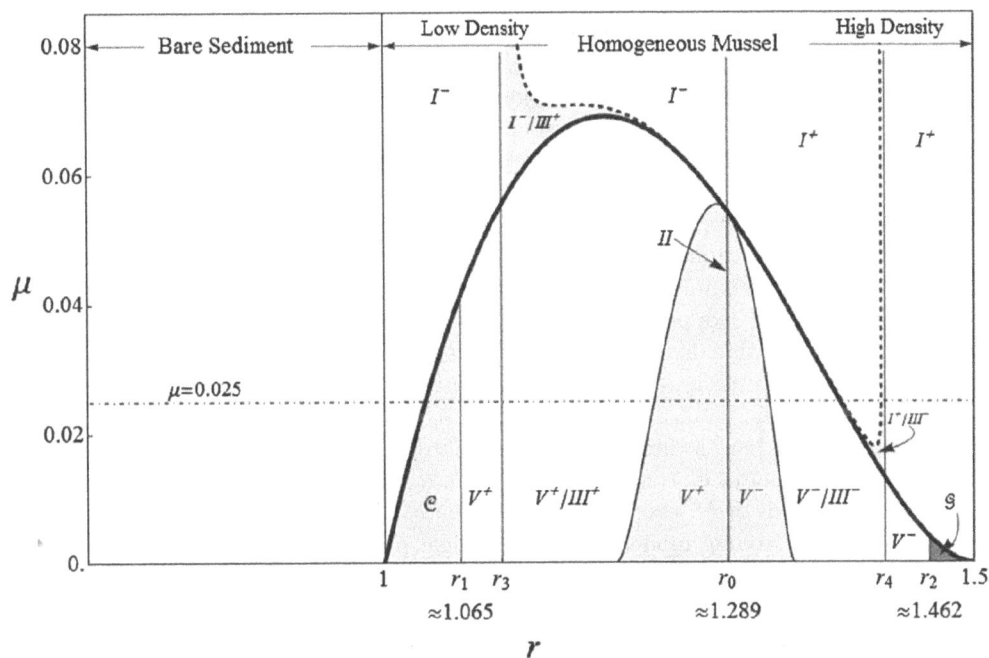

FIGURE 13.16

Stability diagram in the r-μ plane synthesizing the rhombic and hexagonal planform results for our two-dimensional interaction-diffusion model system with $\alpha = 2/3$, denoting the predicted mussel bed patterns. Here we have employed the approximation that $r_0 \cong r_k$, the horizontal line $\mu = 0.025$ has been designated, and the r-axis ranges from $0.75 \leq r \leq 1.50$.

structures would occur where $\sigma > 0$ characterized by isolated patches of vegetation at low density or by isolated patches of bare ground at high density that were a spatial compromise between the periodic patchy vegetation and bare ground or homogeneous vegetation stable states (see Chapter 15). As described in Chapter 12, Wollkind and Kealy co-organized a mini-symposium on diffusion-driven pattern formation in ecological, developmental biological, and behavioral phenomena at the 2012 SIAM Conference on Nonlinear Waves and Coherent Structures held on the Seattle campus of the University of Washington. Wollkind and Cangelosi presented a preliminary report on our mussel-algae model as the closing talk in this session. During that presentation, Michael Ward of the University of British Columbia declared that he too had discovered localized structures to occur in conjunction with such subcriticality for a number of reaction-diffusion systems including the Gray-Scott model (Chen and Ward, [33]). Given the similarity of behavior among all these phenomena, as well as that in Chapter 8, we conjectured with some confidence that isolated clusters of mussels would occur for $1 < r < r_1$ and isolated gaps of bare sediment for $r_2 < r < \alpha^{-1}$ where $0 < \mu < \mu_c$ and identified these regions graphically in Fig. 13.16 using the designations of $\mathcal{C}$ and $\mathcal{G}$, respectively. Finally, associating that region of this figure where $0 < r < 1$ with bare sediment (see Fig. 13.4) completed our morphological identifications.

It remained for us to compare these Turing diffusive instability predictions with relevant experimental or observational evidence and existing numerical simulations involving differential flow instabilities for the associated interaction-dispersion-advection mussel-algae model system. Van de Koppel *et al.* ([260]), Wang *et al.* ([268]), and Liu *et al.* ([122]) studied pattern formation for young beds of blue mussels (*Mytilus edulis*) on intertidal soft-bottom substrates in the Wadden Sea. They reported that within these beds mussels formed regular periodic banded patterns on soft sediment consisting of dense bands alternating with virtually bare sediment containing hardly any mussels at all in the interband regions. Van de Koppel *et al.* ([259]) and Liu *et al.* ([121]) investigated the origin of regular patterns in young beds of *M. edulis* on intertidal flats in the Menai Strait near Bangor, UK. *M. edulis* is a filter feeding mollusk exploiting algal plankton in the benthic marine layer (van de Koppel *et al.*, [259]). They found patterns consisting of regularly spaced clumps at a distance of about 10 cm from each other that formed a coherent labyrinth-like lattice. Van de Koppel *et al.* ([259]) tested in the field and the laboratory their hypothesis that the observed patterns were self-organized and hence would develop spontaneously from homogeneity. For these experiments mussels that were laid out evenly on intertidal flats under wind-sheltered conditions in the field experiment and on homogeneous substrates in the laboratory experiment developed coherent nonrandom spatial patterns that were statistically similar to those field observations. In particular, when mussel densities were decreased, the spatial pattern became more open and the clumps were reduced to isolated clusters as was observed under natural conditions. Liu *et al.* ([121]) presented these experimental results in more detail. Their four photographic reproductions included in Fig. 13.17 consisted of the following sequence of spatial patterns as the initial mussel density was increased: Isolated clusters, open labyrinth clumps, regularly spaced gaps, and dense near homogeneous beds. Specifically the gapped patterns exhibited a honeycomb net-like appearance.

We then compared these observations with our theoretical predictions. Upon examination of Fig. 13.16, we could see that, say for example when $\mu = 0.025$, the predicted sequence as r (or equivalently m_e) increased, transited from isolated clusters to dense homogeneous distributions through intermediate states of pseudo clumps, clumps, bands, pseudo gaps, and gaps. In order to compare these intermediate states with observation, it was necessary for us to recall that the III$^-$ critical point gave rise to mussel patterns that had a sparser central circular region surrounded by denser boundary regions which had their maximum

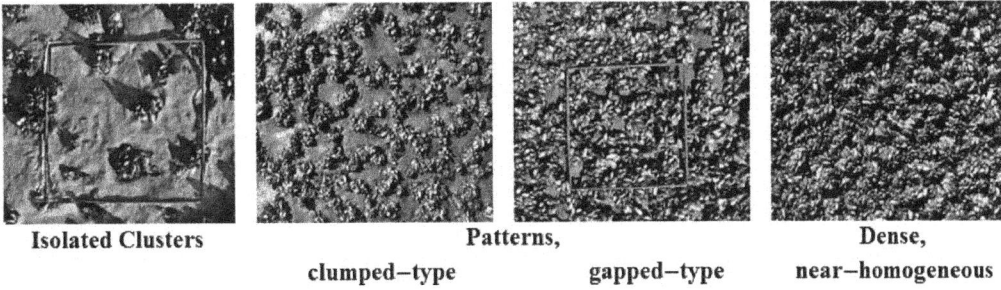

| Isolated Clusters | Patterns, | | Dense, |
| clumped–type | | gapped–type | near–homogeneous |

FIGURE 13.17
Mussel bed patterns observed in the field by Liu *et al.* ([121]). Note that our predicted regions of hexagonal and pseudo clumps or hexagonal and pseudo gaps as initial conditions vary from point to point on the substrate surface can produce such labyrinth-type mussel bed patterns.

density at its hexagonal vertices (see Figs. 2.10 and 7.4) Thus they would have a honeycomb appearance forming web or net-like mussel bed patterning. Finally, for $r = r_0$ where our analysis predicts the occurrence of stable parallel banded patterns, recall the equivalence class of critical point II actually contains two other solutions that represents families of bands making angles of ± 60 degrees with them, for which stable coexistence with either a member of the original family or one another is impossible. Then as initial conditions varied from point to point over a planar environment, such families of bands could give rise to interconnected labyrinth-type bi-continuous patterns (see Figs. 8.1 and 13.18). Therefore, each of the mussel bed patterns described above was represented by a member of this morphological sequence of Fig. 13.17 and each member of this sequence corresponded to a mussel bed pattern described above.

In order to demonstrate that our model could also provide good quantitative agreement when compared with those observed experimental mussel bed patterning, we calculated the predicted dimensionless critical wavelength $\lambda_c = 2\pi/q_c$ where

$$q_c^2(r; \alpha) = (1 - \alpha) \left(1 + \frac{1}{\sqrt{1 - \alpha r}} \right) \text{ for } q_c > 0$$

and then examined its associated dimensional form $\lambda_c^* = \sqrt{D_A/\omega}\lambda_c$ where $D_A = \chi^2$ cm^2/sec. In Fig. 13.19, we plotted $q_c^2(r; 2/3)$ for $1.065 = r_1 < r < r_2 = 1.462$. From this figure, we determined that $0.976 < q_c < 1.558$ in this interval. For the sake of definiteness, again using $r = r_0$ as a reference point, we took $q_c \equiv q_c(r_0; 2/3) = 1.106$, which yielded $\lambda_c = 5.68$. Then, employing our previous results, we found that $\omega = 150/h = 0.04167/\text{sec}$ and hence obtained $\lambda_c^* = 2\sqrt{6}\chi\lambda_c$ cm $= 27.83\chi$ cm. Since, as mentioned above, the observed experimental pattern wavelength was about 10 cm, we wished to determine the χ-value which would yield this result. That is,

$$\lambda_c^* = 27.83\chi \text{ cm } = 10 \text{ cm } \Rightarrow \chi = 0.36.$$

Then, we deduced that

$$D_A = \chi^2 \frac{\text{cm}^2}{\text{sec}} = 0.13 \frac{\text{cm}^2}{\text{sec}} = 0.05 \frac{\text{m}^2}{\text{h}}$$

FIGURE 13.18
Labyrinth-like chemical Turing pattern observed by Ouyang and Swinney ([163]) in their CDIMA/starch system (see Fig. 1.2) as described in Chapter 7.

and noted that this predicted algal lateral dispersal coefficient compared quite favorably with the value of 0.1 cm^2/sec employed by Malchow ([134]) for the same purpose in his phytoplankton-zooplankton interaction-diffusion-advection model system. In this context, motile green algae species of the genus *Chlamydomonas* such as *C. augustae* and *C. reinhardtii* (see Fig. 13.2) exhibit constant isotropic dispersal by virtue of their flagellar swimming motion (see Williams and Bees, [273]). We considered this inverse parameter identification procedure which yielded a prediction in such close agreement with a measured value to be a Rabbit Pulled Out of a Hat!

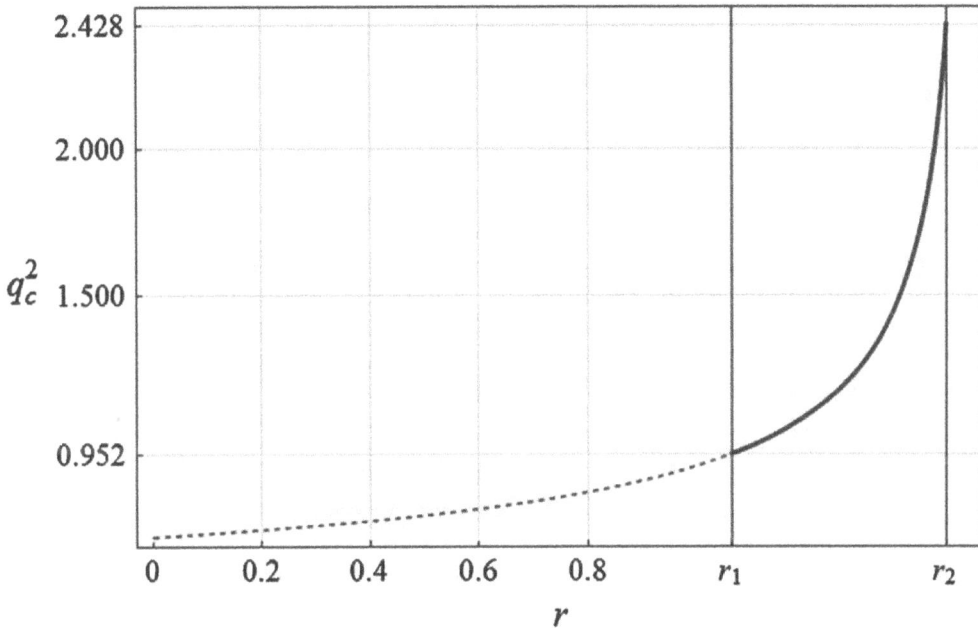

FIGURE 13.19
Plot of q_c^2 versus $r_1 < r < r_2$ for $\alpha = 2/3$.

Next, we compared our model predictions with the numerical simulations of van de Koppel *et al.* ([260]), Wang *et al.* ([268]), and Liu *et al.* ([122]) for their associated interaction-dispersion-advection mussel-algae model system. In particular, these authors set $D_A = 0$ and thus neglected the lateral dispersal of algae, while retaining its advective effect. Introducing $\sqrt{D_M/\omega}$, ω, k_M, and A_{up} as scale factors for space, time, mussel density, and algae concentration, respectively, they obtained a nondimensional system which differed from ours principally in that the Δa term had been eliminated and the dimensionless parameter ν had been replaced by β where $\beta = V/\sqrt{D_M\omega}$. This system had the same critical points as ours and numerical integration of it on a large domain, with either periodic or no-flux boundary conditions, yielded traveling wave patterns that represented migrating band instabilities when $\beta > \beta_c > 0$. The morphological stability results for these numerical analyses of Wang *et al.* ([268]) are summarized in the β-δ plane of Fig. 13.20 where $\delta = \gamma r = ecA_{up}/\omega$, which includes a plot of its marginal stability curve $\beta = \beta_c(\delta)$ with the patterning region consisting of migrating bands lying to its right and homogeneous mussel, to its left. Those bands had a characteristic wavelength related to the critical wavenumber of linear stability theory, as usual, by $\lambda_c = 2\pi/q_c$. Specifically, considering A_{up} and V as variable parameters, taking

$$e = 0.21 \times EX, \; d_M = 0.026 \times EX \; \text{h}^{-1},$$
$$D_M = 0.0005 \times EX \; \text{m}^2\text{h}^{-1} \; \text{with} \; EX = 10^3,$$

and assigning the rest of their parameters our values of van de Koppel *et al.* ([260]), they found that $\gamma = (4/3) \times 10^{-1}$. Then, in addition, for $A_{up} = 1$ gm m^{-3} and $V = 360$ m h^{-1} , Wang *et al.* ([268]) obtained $\delta = 0.140$ and $\beta = 41.5692$, while $\beta_c \cong 13.0$, $q_c \cong 0.065$, and $\lambda_c \cong 96.6644$. Hence, for these values, migrating bands orthogonal to the tidal flow direction were predicted with dimensional wavelength

$$\lambda_c^* = \lambda_c\sqrt{\frac{D_M}{\omega}} \cong 96.6644\sqrt{\frac{0.5 \; \text{m}^2 \; \text{h}^{-1}}{150 \; \text{h}^{-1}}} = 5.58 \; \text{m},$$

compared to the wavelength of the banded mussel bed patterns observed in the Wadden Sea of, on average, approximately 6 m (van de Koppel *et al.*,[260]). We noted that they introduced the factor $EX = 10^3$ primarily to speed up their numerical simulations, especially for the two-dimensional case subsequently handled in more detail by Liu *et al.* ([122]) who, given the behavior described above, employed A_{up} as their bifurcation parameter. This differed from the original analysis of van de Koppel *et al.* ([260]) who implicitly took $EX = 1$, while also employing A_{up} as a bifurcation parameter. Wang *et al.* ([268]) asserted that their predictions were the same as those for the latter. This meant that the q_c and, consequently, the λ_c given above would be unaltered as well. We then re-examined the associated expression for λ_c^* when $EX = 1$ in light of these assertions. Thus

$$\lambda_c^* = \lambda_c\sqrt{\frac{D_M}{\omega}} \cong 96.6644\sqrt{\frac{0.0005 \; \text{m}^2 \; \text{h}^{-1}}{150 \; \text{h}^{-1}}} = 18 \; \text{cm}.$$

In order to obtain the same dimensional critical wavelength for our system, we concluded that

$$\lambda_c^* = 27.83\chi \; \text{cm} = 18 \; \text{cm} \Rightarrow \chi = 0.65$$
$$\Rightarrow D_A = \chi^2 \frac{\text{cm}^2}{\text{sec}} = 0.42\frac{\text{cm}^2}{\text{sec}} = 0.15\frac{\text{m}^2}{\text{h}}.$$

Finally, in the spirit of Wang *et al.* ([268]), we rescaled D_A as well, by taking it equal to $D_A = 0.15 \times EX$ m^2 h^{-1} so that $\mu_0 = D_M/D_A$ would remain invariant under this scaling. Then, employing all of these values, we found they corresponded to

$$\mu = \frac{D_M}{\gamma D_A} = \frac{0.0005 \times EX}{(4/3 \times 10^{-1})(0.15 \times EX)} = 0.025,$$

which was our rationale for using that prototypical value of $\mu = 0.025$ to generate the predicted morphological sequence examined earlier. We definitely considered this evaluation and that prediction to be another Rabbit Pulled Out of a Hat. In this context we observed that, by virtue of $\beta > \beta_c = 13$, Wang *et al.* ([268]) were unable to predict any patterns without significant advection and hence their model could not be used to explain the experimental observations of van de Koppel *et al.* ([259]) and Liu *et al.* ([121]). A question might be raised about a constant advective term being used to model the tidal effects in such mussel-algae systems as opposed to a periodic time varying term. In this context, note that, for the field and laboratory experiments referenced in this chapter, the stationary mussel bed patterning was achieved in 10 hours, which would allow for the assumption of a

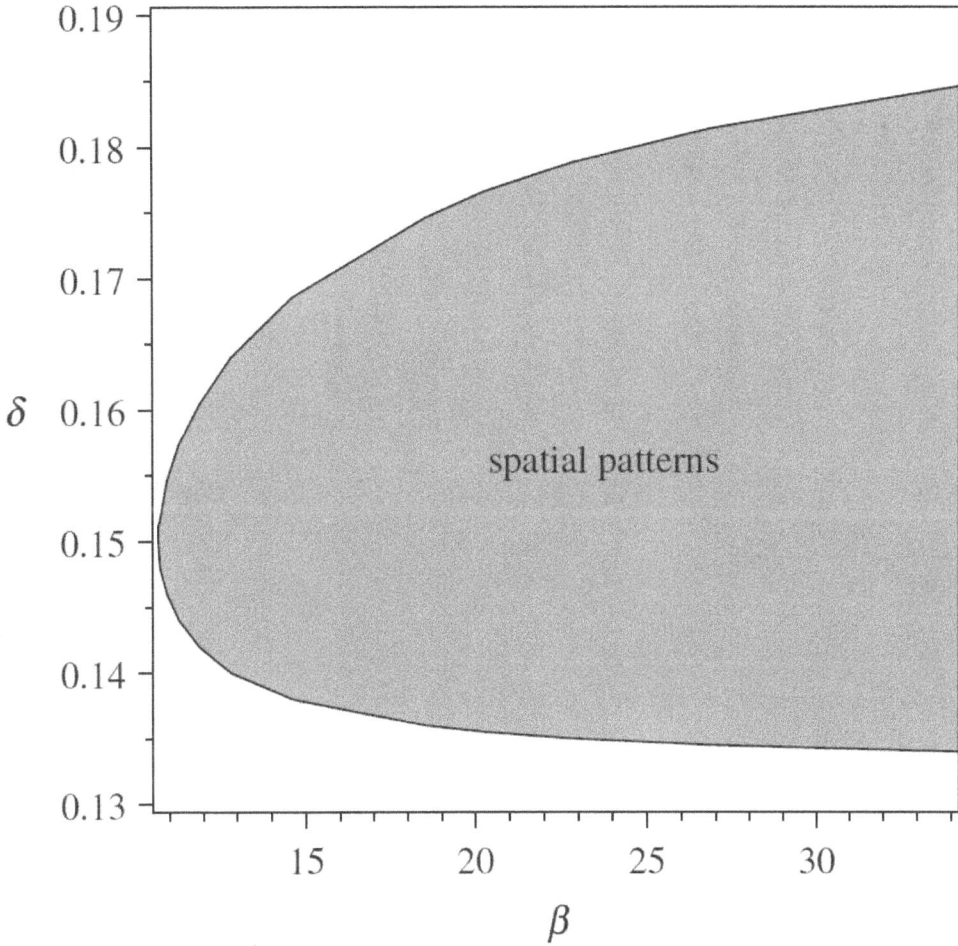

FIGURE 13.20
Wang *et al.*'s ([268]) numerical calculation of stability in (β, δ) space where $\delta = \gamma r$ with $\gamma = (4/3) \times 10^{-1}$.

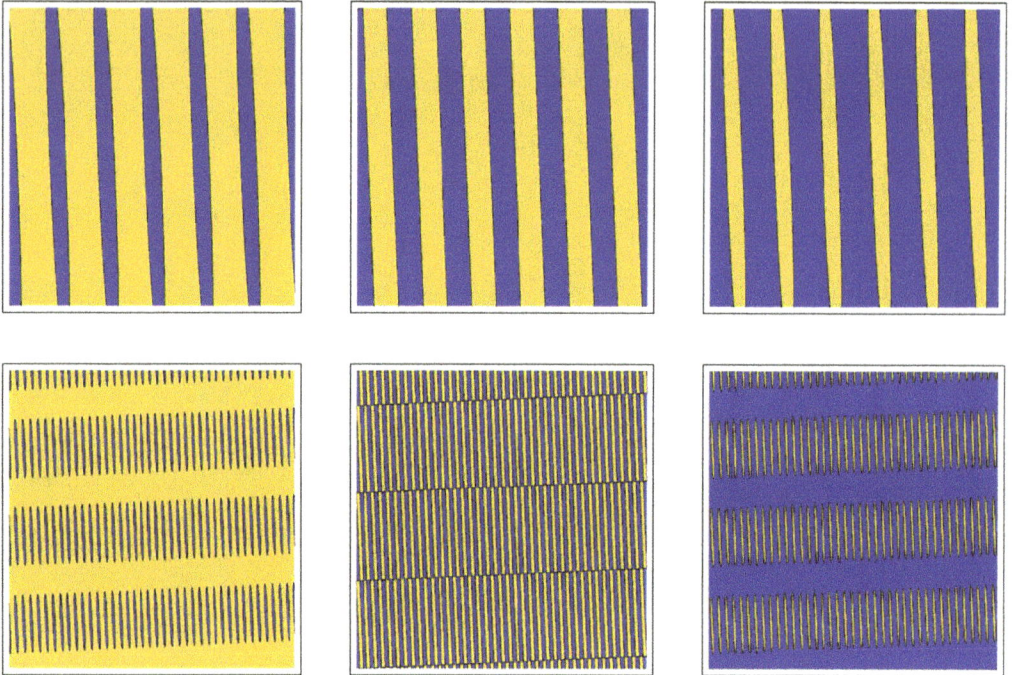

FIGURE 13.21

Rhombic contour plots for $\phi = 0.08725$ (5°) with threshold values from left to right of 1, 0, and, -1. Here, the upper and lower rows have been plotted for length scales consistent with that of the territory sizes employed in the previous rhombic contour plots and ten times this scale, respectively.

constant β as a first approximation.

Recall from Fig. 13.16 that our model could only predict stable bands for $r = r_0$ and for this value of r, side-band structures of zig–zag, and cross-band instabilities can occur (see Chapters 11 and 12) as well as the two other parallel families of bands making angles of $\pm 60°$ with our original family, as related above. Although all these states, which collectively can produce quite complicated labyrinth-like patterns (see Fig. 13.18), only occur for that precise value of r and hence their occurrence may be considered problematic for ecological pattern formation purposes in the $\mu = 0.025$ sequence of morphologies described earlier, the system must pass through such an intermediate labyrinthine state in order to transit from a clumped to a gapped pattern. In spite of the fact that our model does not predict stable bands for $r \neq r_0$, to say that it cannot predict banded mussel patterns of any sort for other values of r would be inaccurate. In order to demonstrate this assertion, it was necessary for us to re-examine our rhombic patterns in more detail when its characteristic angle ϕ took on the representative small values from Table 13.4 corresponding to r in the neighborhood of r_1 or r_2. Toward that end, for $\phi = 0.08725$ (5°), we plotted high, zero, and low threshold versions of that rhombic pattern in Fig. 13.21, which had been portrayed for both small and large window territory sizes with the high threshold versions corresponding to those r's flanking r_1 (see the $\alpha = 0.5$ entry of Table 13.1 in conjunction with Fig. 13.22) and the low threshold ones to those flanking r_2; and noted, in particular, that the large window high-threshold patterns bore a striking resemblance to the banded mussel bank depicted in the aerial photograph of the Wadden Sea in The Netherlands by van de Koppel et al. ([260]).

In concluding, we stated that implicit to our formulation were the continuum mechanical assumptions that the pattern wavelength was large with respect to the average size of the individual mussels but very small with respect to the territorial length-scale characteristic of the sedimentary substrate. Since, under the latter condition, the actual territorial boundary of that sedimentary substrate did not significantly influence the pattern (Graham et al., [73]), it seemed reasonable as a first approximation for us to have considered our interaction-diffusion equations on an unbounded planar spatial domain where m and a satisfied a far-field boundary condition.

Right from the start of the morphological stability problem described in Chapter 2, it became obvious to me that I would always have some anxiety about the outcome of the determination of the Landau coefficients inherent to the various planform expansions associated with nonlinear stability theory and those outlined in this chapter proved to be no exception to this rule. When he initially evaluated the Landau coefficients of our hexagonal planform amplitude-phase equations, Rick found that $a_1 + 4a_2 < 0$. This is a condition under which critical points III$^{\pm}$ do not exist and hence no hexagonal patterns could be predicted for that analysis. Needless to say, he was very despondent about this negative result and Bonni tried to make him feel better by saying don't worry about it since David would no doubt figure a way to Pull a Rabbit Out of a Hat and resolve the situation. That was my motivation for applying the threshold-dependent rhombic pattern formation method of analysis to this problem. I had known since my work with Laura Stephenson that this rhombic methodology could be used to mediate the results of a hexagonal planform analysis but never before considered that the former might be used to replace the latter. Further, although examining in detail high- and low-threshold patterns for the hexagonal planform critical points (see Chapters 7 and 12), I had not really done so for rhombic planform pattern formation. I thought this procedure would only give rise to rhombic arrays of rectangles $V^{(0)}$ of zero threshold (see Fig. 13.14). Thus, when Rick, at my suggestion, first generated the high- and low-threshold rhombic patterns $V^{\pm}$, exhibited in this chapter, and we noticed

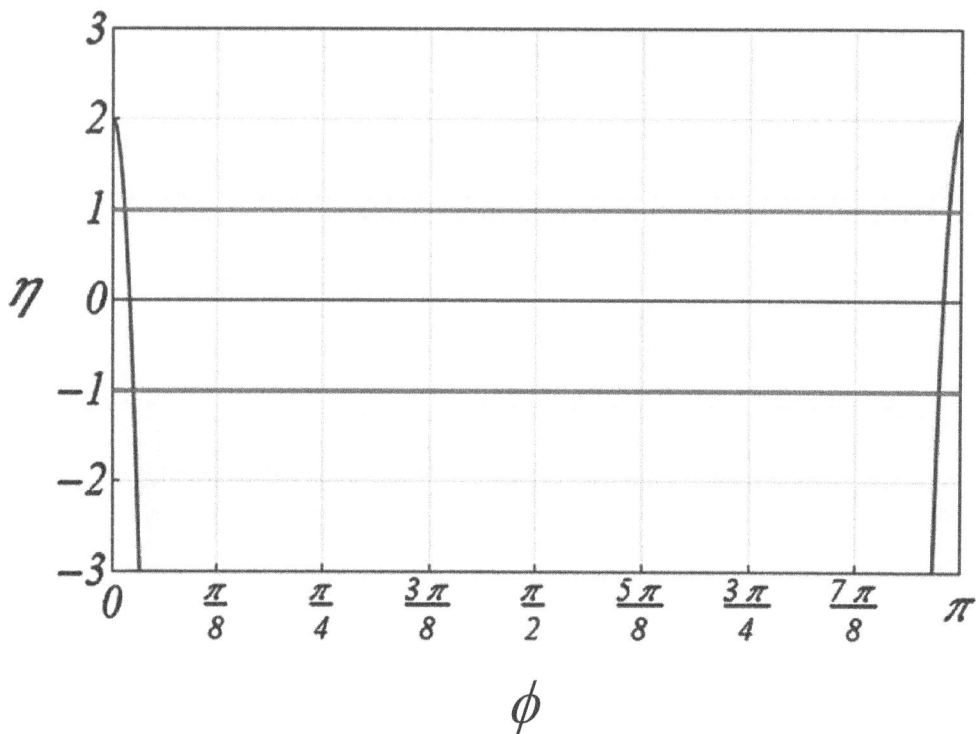

FIGURE 13.22
Plot of η versus ϕ with $\alpha = 1/2$ and $r = 1.05$.

their resemblance to III$^\pm$ hexagonal arrays, a new paradigm of pattern formation had been discovered.

Just as in Chapter 7, it came from a mistake since upon checking his calculations Rick found that indeed there was a range of parameter values with $\alpha = 2/3$ for which $a_1 + 4a_2 > 0$! This was for a very narrow range and then $2a_2 - a_1 < 0$ identically. Hence, most of its pattern formation was generated by the rhombic planform analysis. That was the result presented at the SIAM mini-symposium discussed earlier, which is our reason for labeling it preliminary, and also in the paper submitted to the *Journal of Mathematical Biology* before Rick defended his thesis. Subsequently, we discovered that the hexagonal planform analysis calculations were still wrong, after having Inthira Chaiya determine independently its Landau coefficients by appropriately adapting her thesis analysis (see Chapter 14), so that it could be applied to our system. In particular, although a_0, a_1, and b_1 were correct, a_2 was not, which accounted for our problems with $a_1 + 4a_2$ and $2a_2 - a_1$. Then, we corrected this coefficient and, when a revision was requested for our *J. Math. Biol.* submission, incorporated that modification into its resubmission along with the few minor additions the reviewers suggested such as the reason tidal advection could be considered constant for the differential flow instability model.

We also added the following footnote to our statement about the critical role played by $r = r_0$ in both our planform analyses serving as a partial check on the calculation of their Landau coefficients:

> As a conclusive check on these calculations we evaluated those coefficients for the reaction-diffusion and interaction-diffusion models of Wollkind and Stephenson ([296]) and Kealy and Wollkind ([92]), respectively, and reproduced their computations exactly.

Due to the help provided by them to correct a_2 and produce the results described in this chapter, we included Bonni and Inthira as co-authors for our paper that was accepted by the *J. Math. Biol.* and appeared as Cangelosi *et al.* ([25]). Their inclusion was consistent with my policy of always giving co-authorships to individuals whose contributions were crucial to the publication of a particular paper. These results also comprised the major part of Rick's dissertation, which he successfully defended in his Ph.D. final oral examination and hence became Dr. Richard A. Cangelosi.

We finished the dissemination of this doctoral research by presenting various aspects of it on four different occasions in 2013 and 2014. I gave three of these presentations:

- Nonlinear Stability Anayses of the Sustainability of an Ecological Turing Pattern Formation for an Interaction-Diffusion Mussel-Algae System in a Static Marine Layer, Society of Mathematical Biology, Annual Meeting, Tempe, AZ, June 11, 2013;

- Nonlinear Stability Analyses of Turing Patterns for a Mussel-Algae System, University of Central Florida, Mathematic Department Colloquium, Orlando, FL, March 20, 2014; and WSU Showcase 2014, Pullman, WA, March 28, 2014;

- while Rick gave the other one: Nonlinear Stability Analyses of Turing Patterns for a Mussel-Algae System, PNW MAA, Annual Meeting, Missoula, MT, June 28, 2014.

Just like Bonni the year before, Rick also presented one more talk on this subject at Gonzaga University in Spokane, WA, while successfully interviewing for an Assistant Professorship in their Mathematics Department. Indeed, Rick now starting his eighth year there, has done such a good job of establishing their applied mathematics program and supervising modeling projects for senior theses that he, the same as Bonni, was promoted early to Associate

Professor two years ago and tenured last year. During that time, our ongoing collaboration has continued, as documented by the research to be described in the rest of the chapters of this book.

We close with a discussion of the recent *J. Math. Biol.* publication entitled "Large scale patterns in mussel beds: Stripes or spots," by Jamie Bennett and Jonathan Sherratt ([13]), who performed one- and two-dimensional numerical simulations on an advection-interaction-diffusion mussel-algae system that added our algal lateral dispersal term to the dimensionless differential flow instability model of Wang *et al.* ([268]). Thus, the coefficients of Δm and Δa in their governing equations were unity and $v = D_A/D_M = 1/\mu_0$, respectively, with β treated as the bifurcation parameter. Recall, that for our problem, $\mu_0 = \gamma\mu$ and thus the values $\gamma = (4/3) \times 10^{-1}$ and $\mu = 0.025$ we employed in the comparison with Wang *et al.* ([268]), corresponded to

$$\mu_0 = \left(\frac{4}{3} \times 10^{-1}\right)(0.025) = \left(\frac{4}{3}\right)(0.0025) = \left(\frac{4}{3}\right)\left(\frac{1}{400}\right)$$

$$= \frac{1}{300} = \frac{1}{v} \Rightarrow v = 300.$$

Bennett and Sherratt ([13]) stated that:

> A realistic rate of algal dispersal is difficult to determine, in part due to its obvious simplification of algal movement. Cangelosi *et al.* ([25]) argue the rough estimate $v = 300$, though there is no concrete evidence to support this. In this regard, we assess the effect that algal dispersal has on stability by considering a few different values of v

(they earlier defined $v \gg 1$). In spite of that admonishment, all of Bennett and Sherratt's ([13]) really important 1D and 2D simulations were performed for $v = 300$! In my opinion, we provided very good evidence in support of this estimate but should now be satisfied that they fulfilled Oscar Wilde's adage of "imitation being the sincerest form of flattery"! They also found that stripes could not realistically occur in their model for low values of β, which was consistent with our results for $\beta = 0$ and corroborated this prediction of Cangelosi *et al.* ([25]).

14

Root Suction Driven Vegetative Rhombic Pattern Formation: Rhombic Planform Nonlinear Stability Analysis

---

Inthira Chaiya's research with me played a fundamental role in the full circle nature of my last two years at WSU as a regular Professor of Mathematics before being granted emeritus status, in that, her Ph.D. adviser in the Mathematics Department at Mahidol University was Chontita Rattanakul, who did postdoctoral research with me in conjunction with Adoon (see Chapter 9). When Inthira first came to my office after her arrival on campus, I had already selected her thesis problem which, like Nichaphat's and Bonni's, involved a vegetative pattern formation model in an arid flat environment. This plant-ground water interaction-diffusion system had been constructed from the interaction and diffusion terms of Rietkerk *et al.* ([186]) and von Hardenberg *et al.* ([266]), using Robert May's ([139]) "build a model" methodology. In particular, von Hardenberg *et al.* ([266]) devised a plant-ground water (sometimes called soil water) interaction-diffusion system to model self-organized vegetative pattern formation in arid environments (reviewed by Rietkerk *et al.*,[187]). Here, the positive feedback for an activator consumer (*e.g.*, plants) and the self-diffusivity advantage for an inhibitory limiting resource (*e.g.*, ground water) in that system provided the necessary conditions for the onset of Turing ([256]) pattern formation. Von Hardenberg *et al.*'s ([266]) model also included the effect of plant root suction by adding a cross-diffusion term involving the plant gradient in their ground water flux. Rietkerk *et al.* ([186]) performed numerical simulations using reflecting boundary conditions on a similar interaction-diffusion model system consisting of three partial differential equations describing the spatio-temporal behavior of plant density, soil water, and surface water, respectively, but excluding the root suction cross-diffusion term in their soil water equation. That model had been carefully developed by HilleRisLambers *et al.* ([80]) for a flat semi-arid grazing system.

We wished to formulate an interaction-diffusion model system for $N(X, Y, \tau) \equiv$ plant biomass density (gm/m^2) and $W(X, Y, \tau) \equiv$ ground (soil) water content (mm of depth), where $(X, Y) \equiv$ two-dimensional coordinate system (m, m) and $\tau \equiv$ time (d), based upon the interaction terms of Rietkerk *et al.* ([186]) and the diffusion terms of both von Hardenberg *et al.* ([266]) and Rietkerk *et al.* ([186]), defined on a flat unbounded arid environment. Toward that end, we first introduced the auxiliary dependent variable $O(X, Y, \tau) \equiv$ surface water content (mm of depth) and the coupled interaction-diffusion model system given by

$$\frac{\partial N}{\partial \tau} = Q^{(N)} - \boldsymbol{\nabla}_2 \cdot \boldsymbol{J}^{(N)}, \quad \frac{\partial W}{\partial \tau} = Q^{(W)} - \boldsymbol{\nabla}_2 \cdot \boldsymbol{J}^{(W)}, \quad \frac{\partial O}{\partial \tau} = Q^{(O)} - \boldsymbol{\nabla}_2 \cdot \boldsymbol{J}^{(O)},$$

DOI: 10.1201/9781003195603-14

with source terms (Rietkerk *et al.*, [186])

$$Q^{(N)} = \frac{cg_M WN}{W + k_1} - dN, \ Q^{(W)} = i_M O \frac{N + k_2 f}{N + k_2} - \frac{g_M WN}{W + k_1} - rW,$$

$$Q^{(O)} = R - i_M O \frac{N + k_2 f}{N + k_2};$$

and flux terms (von Hardenberg *et al.*, [266]; Rietkerk *et al.*, [186])

$$\boldsymbol{J}^{(N)} = -D_N \boldsymbol{\nabla}_2 N, \ \boldsymbol{J}^{(W)} = -D_W \boldsymbol{\nabla}_2 (W - \rho N), \ \boldsymbol{J}^{(O)} = -D_O \boldsymbol{\nabla}_2 O;$$

where $\boldsymbol{\nabla}_2 \equiv (\partial/\partial X, \partial/\partial Y)$ and $\boldsymbol{\nabla}_2 \cdot \boldsymbol{J} = (\partial J_1/\partial X, \partial J_2/\partial Y)$ for $\boldsymbol{J} = (J_1, J_2)$. Here, $g_M \equiv$ maximum specific water uptake rate by the plants, $c \equiv$ conversion of water uptake by the plants to plant growth, $d \equiv$ specific loss rate of plant density due to mortality, $k_1 \equiv$ half-saturation constant of specific plant growth and water uptake, $r \equiv$ specific loss rate of soil water due to evaporation and drainage, $R \equiv$ rainfall rate, $i_M \equiv$ maximum specific water infiltration rate, $k_2 \equiv$ saturation constant of water infiltration, $f \equiv$ fraction of maximum specific water infiltration rate in the absence of plants, $D_N \equiv$ dispersal coefficient for plants, $D_W \equiv$ diffusion coefficient for ground water, $D_O \equiv$ diffusion coefficient for surface water, $D_S \equiv$ coefficient of plant root suction, and $\rho \equiv D_S/D_W$.

Fifty years ago, Keller and Segel ([94]) proposed the initiation of slime mold aggregation, viewed as an instability, in a landmark paper with that title. They formulated a mathematical reaction-diffusion model system for the aggregation of the cellular slime mold *Dictyostelium discoideum* involving four dependent variables: Namely, the density of this amoeba; the concentrations of the acrasin and acrasinase produced by it, which mediated its aggregation; and the concentration of an intermediate complex formed by these chemicals in a reversible reaction. Keller and Segel ([94]) then simplified that model to a two-component system involving just the density of the amoeba and the concentration of the acrasin, by making Haldane's assumption that the complex was in chemical equilibrium and the additional assumption that the total concentration of the acrasinase enzyme in both its free and bound state was a constant. This simplified model included flux terms deduced from Fickian self-diffusion of its two components and a chemotaxis term for the amoeba generated by the introduction of cross-diffusion involving the acrasin gradient.

For the sake of model analysis, HilleRisLambers *et al.* ([80]) introduced a quasi *steady-state* approximation by taking $\partial O/\partial \tau \equiv 0$. We next, after Keller and Segel's ([94]) employment of Haldane's assumption in their slime mold problem, simplified our model system by assuming that the surface water was in hydrological equilibrium and making the quasi *stationary* approximation that

$$Q^{(O)} \equiv 0 \Rightarrow R = i_M O \frac{N + k_2 f}{N + k_2}.$$

Then, introducing this into the basic equations by replacing the rate of infiltration term in $Q^{(W)}$ with R, we obtained the final formulation of our interaction-diffusion plant-ground water model system

$$\frac{\partial N}{\partial \tau} = F(N, W) + D_N \nabla_2^2 N, \ \frac{\partial W}{\partial \tau} = G(N, \ W) + D_W (\nabla_2^2 W - \rho \nabla_2^2 N)$$

where

$$\nabla_2^2 \equiv \boldsymbol{\nabla}_2 \cdot \boldsymbol{\nabla}_2, \ F(N, W) = \frac{cg_M WN}{W + k_1} - dN, \ G(N, W) = R - \frac{g_M WN}{W + k_1} - rW.$$

Our main purpose in doing so was to devise a model system of this sort that demonstrated root suction alone could generate the two-dimensional vegetative patterns (*e.g.*, leopard, pearled, and labyrinthine tiger bush) occurring in arid flat environments as described by Rietkerk *et al.* ([187]). In this context, the root suction effect appeared as a cross-diffusion term involving $\nabla^2_2 N$ in the ground water equation. Hence, pattern formation in that event would be driven by cross- rather than self-diffusion. Here again, arid or semi-arid referred to environments where the yearly average rate of rainfall is less than the corresponding rate of evaporation and water was a limiting resource for plant growth.

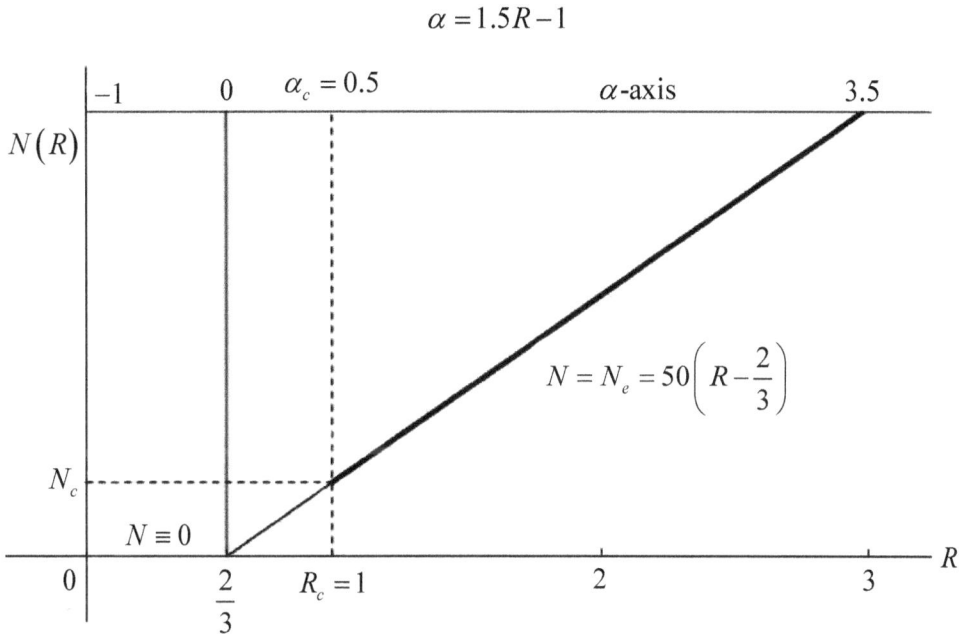

FIGURE 14.1
Plot of the plant component of the equilibrium point versus $0 < R < 3.5$ where R and N are being measured in units of mm/d and gm/m^2, respectively. Here, $\alpha(R) = 1.5R - 1$ and $N = N_e(R) = 50(R - 2/3)$ when $g_M = 0.05$ (mm/d)(gm/m^2), $c = 10$ (gm/m^2)/mm, $d = r = 0.2$/d, and $k_1 = 5$ mm; while $R_c = 1$ or $\alpha_c = 0.5$ corresponds to the N_c such that patterns appear dark where $N > N_c$ and light, where $N < N_c$ (see below).

Since this research was developed in the same manner as Rick's of the previous chapter, we shall replicate that description while particularizing its items to Inthira's problem. There were two equilibrium points of our model system $N \equiv N_0$, $W \equiv W_0$ satisfying

$$F(N_0, W_0) = G(N_0, W_0) = 0$$

given by (see Fig. 14.1)

$$N_0 = 0, \ W_0 = \frac{R}{r};$$

and

$$N_0 = N_e = \left(\frac{c}{d}\right)\left[R - \frac{rk_1}{\delta - 1}\right], \ W_0 = \frac{k_1}{\delta - 1} \text{ with } \delta = \frac{cg_M}{d}.$$

The first corresponding to bare ground existed for all parameter values and was linearly stable for $\delta < 1 + rk_1/R$, while the second only had positive components when $\delta > 1 + rk_1/R$. Thus, there was an exchange of stabilities between these two critical points at $\delta = 1 + rk_1/R$. It was the stability of the community equilibrium point with which we were concerned. Hence, we performed longitudinal, hexagonal, and rhombic planform nonlinear stability analyses on that equilibrium point after introducing the following nondimensional variables and parameters:

$$(x, y) = \left(\frac{d}{D_W}\right)^{1/2} (X, Y), \ t = \tau d, \ n = \frac{N}{N_e}, \ w = \frac{W}{W_e},$$

$$\gamma = \frac{r}{d}, \ \alpha = \frac{(\delta - 1)R - rk_1}{k_1 d}, \ \mu = \frac{D_N}{D_W}, \ \beta = c\rho;$$

which transformed our basic equations into

$$\frac{\partial n}{\partial t} = \Theta(n, w) + \mu \nabla^2 n, \ \frac{\partial w}{\partial t} = \Psi(n, w) + \nabla^2 w - \alpha \beta \nabla^2 n;$$

where $\nabla^2 \equiv \partial^2/\partial x^2 + \partial^2/\partial y^2$ and

$$\Theta(n, w) = \frac{\delta w n}{w + \delta - 1} - n, \ \Psi(n, w) = \alpha + \gamma(1 - w) - \frac{\alpha \delta w n}{w + \delta - 1};$$

while adopting the far field boundary condition:

$$n, w \text{ remain bounded as } x^2 + y^2 \to \infty.$$

We completed our formulation by observing that the bare ground and community equilibrium points now corresponded to

$$n \equiv 0, \ w \equiv 1 + \frac{\alpha}{\gamma}; \text{ and } w = n \equiv 1;$$

respectively, since

$$\Theta\left(0, 1 + \frac{\alpha}{\gamma}\right) = \Psi\left(0, 1 + \frac{\alpha}{\gamma}\right) = \Theta(1, 1) = \Psi(1, 1) = 0.$$

The results of our longitudinal planform linear stability analysis were as follows: To guarantee the onset of an ecological Turing diffusive instability from this uniform steady state which was identically stable to linear homogeneous perturbations, we required:

$$\beta > \beta_0(q^2; \alpha, \gamma, \Delta, \mu) = \left(\frac{\mu}{\alpha \Delta}\right) q^2 + \frac{1}{q^2} + \frac{\mu \gamma}{\alpha \Delta} + \mu,$$

where $\Delta = 1 - \delta^{-1}$ and, as usual, $q = 2\pi/\lambda$, λ being the wavelength of the class of spatially periodic perturbations under investigation. For fixed values of α, γ, Δ, and μ, the marginal stability curve $\beta = \beta_0(q^2; \alpha, \gamma, \Delta, \mu)$ in the first quadrant of the q^2-β plane separated the linearly stable region where $0 \leq \beta < \beta_0(q^2; \alpha, \gamma, \Delta, \mu)$ from this unstable region. The marginal stability curve, the linearly stable region, and the unstable region could be characterized by $\sigma_0 = 0$, $\text{Re}(\sigma_0) < 0$, and $\sigma_0 > 0$, respectively, where σ_0 corresponded to that root of the secular equation having the largest real part. This marginal stability curve had a minimum point at (q_c^2, β_c) given by

$$q_c^2 = \sqrt{\frac{\Delta}{\mu}} \sqrt{\alpha}, \ \beta_c = 2\sqrt{\frac{\mu}{\Delta}} \left(\frac{1}{\sqrt{\alpha}}\right) + \left(\frac{\gamma \mu}{\Delta}\right) \left(\frac{1}{\alpha}\right) + \mu.$$

Hence, when $0 < \beta < \beta_c$ there existed no squared wavenumbers q^2 corresponding to growing disturbances, while when $\beta > \beta_c$ there existed a band of such wavenumbers centered about q_c^2 for which $\sigma_0 > 0$. In particular, taking $q^2 \equiv q_c^2$, as we did in our weakly nonlinear stability analyses, it was possible assuming α, γ, μ, and Δ were fixed, to represent $\sigma_0 = \sigma_c(\beta)$ such that, for β sufficiently close to β_c, σ_c was still real when $\beta < \beta_c$. Then, $\sigma_c(\beta)$ satisfied the conditions that $\sigma_c(\beta) < 0$ when $\beta < \beta_c$, $\sigma_c(\beta_c) = 0$, and $\sigma_c(\beta) > 0$ when $\beta > \beta_c$ (see Fig. 14.2). Hence, the locus $\beta = \beta_c(\alpha; \gamma, \Delta, \mu)$ with $\beta_c(\alpha; \gamma, \Delta, \mu)$ as defined above, was a marginal stability curve in the α-β plane, when γ, Δ, and μ were fixed. We plotted that locus in Fig. 14.3 for $0 < \alpha \leq 3.5$, $\gamma = 1$, $\Delta = 0.6$, and $\mu = 0.01$, corresponding to the typical parameter values (Rietkerk et al., [186]; von Hardenberg et al., [266])

$$g_M = 0.05 \frac{(\text{mm/d})}{(\text{gm/m}^2)}, \quad c = 10 \frac{(\text{gm/m}^2)}{\text{mm}}, \quad d = r = \frac{0.2}{\text{d}}, \quad k_1 = 5 \text{ mm},$$

$$\left(\frac{2}{3}\right) \frac{\text{mm}}{\text{d}} < R \leq 3 \frac{\text{mm}}{\text{d}}, \quad D_N = 0.1 \frac{\text{m}^2}{\text{d}}, \quad D_W = 10 \frac{\text{m}^2}{\text{d}}.$$

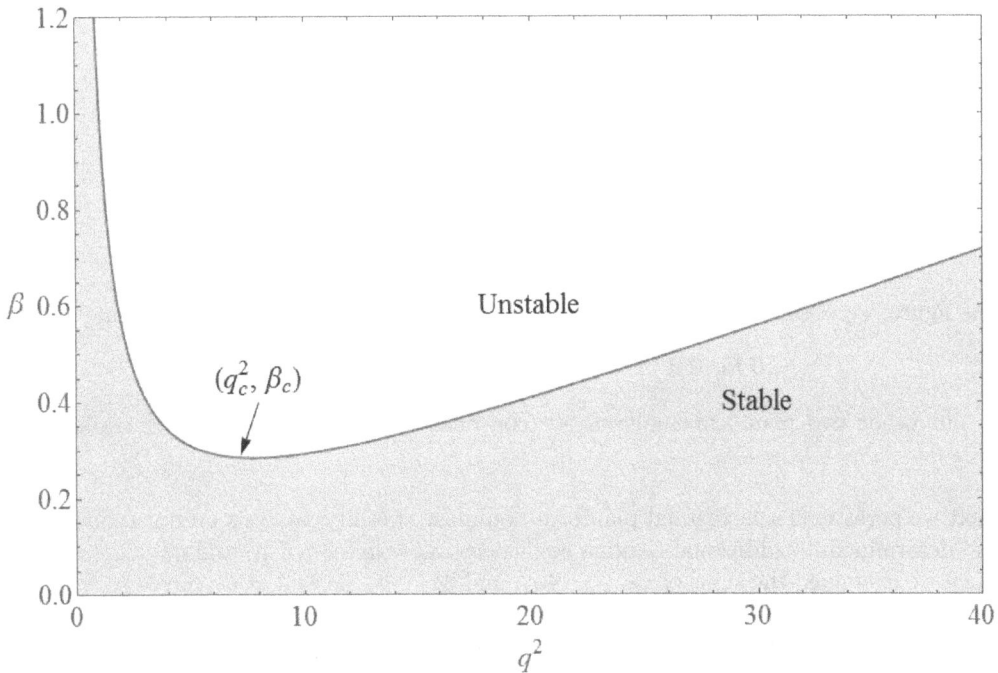

FIGURE 14.2
Plot of $\beta = \beta_0(q^2; \alpha, \gamma, \Delta, \mu)$ for $\alpha = \gamma = 1$, $\Delta = 0.6$, and $\mu = 0.01$.

In this context, we had determined the critical conditions for the onset of diffusive instabilities. To predict the long-time behavior and spatial pattern of such growing disturbances, it was necessary for us to take the nonlinear terms of our governing equations into account. The main nonlinear result of our longitudinal planform analysis was the determination, using the usual Fredholm alternative solvability condition, of its Landau coefficient represented here symbolically by $a_1(\alpha; \gamma, \delta, \mu)$, which is plotted in Fig. 14.4 for the same α-domain and choice of parameter values as used in Fig. 14.3 (note $\Delta = 0.6$ corresponds to $\delta = 2.5$). From

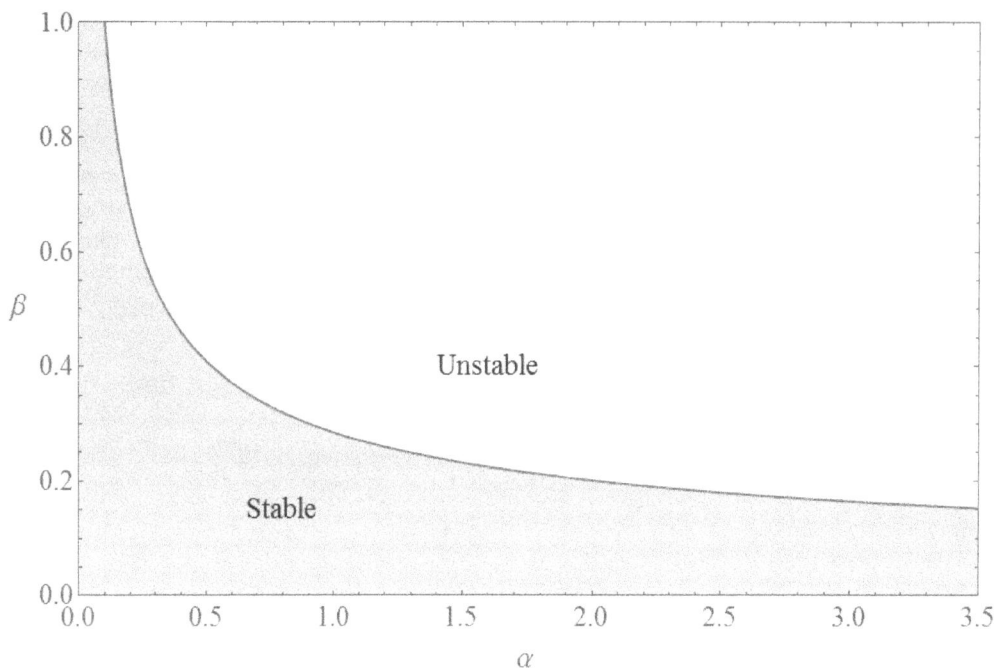

FIGURE 14.3
Plot of $\beta = \beta_c(\alpha; \gamma, \Delta, \mu)$ for $0 < \alpha \leq 3.5$, $\gamma = 1$, $\Delta = 0.6$, and $\mu = 0.01$.

this figure, we observed that $a_1(\alpha; 1, 2.5, 0.01)$ had a zero at $\alpha = 0.172$ such that

$$a_1 < 0 \text{ for } 0 < \alpha < \alpha_1 \equiv 0.172; \ a_1 > 0 \text{ for } \alpha > \alpha_1 \equiv 0.172.$$

In our other two planform analyses, we concentrated on the supercritical regime where $a_1 > 0$.

Next we performed a hexagonal planform nonlinear stability analysis on our model system and determined its additional Landau coefficients $a_0 = a_0(\alpha; \gamma, \delta, \mu)$ and $a_2 = a_2(\alpha; \gamma, \delta, \mu)$. Then we examined the signs of a_0, $a_1 + 4a_2$, and $2a_2 - a_1$, as well as a_2 versus $\alpha_1 < \alpha \leq 3.5$ for $\gamma = 1$, $\delta = 2.5$, and $\mu = 0.01$ in Fig. 14.5 and, observing that they were all identically negative, concluded from Table 8.2 that, although I was stable for $\sigma < 0$, II was not stable since $2a_2 - a_1 < 0$, while III$^{\pm}$ did not exist since $a_1 + 4a_2 < 0$. Hence, we demonstrated that this hexagonal planform analysis did not yield any stable stationary heterogeneous vegetative patterns for our model system. Thus, we decided that if such vegetative patterns were to be predicted for this system it could only happen by means of the rhombic planform analysis.

Hence, we performed that rhombic planform nonlinear stability analysis on our model system and determined its other Landau coefficient $b_1 = b_1(\alpha, \phi; \gamma, \delta, \mu)$ where $\phi \equiv$ rhombic angle. Defining

$$\eta(\alpha, \phi; \gamma, \delta, \mu) = \frac{b_1(\alpha, \phi; \gamma, \delta, \mu)}{a_1(\alpha; \gamma, \delta, \mu)}$$

for $\alpha > \alpha_1$, we catalogued the stability criteria for the rhombic planform critical points:

$$\text{I: } 0 < \beta < \beta_c; \ \text{II: } \beta > \beta_c \text{ and } \eta > 1; \ \text{V: } \beta > \beta_c \text{ and } -1 < \eta < 1.$$

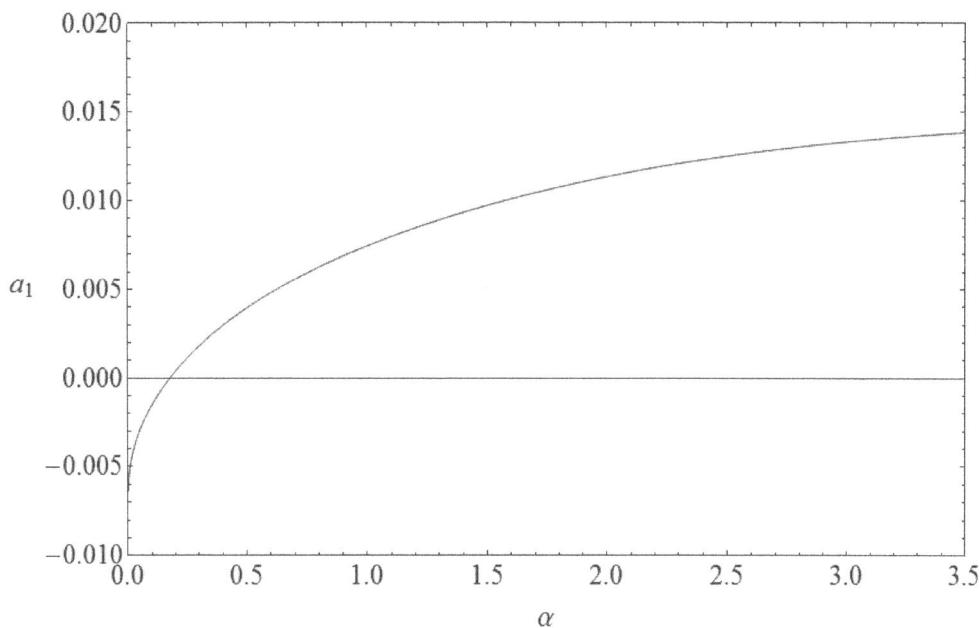

FIGURE 14.4
Plot of $a_1(\alpha; \gamma, \delta, \mu)$ for $0 < \alpha \le 3.5$, $\gamma = 1$, $\delta = 2.5$, and $\mu = 0.01$.

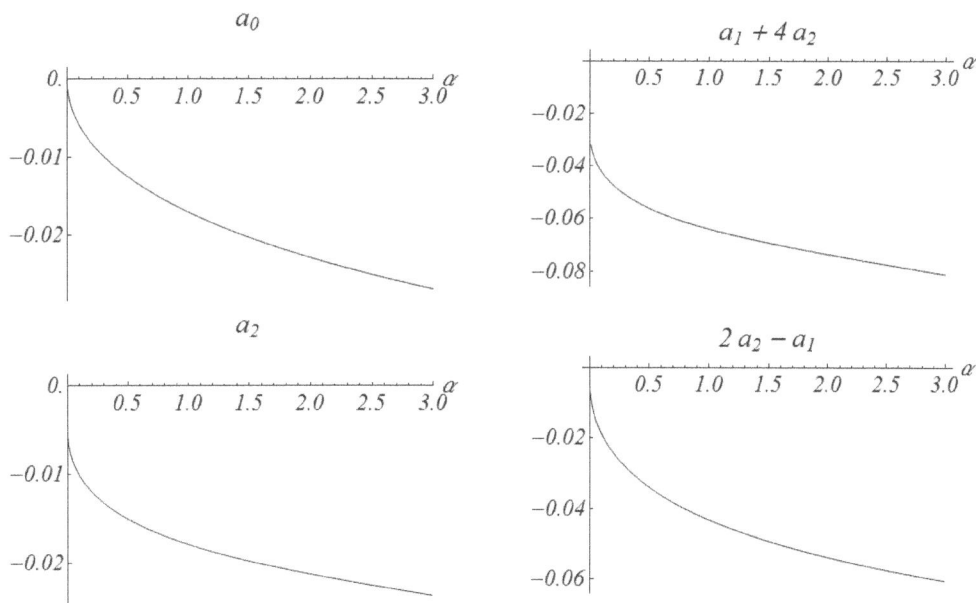

FIGURE 14.5
Plots of a_0, $a_1 + 4a_2$, and $2a_2 - a_1$ versus $\alpha_1 < \alpha \le 3.5$ for $\gamma = 1$, $\delta = 2.5$, and $\mu = 0.01$. Here, for the sake of completeness, an analogous plot of a_2 has been included.

Since critical point II was unstable for our hexagonal planform analysis, stable stationary heterogeneous vegetative patterns could only occur through critical point V, which we identified as a rhombic pattern possessing rhombic angle ϕ and characteristic wavelength $\lambda_c = 2\pi/q_c$ and $\lambda_c^* = (D_W/d)^{1/2}\lambda_c$ in dimensionless and dimensional variables, respectively. Plotting η versus $\phi \in [0,\pi]$ in Fig. 14.6 for a fixed value of α, namely $\alpha = 3$, and with the other parameters taking on their values of Figs. 14.4 and 14.5, we found that there existed an interval of stable rhombic patterns, where $-1 < \eta < 1$, provided $\beta > \beta_c$ given by $\phi \in (\phi_m, \phi_M)$ where $0 < \phi_m < \phi_M < \pi/2$. Again, repeating the process used to produce Fig. 14.6, but for other $\alpha_1 < \alpha \leq 3.5$, we found the same generic behavior as for $\alpha = 3$ and summarized these results for selected values in Table 14.1.

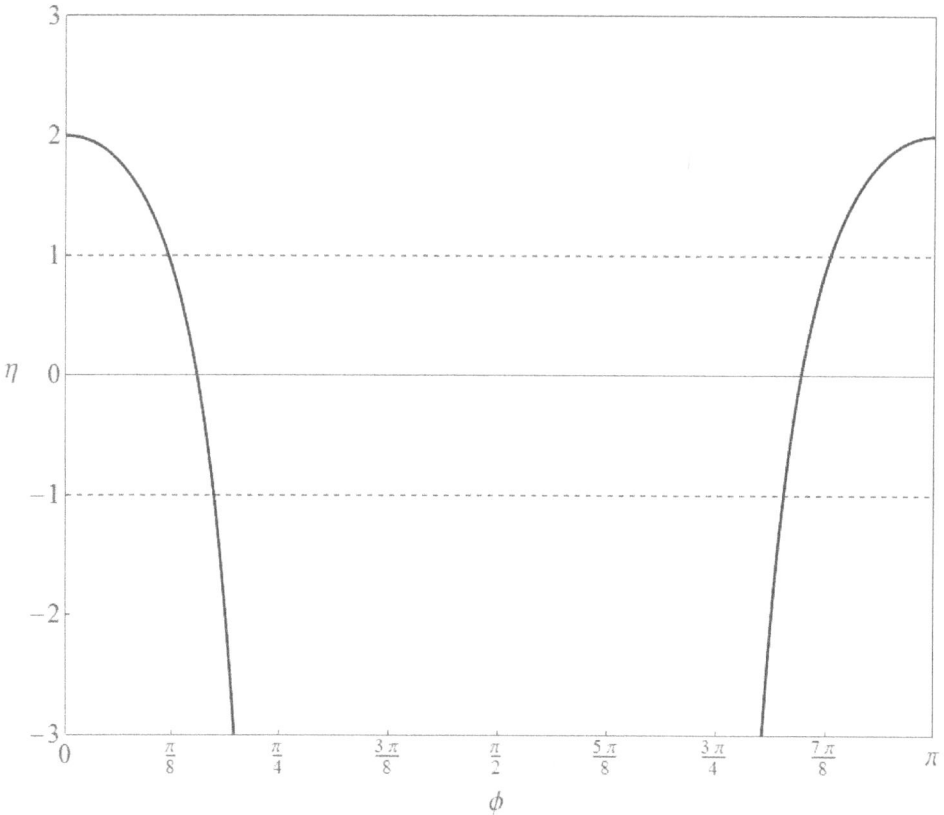

FIGURE 14.6
Plot of η versus $0 \leq \phi \leq \pi$ for $\gamma = 1$, $\delta = 2.5$, $\mu = 0.01$, and $\alpha = 3$. Here, the dashed horizontal lines denote $\eta = \pm 1$. This figure has been plotted for the extended interval $\pi/2 \leq \phi \leq \pi$ in order to demonstrate graphically the symmetry about $\phi = \pi/2$ characteristic of rhombic patterns or $\eta(\alpha, \phi; \gamma, \delta, \mu) = \eta(\alpha, \pi - \phi; \gamma, \delta, \mu)$.

To make a morphological interpretation of these stable rhombic patterns we considered the plots contained in Fig. 14.8 for the representative angle $\phi = 0.5$ of Fig. 14.7. These are threshold contour plots with threshold values of 1, 0, and -1, from left to right, respectively. Here, the spatial variables were measured in units of λ_c and regions exceeding that threshold in each part appeared dark, while those below it appeared light. Hence, the three

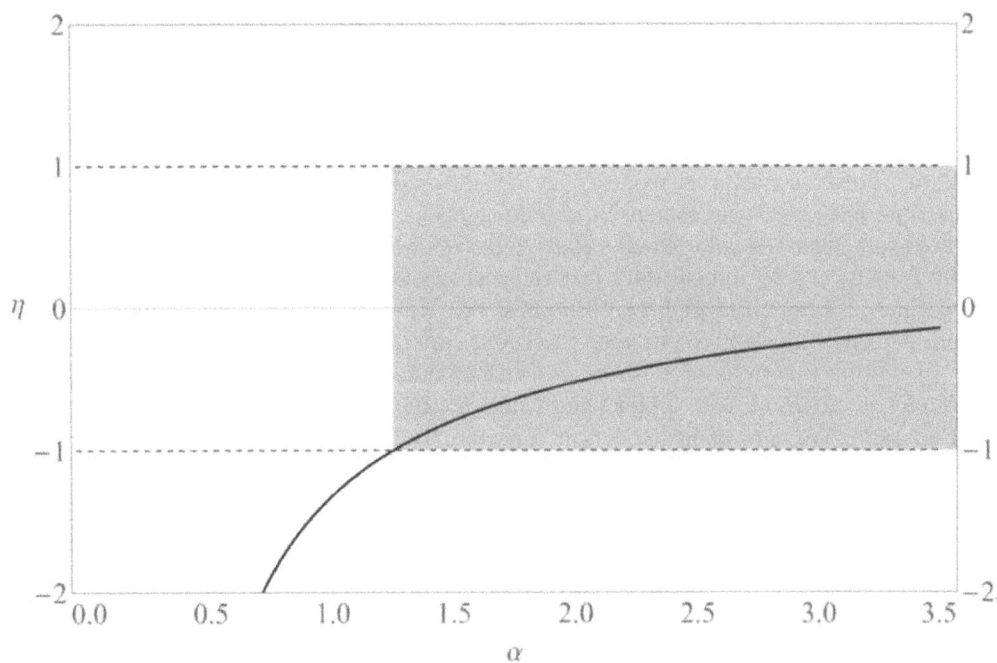

FIGURE 14.7
Plot of η versus $\alpha_1 < \alpha \leq 3.5$ for $\phi = 0.5$ with $\gamma = 1$, $\delta = 2.5$, and $\mu = 0.01$. Here, the α-interval of interest $\alpha \in (1.2520, 3.5]$ is denoted by shading.

TABLE 14.1
Intervals of stable rhombic angles for various values of α with $\gamma = 1$, $\delta = 2.5$, and $\mu = 0.01$.
Note that the α-interval of interest $\alpha \in (1.2520, 3.5]$ for the representative angle $\phi = 0.5$
has been denoted by shading in Fig. 14.7.

α	ϕ_m	ϕ_M	α	ϕ_m	ϕ_M
0.2	0.109632	0.185916	1.8	0.357124	0.521116
0.4	0.243290	0.384288	2.0	0.362099	0.526508
0.5	0.268140	0.416588	2.2	0.366411	0.531150
0.6	0.285492	0.438294	2.4	0.370204	0.535210
0.8	0.308829	0.466429	2.6	0.373583	0.538809
1.0	0.324280	0.484419	2.8	0.376622	0.542033
1.2	0.335538	0.497225	3.0	0.379381	0.544949
1.4	0.344251	0.506968	3.2	0.381903	0.547606
1.6	0.351279	0.514728	3.4	0.385319	0.550045

parts of the contour plot in this figure corresponded to what we have termed higher, zero, and lower threshold patterns, respectively. Traditionally, most pattern formation analyses of this type have used the dimensional homogeneous vegetative solution value N_e as the threshold to trigger the color change from light to dark. Thus all spatial regions characterized by $N = N_e n \geq N_e$ appear dark and those characterized by $N < N_e$, light, where again dark regions correspond to high plant biomass density and light ones to low density or bare ground. This is equivalent to our zero threshold case of Fig. 14.8. We considered α and N_e to be increasing functions of R since, from their definitions, $\alpha(R) = (\delta - 1)N_e(R)/(k_1 c)$ and $N_e(R) = (cd)[R - rk_1(\delta - 1)]$ for fixed values of the other parameters and $\delta > 1 + rk_1/R$ (see Fig. 14.1). We now wished to select a particular $R = R_c$ and adopt the protocol that $N_c = N_e(R_c)$ represented this threshold instead. Then, when $\alpha_1 < \alpha < \alpha_c = \alpha(R_c)$ or, equivalently $N_e < N_c$, a higher threshold pattern of the type depicted in Fig. 14.8 would occur, while, when $\alpha > \alpha_c$ or $N_e > N_c$, a lower threshold type would occur. Given their appearance in Fig. 14.8, we labeled these higher, zero, and lower threshold type rhombic vegetative arrays as pseudo spots, rectangles, and pseudo gaps and denoted them by V^+, V^0, and V^-, respectively.

In order to motivate our specific choice of the proper value to be assigned for R_c and hence, N_c, we first needed to summarize the simulation results of Rietkerk *et al.* ([186]). Their two-dimensional numerical simulations of our original three-component model system with $\rho = 0$ and its other parameters set at values consistent with those we have assigned previously, yielded close-packed vegetative patterns of spots or gaps depending upon whether R was less than or greater than 1 mm/d, respectively (see Fig. 14.9). Motivated by the desire to replicate this behavior and given the similarity in appearance between these two types of patterns and the left- or right-hand parts of Fig. 14.8, respectively, we then selected $R_c = 1$ mm/d $\Rightarrow \alpha_c = 0.5$, since, for our parameter values, $\alpha(R) = 1.5R - 1$ (see Fig. 14.1) where R is being measured in units of mm/d. Thus, from our rhombic threshold argument, we made the prediction that V^+ patterns would occur for $R < R_c$ or $\alpha < \alpha_c = 0.5$ and V^- ones, for $R > R_c$ or $\alpha > 0.5$ when $\beta > \beta_c$. Influenced by that resemblance in appearance just cited, we referred to these periodic rhombic arrays of $V^\pm$ as vegetative pseudo spots or pseudo gaps, respectively. We incorporated these two-dimensional rhombic-planform morphological stability results for $\gamma = 1$, $\delta = 2.5$, and $\mu = 0.01$ in the α-β plane of Fig. 14.10 and identified regions corresponding to the predicted vegetative patterning. Note in this context, that since $N_e(R) = 50(R - 2/3)$, where, in addition, N is being measured in units of gm/m^2 (see

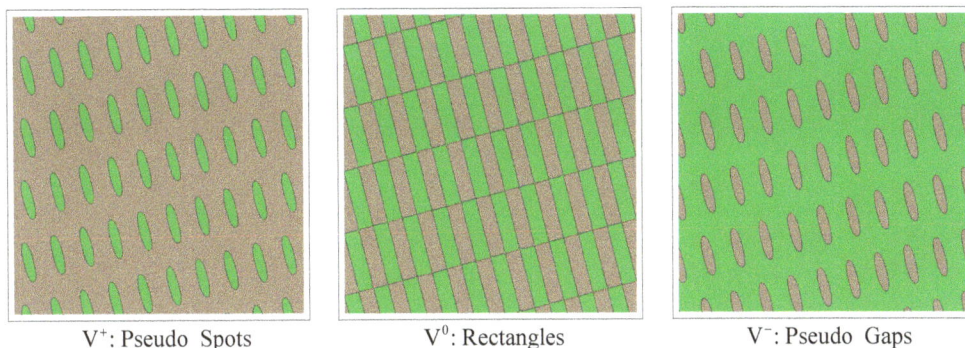

| V⁺: Pseudo Spots | V⁰: Rectangles | V⁻: Pseudo Gaps |

FIGURE 14.8
Rhombic patterns relevant to $\phi = 0.5$ with threshold values from left to right of 1, 0, and -1. Here, the spatial variables are being measured in units of λ_c and regions exceeding that threshold in each part appear dark while those below it appear light.

Fig. 14.1),

$$N_c = N_e(1) = \frac{50 \text{ gm}}{3 \text{ m}^2},$$

while the plant component of the equilibrium point for the model system employed by Rietkerk *et al.* ([186]) was identical with our $N_e(R)$.

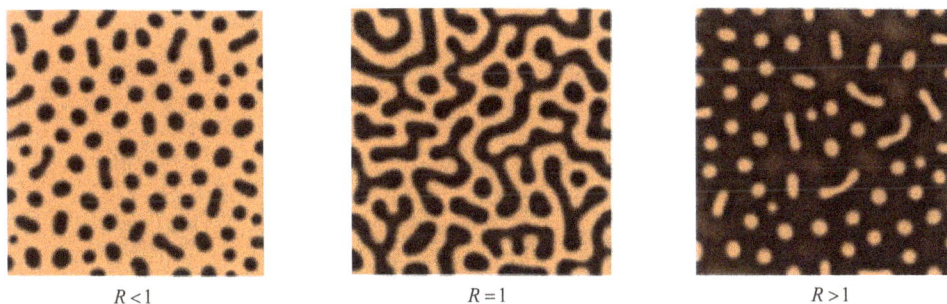

| $R < 1$ | $R = 1$ | $R > 1$ |

FIGURE 14.9
The critical threshold value of $R_c = 1$ was selected to correspond with the simulated patterns adapted from Rietkerk *et al.* ([186]) by Wollkind and Dichone ([284]) for R less than, equal to, or greater than 1, respectively. Here R is being measured in units of mm/d, as in Fig. 14.1.

Specifically, these are identified in Fig. 14.10 where $\beta > \beta_c$, as rhombic arrays of pseudo spots (V^+) for $\alpha_1 < \alpha < \alpha_c$, of rectangles ($V^0$) for $\alpha = \alpha_c$, and of pseudo gaps (V^-) for $\alpha > \alpha_c$, in accordance with our morphological threshold condition. Further, note that the $0 < \beta < \beta_c$ region of Fig. 14.10 which can be identified with a uniform homogeneous vegetative pattern ranges from a relatively sparse distribution ($N_e < N_c = 50/3$ gm/m²) for $0 < \alpha < \alpha_c$ to a relatively dense one ($N_e > N_c = 50/3$ gm/m²) for $\alpha > \alpha_c$ and hence have been designated by $I^\mp$, respectively. Finally, the marginal stability curve denoted by

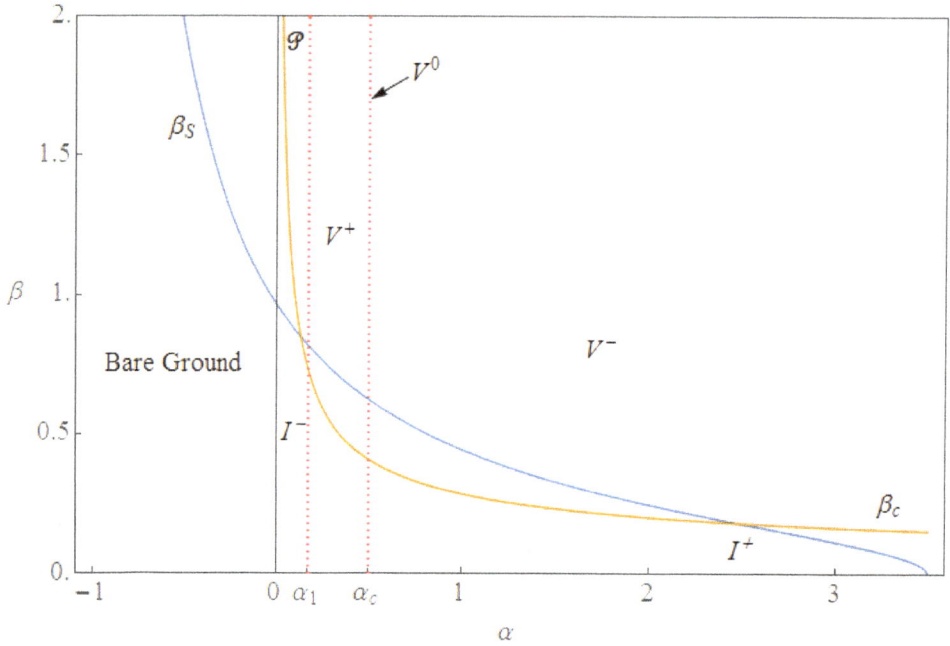

FIGURE 14.10
Stability diagram in the α-β plane for our two-dimensional interaction-diffusion model system with $\gamma = 1$, $\delta = 2.5$, and $\mu = 0.01$, identifying the predicted vegetative patterns. Here, β_c and β_S denote the marginal stability curve and the root suction characteristic as a function of saturation $S = (\alpha + 1)/4.5$ with $m_0 = 0.5$ and $\beta^{(0)} = 0.22$ (see below), respectively.

β_c has a horizontal asymptote given by $\beta = \mu$ while, for $-1 < \alpha < 0$, bare ground or no vegetation is predicted as also indicated in that figure.

So far, with the implicit exception of the above-mentioned I$^-$, we had been concerned with the morphological stability behavior of our model system for $a_1 > 0$. We next considered its behavior for $a_1 < 0$. At the end of Chapter 8, it was shown that a particular partial differential evolution equation containing fourth-order spatial derivatives could be used when $a_1 < 0$ and $\sigma > 0$ to explain the occurrence of dewetting of thin liquid layers by isolated droplet pattern formation for relatively thin layers and by isolated hole formation for relatively thick ones, provided a fifth-order Landau coefficient $a_3 > 0$ (see Fig. 8.8). A comparison of the simulation results of Lejeune $et\ al.$ ([112]) with the weakly nonlinear stability ones of Boonkorkuea $et\ al.$ ([17]) for a strongly related evolution equation describing vegetative pattern formation in arid isotropic environments led to the conjecture that when $a_1 < 0$ localized structures would occur where $\sigma > 0$, characterized by isolated patches of vegetation at low densities that were a spatial compromise between the periodic patchy vegetation and bare ground stable states (see Chapter 11). Chen and Ward ([33]) found local structures occurring in conjunction with such subcriticality for the Gray-Scott reaction-diffusion chemical model system. These occurrences, as explained in Chapter 13, led Cangelosi $et\ al.$ ([25]) to employ the same argument to identify a region of their relevant parameter space with isolated clusters for a mussel-algae interaction-diffusion model system. The resulting morphological sequence deduced from that identification provided close agreement with mussel bed patterning observations, both in the field and laboratory (Wang $et\ al.$, [268]; Liu $et\ al.$, [121]). Given the similarity of behavior among all these phenomena, we conjectured with some confidence that isolated patches of vegetation would occur for $0 < \alpha < \alpha_1$ where $\beta > \beta_c$ and identified that region graphically in Fig. 14.10 using the designation $\mathscr{P}$.

Taking into account, as we shall do in Chapter 15, terms through fifth-order in the expansion and amplitude equation for the longitudinal planform analysis when $a_1 < 0$, by, in particular, considering

$$\frac{dA_1}{dt} \sim \sigma A_1 - a_1 A_1^3 - a_3 A_1^5$$

and calculating a_3, we might enhance our understanding of the morphological stability of this system in the subcritical regime. As demonstrated in that chapter, should $a_3 > 0$ this equation will have three equilibrium points: Namely, 0 and $2a_3 A_e^{\pm 2} = \pm\sqrt{a_1^2 + 4a_3\sigma} - a_1$ such that 0 is globally stable for $\sigma < \sigma_{-1} = -a_1^2/(4a_3) < 0$; A_e^+, for $\sigma > 0$; and in the overlap region $\sigma_{-1} < \sigma < 0$, where either can be stable depending on initial conditions, 0 is stable for $A_1^2(0) < A_e^{-2}$ and A_e^+, for $A_1^2(0) > A_e^{-2}$; while A_e^-, which only exists in that bistability region, is not stable there. Here, the potentially stable critical points 0 and A_e^+ would correspond to I$^-$ and $\mathscr{P}$, respectively.

Determining the generalized marginal curve for our problem with $\sigma \neq 0$, analogous to that of Chapter 12, we would find that

$$\beta_\sigma(\alpha; \gamma, \Delta, \mu) = \frac{2\mu^{1/2}}{\alpha\Delta}[\alpha\Delta + (\gamma + \alpha\Delta)\sigma + \sigma^2]^{1/2} + \frac{\mu\gamma + (1+\mu)\sigma}{\alpha\Delta} + \mu,$$

where $\beta_{\sigma=0}(\alpha; \gamma, \Delta, \mu) = \beta_c(\alpha; \gamma, \Delta, \mu)$. Plotting the marginal curve associated with $\sigma = \sigma_{-1}$,

$$\beta = \beta_{-1}(\alpha; \gamma, \Delta, \mu) = \beta_{\sigma_{-1}}(\alpha; \gamma, \Delta, \mu) < \beta_c(\alpha; \gamma, \Delta, \mu),$$

in Fig. 14.10, we could make the proper morphological identifications. These would differ from those already appearing there only due to the presence of the overlap region

$\beta_{-1} < \beta < \beta_c$ where $\mathscr{P}$, as well as I^- patterns could now occur. Should $a_1^2/(4a_3) \ll 1$, as was the case for Kealy-Dichone *et al.* ([93]) involving hexagonal patterns in that chapter, these marginal curves would almost coincide or $\beta_{-1} \cong \beta_c$, causing the overlap region virtually to disappear and resulting in the exact same identifications as the ones appearing in Fig. 14.10.

Traditionally, morphological sequences of the sort referred to above have been generated from stability diagrams such as Fig. 14.10, by traversing appropriate horizontal or vertical lines in that two-dimensional parameter space (see Cangelosi *et al.*, [25]; Kealy-Dichone *et al.*, [93]). A procedure of this sort is inherently dependent upon the implicit assumption that these two parameters are independent of each other. In the case of Fig. 14.10, however, α and β, being nondimensional measures of rainfall rate and the coefficient of plant root suction, respectively, are actually related. To obtain the proper morphological sequence of vegetative states along a rainfall gradient predicted from Fig. 14.10, it was first necessary for us to deduce that relationship. Toward this end, employing our basic definitions and the parameter values, we found that

$$\beta = c\rho = \frac{cD_S}{D_W} = D_S \frac{\text{gm d}}{\text{mm m}^4}.$$

Note that the units for D_S, as indicated below, are mm m^4/(gm d) consistent with β being a dimensionless parameter. Adopting the root suction characteristic of Roose and Fowler ([188]), we took

$$D_S = \beta^{(0)} f_{m_0}(S) \frac{\text{mm m}^4}{\text{gm d}}$$

where

$$f_{m_0}(S) = (S^{-1/m_0} - 1)^{1-m_0} \text{ for } 0 < m_0 < 1 \text{ and } 0 < S \leq 1.$$

Here, $S \equiv$ the relative water saturation in the soil while the parameters $\beta^{(0)}$ and m_0 were determined from experimental data for different soils. To complete our formulation we let

$$S = \frac{R}{R_M} \text{ for } 0\frac{\text{mm}}{\text{d}} < R \leq R_M$$

where, specifically,

$$R_M = 3\frac{\text{mm}}{\text{d}}.$$

Upon recalling that

$$R = \frac{k_1 d(\alpha + \gamma)}{\delta - 1}$$

and substituting this into that formula for S yielded

$$S = \frac{\alpha + \gamma}{\alpha_M + \gamma} \text{ for } \alpha_0 < \alpha \leq \alpha_M$$

where, specifically,

$$\alpha_0 = -\gamma = -1, \ \alpha_M = 3.5.$$

Finally, selecting $m_0 = 0.5$ after Roose and Fowler ([188]), and incorporating these results into our expression for β, we obtained the one-parameter family of root suction characteristic curves

$$\beta = \beta_S(\alpha) = \beta^{(0)} f_{0.5}\left(\frac{\alpha + 1}{4.5}\right) \text{ for } -1 < \alpha \leq 3.5$$

where
$$f_{0.5}(S) = \frac{(1-S^2)^{1/2}}{S}.$$

We plotted this curve with $\beta^{(0)} = 0.22$ in Fig. 14.10, where the assignment of that parameter had been made both for the purpose of definiteness and to be consistent with our silt loam soil choice for m_0 (Roose and Fowler, [188]). Representing the Turing stability boundary in that figure by $\beta_c(\alpha)$, for ease of exposition, we observed that there existed two points of intersection between it and $\beta_S(\alpha)$ satisfying

$$\beta_S(p_{0,2}) = \beta_c(p_{0,2})$$

where, specifically,
$$p_0 = 0.13217, \quad p_2 = 2.47622.$$

Here, these α-values corresponded to

$$R_0 = 0.75478\frac{\text{mm}}{\text{d}}, \quad R_2 = 2.31748\frac{\text{mm}}{\text{d}},$$

respectively. In this context, we defined

$$p_1 = \alpha_c, \quad R_1 = 0.78133\frac{\text{mm}}{\text{d}}$$

where $\alpha(R_1) = \alpha_1$. The morphological sequence of predicted stable vegetative states along a rainfall gradient obtained upon traversing the curve $\beta = \beta_S(\alpha)$ in the α-β plane of Fig. 14.10 is tabulated in Table 14.2. Note, in general, that

$$p_0 \begin{Bmatrix} < \\ = \\ > \end{Bmatrix} \alpha_1 \text{ if } \beta^{(0)} \begin{Bmatrix} > \\ = \\ < \end{Bmatrix} 0.20.$$

Thus, isolated patches were only predicted for these transit curves provided

$$\beta^{(0)} > 0.20.$$

TABLE 14.2
Morphological stability predictions along a rainfall gradient for $\beta = \beta_S(\alpha)$ in Fig. 14.10. Here, $p_0 = 0.13$, $\alpha_1 = 0.17$, $p_1 = 0.50 = \alpha_c$, $p_2 = 2.48$, $R_0 = 0.75$, $R_1 = 0.78$, and $R_2 = 2.32$.

α-range	R-range (mm/d)	Stable pattern
$-1 < \alpha < 0$	$0 < R < 2/3$	Bare ground
$0 < \alpha < p_0$	$2/3 < R < R_0$	Sparse homogeneous
$p_0 < \alpha < \alpha_1$	$R_0 < R < R_1$	Isolated patches
$\alpha_1 < \alpha < \alpha_c = p_1$	$R_1 < R < R_c = 1$	Pseudo spots
$\alpha = \alpha_c = p_1$	$R = R_c = 1$	Rectangles
$\alpha_c = p_1 < \alpha < p_2$	$R_c = 1 < R < R_2$	Pseudo gaps
$p_2 < \alpha < 3.5$	$R_2 < R < 3$	Dense homogeneous

We next compared our results with those from some recent biological pattern formation studies. We began with the work of Gowda et al. ([72]). These authors examined the standard sequence of patterned states (gaps → labyrinth → spots) generated in a general activator-inhibitor reaction-diffusion system as a bifurcation parameter was varied and then applied their results to the particular von Hardenberg et al. ([266]) plant-ground water model as its precipitation parameter was decreased. They employed both numerical simulation and analytical weakly nonlinear hexagonal-planform bifurcation methods. Gowda *et al.* ([72]) found, for the default set of parameter values von Hardenberg *et al.* ([266]) used in their numerical integration, that, although the simulation method reproduced the latter's standard sequence, the hexagonal planform analysis, as in our problem, failed to predict vegetative patterns. These calculations were performed for $1/\mu = 100$, in accordance with $\mu = 0.01$ (see Fig. 14.11). When those calculations were repeated for $1/\mu = 27$, they found that the same standard sequence of vegetative patterns were produced for both the simulation and weakly nonlinear stability methods with the transition between the two hexagonal states occurring exactly where the second-order Landau coefficient changed sign (see Boonkorkuea *et al.*, [17]). Gowda *et al.* ([72]) concluded that weakly nonlinear stability theory failed to produce the correct results in the first instance because the simulated morphological sequence of vegetative patterns occurred for large amplitudes as the precipitation parameter was decreased.

Given our results, we wished to suggest another possible explanation for this discrepancy: Namely, that a rhombic-planform weakly nonlinear stability analysis might yield a predicted morphological sequence involving pseudo gaps $\rightarrow$ rectangles $\rightarrow$ pseudo spots, as the precipitation parameter was decreased in this case. As supporting evidence for such a conjecture we offered the following obseration. The gap- and spot-type simulated patterns for $1/\mu = 100$, appearing in both Gowda *et al.* ([72]) and von Hardenberg *et al.* ([266]), were much less regular in nature than were the corresponding simulated hexagonal patterns for $1/\mu = 27$ appearing in Gowda *et al.* ([72]). The simulated transition states between these two types of patterns were also different consisting of labyrinths in the former instance but of parallel stripes in the latter case. Since for each value of α we predicted multistable rhombic states with an interval of characteristic angles (see Fig. 14.6 and Table 14.1) and, as initial conditions vary point by point over a flat environment, these states could be selected quite randomly, it was possible to generate simulated patterns resembling those appearing in von Hardenberg *et al.* ([266]) from families of pseudo gaps and pseudo spots including labyrinths from families of rectangles. In this context, the numerically simulated two-dimensional patterns of Rietkerk *et al.* ([186]) used earlier to motivate our choice for R_c also bore a strong resemblance to those of von Hardenberg *et al.* ([266]), including a labyrinthine transition state at $R = 1$ mm/d (see Fig. 14.9). Unlike the von Hardenberg *et al.* ([266]) model, ours is extremely robust to variations in μ. We performed additional rhombic and hexagonal planform nonlinear stability analyses on our system and found identical qualitative behavior for all $0.001 \leq \mu \leq 1$.

We ended this phase of our discussion by restating von Hardenberg *et al.*'s ([266]) claim that the power of model systems such as ours is their predicted sequence of stable states along a rainfall gradient, such as the one summarized in Table 14.2 could be used to motivate an aridity classification scheme which is characterized by the three rainfall thresholds

$$p_0 = 0.13217 < p_1 = 0.5 < p_2 = 2.47622.$$

Here, we were employing the notation of von Hardenberg *et al.* ([266]) for these dimensionless rainfall (precipitation) rate thresholds and, as in Chapter 12, used them to introduce the following possible aridity classes based upon the inherent vegetative states of our system:

- *Dry-subhumid* ($p_2 < \alpha < 3.5$): The only vegetative state the system supports corresponds to a dense homogeneous distribution.

- *Semiarid* ($p_1 < \alpha < p_2$): The only vegetative state the system supports corresponds to pseudo gaps of low threshold type.

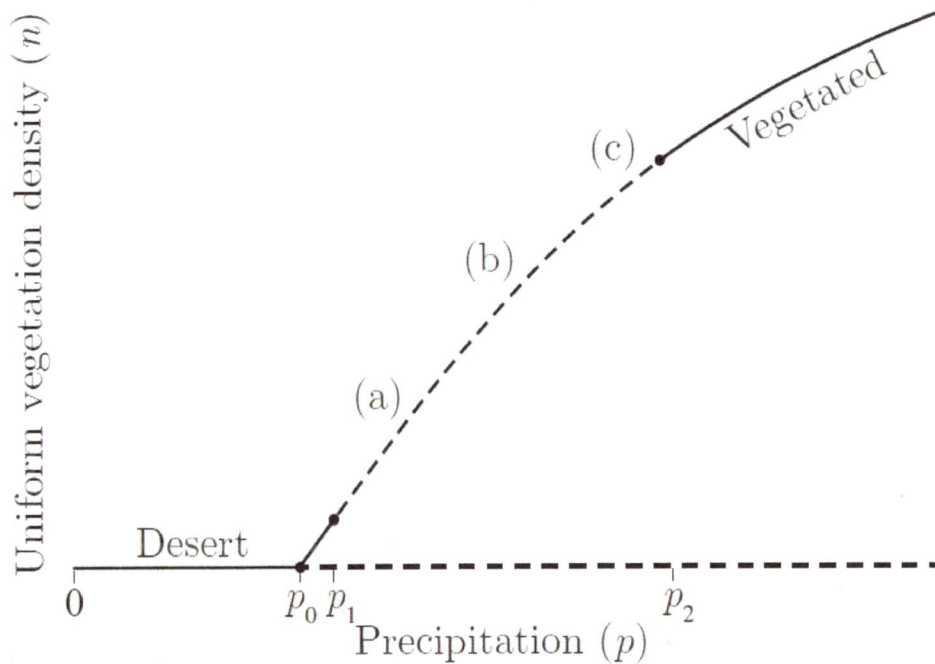

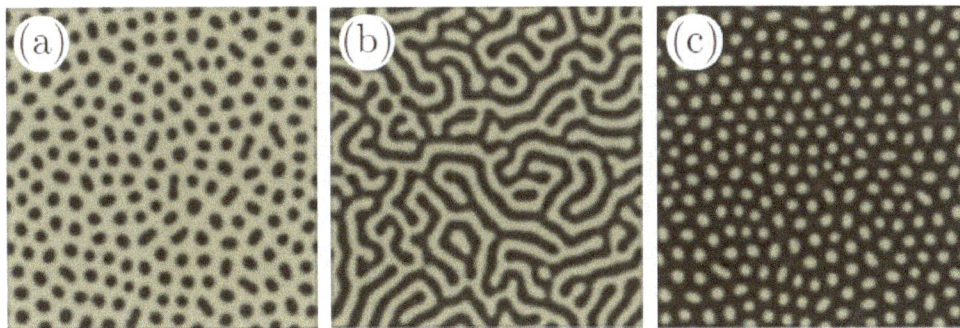

FIGURE 14.11
Simulation results of Gowda *et al.* ([72]) for the von Hardenburg *et al.* ([266]) model with $\mu = 0.01$.

- *Arid* ($p_0 < \alpha < p_1$): The only vegetative states the system supports correspond to either pseudo spots of high threshold type or isolated patches.

- *Hyperarid* ($-1 < \alpha < p_0$): The only possible stable states the system supports correspond to either a sparse homogeneous vegetative distribution or bare ground.

As pointed out by von Hardenberg *et al.* ([266]), the advantage of the proposed aridity classification scheme pertains to the information it contains about dynamical aspects of drylands. Regions whose aridity classes imply the occurrence of upper threshold vegetative patterns, isolated patches, or a sparse homogeneous distribution are vulnerable to desertification. The mere knowledge of that threat, however, allows land managers to reverse this process for those regions by implementing crust disturbance, seed augmentation, or irrigation strategies. Meron *et al.* ([143]) suggested a cycling mechanism between plants and water to account for the formation of bare patches characteristic of vegetative patterning along such a precipitation gradient. Note that a process of this sort occurs in all directions for two-dimensional vegetative patterns (pseudo spots or gaps, rectangles, and isolated patches) but only in two directions for one-dimensional ones (stripes).

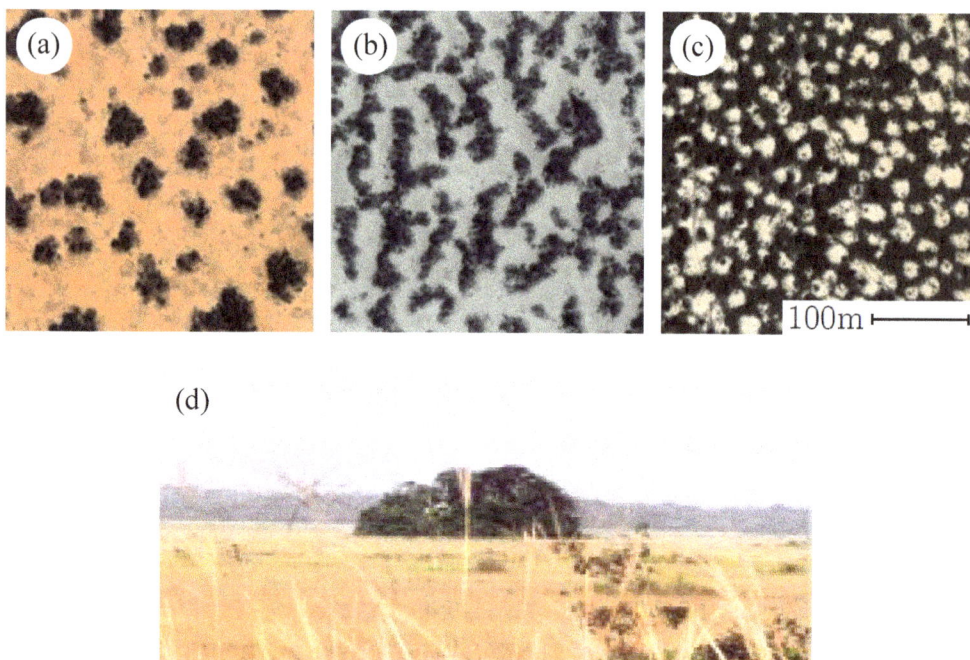

FIGURE 14.12
Images of flat terrain dryland vegetative patterns: (a) Leopard bush spotted patterns in Sudan, (b) Tiger bush labyrinthine patterns in Niger, (c) Pearled bush gapped patterns in Senegal (Gowda *et al.*, [72]), and (d) Localized spots in French Guiana (Lejeune *et al.*, [112]).

It remained for us to compare our theoretical predictions with relevant observational evidence (see Fig. 14.12). As stated in Chapters 11 and 12, striking periodic or localized self-organized vegetative patterns covering widespread areas of arid or semi-arid flat regions of Africa, Australia, the Americas, the Near East, and Asia became noticeable through aerial photography nearly 80 years ago. In this instance, an arid or semi-arid flat region refers to a plateau-like environment that is characterized by an extended dry season where yearly potential evaporation exceeds yearly rainfall and water availability is a limiting factor for plant growth. The periodic patterns reported consisted of spots (leopard bush), labyrinths, or gaps (pearled bush) and the localized ones, isolated spots. These included bushy vegetation punctuated by bare gaps in Senegal, labyrinthine mazes in Niger, and periodic or localized spots of trees or shrubs in Sudan or French Guiana, respectively, as depicted in the photographs comprising Fig. 14.12. Incidentally, labyrinthine patterns have been associated with certain types of tiger bush or banded thicket vegetative distributions found in arid or semiarid flat environments (Couteron et al., [39]). Given the similarity of appearance between the periodic observed patterns of Fig. 14.12a,b,c and the theoretical patterns of Fig. 14.7, we identified our rhombic arrays of pseudo spots (V^+) or pseudo gaps (V^-) with these leopard or pearled bush vegetative patterns, respectively, and our rhombic arrays of rectangles (V^0) with the tiger bush labyrinthine vegetative patterns. In light of those identifications, we then investigated the predicted wavelengths of these vegetative patterns. From $\lambda_c = 2\pi/q_c$, we deduced that

$$\lambda_c = \frac{2\pi \sqrt[4]{\mu/\Delta}}{\sqrt[4]{\alpha}} = \lambda_c(\alpha).$$

Designating the α's associated with $V^\pm$ by $\alpha^\pm$, respectively, it followed from Table 14.2 that

$$\alpha_1 < \alpha^+ < \alpha_c = p_1 < \alpha^- < p_2.$$

Then, we could see from these results that

$$\lambda_1 = \lambda_c(\alpha_1) > \lambda_c^+ = \lambda_c(\alpha^+) > \lambda_c^0 = \lambda_c(\alpha_c) > \lambda_c^- = \lambda_c(\alpha^-) > \lambda_2 = \lambda_c(p_2),$$

and hence concluded that the vegetative distributions of spots in leopard bush had a tendency to be more widely spaced than the labyrinthine components of the mazes, which, in turn, were more widely spaced than the bare patches that regularly punctuate the vegetation cover in pearled bush (Lejeune et al., [113]). Employing our length scale and choice of parameter values, yielded the corresponding dimensional wavelength relationships

$$\lambda_1^* = 25 \text{ m} > \lambda_c^{*+} > \lambda_c^{*0} = 19 \text{ m} > \lambda_c^{*-} > \lambda_2^* = 12.7 \text{ m}$$

consistent with the photographs of Fig. 14.12 and in agreement with Boonkorkuea et al. ([17]), who, interestingly enough, found for the evolution equation of Lejeune et al. ([113]) an identical power law relationship between their pattern wavelength and plant biomass (see Chapter 11), which we considered a minor instance of Pulling a Rabbit Out of a Hat. Finally, it was obvious that our isolated patches $\mathscr{P}$ should be identified with the localized vegetative spots of Fig. 14.12d.

We closed with a more detailed commentary on the role played by cross-diffusion in generating pattern formation instabilities for our two-component model system. Given $(\partial\Theta/\partial n)(1,1) = 0$, our system violated the activator positive feedback necessary condition for the occurrence of a Turing self-diffusive instability which required $(\partial\Theta/\partial n)(1,1) > 0$. Hence, the cross-diffusive effect of plant root suction on ground water generated our instability since, if $\beta = 0$, its community equilibrium point would be identically linearly stable. Indeed, the other requirement of $\mu < 1$ for a Turing self-diffusive instability to occur might also be violated if $\mu = 1$, and a cross-diffusive instability of this type could still be generated although, in our actual parameter range, it was not violated.

Recently, Stancevic *et al.* ([238]) considered a reaction-chemotaxis-diffusion three-component in-host viral dynamics model system for the concentrations of uninfected or infected cells and the virus (see Chapters 16 and 18). They found that the cross-diffusive effect of chemotaxis toward the infected cells by the uninfected ones generated their pattern formation instability in a similar manner as for our two-component system. Since the community equilibrium point of their system was linearly stable in the absence of diffusion and chemotaxis, Stancevic *et al.* ([238]) referred to this as a Turing instability. To distinguish between these two cases, we referred to ours as a Turing cross-diffusive instability instead. The von Hardenberg *et al.* ([266]) two-component nondimensional model system also included the cross-diffusive effect of plant root suction on ground water. Since that system's interaction terms satisfied the activator positive feedback condition for its community equilibrium point, while $1/\mu = 100$ in their default set of parameter values, the presence of this cross-diffusive effect mediated rather than generated their Turing self-diffusive instability. In particular, von Hardenberg *et al.* ([266]) took the coefficient of that term $b = 3$ in this default set. Upon inspection of our system we could see that this coefficient was related to our parameters by

$$b = \alpha\beta.$$

Thus, that assignment would yield the root suction characteristic curve

$$\beta = \frac{3}{\alpha},$$

which, as a decreasing function of α, is in qualitative accord with our formulation for β_S. The three-component model systems of Rietkerk *et al.* ([186]) and HilleRisLanders *et al.* ([80]) by explicitly including surface water were able to generate Turing instabilities where none would have occurred for our simplified two-component version of that model without root suction. Finally, Wang *et al.* ([268]) conducted a definitive analysis of Turing instabilities for their predator-prey model system by including both self- and cross-diffusion terms in the prey and predator equations and performing weakly nonlinear hexagonal-planform bifurcations and numerical simulations on its community equilibrium point. Since the Allee positive feedback effect for the activator prey and the self-diffusivity advantage for the inhibitory predator were satisfied for their specific model, this was an investigation of cross-diffusion mediated rather than generated instabilities.

Implicit to our continuum formulation were the assumptions that the pattern wavelength was relatively large when compared with the mean coverage diameter of an individual plant but quite small when compared with the territorial length scale characteristic of the arid environment, which allowed us to have considered our interaction-diffusion equations on an unbounded spatial domain (Graham *et al.*, [73]). We concluded by noting that, despite the fact these results of our weakly nonlinear stability analyses were only strictly valid in the vicinity of the marginal stability curve and the occurrence of the rhombic vegetative arrays along our specific coefficient of root suction characteristic curve satisfied this particular constraint, recent numerical simulations of pattern formation in a variety of partial differential equation model systems (see Chapter 11) had demonstrated that such theoretical predictions could often be extrapolated to regions of the relevant parameter space substantially removed from the marginal curve (Boonkorkuea *et al.*, [17]; Golovin *et al.*, [69]). We finished by reiterating, for the purpose of emphasis, the fact that all of our rhombic pattern formation results for this model had been generated by the cross-diffusion process of root suction as opposed to the mediating effect such rhombic planform analyses has often had on self-diffusion generated hexagonal Turing patterns (see Chapters 9-11 and 13).

Unlike with the previous chapters in which *all* instances of Pulling Rabbits Out of Hats had been reported upon their occurrence, we have deferred until now the major identifications in the present chapter. The reason for this is that these two instances, taking place before and after Inthira's stay in Pullman, were in the formulation and validation steps of the comprehensive applied mathematical modeling process (Wollkind and Dichone, [284]) and hence, we wanted to group them together. The one in the formulation step was involved with the simplification procedure by which the interaction terms for the three-component system of Rietkerk *et al.* ([186]) were reduced to those in our two-component system. Originally, I selected the interaction terms of our model in an ad hoc manner before realizing that these terms were consistent with the adoption of the hydrological equilibrium assumption for Rietkerk *et al.*'s ([186]) system. The one in the validation step was involved in the specific choice for the root suction characteristic curve of Roose and Fowler ([188]). Realizing what the generic form of such a characteristic curve should be for the limiting cases of small and large rates of rainfall, I originally developed it by using the method of matched asymptotic expansions or matched asymptotes, as Jesse Logan liked to call them (see Chapter 4). In fact, when Inthira returned to Mahidol, the first draft of her dissertation had my ersatz formulation for its characteristic curve. I had tried to find the appropriate reference for such a curve unsuccessfully ever since we began working on her problem. Then, when preparing the paper from her thesis, I made one final attempt to find a reference of this type. I did so by searching for "A model for water uptake by plant roots," which was something I had not tried before. Since this was the exact title of Roose and Fowler ([188]), the first entry for that search naturally was this paper and hence I was in business. Replacing my contribution with their characteristic curve, Rick and I deduced all the results appearing in Fig. 14.10 and Table 14.2; thus completing Inthira's thesis problem and, in the process, returning her favor of helping him correct the a_2-error in his problem. Each of these two fortuitous incidents definitely represented Pulling a Rabbit Out of a Hat! The first one would probably never have occurred if I had not been so intimately familiar with the chemical quasi-equilibrium assumption of Keller and Segel ([94]), as to be related in Chapter 18. The second one provided exactly the formulation for which I had been searching all year and would probably never have occurred had not for my hitting purely by chance on precisely that title of the Roose and Fowler ([188]) paper. The odds of this actually happening were so very small, as to seem almost magical in nature and hence justifying the assertion in Chapter 1 of: "It's Magic, You Dope!"

Inthira's thesis being completed, the dissemination phase could begin. Since she was a co-author of Cangelosi *et al.* ([25]), as related in Chapter 13, there was not as much time pressure in getting another paper published, since that *J. Math. Biol.* contribution could be used as the accepted paper required for Inthira to defend her thesis. Inthira and Chontita wrote the first draft of the paper from her dissertation, I wrote the final manuscript, Rick generated all the figures, and Bonni formatted that paper first in LaTeX for the *J. Math. Biol.* submission and then in the word template for the special issue on Researches in Vegetation of the *American Journal of Plant Sciences* (see Chapter 12). As mentioned in that chapter, the *J. Math. Biol.* rejected our paper. The reasoning of their referees for doing so still boggles my mind. One said that the paper should have used a Ginzburg-Landau formulation and included a side-band instability analysis. The other said hexagonal- planform nonlinear stability analyses of reaction-diffusion systems had to give rise to stable critical points so that there must be a mistake in our calculation of the Landau coefficients for its amplitude-phase equations. Both of these assertions were patently absurd. In the first instance, since there were no stable stripes predicted one could not conduct a side-band instability analysis and in the second, Gowda *et al.*'s ([72]) $\mu = 0.01$ calculations served as a counterexample for the existence of stable critical points for hexagonal-planform nonlinear stability analyses of reaction-diffusion systems. Further, a reviewer can not simply claim calculations of this type are in error without offering some evidence. Finally, neither of the reviewers mentioned the novelty of the cross-diffusion aspect of the rhombic pattern generation nor the threshold contour paradigm of the latter, although the second one claimed that a protocol of this sort to distinguish between vegetated and bare ground states could only be used in chemical pattern formation where there was an actual color change during the reaction but not in ecological pattern formation where there was not. Pioneers in ecosystem modeling analysis such as Ehud Meron ([142]), who had been using biomass density to replicate such patterns with high levels corresponding to dark regions and low levels, light regions, would have been surprised that anyone could be so ignorant of their work. Of course the *Amer. J. Plant Sci.* accepted the paper (see Chapter 12) with some minor revisions (*e.g.*, an explicit statement of the units for D_S) where it appeared as Chaiya *et al.* ([28]).

I also presented the paper entitled "Vegetative rhombic pattern formation driven by root suction for an interaction-diffusion plant-ground water model system in an arid flat environment," on the following five occasions during 2015 and 2016:

- WSU Showcase 2015, Pullman, WA, March 27, 2015;

- NWAPS Meeting, Pullman, WA, May 14, 2015;

- Joint Mathematics Meetings, Seattle, WA, January 7, 2016;

- Data Science Day, Pullman, WA, April 24, 2016; and

- Brown University Applied Mathematics Seminar Series, Providence, RI, May 26, 2016.

It was very well received at all of these venues but especially as part of that prestigious invited seminar series at Brown University, which certainly vindicated our research if we really needed any such endorsement in order to do so. I would also like to cite two of these other occasions as well. During the Joint Meetings in Seattle, I met Donna Chernyk from Springer Nature at the book exposition, who was to serve as our editor for Wollkind and Dichone ([284]), which shall be described in more detail in Chapter 18. Further, the WSU Showcase 2015 played a pivotal role in the full circle nature of this research. During those last two years, I had taught all my six courses, three each year. While explaining my poster at the WSU Showcase 2015, which consisted only of poster presentations, I had an epiphany of sorts. I could keep on teaching those six courses periodically every two years or I could retire and write my textbook with Bonni. I decided right then and there to retire and had all the paperwork completed within a week. My 45-year career as a regular faculty member at WSU was then perfectly symmetric, in that, when it began in the Fall of 1970 I was 27 years old and when it ended in the Spring of 2015 I was 72. "Logic is logic. That's all I say." as Oliver Wendell Holmes ended his famous poem, The Wonderful "One-Hoss Shay."

15

Subcritical Behavior of a Model Interaction-Dispersion Equation: Longitudinal Planform Nonlinear Stability Analysis

Although we previously have conjectured that localized structures can occur in regions of parameter space where $\sigma > 0$ and $a_1 < 0$ for the truncated version of the amplitude equation

$$\frac{dA}{dt} = \sigma A - a_1 A^3 + O(A^5),$$

an examination of this equation's behavior in its subcritical regime has been deferred until now. That was the problem I picked for Mitchell Davis' project when he decided to submit a proposal for funding from the WSU Undergraduate Summer Research Program. Such proposals require that the research be performed under the direction of a faculty member and I agreed to serve in this capacity. Mitch had been a student in my MATH 440 course during the spring semester preceding that summer. So we submitted the proposal and it was denied funding. Most of the support in this program went to students from the humanities and social sciences, since the powers that be felt that compared to the sciences and engineering they had much fewer extramural sources available to them. When I told Bob Dillon about this, Bob, who was a co-principal investigator on an NSF grant for Collaborative Institutional Research to the University of Idaho-WSU Program in Undergraduate Mathematics and Biology (UMB), decided to fund Mitch for that summer, from his grant, to complete a project in biomathematics under my direction. Ironically, the stipend from this grant exceeded the one associated with the summer research program for which he had been denied! The original proposal involved a longitudinal planform analysis of heat conduction in a finite bar with a source term, but for the purposes of the UI-WSU UMB program it was easy to convert that to a related interaction-dispersion partial differential equation for a biological population over a fixed linear territory size with an interaction term of the same form as this particular source term.

Toward that end we considered the following interaction-dispersion partial differential equation boundary value problem for $N = N(s, \tau) \equiv$ population density where $s \equiv$ one-dimensional spatial variable and $\tau \equiv$ time:

$$\frac{\partial N}{\partial \tau} = D_0 \frac{\partial^2 N}{\partial s^2} + R_0 N_e r \left(\frac{N - N_e}{N_e} \right), \ 0 < s < L;$$

$$N(0, \tau) = N(L, \tau) = N_e;$$

with

$$r(\theta) = \theta + \alpha \theta^3 + \gamma \theta^5 + O(\theta^7).$$

Here, $D_0 \equiv$ dispersal constant, $R_0 \equiv$ interaction rate, $N_e \equiv$ equilibrium population density, and $L \equiv$ territory size, while α and γ represented dimensionless interaction coefficients. Note that

$$N(s, \tau) \equiv N_e$$

DOI: 10.1201/9781003195603-15

is an exact solution to this boundary value problem. Introducing the following nondimensional variables and parameter

$$z = \frac{\pi s}{L}, \ t = \frac{D_0 \pi^2 \tau}{L^2}, \ \theta(z,t) = \frac{N(s,\tau) - N_e}{N_e}, \ \beta = \frac{R_0 L^2}{D_0 \pi^2},$$

our original problem transformed into

$$\frac{\partial \theta}{\partial t} - \frac{\partial^2 \theta}{\partial z^2} = \beta r(\theta), \ 0 < z < \pi;$$
$$\theta(0,t) = \theta(\pi,t) = 0.$$

Observe that the exact solution to the dimensional problem corresponds to

$$\theta(z,t) \equiv 0,$$

for our dimensionless one.

This was an extension to fifth-order of a model equation introduced by Wollkind *et al.* ([291]) to illustrate the Stuart ([241])-Watson ([270]) method of weakly nonlinear stability analysis of prototype reaction-diffusion equations. Asymptotic analyses of this sort are very useful for predicting pattern formation in such nonlinear systems, as has been demonstrated in previous chapters. That analysis required the expansion of θ in powers of an unknown amplitude $A(t)$ with spatially dependent coefficients. The pattern formational aspect of this system can be predicted from the long-time behavior of that amplitude which is governed by its Landau ordinary differential equation

$$\frac{dA}{dt} \sim \sigma A - a_1 A^3 - a_3 A^5 = F(A),$$

where σ is the growth rate of linear stability theory and $a_{1,3}$ are the Landau coefficients. That long-time behavior is crucially dependent upon the signs of these Landau coefficients. Wollkind *et al.* ([291]) concentrated on the special case of $r(\theta) = \sin(\theta)$ for the general function employed by Matkowsky ([137]) to develop his two-time method of weakly nonlinear stability theory since their main concern was to compare the results obtained from the application of the Stuart-Watson method with those he deduced. Then, $a_1 > 0$ identically (see below) and it was only necessary to include terms through third-order in $r(\theta)$ to make pattern formation predictions for this problem. In that event, there were two solutions of this reduced system: The first, a homogeneous one that was stable for $\sigma < 0$, and the second, a supercritical re-equilibrated pattern forming one that existed and was stable for $\sigma > 0$ (see Fig. 2.9). These results can be directly applied to our problem for its generalized $r(\theta)$ to fifth-order in the parameter range where $a_1 > 0$. In the range where $a_1 < 0$ and there is so-called subcriticality, the solutions to the reduced problem can grow without bound, and one must take the fifth-order terms into account in order to determine the long-time behavior of the system. Then, we shall show that if there is a parameter range over which the other Landau coefficient $a_3 > 0$, the pattern formation properties of our system can be ascertained without having to resort to considering even higher order terms in $r(\theta)$. This requires the development of a formula for that Landau coefficient and an examination of its sign as a function of α and γ.

Toward this end we sought a Stuart-Watson expansion for the solution of our model equation of the form (Wollkind and Dichone, [284])

$$\theta(z,t) \sim A(t)\sin(z) + A^3(t)[\theta_{31}\sin(z) + \theta_{33}\sin(3z)]$$
$$+ A^5(t)[\theta_{51}\sin(z) + \theta_{53}\sin(3z) + \theta_{55}\sin(5z)].$$

Note that the spatial terms in this expansion satisfied our boundary conditions at $z = 0$ and π, identically. Then, expanding $r(\theta)$ in powers of $A(t)$, employing the relevant trigonometric identities for the resulting products of sine functions contained in its coefficients, and making use of the Landau amplitude equation, we obtained a series of problems, one for each term appearing explicitly in our expansion of the form $A^n(t)\sin(mz)$, given by (Wollkind and Dichone, [284])

$$A(t)\sin(z) : \sigma + 1 = \beta;$$

$$A^3(t)\sin(z) : 3\sigma\theta_{31} - a_1 + \theta_{31} = \beta\left(\theta_{31} + 3\frac{\alpha}{4}\right);$$

$$A^3(t)\sin(3z) : 3\sigma\theta_{33} + 9\theta_{33} = \beta\left(\theta_{33} - \frac{\alpha}{4}\right);$$

$$A^5(t)\sin(z) : 5\sigma\theta_{51} - a_3 - 3a_1\theta_{31} + \theta_{51} = \beta\left(\theta_{51} + \frac{9\alpha\theta_{31}}{4} - \frac{3\alpha\theta_{33}}{4} + \frac{5\gamma}{8}\right).$$

Although there were also two other $A^5(t)$ problems, they have not been catalogued above since only the one proportional to $\sin(z)$ which involves a_3 was required for our purposes. Here, while σ and the θ_{nm} are being considered functions of β, the coefficients $a_{1,3}$ are assumed to be independent of that bifurcation parameter and hence the terminology Landau constants. This assumption is critical for their determination as solvability conditions to be developed below.

We now solved these problems sequentially. Then, from the ones not involving these Landau constants, we obtained in a straight-forward manner that

$$\sigma(\beta) = \beta - 1, \; \theta_{33}(\beta) = \frac{-\alpha\beta}{8(\beta + 3)},$$

while the other two problems yielded

$$2\sigma(\beta)\theta_{31}(\beta) = a_1 + \frac{3\alpha\beta}{4},$$

$$4\sigma(\beta)\theta_{51}(\beta) = a_3 + 3\theta_{31}(\beta)\left(a_1 + \frac{3\alpha\beta}{4}\right) - \frac{3\alpha\beta\theta_{33}(\beta)}{4} + \frac{5\gamma\beta}{8}.$$

(i) Assuming that $\theta_{31}(\beta)$ was well behaved at the critical bifurcation value of $\beta = 1$ and taking the limit of this first relation as $\beta \to 1$, while noting that $\sigma(\beta) = \beta - 1 \to 0$ in this limit, we obtained the solvability condition that

$$a_1 = \frac{-3\alpha}{4}$$

which, upon substitution into this relation, yielded the solution

$$\theta_{31}(\beta) \equiv \theta_{31} = \frac{3\alpha}{8}.$$

Hence, we deduced that

$$a_1 > 0 \text{ for } \alpha < 0 \text{ and } a_1 < 0 \text{ for } \alpha > 0.$$

Thus, as mentioned earlier, when

$$r(\theta) = \sin(\theta) = \theta - \frac{\theta^3}{6} + O(\theta^5) \Rightarrow \alpha = \frac{-1}{6} \Rightarrow a_1 = \frac{1}{8} > 0.$$

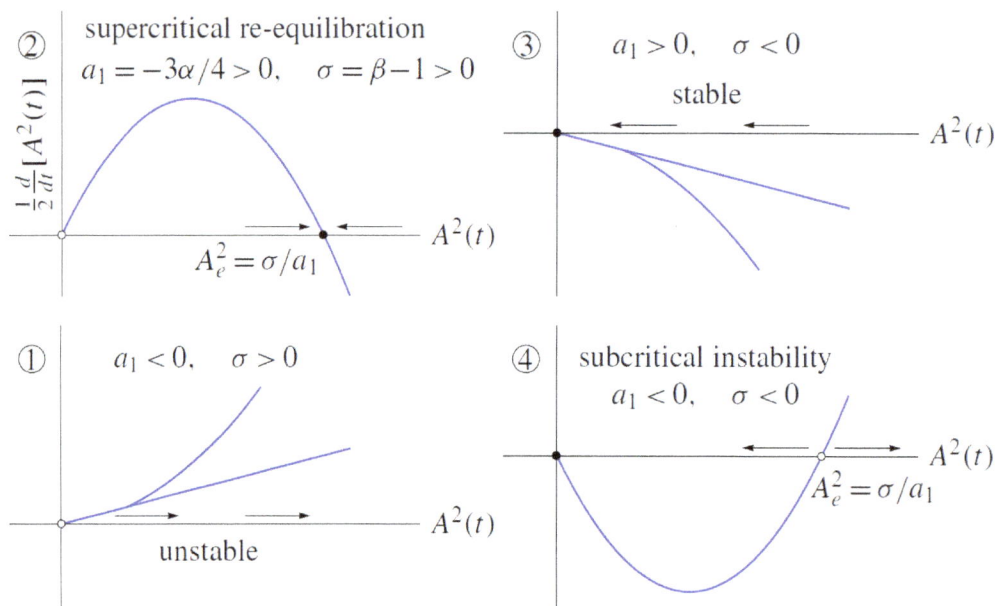

FIGURE 15.1
Plots of $f_3(A^2)$ for the third-order truncated amplitude equation with $\sigma = \beta - 1$ and $a_1 = -3\alpha/4$. Here, the circled numbers correspond to the quadrants in the α-β space of Fig.15.5 with horizontal axis $\beta = 1$ and vertical axis $\alpha = 0$ (see Fig. 2.9).

②　Now, in this case (the circled numbers correspond to those of Fig. 15.1), defining

$$\varepsilon^2 = \frac{\sigma(\beta)}{a_1} > 0 \text{ or } \beta = 1 + \frac{\varepsilon^2}{8},$$

and introducing the rescaled variables

$$\eta = \sigma t, \; \mathcal{A}(\eta) = \frac{A(t)}{\varepsilon},$$

into the reduced amplitude equation through terms of third-order

$$\frac{dA}{dt} = \sigma A - a_1 A^3 + O(A^5),$$

we obtained

$$\frac{d\mathcal{A}}{d\eta} = \mathcal{A} - \mathcal{A}^3 + O(\varepsilon^2),$$

which justified its truncation using a similar procedure to that employed in Chapter 8. Now, multiplying this truncated amplitude equation by $A(t)$ and rewriting it as

$$\frac{1}{2}\frac{dA^2}{dt} = \sigma A^2 - a_1 A^4 = \sigma A^2 \left[1 - \frac{A^2}{\sigma/a_1}\right] = f_3(A^2),$$

we could easily deduce its long-time behavior by means of the four phase-plane plots of $(1/2)dA^2/dt = f_3(A^2)$ that constituted Fig. 15.1 which catalogued the four qualitatively different cases corresponding to the possibility of σ and a_1 being either positive or negative and was equivalent to Fig. 2.9. These served as graphical representations of the cases discussed earlier for the reduced version of our amplitude equation. For the supercritical re-equilibration case of $\sigma, a_1 > 0$, we had

$$\lim_{t\to\infty} A(t) = A_e = \varepsilon,$$

and hence

$$\lim_{t\to\infty} \theta(z,t) \sim \theta_e(z) = \delta \sin(z) \text{ as } \delta \to 0$$

since

$$\lim_{t\to\infty} \theta(z,t) = \varepsilon \sin(z) + \varepsilon^3 [\theta_{31} \sin(z) + \theta_{33}(\beta) \sin(3z)] + O(\varepsilon^5)$$
$$= (\varepsilon + \theta_{31}\varepsilon^3) \sin(z) + \varepsilon^3 \theta_{33}(1) \sin(3z) + O(\varepsilon^5)$$
$$= \delta \sin(z) + \delta^3 \frac{\sin(3z)}{192} + O(\delta^5)$$
$$\sim \delta \sin(z) \text{ as } \delta \to 0$$

where $\delta = \varepsilon + \varepsilon^3 \theta_{31} > 0$. This equilibrium state, plotted in Fig. 15.2, was an arch-type pattern formed from one-cycle of a sine curve with its maximum amplitude δ occurring at $z = \pi/2$.

③　For the stable case of $\sigma < 0, a_1 > 0$, we had

$$\lim_{t\to\infty} A(t) = 0,$$

and hence

$$\lim_{t\to\infty} \theta(z,t) \equiv 0,$$

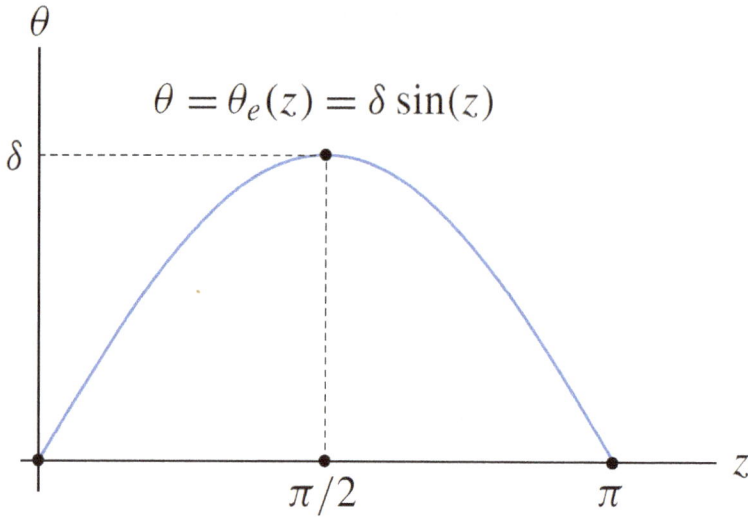

FIGURE 15.2
Plot of the arch solution $\theta_e(z)$ for $0 \le z \le \pi$.

which yielded a uniform homogeneous population distribution.

As an aside, this reduced example through third order terms in $r(\theta) \sim \theta + \alpha\theta^3$ was the model prototype problem I chose to explain weakly nonlinear stability theory to my classes. Segel felt that this procedure was too complicated to present even in graduate courses but I argued this example allowed such a presentation and proceeded to do so in all my classes at WSU. When $a_1 < 0$ for $\alpha > 0$ (see ① and ④ in Fig. 15.1), I told these classes that the truncated problem through third-order yielded unbounded solutions which invalidated the truncation procedure. That is the way I taught this topic until 2014 when it occurred to me in that event, the possibility of extending the expansion to fifth order could be investigated by introducing a homework problem examining the behavior of such an amplitude equation through terms of fifth-order. This was the genesis of the problem I selected for Mitch's project and am now amazed that it took me so long to Pull this particular Rabbit Out of a Hat. That shows most instructors have an inherent tendency to keep teaching things in exactly the same manner as they had before and it takes a major jolt, akin to an epiphany, for them to introduce innovative changes into their methodology. This topic also served as a problem on the written portion of Bonni's preliminary doctoral examination, after the oral part of which, we first began seriously to consider its completion as a future research project.

(ii) We next proceeded to analyze the second Landau constant relation involving a_3 and θ_{51}, in an analogous manner to that just employed, to evaluate a_1 and θ_{31}. Thus, assuming $\theta_{51}(\beta)$ to be well-behaved at $\beta = 1$ and taking the limit of this relation as $\beta \to 1$, we obtained the solvability condition that

$$a_3 = \frac{-5\gamma}{8} - 3\theta_{31}\left(a_1 + \frac{3\alpha}{4}\right) + \frac{3\alpha\theta_{33}(1)}{4} = \frac{-5\gamma}{8} - \frac{3\alpha^2}{128},$$

which, upon substitution into this relation, yielded the solution

$$\theta_{51}(\beta) = \frac{5\gamma}{32} + \frac{9\alpha\theta_{31}}{16} + \frac{3\alpha^2(4\beta+3)}{512(\beta+3)}.$$

Observe that, by virtue of the value of a_1, a_3 is independent of θ_{31}. Also, observe that unlike this quantity, θ_{51} is a function of β. Finally, in addition, note that should we have assumed the Stuart-Watson expansion for $\theta(z,t)$ and the Landau equation for dA/dt also contained even powers of $A(t)$ and $\theta(z,t)$, even multiples of z, then the solvability conditions and solutions for their coefficients would have shown them to be zero (Wollkind et al., [291]). Hence our implicit assumption that these quantities only contained odd powers of $A(t)$ and multiples of z was made without loss of generality and follows as a direct consequence of the form of $r(\theta)$.

Having determined its coefficients, we examined the truncated amplitude equation through terms of fifth-order, i.e., $dA/dt = F(A)$, and deferred until after this examination had been completed a justification for that truncation. We sought conditions under which the inclusion of fifth-order terms would re-equilibrate the growing solutions predicted through third-order when $a_1 < 0$. Hence, we assumed a parameter range in which $a_1 < 0$ or $\alpha > 0$. Further, anticipating our results to be demonstrated below, we assumed that $a_3 > 0$, while as always $\sigma \in \mathbb{R}$. This equation had three equilibrium points

$$A(t) \equiv A_e \text{ such that } F(A_e) = 0$$

satisfying either

$$A_e = 0 \text{ or } 2a_3 A_e^{\pm 2} = \pm\sqrt{a_1^2 + 4a_3\sigma} - a_1.$$

Observe that, since they must be real and positive, A_e^{+2} exists for $\sigma \geq \sigma_{-1} = -a_1^2/(4a_3)$, while A_e^{-2} only exists for $\sigma_{-1} \leq \sigma < 0$. Multiplying our truncated amplitude equation by $A(t)$, we obtained

$$\frac{1}{2}\frac{dA^2}{dt} = \sigma A^2 - a_1 A^4 - a_3 A^6 = a_3 A^2(A_e^{-2} - A^2)(A^2 - A_e^{+2}) = f(A^2).$$

Then, we determined the global stability properties of these equilibrium points by plotting $(1/2)dA^2/dt = f(A^2)$ for $\sigma < \sigma_{-1} < 0$, $\sigma_{-1} < \sigma < 0$, and $\sigma > 0$, respectively, in the three phase-plane plots of Fig. 15.3. From that figure, we could see that 0 was globally stable for $\sigma < \sigma_{-1} < 0$; A_e^{+2}, for $\sigma > 0$; and in the overlap region where either could be stable depending on initial conditions 0 was stable for $0 < A^2(0) < A_e^{-2}$ and A_e^{+2}, for $A^2(0) > A_e^{-2}$, while A_e^{-2} which only existed in that bistability region was not stable there.

To justify this truncation procedure, we considered our Landau equation in the form

$$\frac{dA}{dt} = F(A) + O(A^7),$$

defined $\varepsilon^2 = -a_1$, assumed $a_3 = O(1)$ as $\varepsilon \to 0$, and let $\sigma = O(\varepsilon^4)$. Then, $A_e^{+2} = O(\varepsilon^2)$ which implied that $A_e^+ = O(\varepsilon)$. Note that $\alpha = 10^{-2}$, $\gamma = -2$ yielded Landau constants satisfying these conditions. Now, analogous to our approach at third-order, we introduced the rescaled variables

$$\eta = \sigma t, \ \mathcal{A}(\eta) = \frac{A(t)}{A_e^+} \text{ where } \mathcal{A}, \ \frac{d\mathcal{A}}{d\eta} = O(1) \text{ as } \varepsilon \to 0.$$

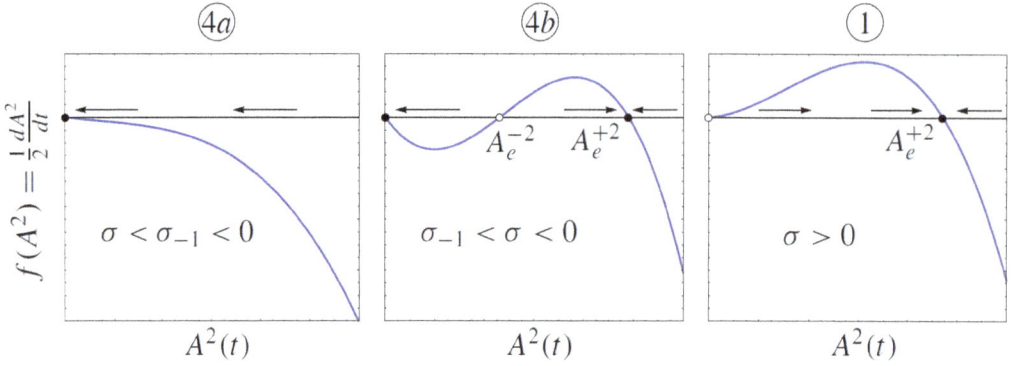

FIGURE 15.3
Plots of $f(A^2)$ for the fifth-order truncated amplitude equation with $a_1 < 0$; $a_3 > 0$; and $\sigma < \sigma_{-1} = -a_1^2/(4a_3) < 0$, $\sigma_{-1} < \sigma < 0$, and $\sigma > 0$, respectively. Here, the circled numbers again correspond to the quadrants in the α-β space of Fig. 15.5.

Since then,

$$\frac{dA}{dt} = \sigma A_e^+ \frac{dA}{d\eta} = O(\varepsilon^5), \ \sigma A = \sigma A_e^+ \mathcal{A} = O(\varepsilon^5), \ a_1 A^3 = a_1 A_e^{+3} \mathcal{A}^3 = O(\varepsilon^5),$$

$$a_3 A^5 = a_3 A_e^{+5} \mathcal{A}^5 = O(\varepsilon^5) \text{ while } O(A^7) = O(A_e^{+7} \mathcal{A}^7) = O(\varepsilon^7),$$

under these conditions, this justified our truncation procedure at fifth-order. Finally, when $\sigma > 0$, we had the same type of equilibrium solution as depicted in Fig. 15.2 except in this case

$$\delta = \varepsilon_0 + \theta_{31}(1)\varepsilon_0^3 + \theta_{51}(1)\varepsilon_0^5 \text{ where } A_e^+ = A_0\varepsilon = \varepsilon_0 \text{ with } A_0 = O(1) \text{ as } \varepsilon \to 0.$$

This result depended upon

$$a_3 > 0 \Rightarrow \gamma < \frac{-3\alpha^2}{80}.$$

Recall that in addition, we had already taken $\alpha > 0$ to guarantee that $a_1 = -3\alpha/4 < 0$. This region is plotted in the fourth quadrant of the α-γ plane of Fig. 15.4. In this context, note from Fig. 15.3 that, unlike the situation depicted in Fig. 15.1 for $\alpha > 0$, all the solutions remain bounded when the fifth-order terms in $r(\theta)$ are retained.

Should there exist a parameter range in a dynamical systems model of a given phenomenon for which the third-order Landau constant $a_1 < 0$ and hence the bifurcation is subcritical, the weakly nonlinear stability analysis must be pushed to fifth order, as originally pointed out by DiPrima *et al.* ([51]). This has been standard operating procedure particularly over the last ten years, when practitioners of the Palermo nonlinear stability theory group began considering fifth-order terms in the Landau equation during their investigation of subcritical bifurcation for a variety of two-component reaction-diffusion systems (Gambino *et al.*, [63, 64]; Tulumello *et al.*, [254]). By necessity, such calculations are long and technically complicated. Thus, when surveying the theory, there is some merit in introducing a simple model equation that preserves all the salient features of a more complex system but considerably reduces the labor involved in determining the Landau constants. This was our rationale for considering the generalized Matkowsky equation under investigation. That

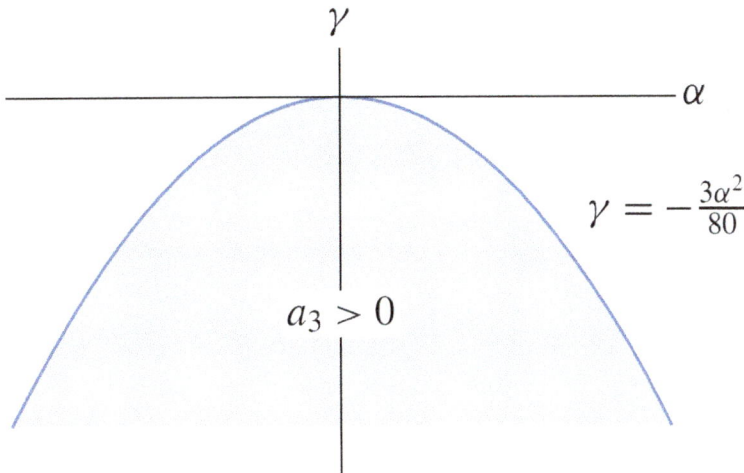

FIGURE 15.4
Plot of the region in the α-γ plane where $a_3 > 0$.

was also the rationale for Drazin and Reid's ([53]) employment of their nondimensionalized version of the Matkowsky equation with $r(\theta) = \sin(\theta)$, in order to develop weakly nonlinear theory relevant to hydrodynamic stability. Matkowsky ([137]) regarded his problem as a mathematical model for temperature distribution in a finite bar with a nonlinear source term, the ends of which were maintained at the ambient, while Drazin and Reid ([53]) offered their corresponding version as a phenomenological model of parallel flow in a channel. Hence, they both envisioned their instabilities to be rate driven by considering the bifurcation parameter $\beta \sim R_0$. For ecological applications, it is often more relevant to envision these instabilities to be territory size driven by considering $\beta \sim L^2$ and then the instability criterion describes the evolution of spatially heterogeneous structure in a specific domain.

Given that the fifth-order extensions, referenced above, primarily concentrated only on the subcritical regime, we next synthesized our fifth-order results of Fig. 15.3 valid for $a_1 < 0$, or equivalently $\alpha > 0$, and $a_3 > 0$, or equivalently $3\alpha^2 + 80\gamma < 0$, with those valid for $a_1 > 0$, or equivalently $\alpha < 0$, and $a_3 > 0$, as well, depicted in the third quadrant of Fig. 15.4. Note, that under these conditions, $A_e^{+2} > 0$ for $\sigma > 0$ and $A_e^{-2} < 0$, identically. If we plot a figure analogous to the supercritical cases of Fig. 15.1 it is obvious that the qualitative morphological behavior of those cases is preserved at fifth order with the only change being now $A_e^2 = A_e^{+2}$. We accomplish this synthesis by means of Fig. 15.5, a bifurcation diagram in α-β space, where the relevant regions associated with these predicted morphological identifications are represented graphically. Since those results also depend on the behavior of σ, while $\sigma = 0$ and $\sigma = \sigma_{-1}$ are the critical loci for that quantity in this regard, it is necessary for us to generate loci equivalent to them in α-β space. In this context, using our previous solvability conditions and definitions, we can deduce the following equivalences:

$$\sigma = \beta - 1 = 0 \iff \beta = 1,$$

$$\sigma = \beta - 1 = \sigma_{-1} = \frac{-a_1^2}{4a_3} = \frac{18\alpha^2}{3\alpha^2 + 80\gamma} \iff \beta = 1 + \frac{18\alpha^2}{3\alpha^2 + 80\gamma};$$

which are plotted in Fig. 15.5. Here, that first locus is a horizontal line parallel to the α-axis which divides our α-β space into the four quadrants formed by it and the β-axis,

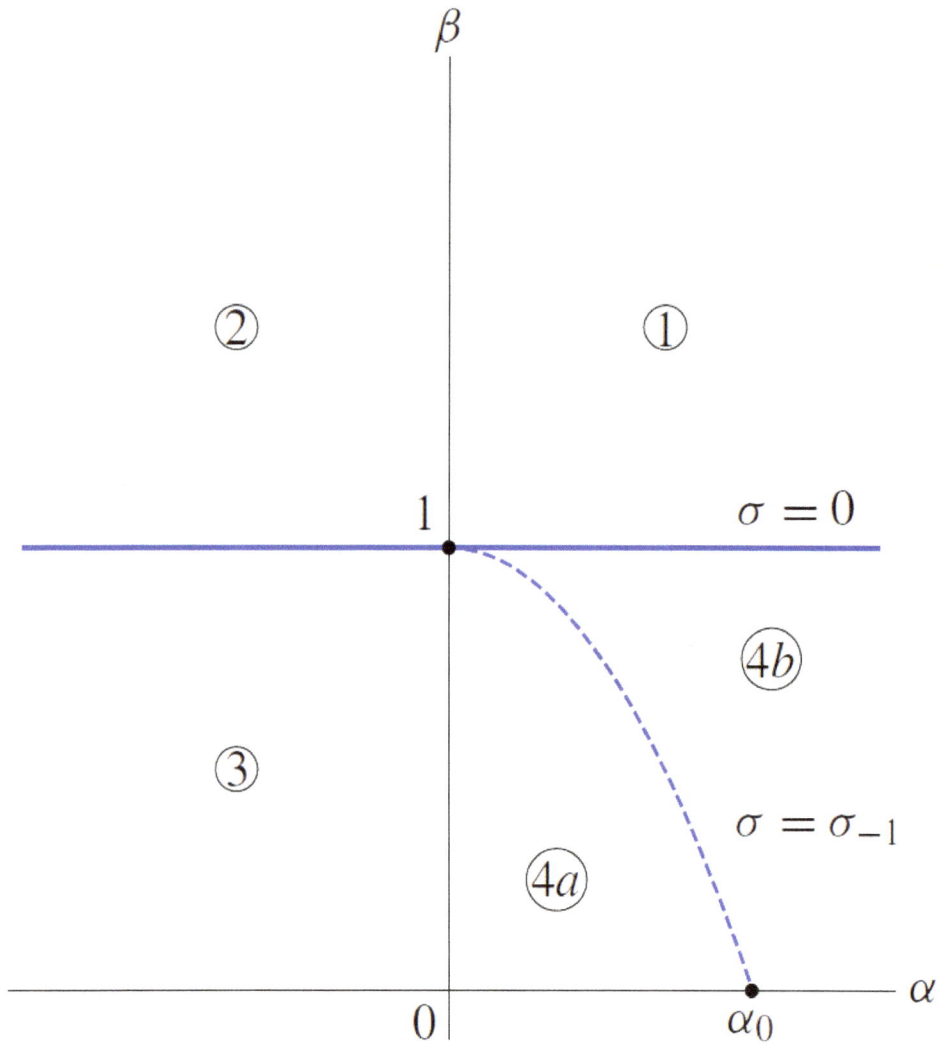

FIGURE 15.5
Bifurcation diagram in α-β space with $\sigma_{-1} = -a_1^2/(4a_3)$, $\sigma = \beta - 1$, $a_1 = -3\alpha/4$, and $a_3 = -5\gamma/8 - 3\alpha^2/128 > 0$ where the circled numbers correspond to the quadrants denoted in Figs. 15.1 and 15.3.

while the second is a concave downward decreasing curve having a horizontal tangent at its β-intercept of 1 and an α-intercept of $\alpha_0 > 0$ where $\alpha_0^2 = -80\gamma/21$ which separates the fourth quadrant of that space into two parts. From an examination of the modification of the supercritical cases of Fig. 15.1 described above and the subcritical cases of Fig. 15.3, we constructed Table 15.1 which catalogued the stable equilibrium points for A^2 in each of the quadrants of Fig. 15.5.

TABLE 15.1
Stable equilibrium points for A^2 in the quadrants of Fig. 15.5.

Quadrant	1	2	3	4a	4b
Stable equilibrium point	A_e^{+2}	A_e^{+2}	0	0	$\left\{ \begin{array}{c} 0 \\ A_e^{+2} \end{array} \right.$

Note that these fifth-order results for our model equation are much more self-consistent than those obtained in the case of its third-order reduction, in that the behavior for the subcritical quadrants 1 and 4a now exactly resemble the behavior for the supercritical quadrants 2 and 3, respectively. In the subcritical quadrant 4b, we have what biologists refer to as metastability, in that the 0 equilibrium point is stable to initially small disturbances but the model will switch to the equilibrium point A_e^{+2} for sufficiently large ones. The existence of such a region of metastability allows our model equation to exhibit outbreak behavior, wherein the maximum population level increases several fold upon a sufficient initial perturbation in amplitude. Returning to our original dimensional formulation, the fact that $A^2 = 0$ represents a globally stable equilibrium point implies that $\lim_{\tau \to \infty} N(s, \tau) = N_e$. Hence, this solution represents a homogeneous population. In many actual biological systems such as the interaction-diffusion plant-ground water one employed by Chaiya et al. ([28]) to model vegetative pattern formation in a flat arid environment (see Chapter 14), the homogeneous patterns in the subcritical parameter range correspond to relatively sparse distributions, while most of those patterns in the supercritical range correspond to much denser distributions where the threshold between these two types of distributions occurs at some N_c. We can induce this sort of behavior in our model equation by adopting the relationship $N_e = N_e(\alpha) = N_c e^{-\alpha}$ which is plotted in Fig. 15.6. Then, from this relation and Table 15.1, in conjunction with Fig. 15.2, we deduced the stable pattern predictions given in Table 15.2 for the quadrants of Fig. 15.5.

TABLE 15.2
Morphological stability predictions for Table 15.1.

Quadrant	Stable pattern
1	Arch
2	Arch
3	Dense Homogeneous
4a	Sparse Homogeneous
4b	$\left\{ \begin{array}{c} \text{Sparse Homogeneous} \\ \text{Arch} \end{array} \right.$

In Chaiya et al. ([28]), it was conjectured that the region of parameter space of subcriticality where $a_1 < 0$ corresponded to isolated vegetative patches when $\sigma > 0$ and low-density homogeneous distributions when $\sigma < 0$, as opposed to the occurrence of periodic patterns

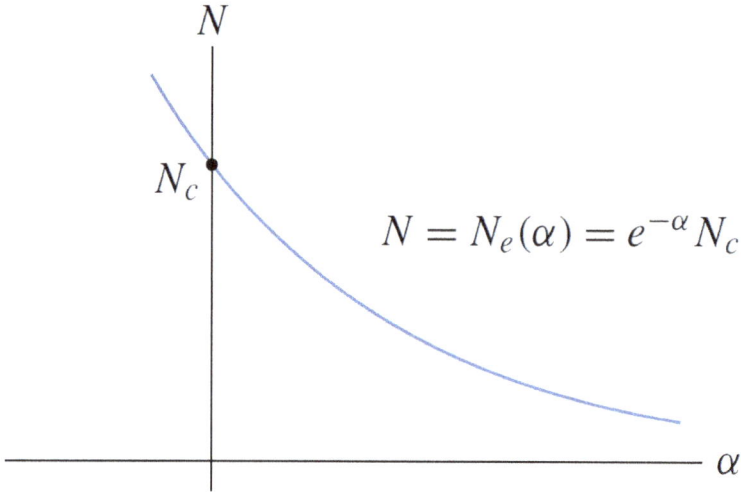

$$N = N_e(\alpha) = e^{-\alpha} N_c$$

FIGURE 15.6
Plot of the population equilibrium density N_e versus α.

for $\sigma > 0$ and high-density homogeneous distributions when $\sigma < 0$ where $a_1 > 0$, which were already predicted by their rhombic-planform two-dimensional nonlinear stability analysis. Such isolated patches are a compromise between periodic patterns and homogeneous stable states that are sparse enough to resemble bare ground. They then associated equilibrium points 0 and A_e^{+2} of quadrants 1 and 4 of Table 15.1 with the sparse homogeneous state and the isolated patch, respectively, that would occur in a postulated fifth-order extension should $a_3 > 0$ for this parameter range. Our fifth-order results, summarized in Table 15.2, represent the first step in a conclusive demonstration of the validity of this conjecture. We conclude by noting that although these results are only strictly asymptotically valid in a neighborhood of the marginal stability curve $\beta = 1$, Boonkorkuea *et al.* ([17]) by comparing their theoretical predictions of this sort with existing numerical simulations of vegetative pattern formation for a model evolution equation, recently showed that the former can often be extrapolated to those regions of parameter space relatively far from the marginal curve (see Chapter 11). These theoretical predictions also associated that region of parameter space where numerical simulation generated isolated patches with $\sigma > 0$ and $a_1 < 0$.

Finally, for the sake of definiteness, we offered a closed-form representation of $r(\theta)$, composed of combinations of common functions, that produced Landau constants consistent in sign with our subcriticality assumptions. Recall the following Maclaurin polynomials truncated through terms of fifth order (Stewart, [240]):

$$\sinh(z) \sim z + \frac{z^3}{6} + \frac{z^5}{120}, \quad \arctan(z) \sim z - \frac{z^3}{3} + \frac{z^5}{5}.$$

Then,

$$4\sinh\left(\frac{\theta}{2}\right) \sim 4\left(\frac{\theta}{2} + \frac{\theta^3}{48} + \frac{\theta^5}{32(120)}\right) = 2\theta + \frac{\theta^3}{12} + \frac{\theta^5}{960},$$

$$2\arctan\left(\frac{\theta}{2}\right) \sim 2\left(\frac{\theta}{2} - \frac{\theta^3}{24} + \frac{\theta^5}{160}\right) = \theta - \frac{\theta^3}{12} + \frac{\theta^5}{80}.$$

Now, defining $r(\theta)$ to be the difference between these two functions we obtain

$$r(\theta) = 2\left[2\sinh\left(\frac{\theta}{2}\right) - \arctan\left(\frac{\theta}{2}\right)\right] \sim \theta + \frac{\theta^3}{6} - \frac{11\theta^5}{960} = \theta + \alpha\theta^3 + \gamma\theta^5$$

which implies

$$\alpha = \frac{1}{6} > 0, \ \gamma = \frac{-11}{960} \text{ such that } 80\gamma + 3\alpha^2 = \frac{-11}{12} + \frac{1}{12} = \frac{-5}{6} < 0.$$

We certainly considered the deduction of this representation to be a Rabbit Pulled Out of a Hat!

This completed Mitch's project and he presented a slide show of these results at the final UI-WSU UMB program meeting at the end of that summer. For the sake of accountability to NSF, Bob encouraged the participants to publish their results in scientific journals relevant to the subject matter of these projects. Hence, I wrote a paper with Mitch, Rick, and Bonni as co-authors reporting the results described in this chapter and submitted it sequentially to the journals where Matkowsky ([137]), Oulton *et al.* ([161]), and Wollkind *et al.* ([291]), all of which involved stability analyses of model reaction-diffusion equations, had appeared: Namely, the *Bulletin of the AMS* (American Mathematical Society), the *American Mathematical Monthly* published by the MAA (Mathematical Association of America), and *SIAM* (Society for Industrial and Applied Mathematics) *Review*. Each of these journals rejected it by editorial decision, without review, because they felt the subject matter of our paper was too specialized, would not be of sufficient interest to a majority of their readership, or was not topical enough, respectively. I then tried the *Journal of Mathematical Analysis* and their editor rejected it out of hand saying the paper did not contain any mathematics! I next submitted it to the *Quarterly of Applied Mathematics* and their reviewer, although liking the paper, felt its content was too similar in spirit to Wollkind *et al.* ([291]) to be publishable by them. Finally, I tried *Involve: A Journal of Mathematics*, created by Kenneth Berenhuat of Wake Forest University, which publishes high-quality manuscripts having at least one-third student authorship being either of undergraduate or graduate level when the research was conducted (Rick was still a graduate student during the relevant period in question) and their reviewer loved the paper, finding it acceptable for publication after our correction of the one typographical error, the latter pointed out [in the introduction of dimensionless variables, θ had been identified as $N - N_e$ instead of $(N - N_e)/N_e$]. Hence, it appeared there as Davis *et al.* ([45]). In addition, Bonni and I each presented a talk based upon this paper in 2017 and 2018. Mine was:

- The behavior of a population interaction-diffusion equation in its subcritical regime, AMS Spring Sectional Meeting, Pullman, WA, April 23, 2017;

while hers was:

- A Stuart-Watson nonlinear stability analysis of a generalized Matkowsky heat equation, Joint Mathematics Meetings, San Diego, CA, January 13, 2018.

Given the title of Bonni's talk, it only seems appropriate to close this chapter by placing our research in the context of that performed by Matkowsky ([137]), since I was present at its inception when we both were attending a series of fluid mechanics stability lectures given by DiPrima and Segel, during my last year at RPI and Bernie's second year as a faculty member in its Department of Mathematics.

One day, Dick was explaining how to truncate the Landau amplitude equation

$$\frac{dA}{dt} = \sigma A - a_1 A^3 + O(A^5) \text{ for } \sigma, a_1 > 0;$$

with which we started this chapter. He said: Consider $\sigma = O(\varepsilon^2)$, $a_1 = O(1)$, and assume that $A = O(\varepsilon)$ since $A_e^2 = \sigma/a_1 = O(\varepsilon^2)$. Then, σA, $a_1 A^3 = O(\varepsilon^3)$, while $O(A^5) = O(\varepsilon^5)$. Thus

$$\frac{dA}{dt} \sim \sigma A - a_1 A^3.$$

Bernie asked: What about the dA/dt term? Wouldn't it also have to be of $O(\varepsilon^3)$ as well, in order for this truncation procedure to be valid? Dick could not seem to understand at exactly what he was driving, although Lee and I realized it immediately. Bernie suggested defining a second time variable $\eta = \varepsilon^2 t$ where $\sigma = \varepsilon^2$ and letting $\mathcal{A}(\eta) = A(t)/\varepsilon$ such that $\mathcal{A}$, $d\mathcal{A}/d\eta = O(1)$. Then, $dA/dt = \varepsilon(d\mathcal{A}/dt) = \varepsilon(d\mathcal{A}/d\eta)(d\eta/dt) = \varepsilon^3(d\mathcal{A}/d\eta) = O(\varepsilon^3)$. Hence, yielding

$$\frac{d\mathcal{A}}{d\eta} \sim \mathcal{A} - a_1 \mathcal{A}^3.$$

It was this insight that allowed Matkowsky ([137]) to develop his two-time method of weakly nonlinear stability theory and its genesis definitely represented a Rabbit Pulled Out of a Hat!

Heretofore, we have been using the $O(A)$ term of our longitudinal planform expansions to examine the linear stability of solutions for all the problems studied in this book. To explain more fully that two-time method Matkowsky ([137]) introduced to analyze such problems, it is necessary for us to reconsider linear stability theory as applied to the model equation being investigated in this chapter. That is, let

$$\theta(z, t; \varepsilon_1) = \theta_e + \varepsilon_1 \theta_1(z, t) + O(\varepsilon_1^2) \text{ where } \theta_e = 0 \text{ and } |\varepsilon_1| \ll 1.$$

Then, upon substituting this into our basic equation, neglecting terms of $O(\varepsilon_1^3)$, and cancelling the resulting common ε_1 factor, we would obtain the following linear perturbation problem for $\theta_1 = \theta_1(z, t)$:

$$\frac{\partial \theta_1}{\partial t} - \frac{\partial^2 \theta_1}{\partial z^2} = \beta \theta_1, \ 0 < z < \pi; \ \theta_1(0, t) = \theta_1(\pi, t) = 0.$$

Seeking a separation of variables normal-mode solution of this equation of the form

$$\theta_1(z, t) = \Theta(z)e^{\sigma t},$$

we find that $\Theta = \Theta(z)$ satisfies the linear eigenvalue problem for σ given by

$$[D^2 - (\sigma - \beta)]\Theta = 0, \ 0 < z < \pi \text{ where } D \equiv \frac{d}{dz}; \Theta(0) = \Theta(\pi) = 0.$$

Assuming the requisite continuity at $z = 0$ and π, we can deduce that

$$D^{2N}\Theta = 0 \text{ at } z = 0 \text{ and } \pi \text{ for } N = 0, 1, 2, 3, \ldots$$

and the eigenvectors from Fourier series analysis are given by (see Wollkind and Dichone, [284])

$$\Theta_{(z)} = \theta_{1m} \sin(mz) \text{ where } \theta_{1m} \neq 0 \text{ for } m = 1, 2, 3, \ldots,$$

with corresponding eigenvalues

$$\sigma_m = \beta - m^2 \text{ for } m = 1, 2, 3, \ldots.$$

Hence, by the superposition principle (see Wollkind and Dichone, [284])

$$\theta_1(z,t) = \sum_{m=1}^{\infty} \theta_{1m} e^{\sigma_m t} \sin(mz).$$

Therefore, $m = 1$ represents its most dangerous mode since $\sigma_m > \sigma_{m+1}$ and thus our stability criterion becomes

$$\sigma_1 = \beta - 1 < 0 \text{ or } \beta = \frac{R_0 L^2}{D_0 \pi^2} < 1,$$

which is identical to that for the $A(t)\sin(z)$ problem. Biologically this means the population disperses faster in the region than it is being produced and the equilibrium solution N_e is linearly stable, while, should $\beta > 1$, the reverse is true and that solution is unstable to linear perturbations.

Matkowsky ([137]) analyzed a model equation which when placed in our form required

$$r(0) = r''(0) = 0, \ r'(0) > 0, \ r'''(0) < 0.$$

Note that $r(\theta) = \sin(\theta)$, as employed by Drazin and Reid ([53]) and Wollkind $et\ al.$ ([291]), satisfied these conditions. Instead of being saddled with the usual deficiencies of linear stability theory which, although producing a solution satisfying a given initial condition at $t = 0$ (see below), is only valid in a boundary layer interval of that time, or nonlinear theory, which, although correctly predicting the long time behavior of the most dangerous mode, cannot satisfy a general initial condition consisting as it does of only this single mode, Matkowsky's ([137]) two-time multiple scales analysis produced a uniformly-valid solution in t, which both satisfied the initial condition and predicted long-time behavior. In particular he posed the initial condition

$$\theta(z,0;\varepsilon_1) \sim \varepsilon_1 h(z) = \varepsilon_1 \sum_{m=1}^{\infty} h_m \sin(mz), \ 0 < z < \pi,$$

and sought a solution of his model equation of the form

$$\theta(z,t;\varepsilon_1) = \zeta(z,t,\eta;\varepsilon_1) \sim \varepsilon_1 \zeta_1(z,t,\eta) + \varepsilon_1^3 \zeta_3(z,t,\eta),$$

with $\eta = \varepsilon_1^2 t$ where $\varepsilon_1^2 = \beta - \beta_c$ for $\beta_c = 1/r'(0)$. Matkowsky ([137]) found that

$$\zeta_1(z,t,\eta) = \mathcal{A}_1(\eta) \sin(z) + \sum_{m=2}^{\infty} \mathcal{A}_m(\eta) e^{\sigma_m(\beta_c)t} \sin(mz)$$

where $\sigma_m(\beta) = r'(0)\beta - m^2$ for $m \geq 1$. Then, the multiple scales uniformity condition at the next order for $\zeta_3(z,t,\eta)$ (see Wollkind and Dichone, [284]), required the following amplitude equation for $\mathcal{A}_1 = \mathcal{A}_1(\eta)$ (Matkowsky, [137]):

$$\frac{d\mathcal{A}_1}{d\eta} = r'(0)\mathcal{A}_1 + \beta_c r'''(0)\frac{\mathcal{A}_1^3}{8}.$$

Now, given that

$$\zeta_1(z, 0, 0) = \sum_{m=1}^{\infty} \mathcal{A}_m(0) \sin(mz),$$

the initial condition will be satisfied provided

$$\mathcal{A}_m(0) = h_m \text{ for } m \geq 1,$$

where, as usual, using Fourier sine series (see Wollkind and Dichone, [284]), the h_m's are related to $h(z)$ by

$$h_m = \frac{2}{\pi} \int_0^{\pi} h(z) \sin(mz) \, dz.$$

Note in this context, that then the amplitude equation Matkowsky deduced after the DiPrima lecture would be equivalent to this one since for $r(\theta) \sim \theta + \alpha\theta^3$:

$$r'(0) = 1, \ r'''(0) = 6\alpha, \ \beta_c = \frac{1}{r'(0)} = 1, \text{ and } \frac{\beta_c r'''(0)}{8} = \frac{3\alpha}{4} = -a_1.$$

Thus, for $a_1 > 0$ or $\alpha < 0$,

$$\lim_{\eta \to \infty} \mathcal{A}_1^2(\eta) = \mathcal{A}_e^2 = \frac{1}{a_1}$$

and hence,

$$\lim_{t \to \infty} \theta(z, t; \varepsilon_1) = \lim_{t, \eta \to \infty} \zeta(z, t, \eta; \varepsilon_1) \sim \varepsilon_1 \lim_{t, \eta \to \infty} \zeta_1(z, t, \eta) = \varepsilon_1 \mathcal{A}_e \sin(z),$$

since $\sigma_m(\beta_c) = 1 - m^2 < 0$ for $m \geq 2$. Observe that, under these conditions,

$$\sigma_1(\beta) = \varepsilon_1^2 \text{ and } \varepsilon_1 \mathcal{A}_e = \sqrt{\frac{\sigma_1(\beta)}{a_1}} \sim \delta,$$

which is consistent with the arch-type equilibrium solution depicted in Fig. 15.2. This completes our sketch of Matkowsky's ([137]) demonstration that the employment of the two-time multiple scales technique of nonlinear stability theory reproduces the Landau equation of the Stuart-Watson method by deduction rather than by assumption and hence preserves its long-time behavior, while allowing satisfaction of an initial condition heretofore restricted to a linear theory approach alone.

There remains one final detail to relate and that concerns the presentation of our work at the 2017 AMS Spring Sectional Meeting. Bob Dillon organized a special session on biological applications at this meeting and invited me to be one of his speakers. That 30-minute talk consisted of a description of the results obtained for our population interaction-dispersion model equation and concluded with its application to the root suction driven vegetative rhombic pattern formation predictions of the plant-ground water interaction-diffusion system, which was the subject of Chapter 14. Thus, the last slides of this presentation were Figs. 14.8, 14.10, and 14.12 from that chapter, which also had been included in Mitch's final UI-WSU UMB program meeting project talk. Hence, it really was an amalgamation of both Davis *et al.* ([45]) and Chaiya *et al.* ([28]). That reversed the order in which these topics have been presented in this book. Clearly, such a talk preserves the natural order from the general to the specific, which the presentations in Chapters 14 and 15 reverse. Such a reversal will occur in any chronological description of my comprehensive applied mathematical modeling results, as per the organization adopted for this book, since I only tend to investigate a general premise after observing the occurrence of the same thing in a number of specific instances. This fulfills the basic definition of inductive reasoning as the derivation of general principles from specific observations. Then, one has to demonstrate the truth of that general premise, as we did in this chapter. That differs from the way pure mathematicians tend to operate: Namely, proving a general result and then applying it to a number of specific examples which satisfy the conditions assumed for this result. That is also the power of model equations from which one can often easily obtain a particular relationship that is difficult to deduce in more complicated systems but once obtained, is simple to show holds in general for those systems. This was Segel's ([212]) rationale for creating model equations, a procedure which has proven itself very useful to me in Pulling Rabbits Out of Hats!

16

Non-Cytopathic Viral-Target Cell Dynamical System Interaction: Dynamical Systems Analysis

While Richard Cangelosi was pursuing his mussel-bed pattern-formation research (see Chapter 13), he began a collaboration with my colleague Elissa Schwartz on an extension of an existing target cell-equine infectious anemia virus (EIAV) interaction model dynamical system relevant to including horse immune responses. Since that research, Schwartz et al. ([207]), was a linear stability analysis of this system's critical points to determine the parameter ranges of interest, followed by numerical simulations over those ranges, Rick wanted to investigate that global behavior using analytical and asymptotic procedures as well. In order to accomplish this, he needed to select appropriate scale factors and introduce dimensionless variables. Knowing my expertise in that area, Rick asked me to help him deduce the optimal scale factors to convert his system into a nondimensional form. Upon my examination of the literature on this subject, I felt that there might be some merit in reconsidering the existing nondimensionalizations of the original model system without explicit immune responses. This literature can be summarized as follows:

Mathematical models have proven valuable in understanding the dynamics of viral infections in vivo within host cells and were originally devised to examine the HIV infection (reviewed by Perelson and Ribeiro, [171]). For interactions of that sort, Anderson and May ([5]) proposed a basic three-component dynamical model system consisting of an uninfected target cell population, an infected such population, and the free virus. This model implied that the propagation of the virus was limited by the availability of susceptible target cells and hence is now characterized as target cell-limited (Phillips, [173]). Assuming a rapid enough time-scale for the free virus dynamics so that a quasi-steady-state approximation could be employed, Tuckwell and Wan ([253]) formally reduced this basic target cell-limited viral model system to a two-component one consisting of the uninfected and infected target cells. They then showed that there were no periodic solutions for the two-component model and that the trajectories of both systems remained quite close. De Leenheer and Smith ([49]) and Prüss et al. ([179]) studied the global stability of the possible equilibrium points for this basic target-cell-limited viral model system and found that its behavior depended upon the size of a particular nondimensional parameter R_0, the basic reproductive number, to be defined below. If $R_0 < 1$, they demonstrated that the virus-free equilibrium point was globally asymptotically stable, while if $R_0 > 1$, this property shifted to the disease-persistence equilibrium point. All of the results cited above used either standard techniques of dynamical systems theory or numerical simulations. Defining $\alpha \equiv$ the ratio of the death rates of the uninfected to the infected target cells, Burg et al. ([22]) classified such viral interactions to be either cytopathic or non-cytopathic depending upon whether $\alpha < 1$ or $\alpha = 1$, respectively. For cytopathic viral interactions, the target cells are killed during the course of infection. Some viruses are intrinsically non-cytopathic because they replicate in a relatively benign manner, while others actively maintain such a state by shutting down all destructive processes, activating nondestructive mechanisms, or inducing alternate

DOI: 10.1201/9781003195603-16

nondamaging replication programs (Plesa *et al.*, [175]).

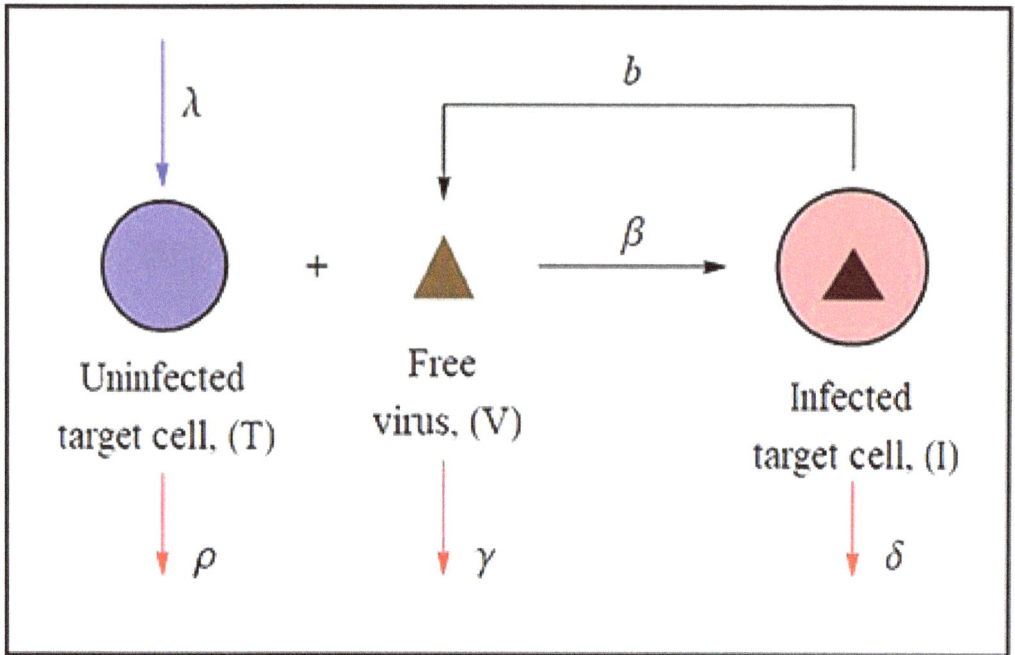

FIGURE 16.1
Schematic diagram of the basic target-cell-limited viral dynamics model illustrating cell-virus interactions. Uninfected target cells (T) can be infected by the free virus (V) to create infected target cells (I). In the case of a non-cytopathic virus, like EIAV, $\alpha = \rho/\delta = 1$ or $\delta = \rho$.

The three-component target cell-virus dynamical model system proposed by Anderson and May ([5]) for $T \equiv$ uninfected target cell population density (cells/ml), $I \equiv$ infected target cell population density (cells/ml), and $V \equiv$ free virus concentration (viral RNA copies/ml) is given by (see Fig. 16.1 and Table 16.1)

$$\frac{dT}{dt} = \lambda - \rho T - \beta TV, \quad \frac{dI}{dt} = \beta TV - \delta I, \quad \frac{dV}{dt} = bI - \gamma V;$$

where $t \equiv$ time (day). Here, it is assumed that the target cells are produced at a constant rate λ and die at a rate ρT; free virus infects target cells at a rate βTV and infected cells die at a rate δI; and free virus particles are produced at a rate bI and cleared at a rate γV. To nondimensionlize this system, a time scale based on the death rate of the uninfected target cell population was selected and hence a dimensionless time $\tau = \rho t$ defined. Then, introducing the nondimensional variables

$$m(\tau) = \frac{T(t)}{T_0}, \quad i(\tau) = \frac{I(t)}{I_0}, \quad v(\tau) = \frac{V(t)}{V_0};$$

where the scale factors T_0, I_0, and V_0 were to be determined; substituting these into the basic equations; and noting that $d/dt = \rho d/d\tau$; this system was transformed into

$$\rho T_0 \frac{dm}{d\tau} = \lambda - \rho T_0 m - \beta T_0 V_0 m v,$$

$$\rho I_0 \frac{di}{d\tau} = \beta T_0 V_0 m v - \delta I_0 i,$$

$$\rho V_0 \frac{dv}{d\tau} = b I_0 i - \gamma V_0 v.$$

Setting the coefficients of all terms in the first transformed equation equal or $\lambda = \rho T_0 = \beta T_0 V_0$ yielded $T_0 = \lambda/\rho$ and $V_0 = \rho/\beta$; while setting the coefficients of the two terms on the right-hand side of the second transformed equation equal or $\beta T_0 V_0 = \delta I_0$ yielded $I_0 = \beta T_0 V_0/\delta = \lambda/\delta$. Now, upon substitution of these values for those scale factors into the transformed equations, the desired nondimensionalized system

$$\frac{dm}{d\tau} = 1 - m - mv, \quad \alpha \frac{di}{d\tau} = mv - i, \quad \varepsilon \frac{dv}{d\tau} = R_0 i - v;$$

was obtained, where

$$\alpha = \frac{\rho}{\delta}, \quad \varepsilon = \frac{\rho}{\gamma}, \quad R_0 = \frac{\lambda \beta b}{\rho \delta \gamma}.$$

The rate parameters contained in these dimensionless quantities are tabulated in Table 16.1.

TABLE 16.1
Units and description of model parameters. Here ml $\equiv$ milliliter and RNA $\equiv$ ribonucleic acid present in a virus that carries its genetic information. Note that the dimensions of $\lambda \beta b$ and $\rho \delta \gamma$ are both $1/\text{day}^3$. Thus the reproductive number R_0 is dimensionless, as stated above.

Rate parameter	Units	Description
λ	cells/(ml × day)	Source rate of uninfected target cells
β	ml/(viral RNA copies × day)	Rate of infection of target cells
b	viral RNA copies/(cells × day)	Free viral production rate
ρ	1/day	Death rate of uninfected target cells
δ	1/day	Death rate of infected target cells
γ	1/day	Free viral clearance rate

When I showed Elissa this nondimensional system, she immediately identified R_0 as the basic reproductive number, which gave her some confidence that my selection of scale factors had been optimal. I also inquired about the possibilities for EIAV infections to have $\alpha = 1$ and $\varepsilon \ll 1$, since in that event it seemed to me we could develop a closed-form asymptotic approximation of the solution to that system. Elissa, who was then conducting the research which would appear as Schwartz et al. ([207]), told me, after consulting her laboratory data relevant to these experimental results, that the EIAV-horse interaction was indeed non-cytopathic and thus $\delta = \rho$ or $\alpha = 1$, while in addition $\varepsilon \ll 1$, as well. She was also very excited about developing an analytic solution for this problem, given that, to her knowledge, this had not yet been done and thus would represent an important contribution to the virology literature. Talk about Pulling a Rabbit Out of a Hat! Hence, in the introduction of Cangelosi et al. ([24]), our *Frontiers in Microbiology* article reporting these results, it was stated that:

We shall consider non-cytopathic retroviral interactions; that is, interactions that satisfy $\alpha = 1$, which is believed to be the case for EIAV (Schwartz et al., [208]). This virus shows many characteristics similar to other retroviruses, including a very rapid replication rate and high levels of antigenic variation. It, however, is unusual among retroviruses in that, most infected animals, after a few episodes of fever and high viral load, progress to a chronic degree phase with low viral load and an absence of clinical disease symptoms. The horses effectively control viral replication through adaptive immune mechanisms. Given that this differs from the retroviruses human immunodeficiency virus (HIV) and simian immunodeficiency virus (SIV), in which the infected develop immunodeficiency and disease, EIAV is especially interesting to study in clinical research as well as by using mathematical models. When adopting the mathematical model system depicted in Fig. 16.1, the viral clearance rate γ implicitly captures these adaptive immune system response mechanisms. Since typically $\varepsilon = \rho/\gamma = O(10^{-3})$, we shall employ a systematic two-time method (Matkowsky, [137]) to deduce a quasi-steady-state asymptotic closed-form analytic solution of that basic target-cell-limited viral dynamics model system. Although such nonlinear problems can be solved numerically, the computation must be performed sequentially for each different set of parameter values. The advantage of this asymptotic approach is that it yields an analytic representation, involving the parameters as well as time, required for least-squares parameter-identification curve-fitting procedures to experimental data. We conclude by applying this approach to an experimental data set for EIAV infection in horses.

Toward this end and operating on the assumptions that the viral-target cell interaction was non-cytopathic, while ε was negligible when compared with terms of $O(1)$, we set $\alpha = 1$ in our nondimensional system of equations and sought a quasi-steady-state solution of it of the form

$$[m, i, v](\tau; \varepsilon) = [m_0, i_0, v_0](\tau) + \boldsymbol{O}(\varepsilon).$$

Upon substituting this solution into that dimensionless system and retaining terms of order $O(1)$, we obtained the differential-algebraic system

$$\frac{dm_0}{d\tau} = 1 - m_0 - m_0 v_0, \quad \frac{di_0}{d\tau} = m_0 v_0 - i_0 \text{ where } v_0 = R_0 i_0.$$

We next constructed the inner (or boundary layer) solution, the outer (or quasi-steady-state) solution, and the uniformly valid additive composite for our basic system. The presence of ε in the last equation of this system suggested it contained interactions that occurred on two widely different time scales, one fast and the other slow. In light of this, we introduced the "transient time" variables

$$\eta = \frac{\tau}{\varepsilon} = \gamma t, \ \mathcal{M}(\eta; \varepsilon) = m(\tau; \varepsilon), \ \mathcal{I}(\eta; \varepsilon) = i(\tau; \varepsilon), \ \mathcal{V}(\eta; \varepsilon) = v(\tau; \varepsilon).$$

Upon substituting these into our basic system and noting that $d/d\eta = \varepsilon d/d\tau$, we obtained the boundary layer equations

$$\frac{d\mathcal{M}}{d\eta} = \varepsilon(1 - \mathcal{M} - \mathcal{MV}), \ \frac{d\mathcal{I}}{d\eta} = \varepsilon(\mathcal{MV} - \mathcal{I}), \ \frac{d\mathcal{V}}{d\eta} = R_0\mathcal{I} - \mathcal{V}.$$

The ratio of the time scales $\varepsilon = \rho/\gamma \ll 1$, was both a consequence of the fact that the virus acts on a fast time scale $\eta = \gamma t$ and the target cells, on a slower time scale $\tau = \rho t$, and a necessary condition for the employment of a quasi-steady-state approach (see Chapter 10). Seeking a solution of these equations of the form

$$[\mathcal{M}, \mathcal{I}, \mathcal{V}](\eta; \varepsilon) = [\mathcal{M}_0, \mathcal{I}_0, \mathcal{V}_0](\eta) + \boldsymbol{O}(\varepsilon),$$

we deduced that

$$\frac{d\mathcal{M}_0}{d\eta} = \frac{d\mathcal{I}_0}{d\eta} = 0, \ \frac{d\mathcal{V}_0}{d\eta} = R_0\mathcal{I}_0 - \mathcal{V}_0;$$

which, upon integration, yielded

$$\mathcal{M}_0(\eta) \equiv m^{(0)}, \ \mathcal{I}_0(\eta) \equiv i^{(0)}, \ \mathcal{V}_0(\eta) = R_0 i^{(0)} + [v^{(0)} - R_0 i^{(0)}]e^{-\eta};$$

where $m^{(0)}$, $i^{(0)}$, and $v^{(0)}$ are the $O(1)$ values as $\varepsilon \to 0$ of the designated initial conditions

$$\mathcal{M}(0; \varepsilon) = m^{(0)}, \ \mathcal{I}(0; \varepsilon) = i^{(0)}, \ \mathcal{V}(0; \varepsilon) = v^{(0)}.$$

We determined the proper initial conditions to impose for the one term outer solution functions by employing the one-term matching rule (see Chapter 4)

$$m_0(0) = \lim_{\eta \to \infty} \mathcal{M}_0(\eta), \ i_0(0) = \lim_{\eta \to \infty} \mathcal{I}_0(\eta), \ v_0(0) = \lim_{\eta \to \infty} \mathcal{V}_0(\eta);$$

which in conjunction with our previous results yielded

$$m_0(0) = m^{(0)}, \ i_0(0) = i^{(0)}, \ v_0(0) = R_0 i^{(0)};$$

where the target cell population density initial values can be normalized to satisfy

$$m^{(0)} + i^{(0)} = 1.$$

Since the target-cell population densities for both their uninfected and infected states have been non dimensionalized by employing the same scale factor λ/ρ given that $\delta = \rho$, this may be accomplished if that common scaling is identified with the initial value of the sum of these densities. Now, taking the sum of the two quasi-steady-state differential equations, we found that $d(m_0 + i_0)/d\tau + (m_0 + i_0) = 1$ with initial condition just determined $m_0(0) + i_0(0) = 1$, which implies that $m_0(\tau) + i_0(\tau) \equiv 1$ or $i_0 = 1 - m_0$. Then, from the quasi-steady-state algebraic relation, it follows that $v_0 = R_0 i_0 = R_0(1 - m_0)$ which, upon substitution into the first differential equation, yielded the Riccati equation for $m_0 = m_0(\tau; R_0)$ given by

$$\frac{dm_0}{d\tau} = 1 - (R_0 + 1)m_0 + R_0 m_0^2 = F(m_0; R_0), \ \tau > 0; \ 0 \leq m_0(0; R_0) = m^{(0)} \leq 1.$$

In order to examine the local versus the global stability behavior of solutions to this equation as a function of R_0, we first determined its equilibrium points $m_e(R_0)$ such that $F(m_e; R_0) = 0$ by factoring

$$F(m_0; R_0) = (R_0 m_0 - 1)(m_0 - 1),$$

which implied that $m_e(R_0) = 1$ or $1/R_0$ were the equilibrium points in question. We next examined the linear stability of these equilibrium points by letting

$$m_0(\tau; \varepsilon_1) = m_e(R_0) + \varepsilon_1 m_1(\tau; \varepsilon_1) + O(\varepsilon_1^2) \text{ where } |\varepsilon_1| \ll 1,$$

substituting this solution into that equation, neglecting terms of $O(\varepsilon_1^2)$, and canceling the common ε_1 factor to obtain

$$\frac{dm_1}{d\tau} = F'(m_e; R_0)m_1 \text{ where } F'(m_e; R_0) = 2R_0 m_e - (R_0 + 1).$$

Then, since

$$F'(1; R_0) = R_0 - 1 \text{ and } F'\left(\frac{1}{R_0}; R_0\right) = 1 - R_0,$$

we deduced that $m_e(R_0) = 1$ was linearly stable for $R_0 < 1$, while $m_e(R_0) = 1/R_0$ was linearly stable for $R_0 > 1$. Note that since $i_e(R_0) = 1 - m_e(R_0)$, these equilibrium points corresponded to $i_e(R_0) = 0$ or $(R_0 - 1)/R_0$, respectively, and thus the second equilibrium point only existed, in this context, where it was stable. Hence, there was an exchange of stability between these two equilibrium points at $R_0 = 1$.

In order to examine the global stability behavior of this equation, we represented $F(m_0; R_0)$ in the form

$$F(m_0; R_0) = F(1; R_0) + F'(1; R_0)(m_0 - 1) + F''(1; R_0)\frac{(m_0 - 1)^2}{2}$$

or, since $F''(1; R_0) = 2R_0$, rewrote that equation as

$$\frac{dm_0}{d\tau} = (R_0 - 1)(m_0 - 1) + R_0(m_0 - 1)^2.$$

Then, making the change of variables

$$y = m_0 - 1,$$

this Riccati equation was transformed into the Bernoulli equation (see Boyce and DiPrima, [18])

$$\frac{dy}{d\tau} + (1 - R_0)y = R_0 y^2,$$

that can be solved by introducing the auxiliary variable $u = 1/y$ for $y \neq 0$ to obtain

$$\frac{1}{y} = \begin{cases} \frac{R_0}{1-R_0} + c_0 e^{(1-R_0)\tau} & \\ c_1 - \tau & \end{cases} \text{ for } \begin{matrix} R_0 \neq 1 \\ R_0 = 1 \end{matrix},$$

while noting that $y \equiv 0$ was a solution to that equation.

Making use of these results and the initial condition, we arrived at the quasi-steady-state approximation for the uninfected target cell population density

$$m_0(\tau; m^{(0)}, R_0) = \begin{cases} 1 + \frac{(1-R_0)(m^{(0)}-1)}{R_0(m^{(0)}-1)+(1-R_0 m^{(0)})e^{(1-R_0)\tau}} & \\ 1 + \frac{m^{(0)}-1}{1+(1-m^{(0)})\tau} & \end{cases} \text{ for } \begin{matrix} R_0 \neq 1 \\ R_0 = 1 \end{matrix}.$$

Note that the corresponding expressions for i_0 and v_0 follow directly from $i_0 = 1 - m_0$ and $v_0 = R_0 i_0 = R_0(1 - m_0)$, respectively. Many similar three component model systems assume that initially the target cells are free of the viral infection. If an assumption of that sort were made for our model by taking $i^{(0)} = 0$ or equivalently $m^{(0)} = 1$, then our solution for $m_0(\tau; m^{(0)}, R_0)$ would yield the unrealistic result that $m_0(\tau; 1, R_0) \equiv 1$. Hence, we approximated that situation by adopting the initial condition $i^{(0)} = a$, or equivalently $m^{(0)} = 1 - a$, instead where the perturbation infected population density a satisfied the condition $0 < a \ll 1$. Specifically, for the relevant plots of Figs. 16.2 and 16.3, we took $a = 0.0001$ which implies that $m^{(0)} = 0.9999$.

Constructing the one-term uniformly valid additive composites defined by (see Chapter 4)

$$m_u^{(0)}(\tau) = m_0(\tau) + \mathcal{M}_0\left(\frac{\tau}{\varepsilon}\right) - m^{(0)},$$

$$i_u^{(0)}(\tau) = i_0(\tau) + \mathcal{I}_0\left(\frac{\tau}{\varepsilon}\right) - i^{(0)},$$

$$v_u^{(0)}(\tau) = v_0(\tau) + \mathcal{V}_0\left(\frac{\tau}{\varepsilon}\right) - R_0 i^{(0)};$$

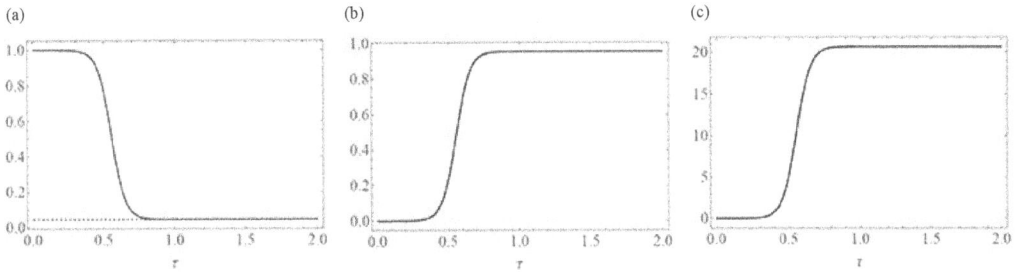

FIGURE 16.2
Plots of the uniformly valid additive composite solutions. (a) Uninfected target cell density, $m_u^{((0))}(\tau)$. (b) Infected target cell density, $i_u^{(0)}(\tau)$. (c) Free virus concentration, $v_u^{(0)}(\tau)$; with the horizontal asymptotes of $1/R_0$, $1-1/R_0$, and R_0-1, respectively. Here, one dimensionless time unit $(\tau = 1)$ is equivalent to 21 days in dimensional time. Parameters used to create the plots are those given in Table 16.2 and correspond to $R_0 = 21.7$ and $\varepsilon = 0.007$.

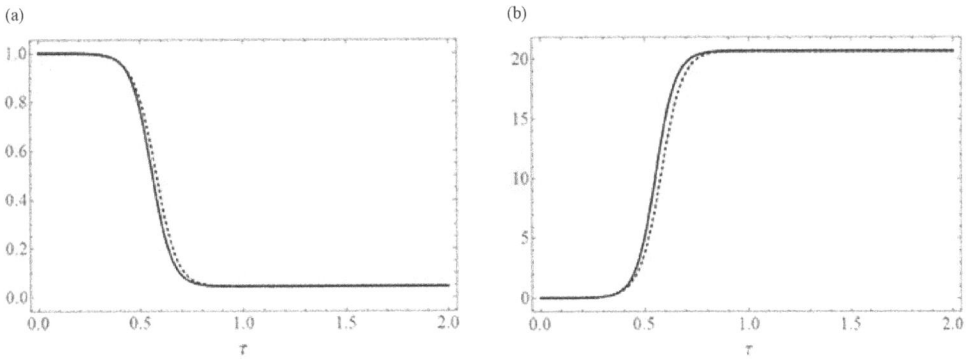

FIGURE 16.3
Comparison of the asymptotic solution of the uninfected target cell population density (solid black curve) in (a) and the EIAV concentration (solid black curve) in (b) with a numerical simulation (dashed curve) of our basic model system. Note that these asymptotic solutions are the same as those of (a) and (c), respectively, in Fig. 16.2. Parameters used to create the plots are those given in Table 16.2 and correspond to $R_0 = 21.7$ and $\varepsilon = 0.007$.

we obtain, from our previous results, that

$$m_u^{(0)}(\tau) = m_0(\tau), \ i_u^{(0)}(\tau) = i_0(\tau), \ v_u^{(0)}(\tau) = v_0(\tau) + [v^{(0)} - R_0 i^{(0)}]e^{-\tau/\varepsilon};$$

where

$$i_0(\tau) = 1 - m_0(\tau) \text{ and } v_0(\tau) = R_0 i_0(\tau) = R_0[1 - m_0(\tau)].$$

Observe, for the target-cell variables, the outer solution is actually uniformly valid to this order.

From the form of $m_0(\tau)$, it is readily seen that when $R_0 = 1$, $m_0(\tau) \to 1$ as $\tau \to \infty$. If $R_0 < 1$, then $m_0(\tau) \to 1$ as $\tau \to \infty$, while if $R_0 > 1$, $m_0(\tau) \to 1/R_0$ as $\tau \to \infty$. This is consistent with the global stability results mentioned earlier. Note further that, except for the exchange of stabilities value of $R_0 = 1$, the linear and global stability behavior with R_0 are actually equivalent for this particular problem.

In Fig. 16.2, we plot the three uniformly valid composite functions $m_u^{(0)}(\tau)$, $i_u^{(0)}(\tau)$, and $v_u^{(0)}(\tau)$. Rate parameter values used are the median ones reported by Schwartz *et al.* ([208]) for EIAV-horse interactions. Specifically, we took the values of these rate parameters as tabulated in Table 16.2.

TABLE 16.2
The rate parameter median values reported by Schwartz *et al.* ([208]) for their EIAV-horse interactions. These values correspond to $R_0 = \lambda\beta b/(\rho^2\gamma) = 21.7$ and $\varepsilon = \rho/\gamma = 0.007$.

Rate parameter	Median value
λ	2019 cells/(ml $\times$ day)
β	3.25×10^{-7} ml/(viral RNA copies $\times$ day)
b	505 viral RNA copies/(cells $\times$ day)
$\rho = \delta$	1/21 per day
γ	6.73 per day

Given that one dimensionless time unit ($\tau = 1$) is equivalent to 21 days in dimensional time, we see that the uninfected target cell population density remains relatively constant for about 7 days ($\tau = 0.33$). This is followed by a period of eight to ten days of rapid infection of that population at the end of which time approximately 95% of it had been infected by EIAV.

In (a) of Fig. 16.3, we provide a comparison of the one-term asymptotic representation for the uninfected target cell density (solid black curve) with a numerical simulation (dashed curve) of the $m(\tau)$ component of our basic system and in (b) of that figure, provide a comparison of the one-term asymptotic representation for the free virus concentration (solid black curve) with a numerical simulation (dashed curve) of the $v(\tau)$ component of our basic system, both using the parameter values given in Table 16.2. Note that these asymptotic solutions are the same as those in (a) and (c), respectively, of Fig. 16.2. In each of these figures, $v^{(0)}$, the initial free virus concentration, was taken to be 450 (viral RNA copies/ml) $\times \beta/\rho = 0.00307$. Observe from Fig. 16.3, the excellent agreement between these analytic asymptotic representations and those numerical simulations. We considered this close correlation to be a Rabbit Pulled Out of a Hat!

Researchers that employ the basic non-cytopathic viral dynamics model system now have an analytic representation involving the parameters that provides a vehicle for least-squares parameter-identification curve-fitting procedures to experimental data. In particular, given a time series uninfected target-cell population density experimental data set $\{(t_n, T_n)\}_{n=1}^{N}$ and our analytic solution for the uninfected target-cell population density in dimensional variables denoted by $T(t; \lambda, \beta, b, \rho, \gamma)$, a parameter identification residual least-squares fit to that data is determined by defining (Torres-Cerna $et\ al.$, [252])

$$E(\lambda, \beta, b, \rho, \gamma) = \sum_{n=1}^{N} [T(t_n; \lambda, \beta, b, \rho, \gamma) - T_n]^2$$

and minimizing this function by solving for λ_c, β_c, b_c, ρ_c, γ_c such that

$$\frac{\partial E}{\partial \lambda}(\lambda_c, \beta_c, b_c, \rho_c, \gamma_c) = \frac{\partial E}{\partial \beta}(\lambda_c, \beta_c, b_c, \rho_c, \gamma_c) = \frac{\partial E}{\partial b}(\lambda_c, \beta_c, b_c, \rho_c, \gamma_c)$$

$$= \frac{\partial E}{\partial \rho}(\lambda_c, \beta_c, b_c, \rho_c, \gamma_c) = \frac{\partial E}{\partial \gamma}(\lambda_c, \beta_c, b_c, \rho_c, \gamma_c) = 0,$$

using the appropriate algorithm. This procedure can be accomplished much more efficiently if one has a closed-form representation for $T(t; \lambda, \beta, b, \rho, \gamma)$, as in our case. Recall that a similar procedure was employed in Chapter 4 to develop the intrinsic growth rate of both mite species as functions of temperature from the Wollkind-Logan temperature-rate equation for arthropods.

We note that for the basic target-cell-limited viral dynamics model system, the deduction of an analytic solution for the quasi-steady-state approximation is crucially dependent on the non-cytopathic condition $\alpha = \rho/\delta = 1$ and we have selected parameter values relevant to this scenario for EIAV. If this were the only non-cytopathic virus, our development restricted to the spread of infection in horse populations might not be representative enough to enlist general interest from virologists. Besides EIAV, however, it has been shown that this non-cytopathic assumption is reasonable for a fairly wide class of important viral interactions in human and other animal populations as well; for example, Hepatitis B and C viruses (Wieland and Chisari, [272]). In addition, non-cytopathic enteroviruses, such as the coxsackie virus B, one of the agents suspected to be responsible for chronic fatigue syndrome (Landay $et\ al.$, [107]), cause persistent infections in their host's cells. Another non-cytopathic virus infecting human populations is the Newcastle disease virus (Carver $et\ al.$, [26]). Finally, Table II in Marcus and Carver ([136]) lists a collection of similar non-cytopathic viruses inducing intrinsic interference among which is the hemadsorption simian virus.

We have been investigating the non-cytopathic interaction of the equine infectious anemia virus infection. While similar to the human immunodeficiency virus, it differs from the latter in not being fatal, partially because the horses' immune systems help effectively to control the virus. Thus, studies of EIAV infection are of importance since they serve as useful prototypes of viral dynamics and immune control, which may have implications in the development of vaccines for HIV and other retroviral infections.

We have assumed that the target-cell populations and the free virus are homogeneously distributed in space. Hence their densities and concentration are functions of time alone. Stancevic $et\ al.$ ([238]) relaxed this assumption by considering a reaction-chemotaxis-diffusion three-component in-host viral dynamics model system for the densities of uninfected or infected target cells and the concentration of a cytopathic virus which were heterogeneous functions of their transverse spatial coordinates, as well as time. They found that the cross-diffusive effect of chemotaxis toward the infected cells by the uninfected ones generated their pattern formation instability in a similar manner as root suction did for the two-component system of Chapter 14. This mechanism was analogous in form to the one referenced in that chapter relevant to the slime-mold model of Keller and Segel ([94]). In our concluding Chapter 18, we shall discuss an extension of that sort for the basic target-cell-limited non-cytopathic viral dynamical model system investigated in this chapter to explain more fully the petechial hemorrhages or minute blood-red spotted patterns that appear on the anemic equine mucous membranes for the chronic degree phase of EIAV infectiousness.

Once our research was done the dissemination phase began. In this case, Rick actually gave a talk on our viral dynamics problem before we prepared a paper on the subject. That talk was entitled "A quasi-steady-state solution for a target-cell limited viral dynamics model with a non-cytopathic effect," by Richard A. Cangelosi and presented at a Spokane Regional Mathematics Colloquium held at Gonzaga University on April 6, 2016. Bonni, who attended this talk, said it was very well received. Given that the experimental results of the definitive study by Schwartz *et al.* ([208]), which appear in Table 16.2 and in our paper Cangelosi *et al.* ([24]) had not yet been completed, Rick used those from his original collaboration with Elissa, namely Schwartz *et al.* ([207]), instead. The relevant rate parameters employed by him for that purpose are tabulated in Table 16.3.

TABLE 16.3
The rate parameter median values reported by Schwartz *et al.* ([207]) for their EIAV-horse interactions. These values correspond to $R_0 = \lambda \beta b/(\rho^2 \gamma) = 17.1$ and $\varepsilon = \rho/\gamma = 0.005$.

Rate parameter	Median value
λ	2000 cells/(ml × day)
β	3.22×10^{-7} ml/(viral RNA copies × day)
b	893 viral RNA copies/(cells × day)
$\rho = \delta$	0.056 per day
γ	11.100 per day

These were also the values of the rate parameters Wollkind and Dichone ([284]) used for the first capstone problem 22.1, in their last chapter, that dealt with this basic viral dynamics model as to be related in our concluding Chapter 18. Rick, in his Gonzaga talk, introduced the concept of the burst size of EIAV, which is defined as the number of RNA viral copies per infected target cell produced by the virus and is equal to b/γ. For the rate parameters of Table 16.3, that would yield the value of 80.45 viral RNA copies/cell, while for those parameters of Table 16.2, it would yield the closely related value of 75.04 viral RNA copies/cell. Since this capstone problem was based on that talk and the latter did not explicitly include an initial viral concentration value $v^{(0)}$, Bonni and I had to select this quantity.

Recall that the uniformly-valid additive composite for the virus was $v_u^{(0)}(\tau) = v_0(\tau) + [v^{(0)} - R_0 i^{(0)}]e^{-\tau/\varepsilon}$. Given that this was the only component of that solution which contained the fast time variable, we wished to select a value for $v^{(0)}$ so that the bracketed term in this solution would not be 0. Therefore, we defined $v^{(0)} = v_1^{(0)} = 1.5 R_0 i^{(0)}$. Let us see how this selection compares with Rick's definition $v_2^{(0)} = 450$ (viral RNA copies/ml) $\times \beta/\rho$. Then, using the parameters from Table 16.3 in conjunction with $i^{(0)} = a = 0.0001$:

$$v_1^{(0)} = 1.5(17.1)(0.0001) = 0.002565$$

and

$$v_2^{(0)} = \frac{(450)(3.33 \times 10^{-7})}{0.056} = 0.002676.$$

What about Pulling this Rabbit Out of a Hat? Or who says, "Great minds don't tend to think alike?"

There was one more thing included in Rick's slide show that did not appear in Cangelosi *et al.* ([24]): Namely, a comparison between our theoretical predictions and experimental measurements. After that paper was published, he realized a comparison with Schwartz *et al.* ([208]) had been omitted and decided it should be included in any future extension of our research, such as the one mentioned above. Toward that end we now compare the quantities tabulated in Table 16.4.

TABLE 16.4
Comparison between experimental measurements of Schwartz *et al.* ([208]) and our theo-retical predictions for the dimensional form of the variables appearing in Fig. 16.3.

Data	R_0	T_e (cells/ml)	V_e (RNA copies/ml)	$\log_{10}(V_e)$
Horse A2202	22.30	1.90×10^3	5.495×10^6	6.74
Horse A2205	21.00	2.02×10^3	1.862×10^6	6.27
Average	21.65	1.96×10^3	3.199×10^6	6.51
Predicted	21.70	1.95×10^3	3.033×10^6	6.48

The entries in that table can be explained as follows: The first two rows represent experi-mental measurements of two selected horses in the study of Schwartz *et al.* ([208]), where T_e and V_e are equilibrium numerical values for the uninfected target cell population density and viral concentration, respectively, in dimensional variables. Each entry in the third row is the arithmetic mean value of the entries in the first two rows, with the exception of its third column entry in which the average of V_e is implicitly meant in the sense of the geometric mean since, for $A, B > 0$,

$$\log_{10}(\sqrt{AB}) = \left(\frac{1}{2}\right) \log_{10}(AB) = \frac{\log_{10}(A) + \log_{10}(B)}{2}.$$

A relationship between the arithmetic and geometric means can be derived as follows:

$$(\sqrt{A} - \sqrt{B})^2 = (\sqrt{A})^2 - 2(\sqrt{A})(\sqrt{B}) + (\sqrt{B})^2 = A + B - 2\sqrt{AB} \geq 0$$

$$\Rightarrow 0 < \min\{AB\} \leq \sqrt{AB} \leq \frac{A + B}{2} \leq \max\{A, B\},$$

with equality holding only if $A = B$.

Finally, we examine the fourth row. Here, $T_e = \lambda/(\rho R_0)$ and $V_e = (R_0 - 1)\rho/\beta$. Hence,

$$T_e = \frac{(2019)(21)}{21.7} \frac{\text{cells}}{\text{ml}} = 1.95 \times 10^3 \frac{\text{cells}}{\text{ml}},$$

$$V_e = \frac{20.7}{(21)(3.25 \times 10^{-7})} \text{ RNA copies/ml} = 3.033 \times 10^6 \text{ RNA copies/ml}$$

$$\Rightarrow \log_{10}(3.033 \times 10^6) = 6 + \log_{10}(3.033) = 6.48.$$

This self-consistency between theory and experiment as exhibited in Table 16.4 is of utmost importance in the biomedical sciences and represents our Pulling another Rabbit Out of a Hat!

About a year and a half after his talk, Rick informed me that he was ready to start writing up these results with Elissa and me as his co-authors and she had suggested we submit our manuscript to the Infectious Diseases section of the journal *Frontiers in Microbiology*. Frontiers journals have an Authors Contribution section, under which Rick stated that: RAC led the project, performed model analysis, ran numerical simulations, and wrote the paper. EJS initiated the project, gathered data for application of the model, and assisted with the interpretation of results and writing the paper. DJW introduced the non-cytopathic assumption, performed model analysis, and assisted with writing the paper. After that paper had been written, we did as Elissa suggested and submitted it to *Frontiers in Microbiology*. Frontiers journals provide an interactive and sequential review process in which referees' reviews are sent directly to the authors by email and they respond to them with modifications of their manuscript in compliance with the referees' suggestions. Even though there were three referees and three rounds of reviews, it was an amazingly efficient process, with our paper being received on November 1, 2017; accepted in revised form on January 10, 2018; and published online on January 31, 2018. The most significant modifications involved the requested listing of other non-cytopathic viruses, a detailed discussion of EIAV in respect to HIV, and the defense of the assumed normalization of the initial condition for the sum of the uninfected and infected target-cell population densities. In addition, the handling editor asked us to explain the least squares parameter identification procedure more fully. Frontiers journals only publish online and being open-access, require an Article Processing Charge (APC) paid before publication which we negotiated to be $1200 with Rick covering $800 of it and Elissa and me, $200 each, that caused a slight delay in our article being published after its acceptance. Further, Frontiers journals actually list the referees and their affiliations in the final version of published articles. In that way, we discovered the identity of our third referee to be Alan Perelson of the Los Alamos National Laboratory, a colleague of Lee Segel and one of the leading experts in virology and immunology.

Since Jeffrey Beall, a librarian at the University of Colorado, Denver, added Frontiers to his blacklist of predatory scientific journal publishers in October of 2015, due to concerns about the quality of their peer review process, it is only fair for us to point out that we found Frontiers' peer review process to be of the highest quality. There are basically two types of methodologies by which scientific journals generate revenue: Namely, open access or subscription. At issue here, in my opinion, was that Beall favored the latter over the former, given that all the journals making his list (which he discontinued on January 15, 2017) were open access. For what it is worth, I like to see articles published with an open access option which allows researchers to download them free of charge. Nothing seems more unreasonable to me than to have my papers in certain subscription journals only available online for a fee. We applaud *Frontiers in Microbiology*'s open access policy and are very glad to have chosen to submit our article to them for possible publication.

After its publication, Rick presented two talks having the same title as our paper, "A quasi-steady-state approximation to the basic target-cell-limited viral dynamics model with a non-cytopathic effect": One, as part of a special session co-organized by us for the 2018 PNW MAA Sectional Meeting held at Seattle University on April 21, 2018; and the other, a virtual presentation at the 3rd International Conference on Applied Microbiology and Beneficial Microbes held in Osaka, Japan, on June 6, 2018. In this context, I presented a preliminary report, as the final talk in our Seattle University 2018 PNW MAA Sectional Meeting special session, on Jeans' criterion for gravitational instabilities with uniform rotation, co-authored by me, my Masters Degree student Kohl Gill, and Bonni, which is the subject of the next chapter.

17

Jeans' Criterion for Gravitational Instabilities with Uniform Rotation: Linear Stability Analysis

I first became aware of Jeans' criterion for gravitational instabilities when refereeing a paper which was extending the model involving a clean gas to one that took cosmic dust into account. That paper started with the relevant linear perturbation equations for this situation. I wanted to see the basic system of governing equations and its exact solution that was being perturbed. Toward this end, I consulted Chandrasekhar ([30]) and realized that Jeans' ([87, 88]) original analysis was flawed, as described below. Strangely enough, adding uniform rotation to the system removed this flaw. That problem has interested me ever since. When Kohl Gill, who was my Masters student, asked me if I had an astrophysical problem he might do for his thesis, Jeans' criterion for gravitational instabilities with uniform rotation immediately came to mind. An MS thesis, unlike a Ph.D. dissertation, could be historical, expository, or critical. Thus, I thought that having Kohl examine the proper development of this problem in a systematic manner would be perfect for the purposes of his thesis, not realizing in so doing he would make a major contribution to the literature.

Consider the governing equations for a self-gravitational adiabatic inviscid fluid of infinite extent undergoing uniform rotation (Chandrasekhar, [30]):

Continuity Equation: $\dfrac{D\rho}{Dt} + \rho \boldsymbol{\nabla} \cdot \boldsymbol{v} = 0;$

Euler's Equation: $\dfrac{D\boldsymbol{v}}{Dt} + 2\boldsymbol{\Omega} \times \boldsymbol{v} + \boldsymbol{\Omega} \times (\boldsymbol{\Omega} \times \boldsymbol{r}) = -\rho^{-1}\wp'(\rho)\boldsymbol{\nabla}\rho + \boldsymbol{g};$

Poisson's Equation: $\boldsymbol{\nabla} \cdot \boldsymbol{g} = -4\pi G_0 \rho.$

Here $t \equiv$ time, $\boldsymbol{r} = (x, y, z) \equiv$ position vector, $\boldsymbol{\Omega} = (0, 0, \Omega_0) \equiv$ uniform rotation vector, $\rho \equiv$ density (mass/unit volume), $\boldsymbol{v} = (u, v, w) \equiv$ velocity vector with respect to the rotating frame, $\boldsymbol{\nabla} = (\partial/\partial x, \partial/\partial y, \partial/\partial z) \equiv$ gradient operator, $D/Dt = \partial/\partial t + \boldsymbol{v} \cdot \boldsymbol{\nabla} \equiv$ material derivative, $\wp(\rho) = p_0(\rho/\rho_0)^{\gamma_0} \equiv$ adiabatic pressure, $\boldsymbol{g} = -\boldsymbol{\nabla}\varphi \equiv$ gravitational acceleration vector with $\varphi \equiv$ self-gravitating potential, and $G_0 \equiv$ universal gravitational constant. The continuity and Euler's equations follow from the conservation of mass and momentum for an inviscid fluid (Lin and Segel, [119]), with the addition of the extra second and third terms on the left-hand side of the latter, which represent the Coriolis effect and centrifugal force, respectively, due to the rotation (Greenspan, [74]). Poisson's equation can be derived from the divergence theorem and Newton's law of gravitation (Binney and Tremaine, [14]) as follows:

Consider this special case of the divergence theorem for $\boldsymbol{g}$ (Segel, [215]):

$$\iint_S \boldsymbol{g} \cdot \boldsymbol{e}_a \, d\sigma = \iiint_R \boldsymbol{\nabla} \cdot \boldsymbol{g} \, d\tau;$$

where $R \equiv$ a small spherical region of radius a, centered about a point in space with bounding surface S possessing unit outward-pointing normal $\boldsymbol{e}_a$. Since by Newton's universal

DOI: 10.1201/9781003195603-17

law of gravitation, $\boldsymbol{g} = -(G_0 M_a / a^2)\boldsymbol{e}_a$ on S, where M_a denotes the mass in this spherical region,

$$\iint_S \boldsymbol{g} \cdot \boldsymbol{e}_a \, d\sigma = -\frac{G_0 M_a}{a^2} \iint_S \boldsymbol{e}_a \cdot \boldsymbol{e}_a \, d\sigma = -\frac{G_0 M_a}{a^2} \iint_S d\sigma$$
$$= -\frac{G_0 M_a}{a^2} 4\pi a^2 = -4\pi G_0 M_a.$$

Given the relationship between M_a and the density ρ

$$M_a = \iiint_R \rho \, d\tau,$$

this special case of the divergence theorem then implies that

$$\iiint_R (\boldsymbol{\nabla} \cdot \boldsymbol{g} + 4\pi G_0 \rho) \, d\tau = 0.$$

Now, employing the integral mean value theorem and cancelling the common spherical volume factor of $(4\pi a^3 / 3)$, this yields (Segel, [215])

$$(\boldsymbol{\nabla} \cdot \boldsymbol{g})^* = -4\pi G_0 \rho^*,$$

where the asterisk denotes the quantity in question being evaluated at an intermediate point in R. Finally, taking the limit as $a \to 0$, results in Poisson's equation for any point in space

$$\boldsymbol{\nabla} \cdot \boldsymbol{g} = -4\pi G_0 \rho.$$

Noting that the cross product (Segel, [215]) is defined by $\boldsymbol{r} \times \boldsymbol{v} \equiv (yw - zv, zu - xw, xv - yu)$:

$$\boldsymbol{\Omega} \times \boldsymbol{v} = \Omega_0(-v, u, 0),$$
$$\boldsymbol{\Omega} \times (\boldsymbol{\Omega} \times \boldsymbol{r}) = (\boldsymbol{\Omega} \cdot \boldsymbol{r})\boldsymbol{\Omega} - (\boldsymbol{\Omega} \cdot \boldsymbol{\Omega})\boldsymbol{r} = -\Omega_0^2(x, y, 0),$$
$$\boldsymbol{\nabla} \cdot \boldsymbol{g} = -\nabla^2 \varphi.$$

Thus, Euler's and Poisson's equations become

Euler's Equation: $\dfrac{D\boldsymbol{v}}{Dt} + 2\Omega_0(-v, u, 0) - \Omega_0^2(x, y, 0) = -\rho^{-1}\wp'(\rho)\boldsymbol{\nabla}\rho - \boldsymbol{\nabla}\varphi;$

Poisson's Equation: $\nabla^2 \varphi = 4\pi G_0 \rho$ where $\nabla^2 \equiv \boldsymbol{\nabla} \cdot \boldsymbol{\nabla}$.

Sir James Jeans ([87, 88]) proposed that a gravitational instability mechanism occurring in the spiral arms of protogalactic nebulae could result in the formation of chains of condensations, which eventually developed into those stars visible in the outer regions of fully evolved galaxies. He suggested that a nonrotating self-gravitating unbounded interstellar cloud of adiabatic gas, which is initially uniform in density and quiescent, should undergo an instability mechanism of this sort when acted on by random infinitesimal perturbations. Jeans deduced a criterion for which such an interstellar cloud would exhibit a gravitational instability, by performing a linear stability analysis on what he assumed to be an exact solution to his governing inviscid gas dynamical model system that was equivalent to our governing equations in the absence of rotation, arriving at the following secular equation satisfied by σ and λ, the growth rate and wavelength, respectively, of his small density fluctuations ([87, 88])

$$\sigma^2 = 4\pi \left(G_0 \rho_0 - \frac{\pi c_0^2}{\lambda^2} \right),$$

where $c_0^2 = \wp'(\rho_0) \equiv$ the speed of sound squared in an adiabatic medium of uniform density ρ_0.

This relation differed from that for the propagation of sound in a homogeneous medium only due to the presence of the gravity term. Then Jeans concluded that there would be instability corresponding to $\sigma^2 > 0$ provided

$$\lambda > \lambda_J = c_0 \sqrt{\frac{\pi}{G_0\rho_0}} \equiv \text{ Jeans' length,}$$

which is known as Jeans' criterion for gravitational instabilities.

The trouble with this derivation was that Jeans first represented his exact static solution to those governing equations symbolically as $\boldsymbol{v} \equiv \boldsymbol{0} = (0,0,0)$, $\rho = \rho_0$, $\varphi = \varphi_0$. Since this analysis was for a nonrotating system with $\Omega_0 = 0$, when he assumed in addition that ρ_0 was uniform to make his perturbation equations constant coefficient, this implicitly required $\nabla\varphi_0 = \boldsymbol{0}$ which implied $\nabla^2\varphi_0 = 0 = 4\pi G_0\rho_0$ or $\rho_0 = 0$ and hence was termed "Jeans swindle" by Binney and Tremaine ([14]). Kiessling ([97]) refuted their claim that Jeans' derivation represented a "swindle" because it could be justified by taking the proper limit of the appropriate cosmological model (see below).

Since spectroscopic evidence (Rubin and Ford, [195]) ultimately showed these nebulae to be rotating, Subrahmanyan Chandrasekhar ([30]) considered the effect of adding uniform rotation to Jeans' governing system of perturbation equations and repeated the latter's analysis, thereby demonstrating that its stability behavior was slightly more complicated, in that it involved an extra instability condition, as well as Jeans' criterion. Chandrasekhar's perturbation analysis suffered from the same deficiency as Jeans, in that he did not develop a parameter relationship for his implicit exact solution and thus treated ρ_0 and Ω_0 as independent. We shall demonstrate that the proper relationship between these parameters eliminates this extra condition and only yields Jeans' instability criterion. Many subsequent linear stability analyses of similar problems influenced by the methodology of these works have treated their associated perturbation systems independently of the actual exact solution of the governing equations and thus replicate this deficiency, including recent studies and reviews of gravitational instabilities (Stahler and Palla, [237]). Hence, we believed there was some merit in performing a systematic linear stability analysis of the relevant exact solution for Chandrasekhar's problem and Kohl and I completed an investigation of this sort. There exists an exact static homogeneous density solution of our basic equations of the form

$$\boldsymbol{v} \equiv \boldsymbol{0} = (0,0,0), \ \rho \equiv \rho_0, \ \varphi = \varphi_0; \text{ where } \varphi_0 \text{ satisfies}$$
$$\nabla\varphi_0 = \Omega_0^2(x,y,0), \ \nabla^2\varphi_0 = 4\pi G_0\rho_0$$
$$\Rightarrow \varphi_0(x,y) = \frac{\Omega_0^2(x^2+y^2)}{2} \text{ with } \Omega_0^2 = 2\pi G_0\rho_0 > 0.$$

Chandrasekhar ([30]), implicitly treating the static homogeneous density state as an exact solution, failed to detect this latter relationship between ρ_0 and Ω_0. Hence, he erroneously considered these two quantities as unrelated independent parameters, as stated above. This was a mistake that Kohl noticed almost immediately, which had not been explicitly pointed out by previous researchers and eluded me as well. In the process, he Pulled a Rabbit Out of a Hat and made a major contribution to the gravitational instability literature with his Masters thesis work! In what follows, we shall sketch that systematic linear stability analysis of this exact solution. Our analysis differs from the corresponding one of Chandrasekhar

([30]) only in that we shall be employing this relationship between ρ_0 and Ω_0, but as Robert Frost ended his famous poem *The Road Not Taken*: "And that has made all the difference."

We sought a linear perturbation solution of our basic equations of the form

$$\boldsymbol{v} = \varepsilon \boldsymbol{v}_1 + \boldsymbol{O}(\varepsilon^2), \ \rho = \rho_0[1 + \varepsilon s + O(\varepsilon^2)],$$

$$\varphi = \varphi_0 + \varepsilon \varphi_1 + O(\varepsilon^2) \text{ where } \boldsymbol{v}_1 = (u_1, v_1, w_1);$$

with $|\varepsilon| \ll 1$; substituted that solution into those equations; neglected terms of $O(\varepsilon^2)$; canceled the resulting ε common factor; and hence deduced that the perturbation quantities to the static homogeneous density state satisfied

$$\frac{\partial s}{\partial t} + \frac{\partial u_1}{\partial x} + \frac{\partial v_1}{\partial y} + \frac{\partial w_1}{\partial z} = 0;$$

$$\frac{\partial u_1}{\partial t} - 2\Omega_0 v_1 + c_0^2 \frac{\partial s}{\partial x} + \frac{\partial \varphi_1}{\partial x} = 0 \text{ where } c_0^2 = \wp'(\rho_0) = \gamma_0 \frac{p_0}{\rho_0} > 0;$$

$$\frac{\partial v_1}{\partial t} + 2\Omega_0 u_1 + c_0^2 \frac{\partial s}{\partial y} + \frac{\partial \varphi_1}{\partial y} = 0;$$

$$\frac{\partial w_1}{\partial t} + c_0^2 \frac{\partial s}{\partial z} + \frac{\partial \varphi_1}{\partial z} = 0;$$

$$2\Omega_0^2 s - \nabla^2 \varphi_1 = 0.$$

Then assuming a normal mode solution for these perturbation quantities of the form

$$[u_1, v_1, w_1, s, \varphi_1](x, y, z, t) = [A, B, C, E, F] e^{i(k_1 x + k_2 y + k_3 z) + \sigma t};$$

where $|A|^2 + |B|^2 + |C|^2 + |E|^2 + |F|^2 \neq 0$, $i = \sqrt{-1}$, and $k_{1,2,3} \in \mathbb{R}$ to satisfy the implicit far-field boundedness property for those quantities; and substituting them into our perturbation system, we obtained the following equations for $[A, B, C, E, F]$ upon cancellation of the exponential common factor:

$$ik_1 A + ik_2 B + ik_3 C + \sigma E = 0;$$

$$\sigma A - 2\Omega_0 B + ic_0^2 k_1 E + ik_1 F = 0;$$

$$2\Omega_0 A + \sigma B + ic_0^2 k_2 E + ik_2 F = 0;$$

$$\sigma C + ic_0^2 k_3 E + ik_3 F = 0;$$

$$2\Omega_0^2 E + k^2 F = 0 \text{ where } k^2 = k_1^2 + k_2^2 + k_3^2.$$

Employing the Pat Monroe algorithm (see Chapter 3) by setting the determinant of the 5×5 coefficient matrix for this linear homogeneous system of constants equal to zero to satisfy their nontriviality property, we obtained

$$k^2[\sigma^4 + (c_0^2 k^2 + 2\Omega_0^2)\sigma^2] + 4\Omega_0^2(c_0^2 k^2 - 2\Omega_0^2)k_3^2 = 0.$$

Defining the wavenumber vector $\boldsymbol{k} = (k_1, k_2, k_3)$, its dot product with $\boldsymbol{\Omega}$ satisfied

$$\boldsymbol{k} \cdot \boldsymbol{\Omega} = k_3 \Omega_0 = |\boldsymbol{k}||\boldsymbol{\Omega}| \cos(\theta) = k \Omega_0 \cos(\theta),$$

θ being the azimuthal angle between $\boldsymbol{k}$ and $\boldsymbol{\Omega}$; which implied that

$$k_3 = k \cos(\theta).$$

Then, substitution of this relationship for k_3 into our previous result and cancellation of k^2, yielded the secular equation

$$\sigma^4 + (c_0^2 k^2 + 2\Omega_0^2)\sigma^2 + 4\Omega_0^2(c_0^2 k^2 - 2\Omega_0^2)\cos^2(\theta) = 0.$$

Since this secular equation was a quadratic in σ^2, we first demonstrated that $\sigma^2 \in \mathbb{R}$ by showing that its discriminant

$$\mathcal{D} = (c_0^2 k^2 + 2\Omega_0^2)^2 - 16\Omega_0^2(c_0^2 k^2 - 2\Omega_0^2)\cos^2(\theta) \geq 0.$$

We considered the two cases of $c_0^2 k^2 - 2\Omega_0^2 \leq 0$ and $c_0^2 k^2 - 2\Omega_0^2 > 0$ separately. For the former case, it was obvious while for the latter one it could be shown by Kohl's observation that

$$\mathcal{D} \geq (c_0^2 k^2 + 2\Omega_0^2)^2 - 16\Omega_0^2(c_0^2 k^2 - 2\Omega_0^2) = (c_0^2 k^2 - 6\Omega_0^2)^2,$$

which definitely represented the Pulling of a Rabbit Out of a Hat. Then for $\theta = \pi/2$, we concluded from our secular equation that

$$\sigma^2 = 0 \text{ or } \sigma^2 = -(c_0^2 k^2 + 2\Omega_0^2) < 0;$$

while for $\theta \neq \pi/2$, we recalled that the stability criteria governing such quadratics: Namely (again, see Chapter 3):

$$\omega^2 + a\omega + b = 0 \text{ with } \mathcal{D} = a^2 - 4b \geq 0; \ \omega < 0 \text{ if and only if } a, b > 0;$$

implied that

$$\sigma^2 < 0 \text{ if and only if } c_0^2 k^2 - 2\Omega_0^2 > 0.$$

Making an interpretation of these results, we deduced that there would only be $\sigma^2 > 0$ and hence unstable behavior provided

$$c_0^2 k^2 - 2\Omega_0^2 < 0,$$

which was equivalent to Jeans' original gravitational instability criterion

$$\lambda > \lambda_J = c_0 \sqrt{\frac{\pi}{G_0 \rho_0}} \equiv \text{ Jeans' length, since } k = \frac{2\pi}{\lambda} \text{ and } \Omega_0^2 = 2\pi G_0 \rho_0.$$

In writing his criterion, Jeans implicitly had to assume that $\rho_0 > 0$, which seems plausible, in that $\rho_0 = 0$ corresponds to a completely empty space (Scheffler and Elsásser, [203]). When gravity is taken into account in the absence of rotation however, such an assumption was not strictly compatible with the equations of hydrostatic equilibrium as we have seen. Thus, Jeans' uniform density solution, as mentioned earlier, was not exact. The problem under examination demonstrates that adding rotation to the system, as Chandrasekhar did, and again performing a standard linear stability analysis of its exact static solution yields Jeans' instability criterion, but in a systematic manner and such a model also has the added advantage of being more astrophysically realistic. Talk about Pulling a Rabbit Out of a Hat! Jeans got the right answer for the wrong reason, as was shown in Kiessling ([97]) by taking the proper limit of the appropriate cosmological model to fix that analysis (see the discussion below). In his review of hydrodynamic stability theory, Lee Segel ([212]) stated that, "Anyone can get the right answer for the right reason. It takes a genius or a physicist to get the right answer for the wrong reason." In this context, Sir James Jeans was both!

The formula for λ_J is of fundamental importance in astrophysics and cosmology, where many significant deductions concerning the formation of galaxies and stars have been based

FIGURE 17.1
Galaxy Evolution Explorer image of the Andromeda galaxy M31, courtesy NASA/JPL Caltech.

upon it. In particular, Jeans' interpretation of the criterion, now bearing his name, was that a gas cloud of characteristic dimension much greater than λ_J would tend to form condensations with mean distance of separation comparable to λ_J that then developed into those protostars observable in the outer arms of spiral galaxies, such as Andromeda M31 (see Fig. 17.1). Sekimura *et al.* ([220]) have demonstrated that, for a secular equation similar in form to Jeans' of $\sigma = \sqrt{2\Omega_0^2 - c^2 k^2}$, λ_J actually corresponds to the so-called critical wavelength λ_c of linear stability theory associated with $\sigma = 0$, while nonlinear stability analyses of physical phenomena involving related secular equations have shown that the observed wavelengths are determined to a close approximation by that λ_c rather than by the dominant wavelength λ_d at which σ achieves its maximum value from linear theory (see Fig. 17.2). Hence Jeans' interpretation, although unusual for linear stability theory (where it is often presumed that such a disturbance associated with the largest growth rate predominates), both anticipated and is consistent with these nonlinear results, since, by the time perturbations have grown enough for the effect of the maximum growth rate to be observed, the neglected nonlinearities may have rendered that linear analysis inaccurate (Segel and Stoeckly, [214]). In this context, note that for a typical value of $\theta \neq \pi/2$, namely $\theta = 0$, we can factor our secular equation to obtain the roots $\sigma^2 = -4\Omega_0^2$ and $\sigma^2 = 2\Omega_0^2 - c_0^2 k^2$. Observe that this last condition, which yields our instability, is equivalent to Jeans' secular equation. For other $\theta \neq \pi/2$, the root of our secular equation which yields the instability behaves in a similar manner.

Thus, using the formula for Jeans' length λ_J with the parameters c_0 and ρ_0 assigned the values

$$c_0 = \left(\frac{2}{3}\right) \times 10^4 \frac{\text{cm}}{\text{sec}} \text{ and } \rho_0 = 10^{-22} \frac{\text{gm}}{\text{cm}^3}$$

employed by Jeans ([88]) for this purpose, but when the polytropic index $\gamma_0 = 4/3$ (Bonnor, [16]) while taking

$$G_0 = 6.67 \times 10^{-8} \frac{\text{cm}^3}{\text{gm sec}^2}$$

in cgs units, yields a Rabbit Pulled Out of a Hat of

$$\lambda_J = 4.58 \times 10^{18} \text{ cm } = 1.48 \text{ pc}$$

where 1 pc $\equiv 3.09 \times 10^{18}$ cm, which compares quite favorably with the mean distance between actual adjacent condensations originally formed in the outer arms of Andromeda since, in those parts of M31, the averaged observed distance between protostars in such chains is about 1.4 pc or somewhat more if allowances are made for foreshortening ([88])! Here pc $\equiv$ parsec, which is an acronym for *par*allax *sec*ond, that represents the distance at which the mean radius of the Earth's orbit subtends an angle of one second of arc and is used in the parallax measurements of large distances to astronomical objects outside the Solar System. Hence, it is a unit of length and not time, which was famously misused in the 1977 Star Wars movie when Hans Solo claimed his Millennium Falcon space ship, "made the Kessel Run in less than 12 parsecs," although George Lucas later tried to justify this error by saying Solo instead was referring to the shorter route under the standard run distance that he was able to travel by skirting the nearby Maw black hole cluster.

Using the definition for Jeans' mass, as that contained in a sphere of diameter λ_J (Binney and Tremaine, [14]), we found that

$$M_J = \frac{4\pi}{3} \left(\frac{\lambda_J}{2}\right)^3 \rho_0 = \frac{\pi}{6} \lambda_J^3 \rho_0 = 5 \times 10^{30} \text{ kgm},$$

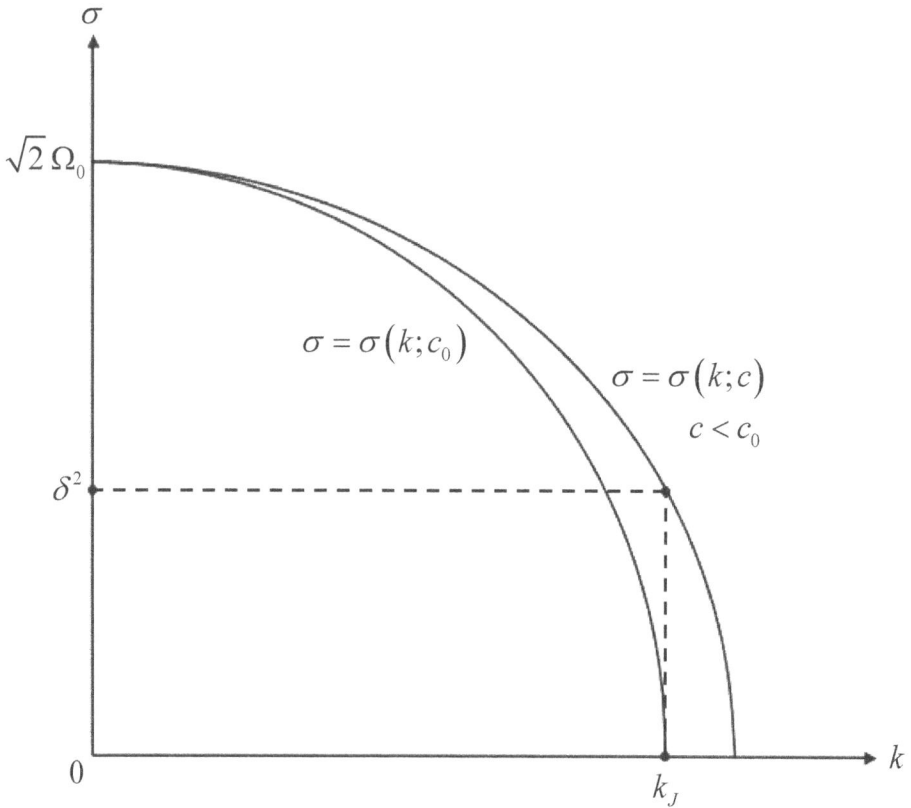

FIGURE 17.2

Schematic plots in the k-σ plane depicting the methodology employed by Sekimura *et al.* ([220]) applied to the Jeans' secular equation $\sigma = \sigma(k; c) = \sqrt{2\Omega_0^2 - c^2 k^2}$. That curve is plotted for both a general speed of sound c and our specific speed $c_0 > c$ in this figure, where $k_J = 2\pi/\lambda_J$ is such that $\sigma(k_J; c_0) = 0$. In a weakly nonlinear stability analysis, one takes the disturbance wavenumber $k \equiv k_J$ and its growth rate to be equal to $\sigma_J(c) = \sigma(k_J; c) = \delta^2 > 0$, where c is close enough to c_0 so that $\delta^2 \ll 1$ (see Fig. 8.3a). Then, in $\lim_{c \to c_0} \sigma_J(c) = 0$, which is a requirement for the application of weakly nonlinear stability theory and any re-equilibrated pattern will exhibit a wavelength of λ_J. Here, $c^2 = \gamma p_0/\rho_0$ and hence the operation $\lim_{c \to c_0}$ is equivalent to $\lim_{\gamma \to \gamma_0}$.

which represents the mass surrounding each of these condensations as compared with the mass of the sun $M_\odot = 2 \times 10^{30}$ kgm or exactly two and one-half solar masses. This differed slightly from Jeans ([88]) original critical mass defined by

$$M_c = \lambda_J^3 \rho_0 = 9.6 \times 10^{30} \text{ kgm such that for } M > M_c,$$

a gas cloud of mass M would undergo gravitational collapse. In this context Fig. 17.3 depicts a Hertzsprung-Russell-type diagram which plots stellar mass in units of $M_\odot$ on a log scale versus star color for Population I stars found in the outer arms of spiral galaxies, such as the Milky Way. Here, these stars are on the so-called "main sequence" of stellar evolution. The form of this figure was motivated by one appearing as an illustration in the original *Life* magazine issue containing the last part "The Starry Universe" of Lincoln Barnett's "The World We Live In," that was serialized in *Life* from 1952 to 1954, each of its 13 parts coming out approximately every two months. For some inexplicable reason, when "The World We Live In" was published in book form, the *Life* editorial staff chose to omit this illustration, which also included Population II stars found in the interior of the Milky Way, much to my consternation, since it has always fascinated me.

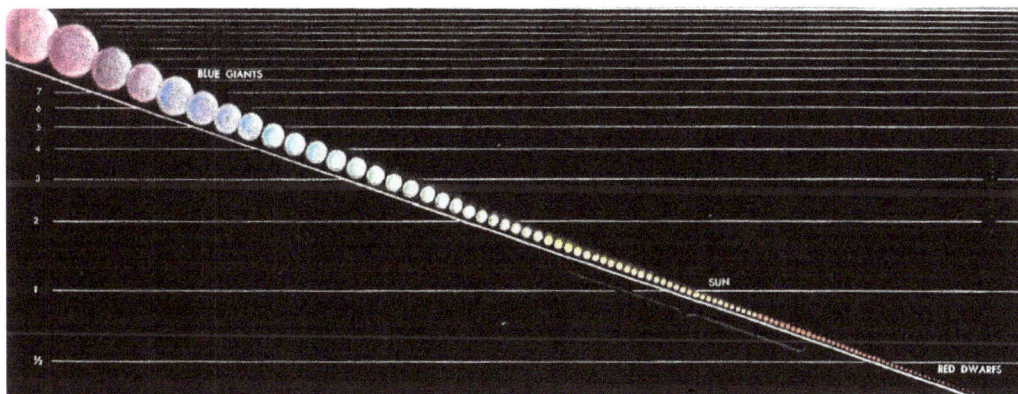

FIGURE 17.3
Plot of size versus color for Population I stars found in the outer arms of spiral galaxies. Here, the scale on the vertical axis indicates size in terms of solar masses and that on the horizontal, color from blue at the left to red at the right with yellow between them. The spiral arm stars show a simple color-size relationship: The largest ones are blue giants and the smallest, red dwarfs; while the sun-types of intermediate size are yellow dwarfs indicated by the bracket.

Given the small size of ρ_0, Chandrasekhar ([30]) was one of those individuals who regarded Jeans' analysis as a close approximation to reality (Scheffler and Elsässer, [203]). Although he oriented his axes so that $\boldsymbol{\Omega} = (0, \Omega_y, \Omega_z)$ with $|\boldsymbol{\Omega}| = \Omega_0$ and $\boldsymbol{k} = (0, 0, k)$, using our more general orientation Chandrasekhar ([30]), in effect, considered uniform rotation Ω_0 in his perturbation equations through the Coriolis force terms of the second and third perturbation equations in order to make the model more realistic, while retaining the coefficient $4\pi\rho_0 G_0$ for s in the fifth one. In so doing, he implicitly assumed that Ω_0 and ρ_0 were independent rather than related parameters. Chandrasekhar ([30]) also took $\sigma = i\omega$ and plotted $\omega^2 = -\sigma^2$ versus k for $\theta = 0$, $\pi/4$, $\pi/2$ and $\Lambda^2 \equiv \Omega_0^2/(\pi G_0 \rho_0) = 0.5$, 1.0, 2.0. Besides Jeans' criterion for $\theta \neq \pi/2$, this yielded an extra extraneous instability criterion for the case of

$\theta = \pi/2$ (see Fig. 17.4):

$$c_0^2 k^2 < 4(\pi G_0 \rho_0 - \Omega_0^2) \text{ should } \Omega_0^2 < \pi G_0 \rho_0.$$

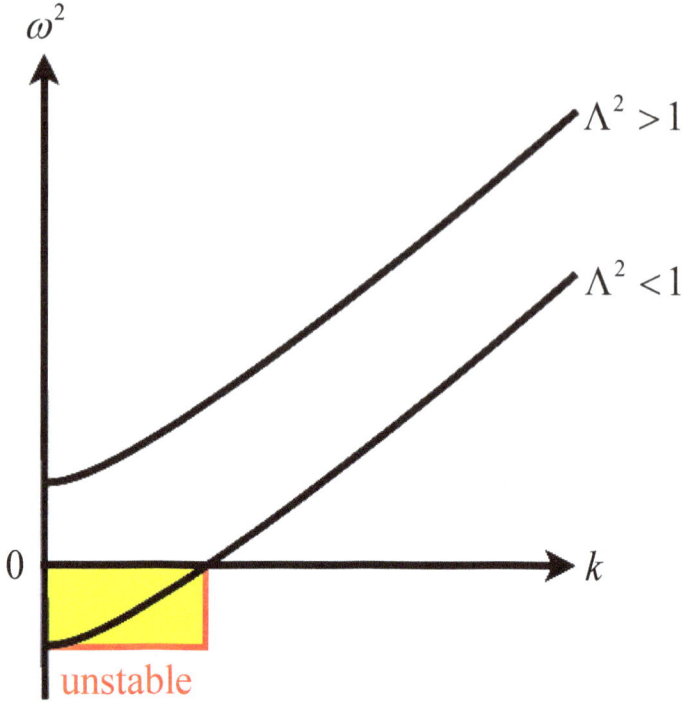

FIGURE 17.4

Plots of $\omega^2 = -\sigma^2 = c_0^2 k^2 + 4(\Omega_0^2 - \pi G \rho_0)$ versus k for Chandrasekhar's dispersion relation when $\theta = \pi/2$ and $\Lambda^2 < 1.0$ or $\Lambda^2 > 1.0$, the lower or upper curves, respectively. Note that $\Lambda^2 = 0.5$ and 2.0 serve as representative values for these two cases, separated by $\Lambda^2 = 1.0$. Here, Chandrasekhar ([30]) was employing Poincaré's ([176]) original estimate of $0 < \Lambda^2 \leq 2.0$.

In point of fact, $\Lambda^2 = 0.5$ is a representative value of that quantity for this instability condition, while $\Lambda^2 = 2.0$, his upper bound, actually corresponds to its value, as per our formula, relating these parameters which implies

$$\Omega_0 = \sqrt{2\pi \rho_0 G_0}.$$

Let us examine the plausibility of this relation, which violates the extra instability criterion identically. In conjunction with the values for ρ_0 and G_0, it yields the uniform rotation

$$\Omega_0 = \frac{6.47 \times 10^{-15}}{\text{sec}}$$

and the corresponding rotational velocity

$$V_0 = R_0 \Omega_0 = 200 \frac{\text{km}}{\text{sec}},$$

for the reference radial distance of

$$R_0 = 1 \text{ kpc } = 10^3 \text{ pc } = 3.09 \times 10^{21} \text{ cm } = 3.09 \times 10^{16} \text{ km},$$

both of which were consistent with the spectroscopic measurements of the Andromeda neb-ula and the observational data of the spiral Milky Way galaxy, in that they nearly coincided with the coordinates of the point of intersection of the rotational curves for these galaxies plotted in Fig. 17.5. We considered such a fortuitous occurrence to be just another Rabbit Pulled Out of a Hat!

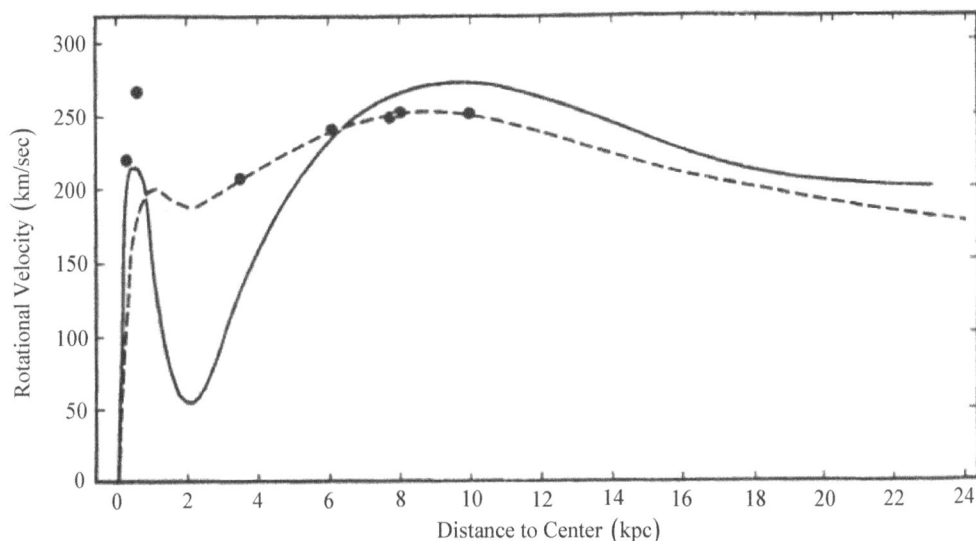

FIGURE 17.5
Comparison of rotational velocities V_0 (km/sec) of Andromeda M31 and the Milky Way Galaxy, as functions of radial distance R_0 (kpc) to their centers from Rubin and Ford ([195]) where the solid and dashed curves represent these plots for M31 and the Galaxy (Schmidt, [204]), respectively. The filled circles are the 7 observed rotational velocities for the Galaxy adopted by Roogour and Oort ([189]) to make the features of the latter conform with those of Andromeda. *Reproduced by permission of the American Astronomical Society which holds the copyright.*

In conclusion, our development presented a systematic linear stability analysis of Chan-drasekhar's ([30]) gravitational instability model in the presence of uniform rotation. We closed by noting that Binney and Tremaine ([14]) considered this gravitational instability model in a cylindrical rotating system as a problem in Chapter 5 of their book "Galac-tic Dynamics." They observed that rotation allowed the Jeans' instability to be analyzed exactly. Since the first part of their problem was to find the condition on Ω_0 so that the homogeneous quiescent gas would be in equilibrium, Binney and Tremaine ([14]) did not examine the plausibility of this condition. Further, the last part of their problem was to show, upon finding the resulting secular equation from its linear stability analysis, that waves propagating perpendicular to the rotation vector were always stable, while those propagating parallel to it were unstable if and only if the usual Jeans' criterion without rotation was satisfied. Although the latter conclusion for $\theta = 0$ agrees with our predictions, the former does not since, when $\theta = \pi/2$, we predicted $\sigma^2 = 0$, as well as those $\sigma^2 < 0$ which only implies a condition of neutral stability. Our results demonstrate that the best way to test the validity of a model for a natural science phenomenon is to compare its theoretical predictions with observable data of this phenomenon. Sir Arthur Conan Doyle characterized that philosophy probably as well as anyone by a Sherlock Holmes quote from A Scandal in Bohemia in his 1891 collection entitled *The Adventures of Sherlock Holmes*:

> It is a capital mistake to theorize before one has data. Insensibly one begins to twist facts to suit theories, instead of theories to suit facts.

Kohl defended his Masters thesis entitled "Jeans' instability" on April 8, 2014, in a talk that outlined his derivation of Jeans' criterion with uniform rotation along the lines of this chapter's development through the determination of the formula for Jeans' length λ_J. Incorporated into our present exposition were additional materials included in capstone problem 22.2 from Wollkind and Dichone ([284]), two presentations at professional meetings, and Gill *et al.* ([66]) in *Involve*. Let us explain the reason for a five-year delay between his thesis defense and that paper. In preparing problem 22.2 for our Springer textbook we re-examined Chandrasekhar's ([30]) derivation of Jeans' criterion with uniform rotation and realized that he had treated ρ_0 and Ω_0 as independent parameters, which convinced us of the significance of Kohl's results. Hence, given our experience with the publication of Davis *et al.* ([45]), we decided to submit a paper on these results with Kohl as our first author to *Involve*. It took almost a year for this paper to be refereed and that review's primary objection would have been almost impossible to anticipate in advance.

This objection was simply that we had mis-characterized the contribution of Kiessling ([97]) and, in the process, did not give it the proper credit for justifying "Jeans swindle," a term the reviewer stated Binney and Tremaine ([14]) had popularized in error. Kiessling's paper claimed to have justified "Jeans swindle" by taking the limit of a cosmological model as its cosmological constant went to zero. We had referenced this paper in relation to "Jeans swindle," in that it did a good job of tracing the history of what was considered a controversy by us. The reviewer felt such a reference implied an endorsement rather than a refutation of the term "Jeans swindle" and wanted us to correct that misconception. In the same vein, the referee also objected to our assertion that adding uniform rotation to the system rectified the flaw in Jeans' analysis because it still was invalid in the limit as $\rho_0 \to 0$ since then $\Omega_0 \to 0$ and said only an approach, which yielded Jeans' criterion in this limit as Kiessling's ([97]) supposedly did, could make that claim. There also was a laundry list of complaints, which can be catalogued as follows: The equations should all be numbered; the derivation of Poisson's equation should be removed; a complete description of Chandrasekhar's dispersion relation in comparison with our secular equation which had been included should be eliminated; some evidence that Chandrasekhar had not corrected nor anyone else noticed his error before we did should be offered; and a defense that patterns of critical wavelength λ_J would occur rather than those of dominant wavelength λ_d should be provided.

Bonni thought this review unreasonable and felt we should withdraw the paper, while Kohl found the model analysis contained in Kiessling ([97]) to be unconvincing, and I considered it to be tangential to the goal of our paper which was, after all, to correct Chandrasekhar's analysis and not Jeans, since Chandrasekhar had already done that by introducing uniform rotation. The items included in the reviewer's laundry list could be handled easily, in my opinion, so I began a revision of our original manuscript by making these modifications: All the equations were numbered; the derivation of Poisson's equation and the analysis of Chandrasekhar's dispersion relation were removed; and descriptions of Chandrasekhar's ([30]) dispersion relation plots, of Binney and Tremaine's ([14]) problem in their Chapter 5, and of a new Fig. 17.2 were added. Finally, the term "Jeans' swindle" was attributed to Binney and Tremaine ([14]), when first introduced, and the reference to Kiessling ([97]) deferred until the end of the revision with:

> Kiessling ([97]) disputes the claim of Binney and Tremaine ([14]) that Jeans' derivation represents a swindle because it can be justified by taking the proper limit of the appropriate cosmological model. This argument is tangential to our development which presents a systematic linear stability analysis of Chandrasekhar's ([30]) gravitational instability model in the presence of uniform rotation. In any event, Jeans' derivation was based on a linear stability analysis of a standard gas dynamical model at a time when the cosmological model used in that argument had yet to be developed.

We then submitted that revision to *Involve*. Having heard nothing for almost six months, I inquired about its status and very soon received the reviewer's comments. The referee did feel the revision was a marked improvement on the original manuscript in all but one respect: Namely, the Kiessling ([97]) matter, which the reviewer wanted brought up as soon as the attribution of "Jeans swindle" to Binney and Tremaine ([14]) was introduced with a statement that Kiessling ([97]) refuted rather than disputed it, which we did with:

> Kiessling ([97]) refuted their claim that Jeans' derivation represented a swindle because it could be justified by taking the proper limit of the appropriate cosmological model.

Before Segel's quote, this individual wanted us to state that, "Jeans got the right answer for the wrong reason, as was shown in Kiessling ([97]), by taking the proper limit of the appropriate cosmological model to fix that analysis." No other mentions were to be included.

Since the initial review contained an endorsement of Kiessling ([97]) that the referee felt we had not taken seriously enough in making the first revision, this individual decided to include a copy of that first review with the second one, while reiterating its relevant portion to convince us of the importance of Kiessling's contribution and the necessity of our including those modifications exactly as suggested above in the next revision if we wished to have the paper published in *Involve*. To let this review speak for itself, we now reproduce that relevant portion of it below, where our statement to be contrasted was "Jeans got the right answer for the wrong reason":

> It's instructive to contrast this statement with the following statement, taken from the Fields Medal-winning paper by Mouhot and Villani on the Landau damping of density disturbances of spatially uniform gases with electrical or gravitational interactions: "[T]he Jeans swindle [is] a trick considered as efficient but logically absurd. In 2003, Kiessling re-opened the case and acquitted Jeans, on the basis that his 'swindle' can be justified by a simple limit procedure, similar to the one presented above;..."

In explanation of this commentary, the reviewer was referring here to the paper Mouhot and Villani ([151]) "On Landau damping," the title of which, in the context of our astrophysical model, is a reference to its nonlinear global behavior in the parameter range where $\lambda < \lambda_J$. Since there is no Nobel Prize offered in Mathematics, the Fields Medal awarded to mathematicians under the age of 40 is often considered to be the equivalence in prestige of such a prize and it was Cédric Villani who had been awarded a 2010 Fields Medal for his proofs of nonlinear Landau damping and convergence to equilibrium in the Boltzmann equation, a fundamental and initially very controversial theory of gases in classical physics. It is helpful for us now to provide the background of this partial quote from Mouhot and Villani

([151]). That material comes from a section of their paper previewing its main mathematical modeling results and is preceded by a discussion of the occurrence of a screening process in which a limiting procedure on Poisson's equation is involved for electrical interactions that does not occur with gravitational ones. Whenever a quote is followed by "...," it is instructive to examine the complete relevant quote with its omitted portion retained, which is what we do next by reproducing the entire quote from Mouhot and Villani ([151]) below:

> In the case of galactic dynamics there is no screening; however it is customary to remove the zeroth-order term of the density. This is known as the Jeans swindle, a trick considered as efficient but logically absurd. In 2003, Kiessling reopened the case and acquitted Jeans, on the basis that his "swindle" can be justified by a simple limit procedure, similar to the one presented above; *however, the physical basis for the limit is less transparent and subject to debate. For our purposes, it does not matter much: since anyway periodic boundary conditions are not realistic in a cosmological setting, we may just as well say that we adopt the Jeans swindle as a simple phenomenological model.*

Note that the omitted material, italicized by us, was a Rabbit Pulled Out of a Hat, changing the whole meaning of the paragraph, which is now no longer the ringing endorsement of Kiessling ([97]) it seemed to be, as was intended by the referee in offering only a partial quotation. There is a word for this and that word is "disingenuous" which has the definition of not being candid or sincere, typically by pretending to know less about something than one really does. Nevertheless, having learned my lesson as related in Chapter 6, we made the suggested modifications exactly as described above and that paper was accepted for publication in *Involve* on the day it was received.

Our dissemination continued with two presentations at the following professional meetings:

- A systematic development of Jeans' criterion with rotation for gravitational instabilities, PNW MAA Section Meeting, Seattle, WA, April 21, 2018;

- An exact derivation of Jeans' criterion with rotation for gravitational instabilities, PNW SIAM Meeting, Seattle, WA, October 19, 2019.

The first presentation, described as a preliminary report at the end of Chapter 16, had the title of our *Involve* paper and was based upon its initial manuscript submitted to that journal. As such, it included the derivation of Poisson's equation and Figure 17.1 of this chapter. The second presentation, which occurred after the publication of our *Involve* paper, excluded the Poisson's equation derivation but included all the five figures appearing in this chapter, plus the concept of Jeans' mass M_J, that was not in the paper because, when we tried to add it, the deadline for making such modifications had already passed. I originally submitted this presentation to the 2019 PNW SIAM Meeting as a contributed paper. When making such a submission, SIAM requests that authors suggest which, if any, of the accepted minisymposia for a meeting might be suitable for their contribution. Then the organizers of such suggested minisymposia decide if they want to add that contribution to their minisymposium. In this case, we suggested the minsymposium entitled Fluid Mechanics: Systems and Models organized by Ralph Showalter, who was mentioned in Chapter 8, now at Oregon State University, and also on the organizing committee of the meeting, and he accepted our contribution for inclusion in the third part of his minisymposium. That meant we now had 30 instead of 15 minutes alloted for our presentation. I made full use of my time and helped Ralph by chairing this part of the minisymposium, introducing each talk.

Incidentally both of these meetings were hosted by the Mathematics Department at Seattle University and I have nothing but praise for the way they carried out their responsibilities in this regard. First of all, the registration fee was only $80 and for that they provided refreshments, a boxed lunch, and a buffet dinner. In comparison the 2016 Joint Mathematics Meetings, which were held in Seattle at the Washington State Convention Center, charged a $480 registration fee for which absolutely nothing was provided at all! Finally, each meeting room was equipped with state of the art presentation facilities including easy to follow instructions for accessing them and individuals offering technical assistance if it were needed.

I shall finish this chapter with some comments about Kohl Gill. He is simply one of the most impressive and enthusiastic graduate students with whom I have ever had the privilege of working. I first met Kohl when he enrolled in my graduate classes. Soon thereafter, he began coming to my office and we engaged in numerous discussions both scientific and philosophical. I started using him as a sounding board for my research ideas, a procedure described in Chapter 1, the importance of which is difficult to overemphasize. In addition, one can not work in a university environment without conflicts and intrigues occurring. Kohl always seemed to react to these things exactly as I did and supportive behavior of this sort can be very helpful for one's mental health. Further, he liked to collect scientific books, another thing Kohl and I had in common. There is a book by Louis Brand ([20]) entitled "Vector and Tensor Analysis," published by John Wiley, which I had used when deriving boundary conditions for surfaces of discontinuity that was out of print. Kohl kept checking for it and finally two copies became available, which he conjectured came from estate sales. We each bought one of them. This may be a trivial example of our interaction but nonetheless one of significance. During the intervening years between his thesis defense and the dissemination of those results, I always felt a vague sense of guilt, which was finally assuaged by his appreciation of Bonni's and my efforts in getting that research published in *Involve* and presented at the two professional meetings described above. This made that long wait seem worthwhile!

All the slide shows for these two presentations, as well as both figures for that *Involve* paper were created by Bonni, who performed a yeoman service in this respect, as she has done for our joint research ever since we began collaborating, which is something that will be discussed further in the concluding next chapter.

18

Conclusions

The purpose of this chapter is to synthesize a number of topics introduced previously, before it concludes by presenting a famous fable for Ph.D. graduate students and their thesis advisors. We shall begin with some additional information on Lee Segel's initial forays into mathematical biology.

During my postdoctoral period, Lee spent a year at the Sloan-Kettering Institute in New York City working with Evelyn Keller on what would become the Keller-Segel ([94]) paper modeling slime mold aggregation (see Chapter 14). One day while visiting him, he asked me if I knew of an algorithm that guaranteed linear stability for problems satisfying quadratic secular equations. Having just encountered such a situation (see Chapter 3), I was able to answer that question immediately. It also was the first time I knew something Lee did not and that is a sobering occurrence for Ph.D. students. Further, being there at the inception of his slime mold research, as it were, I became very interested in this problem. Having a preprint of the paper provided by Lee, I reworked all its analysis as a prelude to presenting it as a phenomenological example in the second semester of my WSU continuum mechanics class. While I was preparing those lessons, the chair of the department Calvin Long, in passing by my office, asked me what I was doing. After I told him, he wanted to know if this would result in any published research. When I responded with, "Not immediately," Cal, visibly upset, said untenured faculty could not afford to waste so much time on their classes. Such an attitude, it seems to me, is extremely shortsighted. Eventually this scholarship did lead to publications (see Chapters 7 and 14), as well as a chapter in Wollkind and Dichone ([284]), that probably would never have been completed without intimate knowledge of Keller and Segel ([94]). In this context, there was a figure in its appendix that I had the occasion to ask Lee about years later (see Fig. 18.1). It is a slightly different version of our Fig. 2.10 but in which a maximum and minimum point, denoted by A and B, respectively, are identified. Although point A is correct, the minimum does not occur at point B, which is a midpoint of the boundary between vertices. Rather, it actually occurs at the vertices themselves. Lee said he didn't know that at the time. This is when Segel proclaimed, "Good doctoral students always think they can improve upon their thesis advisers' results, isn't that right, David?," as related in Chapter 7.

Segel and Jackson ([214]) considered the first occurrence of diffusive instabilities in an ecological context. They examined a general predator–prey interaction–dispersion model system and demonstrated by means of linear stability theory that spatially uniform steady-states, which would have been stable for homogeneously distributed populations, could be destabilized through the introduction of dispersal effects. As usual, the occurrence of these ecological diffusive instabilities depended on the activator prey species exhibiting a positive feedback or Allee effect at equilibrium and having a much lower motility than the inhibitory predaceous one. Then, Segel and Levin ([217]) performed a weakly nonlinear stability analysis on a certain one-dimensional interaction–dispersion system of this type and showed that a new stable nonuniform stationary pattern would emerge following the destabilization of its spatially uniform steady-state. They employed the iterative method of

DOI: 10.1201/9781003195603-18

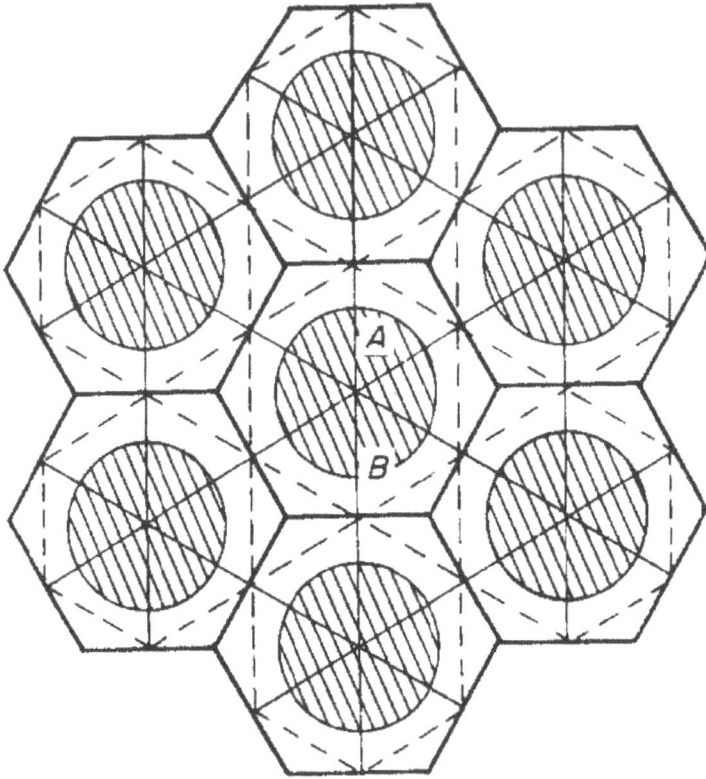

FIGURE 18.1

Superposition of three sinusoidal perturbations 120° apart, each of which has its maximum value of 1 along a light solid line and minimum value of −1 along a dashed one, gives a hexagonal pattern. At A, three maxima coincide to give a resultant largest value of 3. At B, a maximum coincides with two minima giving a value of −1, which is below the mean of 0. Reinforcement and cancellation of this kind lead to values which are greater than the mean in the shaded "cloud" regions and less than the mean in the unshaded regions. Compare with Fig. 2.10a, where its point O corresponds to point A in this figure. Here, point B really represents the maximum on the boundary, while the minimum there, of −1.50 occurring at the vertices, is actually the resultant smallest value of the whole configuration, as shown in Fig. 2.10b.

nonlinear stability theory that besides the alloy solidification problem Segel had offered to me as a possible thesis topic. Levin and Segel ([117]) suggested that diffusive instabilities might explain instances of spatial clumpings for natural communities restricted to two component variables, such as a zooplankton-phytoplankton ecosystem. These results, reviewed by Okubo and Levin ([158]), helped Lee change his mind about mathematical modeling of ecological phenomena not being as important currently as similar endeavors related to developmental biology such as for slime mold.

After I retired, Bonni and I began writing our textbook *Comprehensive Applied Mathematical Modeling in the Natural and Engineering Sciences: Theoretical Predictions Compared with Data*. The basic theme of such modeling is that its theoretical predictions are compared with observational or experimental data from the phenomenon under investigation. The book's main purpose was to demonstrate this process by introducing various case studies. These were arranged in increasing order of complexity of the mathematical methods from advanced calculus and differential equations used to analyze those models, with the methods being developed in parallel to the models by so-called "pastoral interludes" to preserve the goal of a self-contained presentation. Further, these models were deduced from scientific first principles. To reinforce and supplement the material introduced, problems were provided involving closely related case studies. This type of mathematical modeling not only includes the model of a phenomenon but also mappings back and forth to the phenomenon for the purpose of formulating and validating the model and these were emphasized as well. This book concentrated on both the derivation of the governing model equations using continuum mechanical and calculus of variational methods and the properties of the special functions appearing in the solutions of these models, also from first principles. Further, there was a concentration on the simplification procedures required to make model equations tractable without sacrificing their ability to generate theoretical predictions consistent with data. These included the selection of scale factors and introduction of nondimensional variabes and the application of perturbation techniques and asymptotic methods characteristic of linear and nonlinear stability theory, all of which were presented from first principles to make the book as self-contained as possible. Those nonlinear stability analyses were employed to allow an identification of regions in the relevant parameter space resulting from investigations of the model with stable patterns that are useful for comparison with data and difficult to accomplish using numerical simulation methods alone. The book contained 22 chapters and was 607 pages in length. I wrote the initial draft of these chapters in Microsoft Word and produced the required figures, while Bonni reformatted them in LaTeX, editing the text when necessary and embedding the accompanying figures in final form, which she generated. It concluded with six prospective capstone problems from areas of current interest, the first three of which involved virology, cosmology, and chemically reactive fluid flow that are the subjects of our Chapters 16, 17, and 3, respectively.

When thirteen and one-half chapters of this textbook were completed in its initial Word form, I attended the 2016 Joint Mathematics Meetings, as described in earlier chapters, with the intention of finding a publisher. After speaking to a variety of people from various publishing houses without much success, I finally went to the Springer booth at the book fair and met Donna Chernyk, who actually was an editor. My book packet contained a complete prospective Table of Contents, as well as the completed chapters and, unlike the previous individuals that talked with me, she went through them very carefully and showed real interest in what she considered a unique textbook. Further, whereas the other publishers only wanted sample chapters before making a publication decision, Donna wanted the whole book saying, "You wouldn't submit a partial paper to a journal for publication, would you?" and wished to know when we could send her a complete copy. I submitted a Springer book proposal form to her after the meeting and we remained in touch by email. Finally, the book was completed and we submitted it to her. During the review process she issued us a contract and after a series of reviews, resulting in some modifications, the book was accepted and went into production. Before it was printed, Bonni got to meet Donna at the next Joint Mathematics Meetings and they got along famously. Donna constructed a mock-up copy of the book with a paper cover, which she said generated a lot of interest and gave it to Bonni as a souvenir after the meeting was over. Then, the book was finally published later that spring and has done very well. In many respects Pulling Rabbits Out of Hats is a companion volume to that Springer textbook, Wollkind and Dichone ([284]), which itself was an amalgamation of material included in the five graduate courses I created during my 45 years at WSU and that Bonni took from me while there.

Since my retirement, besides Davis *et al.* ([45]), Gill *et al.* ([66]), and these books, Bonni and I have been engaged in two other joint research projects; one, already completed and the paper from it published, while the other is an ongoing effort. We shall describe them in the rest of this chapter, given that they both are closely related to subject matter introduced previously. The first project was concerned with a DMS (Department of Mathematical Sciences) NSF (National Science Foundation) CURM (Center for Undergraduate Research in Mathematics) Grant to Bonni. Earlier, she had asked me to help her choose a problem for such a CUMR proposal. I had suggested a rhombic planform nonlinear stability analysis of the nondimensional spatio-temporal model evolution equation for $h(x, y, t)$, a dimensionless deviation of a thin solid surface from its mean planar position, where $t \equiv$ time and $(x, y) \equiv$ a transverse laboratory Cartesian coordinate system:

$$h_t + 4\nabla^2 h + 2\nabla^4 h + \beta[\sinh(2h) - \alpha h^2] = 0; \ \alpha \in \mathbb{R}, \ \beta > 0.$$

Here, α and β are dimensionless combinations of experimental and material parameters. Then, consistent with that nonlinear stability analysis, retaining only terms through third-order in

$$\sinh(2h) \sim (2h) + \frac{(2h)^3}{6} = 2h + \frac{4h^3}{3},$$

this equation is reduced to its modified Swift-Hohenberg ([243]) truncated form

$$h_t + 4\nabla^2 h + 2\nabla^4 h + \beta\left(2h - \alpha h^2 + \frac{4h^3}{3}\right) = 0.$$

Recall that Chapter 10 dealt with nonlinear stability analyses of such a model evolution equation. Bonni had two of her students, Sydney Schmidt and Stephanie Kolden, perform that analysis on this modified Swift-Hohenberg model equation and write up a report of their project. That report, while technically correct, did not contain any phenomenological applications of those theoretical results. In order to rectify this deficiency, I returned to the isotropic Kuramoto ([104])-Sivashinsky ([230]) equation of Chapter 9 for a thin solid film of dimensional thickness $H(r_1, r_2, \tau)$ undergoing normal-incidence ion-bombardment induced erosion where (r_1, r_2) represents a transverse laboratory Cartesian coordinate system and τ is time:

$$H_\tau + \nu\nabla_2^2 H + D\nabla_2^4 H + J_0 Y_0(H) = \frac{\lambda_0}{2}|\boldsymbol{\nabla}_2 H|^2 + \eta.$$

Recall $\nu \equiv$ the absolute value of the coefficient of negative capillarity; $D \equiv$ the effective surface diffusion coefficient; $J_0 \equiv$ the deterministic component of the ion flux; $Y_0(H) \equiv$ the sputtering yield; $\lambda_0 \equiv$ the tilt-dependent coefficient of the erosion rate; $\eta \equiv$ a noise term resulting from the stochastic component of the ion flux; $\boldsymbol{\nabla}_2 \equiv (\partial/\partial r_1, \partial/\partial r_2)$; $\nabla_2^2 \equiv \boldsymbol{\nabla}_2 \cdot \boldsymbol{\nabla}_2$;

and $\nabla_2^4 \equiv (\nabla_2^2)^2$.

Then, selecting $\ell_0 = (2D/\nu)^{1/2}$ and $\tau_0 = 8D/\nu^2$ as scale factors for length and time, respectively; introducing the dimensionless variables

$$(x,y) = \frac{(r_1, r_2)}{\ell_0}, \ t = \frac{\tau}{\tau_0}, \ h = \frac{H - H_0 + w_n\tau}{\ell_0} \text{ where } w_n = J_0 h_0 \ell_0^2;$$

adopting the sputtering yield relation, consistent with that of Makeev and Barabási ([132]),

$$Y_0(H) = h_0\ell_0^2 \left[1 + \left(\frac{h_0}{\ell_0}\right)\sinh(2h)\right] \Rightarrow J_0 Y_0(H) = w_n \left[1 + \left(\frac{h_0}{\ell_0}\right)\sinh(2h)\right];$$

employing the magnitude of the gradient approximation (Siegmann and Rubenfeld, [228])

$$|\nabla_2 H| \cong \frac{\Delta H}{\Delta r} = \frac{|H - (H_0 - w_n\tau)|}{\ell_0} = |h| = \sqrt{h^2},$$

so that its source term will ultimately be of a modified Swift-Hohenberg ([243]) form, in order to simplify our subsequent analyses (Wollkind *et al.*, [279]); and, after Kahng *et al.* ([91]), taking

$$\eta \equiv 0;$$

we obtained our original model equation

$$h_t + 4\nabla^2 h + 2\nabla^4 h + \beta[\sinh(2h) - \alpha h^2] = 0$$

where

$$\alpha = \frac{\lambda_0\ell_0}{2w_n h_0} = \frac{\lambda_0}{2J_0\ell_0 h_0^2} \in \mathbb{R}, \ \beta = \frac{\tau_0 w_n h_0}{\ell_0^2} = \tau_0 J_0 h_0^2 = \frac{8DJ_0 h_0^2}{\nu^2} > 0.$$

Finally, retention of terms through third-order in $\sinh(2h)$ again yielded the modified Swift-Hohenberg truncation of that equation (reviewed by Cross and Hohenberg, [42])

$$h_t + 4\nabla^2 h + 2\nabla^4 h + \beta\left(2h - \alpha h^2 + \frac{4h^3}{3}\right) = 0.$$

Hence, the magnitude of the gradient approximation represented a Rabbit Pulled Out of a Hat! Here, $h \equiv$ the nondimensional deviation of the interface from its dimensional mean planar position given by $H_0 - w_n\tau_0 t$ where $w_n \equiv$ the normal velocity of erosion and $H_0 \equiv$ initial dimensional uniform thickness of the solid layer; $h_0 \equiv$ the dimensional maximal interfacial deviation from that mean position; $\nabla \equiv (\partial/\partial x, \partial/\partial y)$; $\nabla^2 \equiv \nabla \cdot \nabla$; and $\nabla^4 \equiv (\nabla^2)^2$. Finally, we adopted the constitutive relation $D = D(T_S) = K(T_S) + D_0$ where $K(T_S) \equiv$ thermal surface diffusion coefficient $= K_0 \exp(-T_0/T_S)$ for $K_0, T_0 \equiv$ positive characteristic values and $T_S \equiv$ substrate temperature.

We next summarize the theoretical rhombic planform nonlinear stability analysis results obtained by Schmidt and Kolden ([205]). They found that the coefficients of their amplitude equations satisfied

$$\sigma_0(\beta) = 2(1 - \beta), \ a_1(\alpha) = 1 - \frac{19\alpha^2}{36}, \ b_1(\alpha, \varphi) = 2 - \frac{\alpha^2}{2}\frac{3 + 16\cos^4(\varphi)}{[4\cos^2(\varphi) - 1]^2};$$

where $\varphi \equiv$ the rhombic angle, and thus determined that

$a_1 > 0$ for $\alpha_1 < \alpha < \alpha_2$, $a_1 < 0$ for $\alpha < \alpha_1$ or $\alpha > \alpha_2$;

$$\text{with } \alpha_{2,1} = \pm\frac{6}{\sqrt{19}} = \pm 1.376.$$

Further, Schmidt and Kolden ([205]) predicted that when $0 < \beta < 1$ stable ripples (II) would only occur for $\alpha = 0$ or equivalently, $\lambda_0 = 0$; and for $0 < \alpha < \sqrt{108/73}$ there existed two φ-intervals (φ_m, φ_M) and $(\varphi_\ell, \varphi_r)$, flanking $\varphi = \pi/3$, with

$$0 < \varphi_m(\alpha) < \varphi_M(\alpha) < \frac{\pi}{3} < \varphi_\ell(\alpha) < \varphi_r(\alpha) \leq \frac{\pi}{2},$$

where rhombic patterns of these characteristic angles were stable versus ripples, while for $\sqrt{108/73} < \alpha < 6/\sqrt{19}$ only the (φ_m, φ_M) interval existed. These values are tabulated in Table 18.1.

TABLE 18.1
The φ-range for stable rhombic patterns of these characteristic angles versus α.

α	φ_m	φ_M	φ_ℓ	φ_r
0.1	1.006	1.023	1.071	1.088
0.2	0.963	0.999	1.094	1.127
0.3	0.919	0.974	1.118	1.167
0.4	0.873	0.947	1.142	1.206
0.5	0.826	0.918	1.167	1.245
0.6	0.775	0.887	1.194	1.286
0.7	0.722	0.853	1.223	1.329
0.8	0.664	0.813	1.255	1.376
0.9	0.602	0.768	1.292	1.432
1.0	0.534	0.712	1.377	1.523
1.1	0.456	0.641	1.396	1.571
1.2	0.363	0.544	1.504	1.571
1.3	0.238	0.385	None	None

Thus, since $h_{20} = \alpha\beta/[4(2 - \beta)]$, for $0 < \alpha < \alpha_2 = 6/\sqrt{19}$ ($\lambda_0 > 0$), islands (V$^+$) of the allowable characteristic angles were the only stable patterns predicted and, for $-6/\sqrt{19} = \alpha_1 < \alpha < 0$ ($\lambda_0 < 0$), holes (V$^-$) of these characteristic angles were the only stable patterns predicted. In addition, square patterns (S$^\pm$) of rhombic angle $\varphi = \pi/2$ could only occur for $1.014 = 6/\sqrt{35} < |\alpha| < \sqrt{108/73} = 1.216$. When $\beta > 1$, it was predicted that only the smooth planar surface (I) could occur. These morphological nonlinear stability predictions are represented diagrammatically in the α-β plane of Fig. 18.2.

Then employing the parameter values appropriate for normal-incidence ion sputtering

$$\nu = 2 \times 10^{-15} \frac{cm^2}{sec}, \ D = 0.8 \times 10^{-27} \frac{cm^4}{sec},$$
$$2\lambda_0 = 0.01 \frac{nm}{sec}, \ h_0 = 0.1 \ nm \ (nm \equiv 10^{-9} \ m);$$

consistent with those of Chason *et al.* ([31]) and Facsko *et al.* ([58]) for normal-incidence ion sputtering at $T_S = 423$ K, and applying these theoretical rhombic planform nonlinear stability analysis results obtained by Schmidt and Kolden ([205]) to the experimental outcomes described in Chapter 9, we obtained the identical interpretations as found in that chapter and thus were able to rectify the phenomenological deficiency of their report: Another Rabbit Pulled Out of a Hat!

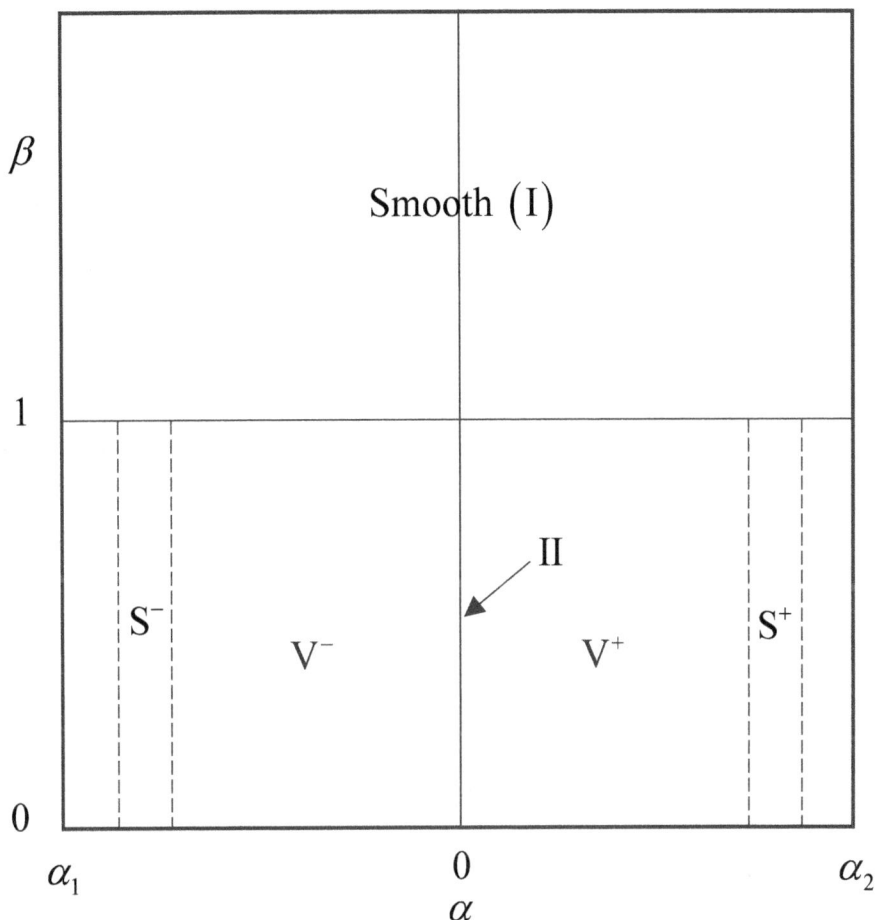

FIGURE 18.2

Morphological nonlinear stability diagram in the α-β plane for our modified Swift- Hohenberg type equation, identifying the predicted ion-sputtered erosion patterns. Here, the regions between the dotted lines indicate parameter ranges where stable rhombic patterns ($V^{\pm}$) of angle $\varphi = \pi/2$ or squares ($S^{\pm}$) can occur, while the region of stable ripples (II) is identified by an arrow and that of a stable smooth surface (I) by $\beta > 1$. In this context, V^+ and V^- patterns represent uniform distributions of islands and holes, respectively, while the S^+ and S^- patterns having rhombic angle $\varphi = \pi/2$ represent checkerboard arrays of mounds and pits, respectively.

Finally, Bonni and I submitted a paper entitled "Rhombic planform nonlinear stability analysis of an ion-sputtering evolution equation," with Sydney and Stephanie as our co-authors, to *Involve* for the purpose of NSF accountability. That paper which synthesized their theoretical results with our experimental interpretations and its six Bonni-generated figures had the following abstract:

A damped Kuramoto-Sivashinsky equation describing the deviation of an interface from its mean planar position during normal-incidence ion-sputtered erosion of a semiconductor or metallic solid surface is derived and the magnitude of the gradient in its source term approximated so that it will be of a modified Swift-Hohenberg form. Next, one-dimensional longitudinal and two-dimensional rhombic planform nonlinear stability analyses of the zero deviation solution to this equation are performed, the former being a special case of the latter. The predicted theoretical morphological stability results of these analyses are then shown to be in very good qualitative and quantitative agreement with relevant experimental evidence involving the occurrence of smooth surfaces, ripples, checkerboard arrays of pits, and uniform distributions of islands or holes once the concept of lower and higher threshold rhombic patterns is introduced based on mean interfacial position.

Indeed, it was in analyzing this model that I got the idea of basing the lower and higher threshold rhombic patterns originally deduced in Chapter 13 on the mean position of the interface as explained in Chapter 10 for both that problem and the one of Chapter 9. If that seems complicated, it was. Sometimes, you have to encounter the same Rabbit on a number of occasions before deciding how to Pull that Rabbit Out of the Hat and then, in retrospect, applying this principle to all of these problems simultaneously. As Segel said, "Once you have the right formulation everything falls into place automatically," or, to paraphrase David Oulton, "Works out just like a champ(ion)."

The two reviewers for this paper were much more prompt and reasonable than the reviewer of Gill *et al.* ([66]) had been. The first reviewer, Richard Cangelosi, who identified himself to me, wanted us to include a comparison of this paper with Pansuwan *et al.* ([166]) described in Chapter 9. That wasn't too hard for us to do and we did so by adding the following paragraph to our discussion:

Our results differ from those obtained by Pansuwan *et al.* ([166]) from their analysis of a related model equation for solid surface erosion caused by ion-sputtering that did not include the simplifying approximation for the magnitude of its gradient. That analysis basically employed the more complicated hexagonal planform method of weakly nonlinear stability theory (reviewed in Chapter 17 of Wollkind and Dichone, [284]) to study pattern formation in this phenomenon and used the easier to implement rhombic planform one to mediate these results by determining the parameter range for stable square patterns instead of generating all its pattern formation predictions, as we do, from our rhombic planform analysis. The reason for this was that since the Cangelosi *et al.* ([25]) paradigm for lower- and higher threshold patterns based upon the mean position of the interface, as we do here, had yet to be developed, the predicted zero-threshold results of their rhombic planform analysis could not be used to compare with experimental and simulated patterns. Our goal being to employ the simplest reasonable model and method of analysis that produce results in agreement with experimental data, we chose to perform that threshold-dependent rhombic planform analysis on the evolution equation developed in Section 1 rather than the corresponding more complex one of Pansuwan *et al.* ([166]). Further, since Facsko *et al.* ([57]) was published simultaneously with Pansuwan *et al.* ([166]), the latter authors used the existing stochastic model of Cuerno *et al.* ([44]) for solid surface erosion via normal-incidence ion sputtering to estimate the value of $\lambda_0/2$ which they took to be equal to 0.01 nm/sec, as opposed to our taking this as the value for $2\lambda_0$. Note that for the value of this parameter employed by Pansuwan *et al.* ([166]) the experiments of Facsko *et al.* ([58]) yielded $\alpha \cong 1.788$, which being greater than α_2 would lie outside our predicted pattern formation range of $0 < \alpha < \alpha_2 = 1.376$.

The second reviewer's most important suggested modification was for us to add a statement about the equivalence class of patterns associated with critical point II and which member of that class would actually be selected. We did so by means of an addition to the caption of our figure corresponding to Fig. 18.2, which explained that the other member of equivalence class II compared to the one of vertical ripples we selected to represent this class simply rotated its deviation function through an angle φ when compared to the latter. Thus, given the isotropic nature of our model evolution equation defined on an unbounded domain, there was no preferred direction and hence both these families of ripples were equally likely to occur (Sekimura *et al.*, [220]) with arbitrary initial conditions determining which of those orientations would actually be selected (Segel, [212]), *e.g.*, if $\varphi = \pi/2$ were selected these vertical ripples would be replaced by horizontal ones instead.

Unlike the case with some of the suggested modifications for Gill *et al.* ([66]), the addition of both these modifications improved our paper, which is the actual purpose of the peer review process when used appropriately. Having made these modifications, our revised manuscript was accepted for publication by *Involve* and recently appeared as Schmidt *et al.*, [206].

Our second ongoing project, in collaboration with Rick Cangelosi, is concerned with relaxing the assumption introduced in Chapter 16 that the EIAV-target cell interaction involved homogeneous populations uniformly distributed in space. After Stancevic *et al.* ([238]), we are now assuming the densities of uninfected or infected target cells and the concentration of our non-cytopathic virus to be heterogeneous functions of their transverse spatial coordinates (r_1, r_2) on planar anemic equine mucous membranes, as well as of time τ. That is, we consider:

$$T = T(r_1, r_2, \tau), \ I = I(r_1, r_2, \tau), \ V = V(r_1, r_2, \tau);$$

$$\frac{\partial T}{\partial \tau} = \lambda - \rho T - \beta TV + D_1 \nabla_2^2 T - \chi_1 \boldsymbol{\nabla}_2 \cdot (T \boldsymbol{\nabla}_2 I)$$

$$\text{where } \boldsymbol{\nabla}_2 \equiv \left(\frac{\partial}{\partial r_1}, \frac{\partial}{\partial r_2} \right) \text{ and } \nabla_2^2 = \boldsymbol{\nabla}_2 \cdot \boldsymbol{\nabla}_2,$$

$$\frac{\partial I}{\partial \tau} = \beta TV - \delta I + D_2 \nabla_2^2 I \text{ with } D_2 = D_1 \text{ and } \delta = \rho,$$

$$\frac{\partial V}{\partial \tau} = bI - \gamma V + D_3 \nabla_2^2 V.$$

Here, the dependent variables, as well as the coefficients of the interaction terms, are as defined in Chapter 16, while it is assumed that the uninfected target cells, infected target cells, and free virus diffuse with diffusion coefficients D_1, D_2, and D_3, respectively, all of which have units of mm^2/day, where for simplicity that motility is identified with Brownian motion. Further, we are assuming that the uninfected and infected target cells both diffuse at the same rate or $D_2 = D_1$ (Stancevic *et al.*, [238]). In addition, uninfected target cells are chemotactically attracted to, and move toward, concentration gradients in the infected target cells with coefficient χ_1 ($[mm^2 \times ml]/[cell \times day]$), since the latter recruit the former by signaling them through the release of chemokines (chemotactic cytokines)

producing inflammation at the site of the infection (Covaleda *et al.*, [40]; Sokol and Luster, [233]) which appears as petechial hemorrhages or minute blood-red spotted patterns on
the anemic equine mucous membranes for the chronic degree phase of EIAV infectiousness
(fs_equine_infectious_anemia.pdf; APHIS Factsheet, 2008).

Introducing the nondimensional variables

$$(x,y) = \frac{(r_1, r_2)}{\sqrt{D_1/\rho}}, \ t = \rho\tau; \ m = \frac{\rho T}{\lambda}, \ i = \frac{\rho I}{\lambda}, \ v = \frac{\beta V}{\rho};$$

our equations become

$$\frac{\partial m}{\partial t} = 1 - m - mv + \nabla^2 m - \chi \boldsymbol{\nabla} \boldsymbol{\cdot} (m \boldsymbol{\nabla} i) \text{ where } \chi = \frac{\lambda \chi_1}{\rho D_1},$$

$$\frac{\partial i}{\partial t} = mv - i + \nabla^2 i,$$

$$\varepsilon \frac{\partial v}{\partial t} = R_0 i - v + \varepsilon \mu \nabla^2 v \text{ where } \varepsilon = \frac{\rho}{\gamma}, \ R_0 = \frac{\lambda \beta b}{\rho^2 \gamma}, \text{ and } \mu = \frac{D_3}{D_1}.$$

Let (see Table 16.2)

$$\rho = \delta = \frac{1}{21} \text{ per day}, \ \gamma = 6.73 \text{ per day};$$

and (Stancevic *et al.*, 2013)

$$D_1 = 1.1 \frac{\mu m^2}{\text{sec}} = 0.09504 \frac{mm^2}{\text{day}}, \ D_3 = 0.0088 \frac{\mu m^2}{\text{sec}} = 0.00076 \frac{mm^2}{\text{day}};$$

where $\mu m \equiv 10^{-6} \text{ m} = 10^{-3} \text{ mm}$ and 1 day = 86,400 sec. Then, as in Chapter 16,

$$\varepsilon = 0.007 \text{ while } \mu = 0.008.$$

The fact that our small parameter ε of Chapter 16 appears in the viral equation diffusion
term times μ, of the same order of magnitude, represented a Rabbit Pulled Out of a Hat!
This more than allowed us again to employ a quasi-equilibrium approximation and obtain
the algebraic relation $v = R_0 i$ (see below). Then our basic equations reduced to the two-
component system

$$\frac{\partial m}{\partial t} = 1 - m - R_0 mi + \nabla^2 m - \chi \boldsymbol{\nabla} \boldsymbol{\cdot} (m \boldsymbol{\nabla} i), \ \frac{\partial i}{\partial t} = R_0 mi - i + \nabla^2 i;$$

where $\boldsymbol{\nabla} \equiv (\partial/\partial x, \partial/\partial y)$ and $\nabla^2 \equiv \boldsymbol{\nabla} \boldsymbol{\cdot} \boldsymbol{\nabla}$; which contains the basic reproductive number
R_0 and the nondimensional chemotaxis coefficient χ. Such a system is intrinsically much
simpler to analyze than the three-component one treated by Stancevic *et al.* ([238]). The
reason for this is that the secular equation produced by a linear stability analysis of our
system is a quadratic as opposed to the associated cubic equation which had to be examined by the latter authors since the Routh-Hurwitz criterion (see Chapters 1 and 3) for such
cubics is much more complex than the relatively simple one for a quadratic (Uspensky, [258]).

Recall from Chapter 16, there are two equilibrium solutions to this system in the absence
of spatial effects, $m \equiv m_e$, $i \equiv i_e$ where $m_e + i_e = 1$, given by $m_e = 1$, $i_e = 0$ or $m_e = 1/R_0$,
$i_e = 1 - 1/R_0$ such that the uninfected state is linearly stable for $0 < R_0 < 1$ and the
community equilibrium point, for $R_0 > 1$. It is the linear stability of these equilibrium
points to our full spatially dependent system with which we are concerned in what follows.

We begin by considering a one-dimensional linear stability solution of this system of the form

$$[m, i](x, y, t) = [m_e, i_e] + \varepsilon_1 [m_{11}, i_{11}] \cos(qx)e^{\sigma t} + \boldsymbol{O}(\varepsilon_1^2)$$

where

$$|\varepsilon_1| \ll 1 \text{ and } |m_{11}|^2 + |i_{11}|^2 \neq 0.$$

Noting that

$$mi = m_e i_e + \varepsilon_1 (m_e i_{11} + i_e m_{11}) \cos(qx)e^{\sigma t} + O(\varepsilon_1^2),$$

this yields a set of linear homogeneous algebraic equations in the constants m_{11} and i_{11} after substitution into that system, neglect of terms of $O(\varepsilon_1^2)$, and cancellation of the common factor:

$$(\sigma + 1 + R_0 i_e + q^2)m_{11} + m_e(R_0 - \chi q^2)i_{11} = 0,$$
$$-R_0 i_e m_{11} + (\sigma + 1 - R_0 m_e + q^2)i_{11} = 0;$$

which, upon imposition of the vanishing of the determinant of the matrix of its coefficients to guarantee the nontriviality property for these constants, results in the following quadratic in σ:

$$(\sigma + 1 + R_0 i_e + q^2)(\sigma + 1 - R_0 m_e + q^2) + R_0 i_e m_e (R_0 - \chi q^2) = 0.$$

Since Turing diffusive instabilities are defined to be those solutions that are stable in the absence of spatial effects but can become unstable if those effects are taken into account, we examine the behavior of this equation for both our equilibrium points sequentially.

For $m_e = 1$ and $i_e = 0$, we obtain

$$(\sigma + 1 + q^2)(\sigma + 1 - R_0 + q^2) = 0 \Rightarrow \sigma_1 = -1 - q^2 \text{ and } \sigma_2 = R_0 - 1 - q^2.$$

Thus since $\sigma_{1,2} < 0$ when $R_0 < 1$, there can be no diffusive instabilities for this solution.

For $m_e = 1/R_0$ and $i_e = (R_0 - 1)/R_0 > 0$, we obtain

$$\sigma^2 + (2q^2 + R_0)\sigma + (R_0 + q^2)q^2 + (R_0 - 1)\left(1 - \frac{\chi q^2}{R_0}\right) = 0,$$

which implies that $\text{Re}(\sigma_{1,2}) < 0$ if and only if

$$(R_0 + q^2)q^2 + (R_0 - 1)\left(1 - \frac{\chi q^2}{R_0}\right) > 0.$$

Hence, there will be a Turing diffusive instability for this solution provided

$$\chi > \chi_0(q^2; R_0) = \frac{R_0}{R_0 - 1}\left[R_0 + q^2 + \frac{R_0 - 1}{q^2}\right] \text{ where } R_0 > 1$$

and marginal stability when $\chi = \chi_0(q^2; R_0)$.

Since

$$\text{both } \lim_{q^2 \to 0} \chi_0(q^2; R_0), \ \lim_{q^2 \to \infty} \chi_0(q^2; R_0) \to \infty \text{ and } \chi_0(q^2; R_0) > 0,$$

this curve must have an absolute minimum at its critical point (q_c^2, χ_c) which satisfies

$$\frac{d\chi_0(q_c^2; R_0)}{dq^2} = 0 \Rightarrow q_c^2(R_0) = \sqrt{R_0 - 1} \text{ and } \chi_c(R_0) = \chi_0(q_c^2; R_0).$$

Hence for $0 < \chi < \chi_c$ there exists no q^2 associated with growing modes, while for $\chi > \chi_c$ there exists a band of such wavenumbers squared centered about $q^2 = q_c^2$. Therefore, this equilibrium point is linearly stable for $0 < \chi < \chi_c$, unstable for $\chi > \chi_c$, and neutrally stable for $\chi = \chi_c$. That marginal stability function is given by

$$\chi_c(R_0) = \frac{R_0}{R_0 - 1}[R_0 + 2\sqrt{R_0 - 1}] = \frac{R_0^2}{R_0 - 1} + \frac{2R_0}{\sqrt{R_0 - 1}}$$

and plotted in the R_0-χ plane of Fig. 18.3. Again since

$$\text{both } \lim_{R_0 \to 1} \chi_c(R_0), \ \lim_{R_0 \to \infty} \chi_c(R_0) \to \infty \text{ and } \chi_c(R_0) > 0,$$

that curve must have an absolute minimum at its critical point which is determined as follows:

$$\chi_c'(R_0) = \frac{2R_0}{R_0 - 1} - \frac{R_0^2}{(R_0 - 1)^2} - \frac{R_0}{(R_0 - 1)^{3/2}} + \frac{2}{(R_0 - 1)^{1/2}} = \frac{F(R_0)}{(R_0 - 1)^2}$$

where

$$F(R_0) = 2R_0(R_0 - 1) - R_0^2 - R_0\sqrt{R_0 - 1} + 2(R_0 - 1)^{3/2} = [R_0 + \sqrt{R_0 - 1}](R_0 - 2).$$

Observing that $F(R_0) < 0$ for $1 < R_0 < 2$, $F(2) = 0$, $F(R_0) > 0$ for $R_0 > 2$, and $\chi_c(2) = 8$, this minimum point is located at $(2,8)$ as indicated in Fig. 18.3. Note that $i_e = 0.5$ at $R_0 = 2$, while $0 < i_e < 0.5$ for $1 < R_0 < 2$ and $0.5 < i_e < 1$ for $R_0 > 2$ is a Rabbit Pulled Out of a Hat!

Finally, let us provide a justification for our quasi-equilibrium approximation to the basic governing partial differential equations for this EIAV-target cell interaction-diffusion-chemotaxis model. Performing a linear stability analysis of the equilibrium solution

$$m \equiv m_e, \ i \equiv i_e, \ v \equiv v_e \text{ where } i_e = 1 - m_e \text{ and } v_e = R_0 i_e$$

to that unreduced model:

$$\frac{\partial m}{\partial t} = 1 - m - mv + \nabla^2 m - \chi \nabla \cdot (m\nabla i), \ \frac{\partial i}{\partial t} = mv - i + \nabla^2 i,$$

$$\varepsilon \frac{\partial v}{\partial t} = R_0 i - v + \varepsilon \mu \nabla^2 v \text{ where } \varepsilon = \frac{\rho}{\gamma} \text{ and } \mu = \frac{D_3}{D_1};$$

we obtain the linear homogeneous algebraic system for the perturbation constants m_{11}, i_{11}, v_{11}:

$$(\sigma + 1 + v_e + q^2)m_{11} - m_e\chi q^2 i_{11} + m_e v_{11} = 0,$$
$$-v_e m_{11} + (\sigma + 1 + q^2)i_{11} - m_e v_{11} = 0,$$
$$0\, m_{11} - R_0 i_{11} + (\varepsilon\sigma + 1 + \varepsilon\mu q^2)v_{11} = 0;$$

where $|m_{11}|^2 + |i_{11}|^2 + |v_{11}|^2 \neq 0$, which, again upon imposition of the vanishing of the determinant of the matrix of its coefficients to guarantee this nontriviality property for these constants or the Pat Munroe algorithm, yields the following cubic in σ:

$$R_0[(\sigma + 1 + v_e + q^2)(\varepsilon\sigma + 1 + \varepsilon\mu q^2) + m_e v_e] +$$
$$(\varepsilon\sigma + 1 + \varepsilon\mu q^2)[(\sigma + 1 + v_e + q^2)(\sigma + 1 + q^2) - m_e v_e\chi q^2] = 0.$$

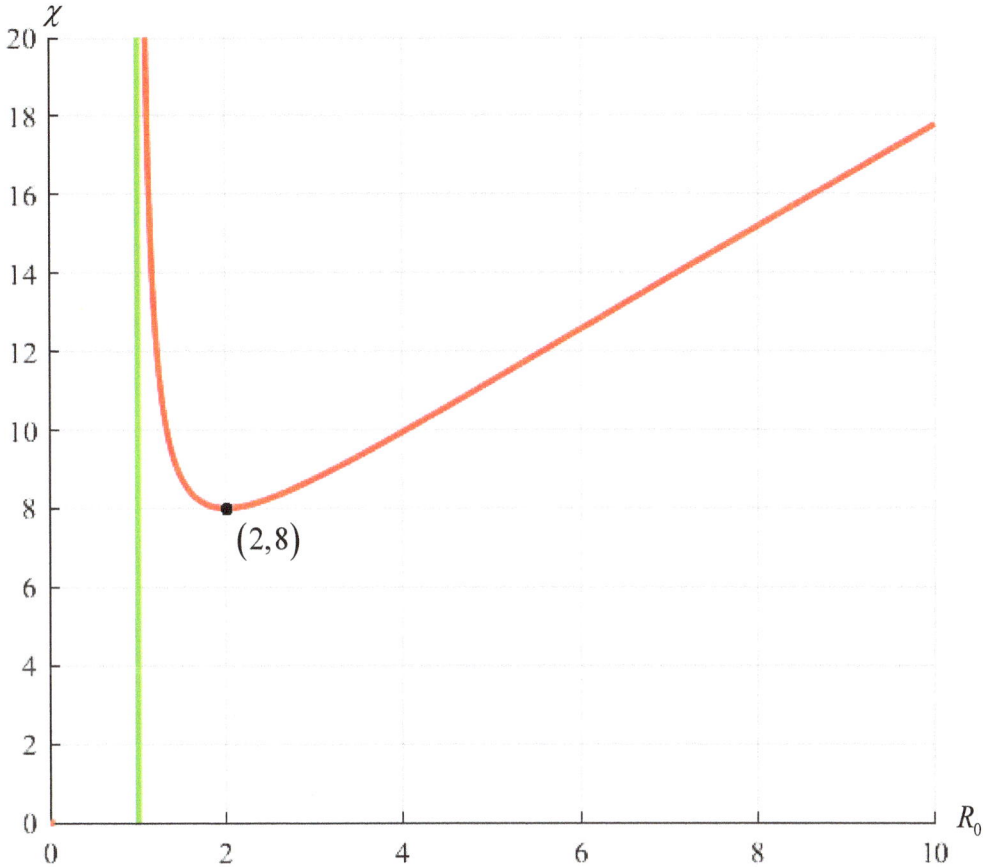

FIGURE 18.3
Plot designating the uninfected region for $0 < R_0 < 1$ and the marginal stability curve (red) $\chi = \chi_c(R_0)$ in the R_0-χ plane for $R_0 > 1$, which has the vertical asymptote (green line) $R_0 = 1$ and represents the infected region when $\chi > \chi_c$. Here, the patterning in that region is caused by the following mechanism: m-target cells are chemotactically attracted to, and move toward, concentration gradients in the i-target cells since the latter recruit the former by signaling them through the release of chemokines (chemotactic cytokines) producing inflammation at the site of the infection which appears as petechial hemorrhages or minute blood-red spotted patterns on the anemic equine mucous membranes for the chronic degree phase of EIAV infectiousness. Further, the homogenous region where $\chi < \chi_c$ is characterized by $i_e = 1 - 1/R_0 > 0$ which ranges from sparse ($1 < R_0 < 2$) to dense ($R_0 > 2$) as R_0 increases over its domain of definition for this community equilibrium point. Unlike Stancevic *et al.* ([238]), our marginal stability curve $\chi = \chi_c(R_0)$ was determined in closed form rather than obtained numerically, which is the advantage of our adopting a quasi-equilibrium system.

From this cubic, we can deduce that there are two roots $\sigma_{1,2} = O(1)$ as $\varepsilon \to 0$ which reduce when $\varepsilon = 0$ to those just examined for the linear stability analysis of the quasi-equilibrium system, while its third root $\sigma_3 \sim -1/\varepsilon$ as $\varepsilon \to 0$. Thus, with no loss of generality, should ε be small, we may consider the full system in the limit as $\varepsilon \to 0$ for our diffusive-instabilies analysis since the root neglected by that quasi-equilibrium approximation is then highly stabilizing. In this instance, recall that $\varepsilon = 0.007$ and $\mu = 0.008$. Hence, not only is ε small but μ is virtually of the same size as well, justifying the employment of the quasi-equilibrium approximation on the unreduced model, as mentioned earlier, which replaces its third governing partial differential equation by the algebraic relation $v = R_0 i$. Indeed, only $\mu = O(1)$ is required. Then substitution of this relation into its first two partial differential equations yields our quasi-equilibrium EIAV-target cell interaction-diffusion-chemotactic model system just analyzed for linear diffusive instabilities.

It remains for us to consider a rhombic planform solution to our quasi-equilibrium EIAV-target cell interaction-diffusion-chemotaxis model system involving $m = m(x, y, t)$ and $i = i(x, y, t)$ of the same form as that employed by Schmidt *et al.* [205] sketched earlier in this chapter. In particular, we shall take $\sigma_0 = \sigma_0(\chi; R_0)$ to be that root of our secular equation causing the diffusive instability when $\chi > \chi_c(R_0)$ and χ sufficiently close to χ_c, so that σ_0 will still be real when $\chi < \chi_c$. Then imposing the solvability conditions for our system as $\chi \to \chi_c$, we shall calculate the Landau coefficients $a_1 = a_1(R_0)$ and $b_1(R_0, \varphi)$, investigate where $a_1(R_0) > 0$, and in that range determine the characteristic rhombic angles φ predicted by this analysis. Further, observe that in this predicted periodic patterned region, the distance between adjacent patterns as measured by the critical wavelength $\lambda_c = 2\pi/\sqrt[4]{R_0-1}$ decreases as R_0 increases consistent with a morphological sequence ranging from sparse to dense. For our purposes, such patterns can be identified with the petechial hemorrhages occurring on the anemic equine mucous membranes during the chronic degree phase of EIAV infectiousness. When Rick and I first talked about extending our EIAV-target cell dynamical system to this chemotaxis-diffusion model, he was reluctant to do so until an application could be found for any potential predicted spatial patterns. This being the case, my finding that petechial hemorrhages occur on equine mucous membranes during EIAV was a Rabbit Pulled Out of a Hat! Here, as in the root suction problem of Chapter 14, the cross-diffusion effect of chemotaxis drives the pattern formation process and this is another example where my intimate knowledge of the Keller and Segel ([94]) results helped me formulate a problem. So much for Cal Long's fifty-year old admonishment. That is the trouble with short-range, as opposed to long-range, planning. As my late colleague Ed Pate used to say, "The trouble with university administrators is that their idea of long-range planning is where do we go for lunch." Incidentally, while we were working on Wollkind and Dichone ([284]), he kept urging me to write a monograph on pattern formation, which this book serves as, besides being a semi-autobiographical annotated account of my research career.

We have been identifying incidents of Pulling Rabbits Out of Hats. We conclude with the fable entitled "The Rabbit, the Fox, and the Wolf," that, rather than identifying what Rabbit is being Pulled Out of a Hat, as in the rest of the book, identifies instead what the Rabbit Pulls Out of the Hat and with which I ended my July 7, 2003, presentation at the Banff Workshop, organized by Simon Levin in honor of Lee Segel's 70th birthday.

The Rabbit, the Fox, and the Wolf: A Fable

One sunny day, a Rabbit came out of her burrow in the ground to enjoy the weather. The day was so nice that the Rabbit became careless, so a Fox sneaked up and caught her.

"I am going to eat you for lunch!" said the Fox.

"Wait!" replied the Rabbit. "You should at least wait a few days. I am just finishing writing my Ph.D. thesis."

"Hah! That's a stupid excuse. What is the title of your thesis, anyway?"

"I am writing my thesis on 'The Superiority of Rabbits over Foxes and Wolves'."

"Are you crazy? I should eat you right now! Everybody knows that a Fox will always defeat a Rabbit."

"Not really; not according to my research. If you like, you can come to my burrow and read it for yourself. If you are not convinced, then you can go ahead and eat me for lunch."

Since the Fox was curious and had nothing better to do, he went with the Rabbit into her burrow and never came back out.

A few days later, the Rabbit was again taking a break from writing, and sure enough, a Wolf came out of the bushes and was ready to eat her for dinner.

"Wait!" exclaimed the Rabbit. "You cannot eat me right now."

"And why might that be?" "I am almost finished writing my Ph.D. thesis on 'The Superiority of Rabbits over Foxes and Wolves'."

The Wolf laughed so hard that he almost lost his grip on the Rabbit.

"Come read it for yourself and you can eat me for dinner if you disagree with my conclusions."

So the Wolf went into the Rabbit's burrow and he never came back out either.

The Rabbit finished writing her thesis and was out celebrating in the lettuce fields. Another rabbit came by and asked, "What's up? You seem to be very happy."

"Yup, I just finished writing my dissertation."

"Congratulations! What is it about?"

"It is entitled 'The Superiority of Rabbits over Foxes and Wolves'."

"Are you sure? That doesn't sound quite right."

"Oh yes, you should come to my burrow and read it for yourself."

So they went together to the Rabbit's burrow. As they entered, the friend saw the typical graduate student abode, albeit rather messy after writing a thesis. The computer with the controversial dissertation was in the middle. To its right there was a pile of fox bones; to its left, a pile of wolf bones; and, in front of it, a Lion.

The moral of the fable is this: The title of a dissertation doesn't matter. The research that went into it is irrelevant. The only thing of importance is the identity of one's thesis advisor and my thesis advisor was Lee Segel.

Bibliography

[1] Ackemann, T., Lange, W. (2001) Optical pattern formation in alkali metal vapors. Appl. Phys. B, 72, 21–34.

[2] Ackemann, T., Logvin, Y.A., Heur, A., Lange, W. (1995) Transition between positive and negative hexagons in optical pattern formation. Phys. Rev. Lett., 75, 3450–3453.

[3] Ahlers, G. (1974) Low-temperature studies of the Rayleigh-Bénard instability and turbulence, Phys. Rev. Lett. 33, 1185–1188.

[4] Alexander, J.I.D., Wollkind, D.J., Sekerka, R.F. (1986) The effect of latent heat on weakly nonlinear morphological stability. J. Crystal Growth 79, 849–865.

[5] Anderson, R.M., May, R.M. (1992) Infectious diseases of humans: Dynamics and control. Oxford University Press, Oxford.

[6] Andrews, R.C. (1951) Nature's ways. Crown, New York.

[7] Aranson, I.S., Gorshkov, K.A., Lomov, A.S., Rabinovich, M.I. (1990) Stable particle-like solutions of multidimensional nonlinear fields. Physica D, 43, 435–453.

[8] Arrowsmith, D.K., Place, C.M. (1982) Ordinary differential equations. Chapman and Hall, London.

[9] Bazykin, A.D. (1976) Structural and dynamical stability of model predator-prey systems. Int. Inst. Appl. Syst. Analysis, Laxenburg, Austria.

[10] Bdzil, J.B., Frirsch, H.L. (1980) Chemically driven convection. J. Chem. Phys. 72, 1875–1886.

[11] Bellman, R., Kalaba R. (eds) (1964) Selected papers on mathematical trends in control theory. Dover, New York.

[12] Bénard, H. (1901) Les tourbillons cellulaires dans une nappe liquide transportant de la chaleur par convection en regime permanent. Annales de Chimie et de Physique 23, 62–144.

[13] Bennett, J.J.R., Sherratt, R.A. (2019) Large scale patterns in mussel beds: Spots or stipes? J. Math. Biol. 78, 315–335.

[14] Binney, J., Tremaine, S. (1987) Galactic dynamics, Princeton University Press, Princeton.

[15] Bischof, J., Scherer, D., Herminghaus, S., Leiderer, P. (1996) Dewetting modes of thin metallic films: Nucleation of holes and spinodal dewetting. Phys. Rev. Lett. 77, 1536–1539.

[16] Bonnor, W.B. (1957) Jeans' formula for gravitational instability. M.N.R.A.S. 117, 104–117.

[17] Boonkorkuea, N., Lenbury, Y., Alvarado, F.J., Wollkind, D.J. (2010) Nonlinear stability analyses of vegetative pattern formation in an arid environment. J. Biol.Dyn. 4, 346–380.

[18] Boyce, W.E., DiPrima, R.C. (2012) Elementary differential equations and boundary value problems. Wiley, New York.

[19] Boyle, NR, Morgan, JA (2009) Flux balance analysis of primary metabolism in *Chlamydomonas reinhardtii*. BMC Systems Biology. 3–4, 1–14.

[20] Brand, L. (1947) Vector and tensor anaslysis.Wiley, New York.

[21] Buckmaster, J.D., Nachman, A. (1978) The buckling and stretching of a viscida, II. Effects of surface tension. Quart. J. Mech. Appl. Math. 31, 157–168.

[22] Burg, D., Rong, L., Neumann, A.U., Dahari, H. (2009) Mathematical modeling of viral kinetics under immune control during primary HIV-1 infection. J. Theor. Biol. 259, 751–759.

[23] Cahn, J.W., Hilliard, J.E. (1958) Free energy of a nonuniform system. I. Interfacial free energy. J. Chem. Phys. 28, 258–267.

[24] Cangelosi, R.A., Schwartz, E.J., Wollkind, D.J. (2018) A quasi-steady-state approximation to the basic target-cell-limited viral dynamics model with a non-cytopathic effect. Frontiers in Microbiology 9, 54:1–6.

[25] Cangelosi, R.A., Wollkind, D.J., Kealy-Dichone, B.J., Chaiya, I. (2015) Nonlinear stability analyses of Turing patterns for a mussel-algae model. J. Math. Biol. 70, 1249–1294.

[26] Carver, D.H., Marcus, P.I., Seto, D.S.Y. (1967) Intrinsic interference: a unique interference system used in assaying non-cytopathic viruses. Archiv für die Gesamte Virusforschung 22, 55–60.

[27] Caughley, G. (1976) Plant-herbivore systems. In: May, R (ed) Theoretical ecology: Principles and applications. W.B. Saunders, Philadelphia, pp. 94–113.

[28] Chaiya, I., Wollkind, D.J., Cangelosi, R.A., Kealy-Dichone, B.J., Rattanakul, C. (2015) Vegetative rhombic pattern formation driven by root suction for an interaction-diffusion plant-ground water model system in an arid at environment. Am J. Plant Sci. 6, 1278–1300.

[29] Chandra, K. (1938) Instability of fluids heated from below, Proc. Roy. Soc. A 164, 231–242.

[30] Chandrasekhar, S. (1961) Hydrodynamic and hydromagnetic stability. Clarendon Press, Oxford.

[31] Chason, E., Mayer, T.M., Kellerman, B.K., McIlroy, D.T., Howard, A.J. (1994) Roughening instability and evolution of Ge(001) surface during ion sputtering. Phys. Rev. Lett. 72, 3040–3043.

[32] Chen, H.S., Jackson, K.A. (1971) Stability of a melting interface. Journal of Crystal Growth 8, 184–190.

[33] Chen, W., Ward, M.J. (2011) The stability and dynamics of localized spot patterns in the two-dimensional Gray-Scott model. SIAM J. Dyn. Syst. 10, 586–666.

[34] Clos-Arceduc, M. (1964) Estude sur photographies aériennes d'une formation végétale sahéleinne: In brousse tigrée. Bulletin de L'InstitutFrancais D'Afrique Noire. Serie A 18, 677–684.

[35] Collings, J.B., Wollkind, D.J. (1990a) A global analysis of a temperature-dependent model system for a mite predator-prey interaction. SIAM J. Appl. Math. 50, 1348–1372.

[36] Collings, J.B., Wollkind, D.J. (1990b) Metastability, hysteresis, and outbreaks in a temperature-dependent model for a mite predator-prey interaction. Mathl. Comput. Modelling. 13, 91–103.

[37] Collings, J.B., Wollkind, D.J., and Moody, M.E. (1990) Outbreaks and oscillations in a temperature dependent model for a mite predator-prey interaction. Theor. Pop. Biol. 38: 159–190.

[38] Couteron, P., Lejeune, O. (2001) Periodic spotted patterns in semi-arid vegetation explained by a propagation-inhibition model. J. Ecol. 89, 616–628.

[39] Couteron, P., Mahamane, A., Ouedraogo, P., Seghieri, J. (2000) Difference between banded thickets (tiger bush) at two sites in West Africa. Journal of Vegetative Science 11, 321–328.

[40] Covaleda, L., Fuller, F.J., Payne, S.L. (2010) EIAV S2 enhances pro-inflammatory cytokine and chemokine response in infected macrophages. Virology 397, 217–223.

[41] Covey, C.C., Schubert, G. (1981) Mesoscale convection in the clouds of Venus. Nature 290, 17–20.

[42] Cross, M.C., Hohenberg, P.C. (1993) Pattern formation outside of equilibrium. Rev. Mod. Phys. 65, 851–1112.

[43] Cuerno, R., Barabási, A-L. (1995) Dynamic scaling of ion-sputtered surfaces. Phys. Rev. Lett. 76, 4746–4749.

[44] Cuerno, R., Makse, H.A., Tomassone, S., Harrington, S.T., Stanley, H.E. (1995) Stochastic model for surface erosion via ion sputtering: Dynamical evolution from ripple morphology to rough morphology. Phys. Rev. Lett. 75, 4464–4467.

[45] Davis, M.G., Wollkind, D.J., Cangelosi, R.A., Kealy-Dichone, B.J. (2018) The behavior of a population interaction-diffusion equation in its subcritical regime. Involve 11, 297–310.

[46] Davis, S.H. (1987) Thermocapillary instabilities. Ann. Rev. Fluid Mech. 19, 403–435.

[47] Davis, S.H., Homsy, G.M. (1980) Energy stability theory for free surface problems: buoyancy-thermocapillary layers. J. of Fluid Mech. 98, 527–553.

[48] Deblauwe, V., Couteron, P., Lejeune, O., Bogaert, J., Barbier, N. (2011) Environmental modulation of self-organized periodic vegetative patterns in Sudan. Ecography 34, 990–1001.

[49] De Leenheer, P., Smith, H.L. (2003) Virus dynamics: a global analysis. SIAM J. Appl. Math. 13, 1313–1327.

[50] DiPrima, R.C. (1967) Vector eigenfunction expansions for the growth of Taylor vortices in the flow between rotating cylinders. In: Ames, WF (ed) Nonlinear partial differential equations. Academic Press, New York, pp. 19–42.

[51] DiPrima, R.C., Echaus, W., Segel, L.A. (1971) Nonlinear wavenumber interaction in near–critical two-dimensional flows. J. Fluid Mech. 49, 705–744.

[52] Doedel, E.J. (1984) The computer-aided bifurcation analysis of predator-prey models. J. Math. Biol. 20, 1–14.

[53] Drazin, P.G., Reid, W.H. (1981) Hydrodynamic stability, Cambridge University Press, Cambridge.

[54] Edelstein-Keshet L. (2005) Mathematical models in biology. SIAM, Philadelphia.

[55] Everleigh, E.S., Chant, D.A. (1982a) Experimental studies on acarine predator-prey interactions: The effect of predator density on prey consumption, predator searching frequency, and functional response to prey density (Acarina: Phytoseiidae). Canadian J. Zool. 60, 611–629.

[56] Everleigh, E.S., Chant, D.A. (1982b) Experimental studies on acarine predator-prey interactions: The effect of predator density on immature survival, adult fecundity and emigration rates, and numerical response to prey density (Acarina: Phytoseiidae). Canadian J. Zool. 60, 630–638.

[57] Facsko, S., Bobek, T., Stahl, A., Kurz, H., Dekorsy, T. (2004) Dissipative continuum model for self-organized pattern formation during ion beam sputtering. Phys. Rev. B 69, 153412–1 – 153412–4.

[58] Facsko, S., Dekorsy, T., Koerdt, C., Trappe, C., Kurz, H., Vogt, A., Hartnagel, H.L. (1999) Formation of ordered nanoscale semiconductor dots by ion sputtering. Science 285, 1551–1553.

[59] Ferm, E.N., Wollkind, D.J. (1982) Onset of Rayleigh-B'enard-Marangoni convection: Comparison between theory and experiment. J. Non-Equilib. Thermodyn. 3, 169–190.

[60] Firth, W.J., Scroggie, A.J. (1994) Spontaneous pattern formation in an absorptive system, Europhys. Lett 26, 521–526.

[61] Fletcher, R.C. (1977) Folding of a single viscous layer: Exact infinitesimal amplitude solution, Tectono-physics 39, 593–606.

[62] Fransz, H.G. (1974) The functional response to prey density in an acarine system (simulation monographs). Pudoc, Wageningen.

[63] Gambino, G., Greco, A.M., Lombardo, M.C., Sammartino, M. (2010) A subcritical bifurcation for a nonlinear reaction-diffusion system. In: Greco, A.M., Rionero, S., Ruggeri, T. (eds) Waves and stability in continuous media. World Scientific Publishing Comp., Singapore, pp. 163–172.

[64] Gambino, G., Lombardo, M.C., Sammartino, M. (2012) Turing instability and traveling fronts for a nonlinear reaction-diffusion system with cross-diffusion. Math. Comput. Simulat. 82, 1112–1132.

[65] Geddes, J.B., Indik, R.A., Moloney, J.V., Firth, W.J. (1994) Hexagons and squares in a passive nonlinear optical system, Phys. Rev. A. 50, 3471–3485.

[66] Gill, K., Wollkind, D.J., Dichone, B.J. (2019) A systematic development of Jeans' criterion with rotation for gravitational instabilities. Involve 12, 1099–1108.

[67] Gmitro, J.I., Scriven, L.E. (1966) A physicochemical basis for pattern and rhythm. In: Warren, KB (ed.) Intracellular Transport, Academic Press, New York, pp. 221–255.

[68] Goldstein, S. (1931) On the stability of superposed streams of fluids of different densities, Proc. Roy. Soc. London Ser. A 132, 524–548.

[69] Golovin, A.A., Matkowsky, B.J., Volpert, V.A. (2008) Turing pattern formation in the Brusselator model with superdiffusion. SIAM J. Appl. Math. 69, 251–272.

[70] Golovin, A.A., Nepomnyashchy, A.A., Pismen, L.M. (1995) Pattern formation in large-scale Marangoni convection with deformable interface. Physica D 81, 117–147.

[71] Goody, R.M. (1956) The influence of radiative transfer on cellular convection. J. Fluid Mech. 1, 424–435.

[72] Gowda, K., Riecke, H., Silber, M. (2014) Transitions between patterned states in vegetation models for semiarid ecosystems, Phys. Rev. E 89, 022701-1–022701-8.

[73] Graham, M.D., Kevrekidis, I.G., Asakura, K., Lauterbach, J., Krischer, K., Rotermund, H-H., Ertl, G. (1994) Effects of boundaries on pattern formation: Catalytic oxidation of CO on platinum. Science 264, 80–82.

[74] Greenspan, H.P. (1968) The theory of rotating fluids, Cambridge University Press, Cambridge.

[75] Gunaratne, G.H., Ouyang, Q., Swinney, H.L. (1994) Pattern formation in the presence of symmetries. Phys. Rev. E 50, 2802–2820.

[76] Hainzl, J. (1988) Stability and Hopf bifurcation for a predator-prey system with several parameters. SIAM J. Appl. Math. 48, 170–190.

[77] Hassell, M.P. (1978) The dynamics of arthropod predator-prey systems. Princeton University Press, Princeton, NJ.

[78] Hastings, A., Serradilla, J.M., Ayala, F.J. (1981) Boundary-layer model for the population dynamics of single species. Proc. Natl. Acad. Sci. 78, 1972–1975. [Note that the θ model mentioned in its abstract is $n_{t+1} = rn_t(1 - n_t^\theta)$ for r, $\theta > 0$ with $\theta = 1$ corresponding to the logistic model].

[79] Herminghaus, S., Jacobs, K., Mecke, K., Bischof, J., Fery, A., Ibn-Elhaj, M., Schlagowski, S. (1998) Spinodal dewetting in liquid crystal and liquid metal films. Science 282 , 916–919.

[80] HilleRisLambers R., Rietkerk M., van den Bosch F., Prins H.T.H., de Kroon H. (2001) Vegetation pattern formation in semi-arid grazing systems. Ecology 82, 50–61

[81] Holling, C.S. (1965) The functional response of predators to prey density and its role in mimicry and population regulation. Mem. Entomol. Soc. Can. 45, 1–60.

[82] Holling, C.S. (1973) Resilience and stability of ecological systems. Ann. Rev. Ecol. Sys. 4, 1–23.

[83] Hoyt, S.C. (1969) Integrated chemical control of insects and biological control of mites on apples in Washington. J. Econ. Entomol. 62, 74–86.

[84] Huffaker, C.B.K., Shea, K.P., Herman, S.G. (1963) Experimental studies on predation (III), complex dispersion and levels of food in an acarine predator-prey interaction. Hilgardia 34, 305–330.

[85] Ince, E.L. (1956) Ordinary differential equations. Dover, New York City.

[86] Jameel, A.T., Sharma, A. (1994) Morphological phase separation in thin liquid films II. Equilibrium contact angles of nanodrops coexisting with thin films. J. Colloid Interface Sci. 164, 416–427.

[87] Jeans, J.H. (1902) The stability of a spherical nebula. Phil. Trans. Roy. Soc. of London 199, 1-53.

[88] Jeans, J.H. (1928) Astronomy and cosmogony, Cambridge University Press, Cambridge.

[89] Jeffreys, H. (1928) Some cases of instability in fluid motion. Proc. Roy. Soc. A 118, 195–208.

[90] Kadar, M., Parisi, G., Zhang, Y-C. (1986) Dynamic scaling of growing interfaces. Phys. Rev. Lett. 56, 889–892.

[91] Kahng, B., Jeong, H., Barabási, A-L. (2001) Quantum dot and hole formation in sputter erosion. Appl. Phys. Lett. 78, 805–807.

[92] Kealy, B.J., Wollkind, D.J. (2012) A nonlinear stability analysis of vegetative Turing pattern formation for an interaction-diffusion plant-surface water system in an arid flat environment. Bull. Math. Biol. 74, 803–833.

[93] Kealy-Dichone, B.J., Wollkind, D.J., Cangelosi, R.A. (2015) Rhombic analysis extension of a plant-surface water interaction-diffusion model for hexagonal pattern formation in an arid at environment. Am. J. Plant Sci. 6, 1256–1277.

[94] Keller, E.F., Segel, L.A. (1970) Initiation of slime mold aggregation viewed as an instability. Journal of Theoretical Biology 26, 399–415.

[95] Khanna, R., Sharma, A. (1997) Pattern formation in spontaneous dewetting of thin apolar films. J. Colloid Interface Sci. 195, 42–50.

[96] Kheshgi, H.S., Scriven, L.E. (1991) Dewetting: Nucleation and growth of dry regions. Chem. Engrng. Sci. 46, 519–526.

[97] Kiessling, M.K-H. (2003) The "Jeans swindle": A true story-mathematically speaking. Adv. Appl.Math. 31, 132–149.

[98] Klausmeier, C.A. (1999) Regular and irregular patterns in semiarid vegetation. Science 284, 1826–1828.

[99] Kolmogorov, A.N. (1936) Sulla teoria di Volterra della Lotka per l'esistenza. Gior. Instituto Ital. Attuari 7, 74–80.

[100] Kondo, S., Asai, R. (1995) A reaction-diffusion wave on the skin of the marine angelfish Pomacanthus. Nature 376, 765–768.

[101] Kondo, A., Miura, T. (2010) Reaction-diffusion model for understanding biological pattern formation. Science 329, 1616–1620.

[102] Koschmieder, E.L. (1967) On convection under an air surface. J. Fluid Mech. 30, 9–15.

[103] Koschmieder, E.L. (1993) Bénard cells and Taylor vortices. Cambridge, London.

[104] Kuramoto, Y. (1978) Diffusion-induced chaos in reaction systems. Suppl. Prog. Theor. Phys. 64, 346–367.

[105] Kuske, R., Matkowsky, B.J. (1994) On roll, square, and hexagonal cellular flames. European Journal of Applied Mathematics 5, 65–93.

[106] Landau, L.D. (1944) On the problem of turbulence. Doklady Akademii Nauk SSR 44, 339–342.

[107] Landay, A.L., Jessop, C., Lennette, E.T. (1991) Chronic fatigue syndrome: clinical conditions associated with immune activation. Lanset 338, 707–712.

[108] Langer, J.S. (1980) Instabilities and pattern formation in crystal growth. Reviews of Modern Physics 52, 1–28.

[109] Lefever, R., Lejeune, O. (1997) On the origin of tiger bush. Bull. Math. Biol. 59, 263–294.

[110] Lefever, R., Lejeune, O., Couteron, P. (2001) A case study of tiger bush in sub-Saharan Sahel. In: Maini, P.K., Othmer, H.G. (eds). Mathematical models for biological pattern formation, Springer-Verlag, New York, pp. 83–112.

[111] Lejeune, O. Tlidi, M. (1999) A model for the explanation of vegetation stripes (tiger bush). J. Veg. Sci., 10, 201–208.

[112] Lejeune, O., Tlidi, M., Couteron, P. (2002) Localized vegetation patches: A self-organized response to resource scarcity. Phys. Rev. E 66, 010901–1–010901–4.

[113] Lejeune, O., Tlidi, M., Lefever, R. (2004) Vegetative spots and stripes: Dissipative structures in arid landscapes. Int. J. Quantum Chem. 98, 261–271.

[114] Lengyel, I., Epstein, I.R. (1991) Modeling of Turing structures in the chlorite-iodide-malonic acid-starch reaction system, Science 251, 650–652.

[115] Lengyel, I., Epstein, I.R. (1992) A chemical approach to designing Turing patterns in reaction-diffusion systems. Natl. Acad. Sci. USA 89, 3977–3979.

[116] Leslie, P.H. (1948). Some further notes on the use of matrices in population mathematics. Biometrica 35, 213–245.

[117] Levin S.A., Segel L.A. (1976) Hypothesis for origin of planktonic patchiness. Nature 259, 659.

[118] Lighthill, M.J. (1960) Dynamics of dissociating gases Part 2. Quasi-equilibrium transfer theory. J. Fluid Mech. 8, 161–182.

[119] Lin, C.C., Segel, L.A. (1974) Mathematics applied to deterministic problems in the natural sciences. MacMillan, New York.

[120] Linsenmaier, W. (1972) Insects of the world. Translated from the German by L. Chadwick. McGraw-Hill, New York.

[121] Liu Q-X., Doelman, A., Rottschafer V., de Jager M., Herman P.M.J., Rietkerk M., van de Koppel J. (2013) Phase separation explains a new class of self-organized spatial patterns in ecological systems. PNAS, 110, 11905–11910.

[122] Liu Q-X., Weerman E.J., Herman P.M.J., Olff H., van de Koppel J. (2012) Alternative mechanisms alter the emergent properties of self-organization in mussel beds. Proc Royal Soc Lond B 279, 2744–2753.

[123] Logan, J.A. (1977) Population model of the association of Tetranychus mcdanieli (Acarina: Tetranychidae) with Metasieulus occidentalis (Acarina: Phytoseiidae) in the apple ecosystem. Ph.D. Dissertation. Washington State University, Pullman.

[124] Logan, J.A. (1982) Recent advances and new directions in Phytoseiid population models. In: Hoyt, SC (ed) Recent advances in the knowledge of the Phytoseiidae. Division of Agricultural Sciences Publications, University of California, Berkeley, pp. 49–71.

[125] Logan, J.A., Hilbert, D.W. (1983) Modeling the effect of temperature on arthropod population systems. In: Lauenroth, W.K., Skogerboe, G.V., Flug, M. (eds) Analysis of ecological systems: State of the art in ecological modeling. Elsevier, Amsterdam, pp. 113–122.

[126] Logan, J.A., Wollkind, D.J., Hoyt, S.C., Tanigoshi, L.K. (1976) An analytic model for description of temperature dependent rate phenomena in arthropods. Envir. Entomol. 5, 1130–1140.

[127] Lotka, A.J. (1925) Elements of physical biology. Williams and Wilkens, Baltimore.

[128] Lugiato, L.A., Lefever, R. (1987) Spatial dissipative structures in passive optical systems. Phys. Rev. Lett., 58, 2209–2211.

[129] Lugiato, L.A., Oldano, C. (1988) Stationary spatial patterns in passive optical systems, Phys. Rev. A 37, 3896–3908.

[130] Macfaydyen, W.A. (1950) Soil and vegetation in British Somaliland. Nature 165, 121.

[131] Makeev, M.A., Barabási, A-L. (1997) Ion-induced effective surface diffusion in ion sputtering. Appl. Phys. Lett. 71, 2800–2802.

[132] Makeev, M.A., Barabási, A-L. (1998) Secondary ion yield changes on rippled interfaces. Appl. Phys. Lett. 72, 906–908.

[133] Makeev, M.A., Cuerno, R., Barabási, A-L. (2002) Morphology of ion-sputtered surfaces. Nucl. Instr. and Meth. in Phys. Res. B 197, 185–227.

[134] Malchow H. (1996) Nonlinear plankton dynamics and pattern formation in an eco-hydrodynamic model system. J Mar Syst 7, 193–202.

[135] Mandel, P., Georgio, M., Erneux, T. (1993) Transverse effects in coherently driven nonlinear cavities, Phys. Rev. A 47, 4277–4286.

[136] Marcus, P.I., Carver, D.H. (1967) Hemadsorption-negative plaque test for viruses inducing intrinsic interference. In: Fundamental techniques in virology. Habel, K., Salzmann, N.P. (eds): Academic Press: New York and London, pp.161–183.

[137] Matkowsky, B.J. (1970) A simple nonlinear dynamic stability problem. Bull. Amer. Math. Soc. 76, 620–625.

[138] Maxwell, J.C. (1868) On governors. Proceedings of the Royal Society of London 16, 270–283.

[139] May, R.M. (1973) Stability and complexity in model ecosystems. Princeton University Press, Princeton.

[140] McKendrick, A. (1926) Applications of mathematics of medical problems. Proceedings of the Edinburgh Mathematical Society 44, 98–133.

[141] Meinhardt, H. (1995) Dynamics of stripe formation. Nature 376, 722–723.

[142] Meron, E. (2015) Nonlinear physics of ecosystems. CRC Press, Baco Raton, Florida.

[143] Meron, E., Gilad, E., von Hardenberg, J., Shachak, M., Zarmi, Y. (2004) Vegetation patterns along a rainfall gradient. Chaos, Solitons and Fractals, 367–376.

[144] Meyer, J.R. (1994) Black jaguars in Belize?: A survey of melanism in the jaguar, Panthera onca. Biological-Diversity. Info/Black_Jaguar.htm.

[145] Michely, T., Comsa, G. (1991) Generation and nucleation of adatoms during ion bombardment of Pt (111). Phys. Rev. B 44, 8411–8414.

[146] Mitlin, V.S. (1993) Dewetting of solid surface: Analogy with spinodal decomposition. J. Colloid Interface Sci.156, 491–497.

[147] Mitlin, V.S., Petviashvili, P.V. (1994) Nonlinear dynamics of dewetting: kinetically stable structures. Phys. Lett. A 192, 323–326.

[148] Moloney, J.V., Newell, A.C. (2004) Nonlinear optics. Westview Press, Boulder, CO.

[149] Morgan, D.S., Doelman, A., Kaper, T.J. (2000) Stationary periodic patterns in the 1-D Gray-Scott model. Methods Appl. Anal. 7, 105–150.

[150] Morris, L.R., Winegard, W.C. (1969) The development of cells during the solidification of a dilute Pb-Sb alloy. Journal of Crystal Growth 5, 361–375.

[151] Mouhot, C., Villani, C. (2011) On Landau damping. Acta Math. 207, 29–201.

[152] Mullins, W.W., Sekerka, R.F. (1964) Stability of a planar interface during solidification of a dilute binary alloy. Journal of Applied Physics 35, 444–45.

[153] Murray, J.D. (1965) On the mathematics of fluidization. Part I. Fundamental equations and wave propagation. J. Fluid Mech. 21, 465–493.

[154] Murray, J.D. (2003) Mathematical biology II: Spatial models and biomedical application. Springer, Berlin.

[155] Nathenson, M. (1990) Convection in thick layers with a linear temperature gradient. Bull. Amer. Phys. Soc. 35, 2273.

[156] Newell, A.C., Whitehead, J.A. (1969) Finite bandwidth, finite amplitude convection. J. Fluid Mech. 38, 279–303.

[157] Normand, C., Pomeau, V., Verlarde, M.G. (1977) Convective instability: a physicists approach. Reviews of Modern Physics 49, 581–624.

[158] Okubo A., Levin S.A. (2001) Diffusion and ecological problems: modern perspectives. Springer, New York.

[159] Oron, A., Bankoff, S.G. (1999) Dewetting of a heated surface by an evaporating liquid film under conjoining/disjoining pressures. J. Colloid Interface Sci. 218, 152–166.

[160] Oulton, D.B., Wollkind, D.J. (1982) A three-dimensional nonlinear stability analysis of the solidification of a dilute binary alloy. Old Dominion University Research Foundation, Norfolk.

[161] Oulton, D.B., Wollkind, D.J., Maurer, R.N. (1979) A stability analysis of a prototype moving boundary problem in heat flow and diffusion. American Mathematical Monthly 3, 175–186.

[162] Ouyang, Q., Li, R., Li, G., Swinney, H.L. (1995). Dependence of Turing pattern wavelength on diffusion rate. J. Chem. Phys. 102, 2551–2555.

[163] Ouyang, Q., Swinney, H.L. (1991). Transition to chemical turbulence, Chaos 1, 411–420.

[164] Ouyang, Q., Swinney, H.L. (1995). Onset and beyond Turing pattern formation. In: Kapral, R., Showalter, K. (eds.) Chemical waves and patterns, Kluwer, Dortrecht, pp. 263–295.

[165] Palmer, H.J., Berg, J.C. (1971) Convective instability in liquid pools heated from below. J. Fluid Mech. 47, 779–787.

[166] Pansuwan, A., Rattanakul, C., Lenbury, Y., Wollkind, D.J., Harrison, L., Rajapakse, I., Cooper, K. (2005) Nonlinear stability analyses of pattern formation on solid surfaces during ion-sputtered erosion. Mathl. Comput. Modelling 41, 939–964.

[167] Park, S., Kahng, B., Jeong, H., Barabási, A-L. (1999) Dynamics of ripple formation in sputter erosion. Phys. Rev. Lett. 83, 3486–3489.

[168] Pearson, J.E. (1992) Pattern formation in a (2+1) activator-inhibitor-immobilizer system. Physica A 188, 178–189.

[169] Pearson, J.E. (1993) Complex patterns in a simple system. Science. 261, 189–192.

[170] Pearson, J.R.A. (1980) On convection cells induced by surface tension. J. Fluid Mech. 4, 489–500.

[171] Perelson, A.S., Ribeiro, R.M. (2013) Modeling the within-host dynamics of HIV infection. BMC Biology 11, 96:1–11.

[172] Peterson, R.T. (1951) Wildlife in Color. Houghton Mifflin, Boston.

[173] Phillips, A.N. (1996) Reduction of HIV concentration during acute infection: independence from a particular immune response. Science 271, 497–499.

[174] Plant, R.E., Mangel, M. (1987) Modeling and simulation in agricultural pest management. SIAM Rev. 29: 235–261.

[175] Plesa, G., McKenna, P.M., Schnell, M.J., Eisenlohr, L.C. (2006) Immunogenicity of cytopathic and noncytopathic viral vectors. J. Virol. 80, 6259–6266.

[176] Poincaré, H. (1903) Figures d'equilibre d'une masse fluide. C. Naude, Paris.

[177] Prager, W. (1961) Introduction to Mechanics of Continua. Ginn, Boston.

[178] Prigogine, I., Nicolis, G.. (1967) On symmetry-breaking instabilities in dissipative systems. J. Chem. Phys. 46, 3542–3550.

[179] Prüss, J., Zacher, R., Schnaubelt, R. (2008) Global asymptotic stability of equilibria in models for virus dynamics. Math. Model. Nat. Phenom. 3, 126–142.

[180] Ramberg, H. (1963) Fluid dynamics of viscous buckling applicable to folding of layered rocks. Bull. Amer. Assoc. Petrol. Geol. 47, 484–505.

[181] Ramberg, H. (1970a) Folding of laterally compressed multilayer in the field of gravity, I. Phys. Earth Planet. Interiors 2, 203–232.

[182] Ramberg, H. (1970b) Folding of laterally compressed multilayer in the field of gravity, II. Phys. Earth Planet. Interiors 4, 83–120.

[183] Rayleigh, Lord. (1916) On convection in a horizontal layer of fluid, when the higher temperature is on the underside. Phil. Mag. 32, 529–546.

[184] Reiter, G. (1995) Dewetting of thin polymer films. Phys. Rev. Lett. 68, 75–78.

[185] Reiter, G. (1998) The artistic side of intermolecular forces. Science 282, 888–889.

[186] Rietkerk, M., Boerlijst, M.C., van Langevelde, F., HilleRisLambers, R., van de Koppel, J., Kumar, L., Prins, H.H.T., de Roos, A.M. (2002) Self-organization of vegetation in arid ecosystems. Am. Nat. 160, 524–530.

[187] Rietkerk, M., Dekker, S.C., de Ruiter, P.C., van de Koppel, J. (2004) Self-organized patchiness and catastrophic shift in ecosystems, Science 305, 1926–1929.

[188] Roose, T., Fowler, A.C. (2004) A model for water uptake by plant roots, J. theor. Biol. 228, 155–171.

[189] Rougoor, G.W., Oort, J.H. (1960) Distribution and motion of interstellar hydrogen in the galactic system with particular reference to the region 3 kiloparsecs of the center. PNAS 46, 1–13.

[190] Rosenzweig, M.L. (1971) Paradox of enrichment: Destabilization of exploitation ecosystems in ecological time. Science 171, 385–387.

[191] Rosenzweig, M.L. (1972) Technical comment on the stability of enriched aquatic ecosystems. Science 175, 564–565.

[192] Rosenzweig, M.L. (1973) Exploitation in three trophic levels. Amer. Natur. 107, 275–294.

[193] Rovinsky, A.B., Menzinger, M. (1992a) Interaction of Hopf and Turing bifurcations in chemical systems. Phys. Rev. A 46, 6315–6322.

[194] Rovinsky, A.B., Menzinger, M. (1992b) Chemical instability induced by a differential flow. Phys. Rev. Lett. 69, 1193–1196.

[195] Rubin, V.C., Ford, W.K.J. (1970) Rotation of the Andromeda Nebula from a spectroscopic survey of emission. Astrophysical Journal. 159, 379–403.

[196] Rusponi, S., Boragno, C., Valbusa, U. (1997) Ripple structure on Ag (110) surface induced by ion sputtering. Phys. Rev. Lett. 76, 2795–2798.

[197] Rusponi, S., Costantini, G., Boragno, C., Valbusa, U. (1998) Scaling laws of ripple morphology on Cu (110). Phys. Rev. Lett. 81, 4184–4187.

[198] Rusponi, S., Costantini, G., Buatier de Mongeot, F., Boragno, C., Valbusa, U. (1999) Patterning a surface on the nanometric scale by ion sputtering. Appl. Phys. Lett. 75, 3318–3320.

[199] Sáez, E., González-Olivares, E. (1999) Dynamics of a predator-prey model. SIAM J. Appl. Math. 59, 1867–1878.

[200] Scanlon, J.W., Segel, L.A. (1973) Some effects of suspended particles on the onset of Bénard convection, Phys. Fluids 16, 1573–1578.

[201] Schechter, R.S., Prigogine, I., Hamm, J.R. (1972) Thermal diffusion and convective stability. Physics of Fluids 15, 379–386.iii

[202] Scheuer, J., Orenstein, M. (1999) Optical vortices crystals: Spontaneous generation in nonlinear semiconductor materials. Science 285, 230–233.

[203] Scheffler, H., Elsässer, H. (1988) Physics of the galaxy and interstellar matter, Springer-Verlag, Berlin.

[204] Schmidt, M. (1965) Rotation parameters and distribution of mass in the galaxy. In: Galactic structure. Blaauw, A, Schmidt, M (eds) University of Chicago Press, Chicago, pp. 513–530.

[205] Schmidt, S., Kolden, S. (2020) Analysis of an evolution equation for a rhombic planform. Student Report, Department of Mathematics, Gonzaga University, Spokane (WA).

[206] Schmidt, S., Kolden, S., Dichone, B.J., Wollkind, D.J. (2021) Rhombic planform nonlinear stability analysis of an ion-sputtering evolution equation. Involve, 14, 119–142.

[207] Schwartz E.J., Pawelek K.A., Harrington K., Cangelosi R., Madrid S. (2013) Immune control of equine infectious anemia virus infection by cell-mediated and humoral responses. Applied Mathematics 4, 171–177.

[208] Schwartz, E.J., Vaidya, N.K., Dorman, K., Carpender, S., Mealey, R.H. (2018) Dynamics of lentiviral infection in vivo in the absence of adaptive host immune responses. Virology 513, 108–113.

[209] Scroggie, A.J., Firth, W.J. (1996) Pattern formation in an alkali-metal vapor with a feedback mirror. Phys. Rev. A, 53, 2752–2764.

[210] Scroggie, A.J., Firth, W.J., MacDonald, G.S., Tlidi, M., Lefever, R., Lugiato, L.A. (1994) Pattern formation in a passive Kerr cavity, Chaos, Solitons & Fractals 4, 1323–1354.

[211] Segel, L.A. (1965) The nonlinear interaction of a finite number of disturbances in a layer of fluid heated from below. J. Fluid Mech. 21, 359–384.

[212] Segel, LA (1966) Nonlinear hydrodynamic stability theory and its applications to thermal convection and curved flows. In: Non-equilibrium thermodynamics, variational techniques, and stability. Donnelly, R.J., Herman, R., Prigogine, I. (eds) University of Chicago Press, Chicago, pp. 164–197.

[213] Segel, L.A. (1969) Distant side-walls cause slow amplitude modulation of cellular convection. J. Fluid Mech. 38, 203–224.

[214] Segel, L.A. (1972) Simplification and scaling. SIAM Review 14, 547–571.

[215] Segel, L.A. (1977) Mathematics applied to continuum mechanics. Macmillan, New York.

[216] Segel L.A., Jackson J.L. (1972) Dissipative structure: an explanation and an ecological example. J. Theoret. Biol. 37, 545–559.

[217] Segel L.A., Levin S.A. (1976) Application of nonlinear stability theory to the study of the effects of diffusion on predator-prey interaction. In: Piccirelli RA (ed) Topics in statistical mechanics and biophysics: a memorial to Julius L. Jackson. AIP Conf. Proc. No. 27. Am. Int. Phys., New York, pp 123–152.d

[218] Segel, L.A., Stoeckly, B. (1972) Instability of a layer of chemotactic cells, attractant and degrading enzyme. J. Theoret. Biol. 37, 561–585.

[219] Segel, L.A., Stuart, J.T. (1962) On the question of the preferred mode in cellular thermal convection. J. Fluid Mech. 13, 289–306.

[220] Sekimura, T., Zhu, M., Cook, J., Maini, P.K., Murray, J.D. (1999) Pattern formation of scale cells in Lepidoptera by differential origin-dependent cell adhesion. Bull. Math. Biol. 61, 807–827.

[221] Sharma, A., Jameel, A.T. (1993) Nonlinear stability, rupture, and morphological phase separation of thin fluid films on apolar and polar substrates. J. Colloid Interface Sci.161, 190–208.

[222] Sharma, A., Khanna, R. (1998) Pattern formation in unstable thin liquid films. Phys. Rev. Lett. 81, 3463–3466.

[223] Sharma, A., Ruckenstein, E. (1986) An analytical nonlinear theory of thin film rupture and its application to wetting films. J. Colloid Interface Sci. 113, 456–479.

[224] Sharp, R.M., Hellawell, A. (1970) Solute redistribution at non-planar, solid-liquid growth fronts, IV. Ripening of cells and dendrites behind the growth front. Journal of Crystal Growth 11, 77–91.

[225] Sherratt, J.A. (2005) An analysis of vegetative stripe formation in semi-arid landscape. J. Math Biol. 51, 183–197.

[226] Sherratt, J.A., Lord, G.J. (2007) Nonlinear dynamics and pattern bifurcations in a model for vegetation stripes in semi-arid environments. Theor. Pop. Biol. 71, 1–11.

[227] Showalter, RE (2003) Diffusion in deforming porous media. Dynamics of Continuous, Discrete, and Impulsive Systems: Series A 10, 661–678.

[228] Siegmann, W.L., Rubenfeld, L.A. (1975) A nonlinear model for double-diffusive convection. SIAM J. Applied Mathematics 29, pp. 540–557.

[229] Silverman, HG, Roberto, FF (2007) Understanding marine mussel adhesion. Mar Biotechnol 9, 661–681.

[230] Sivashinsky, G.I. (1977) Nonlinear analysis of hydrodynamic instability in laminar flames, Part I. Derivation of basic equations, Acta Astronautica 4, 1177–1206.

[231] Smith, R.B. (1979) The folding of a strongly non-Newtonian layer. Amer. J. Sci., 274, 1029–1043.

[232] Snow, C.P. (1959) The Two Cultures. Cambridge University Press, Cambridge.

[233] Sokol C.L., Luster A.D. (2015) The chemokine system in innate immunity. Cold Spring Harb 588 Perspect Biol. 7(5). doi: 10.1101/cshperspect.a016303.

[234] Spiegel, E.A., Veronis, G. (1960) On the Boussinesq approximation for a compressible fluid. Astrophys. J. 131, 442–447.

[235] Springholz, G., Holy, V., Pinczolits, M., Bauer, G. (1998) Self-organized growth of three-dimensional quantum-dot crystals with fcc-like stacking and a tunable lattice constant. Science 282, 734–737.

[236] Sriranganathan, R., Wollkind, D.J., Oulton, D.B. (1983) A theoretical investigation of the development of interfacial cells during the solidification of a dilute binary alloy: Comparison with the experiments of Morris and Winegard. Journal of Crystal Growth 62, 265–283.

[237] Stahler, S., Palla, F. (2004) The formation of stars, Wiley-VCH, Weinheim.

[238] Stancevic, O., Angstmann, C.N., Murray, J.M., Henry, B.I. (2013) Turing patterns from dynamics of early HIV infection, Bull. Math. Biol. 75, 774–795.

[239] Stephenson, L.E., Wollkind, D.J. (1995). Weakly nonlinear stability analyses of one-dimensional Turing pattern formation in activator-inhibitor/immobilizer model systems. J. Math. Biol. 33, 771–815.

[240] Stewart, J. (2007) Essential calculus: Early transcendentals. Thomson, San Diego.

[241] Stuart, J.T. (1960) On the nonlinear mechanics of wave disturbances in stable and unstable flows, 1.The basic behavior in plane Poiseuille flow. J. Fluid Mech. 9, 353–370.

[242] Sutton, O.G. (1950) On the stability of a fluid heated from below, Proc. Roy. Soc. A 204, 297–309.

[243] Swift, J., Hohenberg, P.C. (1977) Hydrodynamic fluctuations at the convective instability. Phys. Rev. A. 15, 319–328.

[244] Takahashi, T., Kamio, A., Nguyen, A.T. (1974) Morphology of the solid-liquid interface and redistribution of solutes in unidirectional solidified aluminum alloys. Journal of Crystal Growth 24/25 477–483.

[245] Taylor, G.I. (1931) Effect of variation in density on the stability of superposed streams of fluid, Proc. Roy. Soc. London Ser. A , 132, 499–523.

[246] Thess, A., Orszag, S.A. (1995) Surface-tension-driven Bénard convection at infinite Prandtl number. J. Fluid Mech. 283, 201–230.

[247] Thorne, K.S. (1994) Black holes & time warps: Einstein's outrageous legacy. Norton, New York.

[248] Tian, E.M., Wollkind, D.J. (2003a) A nonlinear stability analysis of pattern formation in thin liquid films. Interfaces and Free Boundaries 5, 1–25.

[249] Tian, E.M., Wollkind, D.J. (2003b) A nonlinear stability analysis of pattern formation in isothermal thin liquid films. Dynamics of Continuous, Discrete, and Impulsive Systems: Series A 10, 759–782.

[250] Tiller, W.A., Rutter, J.W. (1956) The effect of growth conditions upon the solidification of a binary alloy. Canadian Journal of Physics 34, 96–121.

[251] Tiller, W.A., Sekerka, R.F. (1964) Redistribution of solute during phase transformations. Journal of Applied Physics 35, 2726–2829.

[252] Torres-Cerna, C.E., Alanis, A.Y., Poblete-Castro, T., Bermejo-Jambrina, M., Hernandez-Vargas, E.A. (2016). A comparative study of differential evolution algorithms for parameter fitting procedures. IEEE CEC 1:4662.

[253] Tuckwell, H.C., Wan, F.Y.M. (2004). On the behaviour of solutions in viral dynamical models. BioSystems 73, 157–161.

[254] Tulumello, E., Lombardo, M.C., Sammartino, M. (2014) Cross-diffusion driven instability in a predator-prey system with cross-diffusion. Acta. Appl. Math. 132, 621–633.

[255] Turcotte, D.L., Oxburgh, E.R.(1969) Continental drift. Physics Today 22, 30–41.

[256] Turing, A.M. (1952) The chemical basis of morphogenesis. Phil. Trans. Roy. Soc. London B 237, 37–72.

[257] Ursino, N. (2005) The inuence of soil properties on the formation of unstable vegetation patterns on hillsides of semiarid catchments. Advances in Water Resources 28, 956–963.

[258] Uspensky, J.V. (1948) Theory of equations. McGraw Hill, New York.

[259] van de Koppel J., Gascoigne J.C., Theraulaz G., Rietkerk M., Mooij W.M., Herman P.M.J. (2008) Experimental evidence for spatial self-organization in mussel bed ecosystems. Science 322, 739–742.

[260] van de Koppel J., Rietkerk M., Dankers N., Herman P.M.J. (2005) Self-dependent feedback and regular spatial patterns in young mussel beds. Am Nat 165, E66–E77.

[261] van der Stelt, S., Doelman, A., Hek, G., Rademacher, J.D.M. (2013) Rise and fall of periodic patterns for a generalized Klausmeier-Gray-Scott model. J. Nonlinear Sci. 23, 39–95.

[262] Van Dyke, M.D. (1975) Perturbation methods in fluid mechanics. Parabolic Press. Palo Alto.

[263] Verhoeven, J.D., Gibson, E.D. (1971) Interface stability of the melting solid-liquid interface: I. Sn-Sb alloys. Journal of Crystal Growth 10, 2 9–38.

[264] Volterra, V. (1926) Variazioni e fluttuazioni del numero d'individui in specie animali conviventi. Mem. Acad. Lincei. 2, 31–113.

[265] von Foerster, H. (1959) Some remarks on changing populations. In: Strohlmann Jr., F. (ed) The kinetics of cellular proliferation, Grum and Stratton, New York, pp. 382–407.

[266] von Hardenberg, J., Meron, E., Shachak, M., Zarmi, Y. (2001) Diversity of vegetation patterns and desertification. Phs. Rev. Letts. 87, 198101–1–198101–4.

[267] Walgraef, D. (1997) Spatio-temporal pattern formation, Springer, New York.

[268] Wang, R-H., Liu, Q-X., Sun, G-Q., Jin, A., van de Koppel, J. (2009) Nonlinear dynamic and pattern bifurcations in a model for spatial patterns in young mussel beds. J. R. Soc. Interface. 6, 705-718.

[269] Watkinson, A.J., Alexander, J.I.D. (1979) The importance of geometric systems in multilayer folding. J. Struct. Geol. 1, 95.

[270] Watson, J. (1960) On the nonlinear mechanics of wave disturbances in stable and unstable flows, 2. The development of a solution for plane Poiseuille flow and plane Couette flow. J. Fluid Mech. 9, 371–389.

[271] Weast, R.C. (1968) Handbook of chemistry and physics, 49th ed. Chem. Rubber Co., Cleveland.

[272] Wieland, S.F., and Chisari, F. (2005). Stealth and cunning: hepatitis B and hepatitis C viruses. J. Virol. 79, 9369–9380.

[273] Williams C.R., Bees, M.A. (2011) A tale of three taxes: photo-gyro-gravitactic bioconvection. J. Exp. Biol. 214, 2398–2408.

[274] Williams, M.B., Davis, S.H. (1982) Nonlinear theory of film rupture. J. Colloid Interface Sci. 90, 220–228.

[275] Wollkind, D.J. (1976) Exploitation in three trophic levels: An extension allowing intraspecies carnivore interaction. Amer. Natur. 110, 431–447.

[276] Wollkind, D.J. (1977) Singular perturbation techniques: Comparison of the method of matched asymptotic expansions with that of multiple scales. SIAM Review 19, 502–516.

[277] Wollkind, D.J. (2001) Rhombic and hexagonal weakly nonlinear stability analyses: theory and application. In: Debnath, L (ed) Nonlinear instability analyses II, WIT Press, Southampton, pp. 221–272.

[278] Wollkind, D.J., Alexander, J.I.D. (1982) Kelvin-Helmholtz instability in a layered Newtonian fluid model of the geological phenomenon of rock folding. SIAM J. Appl. Math. 42, 1276–1295.

[279] Wollkind, D.J., Alvarado, F.J., Edmeade, D. (2008) Non-linear stability analyses of optical pattern formation in an atomic sodium ring cavity. IMA J. Appl. Math. 73, 903–935.

[280] Wollkind, D.J., Bdzil, J. (1971) Comments on "chemical instabilities". Physics of Fluids 14, 1813–1814.

[281] Wollkind, D.J., Collings, J.B., Barba, M.C.B. (1991) Diffusive instabilities in a one-dimensional temperature-dependent model system for a mite predator-prey interaction on fruit trees: dispersal motility and aggregative preytaxis effects. J. Math. Biol. 29, 339–362.

[282] Wollkind, D.J., Collings, J.B., Logan J.A. (1988a) Metastability in a temperature-dependent model system for a predator-prey outbreak interaction on fruit trees. Bull. Math. Biol. 50, 379–409.

[283] Wollkind, D.J., Collings, J.B., Logan J.A. (1988b) Temperature-mediated stability of the interaction between spider mites and predatory mites in orchards. Exp. Appl. Acarol. 5, 265–292.

[284] Wollkind, D.J., Dichone, B.J. (2017) Comprehensive applied mathematical modeling in the natural and engineering sciences. Springer, Cham (Switzerland).

[285] Wollkind, D.J., Frisch, H.L. (1970) Linear perturbation analysis of the stability of a dissociating fluid. Physics of Fluids 13, 52–61.

[286] Wollkind, D.J., Frisch, H.L. (1971a) Chemical instabilities I: A heated horizontal layer of a dissociating fluid. Physics of Fluids 14, 13–18.

[287] Wollkind, D.J., Frisch, H.L. (1971b) Chemical Instabilities III: Nonlinear stability analysis of a heated horizontal layer of a dissociating fluid. Physics of Fluids 14, 482–487.

[288] Wollkind, D.J., Logan, J.A. (1976) The use of singular perturbation techniques as a tool for modeling ecosystems, In O'Malley, Jr RE (editor) Asymptotic methods and singular perturbations, AMS, Providence, pp. 151–152.

[289] Wollkind, D.J., Logan, J.A. (1978) Temperature-dependent predator-prey mite ecosystem on apple tree foliage. J. Math. Biol. 6, 265–283.

[290] Wollkind, D.J., Logan, J.A., Berryman, A.A. (1978) Asymptotic methods for modeling biological processes. Res. Pop. Ecol. 29, 79–90.

[291] Wollkind, D.J., Manoranjan, V.S., Zhang, L-M. (1994) Weakly nonlinear stability analyses of prototype reaction-diffusion equations. SIAM Rev. 36, 176–214.

[292] Wollkind, D.J., Oulton, D.B., Sriranganathan, R. (1984a) A nonlinear stability analysis of a model equation for alloy solidification. Journal de Physique 45, 505–516.

[293] Wollkind, D.J., Raissi, S. (1974) A nonlinear stability analysis of the melting of a dilute binary alloy. Journal of Crystal Growth 26, 277–293.

[294] Wollkind, D.J., Segel, L.A. (1970) A nonlinear stability analysis of the freezing of a dilute binary alloy. Philosophical Transactions of the Royal Society of London Series A 268, 351–380.

[295] Wollkind, D.J., Sriranganathan, R., Oulton, D.B. (1984b) Interfacial patterns during plane front alloy solidification. Physica D 12, 215–240.

[296] Wollkind, D.J., Stephenson, L.E. (2000) Chemical Turing pattern formation analyses: Comparison of theory with experiment. SIAM J. Appl. Math. 61, 387–431.

[297] Wollkind, D.J., Stephenson, L.E. (2001) Chemical Turing patterns: A model system of a paradigm for morphogenesis. In: Maini, P.K., Othmer, H.G. (eds). Mathematical models for biological pattern formation, Springer-Verlag, New York, pp. 113–142.

[298] Wollkind, D.J., Vislocky, M. (1990) An interfacial model equation for the bifurcation of solidification patterns during LPEE processes. Earth-Science Reviews 29, 349–368.

[299] Wollkind, D.J., Zhang, L-M. (1994a) The effect of suspended particles on Rayleigh-Bénard convection I. A nonlinear stability analysis of a thermal equilibrium model. Mathl. Comput. Modelling 19, 11–42.

[300] Wollkind, D.J., Zhang, L-M. (1994b) The effect of suspended particles on Rayleigh-Bénard convection II. A nonlinear stability analysis of a thermal disequilibrium model. Mathl. Comput. Modelling 19, 43–74.

[301] Wollkind, D.J., Zhang, L-M. (1997) A nonlinear stability analysis of a prototype problem in Rayleigh-Bénard convection in planetary atmospheres. In: Debnath, L., Choudhury, S.R. (eds) Nonlinear instability analysis. Computational Mechanics Publications, Southhampton, pp.123–144.

[302] Xie, R., Karim, A., Douglas, J.F., Han, C.C., Weiss, R.A. (1998) Spinodal dewetting of thin polymer films. Phys. Rev. Lett. 81, 1251–1254.

[303] Yih, C-S. (1965) Dynamics of nonhomogeneous fluids, Macmillan, New York.

[304] Zeren, R.W., Reynolds, W.C. (1972) Thermal instabilities in two-fluid horizontal layers. J. Fluid Mech. 53, 305–327.

Index

For Product Safety Concerns and Information please contact our EU
representative GPSR@taylorandfrancis.com
Taylor & Francis Verlag GmbH, Kaufingerstraße 24, 80331 München, Germany